Y0-CAS-540

U.S. Customary and Metric Comparisons	To Change From	To	Multiply By
Length			
1 meter =	meters	inches	39.37
39.37 inches	inches	meters	0.0254
1 meter =	meters	feet	3.2808
3.28 feet	feet	meters	0.3048
1 meter =	meters	yards	1.0936
1.09 yards	yards	meters	0.9144
1 centimeter =	centimeters	inches	0.3936
0.39 inch	inches	centimeters	2.54
1 millimeter =	millimeters	inches	0.03937
0.04 inch	inches	millimeters	25.4
1 kilometer =	kilometers	miles	0.6214
0.62 mile	miles	kilometers	1.6093
Weight or Mass			
1 gram =	grams	ounces	0.0353
0.04 ounce	ounces	grams	28.3495
1 kilogram =	kilograms	pounds	2.2046
2.2 pounds	pounds	kilograms	0.4536
Liquid Capacity			
1 liter =	liters	quarts	1.0567
1.06 quarts	quarts	liters	0.9463
Capacity or Volume			
1 cubic inch =	cubic inches	cubic centimeters	16.387
16.387 cubic centimeters	cubic centimeters	cubic inches	0.0610
1 cubic inch =	cubic inches	liters	0.01639
0.01639 liters	liters	cubic inches	61.0128
1 cubic foot =	cubic feet	cubic meters	0.0283
0.0283 cubic meter	cubic meters	cubic feet	35.3357
1 teaspoon =	teaspoons	milliliters	4.93
4.93 milliliters	milliliters	teaspoons	0.2028
1 tablespoon =	tablespoons	milliliters	14.97
14.97 milliliters	milliliters	tablespoons	0.0668
1 fluid ounce =	fluid ounces	milliliters	29.57
29.57 milliliters	milliliters	fluid ounces	0.0338
1 cup = 0.24 liters	cups	liters	0.24
	liters	cups	4.1667
1 pint = 0.47 liters	pints	liters	0.47
	liters	pints	2.1277
1 gallon =	gallons	cubic meters	0.00379
0.00379 cubic meters	cubic meters	gallons	263.85

Special Algebra Patterns for Factoring

$a^2 + 2ab + b^2 = (a + b)^2$

$a^2 - b^2 = (a + b)(a - b)$

$a^3 + b^3 = (a + b)(a^2 - ab + b^2)$

$a^3 - b^3 = (a - b)(a^2 + ab + b^2)$

Symbols

$+$	Add
$-$	Subtract
$\times, \cdot, *, (\ )(\ )$	Multiply
$\div, \overline{}, /, -$	Divide
$=$	Equal to
$\approx$	Approximately equal to
$\neq$	Not equal to
$\%$	Percent
$>$	Greater than
$<$	Less than
$\geq$	Greater than or equal to
$\leq$	Less than or equal to
$\sqrt{}$	Radical sign or square root
$(\ ),[\ ],\{\ \},-$	Grouping symbols
$|\ |$	Absolute value
$f(x)$	Function notation, read "f of x"
$\overleftrightarrow{AB}$	Line AB
$\overline{AB}$	Line segment AB
$\overrightarrow{AB}$	Ray AB
$\simeq, \cong$	Congruent to
$\sim$	Similar to (geometric figures)
$\angle$	Angle
$\perp$	Perpendicular
$\triangle$	Triangle
$\bigcirc$	Circle
$\llcorner$	Right angle
Δ	Delta, change, used with slope
$\{\ldots|\ldots\}$	Such that, used with set notation
Σ	Summation
x_1	Subscript (1)
$\{\ \}, \phi$	Empty or null set
$\in$	Is an element of
$\cup$	Union (of sets)
$\cap$	Intersection (of sets)
π	Constant—Pi (ratio of diameter to circumference of circle, approximately 3.14)
e	Constant—natural exponential; from $\left(1 + \dfrac{1}{n}\right)^n$ where n → ∞, approximately 2.72
i	The square root of -1; $\sqrt{-1}$
∞	Infinity
$\therefore$	Therefore
$\exists$	There exists
$\forall$	For every

College Mathematics for Technology

SIXTH
EDITION

CHERYL CLEAVES

Southwest Tennessee Community College

MARGIE HOBBS

The University of Mississippi

PEARSON

Prentice
Hall

Upper Saddle River, New Jersey
Columbus, Ohio

Library of Congress Cataloging-in-Publication Data

Cleaves, Cheryl S.
    College mathematics for technology.— 6th ed. / Cheryl Cleaves, Margie Hobbs.
        p. cm.
    Includes index.
    ISBN 0-13-048693-0
    1. Mathematics. I. Hobbs, Margie J. II. Title.

QA39.2.C58 2004
512.1—dc21

2003045616

Editor in Chief: Stephen Helba
Senior Acquisitions Editor: Gary Bauer
Editorial Assistant: Natasha Holden
Developmental Editor: Ohlinger Publishing Services
Production Editor: Louise N. Sette
Production Supervision: Carlisle Publishers Services
Design Coordinator: Diane Ernsberger
Cover Designer: Bryan Huber
Cover Art: Corbis
Production Manager: Brian Fox
Marketing Manager: Leigh Ann Sims

This book was set in Palatino and Myriad by Carlisle Communications, Ltd. It was printed and bound by Courier Kendallville, Inc. The cover was printed by Phoenix Color Corp.

Portions of this book were previously published as *Basic Mathematics for Trades and Technologies* by Cleaves, Hobbs, and Dudenhefer (© 1990), *Introduction to Technical Mathematics* by Cleaves, Hobbs, and Dudenhefer (© 1988), and *Vocational-Technical Mathematics Simplified* by Cleaves, Hobbs, and Dudenhefer (© 1987).

Pearson Education Ltd.
Pearson Education Singapore Pte. Ltd.
Pearson Education Canada, Ltd.
Pearson Education—Japan

Pearson Education Australia Pty. Limited
Pearson Education North Asia Ltd.
Pearson Educación de Mexico, S.A. de C.V.
Pearson Education Malaysia Pte. Ltd.

10 9 8 7 6 5 4 3
ISBN 0-13-048693-0

To Sarah
My mother and my first teacher and mentor

To Holly
My daughter and the light of my life

Preface

In *College Mathematics for Technology*, Sixth Edition, we have preserved all the features that made the first five editions one of the most appropriate texts on the market for a comprehensive study of mathematics in career programs. We continue to use real-life situations as a context for applied problems.

Changes in the Sixth Edition

The sixth edition incorporates many valuable suggestions made by users of earlier editions. We have placed greater emphasis on problem solving, included new material, and rearranged some topics. For example, we have expanded topics to include additional approaches such as the completing-the-square method for solving quadratic equations. Functions now include a more in-depth presentation of range and domain. Topics have been rearranged to more clearly distinguish between linear and nonlinear equations and functions. Exponential and logarithmic expressions have been moved to a chapter to follow quadratic and higher-degree equations. Investment and financial formulas are included to emphasize the usefulness of exponential and logarithmic equations.

Our goal is to present a systematic framework for successful learning in mathematics that will strengthen students' *mathematical sense* and give students a greater appreciation for the power of mathematics in everyday life and in the workplace. The new material in this edition has been added to broaden the usefulness of the text to provide the mathematics for the general education core. Many of the explanations have been enhanced with carefully constructed visualizations. Exercises have been updated and new ones added.

Commitment to Improving Mathematics Education

The authors have been and continue to be active in the development, revision, and implementation of the standards (*Crossroads*) of the American Mathematical Association of Two-Year Colleges (AMATYC). We enthusiastically promote the standards and guidelines encouraged by AMATYC, NCTM, MAA, and the SCANS document. The Instructor's Resource Manual gives suggestions and activities for implementing the standards.

Calculator Usage

Calculator tips appropriate for both scientific and graphing calculators are periodically included. These generic tips guide students to use critical thinking to determine how their calculator operates without referring to a user's manual.

We continue to emphasize the calculator as a tool that *facilitates* learning and understanding. We include assessment strategies throughout the text and supplementary materials that enable students to test their understanding of a concept independently of their calculator.

Study Strategies and Reference Features

In our experiences as instructors, we are all too aware of the need for students to develop good study habits and good independent learning skills. Students find a good reference text invaluable as they need to review mathematical concepts. Many students have praised the usefulness of this text as a reference standard. For a detailed description of the features of the text and our suggestions for students, refer to the *To the Student* portion of the preface.

Additional Resources

Several additional resources are available with the adoption of the text. These resources include the Instructor's Resource Manual, a Test Item File and a computerized test item file (TestGen), a Student Solutions Manual, a "How to Study Mathematics" booklet, StudyWizard software (packaged with the text), a Companion Website, a Premium Website, and online course material for WebCT, Blackboard, and CourseCompass. Go to www.prenhall.com/cleaves or contact your Prentice Hall representative for more information.

Acknowledgments

A project such as this does not come together without help from many people. Our first avenue for input is through our students and fellow instructors at the Southwest Tennessee Community College and The University of Mississippi. We also receive input from faculty at other colleges who use the text and from our many AMATYC colleagues. Their comments and suggestions have been invaluable. In addition, we appreciate the assistance we received in ensuring the accuracy of the text. We thank Julie Anderson and Emily Atchley who spent many hours working every problem in the text and Kim Denley who worked many of the problems. However, we take full responsibility for any misprints or errors that may remain in the first printing of this edition.

Supplements for any text are a vital part of the educational support provided to teachers and students. We thank Jimmy Van Alphen who prepared the Test Item File, Jim and Renee Smith who organized the Student Solutions Manual and the solutions for the Instructor's Resource Manual and Marcus Rasco who revised the contents of the Premium and Companion websites.

We wish to express thanks to all the people who helped make this edition a reality. In particular, we thank Gary Bauer, Senior Acquisitions Editor, and Steve Helba, Editor in Chief, whose belief in our work and support of our ideas have been a major factor in the success of this book. We thank Louise Sette, Prentice Hall production editor. We also thank Monica Ohlinger and Megan Becker of Ohlinger Publishing Services and Emily Autumn of Carlisle Publishers Services.

The teaching of mathematics over time produces a wealth of knowledge about instructional strategies and specific content. We are grateful for the many valuable suggestions received in these areas. We wish to thank the following individuals:

Virginia Dewey, York Technical College
Bill Harris, Mountain Empire Community College
Rose Kavanaugh, Ozarks Technical Community College (MO)
Robin G. Moore, North Arkansas College
Toni Parese, Southern Maine Technical College
Pascal Renault, John Tyler Community College (VA)
Fred Toxopeus, Kalamazoo Valley Community College
Jimmie A. VanAlphen, Ozarks Technical Community College (MO)
Terri Wright, New Hampshire Technical College-Manchester

Cheryl Cleaves
Margie Hobbs

To the Student

The mathematics you learn from this book will serve you well and will help you advance your career goals. We have given much thought to the best way to teach mathematics and have done extensive research on how students learn. We have provided a wide variety of features and resources so that you can customize your study to your needs and circumstances. The following features are key to helping you learn the mathematics in this text.

Table of Contents.   The table of contents is your "roadmap" to this text. Study it carefully to determine how the topics are arranged. This will aid you in relating topics to each other.

Glossary/Index.   An extensive glossary/index is an important part of every mathematics book. Use the index to cross-reference topics and to locate other topics that relate to the topic you are studying.

Learning Outcomes.   A learning outcome is what you should be able to do when you master a concept. These outcomes can guide you through your study plan. The chapter opening page lists the learning outcomes for the chapter. Each section begins with a statement of learning outcomes that shows you what you should look for and learn in that section. If you read and think about these outcomes before you begin the section, you will know what to look for as you work through the section. Self-Study Exercises are organized by learning outcomes and the Chapter Summary lists the learning outcomes for your review.

Six-Step Approach to Problem Solving.   Successful problem solvers use a systematic, logical approach. We use a six-step approach to problem solving. This approach gives you a system for solving a variety of math problems. You will learn how to organize the information given and how to develop a logical plan for solving the problem. You are asked to analyze and compare and to estimate as you solve problems. Estimation helps you decide whether your answer is reasonable. You will learn to interpret the results of your calculations within the problem's context, a skill you will use on your job.

Tip Boxes.   These boxes give helpful hints for doing mathematics, and they draw your attention to important observations and connections that you may have missed in an example.

Learning Strategies.   Strategies that help you build a framework for successful learning are found in each chapter. The strategies show ways to manage your learning

Learning Outcome

Tip box

Learning Strategy

Example with explanatory comments

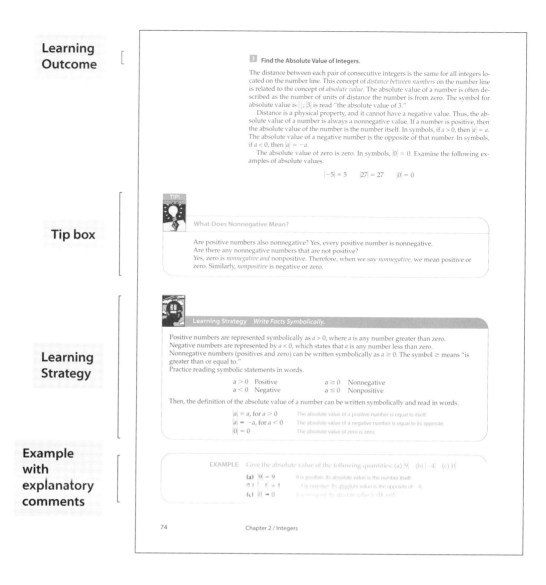

of mathematics that you may not have thought of before. Many of the strategies have to do with your *mathematical sense* and give you a greater appreciation for the power of mathematics in your workplace and your life. Many of these strategies are also useful in other areas of study.

Using Your Calculator.   Calculators are useful in all levels of mathematics. Some tips introduce easy-to-follow calculator strategies. The tips show you how to analyze the procedure and set up a problem for a calculator solution; a sample series of keystrokes is often included. In addition, the tips help you determine how your type of calculator operates for various mathematical processes.

Use of Color in the Text.   As you read the text and work through the examples, notice the items shaded with color or gray. These will help you follow the logic of working through the example. Color also highlights important items and boxed features such as the Tips, Learning Strategies, and rules, procedures, and formulas.

Self-Study Exercises.   These practice sets are keyed to the learning outcomes and appear at the end of each section. Use these exercises to check your understanding of the section. The answers to every exercise are at the end of the text, so you can get immediate feedback on whether you understand the concepts.

Assignment Exercises.   An extensive set of exercises appears at the end of each chapter, so you can review all the learning outcomes presented in the chapter. These

viii   Preface

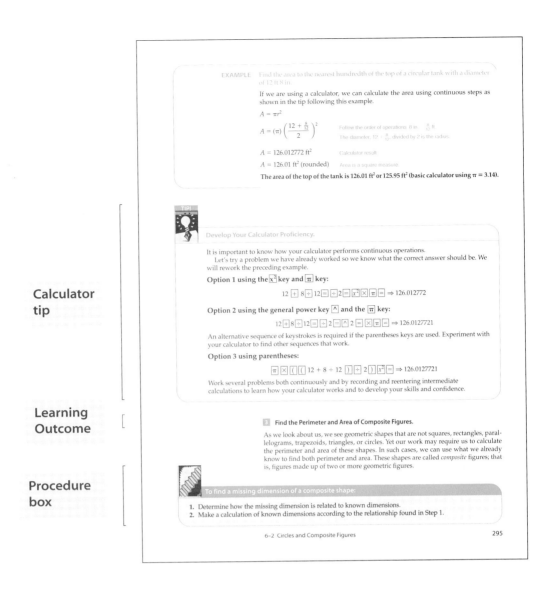

exercises, organized by section, may be assigned as homework, or you may want to work them on your own for additional practice. Challenge problems are at the end of the Assignment Exercises. Answers to the odd-numbered exercises are given at the end of the text, and worked-out solutions appear in a separate Student Solutions Manual available for purchase. Your instructor has the solutions to the even-numbered exercises in the Instructor's Resource Manual.

Concepts Analysis. Too often we focus on the *how to* and overlook the *why* of mathematical concepts. The Concepts Analysis questions further your understanding of a concept and help you see the connections between concepts. Some concepts questions present incorrect solutions to exercises to give you practice in analyzing and correcting errors. Error analysis also reinforces your understanding of concepts. As an added bonus, these exercises strengthen your writing skills. Suggested responses (answers) are found in the Instructor's Resource Manual.

Chapter Summary. Each chapter includes a summary in the form of a two-column chart. The first column lists the learning outcomes of the chapter. The second column gives the procedures and examples for each outcome. Page references are included to facilitate your preview or review of the chapter.

Trial Test. The trial test at the end of each chapter lets you check your understanding of the chapter learning outcomes. You should be able to work each problem without referring to any examples in your text or your notes. Take this test before you take

Six-Step Problem Solving Example

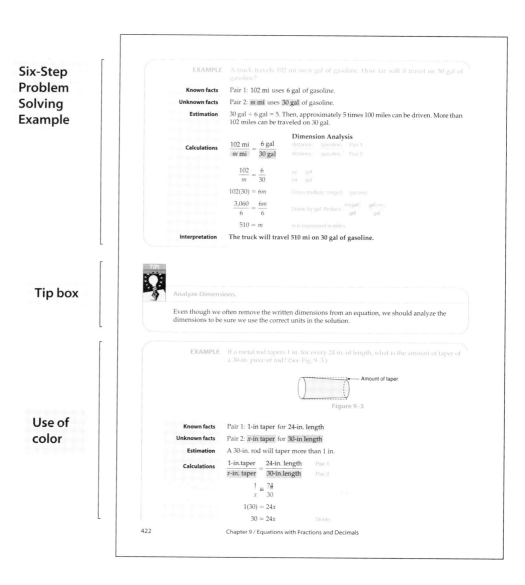

EXAMPLE A truck travels 102 mi on 6 gal of gasoline. How far will it travel on 30 gal of gasoline?

Known facts Pair 1: 102 mi uses 6 gal of gasoline.

Unknown facts Pair 2: m mi uses 30 gal of gasoline.

Estimation 30 gal ÷ 6 gal = 5. Then, approximately 5 times 100 miles can be driven. More than 102 miles can be traveled on 30 gal.

Calculations

Dimension Analysis

$$\frac{102 \text{ mi}}{m \text{ mi}} = \frac{6 \text{ gal}}{30 \text{ gal}}$$ distance ← gasoline Pair 1
distance ← gasoline Pair 2

$$\frac{102}{m} = \frac{6}{30}$$ mi gal
mi gal

$$102(30) = 6m$$ Cross multiply: mi(gal) gal (mi)

$$\frac{3{,}060}{6} = \frac{6m}{6}$$ Divide by gal. Reduce: $\frac{\text{mi.gal}}{\text{gal}}$ $\frac{\text{gal.mi}}{\text{gal}}$

$$510 = m$$ m is expressed in miles

Interpretation The truck will travel 510 mi on 30 gal of gasoline.

Tip box

TIP!

Analyze Dimensions.

Even though we often remove the written dimensions from an equation, we should analyze the dimensions to be sure we use the correct units in the solution.

Use of color

EXAMPLE If a metal rod tapers 1 in. for every 24 in. of length, what is the amount of taper of a 30-in. piece of rod? (See Fig. 9–3.)

— Amount of taper

Figure 9–3

Known facts Pair 1: 1-in taper for 24-in. length

Unknown facts Pair 2: x-in taper for 30-in length

Estimation A 30-in. rod will taper more than 1 in.

Calculations

$$\frac{1\text{-in.taper}}{x\text{-in. taper}} = \frac{24\text{-in. length}}{30\text{-in.length}}$$ Pair 1
Pair 2

$$\frac{1}{x} = \frac{24}{30}$$

$$1(30) = 24x$$

$$30 = 24x$$ Divide.

422 Chapter 9 / Equations with Fractions and Decimals

---

the class test to check and verify your understanding of the chapter material. Answers to the odd-numbered exercises appear at the end of the text, and their solutions appear in a separate Student Solutions Manual available for purchase. Your instructor has the solutions to the even-numbered exercises in the Instructor's Resource Manual.

Student Solutions Manual. This manual can be purchased at your college bookstore or from online bookstores. It gives you extra *learning insurance* to help you master learning outcomes in the text. The manual contains worked-out solutions to the odd-numbered exercises in the Assignment Exercises and the Chapter Trial Test for each chapter of the text. Answers to these exercises appear in the back of your text, but using the manual to study the worked-out solutions may reinforce your problem-solving skills and your understanding of the concepts.

How to Study Mathematics. Your instructor can obtain free copies of this booklet, which describes various learning techniques you can use to learn mathematics.

StudyWizard Software. This software, packaged with the text, provides additional practice with the math concepts presented in the text. Each question contains a reference to the section and outcome in the text where the concept appears, making it easier to find the sections you want to review. Immediate feedback is provided for all questions, allowing you to strengthen your skills and test your knowledge of the concepts. The glossary included on the software allows you to review the terms and concepts presented in the text.

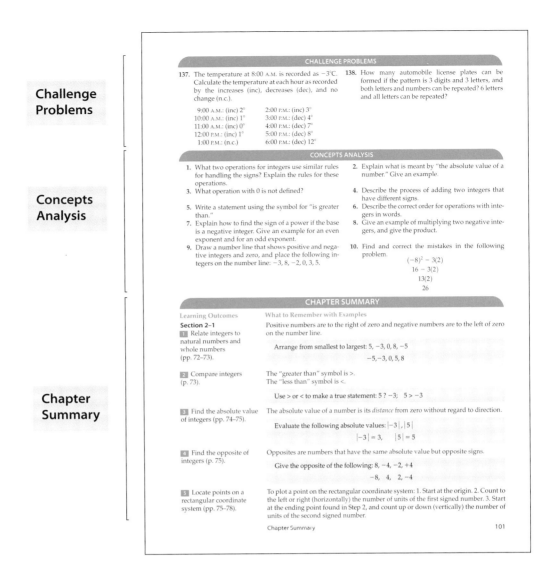

Challenge Problems

Concepts Analysis

Chapter Summary

CHALLENGE PROBLEMS

137. The temperature at 8:00 A.M. is recorded as −3°C. Calculate the temperature at each hour as recorded by the increases (inc), decreases (dec), and no change (n.c.).

9:00 A.M.: (inc) 2°	2:00 P.M.: (inc) 3°
10:00 A.M.: (inc) 1°	3:00 P.M.: (dec) 4°
11:00 A.M.: (inc) 0°	4:00 P.M.: (dec) 7°
12:00 P.M.: (inc) 1°	5:00 P.M.: (dec) 8°
1:00 P.M.: (n.c.)	6:00 P.M.: (dec) 12°

138. How many automobile license plates can be formed if the pattern is 3 digits and 3 letters, and both letters and numbers can be repeated? 6 letters and all letters can be repeated?

CONCEPTS ANALYSIS

1. What two operations for integers use similar rules for handling the signs? Explain the rules for these operations.
2. Explain what is meant by "the absolute value of a number." Give an example.
3. What operation with 0 is not defined?
4. Describe the process of adding two integers that have different signs.
5. Write a statement using the symbol for "is greater than."
6. Describe the correct order for operations with integers in words.
7. Explain how to find the sign of a power if the base is a negative integer. Give an example for an even exponent and for an odd exponent.
8. Give an example of multiplying two negative integers, and give the product.
9. Draw a number line that shows positive and negative integers and zero, and place the following integers on the number line: −3, 8, −2, 0, 3, 5.
10. Find and correct the mistakes in the following problem.
$$(-8)^2 - 3(2)$$
$$16 - 3(2)$$
$$13(2)$$
$$26$$

CHAPTER SUMMARY

Learning Outcomes	What to Remember with Examples
Section 2–1	
1 Relate integers to natural numbers and whole numbers (pp. 72–73).	Positive numbers are to the right of zero and negative numbers are to the left of zero on the number line.
	Arrange from smallest to largest: 5, −3, 0, 8, −5
	−5, −3, 0, 5, 8
2 Compare integers (p. 73).	The "greater than" symbol is >. The "less than" symbol is <.
	Use > or < to make a true statement: 5 ? −3; 5 > −3
3 Find the absolute value of integers (pp. 74–75).	The absolute value of a number is its *distance* from zero without regard to direction.
	Evaluate the following absolute values: $|-3|$, $|5|$
	$|-3| = 3$, $|5| = 5$
4 Find the opposite of integers (p. 75).	Opposites are numbers that have the same absolute value but opposite signs.
	Give the opposite of the following: 8, −4, −2, +4
	−8, 4, 2, −4
5 Locate points on a rectangular coordinate system (pp. 75–78).	To plot a point on the rectangular coordinate system: 1. Start at the origin. 2. Count to the left or right (horizontally) the number of units of the first signed number. 3. Start at the ending point found in Step 2, and count up or down (vertically) the number of units of the second signed number.

Chapter Summary   101

Companion Website.   This free website, available at www.prenhall.com/cleaves, provides even more practice with the math concepts presented in the form of short quizzes for each section of the text. These quizzes are immediately graded, and you have the opportunity to send the results to your instructor via email.

Online Course Material.   Passcodes for WebCT, Blackboard, and CourseCompass, as well as the Premium Website, are available for purchase. The system you use will depend on the software available at your school. These online resources provide numerous multiple-choice questions, including practice and chapter tests to review and check your comprehension. Short-answer quizzes, discussion questions, and review material complete the online packages.

We wish you much success in your study of mathematics. Many of the improvements for this book were suggested by students such as yourself. If you have suggestions for improving the presentation, please give them to your instructor or email the authors at ccleaves@bellsouth.net or mhobbs@watervalley.net.

Cheryl Cleaves
Margie Hobbs

Contents

Whole Numbers and Decimals

1

When we study a subject for the first time or review in some detail a subject we studied a while ago, we begin with the basics. Often, as we examine the basics of a subject, we discover—or rediscover—many pieces of useful information. In this sense, mathematics is no different from any other subject. We begin with a study of whole numbers and decimals and the basic operations that we perform with them.

1–1 | *Whole Numbers, Decimals, and the Place-Value System*

Learning Outcomes

1 Identify place values in whole numbers.

2 Read and write whole numbers in words, standard notation, and expanded notation.

3 Compare whole numbers.

4 Identify place values in decimal numbers.

5 Read and write decimal numbers.

6 Write fractions with power-of-10 denominators as decimal numbers.

7 Compare decimal numbers.

8 Round a whole number or a decimal number to a place value.

9 Round a whole number or a decimal number to a number with one nonzero digit.

Our system of numbers, the *decimal-number system*, uses 10 figures called *digits*: 0, 1, 2, 3, 4, 5, 6, 7, 8, 9. A *whole number* is made up of one or more digits. When a number contains two or more digits, each digit must be in the correct place for the number to have the value we intend it to have.

1 Identify Place Values in Whole Numbers.

Each place a digit occupies in a number has a value called a *place value*. Each place value *increases* as we move from *right* to *left* and each increase is 10 *times* the value of the place to the right. For example, the tens place is 10 times the ones place, the hundreds place is 10 times the tens place, and so on (Fig. 1–1).

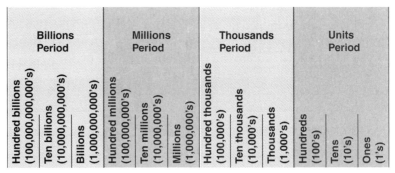

Billions Period			Millions Period			Thousands Period			Units Period		
Hundred billions (100,000,000,000's)	Ten billions (10,000,000,000's)	Billions (1,000,000,000's)	Hundred millions (100,000,000's)	Ten millions (10,000,000's)	Millions (1,000,000's)	Hundred thousands (100,000's)	Ten thousands (10,000's)	Thousands (1,000's)	Hundreds (100's)	Tens (10's)	Ones (1's)

Figure 1–1 Whole-number place values and periods.

The place values are arranged in *periods*, or groups of three. The first group of three is called *units*, the second group of three is called *thousands*, the third group is called *millions*, and the fourth group is called *billions*. Commas are used to mark off these periods. The commas make larger numbers easier to read because we can locate specific place values and interpret the number more easily. Each group of three digits has a hundreds place, a tens place, and a ones place.

In four-digit numbers, the comma separating the units period from the thousands period is optional. Thus, 4,575 and 4575 are both acceptable.

To identify the place value of a digit:

1. Mentally position the number on the place-value chart so that the last digit on the right aligns under the ones place.
2. Identify the place value of each digit according to its position on the chart.

EXAMPLE In the number 2,472,694,500, identify the place value of the digit 7 (Fig. 1–2).

Billions			Millions			Thousands			Units		
Hundred billions	Ten billions	Billions	Hundred millions	Ten millions	Millions	Hundred thousands	Ten thousands	Thousands	Hundreds	Tens	Ones
	2	4	7	2	6	9	4	5	0	0	

Position number on chart.
Identify the place value of 7.

Figure 1–2 Whole-number place-value chart.

┌─ ten-millions place

2,	4 7 2,	694,	500
Billions	Millions	Thousands	Units

7 is in the ten-millions place.

---

2 **Read and Write Whole Numbers in Words, Standard Notation, and Expanded Notation.**

To read whole numbers:

1. Mentally position the number on the place-value chart so that the last digit on the right aligns under the ones place.
2. Examine the number from right to left, separating each period with commas.
3. Identify the leftmost period.
4. From the left, read the numbers in each period and the period name. (The *units* period is not usually read. A period containing all zeros is not usually read.)

EXAMPLE Show how (a) 7543026129 and (b) 2000125 would be read by writing them in words.

Billions			Millions			Thousands			Units		
Hundred billions	Ten billions	Billions	Hundred millions	Ten millions	Millions	Hundred thousands	Ten thousands	Thousands	Hundreds	Tens	Ones
a.		7,	5	4	3,	0	2	6,	1	2	9
b.					2,	0	0	0,	1	2	5

Figure 1 – 3 Period names.

Mentally align the digits on the place-value chart (Fig. 1–3).

Starting at the right, separate each group of three digits with commas.

Identify the leftmost period.

Read numbers in each period and the period name.

Since in example (b) the thousands period contains all zeros, it is *not* read.

(a) **Seven** *billion*, **five hundred forty-three** *million*, **twenty-six** *thousand*, **one hundred twenty-nine**.
(b) **Two** *million*, **one hundred twenty-five**.

TIP!

Special Conventions with Reading and Writing Numbers

- A group name is inserted before each comma.
- The word *and* should not be used when reading whole numbers.
- The numbers from 21 to 99 (except 30, 40, 50, and so on) use a hyphen when they are written (forty-three, twenty-six, and so on).

There are times when we need to write spoken numbers. Numbers written with digits in the appropriate place-value positions are numbers in *standard notation*.

To write whole numbers in standard notation:

1. Write the period names in order from left to right, starting with the first period in the number.
2. Place the digits in each period. Leave blanks for zeros if necessary.
3. Insert zeros as needed for each period. Each period except the leftmost group must have three digits.
4. Separate the periods with commas as needed.

EXAMPLE   The exact cost of a group of airplanes is eight million, two hundred four thousand, twelve dollars. Write this amount as a number in standard notation.

Millions	Thousands	Units	
8	204	_12	Write period names in order.
8	204	012	Place digits in each period.
8,	204,	012	Insert zeros as needed.
			Separate periods with commas.

The cost is $8,204,012.

Learning Strategy   *Read and Illustrate Rules.*

Sometimes the language of mathematics and the details in procedures and rules seem overwhelming. Here are some suggestions that may help.

- Illustrate the rule with an example.
- Make your example with reasonable numbers.
- Reword the rule into your own words.
- Discuss your rule and illustration with a classmate or study partner.
- Be sure you understand the concept of the rule or procedure rather than memorizing meaningless words and steps.

To emphasize the value of a number, we may choose to write a number in *expanded notation.* Expanded notation shows the sum of each digit times its place value.

To write a whole number in expanded notation:

Write 516 in expanded notation.

1. Write each digit times its place value.

5×100
1×10
6×1

2. Write the expanded digits as indicated addition.

$(5 \times 100) + (1 \times 10) + (6 \times 1)$

EXAMPLE   A bank teller only has three types of currency (bills) in the cash drawer: $100, $10, and $1. Using the fewest number of bills, how many bills of each type are needed to cash a check for $243?

$(2 \times 100) + (4 \times 10) + (3 \times 1)$ Mentally visualize 243 in expanded notation.

Each place value in expanded notation represents a type of currency. Interpret the expanded notation as it relates to the type of money.

The cash includes two $100 bills, four $10 bills, and three $1 bills.

3 Compare Whole Numbers.

Whole numbers can be arranged on a *number line* to show a visual representation of the relationship of numbers by size. The most common arrangement is to begin with zero and place numbers on the line from left to right as they get larger.

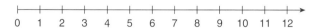

All numbers have a place on the number line and the numbers continue indefinitely without end. A term that is often used to describe this concept is *infinity* and the symbol is ∞.

Whole numbers can be compared by size by determining which of the two numbers is larger or smaller. If two numbers are positioned on a number line, the smaller number is positioned to the left of the larger number. The order relationship can be written in a mathematical statement called an *inequality*. An inequality shows that two numbers are not equal; that is, one is larger than the other. Symbols for showing inequalities are the *less than* symbol < and the *greater than* symbol >.

$5 < 7$ Five is less than seven.
$7 > 5$ Seven is greater than five.

To compare whole numbers:

1. Mentally position the numbers on a number line.
2. Select the number that is farther to the left to be the smaller number.
3. Write an inequality using the *less than* symbol.

smaller number < larger number

or

Write an inequality using the *greater than* symbol.

larger number > smaller number

TIP!

Which Way Does the Inequality Symbol Point?

In using an inequality symbol to relate two numbers, the point of the *less than* symbol is directed toward the *left* like the arrowhead on the *left* of the number line. The point of the *greater than* symbol is directed toward the *right* like the arrowhead on the *right* of the number line.

$$3 < 5 \quad 6 < 14 \quad 4 > 2 \quad 9 > 7$$

EXAMPLE   Write an inequality comparing the numbers 12 and 19:

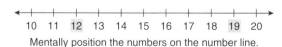

Mentally position the numbers on the number line.

12 is the smaller number.   12 is to the *left* of 19.
$12 < 19$ or $19 > 12$ Use appropriate inequality symbol.

Numbers are used to show *how many* and to show *order*. *Cardinal numbers* show *how many* and *ordinal numbers* show *order* or position (such as first, second, third, fourth, etc.). For example, in the statement "three students are doing a presentation," three is a cardinal number (showing how many). In the statement "Margaret is the third tallest student in the class," third is an ordinal number (showing order).

4 Identify Place Values in Decimal Numbers.

Numbers that are parts of a whole number are called *fractions*. In fraction notation, we write one number over another number.

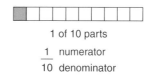

1 of 10 parts

1 numerator
10 denominator

The bottom number, the *denominator*, represents the number of parts that a whole unit contains. The top number, the *numerator*, represents the number of parts being considered.

A special type of fraction is called a *decimal fraction*. Other fractions will be examined in Chapter 3.

A decimal fraction is a fraction whose denominator is 10 or some power of 10, such as 100 or 1,000. Often the terms decimal fraction, decimal number, and decimal are used interchangeably. In fraction notation, 3 out of 10 parts is written as $\frac{3}{10}$. In decimal notation, the denominator 10 is not written but is implied by position on the place-value chart (Fig. 1–4). A decimal point (.) separates whole-number amounts on the left and fractional parts on the right. The fraction $\frac{3}{10}$ can be written in decimal notation as 0.3.

3 out of 10

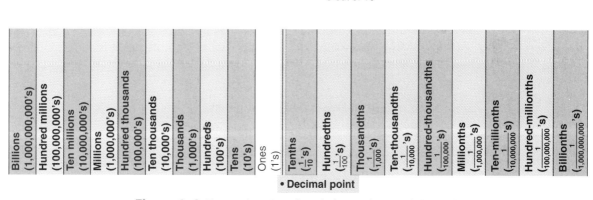

Figure 1–4 Place-value chart for whole numbers and decimals.

To extend the place-value chart to include parts of whole amounts, we place a decimal point (.) after the ones place. The place on the right of the ones place is called the *tenths place*. The *decimal point* is placed between the ones place and tenths place to distinguish between whole amounts and fractional amounts.

TIP!

Use of Commas and Periods in Numbers

The use of a period to separate the whole-number places from the decimal places is not a universally accepted notation. Some cultures use a comma instead. For example, our notation for writing 32,495.8 may be written as 32.495,8 or 32 495,8.

To identify the place value of digits in decimal fractions:

1. Mentally position the decimal number on the decimal place-value chart so that the decimal point of the number aligns with the decimal point on the chart.
2. Identify the place value of each digit according to its position on the chart.

EXAMPLE   Identify the place value of each digit in 32.4675 using the rule.

Align the number 32.4675 with the place-value chart (Fig. 1–5).

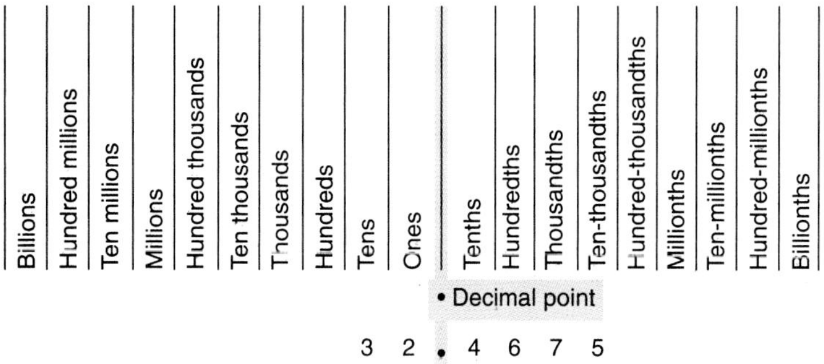

Billions	Hundred millions	Ten millions	Millions	Hundred thousands	Ten thousands	Thousands	Hundreds	Tens	Ones		Tenths	Hundredths	Thousandths	Ten-thousandths	Hundred-thousandths	Millionths	Ten-millionths	Hundred-millionths	Billionths

• Decimal point

3   2   ⟨•⟩   4   6   7   5

Figure 1–5   Place value chart for decimal numbers.

3 is in the tens place.
2 is in the ones place.
4 is in the tenths place.
6 is in the hundredths place.
7 is in the thousandths place.
5 is in the ten-thousandths place.

The place-value chart can be extended on the right side for smaller fractions in the same manner that it is extended on the left for larger numbers.

5   Read and Write Decimal Numbers.

Examine the place-value charts in Fig. 1–4 and Fig. 1–5. Notice the *th* on the end of each place value to the right of the decimal point. When reading decimals, this *th* indicates a decimal number. A decimal place value with two words is hyphenated (ten-thousandths, hundred-millionths). The word *and* indicates the decimal point.

To read decimal numbers:

1. Mentally align the number on the decimal place-value chart so that the decimal point of the number is directly under the decimal point on the chart.
2. Read the whole-number part.
3. Use *and* for the decimal point only if there is a whole-number part.
4. Read the decimal part the same way you read a whole-number part.
5. End by reading the *place value* of the rightmost digit in the decimal part.

EXAMPLE   Read 52.386 by writing it in words.

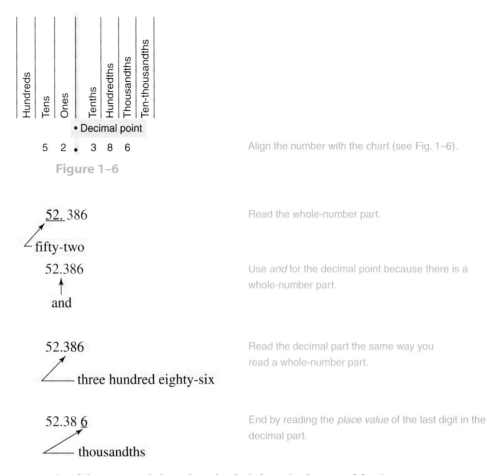

Figure 1–6

Align the number with the chart (see Fig. 1–6).

52. 386

fifty-two

Read the whole-number part.

52.386

and

Use *and* for the decimal point because there is a whole-number part.

52.386

three hundred eighty-six

Read the decimal part the same way you read a whole-number part.

52.38 6

thousandths

End by reading the *place value* of the last digit in the decimal part.

52.386 is "fifty-two and three hundred eighty-six thousandths."

Learning Strategy   *Breaking a Process into Steps*

Sometimes processes that involve several steps can be overwhelming. Some strategies for handling multi-stepped processes are:

• Examine each step of the process to identify processes that are already familiar.
  Look at Steps 1 and 2 of reading a decimal. They refer to skills that have already been learned. Step 3 gives specific instructions for connecting the steps. Step 4 applies the same skill as Step 2 to a different part of the number. Step 5 gives specific instructions for finishing the problem.
• Be sure that you can perform each step independently.
  Since both Steps 2 and 4 involve reading whole numbers, review reading whole numbers if necessary.
• Examine how the steps are connected.
  Steps 1 and 2 focus on the whole-number part of the number, and Steps 4 and 5 focus on the decimal part of the number. Step 3 connects the two parts.

EXAMPLE Read 0.0162 by writing it in words.

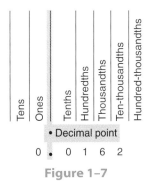

Align the number with the chart (see Fig. 1–7). There is no whole-number part to read.

Figure 1–7

0.0 <u>162</u>
 ↑
one hundred sixty-two

Read the decimal part like a whole number.

0.016 <u>2</u>
 ↑
ten-thousandths

End by reading the *place value* of the last digit in the decimal part. Use the hyphen in ten-thousandths.

0.0162 is "one hundred sixty-two ten-thousandths."

Informal Use of the Word *Point*

Informally, the decimal point is sometimes read as "point." Thus, 3.6 is read "three point six." The decimal 0.0162 can be read as "point zero one six two" or "zero point zero one six two." This informal process is often used in communication to ensure that numbers are not miscommunicated.

Unwritten Decimals

When we write whole numbers, using numerals, we usually omit the decimal point; the decimal point is understood to be at the end of the whole number. Therefore, any whole number, such as 32, can be written without a decimal (32) or with a decimal (32.).

6 Write Fractions with Power-of-10 Denominators as Decimal Numbers.

Fractions like $\frac{1}{10}$ and $\frac{75}{100}$ have denominators that are powers-of-10. Any fraction whose denominator is 10, 100, 1,000, 10,000, and so on, can be written as a decimal number without making any calculations. The decimal number will have the same number of digits in its decimal part as the fraction has zeros in its power-of-ten denominator.

To write a fraction whose denominator is 10, 100, 1000, 10,000, and so on, as a decimal:

1. Use the denominator to find the number of decimal places.

    $10 \rightarrow 1$ place
    $100 \rightarrow 2$ places
    $1{,}000 \rightarrow 3$ places
    $10{,}000 \rightarrow 4$ places

    Write $\frac{17}{1{,}000}$ as a decimal.

    0._ _ _   Three decimal places are needed.

2. Place the numerator so that the last digit is in the farthest place on the right.

    0._ 1 7

3. Fill in any blank spaces with zeros.

    0.017

EXAMPLE   Write $\frac{3}{10}, \frac{25}{100}, \frac{425}{100}$, and $\frac{3}{1{,}000}$ as decimal numbers.

$\frac{3}{10}$ is written **0.3**.   One decimal place

$\frac{25}{100}$ is written **0.25**.   Two decimal places

$\frac{425}{100}$ is written **4.25**.   Two decimal places

$\frac{3}{1{,}000}$ is written **0.003**.   Three decimal places

---

TIP!

Do Ending Zeros Change the Value of a Decimal Number?

When we attach zeros on the *right* end of a decimal number, we do not change the value of the number.

$$0.5 = 0.50 = 0.500 \qquad \frac{5}{10} = \frac{50}{100} = \frac{500}{1{,}000}$$

See equivalent fractions on page 119.

---

7   **Compare Decimal Numbers.**

As with whole numbers, we often need to compare decimals by size. To make valid comparisons, we must compare like amounts. Whole numbers compare with whole numbers, tenths compare with tenths, thousandths compare with thousandths, and so on. One way to compare decimals is to make the decimal parts have the same number of places by attaching zeros at the right end. For instance, to compare 0.7 and 0.68, we change 0.7 to 0.70. Now, which is larger, 0.70 or 0.68? 0.70 is larger. We can also compare digits place by place.

To compare decimal numbers:

1. Compare whole-number parts.
2. If the whole-number parts are equal, compare digits place by place, starting at the tenths place and moving to the right.
3. Stop when two digits in the same place are different.
4. The digit that is larger determines the larger decimal number.

EXAMPLE   Compare the numbers 32.47 and 32.48 to see which is larger.

32.4 7   Look at the whole-number parts: They are the same.
32.4 8   Look at the tenths place for each number. Both numbers have a 4 in the tenths place.
          Look at the hundredths place. They are different and 8 is larger than 7.

32.48 is the larger number.

EXAMPLE   Write two inequalities for the numbers 0.4 and 0.07.
Since the whole-number parts are the same (0), we compare the digits in the tenths place. 0.4 is larger because 4 is larger than 0.

$$0.4 > 0.07 \quad \text{or} \quad 0.07 < 0.4$$

Another procedure for comparing decimals is to affix an appropriate number of zeros to make each number have the same number of decimal places.

$$0.4 \; = \; 0.40 \qquad \frac{4}{10} = \frac{40}{100}$$

Now, compare 0.40 and 0.07. The larger number is 0.40. Then, 0.40 > 0.07.

TIP!

Common Denominators in Decimals

The denominator of an equivalent decimal fraction is determined by the number of decimal places in a number. Decimal fractions have a common denominator if they have the same number of digits to the right of the decimal point. See common denominators of fractions on page 127.

8   **Round a Whole Number or a Decimal Number to a Place Value.**

Rounding a number means finding the closest *approximate number* to a given number. For example, if 37 is rounded to the nearest ten, is 37 closer to 30 or 40? Locate 37 on the number line.

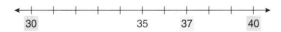

37 is closer to 40 than 30. Thus, 40 is a better approximation to the nearest ten for 37. Another way to say this is that 37 rounded to the nearest ten is 40.

When rounding a number to a certain place value, we must make sure that we are as accurate as our employer wants us to be. Generally, the size of the number and its use dictates the decimal place to which it should be rounded.

Learning Strategy   *Procedures Do Not Substitute for Understanding Concepts.*

Once you understand that rounding means to find the closest approximate number to a given number, we can use procedures or rules to facilitate rounding. Procedures are meaningless and are easily confused or forgotten if you do not understand the concept.

To round a whole or decimal number to a given place value:

1. Locate the digit that occupies the rounding place. Then examine the digit to the immediate right.
2. If the digit to the right of the rounding place is 0, 1, 2, 3, or 4, do not change the digit in the rounding place. If the digit to the right of the rounding place is 5, 6, 7, 8, or 9, add 1 to the digit in the rounding place.
3. Replace all digits to the *right* of the digit in the rounding place with zeros if they are to the left of the decimal point. Drop digits to the right of the digit in the rounding place *and also* to the right of the decimal point.

EXAMPLE   Oregon has a land area of 96,187 square miles. What would be a reasonable approximate number for this land area?

96,187 rounds to the following approximate numbers:

  96,190 to the nearest ten
  96,200 to the nearest hundred
  96,000 to the nearest thousand
100,000 to the nearest ten-thousand

Deciding to which place to round a value is a judgment depending on what use you will make of the rounded or approximate value.

Both 96,000 and 100,000 are reasonable approximations.

Learning Strategy   *How Do I Choose a Rounding Place?*

If the directions do not give a specific rounding place, choose a place that is reasonable for the context of the problem. We illustrate this with two different situations:

If we are comparing the land area of Oregon (96,187 square miles) with the land area of Wyoming (96,988 square miles), Oregon has approximately 96,000 square miles, while Wyoming has approximately 97,000 square miles.

If we are comparing the land area of Oregon (96,187 square miles) with the land area of Texas (262,015 square miles), Oregon has approximately 100,000 square miles, while Texas has approximately 300,000 square miles.

EXAMPLE   Round 46.897 to the hundredths place.

46.8⑨7   9 is in the hundredths place.
46.8⑨7   The next digit to the right is 7, so add 1 to 9. (9 + 1 = 10 and 89 + 1 = 90.)
          Drop the 7 in the thousandths place because it is to the right of the rounding place and also to the right of the decimal.

46.90

Nine Plus One Still Equals Ten

When the digit in the rounding place is 9 and must be rounded up, it becomes 10. The 0 replaces the 9 and 1 is carried to the next place to the left.

EXAMPLE   Round 32.6 to the tens place.

③2.6   3 is in the tens place.

③2.6   Leave 3 unchanged because 2, the next digit to the right, is less than 5. Replace 2 with a

30   zero because it is to the left of the decimal. Drop the 6 because it is to the right of the decimal.

EXAMPLE   Round $293.48 to the nearest dollar.

$29③.48   When we round to the nearest dollar, we are rounding to the *ones* place.

$293

EXAMPLE   Round $71.8986 to the nearest cent.

$71.8⑨86   One cent is 1 hundredth of a dollar, so to round to the nearest cent is to round to the
              hundredths place.

$71.90

---

9 **Round a Whole Number or a Decimal Number to a Number with One Nonzero Digit.**

We can also round so that the rounded number has only one digit that is not a zero. The only difference in this type of rounding and rounding to a specified place value is that the first nonzero digit determines the rounding place.

Look at the following numbers. Notice that the *first* digit on the left is a *nonzero digit* (not a zero); the other digits are all zeros.

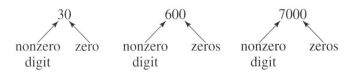

30   600   7000
nonzero zero   nonzero zeros   nonzero zeros
digit   digit   digit

Learning Strategy   *Cumbersome Terminology*

Why do we use the terminology *one nonzero digit* or *first nonzero digit* instead of just saying "the first digit on the left"? All whole numbers begin with a digit that is not a zero.

In a decimal number like 0.00738, the *first nonzero digit*, 7, is not the first digit on the left.

1. Starting at the *left*, find the *first* digit that is not zero. This digit will be a 1, 2, 3, 4, 5, 6, 7, 8, or 9.
2. Round the number to the place value of the first nonzero digit.
3. Replace all digits to the right of the rounding place with zeros up to the decimal point. Drop any digits that are to the right of the digit in the rounding place *and* that also follow the decimal point.

EXAMPLE   Round 78.4 to one nonzero digit.

78.4 ↑	Find the first nonzero digit from the left.
⑦8.4	The first nonzero digit is 7. Round to the place value of the first nonzero digit.
⑦8̲.4	Add 1 to 7 because the next digit to the right is 5 or more.
8_.	Replace each digit between the rounded digit and the decimal point with a zero. Drop all digits after the decimal point.
80	

EXAMPLE   Round 0.083 to one nonzero digit.

0.0⑧3̲	Locate the first nonzero digit. Round to hundredths.
0.08	

TIP!

Exact Amount Versus Approximate Amount

When an amount has been rounded, it is no longer an exact amount. The rounded amount is now an *approximate amount*.

SELF-STUDY EXERCISES 1–1

1 In the number 2,304,976,186, identify the place value of these digits:

1. 3   **2.** 7   **3.** 1   **4.** 0   **5.** 2

In the number 8,972,069,143, identify the place value of these digits:

6. 0   **7.** 4   **8.** 7   **9.** 8   **10.** 6

Show how these numbers are read by writing them as words.

11. 6704   **12.** 89021   **13.** 662900714   **14.** 3000101   **15.** 15407294376   **16.** 150

Write these words as numbers in standard notation. Use commas when necessary.

17. Seven billion, four hundred

18. One million, six hundred twenty-seven thousand, one hundred six

19. Fifty-eight thousand, two hundred one

20. In a telephone conversation, a contractor submitted the following bid for a job: "one thousand six dollars." Write this bid as a number.

Write in expanded notation.

21. 718   **22.** 42   **23.** 1,983   **24.** 8,021   **25.** 52,010   **26.** 700

Write as ordinary numbers.

27. $(5 \times 100) + (3 \times 10) + (7 \times 1)$
28. $(9 \times 1,000) + (0 \times 100) + (0 \times 10) + (8 \times 1)$
29. $(6 \times 10) + (5 \times 1)$
30. $(4 \times 10,000) + (0 \times 1,000) + (2 \times 100) + (1 \times 10) + (7 \times 1)$

3 Write two inequalities to compare each pair of numbers.

31. 6 and 8   **32.** 42 and 32
33. 196 and 148   **34.** 2,802 and 2,517
35. 7,809 and 8,902   **36.** 44,000 and 42,999

4

37. What is the place value of 7 in 32.407?   **38.** What is the place value of 8 in 28.396?
39. What is the place value of 4 in 3.00254?   **40.** What is the place value of 3 in 457.2096532?
41. What is the place value of the 7 in 0.0387?   **42.** What is the place value of 5 in 0.0235?

In the following problems, state what digit is in the place indicated.

43. Tens: 46.3079   **44.** Tenths: 2.0358
45. Thousandths: 520.0765   **46.** Ten-millionths: 0.002178356
47. Hundredths: 402.3786   **48.** Thousands: 8,023, 596.9

5 Write the words for the decimal numbers.

49. 21.387   **50.** 420.059   **51.** 0.89   **52.** 0.0568   **53.** 30.02379   **54.** 21.205085

Write the digits for the word numbers.

55. Three and forty-two hundredths

56. Seventy-eight and one hundred ninety-five thousandths

57. Five hundred and five ten-thousandths

58. Seventy-five thousand thirty-four hundred-thousandths

6 Write the fractions as decimal numbers.

59. $\dfrac{5}{10}$ **60.** $\dfrac{23}{100}$ **61.** $\dfrac{7}{100}$ **62.** $\dfrac{683}{100}$ **63.** $\dfrac{79}{1000}$

64. $\dfrac{468}{1000}$ **65.** $\dfrac{587}{100}$ **66.** $\dfrac{108}{1000}$ **67.** $\dfrac{603}{100}$ **68.** $\dfrac{400}{100}$

7 Compare the number pairs and identify the larger number.

69. 3.72, 3.68   **70.** 7.08, 7.06   **71.** 0.23, 0.3   **72.** 0.56, 0.5   **73.** 2.75, 2.65   **74.** 0.157, 0.2

Compare the number pairs and identify the smaller number.

75. 8.9, 8.88   **76.** 0.25, 0.3   **77.** 0.913, 0.92
78. 0.761, 0.76   **79.** 5.983, 6.98   **80.** 1.972, 1.9735

Arrange the numbers in order from smallest to largest.

81. 0.23, 0.179, 0.314   **82.** 1.9, 1.87, 1.92   **83.** 72.1, 72.07, 73

84. Two micrometer readings are recorded as 0.837 in. and 0.81 in. Which is larger?

85. A micrometer reading for a part is 3.85 in. The specifications call for a dimension of 3.8 in. Which is larger, the micrometer reading or the specification?

86. A washer has an inside diameter of 0.33 in. Will it fit a bolt that has a diameter of 0.325 in.?

87. Aluminum sheeting can be purchased in thicknesses of 0.04 in. or 0.035 in. Which sheeting is thicker?

88. If No. 14 copper wire has a diameter of 0.064 in., and No. 10 wire has a diameter of 0.09 in., which has the larger diameter?

89. A biologist recorded the weights of two endangered animals as 14.8 kilograms and 16.2 kilograms. Which weight is greater?

Write two inequalities for each pair of numbers.

90. 4.2 and 3.8

91. 1.68 and 1.6

92. 7.09 and 7.21

93. 52.97 and 98.4

94. 2.742 and 27.42

95. 15,000 and 13,481.2

96. Cleveland is 119 miles from Toledo, Ohio, and Detroit is 62 miles from Toledo. Write two inequalities that compare these distances. Write a statement about the relative distances of Detroit and Cleveland from Toledo. Compare your statement with those of your classmates.

97. The average July temperature for Laredo, Texas, is 96°F, and the average July temperature for Eugene, Oregon, is 82°F. Write two inequalities to compare the temperatures in these two cities. Write a statement in words to illustrate one of your inequalities.

98. Spokane, Washington, has an average January temperature of 26°F and an average July temperature of 70°F. Write two inequalities to compare Spokane's winter and summer temperatures.

99. San Diego's average January temperature is 57°F, and its average July temperature is 71°F. Write two inequalities to compare San Diego's winter and summer temperatures.

100. The Mexico City earthquake measured 8.1 on the Richter scale, and the 1906 San Francisco earthquake measured 8.3 on the Richter scale. Which earthquake had the greater Richter scale rating?

101. In 1980, the average annual population growth rate for the United States was 1.14, and in 1990, it was 1.02. From these two figures, would you say the growth rate for the United States appears to be rising or falling?

102. For the 1990–2000 decade, the average annual growth rate was 1.7 for Asia and 2.0 for the developing world. Is Asia growing faster or slower than the developing world?

103. If you roll a pair of dice, the probability of rolling a sum of 5 is 0.111 and the probability of rolling a sum of 6 is 0.139. Which sum has the greater probability of being rolled?

104. A 100-watt bulb that burns continuously for two minutes uses 0.003 kilowatt-hours of electricity, and a 800-watt toaster uses 0.026 kilowatt-hours for a piece of toast. Which uses the greater number of kilowatt-hours?

105. To change centimeters to inches, multiply by 0.394, and to change kilometers to miles, multiply by 0.621. Which factor is larger?

8 Round to the place value indicated.

106. Nearest hundred: 468

107. Nearest hundred: 6,248

108. Nearest thousand: 8,263

109. Nearest ten thousand: 429,207

110. Nearest thousand: 39,748

111. Nearest ten thousand: 39,748

112. Nearest million: 285,487,412

113. Nearest ten: 468

114. Nearest billion: 82,629,426,021

115. Nearest ten million: 297,384,726

The *Rand McNally Road Atlas and Travel Guide* gives the altitudes of the following cities:

Corpus Christi, Texas   35 feet above sea level
Denver, Colorado   5,280 feet above sea level
Jacksonville, Florida   19 feet above sea level

Indianapolis, Indiana   717 feet above sea level
Salt Lake City, Utah   4,260 feet above sea level

116. Use your knowledge of rounding to compare the approximate altitudes of Corpus Christi and Jacksonville. Indicate the rounding place you used.

117. To what place should you round to compare the approximate altitudes of Denver and Salt Lake City? Round to show the comparison of approximate altitudes.

118. To what place should you round to show the approximate comparison of altitudes of Indianapolis and Salt Lake City? Round to show the comparison of approximate altitudes.

119. Discuss your decisions for selecting a place value for rounding when comparing the approximate altitudes of the three pairs of cities in Exercises 116–118.

Round to the nearest whole number.

120. 42.7   **121.** 367.43   **122.** 7.983   **123.** 103.06   **124.** 2.9

Round to the nearest tenth.

125. 8.05   **126.** 12.936   **127.** 42.574   **128.** 83.23   **129.** 5.997

Round to the nearest hundredth.

130. 7.036   **131.** 42.0657   **132.** 0.79239   **133.** 3.198051   **134.** 7.7735

Round to the nearest thousandth.

135. 0.21732   **136.** 0.01962   **137.** 1.508567   **138.** 4.2378135   **139.** 7.00394

Round to the nearest dollar.

140. $219.46   **141.** $82.93   **142.** $507.06   **143.** $2.83   **144.** $5.96

Round to the nearest cent.

145. $8.237   **146.** $0.291   **147.** $0.528   **148.** $5.796   **149.** $238.9238

150. A micrometer measure is listed as 0.7835 in. Round this measure to the nearest thousandth.

151. To the nearest tenth, what is the current of a 2.836-amp motor?

152. The diameter of an object is measured as 3.817 in. If specifications call for decimals to be expressed in hundredths, write this measure according to the specifications.

153. The average of four numbers is calculated to be 86.9735 but should be rounded to the nearest whole number. What is the rounded number?

9 Round to numbers with one nonzero digit.

154. 483   **155.** 7.89   **156.** 62.5   **157.** 0.537   **158.** 0.0086

159. 0.095   **160.** 3.07   **161.** 1.52   **162.** 83.09   **163.** 52.8

164. If round steak costs $2.78 per pound, what is the cost per pound to the nearest dollar?

165. An estimate calls for converting 23.077 to an approximate number. What is the approximate number rounded to one nonzero digit?

166. Monthly rainfall (in inches) was identified for these months: January, 0.355; March, 1.785; May, 0.45; July, 1.409; September, 0.07; and December, 2.018. Identify the months with the most similar rainfall by rounding to one nonzero digit.

167. The average response times (in seconds) for drivers braking when they first see a road hazard were measured as follows: driver A, 0.0275; driver B, 0.0264; driver C, 0.0234; driver D, 0.0284; and driver E, 0.0379. Round each response time to one nonzero digit to identify the drivers whose response time was most similar.

1-2 | Adding Whole Numbers and Decimal Numbers

Learning Outcomes

1 Add whole numbers.

2 Add decimal numbers.

3 Estimate and check addition.

Our purpose in this section is to understand addition and the basic properties that give us flexibility in the way we add. These properties will be used again and again throughout our study of mathematics.

1 Add Whole Numbers.

In an addition problem, the numbers being added are called *addends* and the answer is called the *sum* or *total*.

    Two useful properties to know about addition are that it is *commutative* and *associative*. By *commutative*, we mean that it does not matter in what *order* we add numbers. We can add 7 and 6 in any order and still get the same answer:

$$7 + 6 = 13 \qquad 6 + 7 = 13$$

By *associative*, we mean that we can *group* numbers together any way we want when we add and still get the same answer. To add $7 + 4 + 6$, we can group the $4 + 6$ to get 10; then we add the 10 to the 7:

$$7 + (4 + 6) = 7 + 10 = 17$$

Or we can group $7 + 4$ to get 11; then we add the 6 to the 11:

$$(7 + 4) + 6 = 11 + 6 = 17$$

Addition is a *binary operation*; that is, the rules of addition apply to adding *two* numbers at a time. The associative property of addition shows how addition is extended to more than two numbers.

Learning Strategy   *Symbolic Representation of Properties*

If we don't allow ourselves to be intimidated by symbols, we can develop a genuine appreciation for the symbolic representation of properties. Sometimes the simplest property can be very awkward or cumbersome to write in words.

To develop your mathematics communication skills, we suggest that as often as possible, write properties *both* symbolically and in words.

TIP!

Using Symbols to Write Rules and Definitions

Many rules and definitions can be written symbolically. Symbolic representation allows a quick recall of the rule or definition.

Commutative Property of Addition: Two numbers may be added in any order and the sum remains the same.

$$a + b = b + a \text{ where } a \text{ and } b \text{ are numbers.}$$

Associative Property of Addition: Three numbers may be added using different groupings and the sum remains the same.

$$a + (b + c) = (a + b) + c \text{ where } a, b, \text{ and } c \text{ are numbers.}$$

The associative property of addition also allows other possible groupings and extends to more than three numbers.

$$7 + 4 + 6 = 13 + 4 = 17$$
$$3 + 5 + 7 + 9 = 8 + 16 = 24$$

EXAMPLE   Add $7 + 3 + 6 + 8 + 2 + 4$ using the commutative and associative properties.

$$
\begin{array}{l}
\left.\begin{array}{l} 7 \\ 3 \end{array}\right\} 10 \\
 6 \\
\left.\begin{array}{l} 8 \\ 2 \end{array}\right\} 10 \\
\underline{+4} \\
 \mathbf{30}
\end{array}
$$

Group $7 + 3$, $8 + 2$, and $6 + 4$. Then add the three groups ($10 + 10 + 10 = 30$).

Adding zero to any number results in the same number. This property is called the *zero property of addition* and zero is called the *additive identity*.

Zero property of addition:

Adding zero to any number results in the same number.

$$n + 0 = n \quad \text{or} \quad 0 + n = n$$
$$5 + 0 = 5 \quad \text{or} \quad 0 + 5 = 5$$

When adding numbers of two or more digits, the same place values must be aligned under one another so that all the ones are in the far right column, all the tens in the next column, and so on.

To add numbers of two or more digits:

1. Arrange the numbers in columns so that the ones place values are in the same column.
2. Add the ones column, then the tens column, then the hundreds column, and so on, until all the columns have been added. *Carry* whenever the sum of a column is more than one digit. This is also called *regrouping*.

EXAMPLE   Shipping fees are often charged by the total weight of the shipment. Find the total weight of this order: nails, 250 pounds (lb); tacks, 75 lb; brackets, 12 lb; and screws, 8 lb. Arrange in columns and add.

$$
\begin{array}{r}
11 \\
250 \\
75 \\
12 \\
+\ \ 8 \\
\hline
345
\end{array}
$$

The sum of the digits in the right column is 15. Record the 5 in the ones column and *carry* the 1 to the tens column.

Add the tens column. $1 + 5 + 7 + 1 = 13$. Carry the 1.

The total weight is 345 lb.

---

2 Add Decimal Numbers.

When adding, decimal numbers are aligned so that all decimal points fall in the same vertical line. Aligning decimal points has the same effect as using *like* denominators when adding (or subtracting) fractions. Like denominators are discussed in Chapter 3.

To add decimals:

1. Arrange the numbers so that the decimal points are in one vertical line.
2. Add each column.
3. Align the decimal for the sum in the same vertical line.

EXAMPLE Add 42.3 + 17 + 0.36.

```
42.3
17        Note that the decimal in 17 is understood to be at the right end.
0.36

42.30     If we prefer, we may write each number so that all have the same number of decimal places
17.00     by attaching zeros on the right.
0.36
59.66
```

3 Estimate and Check Addition.

Estimating is an increasingly important skill to develop and it depends in part on your *number sense*. You aren't born with number sense. It must be developed. We develop and strengthen our number sense by estimating and doing mental calculations. In making calculations it is important to estimate and check your work.

To estimate the answer to an addition problem:

1. Round each addend to a specific place value or to a number with one nonzero digit.
2. Add the rounded addends.

To check an addition problem:

1. Add the numbers a second time and compare with the first sum.
2. Use a different order or grouping if convenient.

EXAMPLE Elston Home Renovators spent the following amounts on a job: $16,466.15, $23,963.10, and $5,855.20. Estimate the total amount by rounding to thousands. Then find the exact amount and check your answers.

Thousands Place	Estimate	Exact	Check
$16,466.15	$16,000	$16,466.15	$16,466.15
23,963.10	245,000	23,963.10	23,963.10
5,855.20	6,000	5,855.20	5,855.20
	$46,000	$46,284.45	$46,284.45

The estimate and exact answer are close. The exact answer is reasonable.

TIP!

Accuracy of Estimates

The accuracy of estimates varies depending on the place value to which the addends are rounded. The context of the problem will help in determining a reasonable rounding place.

52. A patient's normal body temperature registered at 98.2° and his temperature rose 2.7°. What was his increased body temperature?

53. Your investment portfolio totals $25,915.53 at the beginning of the year and it increases by $2,418.48 over the one-year period. What is the value of your portfolio at the end of the year?

54. eBay is an Internet auction house whose stock is traded publicly. When it was first traded, its stock sold at $8.4375 a share, then immediately increased by $215.8425 a share. What was the new price per share?

55. eBay had 1.2 million users in fall 2003, and by spring 2004, that number had increased by 2.6 million users. How many people were using eBay in early 2004?

56. If your college tuition is $1,564.50 and books cost $624.95, what is the total cost of your tuition and books?

57. Carlee's car odometer registered 36,263.7, then she drove 638.2 miles. What should her odometer have registered at the end of the trip?

58. Lucia purchased groceries costing $5.15, $1.84, $0.59, $7.21, and $1.47 for a party. What is the total cost of her purchases?

59. Caroline sets up a booth once a month at a flea market. For the most recent market, she accepted checks in the amounts of $85.50, $21.75, $13.83, $62.37, and $121.52. What is the total of these checks?

Estimate the sum by rounding to hundreds. Then find the exact sum and check.

60.
```
    6,367.32
   24,036.95
+     582.14
```

61.
```
      532.73
    1,823.14
+   2,384.93
```

62.
```
    2,546.28
    3,626.13
      831.86
+   6,153.15
```

63. On a trip from Albuquerque, New Mexico, to Blowing Rock, North Carolina, Lynette and Tosha purchased 14.2, 12.9, 15.3, 9.2, 8.6, and 12.1 gallons of fuel. Estimate the amount of fuel they purchased by rounding to the nearest ten; then find the exact amount of fuel and compare the estimate with the exact amount.

64. A dietitian purchased apples for $21.87, lettuce for $12.18, steak for $45.97, and tomatoes for $32.15. What is the total of the purchases?

3 Estimate the sum by rounding to hundreds. Then find the exact sum and check.

65.
```
    4,256.65
    3,892.10
      576.46
+   8,293.00
```

66.
```
    34.07
    15.962
     5.81
+    0.523
```

67.
```
    52,843
    17,497
    13,052
+      821
```

68.
```
      0.935
     12.4
    152.07
+    18
```

69.
```
    24,003
     5,874
   319,467
+   52,855
```

70.
```
    24.381
     1.1
    17.92
+   38
```

71. Palmer Associates provided the following prices for items needed to build a sidewalk: concrete, $2,583.45; wire, $43.25; frame material, $18.90; labor, $798. Estimate the cost by rounding each amount to the nearest ten. Find the exact total.

72. Antonio's expenses for one semester are: food, $1,500; lodging, $1,285; books, $288; supplies, $130; transportation, $162. Estimate his expenses by rounding each amount to the nearest hundred. Calculate the exact amount.

Estimate the sums by rounding addends to numbers with one nonzero digit. Then find the exact sum and check.

73.
```
      940
+   8,299
```

74.
```
    478.125
+   146.055
```

75.
```
    1,901
+   6,548
```

76.
```
    149.25
+   652.14
```

77.
```
    16,259
+   36,542
```

78.
```
    32,501
+   16,740
```

79. A hardware store filled an order for nails: 25 lb, $2\frac{1}{2}$-in. common; 16 lb, 4-in. common; 12 lb, 2-in. siding; 24 lb, $2\frac{1}{2}$-in. floor brads; 48 lb, 2-in. roofing; and 34 lb, $2\frac{1}{2}$-in. finish. Find the total weight.

80. If four containers have a capacity of 12 gal, 27 gal, 55 gal, and 21 gal, can 100 gal of fuel be stored in these containers? (Find the total capacity of the containers first.)

81. A printer has three printing jobs that require the following numbers of sheets of paper: 185, 83, and 211. Will one ream of paper (500 sheets) be enough to finish the three jobs?

82. How many feet of fencing are needed to enclose the area shown in Fig. 1–8?

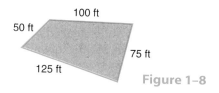

Figure 1–8

Estimate the sums by rounding addends to numbers with one nonzero digit. Then use a calculator to find the exact sum.

83. $47,287 + 33,409 + 81,496 + 28,594$

84. $387,483 + 879,583 + 592,801$

85. $31,592 + 8,584.6 + 13,215.05 + 968$

86. $1,328,591 + 35,803,502 + 10,387,921$

87. Mario's Restaurant had the following daily sales: Sunday, $3,842.95; Monday, $1,285.68; Tuesday, $1,195.57; Wednesday, $1,843.76; Thursday, $1,526.47; Friday, $2,984.89; and Saturday, $4,359.72. Find the total sales for the week.

88. A pharmacy tracked sales of its prescription drugs for 3 months and recorded $82,915.21, $78,529.97, and $97,095.00. What was the sales total for the 3 months?

1–3 | *Subtracting Whole Numbers and Decimals*

Learning Outcomes

1 Subtract whole numbers.

2 Subtract decimal numbers.

3 Estimate and check subtraction.

Subtraction is the *inverse operation* of addition. In addition, we add numbers to get their total (such as $5 + 4 = 9$), but to solve the subtraction problem $9 - 5 = ?$ we ask, "What number must be added to 5 to give us 9?" The answer is 4 because 4 added to 5 gives a total of 9. When we subtract two numbers, the answer is called the *difference* or *remainder*. The initial quantity is the *minuend*. The amount being subtracted from the initial quantity is the *subtrahend*.

1 **Subtract Whole Numbers.**

Subtraction is *not* commutative. $8 - 3 = 5$, but $3 - 8$ does not equal 5; that is, $3 - 8 \neq 5$. The symbol $\neq$ is read "is not equal to."

Subtraction is not associative.

EXAMPLE Show that $9 - (5 - 1)$ does not equal $(9 - 5) - 1$.

$$9 - (5 - 1) = 9 - 4 = 5, \quad \text{but} \quad (9 - 5) - 1 = 4 - 1 = 3$$

When subtracting two or more numbers, if no grouping symbols are included, perform subtractions from left to right.

EXAMPLE Subtract $8 - 3 - 1$.

$$8 - 3 - 1 = 5 - 1 = 4 \quad \text{Subtract from left to right.}$$

Subtracting zero from a number results in the same number:

$$n - 0 = n, \quad 7 - 0 = 7$$

Subtraction and Zeros

Subtracting a number from zero is not the same as subtracting zero from a number; that is, $7 - 0 = 7$, but $0 - 7$ does not equal 7.

To subtract numbers of two or more digits:

1. Arrange the numbers in columns, with the minuend at the top and the subtrahend at the bottom.
2. Make sure the ones digits are in a vertical line on the right.
3. Subtract the ones column first, then the tens column, the hundreds column, and so on.
4. To subtract a larger digit from a smaller digit in a column, *borrow* 1 from the digit in the next column to the left. This is the same as borrowing *one* group of 10; thus, add 10 to the digit in the given column, Then, continue subtracting. The concept of *borrowing* is also referred to as *regrouping*.

EXAMPLE   Subtract $9,327 - 3,514$.
Arrange in columns.

$$
\begin{array}{r}
{\scriptstyle 8\,13} \\
9{,}327 \\
-\ 3{,}514 \\
\hline
5{,}813
\end{array}
$$

Arrange in columns.
In the hundreds place 5 is more than 3. Borrow 1 group of 10 from 9. $9 - 1 = 8$, $10 + 3 = 13$.

Words and Phrases That Imply Subtraction

These phrases indicate subtraction in applied problems:

how many are left	how many more
how much less	how much larger
how much smaller	

Also, some applied problems require more than one operation. Many problems that require more than one operation involve parts and a total. If we know the total and all the parts but one, we can add all the known parts and subtract the result from the total to find the missing part.

EXAMPLE The Froehlichs left Memphis and drove 356 miles on the first day of their vacation. They drove 426 miles on the second day. If they are traveling to Albuquerque, which is 1,050 miles from Memphis, how many more miles do they have to drive?

The phrase, *how many more*, indicates subtraction.

$$1,050 \text{ miles} = \text{total miles}$$
$$356 + 426 + \text{miles left to drive} = 1,050 \text{ miles}$$

To find the miles left to drive, add $356 + 426$ and subtract the result from 1,050.

$$356 + 426 = 782 \qquad 1,050 - 782 = 268$$

2 **Subtract Decimal Numbers.**

When we subtract decimals, we align the digits just like when we add decimals.

To subtract decimals:

1. Arrange the numbers so that the decimal points align.
2. Subtract each column beginning at the right.
3. Interpret blank places as zeros.
4. Place the decimal in the difference in the same vertical line.

EXAMPLE Subtract 7.18 from 15.

Take care to align the decimals properly.

$$\begin{array}{r} 15. \\ - \ \ 7.18 \end{array}$$
Because 15 is a whole number, its decimal point is placed after the 5.

$$\begin{array}{r} 15.00 \\ - \ \ 7.18 \\ \hline 7.82 \end{array}$$
To subtract, we put zeros in the tenths and hundredths places of 15 and then borrow.

When a worker machines an object using a blueprint as a guide, a certain amount of variation from the blueprint specification is allowed for the machining process. This variation is called the *tolerance*. Thus, if a blueprint calls for a part to be 9.47 in. with a tolerance of ± 0.05 (read plus or minus five hundredths), this means that the actual part can be 0.05 in. *more* or 0.05 in. *less* than the specification. To find the *largest* acceptable size of the object, we add $9.47 + 0.05 = 9.52$ in. To find the *smallest* acceptable size of the object, we subtract $9.47 - 0.05 = 9.42$ in. **The dimensions 9.52 in. and 9.42 in. are called the *limit dimensions* of the object**. That is, 9.52 in. is the largest acceptable measure, and 9.42 in. is the smallest acceptable measure.

EXAMPLE Find the limit dimensions of an object with a blueprint specification of 8.097 in. and a tolerance of ± 0.005. (This is often written 8.097 ± 0.005.)

8.097 in. = The blueprint specification for the dimension of an object.
± 0.005 = Tolerance of object's dimension.
Smallest dimension for object = Blueprint specification − Tolerance.
Largest dimension for object = Blueprint specification + Tolerance.

Estimation The tolerance of the part is very small—just five-thousandths, so the limit dimensions for the part should be only a few thousandths of an inch smaller or larger than the blueprint specification.

8.097 − 0.005 = 8.092
8.097 + 0.005 = 8.102

The smallest acceptable dimension for the object is 8.092 in. and the largest acceptable dimension is 8.102 in. That is, the limit dimensions are 8.092 in. and 8.102 in.

3 **Estimate and Check Subtraction.**

Estimating a subtraction problem is similar to estimating an addition problem. The numbers in the problem are rounded before the subtraction is performed. The accuracy of the estimate depends on the method used for rounding.

To estimate the difference:

1. Round each number to the desired place value or to a number with one nonzero digit.
2. Subtract the rounded numbers.

To check subtraction, we can use the relationship between addition and subtraction. If 9 − 5 = 4, then 4 + 5 should equal 9.

To check a subtraction problem:

1. Add the subtrahend and difference.
2. Compare the result of Step 1 with the minuend. If the two numbers are equal, the subtraction is correct.

EXAMPLE Estimate by rounding to hundreds, then find the exact difference, and check.

427.45 − 125

	Estimate	Exact	Check
427.45 − 125.00	400 − 100 **300**	427.45 − 125.00 **302.45**	125.00 + 302.45 **427.45**

Find the difference between $53,943.76 and $34,256.45 using a calculator.

One option:

53943 $\boxed{\cdot}$ 76 $\boxed{-}$ 34256 $\boxed{\cdot}$ 45 $\boxed{=}$ $\boxed{\text{ENTER}}$ or $\boxed{\text{EXE}}$ may replace $\boxed{=}$.

Calculator display: **19687.31**

Learning Strategy *What Happens If the Numbers Are Entered in the Wrong Order?*

Rework the previous example by entering the smaller number first.

34256.45 $\boxed{-}$ 53943.76 $\boxed{=}$

The display shows −19687.31 as the difference. Is this correct? No. Why not? The larger number should be entered first.

Since the digit portion of the display is the same as the correct answer, is it acceptable to just ignore the negative sign? No. 19687.31 and − 19687.31 are not equal. Negative numbers are discussed in Chapter 2.

EXAMPLE Two cuts are made from a 72-in. pipe (see Fig. 1–9). The two lengths cut from the pipe are 28 in. and 15 in. How much of the pipe is left after these cuts are made?

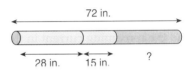

$$72 - 28 - 15 =$$ Subtract from left to right.
$$44 - 15 =$$ $72 - 28 = 44$
$$29$$ $44 - 15 = 29$

Figure 1–9 Lengths cut from pipe.

There are 29 in. of pipe left.

When more than one step is required to solve a problem, there is usually more than one way to solve the problem. In the previous example we could have found the total of the cuts first and then subtracted.

28 in. + 15 in. = 43 in.   Sum of two cuts.
72 in. − 43 in. = 29 in.   Remaining length of pipe.

SELF-STUDY EXERCISES 1–3

1 Subtract.

1. $7 - 2$
2. $9 - 4$
3. $(8 - 3) - 4$
4. $8 - (5 - 4)$
5. $9 - 5 - 2$
6. $(7 - 1) - 3$
7. $8 - (2 - 1)$
8. $(8 - 2) - 1$
9. $9 - 3 - 4$
10. $8 - (5 - 1)$
11. $6 - 3 - 2$
12. $(8 - 6) - 1$
13. $47 - 23$
14. $427 - 26$
15. $3,672 - 2,652$

16. 946
      $- 831$

17. $53,867$
      $-\quad 831$

18. $40,907$
      $- 36,829$

19. $109,806$
      $- 87,927$

20. $52,096$
      $- 38,529$

21. $16,135$
      $- 14,857$

EXAMPLE   Multiply 3(2 + 4).

Multiplying first gives:
3(2 + 4) =
3(2) + 3(4) =
6 + 12 = **18**

Adding first gives:
3(2 + 4) =
3(6) = **18**

EXAMPLE   Multiply 2(6 − 5).

Multiplying first gives:
2(6 − 5) =
2(6) − 2(5) =
12 − 10 = **2**

Subtracting first gives:
2(6 − 5) =
2(1) = **2**

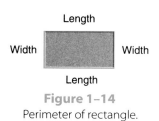

Figure 1–14
Perimeter of rectangle.

The distributive property is found in many formulas. One example is the formula for the perimeter of a rectangle. A *rectangle* is a four-sided geometric shape whose opposite sides are equal in length, and each corner makes a square corner (see Fig. 1–14). The *perimeter* of a rectangle is the distance around the figure. We find perimeters when we fence a rectangular yard, install baseboard in a rectangular room, frame a picture, or outline a flower bed with landscaping timbers.

The *formula* for finding the perimeter of a rectangle is

$$P = 2(l + w) \qquad \text{or} \qquad P = 2l + 2w$$

EXAMPLE   Find the number of feet of fencing needed to enclose a rectangular pasture that is 1,784.6 feet long and 847.3 feet wide.

$P = 2(l + w)$ or $P = 2l + 2w$
$P = 2(1{,}784.6 + 847.3)$ $P = 2(1{,}784.6) + 2(847.3)$
$P = 2(2{,}631.9)$ $P = 3{,}569.2 + 1{,}694.6$
$P = 5{,}263.8$ $P = 5{,}263.8$

The amount of fencing needed is 5,263.8 feet.

4   **Estimate and Check Multiplication.**

To estimate the answer for a multiplication problem:

1. Round both factors to a chosen or specified place value or to one nonzero digit.
2. Then multiply the rounded numbers.

To check a multiplication problem:

1. Multiply the numbers a second time and check the product.
2. Interchange the factors if convenient.

Chapter 1 / Whole Numbers and Decimals

EXAMPLE   Find the approximate and exact cost of 48 flower bulbs if each bulb costs $2.15. Estimate the cost by rounding each factor to a number with one nonzero digit. Then find the exact cost and check your work.

48 = Total number of flower bulbs
$2.15 = Cost of each bulb

Total cost of flower bulbs = Number of bulbs × Cost of each bulb

Estimation

If 50 bulbs were purchased at $2 each, the total cost would be $100. So the exact cost should be close to $100.

48 × $2.15 = $103.20   Number of bulbs × Cost of each bulb

The total cost of 48 flower bulbs is $103.20, which is approximately $100, as noted in the estimation.

The *area* of a geometric shape is the number of square units needed to cover the shape. We use area when we find the amount of carpet needed to cover a rectangular floor, the amount of paint needed for a wall, the amount of asphalt needed to pave a parking lot, the amount of fertilizer needed to treat a yard, or the amount of material needed to produce a rectangular sign. Area is measured in *square units*. One square foot (1 ft^2) indicates a square that measures 1 foot on each side.

5 feet long by 3 feet wide = 15 square feet

To find the area of a rectangle, we multiply the length by the width (see Fig. 1–15). We can state this in symbols using a formula.

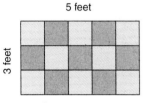

5 feet

3 feet

Figure 1–15
Area of a rectangle.

$A = lw$ Area of a rectangle = length times width

EXAMPLE   Maintenance Consultants needs to apply fertilizer to a customer's lawn. The lawn is 223.4 ft long and 132.8 ft wide. The fertilizer costs $0.004 per square foot to apply. What is the cost of applying the fertilizer?

Lawn is in the shape of a rectangle
Length of the lawn = 223.4 ft
Width of the lawn = 132.8 ft
Fertilizer costs $0.004 per square foot to apply.
Area of a rectangle is the product of the length times the width ($A = lw$)
Total cost = Total number of square feet × Cost per square foot

Estimation

A lawn 200 ft by 100 ft contains 20,000 ft^2, and at a cost of $0.004 per square foot, the total cost is approximately $80. Note: the cost will be higher because we rounded both length and width *down*.

$A = lw$
$A = 223.4(132.8)$
$A = 29,667.52$ square feet

Cost = 29,667.52($0.004)
Cost = $118.67

The total cost of applying fertilizer to the lawn is $118.67, which is higher than the estimated cost of $80, as we anticipated.

When either or both factors of a multiplication problem end in zeros, a shortcut process such as the one in the following example can be used.

> **EXAMPLE**   Multiply 2,600 × 70.
>
> **1.**   2600
>   × 70   Separate the ending zeros from the other digits.
>
> **2.**   2600
>   × 70   Multiply the other digits as if the zeros were not there (26 × 7 = 182).
>   182
>
> **3.**   2600   Attach the zeros to the basic product. Note that the number of zeros affixed to the
>   × 70   basic product is now the same as the sum of the number of zeros at the end of each
>   **182000**   factor.

This process is sometimes necessary whenever a multiplication problem is too long to fit into a calculator. Many calculators have only an eight-digit display window. The problem 26,000,000 × 3,000 would not fit many basic calculators. This shortcut allows us to work the problem with or without a calculator.

> **EXAMPLE**   Multiply 26,000,000 × 3,000.
>
>   26,000,000
>   ×   3,000   Separate ending zeros and multiply 26 × 3.
>   **78,000,000,000**   Attach 9 zeros.

Learning Strategy   *The Mind Is Often Quicker Than the Fingers.*

Don't use your calculator as a crutch. It is a tool! When multiplying 2,500 times 30, you can multiply 25 times 3 mentally. 25 × 3 = 75. Then, attach three zeros to that product.

$$2,500 \times 30 = 75,000$$

This skill does not come automatically. You have to practice it! Like playing a musical instrument or mastering a sport, you don't develop skill by watching. You have to practice!

SELF-STUDY EXERCISES 1–4

1 Multiply the following.

1. 5 × 3   **2.** 5 · 3 · 0 · 6   **3.** (3)(7)(9)(2)   **4.** 7 × 3 × 4   **5.** (3)(2)(0)(8)

6.   32   **7.**   83   **8.**   90   **9.**   503
     × 7   × 7   × 7   × 204

10. What property of multiplication justifies the statement "5(3) = 3(5)"?

11. Jaime Oxnard has 9 wood boxes that he intends to sell for $7 each. If Jaime sells 6 of the boxes, how much money will he make?

12. Explain the associative property of multiplication, and give an example to illustrate the property.

13. Margaret Johnston purchased 3 cases of candy bars. Each case contains 12 bags, and each bag contains 24 pieces of individually wrapped candy. How many individually wrapped candies did Margaret purchase?

14. Chuckee Wright counted 8 unopened boxes of washers. Each box contained 512 washers. What total number of washers will be shown on the inventory sheet?

15. Bill Wepner read 6 books a week over a period of 14 weeks. How many books did Bill read during this time period?

16. Each officer in the Public Safety office wrote, on average, 25 tickets a week. If there are 7 officers, how many tickets were written over a 4-week period?

17. Jossie Moore is planning to sell candy bars in her Smart Shop. She receives 12 boxes, and each box contains 24 candy bars. If Jossie sells the bars for $1 each, how much money will she get for all the bars?

18. Tory Eaton teaches 5 classes, and each class has 46 students. Her sixth class has 19 students enrolled. How many students is Tory teaching?

19. Shinder Blunt worked 40 hours and earned $14 per hour. How much did she earn?

20. Tratec paid $82 each for 290 printers for its stores. How much did they cost?

2 Multiply.

21.
$$\begin{array}{r} 53.4 \\ \times\ 0.29 \end{array}$$

22.
$$\begin{array}{r} 37.7 \\ \times\ 1.5 \end{array}$$

23.
$$\begin{array}{r} 9.27 \\ \times\ 0.35 \end{array}$$

24.
$$\begin{array}{r} 0.215 \\ \times\ 0.27 \end{array}$$

25.
$$\begin{array}{r} 0.271 \\ \times\ 0.32 \end{array}$$

26.
$$\begin{array}{r} 1.93 \\ \times\ 3.7 \end{array}$$

27. 2.78×0.03
28. $73.806(2.305)$
29. $1.9067 \cdot 0.2013$
30. 8.2037×0.602
31. $42(0.73)$
32. $81 \cdot 7.37$
33. 81.7×0.621
34. $72.6(0.532)$
35. $326 \cdot 1.04$
36. 261×2.07
37. $6.35(40)$
38. $9.21 \cdot 20$
39. 93.07×0.01
40. $(76.02)(0.01)$
41. $586.35 \cdot 0.001$
42. 732.65×0.001
43. $(2.0037)(4.6)$
44. $3.0062 \cdot 3.8$
45. 7.04×0.0025
46. $(6.01)(0.0045)$
47. $0.457 \cdot 2.003$
48. 0.387×1.009
49. $(0.05)(0.003)$
50. $0.07 \cdot 0.008$

51. A pipe has an outside diameter of 4.327 in. Find the inside diameter if the pipe wall is 0.185 in. thick.

52. A plastic pipe has an inside diameter of 4.75 in. Find the outside diameter if the pipe wall is 0.25 in. thick.

53. Electrical switches cost $5.70 wholesale. If the retail price is $7.99 each, how much would an electrician save over retail by buying 12 switches at the wholesale price?

54. How much No. 24 electrical wire is needed to make eight pieces each 18.9 in. long?

55. A retailer purchases 15 cases of potato chips at $8.67 per case. If the chips are sold at $12.95 per case, how much profit does the retailer make?

56. A contractor purchases 500 cubic yards (yd^3) of concrete at $21.00 per cubic yard, 24 yd^3 of sand at $5.00 per cubic yard, and 36 yd^3 of fill dirt at $3.00 per cubic yard. What is the total cost of the materials?

57. A micrometer is designed so that the screw thread makes the spindle advance 0.025 in. in one complete turn. How far will the spindle advance in 15 complete turns?

58. A 4 ft × 8 ft sheet of 1-in.-thick steel weighs 1,307 lb and costs $0.30 a pound. Find the cost of five of these sheets.

59. What is the total amperage (A) of a circuit if six 2.5-A devices are connected in parallel?

60. Beverly Vance orders 1,008 bandanas at $3.25 each. How much is the total order?

3 Find the product by multiplying first, and check your answers by adding first.

61. $5(7 + 4)$
62. $3(9 + 11)$
63. $14(3 + 11)$
64. $4(5 - 2)$
65. $1.2(16 - 3)$
66. $7(1.8 - 1)$

Find the perimeter of the rectangles in Figs. 1–16 and 1–17.

67. 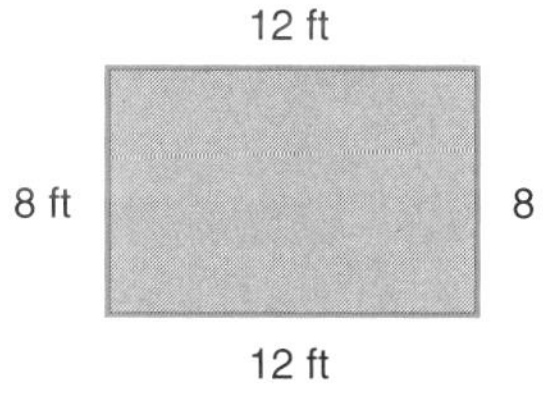
Figure 1–16

68. 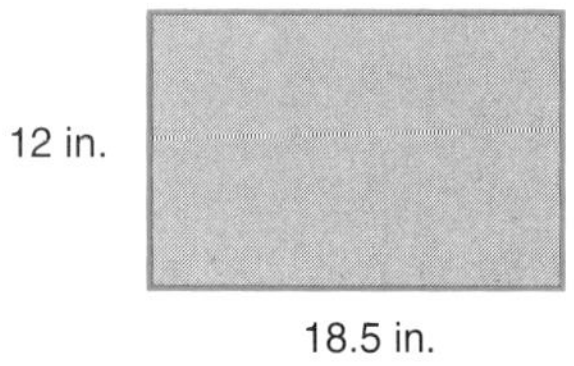
Figure 1–17

69. How much baseboard is required to surround a kitchen if the floor measures 15 ft by 13 ft and no allowance is made for door openings?

70. A dog pen is staked out in the form of a rectangle so that the width is 38 ft and the length is 47 ft. How much fencing must be purchased to build the pen?

4 Complete the multiplication problems as directed.

71. Find the approximate cost of 24 shirts if each shirt costs $32. What is the exact cost of the shirts?

72. Christy Hunsucker receives an invoice for 340 canvas bags that cost $12 each. Approximate then find the exact amount of the invoice.

Find the number of square feet in each rectangle illustrated in Figs. 1–18 and 1–19. Check your work.

73.

8.7 ft

14.2 ft

Figure 1–18

74.

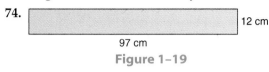

12 cm

97 cm

Figure 1–19

75. Estimate how many square feet of carpet should be purchased to cover a rectangular room that measures 42.3 ft by 32.5 ft. Find the exact area.

76. A rectangular table top is 12 ft long and 8 ft wide. How many square feet of glass are needed to cover it? Check your work.

Use a calculator to find the products.

77. $5896 \times 47.3 \times 12.83$

78. $(52,834)(15)(29.5)$

79. $832 \cdot 703 \cdot 960$

80. $583(294)(627)(0)(8)$

81. Alicia Toliver earns $3,015 per month. What is her annual salary?

82. Alicia Toliver earns $423.20 per week. Calculate her annual salary.

Perform the multiplications.

83. $3,500 \times 42,000,000$

84. $30,800 * 544,000$

85. $70,000(6,000)$

Learning Outcomes

1 Use various symbols to indicate division.

2 Divide whole numbers.

3 Divide decimal numbers.

4 Estimate and check division.

5 Find the numerical average.

1 **Use Various Symbols to Indicate Division.**

Division is the *inverse operation* of multiplication. Since $4 \times 7 = 28$, then 28 divided by 7 is 4 and 28 divided by 4 is 7. The number being divided is called the *dividend*. The number divided by is the *divisor*. The result is the *quotient*.

Division is *not* commutative. $12 \div 6 = 2$, but $6 \div 12$ does not equal 2; that is, $6 \div 12 \neq 2$.

Division is *not* associative. $(12 \div 6) \div 2 = 2 \div 2 = 1$. But $12 \div (6 \div 2) = 12 \div 3 = 4$. That is, $(12 \div 6) \div 2 \neq 12 \div (6 \div 2)$.

1. To use the *divided by* symbol ($\div$), write the dividend first.

$$28 \div 7 = 4 \longleftarrow \text{quotient}$$

dividend ⟶   ⟵ divisor

2. To use the *long-division symbol* ($\overline{)}$), write the dividend under the bar.

$$\begin{array}{r} 4 \longleftarrow \text{quotient} \\ 7\overline{)28} \end{array}$$

divisor ⟶   ⟵ dividend

3. To use the *division bar* or *slash symbol*, write the dividend on top or first.

dividend

$$\frac{28}{7} = 4 \longleftarrow \text{quotient} \quad \text{or} \quad 28/7 = 4 \longleftarrow \text{quotient}$$

divisor

EXAMPLE   Express "2 goes into 8" as a division using the symbol $\div$, the symbol $\overline{)}$, and the bar symbol —. Then solve the problem.

$$8 \div 2 = ? \quad \text{Because } 4 \times 2 = 8, \quad \textbf{8} \div \textbf{2} = \textbf{4}, \quad 2\overline{)8}^{\,\textbf{4}}, \quad \frac{\textbf{8}}{\textbf{2}} = \textbf{4}.$$

When using the "divided by" symbol, the division bar, or the slash as division, the dividend is written first or on top. The divisor is written second or on bottom.

    Technology tools such as the calculator and computer normally use the "divided by" or slash symbol.

    The divisor is written first *only* when using the long division symbol.

    2   **Divide Whole Numbers.**

Long division, like long multiplication, involves using one-digit multiplication facts repeatedly to find the quotient.

    When the quotient is not a whole number, the quotient may have a *whole number part* and a *remainder*. When a dividend has more digits than a divisor, parts of the dividend are called *partial dividends*, and the quotient of a partial dividend and the divisor is called a *partial quotient*.

To divide whole numbers:

1. Beginning with its leftmost digit, identify the first group of digits of the dividend that is larger than or equal to the divisor. This group of digits is the first *partial dividend*.
2. For each partial dividend in turn, beginning with the first:
   a. Divide the partial dividend by the divisor. Write this partial quotient above the rightmost digit of the partial dividend.
   b. Multiply the partial quotient by the divisor. Write the product below the partial dividend, aligning places.
   c. Subtract the product from the partial dividend. Write the difference below the product, aligning places. The difference must be less than the divisor.
   d. Next to the ones place of the difference, write the next digit of the dividend. This is the new partial dividend.
3. When all the digits of the dividend have been used, write the final difference in Step 2c as the remainder (unless the remainder is 0). The whole-number part of the quotient is the number written above the dividend.

EXAMPLE   To divide 881 by 35:

$$35\overline{)88\ 1}$$

The first partial dividend is 88.

$$
\begin{array}{r}
2 \\
35\overline{)00\ 1} \\
70 \\
\hline
18
\end{array}
$$

The partial quotient for 88 ÷ 35 is 2. Multiply 2 × 35 = 70. Then subtract 88 − 70 = 18. The difference 18 is less than the divisor 35.

$$
\begin{array}{r}
2 \\
35\overline{)881} \\
70 \\
\hline
181
\end{array}
$$

1 from the dividend is written next to 18 to form the next partial dividend.

$$
\begin{array}{r}
25 \\
35\overline{)881} \\
70 \\
\hline
181 \\
175 \\
\hline
6
\end{array}
$$

The partial quotient for 181 ÷ 35 is 5. The product of 5 × 35 is 175. The difference of 181 − 175 is 6. The remainder is 6.

881 divided by 35 = **25 R6.**

TIP!

Importance of Placing the First Digit Carefully

The correct placement of the first digit in the quotient is critical. If the first digit is out of place, all digits that follow will be out of place, giving the quotient too few or too many digits.

Chapter 1 / Whole Numbers and Decimals

EXAMPLE To divide 5⟌2,535.

$$\begin{array}{r} 5 \\ 5\overline{\smash{)}2{,}535} \\ \underline{2\,5} \\ 03 \end{array}$$

5 divides into 25 five times. Write 5 over the last digit of the 25. Subtract and bring down the 3.

$$\begin{array}{r} \mathbf{507} \\ 5\overline{\smash{)}2{,}535} \\ \underline{2\,5} \\ 035 \\ \underline{35} \\ 0 \end{array}$$

5 divides into 3 zero times. Write 0 over the 3 of the dividend. Bring down the next digit, which is 5. Divide 5 into the 35: 35 ÷ 5 = 7. Write the 7 over the 5 of the dividend. Multiply 7 × 5 = 35. Subtract.

TIP

What Types of Situations Require Division?

Two types of common situations require division. Both types involve distributing items equally into groups.

1. Distribute a specified total quantity of items so that each group gets a specific equal share. Division determines the number of groups.

    For example, you need to ship 75 crystal vases. With appropriate packaging to avoid breakage, only 5 vases fit in each box. How many boxes are required? You divide the total quantity of vases by the quantity of vases that will fit into one box to determine how many boxes are required:

    75 crystal vases ÷ 5 vases per box = 15 boxes needed

2. Distribute a specified total quantity so that we have a specific number of groups. Division determines each group's equal share.

    For example, how many ounces will each of four cups contain if a carafe of coffee containing 20 ounces is poured equally into the cups? The capacity of the carafe is divided by the number of coffee cups:

    20 ounces ÷ 4 coffee cups = 5 ounces for each cup

When dividing by a one-digit divisor, we can perform the division mentally. This process is called *short division*.

EXAMPLE Divide 7⟌180

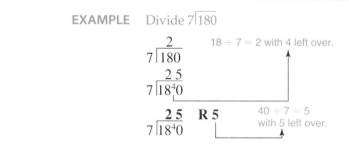

Special properties are associated with division involving zeros. The quotient of zero divided by a nonzero number is zero:

$$0 \div n = 0, \qquad n\overline{)0}, \qquad \frac{0}{n} = 0$$

$$0 \div 5 = 0, \qquad 5\overline{)0}, \qquad \frac{0}{5} = 0$$

The quotient of a number divided by zero is *undefined* or *indeterminate*:

$$n \div 0 \text{ is } undefined, \qquad 0\overline{)\;n\;}^{\,undefined}, \qquad \frac{n}{0} \text{ is undefined}$$

$12 \div 0$ is *undefined*
$0 \div 0$ is *indeterminate*

Dividing any nonzero number by itself yields 1:

$$n \div n = 1, \quad \text{if } n \text{ is not equal to zero;} \quad 12 \div 12 = 1$$

Dividing any number by 1 yields the same number:

$$n \div 1 = n, \quad 5 \div 1 = 5$$

In division, as in subtraction, order is important. For example, $12 \div 4 = 3$; however, $4 \div 12$ is not 3. Therefore, division is *not* commutative. When dividing more than two numbers, as in subtraction, the way we group the numbers is important. That is, division is *not* associative. If there are no grouping symbols, we divide from left to right.

EXAMPLE Divide $16 \div 4 \div 2$.

$16 \div 4 \div 2 = \mathbf{4} \div 2 = \mathbf{2}$ Divide from left to right.

---

3 Divide Decimal Numbers.

Suppose we have 3.6 lb of candy and we want to put 0.4 lb in a package. How many packages would we have? That is, 3.6 lb ÷ 0.4 lb per package = how many packages? Consider the division problem $3.6 \div 0.4$ in long-division form.

$$0.4\overline{)3.6}$$

We will convert the division to an equivalent division with a whole-number divisor.

$$0.4.\overline{)3.6.}^{\;9.}$$

Move the decimal point in the divisor to make the divisor a *whole* number. This changes the value of the divisor, so the decimal point in the dividend must be moved the same number of places to keep the same value of the division problem.

$$\frac{3.6}{0.4} \times \frac{10}{10} = \frac{3.6 \times 10}{0.4 \times 10} = \frac{36}{4} = 9$$

To divide decimal numbers written in long-division form:

1. Move the decimal in the divisor so that it is on the right side of all digits. (By moving the decimal, you are multiplying by 10, 100, 1,000, and so on.)
2. Move the decimal in the dividend to the right as many places as the decimal was moved in the divisor. Attach zeros if necessary. (This is multiplying the dividend by the same number as was used in Step 1.)

Chapter 1 / Whole Numbers and Decimals

3. Write the decimal point in the answer directly above the new position of the decimal in the dividend. (Do this *before* dividing.)
4. Divide as you would in whole numbers.

EXAMPLE   Divide 4.8 ÷ 6.

$$6\overline{)4.8}^{\,.}$$ Insert a decimal point above the decimal point in the dividend.

When the divisor is a whole number, the decimal is understood to be to the right of 6 and is not moved. The decimal is placed in the quotient directly above the decimal in the dividend.

$$6\overline{)4.8}^{\,0.8}$$ Divide.

EXAMPLE   Divide 3.12 ÷ 1.2.

$$1.2\overline{)3.12}$$ Move the decimal in the divisor and dividend.

$$12\overline{)31.2}^{\,.}$$ Write the decimal in the quotient, then divide.

$$\begin{array}{r} 2.6 \\ 12\overline{)31.2} \\ \underline{24} \\ 7\,2 \\ \underline{7\,2} \end{array}$$

To round a quotient to a place value:

1. Divide to one place past the desired rounding place.
2. Attach zeros to the dividend after the decimal if necessary to carry out the division.
3. Round the quotient to the place specified.

EXAMPLE   Divide and round the quotient to the nearest tenth.

$$3.2\overline{)15.27}$$

$$\begin{array}{r} 4.77 \\ 3.2\overline{)15.2\,70} \\ \underline{12\,8} \\ 2\,47 \\ \underline{2\,24} \\ 2\,30 \\ \underline{2\,24} \\ 6 \end{array}$$ Since we are rounding to the nearest tenth, divide to the hundredths place.

4.77 rounds to 4.8.

4 **Estimate and Check Division.**

Estimating a division is similar to estimating other operations. The numbers are rounded *before* the calculation is made.

To estimate division:

1. Round the divisor and dividend to one nonzero digit.
2. Find the first digit of the quotient.
3. Attach a zero in the quotient for each remaining digit in the dividend.

Because division and multiplication are inverse operations, division is checked by multiplication.

To check division:

1. Multiply the divisor by the quotient.
2. Add any remainder to the product in Step 1.
3. The result of Step 2 should equal the dividend.

EXAMPLE Estimate, find the exact answer, and check $913 \div 22$.

Estimate:
$$\frac{40}{20\overline{)900}}$$
20 divides into 90 four whole times. Attach a zero after 4.

Exact:
$$22\overline{)913} \quad \text{or} \quad 22\overline{)913.0}$$
$$\begin{array}{r} 41 \text{ R11} \\ \hline 88 \\ 33 \\ 22 \\ 11 \end{array} \qquad \begin{array}{r} 41.5 \\ \hline 88 \\ 33 \\ 22 \\ 11\,0 \\ \underline{11\,0} \end{array}$$

Check:
$$\begin{array}{r} 41 \\ \times\ 22 \\ \hline 82 \\ 82 \\ \hline 902 \end{array} \quad \begin{array}{r} 902 \\ +\ 11 \\ \hline 913 \end{array} \quad \text{or} \quad \begin{array}{r} 41.5 \\ 2\,2 \\ \hline 83\,0 \\ \underline{830} \\ 913.0 \end{array}$$

The answer checks.

---

5 **Find the Numerical Average.**

We use averages to make comparisons, such as when we compare the average mileage different cars get per gallon of gasoline. There are several types of averages.

In most courses students take, their numerical grade is determined by a process called *numerical averaging*. This average is also called the *arithmetic average* or *mean*. If the grades are 92, 87, 76, 88, 95, and 96, we can find the average grade by adding the grades and dividing by the number of grades. Because we have six grades, we divide the sum by 6.

$$\frac{92 + 87 + 76 + 88 + 95 + 96}{6} = \frac{534}{6} = 89$$

To find the average of a group of numbers or like measures:

1. Add the numbers or like measures.
2. Divide the sum by the number of addends.

EXAMPLE A car involved in an energy efficiency study had the following miles per gallon (mpg) listings for five tanks of gasoline: 21.7, 22.4, 26.9, 23.7, and 22.6 mpg. Find the average miles per gallon for the five tanks of gasoline.

Estimate: The low value is 21.7 and the high value is 26.9. The average will be between the two values.

Exact: $\dfrac{21.7 + 22.4 + 26.9 + 23.7 + 22.6}{5} = \dfrac{117.3}{5} = 23.46$

$= 23.5 \text{ mpg}$ Rounded.

The average is 23.5 mpg.

TIP!

Rounding Answers When Averaging

If the numbers being averaged are expressed to the same place value, the answer is usually rounded to that place value.

SELF-STUDY EXERCISES 1–5

1 Write as divisions using the symbols ÷, $\boxed{}$, and − for each problem.

1. 4 into 8
2. 3 into 9
3. 6 divided into 24
4. 7 divided into 30
5. 6 divided by 2
6. 8 divided by 4

2 Perform the long-division problems.

7. $8\overline{)600}$
8. $9\overline{)207}$
9. $6\overline{)120}$
10. $7\overline{)21}$
11. $3\overline{)48}$
12. $8\overline{)96}$
13. $5\overline{)215}$
14. $37\overline{)1,739}$
15. $26\overline{)312}$
16. $4\overline{)93}$
17. $48\overline{)5,982}$
18. $133\overline{)7,528}$

19. In a building where 46 outlets are installed, 1,472 ft of cable are used. What is the average number of feet of cable used per outlet?

20. Twelve water tanks are constructed in a welding shop at a total contract price of $14,940. What is the price per tank?

21. Five equally spaced holes are drilled in a piece of $\frac{1}{4}$-in. flat metal stock. The centers of the first and last holes are 2 in. from the end (see Fig. 1–20). What is the distance between the centers of any two adjacent holes? (*Caution*: How many equal center-to-center distances are there?)

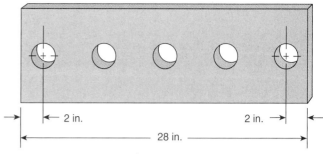

2 in.   2 in.

28 in.

Figure 1–20

3 Divide.

22. $3\overline{)3.78}$

23. $26\overline{)80.34}$

24. $21\overline{)1.323}$

25. $1.2\overline{)342}$

26. $4.8\overline{)28.32}$

27. $7.6\overline{)342}$

28. $6.3\overline{)68.67}$

29. $0.23\overline{)0.0437}$

30. $0.09\overline{)0.0954}$

31. $0.1353 \div 0.41$

32. $8 \div 0.32$

33. $1{,}449 \div 0.07$

34. $8118 \div 0.09$

35. $4066 \div 0.38$

36. $8.28 \div 4.6$

37. A 7.3-ft-long pipe weighs 43.8 lb. What is the weight of 1 ft of pipe?

38. A room requires 770.5 ft^2 of wallpaper, including waste. How many whole single rolls are needed for the job if a roll covers 33.5 ft^2 of surface?

39. The feed per revolution of a drill is 0.012 in. A hole 7.2 in. deep will require how many revolutions of the drill?

40. A piece of channel iron 5.6 ft long is cut into eight pieces. Assuming that there is no waste, what is the length of each piece?

41. If voltage is wattage (W) divided by amperage (A), find the voltage in a 300-W circuit drawing 3.2 A.

42. Exercises 37–41 involve measures. In some instances, the answer is a measure and in others, the answer is an amount telling how many. Explain when the answer will be a measure.

Divide to the nearest hundredths place.

43. $1.3\overline{)25.8}$

44. $2.4\overline{)4.37}$

45. $0.06\overline{)156.07}$

46. $41.7 : 21$

47. $0.23 \div 2.9$

Divide and round the quotient to the place indicated.

48. Nearest tenth: $0.43\overline{)72.8}$

49. Nearest hundredth: $25\overline{)3.897}$

50. Nearest whole number: $4.1\overline{)34.86}$

51. Nearest cent: $5\overline{)\$4.823}$

52. Nearest dollar: $17\overline{)\$24.98}$

53. If 12 lathes cost \$6,895, find the cost of each lathe to the nearest dollar.

54. If 12 electrolytic capacitors cost \$23.75, find the cost of one capacitor to the nearest cent.

55. To find the depth of an American Standard screw thread, the number of threads per inch is divided into 0.6495. Find the depth of thread of a screw that has eight threads per inch. Round to the nearest ten-thousandth.

56. A stack of 40 sheets of fiberglass is 6.9 in. high. What is the thickness of one sheet to the nearest tenth of an inch?

Work the division problems using your calculator. Round the answer to the nearest whole number.

57. $1.58\overline{)85.297}$

58. $4\overline{)1{,}203}$

59. $5.6\overline{)224.178}$

60. $45\overline{)4{,}027{,}500}$

61. $32\overline{)1{,}286{,}400}$

62. $71\overline{)440{,}271}$

63. Three bricklayers laid 3,210 bricks on a job in one day. What was the average number of bricks laid by each bricklayer?

64. A developer divides a tract of land into 14 equally valued parcels. If the tract is valued at \$147,000, what is the value of each parcel?

65. A shipment of 150 machine parts costs \$15,737.50. If this includes a \$25 shipping charge, find the cost of each part.

66. If a shop manager earns \$23,400 annually, what is the monthly salary?

67. A machine operator earns an annual salary of \$29,580. What is the operator's monthly salary?

68. A shipment of 288 headlights is billed out at \$1,329.60, which includes a \$48 freight charge. What is the cost per headlight, excluding freight costs?

69. An invoice for the total cost of refrigerator replacement parts indicates $2,116.80, which includes a $30 freight charge. The catalog price of the parts is $86.95 each. How many parts were ordered?

70. A job requires 95 ft of pipe, which will be cut into 9.3-ft lengths. How many 9.3-ft lengths of pipe will there be?

71. According to a real estate report in the Sunday newspaper, four homes in the Scenic Hills subdivision sold for $86,500, $68,750, $92,780, and $65,990. Find the average selling price of the homes in the subdivision.

72. Suprena Anderson paid the following amounts for her textbooks for the 2003 fall semester at Bolton Technical and Community College: English II, $67.50; general science, $92.75; psychology, $42.50; electronic technology I, $94.25; health, $18.50; and introduction to microcomputers, $42.75. Find the average cost of her textbooks for the semester.

73. The weather report listed the rainfall for the month of July as 0.2 in., 1.4 in., 0.5 in., 1.8 in., 0.6 in., 0.8 in., and 0.2 in. Find the average rain per rainfall to the nearest tenth of an inch.

4 Estimate by rounding to one nonzero digit. Then find the exact answer and round to the nearest hundredth if necessary.

74. $7.2\overline{)904.32}$

75. $122\overline{)384,512}$

76. $221\overline{)824,604}$

77. $7\overline{)59.01}$

78. A landfill job needs 2,294 yd³ of dirt to be delivered. If one truck can carry 18 cubic yards, how many loads will have to be made?

79. If a reel of wire contains 6,362.5 ft of wire, how many 50-ft extension cords can be made?

5 Find the average, then round to the same place value used in the problems.

80. Temperature readings: 82.5°, 76.3°, 79.8°, 84.7°, 80.8°, 78.8°, 80.0°

81. Test scores: 86, 73, 95, 85

82. Weight of cotton bales: 515 lb, 468 lb, 435 lb, 396 lb

83. Monthly income: $873.46, $598.21, $293.85, $546.83, $695.83, $429.86, $955.34, $846.95, $1,025.73, $1,152.89, $957.64, $807.25

84. Average rainfall: 1.25 in., 0.54 in., 0.78 in., 2.35 in., 4.15 in., 1.09 in.

85. Amperes of current: 3.0 A, 2.5 A, 3.5 A, 4.0 A, 4.5 A

1-6 | Exponents, Roots, and Powers of 10

Learning Outcomes

1 Simplify expressions that contain exponents.

2 Square numbers and find the square roots of numbers.

3 Use powers of 10 to multiply and divide.

1 **Simplify Expressions That Contain Exponents.**

The product of repeated factors can be written in shorter form using natural-number exponents. *Natural numbers* are also called *counting numbers* and include all whole numbers except zero. Exponents that are not natural numbers will have a different interpretation. For example, $4 \times 4 \times 4 = 4^3$. The 4 is called the *base* and is the repeated factor. The 3 is the *exponent* and indicates the number of times the factor is repeated. The expression 4^3 is read "four *cubed*" or "four to the third *power*" or "four raised to the third *power*." In expressions where 2 is an exponent, such as 4^2, the expression is usually read as "four *squared*"; however, it may also be read as "four to the second *power*."

The expression 4^3 is written in *exponential notation*. The number 64 is written in *standard notation* and is called the *power*.

To change from Exponential Notation to Standard Notation:

1. Use the base as a factor as many times as indicated by the exponent.
2. Perform the multiplication.

EXAMPLE Identify the base and exponent of the expressions, and write in standard notation.

(a) 5^3 (b) 1.5^2

(a) 5^3 5 is the base; 3 is the exponent.
$5^3 = 5 \times 5 \times 5 = \mathbf{125}$ Standard notation

(b) 1.5^2 1.5 is the base; 2 is the exponent.
$1.5^2 = 1.5 \times 1.5 = \mathbf{2.25}$ Standard notation

Any number with an exponent of 1 is the number itself:

$$a^1 = a, \quad \text{for any base } a \qquad 8^1 = 8 \qquad 2.3^1 = 2.3$$

EXAMPLE Write the exponential expressions in standard notation.

(a) 9^1 (b) 0.13^1

(a) $9^1 = 9$ $0.13^1 = 0.13$

---

2 **Square Numbers and Find the Square Roots of Numbers.**

The result of using a number as a factor 2 times is a *square* number or a *perfect square*. In the expression $7^2 = 49$, the 49 is a perfect square. This terminology evolved from a common formula for finding the area of a square. $A = s^2$, where A represents the area and s represents the length of one side. When we square 3, we write $3^2 = 3 \times 3 = 9$. We say 9 is the square of 3.

EXAMPLE Find the square.

(a) 2   (b) 7   (c) 3.2

(a) 2; $2^2 = 2 \times 2 = \mathbf{4}$
(b) 7; $7^2 = 7 \times 7 = \mathbf{49}$
(c) 3.2; $3.2^2 = 3.2 \times 3.2 = \mathbf{10.24}$

The inverse operation of squaring is taking the square root of a number. The *principal square root* of a perfect square is the number that was used as a factor twice to equal that perfect square. The principal square root of 9 is 3 because 3^2 or $3 \times 3 = 9$.

The *radical sign* $\sqrt{}$ indicates that the square root is to be taken of the number under the bar. This bar serves as a grouping symbol just like parentheses. The number under the bar is called the *radicand*. The entire expression is called a *radical expression*.

radical sign ⟶↓ ↓⟵ bar
$$\sqrt{25} = 5 \leftarrow \text{principal square root}$$
radicand ⟶

1. Select a trial estimate of the square root.
2. Square the estimate.
3. If the square of the estimate is less than the original number, adjust the estimate to a larger number. If the square of the estimate is more than the original number, adjust the estimate to a smaller number.
4. Square the adjusted estimate from Step 3.
5. Continue the adjusting process until the square of the trial estimate is the original number.

EXAMPLE   Find $\sqrt{256}$.

Select 15 as the estimated square root: 15^2 or $15 \times 15 = 225$. The number 225 is less than 256, so the square root of 256 must be larger than 15. We adjust the estimate to 17: 17^2 or $17 \times 17 = 289$. The number 289 is more than 256, so the square root of 256 must be smaller than 17. Now adjust the estimate to 16.

Because $16^2 = 256$, 16 is the square root of 256.

The squares of the numbers 1 through 10 are 1, 4, 9, 16, 25, 36, 49, 64, 81, and 100. These are the only whole numbers from 1 through 100 that are perfect squares.

Learning Strategy   *Memorizing Versus Understanding*

We often try to memorize facts rather than rely on our understanding of a concept. Memorized facts fade with time. For example, do you remember who was the 23rd president of the United States? (By the way, it was Benjamin Harrison.) If we understand the concept of a perfect square, we can generate a list whenever we need it. Of course, the more we use certain facts, the longer our memory retains them.

$11 \times 11 = 121$; $12 \times 12 = 144$; $13 \times 13 = 169$; $14 \times 14 = 196$; $15 \times 15 = 225$
$16 \times 16 = 256$; $17 \times 17 = 289$; $18 \times 18 = 324$; $19 \times 19 = 361$; $20 \times 20 = 400$

3   Use Powers of 10 to Multiply and Divide.

Powers of 10 are numbers whose only nonzero digit is 1. Thus, 10, 100, 1000, and so on are powers of 10 because each value can be written in exponential form with a base of 10.

Compare the number of zeros in standard notation with the exponent in exponential notation.

One million	$1,000,000 = 10^6$	6 zeros
One hundred thousand	$100,000 = 10^5$	5 zeros
Ten thousand	$10,000 = 10^4$	4 zeros
One thousand	$1,000 = 10^3$	3 zeros
One hundred	$100 = 10^2$	2 zeros
Ten	$10 = 10^1$	1 zero
One	$1 = 10^0$	0 zeros

The exponents in powers of 10 indicate the number of zeros used in standard notation.

EXAMPLE   Express as powers of 10.

(a) 10,000,000   (b) 100,000,000   (c) 100,000,000,000

(a) 10,000,000 $= 10^7$
(b) 100,000,000 $= 10^8$
(c) 100,000,000,000 $= 10^{11}$

EXAMPLE   Express in standard notation.

(a) 10^5 (b) 10^{13}

(a) $10^5 = 1\,00,000$
(b) $10^{13} = 10,000,000,000,000$

In some applications, powers of 10 are used to simplify multiplication and division problems. Compare the examples to find the pattern for multiplying a number by a power of 10.

$5 \times 100 = 500$	$5 \times 10^2 = 500$	Decimal point moved two places to the right and two zeros attached.
$27 \times 1,000 = 27,000$	$27 \times 10^3 = 27,000$	Decimal point moved three places to the right and three zeros attached.

To multiply a number by a power of 10:

1. Move the decimal point in the number to the *right* as many places as the 10, 100, 1,000, and so on, has zeros.
2. Attach zeros on the right if necessary.

EXAMPLE   Multiply 237×100.

$237.\underset{\curvearrowright}{00} \times 100 = \mathbf{23,700}$ Attach two zeros to the *right* of the 7 in 237, and insert the appropriate comma.

EXAMPLE   Multiply $36.2 \times 1,000$.

$36.\underset{\curvearrowright}{200} \times 1,000 = \mathbf{36,200}$ Move the decimal point three places to the *right*. Two zeros need to be attached.

Examine these divisions.

$32 \div 10 = 3.2$	Decimal point moved one place to the *left*.
$78.9 \div 100 = 0.789$	Decimal point moved two places to the *left*.
$52,900 \div 1,000 = 52.9$	Decimal point moved three places to the *left*.

To divide a number by a power of 10:

1. Move the decimal to the *left* as many places as the divisor has zeros.
2. Attach zeros to the left if necessary.
3. You may drop zeros to the right of the decimal point if they follow the last nonzero digit of the quotient.

Compare this rule with the rule for multiplying decimal numbers by 10, 100, 1,000, and so on. For multiplication, the decimal shifts to the right; for division, the decimal shifts to the left.

EXAMPLE If 100 lb of floor cleaner costs $63, what is the cost of 1 lb?

Divide $63 by 100.

$63 \div 100 = 0.63 = \$0.63$ Decimal is after 3 in 63. Move decimal two places to the left.

The cost of 1 lb is $0.63.

Some scientific and graphing calculators have specific power keys, such as keys for squares and cubes. They also have a "general power" key that can be used for all powers. Even though the "general power" key can be used to square and cube numbers, the "square" and "cube" keys require fewer keystrokes.

TIP!

Labels on Calculator Keys Are Not Universal.

To show calculator steps, we often identify a common label for a function in a box. The exact label will vary with the specific calculator model. Some calculators also provide many functions on menus instead of on specific keys. Even though we use a key notation in this text to show a calculator function, this function may actually appear on a menu. Check your calculator manual for exact location and labeling of functions.

Common Labels for Power Keys: $\boxed{x^2}$ $\boxed{x^3}$ $\boxed{\wedge}$ $\boxed{x^y}$

Common Labels for Square Root Keys: $\boxed{\sqrt{}}$ $\boxed{\sqrt{x}}$

EXAMPLE Use the calculator to find 35^2.

$35\ \boxed{x^2}\ \boxed{=}$ On some calculators pressing the equal key may not be required.

1,225

EXAMPLE Use the calculator to find 2.3^5.

$2\ \boxed{\cdot}\ 3\ \boxed{\wedge}\ 5\ \boxed{=}$ $\boxed{\wedge}$ and $\boxed{x^y}$ are the most common labels for the general power key.

64.36343

EXAMPLE (a) Evaluate $\sqrt{529}$.

$\boxed{\sqrt{}}\ 529\ \boxed{=}$

23

(b) Evaluate $\sqrt{10.89}$.

$\boxed{\sqrt{}}\ 10\ \boxed{\cdot}\ 89\ \boxed{=}$

3.3

Parentheses Indicate Multiplication, the Distributive Property, or a Grouping.

Parentheses can indicate multiplication or an operation that should be done first. If the parentheses contain an operation, they indicate a grouping. Otherwise, they indicate multiplication. The expression 5(2) indicates multiplication, while (4 + 6) indicates a grouping.

Many calculators have parentheses keys $\boxed{(}\,\boxed{)}$. These keys are used for all types of grouping symbols.

EXAMPLE   Evaluate $5 \times \sqrt{16} - 5 + [15 - (3 \times 2)]$.

$5 \times \sqrt{16} - 5 + [15 - (3 \times 2)]$	Work innermost grouping: 3×2.	P
$5 \times \sqrt{16} - 5 + [15 - 6]$	Work remaining grouping: $15 - 6$.	P
$5 \times \sqrt{16} - 5 + 9$	Find square root: $\sqrt{16} = 4$.	E
$5 \times 4 - 5 + 9$	Multiply: 5×4.	M D
$20 - 5 + 9$	Add and subtract from left to right.	A S
$15 + 9 = \mathbf{24}$		

EXAMPLE   Evaluate $3.2^2 + \sqrt{21 - 5} \times 2$.

$3.2^2 + \sqrt{21 - 5} \times 2$	Do operations within grouping first; the bar of the radical symbol is a grouping symbol: $21 - 5 = 16$.
$3.2^2 + \sqrt{16} \times 2$	Evaluate exponent and square root from left to right: $3.2^2 = 10.24$; $\sqrt{16} = 4$.
$10.24 + 4 \times 2$	Then multiply: $4 \times 2 = 8$.
$10.24 + 8$	Add last.
$10.24 + 8 = \mathbf{18.24}$	

A division bar serves as a grouping symbol in the same way parentheses do.

EXAMPLE   Makesha Lee took six tests and scored 87, 92, 76, 85, 95, 89. Find Makesha's average score.

$\dfrac{87 + 92 + 76 + 85 + 95 + 89}{6} =$	The division bar serves to group the addends. Perform operations that are grouped first.
$\dfrac{524}{6} =$	Perform division.
87.3	

The average score is 87.3.

---

2 **Solve Applied Problems Using Problem-Solving Strategies.**

Problem solving is an important skill in the workplace and in everyday life. In developing good problem-solving skills, it is helpful to use a systematic problem-solving plan. A plan gives you a framework for approaching problems.

Learning Strategy *Develop Problem-Solving Skills with a Positive Attitude.*

It is important to keep a positive attitude when solving problems.

Work with a study partner or group.

Working with a partner or group develops confidence. Each of you verbalizes your understanding of the problem, and you can discuss and correct any misconceptions.

You and your partners can examine several approaches to the problem and discuss their pros and cons. This discussion strengthens your investigative skills and gives you more options than you would probably come up with alone. You also may discover that there may be several *correct* approaches to solving the problem.

Be persistent.

Even the best problem solvers do not always use a correct approach on the first try. The mark of a good problem solver is persistence. Learn from failed attempts. Learn what doesn't work and why. This makes the correct approach easier to understand when you find it.

Don't make the "ruts" too deep.

After pursuing several dead ends, don't continue to the point of frustration. In other words, don't spin your wheels! The harder you work, the deeper the ruts get. Back away from the problem, and when you come back to it, you may think of a new approach you haven't tried before.

Another option is to seek help from others (additional classmates, tutors, your instructor). Just because a problem doesn't "click" right away, doesn't mean that you can't solve problems.

You will encounter many different plans for solving problems. They are all trying to help you organize the problem details so that you can find an appropriate procedure for solving the problem. Our plan, like others, is very structured. As you develop confidence and skill in solving problems you will probably use less structured and more intuitive approaches.

Six-Step Problem-Solving Plan
1. **Unknown Facts.** What facts are missing from the problem? What are you trying to find?
2. **Known Facts.** What relevant facts are known or given? What facts must you bring to the problem from your own background?
3. **Relationships.** How are the known facts and the unknown facts related? What formulas or definitions can you use to establish a model?
4. **Estimation.** What are some of the characteristics of a reasonable solution? For instance, should the answer be more than a certain amount or less than a certain amount?
5. **Calculations.** Perform the operations identified in the relationships.
6. **Interpretation.** What do the results of the calculations represent within the context of the problem? Is the answer reasonable? Have all unknown facts been found? Do the unknown facts check in the context of the problem?

EXAMPLE The 7th Inning buys baseball cards from eight different vendors. In November, the company purchased 8,832 boxes of cards. If an equal number of boxes was purchased from each vendor, how many boxes of cards were supplied by each vendor?

Unknown fact	Number of boxes of cards supplied by each vendor
Known facts	8,832 = Total number of boxes purchased 8 = Total number of vendors An equal number of boxes were purchased from each vendor.
Relationships	Total boxes purchased ÷ Number of vendors = Boxes purchased by each vendor
Estimation	If 8,000 boxes were purchased in equal amounts from eight vendors, then 1,000 were purchased from each vendor. Since more than 8,000 boxes were purchased, then more than 1,000 were purchased from each vendor.
Calculations	$\begin{array}{r} 1{,}104 \\ 8\overline{)8{,}832} \end{array}$
Interpretation	**Each vendor supplied 1,104 boxes of cards.**

In identifying the relationships of a problem or in developing your plan for solving a problem, it is often helpful if you look for key words or phrases that give you clues about the mathematical operations involved in the relationships. Key words give clues as to whether one quantity is added to, subtracted from, or multiplied or divided by another quantity. For example, if a problem tells you that Carol's salary in 2004 exceeds her 2003 salary by $2,500, you know that you should add $2,500 to her 2003 salary to find her 2004 salary.

Table 1–1   Key Words and What They Generally Imply in Word Problems

Addition	Subtraction	Multiplication	Division	Equality
The sum of	Less than	Times	Divide(s)	Equals
Plus/total	Decreased by	Multiplied by	Divided by	Is/was/are
Increased by	Subtracted from	Of	Divided into	Is equal to
More/more than	Difference between	The product of	Half of (divided by 2)	The result is
Added to	Diminished by	Twice (2 times)	Third of (divided by 3)	What is left
Exceeds	Take away	Double (2 times)	Per	What remains
Expands	Reduced by	Triple (3 times)	How big is each part?	The same as
Greater than	Less/minus	Half of ($\frac{1}{2}$ times)	How many parts can be made from	Gives/giving
Gain/profit	Loss	Third of ($\frac{1}{3}$ times)		Makes
Longer	Lower			Leaves
Older	Shrinks			
Heavier	Smaller than			
Wider	Younger			
Taller	Slower			
Larger than	Shorter			

More than one relationship may be needed to find the unknown facts. When this is the case, you usually need to read the problem several times to find all the relationships and plan your solution strategy.

EXAMPLE Carlee Anne McAnally needs to ship 78 crystal vases. With standard packing to prevent damage, 5 vases fit in each available box. How many boxes are required to pack the vases?

Unknown fact   Number of boxes required to ship the vases

Known facts   Total vases to be shipped = 78
Number of vases per box = 5

Relationships   Total boxes needed = Total number of vases ÷ Number per box
Total boxes needed = 78 ÷ 5

Estimation $70 ÷ 5 = 14$ Round down the dividend.
$80 ÷ 5 = 16$ Round up the dividend.
Since 78 is between 70 and 80, the number of boxes needed is between 14 and 16.

Calculation $78 ÷ 5 = 15$ R 3

Interpretation   **16 boxes are needed**; 15 boxes will contain 5 vases each, and 1 box will contain 3 vases. The box with 3 vases will need extra packing.

Learning Strategy   *Using Guess and Check to Solve Problems*

An effective strategy for solving problems involves guessing. Make a guess that you think may be reasonable, and check to see if the answer is correct. If your guess is not correct, decide if it is too high or too low. Make another guess based on what you learned from your first guess. Continue until you find the correct answer.

Let's try guessing in the previous example. We found that we could pack 70 vases in 14 boxes and 80 vases in 16 boxes. Since we need to pack 78 vases, how many vases can we pack with 15 boxes? $15 × 5 = 75$. Still not enough. Therefore, we will need 16 boxes, but the last box will not be full.

You can probably think of other ways to solve this problem. This is good. Any plan that leads to a correct solution is acceptable. Some plans will be more efficient than others, but you develop your problem-solving skills by pursuing a variety of strategies.

SELF-STUDY EXERCISES 1–7

1 Use the order of operations to evaluate each problem.

1. $5^2 + 4 - 3$
2. $4^2 + 6 - 4$
3. $4(3) - 9 ÷ 3$
4. $5 \cdot 2.9 - 4 ÷ 2$
5. $25 ÷ 5 \cdot 4.8$
6. $64 ÷ 4(2)$
7. $48 ÷ 8 × 3$
8. $5 × 12 ÷ 6$
9. $72 ÷ 9 \cdot 3$
10. $15 - 2 × 3$
11. $4^2 - (4)(3) + 6$
12. $17 - 4 × 2$
13. $6 × \sqrt{36} - 2 × 3$
14. $3 \cdot \sqrt{81} - (3)(4)$
15. $4^2 \cdot 3^2 + (4 + 2)(2)$
16. $2^2 \cdot 5^2 + (2 + 1)(3)$
17. $54 - 3^3 - \dfrac{8}{2}$
18. $156 - 2^3 - \dfrac{9}{3}$
19. $32 - 2.05 × 4^2 ÷ 2$
20. $5.2^2 - 3 × 2^2 ÷ 6$
21. $4 + 15 ÷ 3 - \dfrac{0}{7}$
22. $3 + 8 ÷ 4 - \dfrac{0}{5}$
23. $3 - 2 + 3 × 3 - \sqrt{9}$
24. $2 - 1 + (4)(4) - \sqrt{4}$
25. $2^4 × (7 - 2) × 2$
26. $3^4 × (9 - 3) × 3$
27. $124 - 8 \cdot 7 + 12$

Use a calculator to perform the operations.

28. $6 \times 9^2 - \dfrac{12}{4}$ **29.** $7 \times 8^2 - \dfrac{10}{2}$ **30.** $4^3 + 14 - 8$

31. $2^3 + 12 - 7$ **32.** $2 \times \sqrt{16} + (8 - \sqrt{25})$ **33.** $4 \times \sqrt{49} + (9 - \sqrt{64})$

34. $3(2^2 + 1) - 30 \div 3$ **35.** $4(3^2 + 2) - 60 \div 12$ **36.** $3 + 10 \div 5 + 2$

37. $4 + 12 \div 4 + 7$ **38.** $25 - 5^2 \div (7 - 2)$ **39.** $36 - 6^2 \div (8 - 2)$

40. $2(3.1^2 + 2) - \sqrt{7.29}$ **41.** $3^4 - 2 \times 4.6 \div \dfrac{0}{2}$ **42.** $24(3^2 + 1) - 48.25$

43. State the first operation in the order of operations.   **44.** State the last operation in the order of operations.
45. Use the order of operations to average 42, 66, and 93.   **46.** Find the average of 86, 97, 84, and 77.

> 2

47. If you have 348 packages of Halloween candy to rebox for shipment to a discount store and you can pack 12 packages in each box, how many boxes will you need?

48. If American Communications Network (ACN) has an annual payroll of $5,602,186 for its 214 employees, what is the average salary of an ACN employee?

49. In a recent year, 21,960 people visited Bio Fach, Germany's biggest ecologically sound consumer goods trade fair. This figure was up from the 18,090 the previous year and 16,300 two years ago. What was the increase in visitors to Bio Fach over the two-year period?

50. The On-The-Square Card and Gift Shop buys cards from six vendors. In November, the company purchased 6,480 boxes of cards. If the shop purchased an equal number of boxes from each vendor, how many boxes of cards did each vendor supply?

ASSIGNMENT EXERCISES

Section 1–1

Write as decimal numbers.

1. (a) $\dfrac{3}{10}$ (b) $\dfrac{15}{100}$ (c) $\dfrac{4}{100}$

2. (a) $\dfrac{75}{1,000}$ (b) $\dfrac{21}{10}$ (c) $\dfrac{652}{100,000}$

3. What is the place value of 6 in 21.836?   **4.** What is the place value of 5 in 13.0586?
5. In 13.7213, what digit is in the tenths place?   **6.** In 15.02167, what digit is in the ten-thousandths place?

7. Identify the place value of the digit 3:
    (a) 430   (b) 34,789   (c) 3,456,521
8. Identify the place value of the digit 2:
    (a) 2,785,901   (b) 45,923
9. Write 56,109,110 in words.   **10.** Write 61,201 in words.
11. Write one million, two hundred sixty-five thousand, four hundred one in standard notation.   **12.** Write thirty-two thousand, three hundred twenty-one in standard notation.
13. Write the words for 6.803.   **14.** Write the words for 0.0712.
15. The decimal equivalent of $\frac{5}{8}$ is six hundred twenty-five thousandths. Write this decimal using digits.   **16.** The thickness of a sheet of aluminum is forty-thousandths of an inch. Write this thickness as a decimal number.

17. Round to the indicated places.
    (a) 36 to the nearest tens
    (b) 74 to the nearest tens
    (c) 24.237 to the nearest whole number
    (d) $42.98 to the nearest dollar
    (e) 83.052 to tens
    (f) $8.9378 to the nearest cent
    (g) $0.9986 to the nearest cent
    (h) 0.097032 to hundred-thousandths
18. Round to the indicated places.
    (a) Nearest thousand: 65,763
    (b) Nearest thousand: 28,714
    (c) Nearest ten million: 497,283,016
    (d) Nearest hundred: 8,236
    (e) Nearest ten thousand: 248,217
    (f) Nearest hundredth: 7.0893
    (g) Nearest thousandth: 1.078834
    (h) Nearest tenth: 0.09783

19. Round to the indicated places.
  (a) Nearest hundred: 468
  (b) Nearest ten thousand: 49,238
  (c) Nearest tenth: 41.378
  (d) Nearest hundredth: 6.8957
  (e) Nearest ten-thousandth: 23.46097

20. Round to numbers with one nonzero digit.
  (a) 98
  (b) 94
  (c) 25,786
  (d) 34,786
  (e) 12.83
  (f) 0.0736
  (g) 7.93
  (h) 1.876

21. Which of these decimal numbers is larger: 4.783 or 4.79?

22. Which of these decimal numbers is smaller: 0.83 or 0.825?

23. Write these decimal numbers in order of size from smallest to largest: 0.021, 0.0216, 0.02.

24. Two measurements of an object are recorded. If the measures are 4.831 in. and 4.820 in., which is larger?

25. The decimal equivalent of $\frac{7}{8}$ is 0.875. The decimal equivalent of $\frac{6}{7}$ is approximately 0.857. Which fraction is larger?

26. Two parts are machined from the same stock. They measure 1.023 in. and 1.03 in. after machining. Which part has been machined more; that is, which part is now smaller?

Section 1–2

Add using a calculator or as directed by your instructor.

27. **(a)** $6 + 9 + 3 + 5$
  (b) $5 + 1 + 6 + 3 + 3$
  (c) $8 + 5 + 3 + 6 + 2 + 4$
  (d) $7 + 4 + 3 + 2 + 5 + 4$

28. **(a)** $3.47 + 42.32 + 3.82 + 4.09$
  (b) $6.2 + 32.7 + 46.82 + 0.29 + 4.237$
  (c) $86.3 + 9.2 + 70.02 + 3 + 2.7$
  (d) $42 + 3.6 + 2.1 + 7.83$

29. An air conditioner uses 10.4 kW (kilowatts), a stove uses 15.3 kW, a washer uses 2.9 kW, and a dryer uses 6.3 kW. What is the total number of kilowatts used?

30. Add.

	(a)	(b)
	3,456.08	12,467
	+ 2,147.76	+ 24,378

	(c)	(d)
	23,609	$43,045.36
	2,200	5,047.47
	76	87.10
	+ 124	+ 213.08

31. Estimate by rounding to thousands, then find the exact answer. Check your answer
  (a) $16,742.83
     + $12,349.26
  (b) 17,402
     + 18,646

32. A do-it-yourself project requires $57.32 for concrete, $74.26 for fence posts, and $174.85 for fence boards. Estimate the cost by rounding to numbers with one nonzero digit, then find the exact cost. Check your answer.

Section 1–3

33. Subtract.
  (a) $21.34 - 16.73$
  (b) $15.934 - 12.807$
  (c) $9 - 7$
  (d) $5 - 0$
  (e) $8 - 3 - 2 - 3$
  (f) $284.73 - 79.831$
  (g) $345 - 201$
  (h) $13,342 - 1,202$

34. Estimate by rounding to hundreds, then find the exact answer. Check your answer.
  (a) $12,346.87 - $4,468.63
  (b) $3,495 - 3,090$
  (c) $6,767 - 478$
  (d) $293.86 - 148$

35. A blueprint calls for the length of a part to be 8.296 in. with a tolerance of ± 0.005 in. What are the limit dimensions of the part?

36. For a moving sale, a family sold a sofa for $75 and a table for $25. If a newspaper ad for the sale cost $12.75, how much did the family clear on the two items sold?

37. Planning a vacation, the Mendez family selected a scenic route covering 653 miles and a direct route covering 463 miles. Estimate the difference by rounding to one nonzero digit number. Find the exact answer and check.

38. A new foreign car costs $12,677, and a comparable new American car costs $11,859. Estimate the difference by rounding to thousands. Find the exact answer and check.

135. You are designing a playground that has 2,160 yd² of space. If the playground is rectangular, determine how long and how wide it should be if one piece of equipment requires a space at least 15 yd long. Give whole-number solutions.

136. Of all the possible ways to design the playground in Problem 135, which rectangle requires the least amount of fencing? Give whole-number solutions.

137. Use a calculator to evaluate the powers.

    (a) 0.03^2 **(b)** 0.07^2 **(c)** 0.005^2

    (d) 0.009^2 **(e)** 0.02^3 **(f)** 0.004^3

1. Addition and subtraction are inverse operations. Write the following addition problem as a subtraction problem, and find the value of the number represented by the letter n: $1.2 + n = 1.7$.

2. Multiplication and division are inverse operations. Write the following multiplication problem as a division problem, and find the value of the number represented by the letter n: $5 \times n = 4.5$.

3. Squaring and finding square roots are inverse operations. Write the following square root as a squaring problem, and find the value of the number represented by the letter n: $\sqrt{n} = 6$.

4. Give an example that shows subtraction is not associative.

5. Give an example that shows division is not commutative.

6. Give the steps in the order of operations.

Find and explain the mistake, then rework each problem correctly.

7. $2.5 + 4.9$

$$\begin{array}{r} 2.5 \\ +\ 4.9 \\ \hline 6.14 \end{array}$$

8. $2 + 5(4) =$

$\quad\quad 7(4) = 28$

9. $\sqrt{9} = 01$

10. How are the procedures for adding whole numbers and adding decimals related?

11. Without making any calculations, do you think 0.004 is a perfect square? Why or why not?

12. Without making any calculations, do you think 0.008 is a perfect cube? Why or why not?

13. Can you find at least one exception to the generalization that a perfect square decimal has an even number of decimal places? Illustrate your answer.

Learning Outcomes
Section 1–1

What to Remember with Examples

1 Identify place values in whole numbers (p. 3).

Each digit has a place value. Place values are grouped in periods of three digits. Place-value names are shown in Fig. 1–1.

> Identify the place value of each digit in the number 3,628: 3 thousands 6 hundreds 2 tens 8 ones.

2 Read and write whole numbers in words, standard notation, and expanded notation (pp. 4–5).

Standard notation is the usual form of a number, that is, the number written in digits using place values.

Read from the left the numbers in each period and the period name. The name *units* is not read.

> The number 345,230 is written in standard form.
> Write 3,462 in words: three thousand, four hundred sixty-two.

To write numbers in expanded notation, multiply each digit by the power of 10 that represents its place value and add.

> Write 487 in expanded notation.
> $4 \times 100 + 8 \times 10 + 7 \times 1$

| 3 Compare whole numbers (pp. 6–7). | Mentally position the numbers on a number line. The leftmost number is the smaller number. Write an inequality using either the *less than* < or *greater than* > symbol. |

Write two inequalities comparing 5 and 9.
5 < 9 or 9 > 5

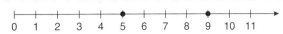

| 4 Identify place values in decimal numbers (pp. 7–8). | Align the decimal point in the number under the decimal point in the place-value chart. |

In 3.75, 3 is in the ones place, 7 is in the tenths place, and 5 is in the hundredths place.

| 5 Read and write decimal numbers (pp. 8–10). | Read the whole-number part, "*and*" for the decimal point, then the decimal part, and the place value of the rightmost digit. |

Write 32.075 in words: thirty-two and seventy-five thousandths.

| 6 Write fractions with power-of-10 denominators as decimal numbers (pp. 10–11). | Write the numerator as the decimal number with the same number of decimal digits as there are zeros in the denominator. |

Express $\dfrac{53}{1,000}$ as a decimal: 0.053.

| 7 Compare decimal numbers (pp. 11–12). | Compare decimal numbers by comparing the digits in the same place beginning from the left of each number. |

Which decimal is larger, 0.23 or 0.225? Both numbers have the same digit, 2, in the tenths place. 0.23 is larger because it has a 3 in the hundredths place, while 0.225 has a 2 in that place.

| 8 Round a whole number or a decimal number to a place value (pp. 12–14). | **Whole number: 1.** Locate the rounding place. **2.** Examine the digit to the right. **3.** Round down if the digit is less than 5. **4.** Round up if the digit is 5 or more. |

Decimal number:

Round as in whole numbers; however, digits on the right side of the digit in the rounding place are dropped rather than replaced with zeros.

Round 3,624 to the nearest tens place: 2 is in the tens place, 4 is the digit to the right; 4 is less than 5, so round down. The rounded value is 3,620.
Round 5.847 to the nearest tenth: 8 is in the tenths place; 4 is less than 5, so leave 8 as is and drop the 4 and 7. The rounded value is 5.8.

| 9 Round a whole number or a decimal number to a number with one nonzero digit (pp. 14–15). | **Whole number: 1.** Locate the first nonzero digit on the left. **2.** Examine the digit to the right. **3.** Round down if the digit is less than 5. **4.** Round up if the digit is 5 or more. |

Decimal number:

To round decimals to one nonzero digit is to round to the first place from the left that is not zero.

Round 4,789 to one nonzero digit: 4 is the first nonzero digit on the left, 7 is the digit to the right; 7 is 5 or more, so round up. The rounded value is 5,000.

Round 0.0372 to one nonzero digit: 3 is the first nonzero digit from the left; 7 is 5 or greater, so change 3 to 4 and drop the remaining digits on the right. The rounded value is 0.04.

Chapter Summary

Section 1–2

1 Add whole numbers (pp. 18–20).

Addition is a binary operation that is commutative and associative.

Zero property of addition: $0 + n = n + 0 = n$

Arrange addends in columns of like places. Add each column beginning with the ones place. Carry when necessary.

$5 + 7 = 7 + 5 = 12$ Commutative property of addition
$(3 + 2) + 6 = 3 + (2 + 6) = 11$ Associative property of addition
$0 + 3 = 3 + 0 = 3$ Zero property of addition

$$
\begin{array}{r}
4{,}824 \quad \text{Addend}\\
+\;\;\; 745 \quad \text{Addend}\\
\hline
5{,}569 \quad \text{Sum}
\end{array}
$$

2 Add decimal numbers (pp. 20–21).

Add decimal numbers by arranging the addends so that the decimal points are aligned.

Add: $43.35 + 3.7 + 0.462$

$$
\begin{array}{r}
43.35 \quad \text{Addend}\\
3.7 \quad \text{Addend}\\
+\; 0.462 \quad \text{Addend}\\
\hline
47.512 \quad \text{Sum}
\end{array}
$$

3 Estimate and check addition (pp. 21–22).

Estimate by rounding the addends before finding the sum. Check addition by adding a second time.

Estimate the sum by rounding to the nearest hundred: $483 + 723$; $500 + 700 = 1{,}200$. Exact sum: $1{,}206$.

Section 1–3

1 Subtract whole numbers (pp. 25–27).

Subtraction is a binary operation that is *not* commutative or associative. Addition and subtraction are inverse operations.

Zero property of subtraction: $n - 0 = n$

Arrange numbers in columns of like places. Subtract each column, beginning with the ones place. Borrow when necessary.

If $5 + 4 = 9$, then $9 - 5 = 4$ or $9 - 4 = 5$.   Inverse operations
$7 - 0 = 7$ Zero property of subtraction

$$
\begin{array}{r}
4{,}227\\
-\;\;\; 745\\
\hline
3{,}482
\end{array}
$$

2 Subtract decimal numbers (pp. 27–28).

Subtract decimal numbers by arranging the minuend and subtrahend so that the decimal points are aligned.

Subtract: $53.824 - 4.0423$

$$
\begin{array}{r}
53.824 \quad \text{Minuend}\\
-\;\; 4.0423 \quad \text{Subtrahend}\\
\hline
49.7817 \quad \text{Difference}
\end{array}
$$

3 Estimate and check subtraction (pp. 28–29).

Estimate by rounding the minuend and subtrahend before finding the difference. Check subtraction by adding the difference and the subtrahend. The result should equal the minuend.

Estimate the difference by rounding to the nearest hundred: $783 - 423$ $800 - 400 = 400$. Exact difference: $783 - 423 = 360$. Check: $360 + 423 = 783$.

Chapter 1 / Whole Numbers and Decimals

Section 1–4

1 Multiply whole-number factors (pp. 31–33).

Multiplication is a binary operation that is commutative and associative.

Zero property of multiplication: $n \times 0 = 0 \times n = 0$

Arrange numbers in columns of like place values. Multiply the multiplicand by each digit in the multiplier. Add the partial products.

$5(7) = 7(5) = 35;$ Commutative property of multiplication
$(3 \times 2) \cdot 6 = 3 \cdot (2 \times 6) = 36:$ Associative property of multiplication
$5 \times 0 = 0 \times 5 = 0$ Zero property of multiplication

259	Multiplicand
× 23	Multiplier
777	Partial product
5 18	Partial product
5,957	Product

2 Multiply decimal factors (pp. 33–35).

Multiply the decimal numbers as in whole numbers, and count the number of decimal digits in both numbers. Place the decimal in the product to the left of the same number of digits, counting from the right.

Multiply: 3.25×0.53

3.25	Two decimal places
× 0.53	Two decimal places
9 75	
1 62 5	
1.72 25	Four decimal places

3 Apply the distributive property (pp. 35–36).

$a(b + c) = ab + ac$
$a(b - c) = ab - ac$

$3(5 + 6) = 3(5) + 3(6)$
$3(11) \quad = 15 + 18$
$\quad\quad 33 = 33$

4 Estimate and check multiplication (pp. 36–38).

Round the factors, then multiply. Check multiplication by multiplying a second time.

Estimate the product by rounding to one nonzero digit: 483×72;
$500 \times 70 = 35,000$. Exact product: 34,776.

Section 1–5

1 Use various symbols to indicate division (pp. 40–41).

The division a divided by b can be written as

$$a \div b \qquad b\overline{)a} \qquad \frac{a}{b} \qquad \text{or} \qquad a/b$$

The divisor, b, cannot be zero.

Write 12 divided by 4 in four ways.

$$12 \div 4 \qquad 4\overline{)12} \qquad \frac{12}{4} \qquad 12/4$$

2 Divide whole numbers (pp. 41–44).

Align the numbers properly in long division.

```
      20 R15
23)475
   46
   15
    0
   15
```

Chapter Summary

| **3** Divide decimal numbers (pp. 44–46). | When dividing by a decimal number, move the decimal in the divisor to the right end; move the decimal in the dividend the same number of places to the right. Place the decimal in the quotient. |

Divide one place past the specified place value. Attach zeros in the dividend if necessary. Round to the specified place.

Divide:

$$
\begin{array}{r}
3.8 \\
2.1\overline{)7.9\,8} \\
\underline{6\,3} \\
1\,6\,8 \\
\underline{1\,6\,8}
\end{array}
$$

Divide and round to tenths:

$$
\begin{array}{r}
1.70 \approx 1.7 \\
15\overline{)25.60} \\
\underline{15} \\
10\,6 \\
\underline{10\,5} \\
10
\end{array}
$$

4 Estimate and check division (p. 46).

To estimate division, round the dividend and divisor before finding the quotient. Find the first digit of the quotient, and add a zero for each remaining digit in the dividend.

To check division, multiply the quotient by the divisor and add the remainder.

Estimate $2{,}934 \div 42$.

$$
\begin{array}{r}
70 \\
40\overline{)3{,}000}
\end{array}
$$

Exact quotient = 69 R36

Check:

$69 \times 42 = 2{,}898$

$2{,}898 + 36 = 2{,}934$

5 Find the numerical average (pp. 46–47).

Find the sum of the values and divide the sum by the number of values.

Find the average of 74, 65, and 85.

$74 + 65 + 85 = 224$ Find the sum of the values.

$224 \div 3 = 74.6$ or 75 (rounded)   Divide by the number of values.

Section 1–6

1 Simplify expressions that contain exponents (pp. 49–50).

To simplify an exponential expression, use the base as a factor the number of times indicated by the exponent.

$a^1 = a$, for any base a

$5^3 = 5 \times 5 \times 5 = 125$ Five is used as a factor 3 times.

$7^1 = 7$ Any number raised to the first power is the number.

2 Square numbers and find the square roots of numbers (pp. 50–51).

Squaring and finding square roots are inverse operations.

$7^2 = 49$

$\sqrt{49} = 7$

Chapter 1 / Whole Numbers and Decimals

3 Use powers of 10 to multiply and divide (pp. 51–53).

To multiply a decimal number by a power of 10, move the decimal to the *right* as many digits as the power of 10 has zeros. Attach zeros if necessary.

To divide a decimal by a power of 10, move the decimal to the *left*. Note that this process is just the opposite of that for multiplication.

> Multiply:
> $27 \times 10^3 = 27{,}000$
> $18 \times 100 = 1{,}800$
> $23.52 \times 1{,}000 = 23{,}520$
>
> Divide:
> $3.52 \div 10 = 0.352$
> $400 \div 10^2 = 4$
> $23{,}000 \div 10^3 = 23$

Section 1–7

1 Apply the order of operations to a series of operations (pp. 54–56).

Order of operations:

1. Parentheses or groupings, innermost first.
2. Exponential operations and roots from left to right.
3. Multiplication and division from left to right.
4. Addition and subtraction from left to right.

> Simplify.
>
> | | |
> |---|---|
> | $3^2 + 5(6 - 4) \div 2$ | Work inside parentheses. |
> | $3^2 + 5(2) \div 2$ | Raise to power. |
> | $9 + 5(2) \div 2$ | Multiply. |
> | $9 + 10 \div 2$ | Divide. |
> | $9 + 5$ | Add. |
> | 14 | |

2 Solve applied problems using problem-solving strategies (pp. 57–59).

Six-Step Problem-Solving Plan

> A shipment of textbooks to a college bookstore is sent in two boxes. One box weighs 9 lb, and the second box weighs 14 pounds more than the first box. What is the total weight of the boxes?

Unknown facts

Weight of second box
Total weight of the two boxes together

Known facts

First box weighs 9 lb.
Second box weighs 14 lb more than first box.

Relationships

$$\text{Weight of first box} = 9$$
$$\text{Weight of second box} = 9 + 14$$
$$\text{Total weight} = 9 + 9 + 14$$

Estimation

Since the second box weighs more than twice as much as the first, the two boxes together weigh approximately 3×9 or 27 lb.

Calculation

$$9 + 9 + 14 = 32$$

Interpretation

The two boxes weigh 32 lb together, and 9 lb and 23 lb separately.

Describe the location of the points in Fig. 2–5 by giving the amount of horizontal and vertical movement from the origin.

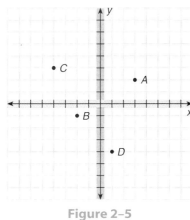

Point A: horizontal movement, +3 ; vertical movement, +2
Point B: horizontal movement, −2 ; vertical movement, −1
Point C: horizontal movement, −4 ; vertical movement, +3
Point D: horizontal movement, +1 ; vertical movement, −4

Figure 2–5

The location of points on a rectangular coordinate system can be written in an abbreviated, symbolic form. *Point notation* uses two signed numbers to show the horizontal and vertical movement from the origin to a given point. Horizontal movement is represented by the *x-coordinate* and is written first. Vertical movement is represented by the *y-coordinate* and is written second. These signed numbers are separated with a comma and enclosed in parentheses.

TIP!

Point Notation

Symbolically, we represent the location of a point as

(horizontal movement, vertical movement)   or (x, y)

where x is the horizontal movement from the origin, indicated by the x-coordinate, and y is the vertical movement from the x-axis, indicated by the y-coordinate.

EXAMPLE Write the points in the preceding example (Fig. 2–5) using point notation.

Point A = (horizontal movement of +3, vertical movement of +2) or **(+3, +2)**
Point B = (horizontal movement of −2, vertical movement of −1) or **(−2, −1)**
Point C = (horizontal movement of −4, vertical movement of +3) or **(−4, +3)**
Point D = (horizontal movement of +1, vertical movement of −4) or **(+1, −4)**

To *plot* a point means to show its location on the rectangular coordinate system.

To plot a point on the rectangular coordinate system:

1. Start at the origin.
2. Count to the left or right the number of units of the first signed number.
3. From the ending point of Step 2, count up or down the number of units from the second signed number.

Chapter 2 / Integers

1 Exercises 1–8 re

1. On which side c
3. How far does th
   direction?
5. What sign (+ or

7. In which directi
   numbers increa:
9. Draw a thermor
   20° above zero,
   10° below zero.

2 Use the "greate

11. 5 ___ 8
15. −5 ___ −9

Rewrite the express
numbers.
19. x is less than y.
21. 3 + 5 is greater
23. k is greater tha

3 Give the value

25. $|23|$
29. $|-13|$

4 Illustrate the n

33. 7
37. Which numbe:

5 Locate the poi

39. R 40. !
43. V 44. !

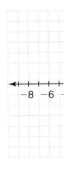

−8 −6

Which Way Do I Go First?

Proper use of point notation depends on knowing which direction applies to which number. Horizontal comes first. Think of other words that imply "horizontal movement" right/left, forward/back. For vertical movement, think up/down, rise/fall.

A memory strategy might be to think of the alphabetical order of some key words.

Alphabetically, *horizontal* comes before *vertical*. *Right* comes before *up*.

EXAMPLE Plot these points: Point $A = (3, 1)$, point $B = (-2, 5)$, point $C = (-3, -2)$, point $D = (1, -3)$.

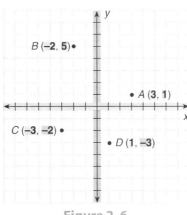

A: **right 3, up 1**
B: **left 2, up 5**
C: **left 3, down 2**
D: **right 1, down 3**

Figure 2–6

Learning Strategy   *Develop Your Spatial Sense*

Moving in the wrong direction is a common mistake when plotting points. To develop your spatial sense, consider the signs of the coordinates of points being plotted. Then, visualize in which part of the graph the point will fall.

Refer to the points $A = (3, 1)$; $B = (-2, 5)$; $C = (-3, -2)$; and $D = (1, -3)$.

Point A (3, 1) moves *right* and *up*, as shown in Fig. 2–7.

Point B (−2, 5) moves *left* and *up*, as shown in Fig. 2–8.

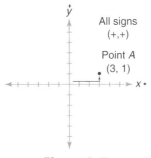

Figure 2–7

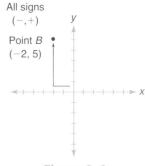

Figure 2–8

Point C (−
in Fig. 2–9.

To add signed numbers with unlike signs:

1. *Subtract* the smaller absolute value from the larger absolute value.
2. Give the sum the sign of the number with the larger absolute value.

EXAMPLE Add $7 + (-12)$.

$7 + (-12) =$ Signs are unlike.

$7 + (-12) = \mathbf{-5}$ Subtract absolute values: $12 - 7 = 5$. Keep the sign of the -12 (negative), the
                   larger absolute value.

TIP!

To Add, You . . . Subtract?

Phrases we used in arithmetic for adding integers, such as "find the sum" or "add," can be
confusing. In addition of integers, sometimes we add absolute values and sometimes we subtract
absolute values. We often use the word "combine" to imply the addition of numbers with either like
or unlike signs.

EXAMPLE Add $-5 + 7$.

$-5 + 7 =$ Signs are unlike.

$-5 + 7 = \mathbf{2}$ Subtract absolute values: $7 - 5 = 2$. Keep the sign of the 7 (positive), the larger
                   absolute value.

Learning Strategy *Rules and Procedures Are Shortcuts.*

Rules and procedures help us perform operations efficiently, but they don't take the place of understanding
the concepts. Let's look again at the examples $7 + (-12)$ and $-5 + 7$ in Figs. 2–18 and 2–19.
  Since we move back more spaces than we move forward, we end on the negative side of zero.

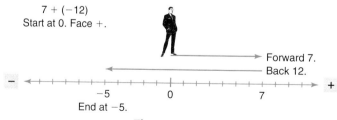

Figure 2–18

Since we move forward more spaces than we move back, we end on the positive side of zero.

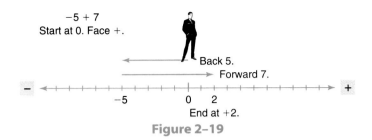

$-5 + 7$
Start at 0. Face $+$.

Back 5.
Forward 7.

-5 0   2

End at $+2$.

Figure 2–19

- The rule for subtracting absolute values when the numbers have *unlike* signs takes into account that the movements are in opposite directions.
- The rule for giving the sum the sign of the number with the larger absolute value takes into account that the number that creates the larger movement determines whether the sum will be positive or negative.

When an addition involves several positive numbers and several negative numbers, it is convenient to group all the positive numbers and all the negative numbers and to add each group separately. We then add the sums of each group. We can do this because of the associative property of addition. The grouping of addends in an addition does not matter. In the next example we combine the commutative and associative properties of addition to add several signed numbers.

EXAMPLE   Add $8 + (-9) + 13 + (-15)$.

$8 + (-9) + 13 + (-15) =$ Signs are unlike. Rearrange and group numbers with like signs.

$[8 + 13] + [-9 + (-15)] =$ The brackets, [ ], show the grouping and separate the operational plus sign from the directional minus sign of -9. Add groups of numbers with like signs.

$21 + (-24) = -3$ Add resulting sums using the rule for adding numbers with unlike signs.

The properties of addition that involve the number zero apply to integers as well as to whole numbers and decimals. When zero is added to a number, the number is not changed. We say that zero is the *additive identity*. The sum of a number and its opposite is zero. We call the opposite of a number the *additive inverse* of that number. We can also write these definitions using symbols.

Zero is the *additive identity* because $a + 0 = a$ for all values of a.
The opposite of a number is the *additive inverse* of the number because $a + (-a) = 0$ for all values of a.

EXAMPLE   Add: (a) $-2 + 0$ (b) $5 + (-5)$ (c) $0 + (-3) + 4$ (d) $5 + (-2) + (-5)$

(a) $-2 + 0 = -2$ Additive identity.
(b) $5 + (-5) = 0$ Additive inverse.
(c) $0 + (-3) + 4 =$ Additive identity.
            $-3 + 4 = 1$ Add numbers with unlike signs.
(d) $5 + (-2) + (-5) =$ Commutative and associative properties.

        $5 + (-5) + (-2) =$ Additive inverse.

            $0 + (-2) = -2$ Additive identity.

SELF-STUDY EXERCISES 2-2

1 Add.

1. $8 + 15$

4. $(-12) + (-17)$

7. $7 + 10 + 12$

10. $-21 + (-38)$

13. $-5 + (-12) + (-36)$

2. $-9 + (-7)$

5. $-24 + (-31)$

8. $-5 + (-8) + (-7)$

11. $-32 + (-16)$

14. $52 + (+92) + (+88)$

3. $-5 + (-8)$

6. $-78 + (-46)$

9. $12 + 87$

12. $(-58) + (-103)$

15. $-107 + (-502) + (-396)$

2 Add.

16. $5 + (-3)$

19. $-8 + 4$

22. $-4 + (+2)$

25. $-4 + 6$

28. $21 + (-14)$

31. $47 + (-82) + 2$

34. $4 + 2 + (-3) + 10$

37. $15 + (-15)$

40. $(-396) + (396)$

43. $-3 + 0$

46. $-8 + 0$

17. $-3 + 5$

20. $9 + (-3)$

23. $-3 + 7$

26. $-18 + 8$

29. $17 + (-4) + 3 + (-1)$

32. $14 + (-6) + 1$

35. $2 + (-1) + 8$

38. $-7 + 7$

41. $503 + (-503)$

44. $0 + (-7)$

47. $0 + (-28)$

18. $8 + (-4)$

21. $3 + (-9)$

24. $4 + (-6)$

27. $32 + (-72)$

30. $-3 + 2 + (-7)$

33. $-7 + (-3) + (-1)$

36. $-2 + 1 + (-8) + 12$

39. $92 + (-92)$

42. $0 + 9$

45. $18 + 0$

48. $12 + 0 + (-12)$

49. Write a sum that shows the additive inverse.

51. Explain why the sum of two positive numbers is positive and the sum of two negative numbers is negative.

53. You open a bank account by depositing $242. You then write checks for $21, $32, and $123. What is your new account balance?

55. Memphis, Tennessee, is $-6{:}00$ GMT and Portland, Oregon, is 2 time zones west of Memphis. Movement westward is represented by negative numbers. Find the GMT for Portland.

50. What is the additive identity?

52. A stock priced at $42 has the following changes in one week: $-3, +8, -6, -7, +2$. What is the value of the stock at the end of the week?

54. Paris, France, is assigned Greenwich Mean Time (GMT) of $+1{:}00$, and Jerusalem, Israel, is 1 time zone east (1 hour) of Paris. If east is represented by adding a positive number, what is the GMT for Jerusalem, Israel?

56. The temperature at the South Pole is recorded as $-38°F$ at midnight. The temperature rises 15° by noon. Determine the temperature reading at noon.

2–3 Subtracting Integers

Learning Outcomes

1 Subtract integers.

2 Subtract with zero or opposites.

3 Combine addition and subtraction.

Before we discuss the subtraction of integers, let's examine two expressions involving the subtraction of integers, $6 - 9$ and $5 - (-3)$, and interpret the meaning of the notation. The first expression, $6 - 9$, means positive 9 is subtracted from positive 6. The sign between the 6 and the 9 means subtraction and is read as "6 subtract 9" or "6 minus 9," just as we did with whole numbers. The minus sign is the operational sign telling us to subtract positive 9 from positive 6. The expression can also be written as $6 - (+9)$.

The second expression, $5 - (-3)$, means that negative 3 is subtracted from positive 5. The first minus sign between 5 and 3 is the operational sign telling us to subtract. The second minus sign is the directional sign to show negative 3.

84 Chapter 2 / Integers

1 Subtract Integers.

To show subtraction on the number line, we face the positive end of the number line and step forward or backward as determined by the sign of the first number. Then we turn and face the negative end of the number line and step forward or backward as determined by the sign of the second number (subtrahend), as shown in Figs. 2–20 and 2–21.

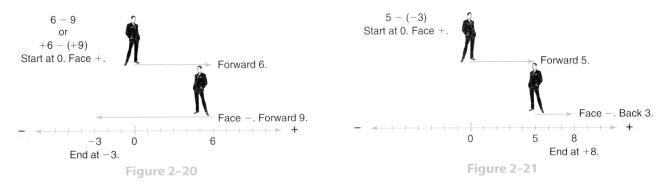

Figure 2–20

Figure 2–21

We can use the relationship between subtraction and addition to subtract integers efficiently.

We can show that subtracting a number is the *same as* adding the opposite of the number by using a number line.

$$6 - 9$$
$$+6 - (+9) \qquad \text{Subtract } +9.$$
$$+6 + (-9) \qquad \text{Add the opposite of } +9 \text{ or add } -9.$$

When we add on the number line, we always face the positive end of the number line, as shown in Figs. 2–22 and 2–23.

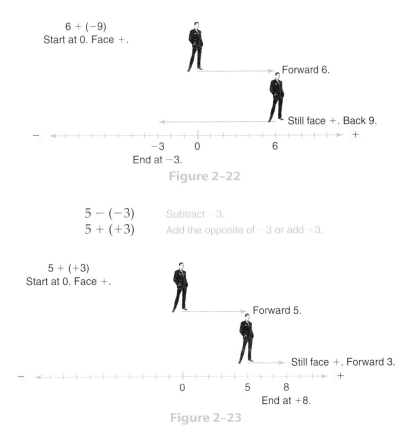

Figure 2–22

$$5 - (-3) \qquad \text{Subtract } -3.$$
$$5 + (+3) \qquad \text{Add the opposite of } -3 \text{ or add } +3.$$

Figure 2–23

Addition and subtraction are related. Look at this relationship in words.

To subtract signed numbers:

1. Change the subtraction sign to addition.
2. Change the sign of the second number (subtrahend) to its opposite.
3. Apply the appropriate rule for adding signed numbers.

EXAMPLE Subtract: (a) $2 - 6$ (b) $-9 - 5$ (c) $12 - (-4)$ (d) $-7 - (-9)$

(a) $\quad 2 - 6 =$ Subtract positive 6 from positive 2.
$\quad\quad 2 - (+6) =$ Write sign of subtrahend.
$\quad\quad 2 + (-6) =$ Add the opposite of $+6$, which is -6, to 2.
$\quad\quad 2 + (-6) = \mathbf{-4}$ Apply the rule for adding numbers with unlike signs.

(b) $\quad -9 - 5 =$ Subtract positive 5 from negative 9.
$\quad\quad -9 - (+5) =$ Write sign of subtrahend.
$\quad\quad -9 + (-5) =$ Add the opposite of $+5$, which is -5, to -9.
$\quad\quad -9 + (-5) = \mathbf{-14}$ Apply the rule for adding numbers with like signs.

(c) $\quad 12 - (-4) =$ Subtract negative 4 from positive 12.
$\quad\quad 12 + (+4) =$ Add the opposite of -4, which is $+4$, to 12.
$\quad\quad 12 + 4 = \mathbf{16}$ Apply the rule for adding numbers with like signs.

(d) $\quad -7 - (-9) =$ Subtract negative 9 from negative 7.
$\quad\quad -7 + (+9) =$ Add the opposite of -9, which is $+9$, to -7.
$\quad\quad -7 + 9 = \mathbf{2}$ Apply the rule for adding numbers with unlike signs.

TIP!

Writing Subtractions as Equivalent Additions

When writing a subtraction as an equivalent addition, you make *two* changes.

1. Change the operation from subtraction to addition.
2. Change the subtrahend to its opposite.

$2 - (+6)$	$-9 - (+5)$	$12 - (-4)$	$-7 - (-9)$
$2 + (-6)$	$-9 + (-5)$	$12 + (+4)$	$-7 + (+9)$
↑ ↑	↑ ↑	↑ ↑	↑ ↑
1. 2.	1. 2.	1. 2.	1. 2.

2 Subtract with Zero or Opposites.

Subtractions involving zero and opposites are interpreted first as equivalent addition problems. Since zero has no opposite, $+0$ and -0 are still 0.

EXAMPLE Evaluate: (a) $8 - 0$ (b) $0 - 15$ (c) $-32 - (-32)$ (d) $-15 - 15$

(a) $\quad 8 - 0 =$ Subtract zero from positive 8.
$\quad\quad 8 + (0) =$ Add zero to positive 8.
$\quad\quad 8 + (0) = \mathbf{8}$ Zero added to any number is that number (additive identity).

Chapter 2 / Integers

(b)	$0 - 15 =$	Subtract positive 15 from zero.
	$0 - (+15) =$	Write sign of subtrahend.
	$0 + (-15) =$	Add the opposite of $+15$, which is -15, to zero.
	$0 + (-15) = \mathbf{-15}$	Any number added to zero (additive identity) is that number.
(c)	$-32 - (-32) =$	Subtract negative 32 from negative 32.
	$-32 + (+32) =$	Add the opposite of -32, which is $+32$, to -32.
	$-32 + 32 = \mathbf{0}$	A number added to its opposite (additive inverse) is zero.
(d)	$-15 - (+15) =$	Subtract positive 15 from negative 15.
	$-15 + (-15) =$	Add the opposite of $+15$, which is -15, to -15.
	$-15 + (-15) = \mathbf{-30}$	Apply the rule for adding numbers with like signs.

3 **Combine Addition and Subtraction.**

When we express the sum and difference of integers with all the appropriate operational and directional signs, we have an expression with both an operational sign and a directional sign between every two numbers. We have four different possibilities when two signs are written together. The following allows us to simplify these four possibilities.

TIP!

Omitting Signs

Writing mathematical expressions that include every operational and directional sign is cumbersome. In general practice, we omit as many signs as possible. When two signs are written between two numbers, we can write a simplified expression with only one sign.

	Double Signs		Single Sign
Like Signs	Plus, Plus: $+3 + (+5)$	Add $+3$ and $+5$.	$3 + 5$
	Minus, Minus: $+3 - (-5)$	Change to addition.	
		$+3 + (+5)$	$3 + 5$
Unlike Signs	Plus, Minus: $+3 + (-5)$	Add $+3$ and -5.	$3 - 5$
	Minus, Plus: $+3 - (+5)$	Change to addition.	$3 - 5$
		$+3 + (-5)$	$3 - 5$

To generalize, *two like signs between integers,* whether both plus or both minus, translate to adding a positive number. Use just one plus sign:

$$+ + \rightarrow + \qquad - - \rightarrow + \qquad \textbf{like signs} \rightarrow +$$

Two unlike signs between integers, either a plus/minus, or a minus/plus, translate to adding a negative number. Use just one minus sign:

$$+ - \rightarrow - \qquad - + \rightarrow - \qquad \textbf{unlike signs} \rightarrow -$$

To add and subtract more than two numbers:

1. Rewrite the problem so that all integers are separated by only one sign.
2. Add a series of signed numbers.

Evaluate: (a) $3 - (-5) - 6$ (b) $-8 + 10 - (-7)$

(a) $3 - (-5) - 6 =$ Rewrite with only one sign between integers.

        $3 + 5 - 6 =$ Add numbers with like signs.

          $8 \quad - 6 = \mathbf{2}$ Apply the rule for adding numbers with unlike signs.

(b) $-8 + 10 - (-7) =$ Rewrite with only one sign between integers.

      $-8 + 10 + 7 =$ Add numbers with like signs.

          $-8 + 17 = \mathbf{9}$ Apply the rule for adding numbers with unlike signs.

The key to solving applied problems with integers is to identify which numbers are positive and which are negative.

Positive key words: profits, gains, money in the bank, temperatures above zero, receipts, income, winnings, and so on.

Negative key words: losses, deficits, checks that cleared the bank, temperatures below zero, drops, declines, payments, and so on.

A landscaping business makes a profit of \$345 one week, has a loss of \$34 the next week, and makes a profit of \$235 the third week. What is the total profit?

The total profit is the sum of the weekly profits and losses.

$\$345 \; - \; \$34 \; + \; \$235$ Interpret profits as positive and losses as negative.
profit   loss   profit

$345 + 235 \; - \; 34$ Rearrange and add positives.

      $580 \; - 34 = 546$ Apply rule for adding numbers with unlike signs.

The total profit for the   Interpret answer.
three weeks was \$546.

In a recent year, 98°F was the highest temperature in Boston, and −2°F was the lowest temperature. What was the temperature range for the city that year? (The range is the difference between the highest and lowest values.)

$98 - (-2) =$ Subtract −2 from 98.

    $98 + 2 = 100$ Rewrite with only one sign between integers, and apply the appropriate rule for adding integers.

The temperature range for   Interpret answer.
Boston that year was 100°F.

SELF-STUDY EXERCISES 2−3

1 Subtract.

1. $-3 - 9$ **2.** $8 - 2$ **3.** $9 - 15$ **4.** $-3 - (-7)$

5. $-11 - 14$ **6.** $-6 - (-3)$ **7.** $5 - (-3)$ **8.** $8 - 11$

9. $-8 - 1$ **10.** $11 - (-2)$ **11.** $(-8) - (-7)$ **12.** $(-15) - (-7)$

Subtract.

13. $15 - 0$ **14.** $0 - 8$ **15.** $-12 - 0$ **16.** $0 - (-8)$
17. $0 - (-7)$ **18.** $10 - 0$ **19.** $28 - (-28)$ **20.** $-46 - 46$
21. $7 - (-7)$ **22.** $-18 - 18$ **23.** $5 - 5$ **24.** $21 - 21$

3 Evaluate.

25. $5 + 8 - 9$ **26.** $6 - 4 + 5$ **27.** $9 + 2 - 5$ **28.** $-1 + 1 - 4$
29. $5 + 3 - 7$ **30.** $7 + 3 - (-4)$ **31.** $-8 - 2 - (-7)$ **32.** $-3 + 4 - 7 - 3$
33. $2 - 4 - 5 - 6 + 8$ **34.** $8 - 3 + 2 - 1 + 7$ **35.** $-5 - 3 + 8 - 2 + 4$ **36.** $6 - (-3) + 5 - 6 - 9$

37. The temperatures for Bowling Green, Kentucky, ranged from 102°F to −5°F. What was the temperature range for the city?

38. New Boston, Texas, registered −8°F as its lowest temperature one year and 99°F as its highest temperature for the same year. What was the temperature range for New Boston?

39. Computing Solutions records a profit of $28,296 one quarter (three months), a loss of $1,896 for the second quarter, a profit of $52,597 for the third quarter, and a profit of $36,057 for the fourth quarter. What is their net profit for the year?

40. Explain the difference between subtracting zero from a number and subtracting a number from zero.

41. Time zones from Greenwich, England, west to the international date line are assigned negative numbers from −1 to −12. If Lima, Peru, is assigned a Greenwich Mean Time (GMT) of −5:00 and Los Angeles, California, which is west of Lima, is assigned −8:00 (GMT), express the time difference between Lima and Los Angeles as a positive or negative number.

42. New Orleans is assigned a GMT of −6:00 and Denver is assigned a GMT of −7:00. Express the time difference from New Orleans to Denver as a positive or negative number.

2-4 | *Multiplying Integers*

Learning Outcomes
1 Multiply two integers.
2 Multiply several integers.
3 Multiply with zero.
4 Evaluate powers of integers.

Multiplying absolute values of integers is exactly the same as multiplying whole numbers. However, the rules for multiplying integers must also include the assignment of the proper sign to the product.

1 **Multiply Two Integers.**

Just like with whole numbers, multiplication of integers is a binary operation that is commutative and associative; that is, two factors can be multiplied at a time in any order. More than two factors can be grouped in any manner. These examples show how we assign signs to the product of two integers:

Like Signs

$4(6) = 24, \quad (-3)(-7) = 21$

Unlike Signs

$-10(2) = -20, \quad 8(-2) = -16$

To multiply two integers:

1. Multiply the absolute values of the numbers as in whole numbers.
2. If the factors have like signs, the sign of the product is positive.
3. If the factors have unlike signs, the sign of the product is negative.

EXAMPLE Multiply: (a) $-12(-2)$ (b) $10 \cdot 3$ (c) $25(-3)$ (d) $-5 * 7$

(a) $-12(-2) = \mathbf{24}$ Like signs give a positive product.
(b) $10 \cdot 3 = \mathbf{30}$ Like signs give a positive product.
(c) $25(-3) = \mathbf{-75}$ Unlike signs give a negative product.
(d) $-5 * 7 = \mathbf{-35}$ Unlike signs give a negative product.

When the number 1 is multiplied by a number, the result is the number. The number 1 is the *multiplicative identity* because $a \cdot 1 = 1 \cdot a = a$ for all values of a.

2 Multiply Several Integers.

Multiplication, like addition, is a binary operation, so when we multiply three or more factors, we multiply two at a time.

EXAMPLE Multiply: (a) $4(-2)(6)$ (b) $-3(4)(-5)$ (c) $-2(-8)(-3)$ (d) $-2(-3)(-4)(-1)$

(a) $4(-2)(6) =$ Multiply the first two factors and apply the rule for factors with unlike signs.

$\qquad -8(6) = \mathbf{-48}$ Multiply and apply the rule for factors with unlike signs.

(b) $-3(4)(-5) =$ Multiply the first two factors and apply the rule for factors with unlike signs.

$\qquad -12(-5) = \mathbf{60}$ Multiply and apply the rule for factors with like signs.

(c) $-2(-8)(-3) =$ Multiply the first two factors and apply the rule for factors with like signs.

$\qquad 16(-3) = \mathbf{-48}$ Multiply and apply the rule for factors with unlike signs.

(d) $-2(-3)(-4)(-1) =$ Multiply the first two and the last two factors.

$\qquad 6(4) = \mathbf{24}$ Multiply and apply the rule for factors with like signs.

If we examine the multiplications in the preceding example more closely, we see that the number of negative factors affects the sign of the answer.

	Number of Negative Factors	Sign of Product
$4(-2)(6)$	1	$-$
$-3(4)(-5)$	2	$+$
$-2(-8)(-3)$	3	$-$
$-2(-3)(-4)(-1)$	4	$+$

To determine the sign of the product when multiplying three or more factors:

1. The sign of the product is *positive* if the number of negative factors is *even*.
2. The sign of the product is *negative* if the number of negative factors is *odd*.

EXAMPLE Multiply: (a) $-2(6)(-1)(-3)$ (b) $(2)(-5)(1)(-3)$

(a) $-2(6)(-1)(-3) = \mathbf{-36}$ Multiply absolute values. The odd number of negative factors makes the product negative.

(b) $(2)(-5)(1)(-3) = \mathbf{30}$ Multiply absolute values. The even number of negative factors makes the product positive.

Chapter 2 / Integers

3 Multiply with Zero.

The *zero property of multiplication* extends to integers. The product of zero and any number is zero: $x \cdot 0 = 0$.

TIP!

Effect of a Zero Factor

If we have two or more factors and one factor is zero, we can immediately write the product as zero without having to work through the steps.

EXAMPLE Multiply: (a) $3(-21)(2)(0)$ (b) $-9(-2)(8)(-1)$

(a) $3(-21)(2)(0) = \mathbf{0}$ Zero is a factor.
(b) $-9(-2)(8)(-1) = \mathbf{-144}$ Zero is not a factor.

Applied problems often require multiplication of integers.

EXAMPLE In a three-week period, a technology stock declined approximately 2 points each week. How many points did the stock decline in the three weeks?

Because there are equal declines each week, we multiply the amount of weekly decline times the number of weeks: $-2(3) = -6$. Thus, **the stock declined (negative) a total of 6 points over the three-week period.**

4 Evaluate Powers of Integers.

Raising a number to a natural-number power is an extension of multiplication, so determining the sign of the result is similar to multiplying several integers. Observe the pattern for determining the sign of the result:

Positive Base	Negative Base
$(+4)^2 = (+4)(+4) = +16$	$(-4)^2 = (-4)(-4) = +16$
$(+4)^3 = (+4)(+4)(+4) = +64$	$(-4)^3 = (-4)(-4)(-4) = -64$

To raise integers to a natural-number power, use the following patterns:

1. A positive number raised to any natural-number power is positive.
2. Zero raised to any natural-number power is zero.
3. A negative number raised to an even natural-number power is positive.
4. A negative number raised to an odd natural-number power is negative.

EXAMPLE Evaluate the powers: (a) 4^3 (b) 0^8 (c) $(-2)^4$ (d) $(-3)^5$

(a) $4^3 = 4(4)(4) = \mathbf{64}$ Base is positive.
(b) $0^8 = \mathbf{0}$ Base is zero.
(c) $(-2)^4 = (-2)(-2)(-2)(-2) = \mathbf{16}$ Base is negative, exponent is even.
(d) $(-3)^5 = (-3)(-3)(-3)(-3)(-3) = \mathbf{-243}$ Base is negative, exponent is odd.

Or, use the distributive principle:

$$8(4 - 6) =$$ Distribute. Multiply each term in the parentheses by the factor 8.

$$8(4) + 8(-6) =$$

$$32 + -48 = -16$$ Add 32 and -48.

Symbols for division are also grouping symbols.

EXAMPLE   Evaluate $\dfrac{3 - 5}{2} - \dfrac{9}{2 + 1} + 4(5)$.

$$\dfrac{3 - 5}{2} - \dfrac{9}{2 + 1} + 4(5) =$$ Perform operations grouped by the fraction bar.

$$\dfrac{-2}{2} - \dfrac{9}{3} + 4(5) =$$ Multiply and divide.

$$-1 - 3 + 20 =$$ Add and subtract last.

$$-4 + 20 = 16$$

EXAMPLE   Evaluate $-12 \div 3 - (2)(-5)$.

$$-12 \div 3 - (2)(-5) =$$

$$-12 \div 3 - (2)(-5) =$$ Multiply and divide first.

$$-4 - (-10) =$$ Change to a single sign between the numbers.

$$-4 + 10 = 6$$ Add integers with unlike signs.

The problem in the preceding example can also be written without parentheses around the 2. The problem is then expressed as

$$-12 \div 3 - 2(-5)$$

We multiply and divide first, but notice how we perform the multiplication.

$$-12 \div 3 - 2(-5) =$$ Consider the minus sign as the sign of the 2. Add the results of the division and the multiplication. $-12 \div 3 = -4$ and $-2(-5) = 10$.

$$-4 + 10 =$$

$$-4 + 10 = 6$$ Add integers with unlike signs.

EXAMPLE   Evaluate $10 - 3(-2)$.

$$10 - 3(-2) =$$ Think of the multiplication as -3 times -2.

$$10 + 6 =$$ Interpret as addition.

$$10 + 6 = 16$$ Add integers with like signs.

Only after parentheses and all multiplications and/or divisions are taken care of do we perform the final additions and/or subtractions from left to right.

EXAMPLE   Evaluate $4 + 5(2 - 8)$.

$$4 + 5(2 - 8) =$$ Perform operations in parentheses.

$$4 + 5(-6) =$$ Multiply.

$$4 - 30 =$$ Add integers with unlike signs.

$$4 - 30 = -26$$

The Order of Operations Is Important.

Note what happens if we proceed *out of order* in the preceding example!

Incorrectly Worked

$4 + 5(2 - 8)$	*Incorrectly* add first instead of last.
$9(2 - 8)$	Perform operation in parentheses second instead of first.
$9(-6)$	Multiply last instead of second.
$9(-6) = -54$	Incorrect solution.

We get a wrong answer. *The order of operations must be followed to arrive at a correct solution.*

EXAMPLE Evaluate $5 + (-2)^3 - 3(4 + 1)$.

$5 + (-2)^3 - 3 \ (4 + 1) =$	Perform operation in parentheses.
$5 + (-2)^3 - 3 \ (5) =$	Raise to a power.
$5 + (-8) - 3(5) =$	Multiply.
$5 + (-8) - 15 =$	Change to one sign only between numbers.
$5 - 8 - 15 =$	First addition of integers.
$-3 - 15 =$	Remaining addition of integers.
-18	

Since negative numbers are such an integral part of our everyday lives, practically all types of calculators, even the basic calculator, can deal with negative values. However, the notation for negatives and the process for entering negatives varies widely from calculator to calculator. We will look at some of the most common options.

Some calculators use the subtraction key for both subtraction and entering negative numbers. Others have a special key for entering a negative sign that is labeled as a negative sign enclosed in parentheses $\boxed{(-)}$. A common option with basic calculators is the *sign-change key* $\boxed{+/-}$. This key is a *toggle key* and changes the sign of the number in the display to the opposite sign.

EXAMPLE Use a calculator to evaluate.

(a) $2 - 7$ (b) $(-7)(2)$ (c) $\dfrac{-4}{2} + 14$

The most common options:

(a) $2 - 7$ $2 \boxed{-} 7 \boxed{=} \Rightarrow -5$ $\boxed{=}$ may be labeled $\boxed{ENTER}$ or $\boxed{EXE}$.

(b) $(-7)(2)$ $\boxed{(-)} 7 \boxed{\times} 2 \boxed{=} \Rightarrow -14$

or

$\boxed{(-)} 7 \boxed{(} 2 \boxed{)} \boxed{=} \Rightarrow -14$

or

$7 \boxed{+/-} \boxed{\times} 2 \boxed{=} \Rightarrow -14$

Some calculators interpret parentheses with no operational symbol preceding as multiplication. The sign-change key is entered after the absolute value of the number.

(c) $\dfrac{-4}{2} + 14$ $\boxed{(-)} 4 \boxed{\div} 2 \boxed{=} \boxed{+} 14 \boxed{=} \Rightarrow 12$ $\boxed{=}$ may not be required if calculator applies the order of operations. A slash $\boxed{/}$ may be used for division on some calculators.

When using fraction terminology or notation to describe division, the numerator is divided by the denominator in all cases.

$$\text{denominator}\,\overline{)\text{numerator}} \qquad \text{numerator} \div \text{denominator} \qquad \frac{\text{numerator}}{\text{denominator}}$$

Another term for a fraction is *rational number*. A rational number can be written in various forms: natural numbers, whole numbers, fractions, and decimals. We use the fraction form to define a rational number.

A rational number is any number that can be written in the form of a fraction, with an integer for the numerator and a nonzero integer for the denominator.

To interpret natural numbers, whole numbers, and integers as rational numbers, consider one whole unit as one part. Thus, 5 can be written in fraction form as $\frac{5}{1}$.

EXAMPLE Answer each question as it relates to $\frac{2}{3}$ (Fig. 3–7).

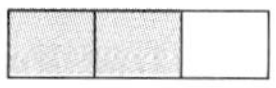

Figure 3–7

(a) One unit is divided into how many parts?
The denominator shows one whole is divided into **three parts**.

(b) How many of these parts are used?
The numerator shows that **two parts** are used.

(c) The numerator of the fraction is _____.
The numerator is the top number, so the numerator of $\frac{2}{3}$ is **2**.

(d) The denominator of the fraction is _____.
The denominator is the bottom number, so the denominator of $\frac{2}{3}$ is **3**.

(e) The fraction can be read as _____ divided by _____.
The numerator is divided by the denominator, so **2** is divided by **3**.

(f) _____ is the divisor of the division.
The divisor is the number we are dividing by, so **3** is the divisor.

(g) _____ is the dividend of the division.
The dividend is the number being divided, so **2** is the dividend.

(h) Is this common fraction proper or improper?
A proper fraction is less than one whole amount. In $\frac{2}{3}$, the whole is divided into three parts and 2 is less than 3. Therefore, $\frac{2}{3}$ is a **proper fraction**.

(i) This fraction represents _____ out of _____ parts.
$\frac{2}{3}$ is *two* **out of** *three* parts.

(j) Does this fraction have a value less than, more than, or equal to 1?
Since two parts is less than the three parts needed for one whole, then $\frac{2}{3}$ is **less than 1**.

Learning Strategy *Learning Definitions.*

A key strategy for learning new definitions is to understand the basic concepts, to connect the new words to other words or concepts you already know, and to use memory devices.

Learning terminology is important in developing your math survival skills. When you forget the details of a procedure (and you probably will if you don't use the procedure often), the terminology will help you use the *index* or *table of contents* to find the topic in the text that will refresh your skills.

An improper fraction can be written as a whole number when the denominator divides evenly into the numerator ($\frac{8}{4} = 2$). When the fraction is more than one unit and the denominator cannot divide evenly into the numerator, the improper fraction can be written as a combination of a whole number and a fractional part, such as $\frac{9}{4} = 2\frac{1}{4}$.

A *mixed number* consists of both a whole number and a fraction. The whole number and fraction are added together. *Example*: $3\frac{2}{5}$ means 3 whole units and $\frac{2}{5}$ of another unit or $3\frac{2}{5} = 3 + \frac{2}{5}$.

Sometimes in evaluating formulas or solving equations we have a fraction, decimal, or mixed number as the numerator or denominator of a fraction. This type of fraction is called a *complex fraction*. A *complex fraction* has a fraction or mixed number in its numerator or in its denominator, or in both. Some examples of complex fractions are

$$\frac{\frac{1}{2}}{2}, \quad \frac{7}{\frac{3}{4}}, \quad \frac{3\frac{1}{3}}{7}, \quad \frac{5}{4\frac{1}{2}}, \quad \frac{\frac{1}{8}}{\frac{5}{16}}, \quad \frac{6\frac{3}{8}}{4\frac{1}{2}}$$

It is very important to interpret these fractions as division.

$\dfrac{\frac{1}{2}}{2}$ is read as $\dfrac{1}{2}$ divided by 2. $\qquad \dfrac{7}{\frac{3}{4}}$ is read as 7 divided by $\dfrac{3}{4}$.

As you learned in Chapter 1, many job-related applications make use of a decimal fraction whose denominator is 10 or a power of 10, such as 100, 1,000, and so on. We summarize our study of rational numbers or fractions in Fig. 3–8.

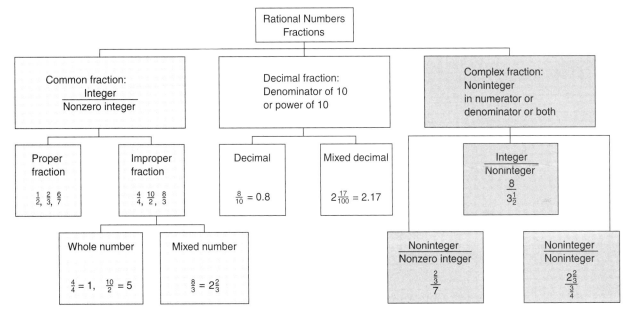

Figure 3–8

SELF-STUDY EXERCISES 3–1

1 Write a fraction to represent the shaded portion in Exercises 1–7 (Figs. 3–9 to 3–15).

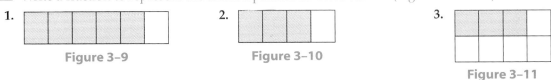

1.

Figure 3–9

2.

Figure 3–10

3.

Figure 3–11

4.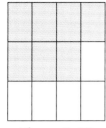

Figure 3–12

5.

Figure 3–13

6.

Figure 3–14

7.

Figure 3–15

Examine each common fraction in Exercises 8–11 and answer the questions a–j.

8. $\dfrac{5}{6}$

9. $\dfrac{8}{8}$

10. $\dfrac{11}{5}$

11. $\dfrac{12}{3}$

(a) One unit has been divided into how many parts?
(b) How many of these parts are used?
(c) The numerator of the fraction is _____.
(d) The denominator of the fraction is _____.
(e) The fraction can be read as _____ divided by _____.
(f) _____ is the divisor of the division.
(g) _____ is the dividend of the division.
(h) Is this common fraction proper or improper?
(i) This fraction represents _____ out of _____ parts.
(j) Does this fraction have a value less than, more than, or equal to 1?

Match the best description (a–f) with the number in each exercise.

_____**12.** $2\dfrac{3}{7}$

_____**13.** 0.689

_____**14.** $\dfrac{8}{3}$

_____**15.** $14\dfrac{3}{10}$

_____**16.** 6.27

_____**17.** 3 7

_____**18.** 0.3

_____**19.** $\dfrac{3}{7}$

_____**20.** $\dfrac{3}{1\frac{1}{2}}$

_____**21.** $\dfrac{1\frac{1}{2}}{3}$

_____**22.** $\dfrac{\frac{3}{4}}{\frac{2}{7}}$

(a) Decimal fraction
(b) Mixed number
(c) Complex fraction
(d) Mixed-decimal fraction
(e) Proper fraction
(f) Improper fraction

3–2 | Multiples, Divisibility, and Factor Pairs

Learning Outcomes

1 Find multiples of a natural number.

2 Determine the divisibility of a number.

3 Find all factor pairs of a natural number.

Fractions indicate division, and multiplication and division are inverse operations. Before we examine computations with fractions, let's look at some relationships involving multiplication and division.

1 **Find Multiples of a Natural Number.**

If we count by 3s, such as 3, 6, 9, 12, 15, 18, we obtain natural numbers that are *multiples* of 3 because each is the product of 3 and a natural number; that is,

$$3 = 3 \times 1, \qquad 6 = 3 \times 2, \qquad 9 = 3 \times 3$$
$$12 = 3 \times 4, \qquad 15 = 3 \times 5, \qquad 18 = 3 \times 6$$

A *multiple* of a natural number is the product of that number and a natural number.

EXAMPLE Show that 2, 4, 6, 8, and 10 are multiples of 2 by writing each as the product of 2 and a natural number.

$$2 = 2 \times 1, \quad 4 = 2 \times 2, \quad 6 = 2 \times 3, \quad 8 = 2 \times 4, \quad 10 = 2 \times 5$$

Natural numbers that are multiples of 2 are *even numbers*. Natural numbers that are not multiples of 2 are *odd numbers*.

EXAMPLE Find five multiples of 16.

$$1 \times 16 = 16, \quad 2 \times 16 = 32, \quad 3 \times 16 = 48, \quad 4 \times 16 = 64, \quad 5 \times 16 = 80$$

16, 32, 48, 64, 80 are multiples of 16.

2 Determine the Divisibility of a Number.

As we continue to develop our number sense, an important skill to acquire is to be able to determine when a number will divide into another number with no remainder. We say that a number is *divisible* by another number if the quotient has no remainder or if the dividend is a multiple of the divisor.

We want to be able to determine divisibility by *inspection*. This means that we can examine the number being divided (dividend) and decide if it is divisible by a divisor without actually having to perform the division.

EXAMPLE Is 35 divisible by 7?

35 is divisible by 7 if $35 \div 7$ has no remainder or if 35 is a multiple of 7.

$$35 \div 7 = 5 \quad \text{or} \quad 35 = 5 \times 7$$

Yes, 35 is divisible by 7.

EXAMPLE A holiday party is planned for 47 first-grade students. The teachers have a gross (144) of assorted party favors. Will each child receive an equal number of favors with none left over?

This problem is really asking whether 144 is divisible by 47.

$144 \div 47 = 3\text{R}3$ or $144 \div 47 = 3.063829787$ By calculator.

$144 \div 47$ has a remainder, and the calculator answer is not a natural number; therefore, 144 is not divisible by 47. **No, there will be 3 party favors left over.**

Rules or tests can help us decide by inspection if certain numbers are divisible by other numbers.

Tests for divisibility:

A number is divisible by

1. 2 if the last digit is an even number (0, 2, 4, 6, or 8).
2. 3 if the sum of its digits is divisible by 3.
3. 4 if the last two digits form a number that is divisible by 4.
4. 5 if the last digit is 0 or 5.
5. 6 if the number is divisible by *both* 2 *and* 3.

In the preceding example $\frac{1}{2}$ is multiplied by a fraction whose value is 1, and 1 times any number does not change the value of that number. Written symbolically,

$$\frac{n}{n} = 1 \text{ when } n \neq 0 \qquad \text{and} \qquad 1 \times n = n$$

Fractions in the same family can be generated by multiplying the fraction by 1 in the form of $\frac{2}{2}$, $\frac{3}{3}$, $\frac{4}{4}$, and so on.

The concept presented in the preceding example and tip is referred to as the *fundamental principle of fractions*. If the numerator and denominator of a fraction are multiplied by the same nonzero number, the value of the fraction remains unchanged.

To change a fraction to an equivalent fraction with a larger denominator:

1. Divide the larger denominator by the original denominator.
2. Multiply the original numerator and denominator by the quotient found in Step 1. That is, multiply by 1 in the form of $\frac{n}{n}$ when $n \neq 0$.

EXAMPLE   Find a fraction equivalent to $\frac{2}{3}$ that has a denominator of 12.

3 times what number is 12? Or 12 divided by 3 is what number? The answer is 4.

$$\frac{2 \times 4}{3 \times 4} \qquad \text{or} \qquad \frac{2}{3} \times \frac{4}{4} = \frac{8}{12} \qquad \text{Apply the fundamental principle of fractions.}$$

EXAMPLE   Change $\frac{5}{8}$ to an equivalent fraction whose denominator is 32.

$$\frac{5}{8} = \frac{?}{32}, \qquad 32 \div 8 = 4 \text{ and } 4 \times 5 = 20 \qquad \text{Apply the fundamental principle of fractions.}$$

$$\frac{5}{8} \times \frac{4}{4} = \frac{20}{32}$$

2 Write Equivalent Fractions with Lowest Denominators.

Because each fraction has an unlimited number of equivalent fractions, we usually work with decimal equivalents or fractions in *lowest terms*.

By *lowest terms*, we mean that no *whole* number other than 1 divides evenly into both the numerator and denominator. Another way of saying this is that the numerator and denominator have no common factors other than 1. When we find an equivalent fraction with smaller numbers and there are no common factors in the numerator and denominator, we have *reduced to lowest terms*.

To change a fraction to an equivalent fraction with a smaller denominator or to reduce a fraction to lowest terms:

1. Find a common factor greater than 1 for the numerator and denominator.
2. Divide both the numerator and denominator by this common factor.
3. Continue until the fraction is in lowest terms or has the desired smaller denominator.

Note: To reduce to lowest terms in the fewest steps, find the greatest common factor (GCF) in Step 1.

EXAMPLE Reduce $\frac{8}{10}$ to lowest terms.

Prime factors of 8: $2 \times 2 \times 2$ or 2^3

Prime factors of 10: 2×5

The greatest common factor (GCF) is 2.

$$\frac{8 \div 2}{10 \div 2} \qquad \text{or} \qquad \frac{8}{10} \div \frac{2}{2} = \frac{4}{5}$$

TIP!

Reducing and the Properties of 1

To reduce the fraction $\frac{8}{10}$, we divide by the whole number 1 in the form of $\frac{2}{2}$. That is, $\frac{8}{10} \div \frac{2}{2} = \frac{4}{5}$. A nonzero number divided by itself is 1, and to divide a number by 1 does not change the value of the number. Symbolically,

$$\frac{n}{n} = 1 \text{ when } n \neq 0 \qquad \text{and} \qquad n \div 1 = n \quad \text{or} \quad \frac{n}{1} = n$$

EXAMPLE Reduce $\frac{18}{24}$ to lowest terms.

Prime factors of 18: $2 \times 3 \times 3$ or 2×3^2
Prime factors of 24: $2 \times 2 \times 2 \times 3$ or $2^3 \times 3$
The GCF is 2×3 or 6.

$$\frac{18 \div 6}{24 \div 6} \qquad \text{or} \qquad \frac{18}{24} \div \frac{6}{6} = \frac{3}{4}$$

TIP!

Do You Have to Use the GCF to Reduce to Lowest Terms?

A fraction can be reduced to lowest terms in the fewest steps by using the *greatest common factor;* however, it can still be reduced correctly using any common factor; this just takes a few more steps.

$$\frac{18 \div 2}{24 \div 2} = \frac{9}{12} \qquad \text{Reduce with common factor 2.}$$

$$\frac{9 \div 3}{12 \div 3} = \frac{3}{4} \qquad \text{Reduce with common factor 3.}$$

Writing Whole Numbers in Mixed-Number Form

When you add mixed numbers and whole numbers, think of the whole number as a mixed number with zero as the numerator in the fraction.

$$16 = 16\frac{0}{24}$$

$$+ \ 1\frac{13}{24} = 1\frac{13}{24}$$

$$17\frac{13}{24}$$

EXAMPLE Find the largest acceptable measurement of a part if the blueprint calls for the part to be 2 in. long and the tolerance is $\pm\frac{1}{8}$ in.

Tolerance is the amount a part can vary from the blueprint specification. To find the largest acceptable measure, we add. Review tolerance in Chapter 1, Section 3, Outcome 2.

$$2 + \frac{1}{8} = 2\frac{1}{8}$$

The largest acceptable measure is $2\frac{1}{8}$ in.

SELF-STUDY EXERCISES 3–8

1 Add; reduce answers to lowest terms and convert any improper fractions to whole or mixed numbers.

1. $\frac{5}{16} + \frac{1}{16}$

2. $\frac{1}{2} + \frac{1}{8} + \frac{3}{4}$

3. $\frac{1}{8} + \frac{1}{2}$

4. $\frac{3}{8} + \frac{5}{32} + \frac{1}{4}$

5. $\frac{5}{16} + \frac{1}{4}$

6. $\frac{15}{16} + \frac{1}{2}$

7. $\frac{3}{32} + \frac{5}{64}$

8. $\frac{7}{8} + \frac{3}{5}$

9. $\frac{3}{4} + \frac{8}{9}$

10. $\frac{7}{8} + \frac{5}{24}$

11. $\frac{3}{5} + \frac{4}{5}$

12. $\frac{5}{7} + \frac{4}{21}$

13. What is the thickness of a countertop made of $\frac{7}{8}$-in. plywood and $\frac{1}{16}$-in. Formica?

14. Three pieces of steel are joined together. What is the total thickness if the pieces are $\frac{1}{2}$ in., $\frac{7}{16}$ in., and $\frac{29}{32}$ in.?

15. Three books are placed side by side. They are $\frac{5}{16}$ in., $\frac{7}{8}$ in., and $\frac{3}{4}$ in. wide. What is the total width of the books if they are polywrapped in one package?

16. Find the outside diameter of a pipe (Fig. 3–19) whose wall is $\frac{1}{2}$ in. thick if its inside diameter is $\frac{7}{8}$ in.

17. What length bolt is needed to fasten two pieces of metal each $\frac{7}{16}$ in. thick if a $\frac{1}{8}$-in. lock washer and a $\frac{1}{4}$-in. nut are used?

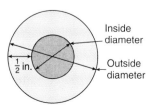

Figure 3–19

2 Add; reduce answers to lowest terms and convert any improper fractions to whole or mixed numbers.

18. $2\frac{3}{5} + 4\frac{1}{5}$

19. $1\frac{5}{8} + 2\frac{1}{2}$

20. $3\frac{3}{4} + 7\frac{3}{16} + 5\frac{7}{8}$

21. $\frac{1}{6} + \frac{7}{9} + \frac{2}{3}$

22. $2\frac{1}{4} + 2\frac{9}{16}$

23. $1\frac{5}{16} + 4\frac{7}{32}$

24. $3\frac{1}{4} + 1\frac{7}{16}$

25. $4\frac{1}{2} + 9$

26. $\frac{2}{3} + 3\frac{4}{5}$

27. $5\frac{1}{8} + 3 + 4\frac{9}{16}$

28. The studs (interior supports) of an outside wall are $5\frac{3}{4}$ in. thick. The inside wallboard is $\frac{7}{8}$ in. thick, and the outside covering is $2\frac{3}{16}$ in. thick. What is the total thickness of the wall?

29. A blueprint calls for a piece of bar stock $3\frac{7}{8}$ in. long. If a tolerance of $\pm\frac{1}{16}$ in. is allowed, what is the longest acceptable measurement for the bar stock?

30. If $4\frac{3}{8}$ gal of water are used to dilute $7\frac{1}{4}$ gal of acid, how many gallons are in the mixture?

31. How much bar stock is needed to make bars of the following lengths: $10\frac{1}{4}$ in., $8\frac{7}{16}$ in., $5\frac{15}{32}$ in.? Disregard waste.

32. Three pieces of I-beam each measuring $7\frac{5}{8}$ in. are needed to complete a job. How much I-beam is needed?

33. If $5\frac{1}{8}$ cups of water are mixed with $\frac{3}{4}$ cup of Kool Aid, how many cups are in the mixture?

3–9 | Subtracting Fractions and Mixed Numbers

Learning Outcomes
1 Subtract fractions.
2 Subtract mixed numbers.

1 Subtract Fractions.

The steps for subtracting fractions and mixed numbers are very similar to the steps for adding fractions and mixed numbers.

To subtract fractions:

1. If the denominators are not the same, find the least common denominator.
2. Change each fraction not expressed in terms of the common denominator to an equivalent fraction having the common denominator.
3. Subtract the numerators.
4. The common denominator will be the denominator of the difference.
5. Reduce the answer to lowest terms.

EXAMPLE Subtract $\dfrac{3}{8} - \dfrac{7}{32}$.

$$\frac{3}{8} = \frac{12}{32} \qquad \text{Change } \tfrac{3}{8} \text{ to an equivalent fraction with a denominator of 32.}$$

$$-\frac{7}{32} = \frac{7}{32} \qquad \text{Subtract numerators and keep the common denominator.}$$

$$\frac{5}{32}$$

2   **Subtract Mixed Numbers.**

To subtract mixed numbers:

1. If the fractional parts of the mixed numbers do not have the same denominator, change them to equivalent fractions with a common denominator.
2. When the fraction in the minuend is larger than the fraction in the subtrahend, go to Step 6.
3. When the fraction in the subtrahend is larger than the fraction in the minuend, borrow (regroup) one whole number from the whole-number part of the minuend. This makes the whole number 1 less.
4. Change the whole number borrowed to an improper fraction with the common denominator. For example, $1 = \frac{3}{3}$, $1 = \frac{8}{8}$, $1 = \frac{n}{n}$, where n is the common denominator.
5. Add the borrowed fraction ($\frac{n}{n}$) to the fraction already in the minuend.
6. Subtract the fractional parts and the whole-number parts.
7. Reduce the answer to lowest terms.

EXAMPLE   Subtract $15\frac{7}{8} - 4\frac{1}{2}$.

$$15\frac{7}{8} = 15\frac{7}{8}$$ LCD is 8. Change $\frac{1}{2}$ to $\frac{4}{8}$

$$-\ 4\frac{1}{2} = \ \ 4\frac{4}{8}$$ Subtract fractions.

$$\overline{\hspace{3em} 11\frac{3}{8}}$$ Subtract whole numbers.

EXAMPLE   Subtract $15\frac{3}{4}$ from $18\frac{1}{2}$.

LCD = 4. Change $\frac{1}{2}$ to $\frac{2}{4}$.

$$18\frac{1}{2} = 18\frac{2}{4} = 17\frac{4}{4} + \frac{2}{4} = 17\frac{6}{4}$$ Borrow. $18 - 1 = 17$, $1 = \frac{4}{4}$, $\frac{4}{4} + \frac{2}{4} = \frac{6}{4}$

$$-\ 15\frac{3}{4} = 15\frac{3}{4} = 15\frac{3}{4} \hspace{2em} = 15\frac{3}{4}$$ Subtract fractions.

$$\overline{\hspace{6em} 2\frac{3}{4}}$$ Subtract whole numbers.

EXAMPLE   A metal block weighing $127\frac{1}{2}$ lb is removed from a flatbed truck with a payload of $433\frac{3}{8}$ lb. How many pounds remain on the truck?

$$433\frac{3}{8} = 433\frac{3}{8} = 432\frac{11}{8}$$ $433 - 1 = 432$, $1 = \frac{8}{8}$, $\frac{8}{8} + \frac{3}{8} = \frac{11}{8}$

$$-\ 127\frac{1}{2} = 127\frac{4}{8} = 127\ \frac{4}{8}$$ Subtract fractions. Subtract whole numbers.

$$\overline{\hspace{6em} 305\ \frac{7}{8}}$$

$305\frac{7}{8}$ lb remains on the truck.

       Chapter 3 / Fractions

Think of Whole Numbers in Mixed-Number Form Before Subtracting.

As in addition, when subtracting whole numbers and mixed numbers, consider the whole number to have zero fractional parts. Then follow the same procedures as before. Borrow when necessary.

EXAMPLE   Subtract 27 from $45\frac{1}{3}$.

$$45\frac{1}{3} = 45\frac{1}{3}$$

$$\underline{- 27 \quad = 27\frac{0}{3}} \qquad \text{Write 27 as a mixed number. Subtract.}$$

$$18\frac{1}{3}$$

EXAMPLE   How many feet of wire are left on a 100-ft roll if $27\frac{1}{4}$ ft are used from the roll?

$$100 = 100\frac{0}{4} = 99\frac{4}{4} \qquad \text{Write 100 as a mixed number. Borrow.}$$

$$\underline{- 27\frac{1}{4} = 27\frac{1}{4} = 27\frac{1}{4}} \qquad \text{Subtract.}$$

$$72\frac{3}{4}$$

$72\frac{3}{4}$ ft of wire are left on the roll.

EXAMPLE   Three lengths measuring $5\frac{1}{4}$ in., $7\frac{3}{8}$ in., and $6\frac{1}{2}$ in. are cut from a 64-in. bar of angle iron. If $\frac{3}{16}$ in. is wasted on each cut, how many inches of angle iron remain?

Visualize the problem by making a sketch (see Fig. 3–20). Then find the total amount of angle iron used. This includes the three lengths and the waste for three cuts.

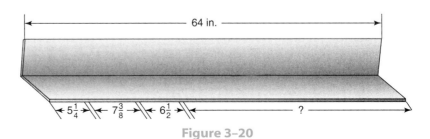

Figure 3–20

Total removed and wasted = 3 lengths + 3 cuts

Three Lengths   Three cuts

$$5\frac{1}{4} + 7\frac{3}{8} + 6\frac{1}{2} + \frac{3}{16} + \frac{3}{16} + \frac{3}{16} =$$

Change fractions to equivalent fractions with LCD of 16 and add 5 + 7 + 6 = 18

4 + 6 + 8 + 3 + 3 + 3 = 27.

$$5\frac{4}{16} + 7\frac{6}{16} + 6\frac{8}{16} + \frac{3}{16} + \frac{3}{16} + \frac{3}{16} = 18\frac{27}{16}$$ Write answer as standard mixed number.

$$18\frac{27}{16} = 18 + \frac{27}{16} = 18 + 1\frac{11}{16} = 19\frac{11}{16}$$ Total amount removed and wasted.

Amount of angle iron remaining = $\dfrac{\text{Beginning}}{\text{length}} - \dfrac{\text{Total iron removed}}{\text{and wasted}}$

$$64 - 19\frac{11}{16} =$$ Borrow and subtract.

$$63\frac{16}{16} - 19\frac{11}{16} = 44\frac{5}{16}$$ Total remaining.

$44\frac{5}{16}$ in. of angle iron remains.

SELF-STUDY EXERCISE 3-11

1 Subtract; reduce when necessary.

1. $\dfrac{7}{8} - \dfrac{5}{8}$
2. $\dfrac{9}{16} - \dfrac{3}{8}$
3. $\dfrac{7}{16} - \dfrac{3}{8}$
4. $\dfrac{5}{8} - \dfrac{1}{2}$
5. $\dfrac{5}{32} - \dfrac{1}{64}$
6. $\dfrac{7}{8} - \dfrac{3}{4}$

2 Subtract; reduce when necessary.

7. $9\dfrac{11}{16} - 5$
8. $23\dfrac{3}{16} - 5\dfrac{7}{16}$
9. $9\dfrac{1}{4} - 4\dfrac{5}{16}$
10. $9\dfrac{1}{32} - 3\dfrac{3}{8}$
11. $14\dfrac{1}{7} - 12\dfrac{3}{7}$
12. $8\dfrac{1}{2} - 6\dfrac{3}{4}$

13. A length of bar stock $16\frac{3}{8}$ in. long is cut so that a piece only $7\frac{9}{16}$ in. long remains. What is the length of the cutoff piece? Disregard waste.
14. A concrete foundation includes $7\frac{7}{8}$ in. of base fill. If the foundation is to be 18 in. thick, how thick must the concrete be?
15. A casting is machined so that $22\frac{1}{5}$ lb of metal remains. If the casting weighed $25\frac{3}{10}$ lb, how many pounds were removed by machine?
16. A bolt $2\frac{5}{8}$ in. long fastens two pieces of wood 1 in. and $1\frac{7}{32}$ in. thick. If a $\frac{3}{32}$-in.-thick lock washer and a $\frac{1}{8}$-in.-thick washer are used, what thickness is the nut if it is even with the end of the bolt? The measure of a bolt length does not include the bolt head.

17. Find the missing length in Fig. 3–21.

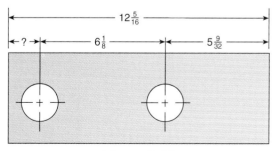

Figure 3–21

Learning Outcomes

1 Multiply fractions.

2 Multiply mixed numbers.

3 Raise a fraction to a power.

When multiplying a fraction by a fraction, we are finding *a part of a part*. For instance, $\frac{1}{2} \times \frac{1}{2}$ is $\frac{1}{2}$ of $\frac{1}{2}$ (Fig. 3–22). The word "of" is the clue that we must multiply to find the part we are looking for.

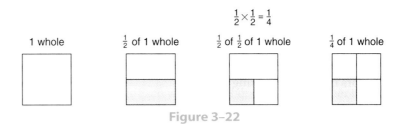

Figure 3–22

Adding or subtracting fractions and mixed numbers requires a common denominator. In multiplying fractions, we do *not* change fractions to equivalent fractions with a common denominator. Look at two more examples of taking a part of a part (Figs. 3–23 and 3–24).

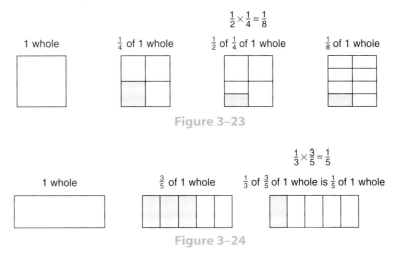

Figure 3–23

Figure 3–24

1 **Multiply Fractions.**

To multiply fractions:

1. Multiply the numerators of the fractions to get the numerator of the product.
2. Multiply the denominators to get the denominator of the product.
3. Reduce the product to lowest terms.

EXAMPLE Find $\frac{1}{2}$ of $\frac{1}{4}$.

$$\frac{1}{2} \times \frac{1}{4} = \frac{1}{8}$$ Multiply numerators.
    Multiply denominators.

Find $2\frac{1}{2} \div 3\frac{1}{3}$.

$2\frac{1}{2} \div 3\frac{1}{3} =$ Change each mixed or whole number to an improper fraction.

$\frac{5}{2} \div \frac{10}{3} =$ Change division to equivalent multiplication.

$\frac{\overset{1}{5}}{2} \times \frac{3}{\underset{2}{10}} =$ Reduce and multiply.

$\frac{3}{4}$

EXAMPLE   Find $5\frac{3}{8} \div 3$.

$5\frac{3}{8} \div 3 =$ Change mixed number and whole number to improper fractions.

$\frac{43}{8} \div \frac{3}{1} =$ Change division to equivalent multiplication.

$\frac{43}{8} \times \frac{1}{3} =$ Multiply.

$\frac{43}{24} = 1\frac{19}{24}$

EXAMPLE   A developer subdivides $5\frac{1}{4}$ acres into lots; each lot is $\frac{7}{10}$ of an acre. How many lots are made?

$$5\frac{1}{4} \div \frac{7}{10} = \frac{21}{4} \div \frac{7}{10} = \frac{\overset{3}{21}}{\underset{2}{4}} \times \frac{\overset{5}{10}}{\underset{1}{7}} = \frac{15}{2} = 7\frac{1}{2}$$

Seven lots are made and each is $\frac{7}{10}$ of an acre. The $\frac{1}{2}$ lot is left over or combined with one of the other lots.

EXAMPLE   How many pieces of cable, each $1\frac{3}{4}$ ft long, can be cut from a reel containing 100 ft of cable? See Fig. 3–27.

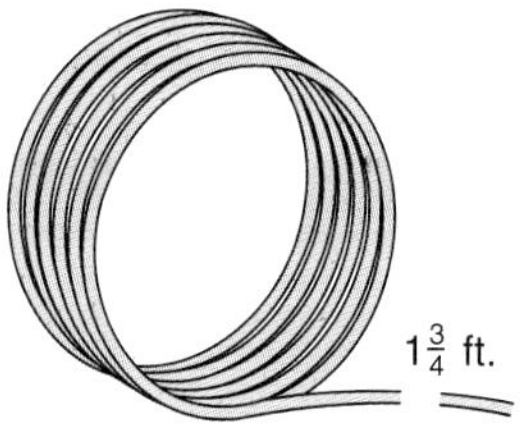

$$100 \div 1\frac{3}{4} = \frac{100}{1} \div \frac{7}{4} = \frac{100}{1} \times \frac{4}{7} = \frac{400}{7} = 57\frac{1}{7}$$

Fifty-seven pieces of cable can be cut to the desired length. The extra $\frac{1}{7}$ of the desired length is considered waste.

$1\frac{3}{4}$ ft.

Figure 3–27

A piece of trophy column stock that is $21\frac{1}{2}$ in. long is cut into four equal trophy columns (Fig. 3–28). If $\frac{1}{16}$ in. is wasted on each cut, find the length of each piece:

Figure 3–28

Use the Six-Step Problem-Solving Strategy.

Unknown facts   Length of cuts to be made.

Known facts   4   Note the number of pieces needed.

3   Note the number of cuts to be made ($4 - 1 = 3$).

$\frac{1}{16}$ in.   Note the waste for each cut.

$21\frac{1}{2}$ in.   Note the length of trophy column stock.

Relationships   Length of each piece = (Total length $-$ 3 cuts $\times$ Amount wasted for each cut) $\div$ 4 pieces of stock needed.

Estimation   If the stock is 20 in. long and cut into 4 pieces and if waste is disregarded, each piece will be 5 in.

Calculation   Three cuts are to be made and each cut wastes $\frac{1}{16}$ in. Find the amount wasted.

$$\frac{1}{16} \times 3 = \frac{3}{16}$$ Total waste.

$$21\frac{1}{2} - \frac{3}{16}$$ Subtract to find the amount of stock that will be left to divide equally into four trophy columns.

$$21\frac{1}{2} = 21\frac{8}{16}$$ Align vertically and use common denominators.

$$-\frac{3}{16} = \frac{3}{16}$$

$$\overline{\qquad 21\frac{5}{16} \text{ in.}}$$ Amount of stock left to be divided.

Find the length of each trophy column.

$$21\frac{5}{16} \div 4 = \frac{341}{16} \div \frac{4}{1} = \frac{341}{16} \times \frac{1}{4} = \frac{341}{64} = 5\frac{21}{64}$$

Interpretation   **Each trophy column is $5\frac{21}{64}$ in. long.**

4 Simplify Complex Fractions.

A *complex fraction* is a fraction in which either the numerator or the denominator or both contain a fraction, or a mixed number. Recall also that fractions indicate division as we read from top to bottom. The large fraction line is read as "divided by." For example,

$$\frac{2\frac{1}{2}}{7} \quad \text{is read} \quad \text{"}2\frac{1}{2}\text{ divided by 7"}$$

$$\frac{4}{1\frac{1}{2}} \quad \text{is read} \quad \text{"4 divided by }1\frac{1}{2}\text{"}$$

If we think of fractions as divisions, then complex fractions are another way of writing divisions.

To simplify a complex fraction:

1. Rewrite the fraction with the divided by symbol ÷.
2. Perform the indicated division.

EXAMPLE   Simplify $\dfrac{6\frac{3}{8}}{4\frac{1}{2}}$.

$$\dfrac{6\frac{3}{8}}{4\frac{1}{2}} =$$ Rewrite using the ÷ symbol.

$$6\frac{3}{8} \div 4\frac{1}{2} =$$ Perform the indicated division. Write mixed numbers as improper fractions.

$$\frac{51}{8} \div \frac{9}{2} =$$ Change division to multiplication.

$$\frac{\overset{17}{\cancel{51}}}{\underset{4}{\cancel{8}}} \times \frac{\overset{1}{\cancel{2}}}{\underset{3}{\cancel{9}}} =$$ Reduce and multiply.

$$\frac{17}{12} = \mathbf{1\frac{5}{12}}$$ Write as a mixed number.

We sometimes use complex fractions to change mixed decimals to fractions.

EXAMPLE   Write $0.33\frac{1}{3}$ as a fraction.

$$0.33\frac{1}{3} = \frac{33\frac{1}{3}}{100} = 33\frac{1}{3} \div 100$$ Count places for digits only; that is, do not count the fraction $\frac{1}{3}$ as a place.

$$33\frac{1}{3} \div 100 = \frac{100}{3} \div \frac{100}{1} = \frac{\overset{1}{\cancel{100}}}{3} \times \frac{1}{\underset{1}{\cancel{100}}} = \mathbf{\frac{1}{3}}$$ Change division to multiplication. Reduce and multiply.

What Place Does $\frac{1}{3}$ Hold If the Decimal Is $0.33\frac{1}{3}$?

When changing decimals to fractions, we divide by the place value of the last digit. Is $\frac{1}{3}$ in $0.33\frac{1}{3}$ in the hundredth's or thousandth's place? Hundredth's. The fraction attaches to the last digit.

$0.12\frac{1}{2}$ is read "twelve and one-half hundredths" $0.008\frac{1}{3}$ is read "eight and one-third thousandths"

1 Give the reciprocal.

1. $\dfrac{3}{7}$

2. $\dfrac{5}{8}$

3. $2\dfrac{1}{5}$

4. $3\dfrac{2}{3}$

5. 7

6. 8

7. 0.9

8. 0.25

9. 1.3

10. 1.8

2 Divide and reduce answers to lowest terms. Convert improper fractions to whole or mixed numbers.

11. $\dfrac{1}{2} \div \dfrac{7}{12}$

12. $\dfrac{4}{5} \div \dfrac{8}{9}$

13. $\dfrac{11}{32} \div \dfrac{3}{8}$

14. $\dfrac{3}{4} \div \dfrac{3}{8}$

15. $\dfrac{7}{10} \div \dfrac{2}{3}$

16. $\dfrac{5}{6} \div \dfrac{2}{5}$

17. $\dfrac{7}{8} \div \dfrac{3}{16}$

18. $\dfrac{1}{5} \div \dfrac{1}{2}$

19. One-half inch is divided into $\frac{1}{16}$-in. segments. How many $\frac{1}{16}$-in. segments are in $\frac{1}{2}$ in.?

20. Seven-sixteenths is divided into $\frac{1}{64}$-in. segments. How many $\frac{1}{64}$-inch segments are in $\frac{7}{16}$ in?

3 Divide and reduce answers to lowest terms. Convert improper fractions to whole or mixed numbers.

21. $8 \div \dfrac{1}{4}$

22. $10 \div \dfrac{3}{4}$

23. $3\dfrac{1}{8} \div \dfrac{1}{4}$

24. $2\dfrac{1}{2} \div 4$

25. $7\dfrac{3}{8} \div \dfrac{1}{2}$

26. $30\dfrac{1}{3} \div 4\dfrac{1}{3}$

27. $1\dfrac{1}{7} \div \dfrac{2}{7}$

28. $3\dfrac{3}{4} \div 1\dfrac{1}{2}$

29. $2\dfrac{2}{9} \div 1\dfrac{2}{3}$

30. $2\dfrac{1}{10} \div 3\dfrac{1}{2}$

31. Lumber is given in rough dimensions. Rough lumber that is 2 in. thick finishes to be $1\frac{5}{8}$ in. How many finished 2 × 4's are in a stack that is $29\frac{1}{4}$ in. high?

32. How many $4\frac{5}{8}$-ft lengths can be cut from a 50-ft length of conduit?

33. A truck will hold 21 yd³ (cubic yards) of gravel. If an earth mover has a shovel capacity of $1\frac{3}{4}$ yd³, how many shovelfuls are needed to fill the truck?

34. Three shelves of equal length are cut from a 72-in. board. If $\frac{1}{8}$ in. is wasted on each cut, what is the maximum length of each shelf? (Two cuts are made to divide the entire board into three equal lengths.)

35. If $\frac{1}{8}$ in. represents 1 ft on a drawing, find the dimensions of a room that measures $2\frac{1}{2}$ in. by $1\frac{7}{8}$ in. on the drawing. (How many $\frac{1}{8}$s are there in $2\frac{1}{2}$; how many $\frac{1}{8}$s are there in $1\frac{7}{8}$?)

36. A segment of I-beam is $10\frac{1}{2}$ ft long. Into how many whole $2\frac{1}{4}$-ft pieces can it be divided? Disregard waste.

37. How many $17\frac{5}{8}$-in. strips of quarter-round molding can be cut from a piece $132\frac{3}{4}$ in. long? Disregard waste.

38. How many $9\frac{1}{4}$-in. drinking straws can be cut from a $216\frac{1}{2}$-in. length of stock? How much stock is left over?

39. It takes $12\frac{1}{2}$ minutes to sand a piece of oak stock. If Isom Tibbs works 100 minutes, how many pieces does he sand?

4 Divide and reduce answers to lowest terms. Convert improper fractions to whole or mixed numbers.

40. $\dfrac{\frac{1}{2}}{\frac{2}{5}}$

41. $\dfrac{\frac{2}{2}}{\frac{2}{3}}$

42. $\dfrac{\frac{7}{8}}{1\frac{1}{4}}$

43. $\dfrac{2\frac{2}{5}}{5}$

44. $\dfrac{3\frac{1}{4}}{9\frac{3}{4}}$

45. $\dfrac{12\frac{1}{2}}{2\frac{1}{2}}$

46. $\dfrac{\frac{7}{8}}{1\frac{1}{4}}$

47. $\dfrac{1\frac{1}{3}}{\frac{1}{6}}$

48. $\dfrac{10}{3\frac{1}{3}}$

49. $\dfrac{33\frac{1}{3}}{100}$

Write as a fraction.

50. $0.16\frac{2}{3}$

51. $0.83\frac{1}{3}$

52. $0.66\frac{2}{3}$

53. $0.37\frac{1}{2}$

54. $0.12\frac{1}{2}$

3–12 | *Signed Fractions and Decimals*

Learning Outcomes

1 Change a signed fraction to an equivalent signed fraction.

2 Perform basic operations with signed fractions.

3 Perform basic operations with signed decimals.

4 Apply the order of operations with signed fractions and decimals.

1 Change a Signed Fraction to an Equivalent Signed Fraction.

A fraction has three signs: the sign of the fraction, the sign of the numerator, and the sign of the denominator. The fraction $\frac{2}{3}$ expressed as a *signed fraction* is $+\frac{+2}{+3}$. When a signed fraction has negative signs, we sometimes change the signed fraction to an equivalent signed fraction.

The rules for operating with integers can be extended to apply to signed fractions.

To find an equivalent signed fraction:

1. Identify the three signs of the fraction (sign of the fraction, sign of the numerator, and the sign of the denominator).
2. Change any two of the three signs to the opposite sign.

EXAMPLE Change $-\dfrac{-2}{-3}$ to three equivalent signed fractions.

$$-\frac{-2}{-3} = +\frac{+2}{-3}$$ Change the signs of the fraction and the numerator.

$$-\frac{-2}{-3} = +\frac{-2}{+3}$$ Change the signs of the fraction and the denominator.

$$-\frac{-2}{-3} = -\frac{+2}{+3}$$ Change the signs of the numerator and the denominator.

TIP!

Changing the signs of a fraction is a manipulation tool that simplifies our work when we perform basic operations with signed fractions.

Why Would We Want to Change the Signs of a Fraction?

When changing any two signs of a fraction, we can accomplish these desirable outcomes.

- Avoid dealing with negatives.

$$-\frac{-3}{+4} = +\frac{+3}{+4} \quad \text{or} \quad \frac{3}{4}$$

- Avoid negative denominators.

$$-\frac{+3}{-4} = +\frac{+3}{+4} \quad \text{or} \quad \frac{3}{4}$$

$$+\frac{-6}{-7} = +\frac{+6}{+7} \quad \text{or} \quad \frac{6}{7}$$

$$-\frac{+5}{-9} = +\frac{+5}{+9} \quad \text{or} \quad \frac{5}{9}$$

- Change subtraction to addition.

$$-\frac{+5}{+6} = +\frac{-5}{+6} \quad \text{or} \quad \frac{-5}{6}$$

$$+\frac{+2}{-5} = +\frac{-2}{+5} \quad \text{or} \quad \frac{-2}{5}$$

- Deal with fewer negatives.

$$-\frac{-5}{-8} = +\frac{-5}{+8} \quad \text{or} \quad \frac{-5}{8}$$

2 **Perform Basic Operations with Signed Fractions.**

To add and subtract signed fractions:

1. Change the signs of the fractions so the denominators are positive and subtractions are expressed as addition.
2. Apply the appropriate rules for adding integers (signed numbers) and for adding fractions.

EXAMPLE   Add $\dfrac{-3}{4} + \dfrac{5}{-8}$.

$\dfrac{-3}{4} + \dfrac{-5}{8}$ Change the signs of the numerator and denominator in the second fraction so that both denominators are positive.

$\dfrac{-6}{8} + \dfrac{-5}{8}$ Change to equivalent fractions with a common denominator.

$\dfrac{-11}{8}$ Add numerators, applying the rule for adding numbers with like signs.

$-1\dfrac{3}{8}$ Change to mixed number. The sign of the mixed number is determined by the rule for dividing numbers with unlike signs.

EXAMPLE   Subtract $\dfrac{-3}{7} - \dfrac{-5}{7}$.

$\dfrac{-3}{7} + \dfrac{5}{7}$ Change subtraction to addition by changing the signs of the second fraction and the numerator.

$\dfrac{2}{7}$ Apply the rule for adding numbers with unlike signs.

To multiply or divide signed fractions:

1. Write divisions as equivalent multiplications.
2. Apply the appropriate rules for multiplying integers (signed numbers) and for multiplying fractions.

EXAMPLE   Multiply $\left(\dfrac{-4}{5}\right)\left(\dfrac{3}{-7}\right)$.

$$\dfrac{-4}{5} \cdot \dfrac{3}{-7}$$ Multiply numerators and denominators.

$$\dfrac{-12}{-35}$$ Apply rule for dividing numbers with like signs.

$$\dfrac{\mathbf{12}}{\mathbf{35}}$$

EXAMPLE   Simplify $\left(\dfrac{-2}{3}\right)^3$.

$$\left(\dfrac{-2}{3}\right)^3$$ Cube numerator and cube denominator.

$$\dfrac{(-2)^3}{3^3}$$ $(-2)(-2)(-2) = -8;\ (3)(3)(3) = 27$

$$\dfrac{-8}{27} \quad \text{or} \quad -\dfrac{8}{27}$$ Manipulate signs if desired.

 1 **Perform Basic Operations with Signed Decimals.**

To perform basic operations with signed decimals:

1. Determine the indicated operations.
2. Apply the appropriate rules for signed numbers and for decimals.

EXAMPLE   Add -5.32 and -3.24.

$$\begin{array}{r} -5.32 \\ -3.24 \\ \hline \mathbf{-8.56} \end{array}$$ Align decimals and use the rule for adding numbers with like signs.

EXAMPLE   Subtract -3.7 from 8.5.

$$8.5 - (-3.7) = 8.5 + 3.7 = \mathbf{12.2}$$ Change subtraction to an equivalent addition and add numbers with like signs.

EXAMPLE   Multiply 3.91 and -7.1.

$$\begin{array}{r} 3.91 \\ \times\ -7.1 \\ \hline 391 \\ 27\ 37 \\ \hline \mathbf{-27.761} \end{array}$$ Apply rule for multiplying numbers with unlike signs.

EXAMPLE   Divide: $(-1.2) \div (-0.4)$.

$$-0.4 \overline{\smash{)}\,-1.2}^{\,\,+3.}$$

Shift decimal points and divide. Use rule for dividing numbers with like signs.

TIP!

The Rules of Signed Numbers Apply to All Types of Numbers.

The rules of signed numbers apply to all types of numbers. First apply the appropriate rule for signed numbers, then perform the necessary calculations.

4   **Apply the Order of Operations with Signed Fractions and Decimals.**

Signed fractions and decimals follow the same order of operations as whole numbers and integers. Review Chapter 1, Section 7 and Chapter 2, Section 6.

EXAMPLE   Evaluate $-\frac{1}{2} + 4\left(\frac{3}{8} - \frac{5}{8}\right)$.

$$-\frac{1}{2} + 4\left(\frac{3}{8} - \frac{5}{8}\right) =$$ Perform operation inside grouping. $\frac{3}{8} - \frac{5}{8} = \frac{-2}{8} = -\frac{1}{4}$.

$$-\frac{1}{2} + 4\left(-\frac{1}{4}\right) =$$ Multiply. $4\left(-\frac{1}{4}\right) = -1$.

$$-\frac{1}{2} - 1 =$$ Change -1 to an equivalent fraction with a denominator of 2.

$$-\frac{1}{2} - \frac{2}{2} =$$ Add like fractions with like signs.

$$-\frac{3}{2} \text{ or } -1\frac{1}{2}$$ Change to signed mixed number if desired.

EXAMPLE   Evaluate $3.2 - 1.7\,(0.2)^3$.

$$3.2 - 1.7(0.2)^3 =$$ Raise 0.2 to the third power. $(0.2)(0.2)(0.2) = 0.008$.

$$3.2 - 1.7(0.008) =$$ Multiply. $-1.7(0.008) = -0.0136$.

$$3.2 - 0.0136 =$$ Subtract (add opposite).

3.1864

SELF-STUDY EXERCISES 3–12

1   Change the fractions to three equivalent signed fractions.

1. $+\dfrac{+5}{+8}$

2. $-\dfrac{3}{4}$

3. $\dfrac{-2}{-5}$

4. $-\dfrac{-7}{-8}$

5. $\dfrac{7}{8}$

2 Perform the operations.

6. $\dfrac{-7}{8} + \dfrac{5}{8}$ 　　**7.** $\dfrac{-4}{5} + \dfrac{-3}{10}$ 　　**8.** $\dfrac{1}{2} - \dfrac{-3}{5}$ 　　**9.** $\dfrac{-3}{5} \times \dfrac{10}{-11}$ 　　**10.** $-\dfrac{5}{8} \div \dfrac{4}{5}$

3 Perform the operations with signed decimals. Round to hundredths if necessary.

11. $5.823 - 32.12$ 　　　　**12.** $-8.32 + 7.21$ 　　　　**13.** $-84.23 - 7.21$ 　　　　**14.** $34.6(-3.2)$
15. $-7.2(8.2)$ 　　　　　**16.** $-83.1(-4.1)$ 　　　　**17.** $83.2 \div (-3)$ 　　　　**18.** $-0.826 \div (-2)$
19. $-3.2 + 7.8$ 　　　　　**20.** $4.23 - 4.2$ 　　　　　**21.** $4.6 \div (-2)$ 　　　　**22.** $-3.8 + (-1.7)$

4 Evaluate.

23. $\dfrac{4}{5} + 3\left(-\dfrac{2}{3} + \dfrac{1}{3}\right)$ 　　　　**24.** $\dfrac{-5}{8} + 7\left(\dfrac{1}{8} - \dfrac{7}{8}\right)$ 　　　　**25.** $\dfrac{1}{2} - 3\left(2\dfrac{3}{4} - \dfrac{1}{4}\right)$

26. $-\dfrac{3}{5} + \dfrac{1}{2}(2 - 8)$ 　　　　**27.** $\left(-\dfrac{2}{3}\right)^2 - \dfrac{1}{2}\left(\dfrac{3}{8}\right)$ 　　　　**28.** $-\dfrac{3}{5} + \left(\dfrac{2}{7}\right)^2$

29. $5.2 + 3.8(-4.1)$ 　　　　**30.** $1.3 - (2.1)^2(1.2)$ 　　　　**31.** $(0.3)^2 + 5.7(-2.1)$

ASSIGNMENT EXERCISES

Section 3–1

Write a fraction that represents the shaded portion of Figs. 3–29 to 3–33.

1.

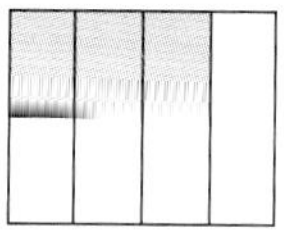

Figure 3–29

2.

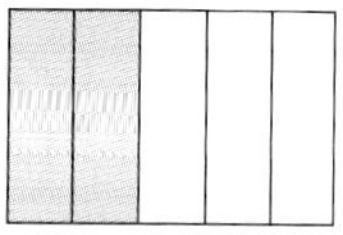

Figure 3–30

3.

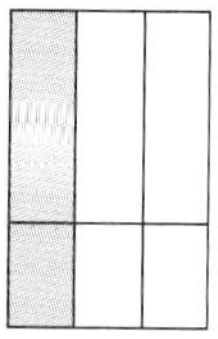

Figure 3–31

4.

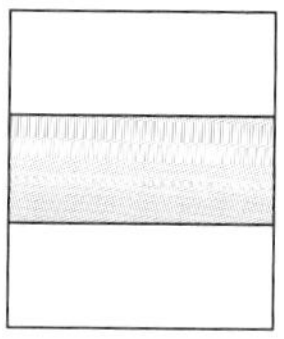

Figure 3–32

5.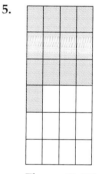

Figure 3–33

Use the numbers in Exercises 6–8 to answer the questions a–j for each number.

6. $\dfrac{3}{7}$

7. $\dfrac{9}{4}$

8. $\dfrac{2}{2}$

(a) What is the numerator of the fraction?
(b) What is the denominator of the fraction?
(c) One unit has been divided into how many parts?
(d) How many of these parts are used?
(e) The fraction can be read as ____ divided by ____.
(f) What is the divisor of the division?
(g) What is the dividend of the division?
(h) Is this fraction proper or improper?
(i) Does this fraction have a value less than, more than, or equal to 1?
(j) This fraction represents ____ out of ____ parts.

Match the best description a–f with each fraction.

____ **9.** $\dfrac{5}{2}$ 　　　____**10.** 0.172 　　　____**11.** $6\dfrac{1}{7}$ 　　　____**12.** 4.59

____**13.** $\dfrac{7}{11}$ 　　　____**14.** $\dfrac{\frac{2}{3}}{\frac{1}{4}}$ 　　　____**15.** $33\dfrac{1}{3}$

(a) Mixed-decimal fraction
(b) Decimal fraction
(c) Complex fraction
(d) Mixed number
(e) Proper fraction
(f) Improper fraction

Write five multiples of each number.

16. 31 **17.** 10 **18.** 9 **19.** 21 **20.** 14
21. 7 **22.** 11 **23.** 8 **24.** 15

Is the number divisible by the given number? Explain.

25. 153 by 3 **26.** 8,234 by 4 **27.** 8,726 by 6 **28.** 5,986 by 5
29. 240 by 10 **30.** 5,845 by 5 **31.** 63,539 by 9 **32.** 52,428 by 8

List all factor pairs for each number, then write the factors in order from smallest to largest.

33. 48 **34.** 50 **35.** 51 **36.** 63 **37.** 74

Section 3-3

Identify each number as *prime* or *composite*. Explain.

38. 17 **39.** 18 **40.** 20 **41.** 21 **42.** 29

Find the prime factorization of each number. Write in factored form and then in exponential notation.

43. 42 **44.** 48 **45.** 98 **46.** 120

Section 3-4

Find the least common multiple.

47. 18 and 40 **48.** 12 and 18 **49.** 12, 18, and 30 **50.** 6, 10, and 12

Find the greatest common factor.

51. 10 and 12 **52.** 12 and 18 **53.** 12, 18, and 30 **54.** 4, 9, and 16

Section 3-5

Find the equivalent fractions using the indicated denominators.

55. $\dfrac{5}{8} = \dfrac{?}{24}$ **56.** $\dfrac{3}{7} = \dfrac{?}{35}$ **57.** $\dfrac{5}{12} = \dfrac{?}{60}$ **58.** $\dfrac{4}{5} = \dfrac{?}{40}$

59. $\dfrac{2}{3} = \dfrac{?}{15}$ **60.** $\dfrac{4}{9} = \dfrac{?}{18}$ **61.** $\dfrac{3}{4} = \dfrac{?}{32}$ **62.** $\dfrac{1}{6} = \dfrac{?}{30}$

63. $\dfrac{1}{5} = \dfrac{?}{55}$ **64.** $\dfrac{7}{8} = \dfrac{?}{64}$ **65.** $\dfrac{4}{5} = \dfrac{?}{20}$ **66.** $\dfrac{1}{6} = \dfrac{?}{15}$

Reduce to lowest terms.

67. $\dfrac{6}{12}$ **68.** $\dfrac{8}{12}$ **69.** $\dfrac{4}{32}$ **70.** $\dfrac{26}{64}$

71. $\dfrac{2}{8}$ **72.** $\dfrac{8}{32}$ **73.** $\dfrac{34}{64}$ **74.** $\dfrac{16}{64}$

75. $\dfrac{12}{32}$ **76.** $\dfrac{45}{90}$ **77.** $\dfrac{6}{8}$ **78.** $\dfrac{75}{100}$

Change the decimals to fractions in lowest terms.

79. 0.7 **80.** 0.83 **81.** 0.95 **82.** 0.25
83. 0.872 **84.** 0.081 **85.** 0.02 **86.** 0.005

Change the fractions to decimals. If necessary, round to the nearest thousandth.

87. $\dfrac{1}{5}$ **88.** $\dfrac{1}{10}$ **89.** $\dfrac{5}{8}$ **90.** $\dfrac{3}{7}$ **91.** $\dfrac{9}{11}$

Section 3-6

Write the improper fractions as whole or mixed numbers.

92. $\dfrac{35}{7}$ **93.** $\dfrac{18}{5}$ **94.** $\dfrac{27}{6}$ **95.** $\dfrac{39}{8}$ **96.** $\dfrac{21}{15}$

97. $\dfrac{43}{8}$ **98.** $\dfrac{22}{7}$ **99.** $\dfrac{175}{2}$ **100.** $\dfrac{135}{3}$ **101.** $\dfrac{18}{12}$

Write the whole or mixed numbers as improper fractions.

102. $6\frac{1}{2}$ **103.** 8 **104.** $10\frac{1}{2}$ **105.** $7\frac{1}{8}$ **106.** $5\frac{7}{12}$ **107.** $9\frac{3}{16}$

108. $7\frac{8}{17}$ **109.** $4\frac{3}{5}$ **110.** $9\frac{1}{9}$ **111.** 12 **112.** $16\frac{2}{3}$ **113.** $5\frac{1}{3}$

Change the whole number to an equivalent fraction using the indicated denominator.

114. $5 = \frac{?}{3}$ **115.** $2 = \frac{?}{10}$ **116.** $6 = \frac{?}{4}$ **117.** $11 = \frac{?}{3}$ **118.** $7 = \frac{?}{5}$

Section 3–7

Find the least common denominator.

119. $\frac{1}{4}, \frac{1}{3}, \frac{1}{5}$ **120.** $\frac{7}{8}, \frac{2}{3}$ **121.** $\frac{3}{4}, \frac{1}{16}$

122. $\frac{1}{12}, \frac{3}{4}$ **123.** $\frac{5}{12}, \frac{3}{10}, \frac{13}{15}$ **124.** $\frac{1}{12}, \frac{3}{8}, \frac{15}{16}$

125. Is a $\frac{5}{8}$-in. wrench larger or smaller than a $\frac{9}{16}$-in. wrench?

126. An alloy contains $\frac{2}{3}$ metal A, and the same quantity of another alloy contains $\frac{3}{5}$ metal A. Which alloy contains more of metal A?

127. Is a $\frac{3}{8}$-in.-thick piece of plasterboard thicker than a $\frac{1}{2}$-in.-thick piece?

128. A $\frac{9}{16}$-in. tube must pass through an opening in a wall. Is a $\frac{3}{4}$-in.-diameter hole large enough?

129. Is a $\frac{19}{32}$-in. wrench larger or smaller than a $\frac{7}{8}$-in. bolt head?

130. A cook top that has a width of $22\frac{1}{4}$ in. needs to fit inside an opening of $22\frac{5}{16}$ in. wide. Will the opening need to be made larger?

Which fraction is smaller?

131. $\frac{5}{8}, \frac{3}{8}$ **132.** $\frac{3}{7}, \frac{2}{7}$ **133.** $\frac{3}{8}, \frac{4}{8}$ **134.** $\frac{5}{9}, \frac{4}{9}$ **135.** $\frac{1}{4}, \frac{3}{16}$

136. $\frac{5}{8}, \frac{11}{16}$ **137.** $\frac{7}{8}, \frac{27}{32}$ **138.** $\frac{7}{64}, \frac{1}{4}$ **139.** $\frac{1}{2}, \frac{9}{19}$

Which common or decimal fraction is larger?

140. $0.127, \frac{1}{8}$ **141.** $0.26, \frac{3}{4}$ **142.** $0.08335, \frac{1}{6}$ **143.** $0.272, \frac{3}{11}$

Section 3–8

Add; reduce sums to lowest terms and convert improper fractions to mixed numbers or whole numbers.

144. $\frac{1}{8} + \frac{5}{16}$ **145.** $\frac{3}{16} + \frac{9}{64}$ **146.** $\frac{3}{14} + \frac{5}{7}$

147. $\frac{3}{5} + \frac{5}{6}$ **148.** $3\frac{7}{8} + 7 + 5\frac{1}{2}$ **149.** $2\frac{7}{16} + 6\frac{5}{32}$

150. $9\frac{7}{8} + 5\frac{3}{4}$ **151.** $3\frac{7}{8} + 5\frac{3}{16} + 1\frac{7}{32}$ **152.** $2\frac{1}{4} + 3\frac{7}{8}$

153. A pipe is cut into two pieces measuring $7\frac{5}{8}$ in. and $10\frac{7}{16}$ in. How long was the pipe before it was cut? Disregard waste.

154. Find the total thickness of a wall if the outside covering is $3\frac{7}{8}$ in. thick, the studs (interior supports) are $3\frac{7}{8}$ in., and the inside covering is $\frac{5}{16}$-in. paneling.

155. If $7\frac{5}{16}$ in. of a piece of square bar stock is turned (machined) so that it is cylindrical and $5\frac{9}{32}$ in. remains square, what is the total length of the original bar stock?

156. Two castings weigh $27\frac{1}{2}$ lb and $20\frac{3}{4}$ lb. What is the total weight of the two castings?

157. Three metal rods measuring $3\frac{1}{8}$ in., $5\frac{3}{32}$ in., and $7\frac{9}{16}$ in. are welded together end to end. How long is the welded rod?

158. In Fig. 3–34, what is the length of side A?

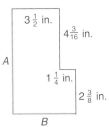

$3\frac{1}{2}$ in.

$4\frac{3}{16}$ in.

A

$1\frac{1}{4}$ in.

$2\frac{3}{8}$ in.

B

Figure 3–34

159. In Fig. 3–34, what is the length of side B?

160. A hollow-wall fastener has a grip range up to $\frac{3}{4}$ in. Is it long enough to fasten three sheets of metal $\frac{5}{16}$ in. thick, $\frac{3}{8}$ in. thick, and $\frac{1}{16}$ in. thick?

161. Figure 3–35 shows $\frac{1}{2}$-in. copper tubing wrapped in insulation. What is the distance across the tubing and insulation?

162. In Exercise 161, what would be the overall distance across the tubing and insulation if $\frac{3}{8}$-in.-ID (inside diameter) tubing were used?

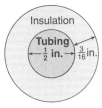

Insulation

Tubing

$\frac{1}{2}$ in. $\frac{3}{16}$ in.

Figure 3–35

Section 3–9

Subtract; reduce to lowest terms when necessary.

163. $\dfrac{5}{9} - \dfrac{2}{9}$

164. $\dfrac{11}{32} - \dfrac{5}{64}$

165. $3\dfrac{5}{8} - 2$

166. $7 - 4\dfrac{3}{8}$

167. $8\dfrac{7}{8} - 2\dfrac{29}{32}$

168. $7 - 2\dfrac{9}{16}$

169. $12\dfrac{11}{16} - 5$

170. $48\dfrac{5}{12} - 12\dfrac{11}{15}$

171. $122\dfrac{1}{2} - 87\dfrac{3}{4}$

172. Pins of $2\frac{3}{8}$ in. and $3\frac{7}{16}$ in. are cut from a drill rod 12 in. long. If $\frac{1}{16}$ in. of waste is allowed for each cut, how many inches of drill rod are left?

173. A bolt 2 in. long fastens a piece of $\frac{7}{8}$-in.-thick wood to a piece of metal. If a $\frac{3}{16}$-in.-thick lock washer, a $\frac{1}{16}$-in. washer, and a $\frac{7}{16}$-in.-thick nut are used, what is the thickness of the metal if the nut is even with the end of the bolt after tightening?

174. Four lengths measuring $6\frac{1}{4}$ in., $9\frac{3}{16}$ in., $7\frac{1}{8}$ in., and $5\frac{9}{32}$ in. are cut from 48 in. of copper tubing. How much copper tubing remains? Disregard waste.

175. A piece of tapered stock has a diameter of $2\frac{5}{16}$ in. at one end and a diameter of $\frac{55}{64}$ in. at the other end. What is the difference in the diameters?

Section 3–10

Multiply and reduce answers to lowest terms. Convert improper fractions to whole or mixed numbers.

176. $\dfrac{3}{5} \times \dfrac{10}{21}$

177. $\dfrac{1}{3} \times \dfrac{7}{8}$

178. $\dfrac{2}{5} \times \dfrac{7}{10}$

179. $\dfrac{7}{9} \times \dfrac{3}{8}$

180. $\dfrac{2}{3} \times \dfrac{5}{8} \times \dfrac{3}{16}$

181. $\dfrac{15}{16} \times \dfrac{4}{5} \times \dfrac{2}{3}$

182. $5 \times \dfrac{3}{4}$

183. $\dfrac{7}{16} \times 18$

184. $\dfrac{3}{16} \times 184$

185. $1\dfrac{1}{2} \times \dfrac{4}{5}$

186. $3\dfrac{1}{3} \times 4\dfrac{1}{2}$

187. $1\dfrac{3}{4} \times 1\dfrac{1}{7}$

188. In a concrete mixture, $\frac{4}{7}$ of the total volume is sand. How much sand is needed for 135 cubic yards (yd^3) of concrete?

189. Concrete blocks are 8 in. high. If a $\frac{3}{8}$-in. mortar joint is used, how high will a wall of 12 courses of concrete blocks be? (*Hint:* There are 12 rows of mortar joints.)

190. An adjusting screw will move $\frac{3}{64}$ in. for each full turn. How far will it move in four full turns?

191. A chef is making a dessert that is $\frac{3}{4}$ the original recipe. How much flour should be used if the original recipe calls for $3\frac{2}{3}$ cups of flour?

192. If an alloy is $\frac{3}{5}$ copper and $\frac{2}{5}$ zinc, how many pounds of each metal are in a casting weighing $112\frac{1}{2}$ lb?

193. A water pipe has an outside diameter of $18\frac{3}{4}$ cm. What is the width of eight pipes that have the same diameter?

Raise the fractions to the indicated power.

194. $\left(\dfrac{1}{5}\right)^2$ **195.** $\left(\dfrac{3}{4}\right)^2$ **196.** $\left(\dfrac{5}{6}\right)^2$ **197.** $\left(\dfrac{4}{9}\right)^2$

198. $\left(\dfrac{1}{7}\right)^3$ **199.** $\left(\dfrac{1}{2}\right)^3$ **200.** $\left(\dfrac{7}{8}\right)^2$ **201.** $\left(\dfrac{9}{10}\right)^2$

Give the reciprocal.

202. $\dfrac{7}{8}$ **203.** 4   **204.** $2\dfrac{3}{5}$ **205.** 0.7   **206.** 1.8

Section 3–11

Divide and reduce answers to lowest terms. Convert improper fractions to whole or mixed numbers.

207. $\dfrac{7}{8} \div \dfrac{3}{4}$ **208.** $\dfrac{4}{9} \div \dfrac{5}{16}$ **209.** $\dfrac{7}{8} \div \dfrac{3}{32}$

210. $8 \div \dfrac{2}{3}$ **211.** $18 \div \dfrac{3}{4}$ **212.** $35 \div \dfrac{5}{16}$

213. $5\dfrac{1}{10} \div 2\dfrac{11}{20}$ **214.** $27\dfrac{2}{3} \div \dfrac{2}{3}$ **215.** $7\dfrac{1}{5} \div 12$

216. On a house plan, $\frac{1}{4}$ in. represents 1 ft. Find the dimensions of a porch that measures $4\frac{1}{8}$ in. by $6\frac{1}{2}$ in. on the plan. (How many $\frac{1}{4}$s are there in $4\frac{1}{8}$; how many $\frac{1}{4}$s are there in $6\frac{1}{2}$?)

217. A pipe that is 12 in. long is cut into four equal parts. If $\frac{3}{16}$ in. is wasted per cut, what is the maximum length of each pipe? (It takes three cuts to divide the entire length into four equal parts.)

218. A stack of $\frac{5}{8}$-in. plywood is $21\frac{7}{8}$ in. high. How many sheets of plywood are in the stack?

219. A rod $1\frac{1}{8}$ yd long is cut into six equal pieces. What is the length of each piece? Disregard waste.

220. If $7\frac{1}{2}$ gal of liquid are distributed equally among five containers, what is the number of gallons per container?

221. Fabric that is $22\frac{1}{2}$ yd long is cut into lengths of $\frac{5}{8}$ yd. How many equal lengths can be made?

Divide and reduce answers to lowest terms. Convert improper fractions to whole or mixed numbers.

222. $\dfrac{\frac{5}{8}}{2\frac{1}{8}}$ **223.** $\dfrac{\frac{1}{3}}{6}$ **224.** $\dfrac{\frac{4}{4}}{5}$

225. $\dfrac{\frac{8}{1}}{1\frac{1}{2}}$ **226.** $\dfrac{3\frac{1}{4}}{5}$ **227.** $\dfrac{2\frac{1}{5}}{8\frac{4}{5}}$

228. $\dfrac{16\frac{2}{3}}{3\frac{1}{3}}$ **229.** $\dfrac{12\frac{1}{2}}{100}$ **230.** $\dfrac{37\frac{1}{2}}{100}$

Section 3–12

Change each fraction to three equivalent signed fractions.

231. $-\dfrac{3}{8}$ **232.** $\dfrac{-5}{9}$ **233.** $\dfrac{-7}{-8}$ **234.** $\dfrac{4}{5}$ **235.** $\dfrac{2}{-5}$

Perform the indicated operations.

236. $\dfrac{-7}{8} + \dfrac{-3}{8}$

237. $\dfrac{5}{9} - \dfrac{-3}{7}$

238. $\dfrac{-5}{8} * \dfrac{-2}{3}$

239. $\dfrac{-4}{5} \div -\dfrac{7}{15}$

240. $3.23 + (-4.61)$

241. 0.27×0.13

242. $-4.36 + (-7.23)$

243. $-12.4 \div 0.2$

244. $\dfrac{-11}{12} - \dfrac{-7}{8}$

245. $-\dfrac{7}{8} + \left(-\dfrac{5}{12}\right)$

246. $1\dfrac{3}{5} \div \left(-7\dfrac{5}{8}\right)$

247. $-2\dfrac{5}{8} \times 4\dfrac{1}{2}$

248. $\dfrac{3}{5} + \dfrac{2}{3}\left(-\dfrac{3}{5}\right)$

249. $-\dfrac{1}{2} + \dfrac{3-5}{3}$

250. $\dfrac{3}{4} - \left(-\dfrac{1}{2}\right)^2$

251. $0.2 - 3.1(-7.6)$

252. $-0.7 - (-7.2 + 5)$

253. $(-0.2)^2 + 5.7$

CHALLENGE PROBLEMS

254. Len Smith has 180 ft of fencing and needs to build two square or rectangular holding yards for his two sheltie dogs. How should he design the two holding yards to get the largest area for each dog?

255. When documents are printed on both sides of paper, we put odd-numbered pages on the front and even-numbered pages on the back. A new unit or chapter is started on the front of a sheet of paper, even if this creates a preceding blank page. Assign page numbers to the document based on these guidelines.

Chapter 1, 15 pages;   Chapter 2, 17 pages;
Chapter 3, 24 pages;   Chapter 4, 15 pages

(a) What page number will start Chapter 2?
(b) How many pages are in the document?
(c) How many blank pages are in the document?

CONCEPTS ANALYSIS

1. What two operations require a common denominator?

2. Explain how to find the reciprocal of a fraction.

3. What steps must be followed to find the reciprocal of a mixed number?

4. What number can be written as any fraction that has the same numerator and denominator?

5. What operation requires the use of the reciprocal of a fraction?

6. Name the operation that has each of the following for an answer: sum, difference, product, quotient.

7. What operation must be used to solve an applied problem if the total and one of the parts are given?

8. What does the denominator of a fraction indicate?

9. What does the numerator of a fraction indicate?

10. What kind of fraction has a value less than 1?

Find, explain, and correct the mistakes in these problems.

11. $\dfrac{5}{8} + \dfrac{1}{8} = \dfrac{6}{16} = \dfrac{3}{8}$

12. $\begin{array}{r} 12 \\ -5\dfrac{3}{4} \\ \hline 7\dfrac{3}{4} \end{array}$

13. $\dfrac{3}{5} \times 2\dfrac{1}{5} = 2\dfrac{3}{25}$

14. $\dfrac{5}{8} \div 4 = \dfrac{5}{8} \times \dfrac{4}{1} = \dfrac{5}{2} = 2\dfrac{1}{2}$

15.
$$12\dfrac{3}{4} = 12\dfrac{6}{8} = 11\dfrac{16}{8}$$
$$\underline{-4\dfrac{7}{8} = -4\dfrac{7}{8} = -4\dfrac{7}{8}}$$
$$7\dfrac{9}{8} = 7 + 1\dfrac{1}{8} = 8\dfrac{1}{8}$$

Change $\frac{5}{8}$ to a decimal:

$$
\begin{array}{r}
0.625 \\
8\overline{)5.000} \\
\underline{4\ 8} \\
20 \\
\underline{16} \\
40 \\
\underline{40}
\end{array}
\qquad \text{Divide by denominator.}
$$

Section 3–6

1 Convert improper fractions to whole or mixed numbers (p. 125).

To convert an improper fraction to a whole or mixed number: **1.** Divide the numerator by the denominator.   **2.** Write any remainder as a fraction with the original denominator as its denominator.

Convert $\frac{18}{6}, \frac{15}{4}$ to whole or mixed numbers.

$$
\frac{18}{6} = 3, \qquad \frac{15}{4} = 3\frac{3}{4}
$$

2 Convert mixed numbers and whole numbers to improper fractions (p. 126).

To convert a mixed number to an improper fraction: **1.** Multiply the whole number by the denominator.   **2.** Add the numerator to the result of Step 1.   **3.** Place the sum from Step 2 over the original denominator.

To convert a whole number to an improper fraction: **1.** Write the whole number as the numerator.   **2.** Write 1 as the denominator.

Change $4\frac{7}{8}$ and 7 to improper fractions.

$$
4\frac{7}{8} = \frac{(8 \times 4) + 7}{8} = \frac{39}{8} \qquad 7 = \frac{7}{1}
$$

Section 3–7

1 Find common denominators (pp. 127–128).

To find the least common denominator, use the process shown in Section 4, Outcome 1 for finding the least common multiple.

Find the least common denominator for $\frac{7}{18}$ and $\frac{5}{24}$.

$$
\begin{aligned}
18 &= 2 \cdot 3 \cdot 3 = 2 \cdot 3^2 \\
24 &= 2 \cdot 2 \cdot 2 \cdot 3 = 2^3 \cdot 3 \\
\text{LCD} &= 2^3 \cdot 3^2 = 2 \cdot 2 \cdot 2 \cdot 3 \cdot 3 = 72
\end{aligned}
$$

The smallest number that can be divided evenly by both 18 and 24 is 72.

2 Compare fractions (pp. 128–129).

To compare fractions:

1. Write the fractions as equivalent fractions with common denominators.
2. Compare the numerators. The larger numerator indicates the larger fraction.
To compare mixed numbers:
1. Compare the whole-number parts, if different.   **2.** If the whole-number parts are equal, write the fractions with common denominators.   **3.** Compare the numerators.

Which fraction is smaller, $\frac{2}{5}$ or $\frac{5}{12}$?

$$
\frac{2}{5} = \frac{24}{60}
$$

$$
\frac{5}{12} = \frac{25}{60}
$$
Since $\frac{24}{60}$ is smaller than $\frac{25}{60}$, $\frac{2}{5}$ is smaller than $\frac{5}{12}$.

Section 3–8

1 Add fractions
(pp. 130–131).

To add fractions: **1.** Find the common denominator.   **2.** Change each fraction to an equivalent fraction with the common denominator.   **3.** Add the numerators and place the sum over the common denominator.   **4.** Reduce the sum if possible.   **5.** Convert improper fractions to whole or mixed numbers if desired.

Add:

$$\frac{1}{7} + \frac{3}{7} + \frac{2}{7} = \frac{6}{7} \qquad \frac{5}{8} + \frac{3}{4} = \frac{5}{8} + \frac{6}{8} = \frac{11}{8} = 1\frac{3}{8}$$

2 Add mixed numbers
(pp. 131–132).

To add mixed numbers: **1.** Convert each fraction to an equivalent fraction with the LCD.   **2.** Place the sum of the numerators over the LCD.   **3.** Add the whole-number parts.   **4.** Write the improper fraction from Step 2 as a whole or mixed number; add the result to the whole number from Step 3.   **5.** Simplify if necessary.

Add $4\frac{3}{4} + 5\frac{2}{8} + 1\frac{1}{2}$.

$$4\frac{3}{4} = 4\frac{6}{8}$$

Convert each fraction to an equivalent fraction with the LCD.

$$5\frac{2}{8} = 5\frac{2}{8}$$

$$1\frac{1}{2} = 1\frac{4}{8}$$

Add whole numbers.
Add fractions.

$$10\frac{12}{8}$$

Simplify. $\dfrac{12}{8} = 1\dfrac{4}{8}$

$$10 + 1\frac{4}{8} = 11\frac{4}{8} = 11\frac{1}{2}$$

Section 3–9

1 Subtract fractions
(p. 133).

To subtract fractions: **1.** Convert each fraction to an equivalent fraction that has the LCD as its denominator.   **2.** Subtract the numerators.   **3.** Place the difference over the LCD.   **4.** Reduce if possible.

Subtract $\frac{5}{8} - \frac{7}{16}$.

$$\frac{5}{8} = \frac{10}{16}$$

Convert each fraction to an equivalent fraction with a common denominator.

$$-\frac{7}{16} = \frac{7}{16}$$

Subtract the numerators.

$$\frac{3}{16}$$

2 Subtract mixed numbers (pp. 134–136).

To subtract mixed numbers: **1.** Write fractions as equivalent fractions with common denominators.   **2.** Borrow from the whole number and add to the fraction if the fraction in the minuend is smaller than the fraction in the subtrahend.   **3.** Subtract the fractions.   **4.** Subtract the whole numbers.   **5.** Simplify if necessary.

Chapter Summary

Subtract $5\frac{3}{8} - 3\frac{9}{16}$.

$$5\frac{3}{8} = 5\frac{6}{16} = 4\frac{22}{16} \qquad \text{Borrow.}$$

$$-\; 3\frac{9}{16} = 3\frac{9}{16} = 3\frac{9}{16} \qquad \begin{array}{l}\text{Subtract whole numbers. Subtract}\\ \text{fractions.}\end{array}$$

$$\overline{\phantom{-\; 3\frac{9}{16} = 3\frac{9}{16} = }\;1\frac{13}{16}}$$

Section 3–10

1 Multiply fractions (pp. 137–139).

To multiply fractions: **1.** Reduce any numerator and denominator that have a common factor. **2.** Multiply the numerators to get the numerator of the product. **3.** Multiply the denominators to get the denominator of the product. **4.** Be sure the product is reduced.

Multiply $\frac{4}{5} \times \frac{7}{10} \times \frac{15}{35}$.

$$\overset{2}{\underset{1}{\cancel{\frac{4}{5}}}} \times \overset{1}{\underset{5}{\cancel{\frac{7}{10}}}} \times \overset{3}{\underset{5}{\cancel{\frac{15}{35}}}} = \frac{6}{25} \qquad \text{Reduce and multiply.}$$

2 Multiply mixed numbers (pp. 139–140).

To multiply fractions, whole numbers, and mixed numbers: **1.** Write whole numbers as fractions with denominators of 1. **2.** Write mixed numbers as improper fractions. **3.** Reduce as much as possible. **4.** Multiply the numerators to get the numerator of the product. **5.** Multiply the denominators to get the denominator of the product. **6.** Write the product as a whole number, mixed number, or fraction in lowest terms.

Multiply $4 \times 3\frac{1}{5} \times \frac{2}{7}$.

$$\frac{4}{1} \times \frac{16}{5} \times \frac{2}{7} = \frac{128}{35} = 3\frac{23}{35}$$

3 Raise a fraction to a power (p. 141).

To raise a fraction or quotient to a power: **1.** Raise the numerator to the power. **2.** Raise the denominator to the power.

Raise $\left(\frac{3}{5}\right)^3$ to the indicated power.

$$\left(\frac{3}{5}\right)^3 = \frac{3^3}{5^3} = \frac{3 \cdot 3 \cdot 3}{5 \cdot 5 \cdot 5} = \frac{27}{125}$$

Section 3–11

1 Find reciprocals (p. 142).

To write the reciprocal of a number: **1.** Express the number as a fraction. **2.** Interchange the numerator and denominator.

Find the reciprocal of $\frac{3}{5}$, 6, $2\frac{3}{4}$, and 0.2.

The reciprocal of $\frac{3}{5}$ is $\frac{5}{3}$; the reciprocal of 6 is $\frac{1}{6}$; the reciprocal of $2\frac{3}{4}$ or $\frac{11}{4}$ is $\frac{4}{11}$.

The reciprocal of 0.2 or $\dfrac{2}{20}$ is $\dfrac{10}{2}$ or 5.

2 Divide fractions (pp. 142–143).

To divide fractions: **1.** Replace the divisor with its reciprocal. **2.** Change the division to multiplication. **3.** Multiply.

Divide $\frac{4}{5} \div \frac{8}{9}$.

$$\frac{4}{5} \div \frac{8}{9} = \frac{4}{5} \times \frac{9}{\underset{2}{\cancel{8}}}^{1} = \frac{9}{10} \qquad \begin{array}{l}\text{Change to an equivalent}\\ \text{multiplication and multiply.}\end{array}$$

3 Divide mixed numbers (pp. 143–145).	To divide mixed numbers: **1.** Write each mixed number as an improper fraction. **2.** Change the division to an equivalent multiplication by using the reciprocal of the divisor. **3.** Simplify if possible. **4.** Multiply.

Divide $4\frac{2}{3} \div 1\frac{1}{6}$.

$$4\frac{2}{3} \div 1\frac{1}{6} = \frac{14}{3} \div \frac{7}{6} = \frac{\overset{2}{\cancel{14}}}{\cancel{3}} \times \frac{\overset{2}{\cancel{6}}}{\cancel{7}} = \frac{4}{1} \text{ or } 4.$$

4 Simplify complex fractions (pp. 145–146).	To simplify a complex fraction: **1.** Write mixed numbers and whole numbers as improper fractions. **2.** Write the division (as indicated by the fraction bar) and then change to an equivalent multiplication. **3.** Simplify if possible. **4.** Multiply. **5.** Convert the improper fraction to a whole or mixed number if necessary.

Simplify:

$$\frac{4\frac{1}{2}}{3\frac{3}{5}} = \frac{\frac{9}{2}}{\frac{18}{5}} = \frac{9}{2} \div \frac{18}{5} = \frac{\overset{1}{\cancel{9}}}{2} \times \frac{5}{\underset{2}{\cancel{18}}} = \frac{5}{4} = 1\frac{1}{4}$$

Section 3–12

1 Change a signed fraction to an equivalent signed fraction (pp. 148–149).	If any two of the three signs of a signed fraction are changed, the fraction's value does not change.

Write three equivalent signed fractions for $-\frac{7}{8}$.

$$-\frac{7}{8} = -\frac{+7}{+8} \quad \text{Equivalent fractions are } +\frac{+7}{-8} \text{ or } -\frac{-7}{-8} \text{ or } +\frac{-7}{+8}$$

2 Perform basic operations with signed fractions (pp. 149–150).	To add or subtract signed fractions: **1.** Write equivalent fractions that have positive integers as denominators and have common denominators. **2.** Add or subtract the numerators using the rules for adding signed numbers. To multiply or divide signed fractions: **1.** Multiply or divide the fractions. **2.** Apply the rules for multiplying or dividing signed numbers.

Add $\frac{-5}{8} + \frac{7}{8}$.

$$\frac{-5}{8} + \frac{7}{8} = \frac{2}{8} = \frac{1}{4}$$

3 Perform basic operations with signed decimals (pp. 150–151).	**1.** Determine the indicated operations. **2.** Apply the appropriate rules for signed numbers and for decimals.

Add $4.37 + (-2.91)$

$$4.37 - 2.91 = 1.46 \qquad \text{Add decimals with unlike signs.}$$

4 Apply the order of operations with signed fractions and decimals (p. 151).	Signed fractions and decimals follow the same order of operations as whole numbers (Chapter 1, Section 7) and integers (Chapter 2, Section 6).

Evaluate $-\frac{1}{4} - \frac{2}{3}\left(-\frac{1}{2}\right)$

$$-\frac{1}{4} - \frac{2}{3}\left(-\frac{1}{2}\right)$$ Multiply using rules for like signs and multiplying fractions.

$$-\frac{1}{4} + \frac{1}{3}$$ Use the rule for adding numbers with unlike signs.

$$-\frac{3}{12} + \frac{4}{12}$$ Change to equivalent fractions with the LCD.

$$\frac{1}{12}$$

Evaluate $(-0.3)^2 - 2.8$

$$(-0.3)^2 - 2.8$$ Square signed decimal.

$$0.09 - 2.8$$ Add decimals with unlike signs.

$$-2.71$$

CHAPTER TRIAL TEST

Represent as fractions.

1. 3 out of 4 people in a survey

2. $7 \div 9$

Convert to mixed or whole numbers.

3. $\frac{9}{3}$

4. $\frac{14}{9}$

Convert to improper fractions.

5. $4\frac{6}{7}$

6. $3\frac{1}{10}$

Write the prime factors in exponential notation.

7. 96

8. 132

Perform the operations. Write your answers in lowest terms.

9. $\frac{5}{6} \times \frac{3}{10}$

10. $\frac{3}{7} \times \frac{2}{9}$

11. $2\frac{2}{9} \times 1\frac{3}{4}$

12. $7 \times \frac{1}{3}$

13. $7\frac{1}{2} \div \frac{5}{9}$

14. $\frac{4\frac{2}{3}}{2\frac{1}{2}}$

15. $\frac{7}{12} + \frac{5}{6}$

16. $\frac{5}{12} \div \frac{5}{6}$

17. $2\frac{3}{7} + 5 + \frac{1}{2}$

18. $\frac{3}{32} + 4 + 1\frac{3}{4}$

19. $\frac{7}{9} - \frac{2}{3}$

20. $6\frac{1}{4} - 2\frac{3}{4}$

21. $\frac{5\frac{2}{3}}{1\frac{1}{9}}$

22. $-\frac{2}{5} + \frac{1}{10}$

23. $-1.3 - (3.1 - 5.4)$

Determine which is larger. Show your work.

24. $\frac{7}{8}, \frac{11}{12}$

25. $\frac{7}{32}, \frac{5}{16}$

Arrange the fractions in order, beginning with the smallest. Show your work.

26. $\frac{5}{7}, \frac{10}{21}, \frac{3}{4}$

Write the decimal equivalent.

27. $\frac{3}{5}$

28. $\frac{5}{8}$

Solve the problems.

29. Two of the seven security employees at the local community college received safety awards from the governor. Represent the part of the total number of employees who received an award as a fraction.

30. A candy-store owner mixes $1\frac{1}{2}$ lb of caramels, $\frac{3}{4}$ lb of chocolates, and $\frac{1}{2}$ lb of candy corn. What is the total weight of the mixed candy?

31. A homemaker has $5\frac{1}{2}$ cups of sugar on hand to make a batch of cookies requiring $1\frac{2}{3}$ cups of sugar. How much sugar is left?

32. If $6\frac{1}{4}$ ft of wire is needed to make one electrical extension cord, how many extension cords can be made from $68\frac{3}{4}$ ft of wire?

33. A costume maker figures one costume requires $2\frac{2}{3}$ yd of red satin. How many yards of red satin are needed to make three costumes?

34. Will a $\frac{5}{8}$-in.-wide drill bit make a hole wide enough to allow a $\frac{1}{2}$-in. (outside diameter) copper tube to pass through?

Percents

4

Learning Outcomes

1 Change any number to its percent equivalent.

2 Change any percent to its numerical equivalent.

The word *percent* means "per hundred" or "for every hundred." Thus, 35 percent means 35 per hundred, or 35 out of every hundred, or $\frac{35}{100}$. Also, 100 percent means 100 out of 100 parts, or $\frac{100}{100}$, or 1 whole quantity. The symbol % is used to represent "percent."

1 **Change Any Number to Its Percent Equivalent.**

We often need to use percents, but the numbers we are given are not expressed as percents. To express a number as a percent, we multiply by 1 in the form of 100%. For example, the fraction $\frac{1}{2}$ can be written as the decimal 0.5. Both $\frac{1}{2}$ and 0.5 change to the same equivalent percent.

$$\frac{1}{2} \cdot 100\% = \frac{1}{\underset{1}{2}} \cdot \frac{\overset{50}{\cancel{100}}\%}{1} = 50\% \qquad 0.5 \cdot 100\% = 50\%$$

To change any number to its percent equivalent:

Multiply by 1 in the form of 100%.

We can use this rule to change any type of number—fraction, decimal, whole number, or mixed number—to a percent equivalent.

Multiplying by 1 Can Take Many Forms.

The multiplicative identity states that $n \times 1 = n$. In fractions, we use the multiplicative identity to change a fraction to an equivalent fraction. We can use the property: $\frac{n}{n} = 1$.

$$\frac{1}{2} \cdot \frac{3}{3} = \frac{3}{6} \qquad \frac{3}{3} = 1.$$

With percents, we use the property of 1 in the form of 100%.

$$\frac{1}{2} \times 100\% = 50\% \qquad 100\% = 1.$$

EXAMPLE Change the fractions to percent equivalents: $\frac{1}{4}, \frac{1}{3}, \frac{3}{8}, \frac{1}{200}$, and $\frac{3}{1,000}$.

$$\frac{1}{4} \cdot 100\% = \frac{1}{\underset{1}{4}} \cdot \frac{\overset{25}{\cancel{100}}\%}{1} = \mathbf{25\%}$$

$$\frac{1}{3} \cdot 100\% = \frac{1}{3} \cdot \frac{100\%}{1} = \frac{100\%}{3} = \mathbf{33\frac{1}{3}\%}$$

$$\frac{3}{8} \cdot 100\% = \frac{3}{\underset{2}{8}} \cdot \frac{\overset{25}{\cancel{100}}\%}{1} = \frac{75\%}{2} = \mathbf{37\frac{1}{2}\%}$$

$$\frac{1}{200} \cdot 100\% = \frac{1}{\cancel{200}} \cdot \frac{\overset{1}{\cancel{100\%}}}{1} = \frac{1}{2}\%$$

$\frac{1}{2}\%$ means $\frac{1}{2}$ of every hundredth, or $\frac{1}{2}$ of 1%.

$$\frac{3}{1,000} \cdot 100\% = \frac{3}{\cancel{1,000}} \cdot \frac{\overset{1}{\cancel{100\%}}}{1} = \frac{3}{10}\%$$

$\frac{3}{10}\%$ means $\frac{3}{10}$ of every hundredth, or $\frac{3}{10}$ of 1%.

EXAMPLE Change the decimals 0.3 and 0.006 to percent equivalents.

$0.3 \cdot 100\% = 030.\% = \mathbf{30\%}$ Use the shortcut procedure to multiply by 100: Move the decimal point two places to the right.

$0.006 \cdot 100\% = 000.6\% = \mathbf{0.6\%}$ 0.6% means 0.6 of every hundredth, or 0.6 of 1%.

EXAMPLE Change the whole numbers 1, 3, and 7 to their percent equivalents.

$1 \cdot 100\% = \mathbf{100\%}$ 100 out of 100 or all of 1 quantity.

$3 \cdot 100\% = \mathbf{300\%}$ 300 out of 100 or 3 whole quantities.

$7 \cdot 100\% = \mathbf{700\%}$ 7 whole quantities, or 7 times a quantity.

EXAMPLE Change the mixed numbers and decimals to their percent equivalents: $1\frac{1}{4}$, $3\frac{2}{3}$, 5.3, and 5.12.

$$1\frac{1}{4} \cdot 100\% = \frac{5}{\cancel{4}} \cdot \frac{\overset{25}{\cancel{100\%}}}{1} = \mathbf{125\%}$$

Write $1\frac{1}{4}$ as an improper fraction and multiply.

$$3\frac{2}{3} \cdot 100\% = \frac{11}{3} \cdot \frac{100\%}{1} = \frac{1,100\%}{3} = \mathbf{366\frac{2}{3}\%}$$

Write $3\frac{2}{3}$ as an improper fraction and multiply.

$$5.3 \cdot 100\% = 530.\% = \mathbf{530\%}$$

Multiply by 100 by moving the decimal two places to the right.

$$5.12 \cdot 100\% = 512.\% = \mathbf{512\%}$$

---

2   **Change Any Percent to Its Numerical Equivalent.**

Percents are a convenient way to express the relationship of any quantity to 100. They are excellent time-savers when we make comparisons or state problems on the job. However, we cannot use percents to solve a problem. Instead, we first convert the percents to fraction-, decimal-, whole-, or mixed-number equivalents.

To change a percent to a numerical equivalent:

Divide by 1 in the form of 100%.

TIP!

Dividing by 1 Can Take Many Forms.

With whole numbers, $\frac{n}{n} = 1$. To reduce fractions, we use this property again as

$$\frac{6}{10} \div \frac{2}{2} = \frac{6 \div 2}{10 \div 2} = \frac{3}{5}$$

With percents we apply this property using 1 in the form of 100%.

The numerical equivalent of a percent can be expressed in fraction or decimal form as shown in the following example.

EXAMPLE   Change the percents to their fraction and decimal equivalents: 75%, 38%, and 5%.

	Fraction equivalent	Decimal equivalent	
75%	75% ÷ 100%	75% ÷ 100%	For fractions, change division to an equivalent multiplication.
	$\dfrac{\overset{3}{\cancel{75}}\%}{1} \cdot \dfrac{1}{\underset{4}{\cancel{100}}\%} = \dfrac{3}{4}$	$.\underset{\curvearrowleft}{75} = \mathbf{0.75}$	For decimals, use the shortcut procedure for dividing by 100%: Move the decimal point two places to the left.
38%	38% ÷ 100%	38% ÷ 100%	
	$\dfrac{\overset{19}{\cancel{38}}\%}{1} \cdot \dfrac{1}{\underset{50}{\cancel{100}}\%} = \dfrac{19}{50}$	$.\underset{\curvearrowleft}{38} = \mathbf{0.38}$	$\dfrac{\%}{\%}$ reduces to 1.
5%	5% ÷ 100%	5% ÷ 100%	
	$\dfrac{\overset{1}{\cancel{5}}\%}{1} \cdot \dfrac{1}{\underset{20}{\cancel{100}}\%} = \dfrac{1}{20}$	$.\underset{\curvearrowleft}{05} = \mathbf{0.05}$	

TIP!

Division Expressed as Multiplication

As with fractions, we see that it is again convenient to change division to an equivalent multiplication. Is *dividing* by 100% the same as *multiplying* by $\dfrac{1}{100\%}$? Yes.

$$\text{A percent} \div 100\% = \text{A percent} \div \frac{100\%}{1} = \text{A percent} \cdot \frac{1}{100\%}$$

Some percents change more conveniently to a fraction equivalent, and others change more conveniently to a decimal equivalent. In solving problems, we normally change the percent to the most convenient equivalent for the problem. In the following examples, both fraction and decimal equivalents are given, and you can judge for yourself when fraction equivalents are more convenient than decimal equivalents, and vice versa.

EXAMPLE Change the percents to their fraction and decimal equivalents: $33\frac{1}{3}\%$, $37\frac{1}{2}\%$.

Fractional equivalent	**Decimal equivalent**
$33\frac{1}{3}\% \div 100\%$	

$$\frac{\overset{1}{\cancel{100\%}}}{3} \cdot \frac{1}{\underset{1}{\cancel{100\%}}} = \frac{1}{3} \qquad 33\frac{1}{3}\% = \mathbf{0.33\frac{1}{3}} \text{ or } 0.33 \text{ (rounded)}$$

A decimal point separates whole quantities from fraction parts. Therefore, there is an *unwritten* decimal between 33 and $\frac{1}{3}$. Because $\frac{1}{3}$ does not change to a terminating decimal equivalent, using the decimal equivalent of $33\frac{1}{3}\%$ will create more extensive calculations. Using a rounded decimal equivalent changes the result from an exact to an approximate amount.

Fractional equivalent	**Decimal equivalent**
$37\frac{1}{2}\% \div 100\%$	

$$\frac{\overset{3}{\cancel{75}}}{2}\% \cdot \frac{1}{\underset{4}{\cancel{100\%}}} = \frac{3}{8} \qquad 37\frac{1}{2}\% = \mathbf{0.37\frac{1}{2}}$$

Decimal equivalents are desirable when we use calculators. However, the mixed decimal $0.37\frac{1}{2}$ would not be adaptable to most calculators because of the fraction that remains. For that reason, when changing mixed-number percents to decimal form, we first change the fractional part of the mixed number to its decimal equivalent.

$$\frac{1}{2} = 0.5 \qquad 2\overline{)1.0}^{\,0.5}$$

Thus, $37\frac{1}{2}\% = 37.5\%$. Then, divide by 100%.

$$37.5\% = \mathbf{0.375}$$

Since no rounding was necessary, the decimal equivalent is an exact amount.

EXAMPLE Change 5.25% to its fractional and decimal equivalents.

Fractional equivalent   **Decimal equivalent**

First, write the percent in fraction form.

$$5.25\% = 5\frac{25}{100}\% = 5\frac{1}{4}\% \qquad 5.25\% = 5.25\% \div 100\% = \mathbf{0.0525}$$

$$5\frac{1}{4}\% = 5\frac{1}{4}\% \div 100\%$$

$$= \frac{21}{4}\% \cdot \frac{1}{100\%}$$

$$= \frac{21}{400}$$

Table 4-1 Common Percent, Fraction, and Decimal Equivalents

Percent	Fraction	Decimal	Percent	Fraction	Decimal
10%	$\frac{1}{10}$	0.1	60%	$\frac{3}{5}$	0.6
20%	$\frac{1}{5}$	0.2	$66\frac{2}{3}\%$	$\frac{2}{3}$	$0.66\frac{2}{3}$ or 0.667[a]
25%	$\frac{1}{4}$	0.25	70%	$\frac{7}{10}$	0.7
30%	$\frac{3}{10}$	0.3	75%	$\frac{3}{4}$	0.75
$33\frac{1}{3}\%$	$\frac{1}{3}$	$0.33\frac{1}{3}$ or 0.333[a]	80%	$\frac{4}{5}$	0.8
40%	$\frac{2}{5}$	0.4	90%	$\frac{9}{10}$	0.9
50%	$\frac{1}{2}$	0.5	100%	$\frac{1}{1}$	1.0

[a] These decimals can be expressed with fractions or rounded decimals. Their fraction equivalents are exact amounts, and rounded decimals are approximate amounts.

To help us remember these equivalents, let's group them differently and suggest some mental calculations as memory aids.

$$50\% = \frac{1}{2} = 0.5$$

This is one of the easiest to remember because of its direct comparison to money. One dollar is 100 cents. One-half dollar is 50 cents. One-half dollar is $0.50 in dollar-and-cent notation.

$$25\% = \frac{1}{4} = 0.25 \qquad 75\% = \frac{3}{4} = 0.75$$

Relating these to money, one-fourth of a dollar is 25 cents or $0.25. Three-fourths of a dollar is three times as much as 25 cents. It is 75 cents or $0.75.

The equivalents of 10% and multiples of 10% are worth remembering.

EXAMPLE Write the percents as decimals. 10%, 20%, 30%, 40%, 50%, 60%, 70%, 80%, 90%, 100%, $33\frac{1}{3}\%$, and $66\frac{2}{3}\%$.

$$10\% = \frac{1}{10} = \textbf{0.1} \qquad\qquad 60\% = \frac{6}{10} = \frac{3}{5} = \textbf{0.6}$$

$$20\% = \frac{2}{10} = \frac{1}{5} = \textbf{0.2} \qquad\qquad 70\% = \frac{7}{10} = \textbf{0.7}$$

$$30\% = \frac{3}{10} = \textbf{0.3} \qquad\qquad 80\% = \frac{8}{10} = \frac{4}{5} = \textbf{0.8}$$

$$40\% = \frac{4}{10} = \frac{2}{5} = \textbf{0.4} \qquad\qquad 90\% = \frac{9}{10} = \textbf{0.9}$$

$$50\% = \frac{5}{10} = \frac{1}{2} = \textbf{0.5} \qquad\qquad 100\% = \frac{1}{1} = \textbf{1}$$

When 3 is divided into 100, the result is $33\frac{1}{3}$.

$$\frac{33\frac{1}{3}}{3\overline{)100}}$$

$$33\frac{1}{3}\% = \frac{1}{3} = 0.33\frac{1}{3}$$

Two times $33\frac{1}{3}$ is $66\frac{2}{3}$.

$$66\frac{2}{3}\% = \frac{2}{3} = 0.66\frac{2}{3}$$

TIP!

Estimating Percents

Find a relationship between the given fraction and some common fraction and percent equivalent that you know from memory. Then, use that common equivalent to estimate the percent equivalent of the given fraction.

- Approximately what percent is $\frac{1}{6}$ of a quantity? $\frac{1}{6} = \frac{1}{2}$ of $\frac{1}{3}$ and $\frac{1}{3}$ equals $33\frac{1}{3}\%$. What is $\frac{1}{2}$ of $33\frac{1}{3}\%$? Thus, $\frac{1}{6}$ of a quantity is approximately 16% or 17%.
- Approximately what percent is $\frac{3}{8}$ of a quantity? $\frac{3}{8}$ is halfway between $\frac{2}{8}$ or $\frac{1}{4}$ and $\frac{4}{8}$ or $\frac{1}{2}$. Then, $\frac{3}{8}$ is halfway between 25% and 50% or $37\frac{1}{2}\%$. So, $37\frac{1}{2}\%$ is the exact equivalent rather than an estimated equivalent.

EXAMPLE   Estimate the percent equivalent of $\frac{3}{20}$ and $\frac{4}{7}$.

Approximately what percent is $\frac{3}{20}$?

$$\frac{1}{20} = \frac{1}{2} \text{ of } \frac{1}{10} = \frac{1}{2} \text{ of } 10\%$$

$$\frac{1}{20} = \frac{1}{2} \text{ of } 10\% = 5\% \qquad \frac{3}{20} \text{ is 3 times } \frac{1}{20}.$$

Then, $\frac{3}{20} = 3 \cdot \frac{1}{20} = 3 \times 5\% = \mathbf{15\%}$ This is an exact equivalent.

Approximately what percent is $\frac{4}{7}$?

$$\frac{3}{7} < \frac{1}{2} \text{ and } \frac{4}{7} > \frac{1}{2} \qquad \text{Compare } \frac{4}{7} \text{ to } \frac{1}{2}.$$

$$\frac{4}{7} > 50\% \qquad \frac{1}{2} = 50\%.$$

55% or **60%** would be a good estimate.

EXAMPLE 20% of what number is 45?

This time we know the rate, 20%, and the percentage, 45. We are looking for the base, as indicated by the key word *of*. The base is the original or whole amount. The base is larger than the percentage when the rate is less than 100%.

Option 1

$$\frac{20}{100} = \frac{45}{B}$$

$$20 \cdot B = 100 \cdot 45$$

$$20 \cdot B = 4{,}500$$

$$B = \frac{4{,}500}{20}$$

$$B = 225$$

Option 2

$$\frac{1}{5} = \frac{45}{B}$$

$$B = 5 \cdot 45$$

$$B = 225$$

Option 3

$$\frac{0.2}{1} = \frac{45}{B}$$

$$45 = 0.2 \cdot B$$

$$\frac{45}{0.2} = B$$

$$225 = B$$

20% of 225 is 45.

In the following example the rate is more than 100% so we expect the base to be less than the percentage.

EXAMPLE 398.18 is 215% of what number?

Here, we are looking for the base, as indicated by the key word *of*. We are given the percentage and the rate.

$$\frac{215}{100} = \frac{398.18}{B}$$

$$215 \cdot B = 39{,}818 \qquad \text{Cross multiply.}$$

$$B = \frac{39{,}818}{215} \qquad \text{Divide.}$$

$$B = 185.2$$

398.18 is 215% of 185.2.

Because the rate is more than 100%, we expected the base to be smaller than the percentage. Thus, the answer is reasonable.

When solving applied problems, our most difficult task is identifying the two given parts and determining which part is missing. Then the proportion can be set up and solved.

EXAMPLE If a type of solder contains 55% tin, how many pounds of tin are needed to make 10 lb of solder?

First, let's be sure we understand the word *solder*. Solder is a mixture of metals. In this problem, the *total amount* or the *base* is the 10 lb of solder, and it is made of tin and other metals.

Known facts   55% of 10 lb of solder is tin.
Rate: percent of tin = 55%
Base: amount of solder = 10 lb

Unknown fact	Percentage or number of pounds of tin $= P$
Relationships	Percentage proportion

$$\frac{R}{100} = \frac{P}{B} \qquad \text{Substitute known values.}$$

$$\frac{55}{100} = \frac{P}{10 \text{ lb}}$$

Estimation	To estimate, 55% is more than $\frac{1}{2}$. $\frac{1}{2}$ of 10 = 5. So the amount of tin should be more than 5 lb.

Calculations	$\dfrac{55}{100} = \dfrac{P}{10}$ Solve for P. Cross multiply.

$$55 \cdot 10 = 100 \cdot P$$

$$550 = 100 \cdot P \qquad \text{Divide by 100.}$$

$$\frac{550}{100} = P$$

$$5\frac{1}{2} = P$$

Interpretation	Thus, $5\frac{1}{2}$ lb of tin are needed to make 10 lb of solder.
	To check the reasonableness of the answer, $5\frac{1}{2}$ lb is a little more than 5.

EXAMPLE	If a 150-horsepower (hp) engine delivers only 105 hp to the driving wheels of a car, what is the efficiency of the engine?

Efficiency means the *percent* the output (105 hp) is of the total amount (150 hp) the engine is capable of delivering. Thus, the base amount is 150 hp, the part or percentage delivered is 105 hp, and the percent of 150 represented by 105 is the rate or efficiency.

Known facts	Base: Total amount of horsepower = 150 hp
	Percentage: Amount of horsepower the engine delivers = 105 hp
Unknown facts	Efficiency or percent of the total horsepower
Relationships	Percentage proportion

$$\frac{R}{100} = \frac{P}{B} \qquad \text{Substitute known values.}$$

$$\frac{R}{100} = \frac{105}{150}$$

Estimation	Because the engine is not operating at full capacity (150 hp), we expect the efficiency to be less than 100%. Since 105 is more than $\frac{1}{2}$ of 150, the percent or rate will be more than 50%. A more precise estimate is

$$\frac{100}{150} = \frac{2}{3} = 66\frac{2}{3}\%$$

Since $105 > 100$ > is read "is greater than."

$$\text{Rate} > 66\frac{2}{3}\%$$

85. A real estate salesperson earns 4% commission. What is the commission on a property that sold for $295,800?

86. A manufacturer gives a 2% discount to customers paying cash. If a parts store pays cash for an order totaling $875.84, what amount is saved? Calculate to the nearest cent.

87. Find the cash price for an order of hospital supplies totaling $3,985.57 if a 3% discount is offered for cash orders. Calculate to the nearest cent.

88. What commission is earned by a salesperson who sells $18,890 in merchandise if a 5% commission is paid on all sales?

89. A manufacturer's representative is paid a salary of $140 per week and 7% commission on all sales over $3,200 per week. The sales for a recent week were $7,412. What is the representative's salary for that week?

90. A computer store manager is paid a salary of $2,153 monthly plus a bonus of 1% of the net earnings of the business. Find the total salary for a month when the net earnings of the business is $105,275.

91. An electrician purchases $650 worth of electrical materials. A finance charge of $1\frac{1}{2}$% per month is added to the bill. What is the finance charge for 1 month?

92. If $10,000 is invested for 3 months at 6% per year, how much interest is earned?

93. A nurse paid $10.24 in monthly interest on a credit-card account that had an average daily balance of $584.87. Find the monthly rate of interest. Round to the nearest hundredth of a percent.

94. Find the interest on a loan of $2,450 at 7% per year for 1 year.

95. Find the interest on a loan of $5,840 at 10% per year for 2 years, 6 months.

96. Find the annual interest rate if a deposit of $5,000 earns $125 in 1 year.

4-3 | Increases and Decreases

Learning Outcomes

1 Find the amount of increase or decrease in percent problems.
2 Find the new amount directly in percent problems.
3 Find the rate or the base in increase or decrease problems.

1 **Find the Amount of Increase or Decrease in Percent Problems.**

Percents are often used in problems dealing with increases or decreases.

TIP!

Relating Increases and Decreases to Percents

Increases and decreases are applications of the percentage formula or percentage proportion.

	Rate	Base	Percentage
Increase	Rate of increase	Original amount	Amount of increase
Decrease	Rate of discount	Original amount	Amount of decrease

New amount for increase $=$ Original amount $+$ Amount of increase
New amount for decrease $=$ Original amount $-$ Amount of decrease

EXAMPLE Medical assistants are to receive a 9% increase in wages per hour. If they were making $9.25 an hour, what is the *amount of increase per hour* (to the nearest cent)? Also, what is the *new wage per hour*?

The original wage per hour is the base, and we want to find the amount of increase (percentage).

Estimation 10% of $9.25 is $0.92. The increase will be less than $0.92. The new wage per hour will be approximately $10.

$$\frac{9}{100} = \frac{P}{9.25} \qquad \text{P represents amount of increase, and 9\% is the rate of increase.}$$

$$9 \cdot 9.25 = 100 \cdot P$$
$$83.25 = 100 \cdot P$$
$$\frac{83.25}{100} = P$$
$$\$0.8325 = P \qquad \text{$\$0.83$ to the nearest cent is the amount of increase.}$$

Interpretation **The medical assistants will receive an $0.83 per hour increase in wages.**

$\$9.25 + \$0.83 = \$10.08$ New amount = Original amount + Amount of increase.

Their new hourly wage will be $10.08.

EXAMPLE Molten iron shrinks 1.2% while cooling. What is the cooled length of a piece of iron if it is cast in a 24-cm pattern?

First, we find the amount of shrinkage (amount of decrease, percentage). The original amount, 24 cm, is the base.

Estimation The cooled piece will be less than 24 cm.

$$\frac{1.2}{100} = \frac{P}{24} \qquad \text{P is the amount of decrease, and 1.2\% is the rate of decrease.}$$

$$1.2 \cdot 24 = 100 \cdot P$$
$$28.8 = 100 \cdot P$$
$$\frac{28.8}{100} = P$$
$$0.288 = P$$

Interpretation The amount of shrinkage is 0.288 cm, so the length of the cooled piece (new amount) is

$24 - 0.288 = \textbf{23.712 cm}$ New amount = Original amount − Amount of decrease.

EXAMPLE Uncut earth is hard, packed soil. As it is dug the volume increases or swells. A contractor figures that there will be a 20% earth swell when a mixture of uncut loam and clay soil is excavated. If 150 cubic yards (yd^3) of uncut earth is to be removed, taking into account the earth swell, how many cubic yards will have to be hauled away?

To find the amount of earth swell (amount of increase, percentage), find 20% of 150 yd^3, the original amount or base.

Estimation 10% of 150 = 15, so 20% of 150 = 30. The amount of earth to be hauled away is estimated to be 150 + 30 = 180 yd^3.

(a) What percent of 10 is 5?

(b) 25% of what number is 3?

(c) 3 is 20% of what number?

2 Solve percent problems using the percentage formula (pp. 176–178).

Use key words like *of* or *total* for the base and *is* or *portion* for the percentage to identify the missing element. Then select the appropriate percentage formula.

$$P = R \cdot B \qquad R = \frac{P}{B} \qquad B = \frac{P}{R}$$

What amount is 5% of $200? *Of* identifies $200 as the base.
The percent or rate is 5%. The missing element is the percentage.

$P = RB$ — Select the appropriate percentage formula and substitute given amounts.
$P = 5\% \,(\$200)$ — Change 5% to fraction or decimal equivalent by dividing by 100%.
$P = 0.05 \,(200)$ — Multiply.
$P = \$10$ — Percentage.

What percent of 6 is 2?
Is suggests 2 is the percentage. *Of* identifies 6 as the base. The rate is missing.

$R = \dfrac{P}{B}$ — Select the appropriate percentage formula and substitute given amounts.

$R = \dfrac{2}{6}$ — Reduce fraction or change to decimal equivalent by dividing.

$R = \dfrac{1}{3}$ — Change to a percent by multiplying by 100%.

$R = 33\dfrac{1}{3}\%$ — Rate $\left(\dfrac{1}{3} \times \dfrac{100\%}{1} = \dfrac{100\%}{3} = 33\dfrac{1}{3}\% \right)$.

12 is 24% of what number?
The percent is 24%. 12 is the part or percentage and is suggested by *is*. *Of* identifies "what number" as the missing base.

$B = \dfrac{P}{R}$ — Select the appropriate percentage formula and substitute given amounts.

$B = \dfrac{12}{24\%}$ — Change 24% to a decimal equivalent by dividing by 100%.

$B = \dfrac{12}{0.24}$ — Divide.

$B = 50$ — Base.

3 Solve percentage problems using the percentage proportion (pp. 178–187).

To solve a proportion, multiply diagonally across the equal sign to find the cross products. Then divide the cross product of the two known factors by the number factor of the other cross product.

$\dfrac{3}{y} = \dfrac{2}{5}$ — Multiply cross products diagonally across the equal sign.

$3 \times 5 = 2 \times y$

$15 = 2 \times y$ — Divide by 2 (the number factor of the other cross product).

$\dfrac{15}{2} = y$ — Change to mixed-number or mixed-decimal equivalent.

$7\dfrac{1}{2} = y$ or $y = 7.5$

208   Chapter 4 / Percents

What amount is 15% of $40? *Of* identifies $40 as the base.
The percent or rate is 15%. The missing element is the percentage.

$$\frac{R}{100} = \frac{P}{B}$$ Identify the rate, base, and percentage and substitute given amounts into the percentage proportion.

$$\frac{15}{100} = \frac{P}{40}$$ Multiply cross products.

$$15 \times 40 = P \times 100$$

$$600 = P \times 100$$ Divide by 100.

$$\frac{600}{100} = P$$

$$\$6 = P$$ Percentage.

LaQuita sold 12 boxes of candy for a school project. If she started with 25 boxes, what percent of the candy did she sell?
The percent is missing. 25 boxes is the total amount or base. 12 is the part or percentage.

$$\frac{R}{100} = \frac{12}{25}$$ Substitute given amounts into the percentage proportion.

$$25 \times R = 12 \times 100$$ Estimation:

$$25 \times R = 1{,}200$$ 12 is approximately $\frac{1}{2}$ of 25.

$$R = \frac{1{,}200}{25}$$ $\frac{1}{2} = 50\%$.

   Interpretation:

$$R = 48\%$$ 48% of the candy was sold.

4 Calculate sales tax and payroll deductions (pp. 187–189).

The rate is the percent of tax or deduction. The base is the amount of the purchase or the gross pay. The percentage is the amount of tax or deduction. Solve using the percentage proportion or percentage formula.

If the sales tax rate is 7%, find the tax and total bill on a purchase of $30.

$$\frac{7}{100} = \frac{P}{30}$$ Estimation:
   10% of $30 = $3
   Sales tax < $3

$$7 \times 30 = 100 \times P$$ Total bill < $30 + $3 or < $33.

$$210 = 100 \times P$$

$$\frac{210}{100} = P$$

$$\$2.10 = P$$
   Interpretation:

$$\$30 + \$2.10 = \$32.10$$ Sales tax is $2.10 and total cost is $32.10.

If 8% of gross pay is deducted for retirement, how much is deducted for retirement from a monthly salary of $1,650? What is the net pay?

$$\frac{8}{100} = \frac{P}{1{,}650}$$ Estimation:
   10% of $1,650 = $165.
   Deduction < $165.

$$8 \times 1{,}650 = 100 \times P$$ $1,650 − $165 = $1,485.
   Net pay > $1,485.

$$13{,}200 = 100 \times P$$

$$\frac{13{,}200}{100} = P$$

$$\$132 = P$$
   Interpretation:

$$\$1{,}650 - \$132 = \$1{,}518$$ The retirement deduction is $132 and the net pay is $1,518.

The rate is the percent of discount or commission. The base is the price or amount of sales. The percentage is the amount of the discount or commission. Solve using the percentage proportion or percentage formula with P as the missing element.

A $45 dress is on sale at 20% off. How much is the discount? What is the sale price?

$$\frac{20}{100} = \frac{P}{45}$$

Estimation:
10% of $45 = $4.50.
20% of $45 = 2 × $4.50.

$$100 \times P = 20 \times 45$$

20% of $45 = $9 discount.
Sale price = $45 − $9 = $36.

$$100 \times P = 900$$

$$P = \frac{900}{100}$$

$$P = \$9$$

Interpretation:

$$\$45 - \$9 = \$36$$

The discount is $9 and the sale price is $36.

Chang sold $5,000 of furniture and earned a 3% commission. How much is his commission?

$$\frac{3}{100} = \frac{P}{5,000}$$

Estimation:
1% of $5,000 = $50.
3% of $5,000 = 3 × $50 = $150.

$$100 \times P = 3 \times 5,000$$

$$100 \times P = 15,000$$

$$P = \frac{15,000}{100}$$

Interpretation:

$$P = \$150$$

The commission is $150.

Sometimes your estimate is the same as the exact answer.

The rate is the percent of interest. The base is the amount of the loan or investment. The percentage is the amount of interest. Solve using the percentage proportion or percentage formula with P as the missing element.

Boris borrowed $2,500 at 8.5% interest for 1 year. How much did he have to repay?

$$\frac{8.5}{100} = \frac{P}{2,500}$$

Estimation:
10% of $2,500 = $250.
Interest < $250.

$$100 \times P = 8.5 \times 2,500$$

Amount to repay < $2,500 + $250 or $2,750.

$$100 \times P = 21,250$$

$$P = \frac{21,250}{100}$$

$$P = \$212.50$$

Interpretation:
$212.50 is paid in interest and $2,712.50 is the total repaid.

$$\$2,500 + \$212.50 = \$2,712.50$$

Interest on an investment is found the same way.

Section 4–3

The original amount is the base. The new amount is the original amount plus the increase or the original amount minus the decrease. Subtract the new amount and the original amount to find the increase or decrease.

Julio made $15.25 an hour but took a 20% pay cut. What was the new hourly pay?

$$\frac{20}{100} = \frac{P}{15.25}$$

Estimation:
10% of $15 is $1.50.
20% of $15 is $3.00.

$$100 \times P = 20 \times 15.25$$
$$100 \times P = 305$$
$$P = \frac{305}{100}$$
$$P = 3.05$$ Hourly pay cut (decrease)
$$\$15.25 - \$3.05 = \$12.20$$ Original amount − decrease = new amount

2 Find the new amount directly in percent problems (pp. 198–201).

Add the percent of increase to 100% or subtract the percent of decrease from 100%. Use this new percent in the percentage proportion or percentage formula.

A project requires 5 lb of galvanized nails. If 15% of the nails will be wasted, how many pounds must be purchased?

$$\frac{115}{100} = \frac{P}{5}$$

$B = 5$ lb, $R = 100\% + 15\% = 115\%$
P is missing.

$$100 \times P = 5 \times 115$$

Estimation:
10% of 5 = 0.5.

$$100 \times P = 575$$

More than 0.5 or $\frac{1}{2}$ lb of nails must be added for waste.
Order > 5.5 lb.

$$P = \frac{575}{100}$$

Interpretation:

$$P = 5.75 \text{ lb}$$ 5.75 lb must be ordered.

3 Find the rate or the base in increase or decrease problems (pp. 201–202).

Subtract the original amount and the new amount to find the amount of increase or decrease. Then use the percentage proportion or percentage formula, with R or B as the missing element.

A 5-in. power edger blade now measures $4\frac{3}{4}$ in. What is the percent of wear?

$$\frac{R}{100} = \frac{\frac{1}{4}}{5}$$

$4\frac{4}{4} - 4\frac{3}{4} = \frac{1}{4}$. The part or percentage is $\frac{1}{4}$. The base or original amount is 5 in.

$$5 \times R = \frac{1}{4} \times 100$$

Estimation:
Wear is much less than 1 in.; 1 in. $= \frac{1}{5}$ original length.

$$5 \times R = 25$$

$\frac{1}{5} = 20\%$

$$R = \frac{25}{5}$$

Wear < 20%. (< is read "is less than.")

$$R = 5\%$$

Interpretation:
The percent of wear is 5%.

A PC has 20% of its hard drive storage capacity filled. If the PC now has 5.12G of storage capacity available, what was the original storage capacity?

Current storage capacity available is 100% − 20% = 80% of the original storage capacity, or base. 5.12G is the part or percentage.

$$\frac{80}{100} = \frac{5.12}{B}$$

Estimation:

$$20\% = \frac{1}{5}$$

$$80 \times B = 5.12 \times 100$$

$$\frac{1}{5} \text{ of } 5 = 1$$

$$80 \times B = 512$$

$\frac{1}{5}$ of original storage capacity > 1

Original storage capacity > 5.12 + 1 or

$$B = \frac{512}{80}$$

Original storage capacity > 6.12G

Interpretation:

$$B = 6.4G$$

Original storage capacity was 6.4G

CHAPTER TRIAL TEST

1. Write $\frac{4}{5}$ as a percent.
2. Write 85% as a fraction.
3. Write 0.3% as a decimal.

Identify the given and missing elements as R (rate), B (base), and P (percent).

4. $10 is what percent of $35?
5. 40% of 10 X-ray technicians is how many?
6. What percent of 24 syringes is 8?
7. 9 is what percent of 27 dogwood trees?
8. How many books is 30% of 40 books?
9. 12% of 50 grass plugs is how many?

Solve

10. 10% of 150 is what number?
11. What number is $6\frac{1}{4}\%$ of 144?
12. What percent of 275 is 33?
13. 45.75 is 15% of what number?
14. 55 is what percent of 11?
15. 250% of what number is 287.5?
16. What percent of 360 is 1.2?
17. 245% of what number is 164.4? Round to the nearest hundredth.
18. 5.4% of 57 is what number?
19. Find the complement of 88%.
20. Find the complement of 13%.
21. $15,000 is invested at 14% per year for 3 months. How much interest is earned on the investment?

22. A casting measuring 48 cm when poured shrinks to 47.4 cm when cooled. What is the percent of decrease?
23. An electronic parts salesperson earned $175 in commission. If the commission is 7% of sales, how much did the salesperson sell?
24. Electronic parts increased 15% in cost during a certain period, amounting to an increase of $65.15 on one order. How much would the order have cost before the increase? Round to the nearest cent.
25. In 2003, an area vocational school had an enrollment of 325 men and 123 women. In 2004, there were 149 women. What was the percent of increase of women students? Round to the nearest hundredth.
26. The payroll for a hobby shop for 1 week is $1,500. If federal income and social security taxes average 28%, how much is withheld from the $1,500?
27. During one period, a bakery rejected 372 items as unfit for sale. In the following period, the bakery rejected only 323 items, a decrease in unfit bakery items. What was the percent of decrease? Round to the nearest hundredth.

28. Materials to landscape a new home cost $643.75. What is the amount of tax if the rate is 6%? Round to the nearest cent.
29. A casting weighed 36.6 kg. After milling, it weighs 34.7 kg. Find the percent of weight loss to the nearest whole percent.
30. After soil was excavated for a project, it swelled 15%. If 275 yd³ were excavated, how many cubic yards of soil were there after excavation?
31. The total bill for machinist supplies was $873.92 before a discount of 12%. How much was the discount? Round to the nearest cent.
32. A business paid a $5.58 finance charge on a monthly balance of $318.76. What was the monthly rate of interest? Round to the nearest hundredth.

Direct Measurement

5

213

3 tons – approximately

Figure 5–6

1 ounce — approximately two small bottles of fingernail polish

Figure 5–7

1 cup — approximately one large coffee cup

Figure 5–8

1 pint – one large single-serving container of milk

Figure 5–9

1 quart

Figure 5–10

Ton: A ton is 2,000 pounds. The word originally meant a weight or measure. The typical American-made SUV weighs 3 tons and may help you visualize this measure (Fig. 5–6). The ton is used for extremely heavy items, such as very large volumes of grain, the weights of huge animals like elephants (4–7 tons), and the weights by which coal and iron are sold.

Capacity or Volume

The U.S. customary system includes units for both liquid and dry capacity measures; however, the dry capacity measures are seldom used. It is more common to express dry measures in terms of weight than in terms of capacity.

Common U.S. customary units of measure for capacity or volume are the ounce, cup, pint, quart, and gallon. Table 5–3 gives the relationships among the liquid measures for capacity or volume. In the U.S. customary system the term *ounce* represents both weight and liquid capacity. The measures are different and have no common relationship. The context of the problem will suggest whether the unit for weight or capacity is meant.

Ounce: A liquid ounce is a volume, not a weight. It is about the volume of two small bottles of fingernail polish or perfume (Fig. 5–7). Liquid ounces are used for bottled medicine, canned or bottled carbonated beverages, baby bottles and formula, and similar-sized quantities.

Cup: A cup is equal to 8 ounces. The term is derived from an old English word for tub, a kind of container. This measure is about the volume of a coffee cup (Fig. 5–8). Cups are most often used for 8-oz quantities in cooking. Measuring cups used for cooking usually divide their contents into cups and portions of cups.

Pint: A pint is two cups. Its name comes from an Old English word meaning the spot that marks a certain level in a measuring device. It is the quantity of a *medium-size* paper container of milk (Fig. 5–9). Liquids like milk and automobile motor additives, ice cream, and produce like strawberries are packaged in pint containers. In some localities, shucked oysters are sold by the pint.

Quart: A quart is two pints, or four cups, or 32 ounces. The term is derived from an Old English word for fourth. It is a fourth of a gallon. Mayonnaise and various citrus juices are often sold in quart containers (Fig. 5–10). Motor oil and ice cream are also packaged in quart containers. Quarts are used to measure liquid quantities in cooking and also the capacities of cookware like mixing bowls, pots, and casserole dishes.

Gallon: A gallon is four quarts. It is based on the "wine gallon" of British origin. Paint, varnish, and stain are often sold in gallon cans (Fig. 5–11). Motor fuel and heating oil are usually sold by the gallon, as are large quantities of liquid propane and various chemicals. Often, statistics on liquid consumption, such as water or alcoholic beverages, are reported in gallons consumed per individual during a specified time period.

Table 5–3	U.S. Customary Units of Liquid Capacity or Volume	
3 teaspoons (t) = 1 tablespoon (T)		2 tablespoons (T) = 1 ounce (oz)
8 ounces (oz) = 1 cup (c)		4 cups (c) = 1 quart (qt)
2 cups (c) = 1 pint (pt)		4 quarts (qt) = 1 gallon (gal)
2 pints (pt) = 1 quart (qt)		

1 gallon – one large can of paint

Figure 5–11

Chapter 5 / Direct Measurement

EXAMPLE Select a reasonable unit of measure for the following:

1. Package of rice	**Pounds or ounces**
2. Height of Doctor Washington	**Feet, or feet and inches**
3. A hippopotamus	**Tons or pounds**
4. Sugar for a cake	**Cups**
5. Expensive French perfume	**Ounces**
6. Distance between Memphis and New Orleans	**Miles**
7. Bottle of suntan lotion	**Ounces**
8. Container of eggnog	**Gallon, quart, or pint**
9. Sherbet in grocery store freezer	**Gallon or pint**
10. Tank of pesticide	**Gallons**
11. Man's dress shirt size	**Inches**
12. Material for draperies	**Yards**

2 **Convert One U.S. Customary Unit of Measure to Another Using Unity Ratios.**

Using the relationship between two units of measure, we can form a ratio in two different ways that has a value of 1. We call this type of ratio a *unity ratio*.

A ratio is a fraction. A unity ratio, then, is a fraction with one unit of measure in the numerator and a different, but equivalent, unit of measure in the denominator. Some examples of unity ratios are

$$\frac{12 \text{ in.}}{1 \text{ ft}}, \quad \frac{1 \text{ ft}}{12 \text{ in.}}, \quad \frac{3 \text{ ft}}{1 \text{ yd}}, \quad \frac{1 \text{ mi}}{5{,}280 \text{ ft}}$$

In each unity ratio, the value of the numerator equals the value of the denominator. A ratio with the numerator and denominator equal has a value of 1. When we convert from one unit of measure to another, we use a unity ratio that contains the original unit and the new unit.

EXAMPLE Write two unity ratios that relate the pair of measures.

(a) ounces and pounds (b) cups and pints

(a) The relationship between ounces and pounds is 1 lb contains 16 oz. The unity ratios involving ounces and pounds are

$$\frac{1 \text{ lb}}{16 \text{ oz}} \quad \text{and} \quad \frac{16 \text{ oz}}{1 \text{ lb}}$$

(b) The relationship between cups and pints is 1 pint contains 2 cups. The unity ratios involving cups and pints are

$$\frac{1 \text{ pint}}{2 \text{ cups}} \quad \text{and} \quad \frac{2 \text{ cups}}{1 \text{ pint}}$$

Unity ratios help us convert from one unit of measure to another.

To change from one U.S. customary unit of measure to another using unity ratios:

1. Set up the original amount as a fraction with the original unit of measure in the numerator.
2. Multiply this fraction by a unity ratio with the original unit in the denominator and the new unit in the numerator.
3. Reduce like units of measure and all numbers wherever possible.

EXAMPLE Find the number of inches in 5 ft.

Multiply 5 ft by a unity ratio that contains both inches and feet.
Because 5 ft is a whole number, we write it with 1 as the denominator.

$$\frac{5\text{ ft}}{1}\left(\frac{\quad}{\quad}\right)$$

Place the original unit with the 5 in the *numerator* of the first fraction.

$$\frac{5\text{ ft}}{1}\left(\frac{\quad}{\text{ft}}\right)$$

We are changing *from* (feet), so we place ft in the *denominator* of the unity ratio, which is shown in parentheses. This allows us to reduce the units later.

$$\frac{5\text{ ft}}{1}\left(\frac{\text{in.}}{\text{ft}}\right)$$

To change *to* inches, place inches in the *numerator* of the unity ratio.

$$\frac{5\text{ ft}}{1}\left(\frac{12\text{ in.}}{1\text{ ft}}\right)=60\text{ in.}$$

Place in the unity ratio the numerical values that make these two units of measure equivalent (1 ft = 12 in.). Complete the calculation, reducing wherever possible.

5 ft = 60 in.

EXAMPLE How many pints are in 4.5 quarts?

$$\frac{4.5\text{ qt}}{1}\left(\frac{2\text{ pt}}{1\text{ qt}}\right)=4.5\,(2\text{ pt})=9\text{ pt}$$

From quart (denominator) to pint (numerator).

4.5 qt = 9 pt.

When working with units of measure, it is very important to include the measuring unit in our analysis. Measurements are also referred to as *dimensions*, and the systematic examination of the appropriate measuring units of a solution is referred to as *dimension analysis*.

Sometimes it is necessary to convert a U.S. customary unit to a unit that is *not* the next larger or smaller unit of measure.

TIP!

Changing to Any Larger or Smaller Unit

To change from a U.S. customary unit to one other than the next larger or smaller unit, proceed as before, but multiply the original amount by as many unity ratios as needed to attain the new U.S. customary unit.
For instance, to change from yards to inches: yards → feet → inches.
To change from gallons to ounces: gallons → quarts → pints → cups → ounces.

EXAMPLE How many inches are in $2\frac{1}{3}$ yd?

$$2\frac{1}{3}\text{ yd}=\frac{7}{3}\text{ yd}$$

Write $2\frac{1}{3}$ as an improper fraction.

$$\frac{7\text{ yd}}{3}\left(\frac{\text{ft}}{\text{yd}}\right)\left(\frac{\text{in.}}{\text{ft}}\right)$$

Multiply the improper fraction by two unity ratios. To change from yards to inches, first change from yards to feet, then from feet to inches. Place the original unit in the *numerator* of the improper fraction, $\frac{7}{3}$.

$$\frac{7\text{ yd}}{3}\left(\frac{3\text{ ft}}{1\text{ yd}}\right)\left(\frac{12\text{ in.}}{1\text{ ft}}\right)=84\text{ in.}$$

Insert the appropriate numerical values for each unity ratio (3 ft = 1 yd; 12 in. = 1 ft), multiply, reducing wherever possible.

$2\frac{1}{3}$ **yd = 84 in.**

Alternative method

If we use the relationship for inches and yards, 36 in. = 1 yd, we need only one unity ratio for the calculation. That is, we convert $2\frac{1}{3}$ yd ($\frac{7}{3}$ yd) to inches as follows:

$$\frac{7 \text{ yd}}{3}\left(\frac{36 \text{ in.}}{1 \text{ yd}}\right)$$

Set up the unity ratio using 36 in. = 1 yd.

$$\frac{7 \text{ yd}}{\underset{1}{3}}\left(\frac{\overset{12}{36} \text{ in.}}{1 \text{ yd}}\right) = 84 \text{ in.}$$

Reduce units and numbers; then multiply.

$$2\frac{1}{3} \text{ yd} = 84 \text{ in.}$$

Learning Strategy *Focus on One Thing at a Time.*

Sometimes the steps in a multi-stepped problem can be overwhelming. It is often helpful to focus on one aspect of the problem at a time. In the example changing yards to inches, focus first on just the units of measure or dimensions.

$$\frac{\text{yd}}{1}\left(\frac{\text{ft}}{\text{yd}}\right)\left(\frac{\text{in.}}{\text{ft}}\right)$$

Reduce as appropriate. Yards reduce to 1. Feet reduce to 1. The only measuring unit left is inches. Therefore, the result will be in inches.

Next, focus on the numbers.

$$\frac{7\text{yd}}{3}\left(\frac{3 \text{ ft}}{1 \text{ yd}}\right)\left(\frac{12 \text{ in.}}{1 \text{ ft}}\right) \to \frac{7}{3}\left(\frac{3}{1}\right)\left(\frac{12}{1}\right) = 84$$

Putting the number and unit together, you have 84 in.

EXAMPLE Find the number of ounces in 2 gallons.

To change from gallons to ounces, we use the following conversions:

gallons → quarts → pints → cups → ounces

$$\frac{2 \text{ gal}}{1}\left(\frac{\text{qt}}{\text{gal}}\right)\left(\frac{\text{pt}}{\text{qt}}\right)\left(\frac{\text{c}}{\text{pt}}\right)\left(\frac{\text{oz}}{\text{c}}\right) =$$

Set up ratios, and insert relationship numbers. Reduce measures.

$$\frac{2 \text{ gal}}{1}\left(\frac{4 \text{ qt}}{1 \text{ gal}}\right)\left(\frac{2 \text{ pt}}{1 \text{ qt}}\right)\left(\frac{2 \text{ c}}{1 \text{ pt}}\right)\left(\frac{8 \text{ oz}}{1 \text{ c}}\right) =$$

$$2(4)(2)(2)(8) \text{ oz} = 256 \text{ oz}$$

Multiply.

2 gal = 256 oz.

Alternative method

We can work this same example with fewer unity ratios if we make some preliminary calculations using the relationships. Multiply 4 (cups) by 8 (ounces per cup) to find the number of ounces in a quart; that is, $4 \times 8 = 32$ oz.

$$\frac{2 \text{ gal}}{1}\left(\frac{4 \text{ qt}}{1 \text{ gal}}\right)\left(\frac{32 \text{ oz}}{1 \text{ qt}}\right)$$

Set up unity ratios.

$$\frac{2 \text{ gal}}{1}\left(\frac{4 \text{ qt}}{1 \text{ gal}}\right)\left(\frac{32 \text{ oz}}{1 \text{ qt}}\right) = 256 \text{ oz}$$

Reduce and multiply.

2 gal = 256 oz.

When estimating unit conversions, first see if the new unit is larger or smaller than the original unit.

Larger to Smaller

Each larger unit can be divided into smaller units. Larger-to-smaller conversions mean *more* smaller units. *More* implies multiplication.

    To convert a U.S. customary unit to a desired *smaller* unit: *Multiply* the number of larger units by the number of smaller units that equals 1 larger unit.

2 yd = _____ ft	The smaller unit is feet: 3 ft = 1 yd.
2×3 ft = 6 ft	Multiply number of yards by 3 ft.
2 yd = 6 ft	Dimension analysis: $\dfrac{2 \text{ yd}}{1} \times \dfrac{3 \text{ ft}}{1 \text{ yd}} = 6 \text{ ft}$

The key word clues are:

Larger to smaller unit → obtain more units → multiply

Smaller to Larger

Several small units combine to make one large unit. Thus, smaller-to-larger conversions mean *fewer* large units. *Fewer* implies division.

    To convert a U.S. customary unit to a *larger* unit: *Divide* the original unit by the number of smaller units that equals 1 desired larger unit.

12 ft = _____ yd	The larger unit is yards: 1 yd = 3 ft.
$12 \div 3 = 4$	Divide by 3 ft to get yards.
12 ft = 4 yd	Dimension analysis: $\dfrac{12 \text{ ft}}{1} \times \dfrac{1 \text{ yd}}{3 \text{ ft}} = 4 \text{ yd}$

Using key word clues:

Smaller to larger unit → fewer units → divide

Estimation can catch errors in setting up the problem, but it will not likely catch calculation errors.

3 **Convert from One U.S. Customary Unit of Measure to Another Using Conversion Factors.**

    Unity ratios can be used to develop conversion factors. With conversion factors, you *always* multiply to change from one measuring unit to another.

To develop a conversion factor for converting from one measure to another:

1. Write a unity ratio that changes the given unit to the new unit.
2. Change the fraction (or ratio) to its decimal equivalent by dividing the numerator by the denominator.

EXAMPLE   Develop two conversion factors relating pounds and ounces.

<table>
<tr><td align="center">Pounds to Ounces</td><td align="center">Ounces to Pounds</td></tr>
<tr><td align="center">$\dfrac{\text{pounds}}{1}\left(\dfrac{\text{ounces}}{\text{pounds}}\right)$</td><td align="center">$\dfrac{\text{ounces}}{1}\left(\dfrac{\text{pounds}}{\text{ounces}}\right)$</td></tr>
<tr><td align="center">$\dfrac{\text{pounds}}{1}\left(\dfrac{16 \text{ ounces}}{1 \text{ pound}}\right)$</td><td align="center">$\dfrac{\text{ounces}}{1}\left(\dfrac{1 \text{ pound}}{16 \text{ ounces}}\right)$</td></tr>
</table>

$$\frac{16}{1} = 16 \qquad\qquad \frac{1}{16} \text{ or } 1 \div 16 = 0.0625$$

pounds × 16 = ounces   ounces × 0.0625 = pounds

To change from one U.S. customary unit of measure to another using conversion factors:

1. Select the appropriate conversion factor.
2. Multiply the original measure by the conversion factor.

EXAMPLE   Use a conversion factor to convert 54 ounces to pounds.

ounces × 0.0625 = pounds   Conversion factor for ounces to pounds.
56 × 0.0625 = 3.5 pounds   Multiply.

56 ounces is 3.5 pounds.

U.S. Customary Conversion Factors

		TO CHANGE	
	From	To	Multiply By
Length or Distance			
12 inches (in.) = 1 foot (ft)	feet	inches	12
	inches	feet	0.0833333
3 feet (ft) = 1 yard (yd)	yards	feet	3
	feet	yards	0.3333333
36 inches (in.) = 1 yard (yd)	yards	inches	36
	inches	yards	0.0277778
5,280 feet (ft) = 1 mile (mi)	miles	feet	5280
	feet	miles	0.0001894
Weight or Mass			
16 ounces (oz) = 1 pound (lb)	pounds	ounces	16
	ounces	pounds	0.0625
2,000 pounds (lb) = 1 ton (T)	pounds	tons	2000
	tons	pounds	0.0005
Liquid Capacity or Volume			
8 ounces (oz) = 1 cup (c)	cups	ounces	8
	ounces	cups	0.125
2 cups (c) = 1 pint (pt)	cups	pints	2
	pints	cups	0.5
2 pints (pt) = 1 quart (qt)	quarts	pints	2
	pints	quarts	0.5
4 quarts (qt) = 1 gallon (gal)	gallons	quarts	4
	quarts	gallons	0.25

Additional conversion factors are found on the inside covers of the text.

1 Add U.S. customary measures.

2 Subtract U.S. customary measures.

1 Add U.S. Customary Measures.

We can add U.S. customary measures *only* when their units are the same. Measures with the same units are *like* measures. Measures with different units are *unlike* measures.

To add unlike U.S. customary measures:

1. Convert each measure to a measure with a common U.S. customary unit.
2. Add.

EXAMPLE Add 3 ft + 2 in.

Because 2 in. is a fraction of a foot, we can avoid working with fractions by converting the larger unit (feet) to the smaller unit (inches).

$$3 \text{ ft} = \frac{3 \text{ ft}}{1} \left(\frac{12 \text{ in.}}{1 \text{ ft}} \right) = 36 \text{ in.}$$ Convert ft to in., and add like measures.

36 in. + 2 in. = 38 in.

Learning Strategy *Relate to Previous Processes.*

We are using a process that requires us to add *like quantities*. When have we dealt with *likes* quantities before? We did so when we used *like* signs in adding integers and *like* denominators when adding fractions.

What did we do when we had *unlike* quantities? With integers having *unlike* signs, we use a specified rule. With fractions having *unlike* denominators, we change to fractions with a common denominator. Which process will we pattern after here?

In many instances, unlike units can be changed to a common unit. To add 3 ft + 2 in., we change 3 ft to inches. Are there situations when we cannot change to a common unit? Yes, but sometimes in our investigation of possible solutions, we may temporarily consider adding unlike types of measures.

Can we add 3 ft + 2 lb? No. Why not? Feet and pounds measure different types of units: length and weight.

To add mixed U.S. customary measures:

1. Align the measures vertically so the common units are written in the same vertical column.
2. Add.
3. Express the sum in standard notation.

EXAMPLE   Add and write the answer in standard form: 6 lb 7 oz and 3 lb 13 oz.

$$\begin{array}{r} 6\text{ lb }\ 7\text{ oz} \\ +\ 3\text{ lb }13\text{ oz} \\ \hline 9\text{ lb }20\text{ oz} \end{array}$$ Write in standard notation, 20 oz = 1 lb 4 oz.

Thus, 9 lb 20 oz = 10 lb 4 oz.

2 **Subtract U.S. Customary Measures.**

As in addition, to subtract measures we find a common unit of measure.

To subtract unlike U.S. customary units:

1. Convert to a common U.S. customary unit.
2. Subtract.

EXAMPLE   Subtract 15 in. from 2 ft.

Changing 15 in. to feet gives us a mixed number, so it is more convenient to convert 2 ft to inches.

2 ft = 24 in.   Convert ft to in. and subtract.
24 in. − 15 in. = **9 in**.

To subtract mixed U.S. customary measures:

1. Align the measures vertically so that the common units are written in the same vertical column.
2. Subtract.

EXAMPLE   Subtract 5 ft 3 in. from 7 ft 4 in.

$$\begin{array}{r} 7\text{ ft }4\text{ in.} \\ -\ 5\text{ ft }3\text{ in.} \\ \hline \mathbf{2\text{ ft }1\text{ in.}} \end{array}$$ Align like measures in a vertical line, then subtract.

When we subtract mixed measures we use our knowledge of borrowing to subtract a larger unit from a smaller one.

To subtract a larger U.S. customary unit from a smaller unit in mixed measures:

1. Align the common measures in vertical columns.
2. Borrow one unit from the next larger unit of measure in the minuend, convert to the equivalent smaller unit, and add it to the smaller unit.
3. Subtract the measures.

Area measure is a measure of a surface instead of a length. Area indicates the number of squares required to cover a surface.

$$\text{length} \times \text{length} = \text{length}^2 \text{ or area}$$

To multiply a length measure by a like length measure:

1. Multiply the numbers associated with each like unit of measure.
2. The product is a square unit of measure.

EXAMPLE   A desktop is 2 ft × 3 ft (Fig. 5–13). What is the number of square feet in the surface?

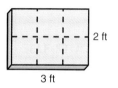

2 ft

3 ft

$2 \text{ ft} \times 3 \text{ ft} = \mathbf{6 \text{ ft}^2}$ Multiply numbers. Product is ft².

Figure 5–13

3 **Divide a U.S. Customary Measure by a Number.**

We are frequently required to divide measures by a number. For example, if a recipe calls for 2 gal 2 qt of liquid and the recipe is halved (divided by 2), how much liquid should be used? To solve this problem, we divide the measure by 2.

To divide a U.S. customary measure by a number that divides evenly into each measure:

1. Divide the numbers associated with each unit of measure by the given number.
2. Write the resulting measure in standard notation.

EXAMPLE   How much milk is needed for a half-recipe if the original recipe calls for 2 gal 2 qt?

$2\overline{)2 \text{ gal } 2 \text{ qt}}$ Divide each measure by 2.

$\dfrac{1 \text{ gal } 1 \text{ qt}}{2\overline{)2 \text{ gal } 2 \text{ qt}}}$

The half-recipe requires 1 gal 1 qt of milk.

If a given number does not divide evenly into each measure, there will be a remainder. For instance, if we divide 5 gal by 2, the 2 divides into the 5 gal 2 times with 1 gal left over as a remainder.

To divide U.S. customary measures by a number that does not divide evenly into each measure:

1. Set up the problem and proceed as in long division.
2. When a remainder occurs after subtraction, convert the remainder to the same unit used in the next smaller measure, and add it to the quantity in the next smaller measure.
3. Divide the given number into this next smaller unit.
4. If a remainder occurs when the smallest unit is divided, express the remainder as a fractional part of the smallest unit.

EXAMPLE Divide 5 gal 3 qt 1 pt by 3.

$$
\begin{array}{r}
\text{1 gal} \quad\quad \text{3 qt} \quad 1\frac{2}{3}\text{pt} \\
3\overline{\big|\text{5 gal} \quad\quad \text{3 qt} \quad \text{1 pt}} \\
\underline{\text{3 gal}} \quad\quad\quad\quad\quad\quad\quad \\
\text{2 gal} = \underline{\text{8 qt}} \quad\quad\quad\quad \\
\text{11 qt} \quad\quad\quad \\
\underline{\text{9 qt}} \quad\quad\quad \\
\text{2 qt} = \underline{\text{4 pt}} \\
\text{5 pt} \\
\underline{\text{3 pt}} \\
\text{2 pt}
\end{array}
$$

5 gal ÷ 3 = 1 gal, remainder 2 gal

2 gal = 8 qt

3 qt + 8 qt = 11 qt

11 qt ÷ 3 = 3 qt, remainder 2 qt

2 qt = 4 pt

1 pt + 4 pt = 5 pt

5 pt ÷ 3 = 1 pt, remainder 2 pt

Write the final remainder, **2**, as a fraction $\frac{2}{3}$ pints, and add to 1 pint to get $1\frac{2}{3}$ pints.

Thus, 1 gal 3 qt $1\frac{2}{3}$ pt is the solution.

---

4 **Divide a U.S. Customary Measure by a Measure.**

If tubing is manufactured in lengths of 8 ft 4 in. and a part is 10 in. long, how many parts can be cut from the length of tubing if we do not account for waste? To solve such a problem, we express both measures in the same unit, just as in adding and subtracting measures. We generally convert to the *smallest* unit used in the example.

To divide a U.S. customary measure by a U.S. customary measure:

1. Convert both measures to the same unit if they are different.
2. Write the division as a fraction, including the common unit in the numerator and the denominator.
3. Reduce the units and divide the numbers.

EXAMPLE If tubing is manufactured in lengths of 8 ft 4 in. and a part is 10 in. long, how many parts can be cut from the length of tubing if we do not account for waste? Divide 8 ft 4 in. by 10 in.

$$8\ \text{ft} = \frac{8\ \cancel{\text{ft}}}{1}\left(\frac{12\ \text{in.}}{1\ \cancel{\text{ft}}}\right) = 96\ \text{in.}$$

Convert 8 ft to inches. Then add to 4 in.

$$8\ \text{ft}\ 4\ \text{in.} = 96\ \text{in.} + 4\ \text{in.} = 100\ \text{in.}$$

Write the mixed measure as a measure with one measuring unit.

$$100\ \text{in.} \div 10\ \text{in.}$$

Divide the total length by the length of each part.

If we write this division in fraction form, we can see more easily that the common units reduce. In other words, the answer will be a number (not a measure) telling *how many equal parts* can be cut from the tubing.

$$\frac{100\ \cancel{\text{in.}}}{10\ \cancel{\text{in.}}} = 10$$

Therefore, 10 parts of equal length can be cut from the tubing.

Change from One U.S. Customary Rate Measure to Another.

A *rate measure* is a ratio of two different kinds of measures. A rate measure is often referred to as a *rate*. Some examples of rates are 55 miles per hour or 20 cents per mile. In each of these rates, the word *per* means *divided by*.

The rate 55 miles per hour means 55 miles ÷ 1 hour or $\frac{55 \text{ mi}}{1 \text{ hr}}$. The rate 20 cents per mile means 20 cents ÷ 1 mile or $\frac{20 \text{ cents}}{1 \text{ mi}}$.

Many rate measures involve measures of time. The units we use to measure time are universally accepted. The basic units of time are the year, month, week, day, hour, minute, and second. Table 5–4 gives the relationships among the units for time. These units are often used when working with rates.

Table 5–4 Units of Time	
1 year (yr) = 12 months (mo)	1 minute (min) = 60 seconds (sec)[b]
1 year (yr) = 365 days (da)	1 millisecond (ms) = $\dfrac{1}{1,000}$ second
1 week (wk) = 7 days (da)	
1 day (da) = 24 hours (hr)	
1 hour (hr) = 60 minutes (min)[a]	1 nano second (ns) = $\dfrac{1}{1,000,000,000}$ second

[a] The symbol ′ means feet (3′ = 3 ft) or minutes (60′ = 60 minutes).
[b] The symbol ″ means inches (8″ = 8 in.) or seconds (60″ = 60 seconds).

To convert one U.S. customary rate measure to another:

1. Compare the **units** of both numerators and both denominators to determine which units will change.
2. Multiply each unit that changes by a unity ratio containing the new unit so that the unit to be changed will reduce.

EXAMPLE   Change $8\,\dfrac{\text{pt}}{\text{min}}$ to $\dfrac{\text{qt}}{\text{min}}$.

Estimation   Pints to quarts is *smaller* to *larger*, so there will be fewer quarts.

Examine the rates:

   Numerators—pints change to quarts
   Denominators—no change

$$\frac{\cancel{\text{pt}}}{\text{min}}\left(\frac{\text{qt}}{\cancel{\text{pt}}}\right) = \frac{\text{qt}}{\text{min}}$$ Develop a unity ratio with pints in the denominator.

$$\frac{\overset{4}{\cancel{8}}\,\cancel{\text{pt}}}{\text{min}}\left(\frac{1 \text{ qt}}{\underset{1}{\cancel{2}}\,\cancel{\text{pt}}}\right) = \frac{4 \text{ qt}}{\text{min}}$$ Insert numbers in the unity ratio and reduce. Multiply.

Interpretation   Thus, $8\,\dfrac{\text{pt}}{\text{min}}$ equals $4\,\dfrac{\text{qt}}{\text{min}}$.

A separate unity ratio is used for each unit in the rate that changes. For example, when both the numerator and denominator of a rate measure change, we multiply by at least two unity ratios to make the conversion: First, we multiply by a unity ratio that changes the unit in the numerator of the original rate measure. Then, we multiply by a unity ratio that changes the unit in the denominator of the original rate measure.

Chapter 5 / Direct Measurement

EXAMPLE   Change 60 miles per hour to feet per second.

$$60 \text{ miles per hour} = 60\,\frac{\text{mi}}{\text{hr}}$$ Write rate as a fraction.

$$\text{feet per second} = \frac{\text{ft}}{\text{sec}}$$

Numerators—miles change to feet   Examine the changes in the measures.
Denominators—hours change to seconds

$$60\,\frac{\text{mi}}{\text{hr}}\left(\frac{5{,}280 \text{ ft}}{1 \text{ mi}}\right)\left(\frac{1 \text{ hr}}{60 \text{ min}}\right)\left(\frac{1 \text{ min}}{60 \text{ sec}}\right)$$

(miles to feet) (hours to minutes) (minutes to seconds)

Develop appropriate unity ratios, and reduce.

$$\frac{5{,}280 \text{ ft}}{60 \text{ sec}} = 88\,\frac{\text{ft}}{\text{sec}}$$ Divide.

At 60 miles per hour you are traveling 88 feet per second.

SELF-STUDY EXERCISES 5–3

1 Multiply and write the answers for mixed measures in standard notation.

1. 12 mi
   5

2. 18 gal
   6

3. 21 lb
   12

4. 3 qt 1 pt
   4

5. 7 lb 3 oz
   8

6. 5 gal 2 qt
   7

7. 8 gal 3 qt
   5

8. 7 ft 3 in.
   8

9. Tuna is packed in 1-lb 8-oz cans. If a case contains 24 cans, how much does a case weigh?

10. A car used 1 qt 1 pt of oil each month for 5 months. Find the total amount of oil used.

2 Multiply.

11. 5 in. × 7 in.

12. 12 ft × 9 ft

13. 15 yd × 12 yd

14. 4 mi × 27 mi

15. A room is to be covered with square linoleum tiles that are 1 ft by 1 ft. If the room is 18 ft by 21 ft, how many tiles (square feet) are needed?

16. A horticulturist stores a stock solution of fertilizer in two tanks, each with a capacity of 23 gal 9 oz. How much liquid fertilizer is needed to fill both tanks?

17. Latonya has three containers, each containing 1 qt 3 pt of photographic solution. How much total photographic solution is in all three containers?

18. A package of nails weighs 1 lb 4 oz. How much would five packages weigh?

19. If a history textbook weighs 2 lb 8 oz, how much would three of the same books weigh?

3 Divide.

20. 12 gal ÷ 2

21. 3 days 6 hr ÷ 2

22. 20 yd 2 ft 6 in. ÷ 2

23. 4 yd 1 ft 9 in. ÷ 3

24. 4 gal 3 qt 1 pt ÷ 6

25. 60 gal ÷ 9 (express in gallons only)

26. 5 hr ÷ 3 (express in hours and minutes)

27. If eight trucks require 42 qt of oil for each to get a complete oil change, how many quarts are required for each truck?

28. Sixty feet of wire are required to complete eight jobs. If each job requires an equal amount of wire, find the amount of wire required for one job. (Express in feet and inches.)

29. A vat holding 10 gal 2 qt of defoliant is emptied equally into three tanks. How many gallons and quarts are in each tank?

30. How many pieces of $\frac{1}{2}$-in. OD (outside diameter) plastic pipe 8 in. long can be cut from a piece 72 in. long?

31. A roll of No. 14 electrical cable 150 ft long is divided into 30 equal sections. How long is each section?

32. A greenhouse attendant has a container with 6 gal 2 qt 10 oz of potassium nitrate solution that will be stored in two smaller containers. How much solution will be stored in each smaller container?

33. For a family picnic, Mr. Sonnier prepared 96 lb 12 oz of boiled crawfish. He brought the crawfish to the picnic site in four containers containing equal amounts. How much did the crawfish in each container weigh?

4 Divide.

34. 36 ft ÷ 12 ft

35. 51 in. ÷ 3 in.

36. 2 ft 8 in. ÷ 4 in.

37. 6 lb 12 oz ÷ 6 oz

38. 2 ft 6 in. ÷ 10 in.

39. How many 6-in. pieces can be cut from 48 in. of pipe?

40. How many 2-lb boxes can be filled from 18 lb of nails?

41. How many 15-oz cans are in a case if the case weighs 22 lb 8 oz?

42. How many 8-dollar tickets can be purchased for 72 dollars?

43. How many pieces of wood 5 in. long can be cut from a piece 45 in. long?

5 Change to the indicated rate measure.

44. $\dfrac{60 \text{ qt}}{\text{sec}} = \underline{\hspace{1cm}} \dfrac{\text{gal}}{\text{sec}}$

45. $\dfrac{45 \text{ lb}}{\text{hr}} = \underline{\hspace{1cm}} \dfrac{\text{lb}}{\text{min}}$

46. $\dfrac{3 \text{ mi}}{\text{hr}} = \underline{\hspace{1cm}} \dfrac{\text{ft}}{\text{hr}}$

47. $\dfrac{144 \text{ lb}}{\text{min}} = \underline{\hspace{1cm}} \dfrac{\text{oz}}{\text{min}}$

48. $\dfrac{30 \text{ gal}}{\text{min}} = \underline{\hspace{1cm}} \dfrac{\text{qt}}{\text{sec}}$

49. $\dfrac{30 \text{ lb}}{\text{day}} = \underline{\hspace{1cm}} \dfrac{\text{oz}}{\text{hr}}$

50. $\dfrac{8 \text{ in}}{\text{sec}} = \underline{\hspace{1cm}} \dfrac{\text{ft}}{\text{min}}$

51. $\dfrac{5 \text{ mi}}{\text{min}} = \underline{\hspace{1cm}} \dfrac{\text{ft}}{\text{sec}}$

52. A car traveling at the rate of 30 mph (miles per hour) is traveling how many feet per second?

53. A pump that can pump $45 \frac{\text{gal}}{\text{hr}}$ can pump how many quarts per minute?

54. A pump can dispose of sludge at the rate of $3{,}200 \frac{\text{lb}}{\text{hr}}$. How many pounds can be disposed of per minute? Round to the nearest tenth.

55. If water flows through a pipe at the rate of $50 \frac{\text{gal}}{\text{min}}$, how many gallons will flow per second? Round to the nearest tenth.

5–4 | *Introduction to the Metric System*

Learning Outcomes

1 Identify uses of metric measures of length, weight, and capacity.

2 Convert from one metric unit of measure to another.

3 Make calculations with metric measures.

The *metric system* is an international system of measurement that uses standard units and powers-of-10 prefixes to indicate other units of measure.

In the metric system, or the International System of Units (SI), there is a standard unit for each type of measurement. The *meter* is used for length or distance, the *gram* is used for weight, and the *liter* is used for capacity or volume. A prefix is affixed to the standard unit to indicate a measure greater than the standard unit or less than the standard unit.

1 Identify Uses of Metric Measures of Length, Weight, and Capacity.

The most common prefixes for units *smaller* than the standard unit are:

deci- $\dfrac{1}{10}$ of **centi-** $\dfrac{1}{100}$ of **milli-** $\dfrac{1}{1{,}000}$ of

The most common prefixes for units *larger* than the standard unit are

deka- 10 times   **hecto-** 100 times   **kilo-** 1,000 times

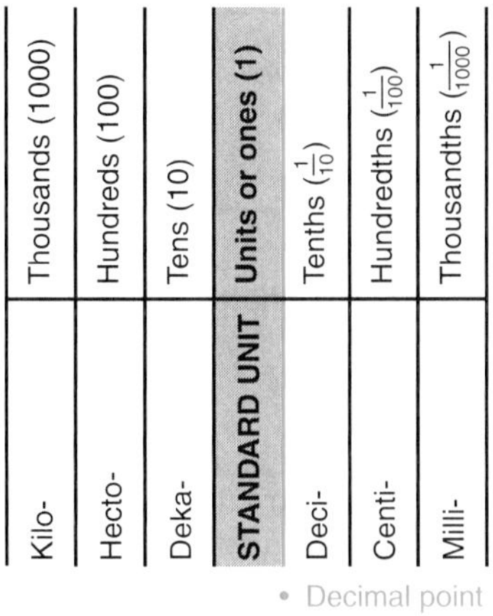

Thousands (1000)	Hundreds (100)	Tens (10)	Units or ones (1)	Tenths $(\frac{1}{10})$	Hundredths $(\frac{1}{100})$	Thousandths $(\frac{1}{1000})$
Kilo-	Hecto-	Deka-	**STANDARD UNIT**	Deci-	Centi-	Milli-

• Decimal point

Figure 5–14

We can compare our decimal place-value chart with the prefixes (Fig. 5–14). The standard unit (whether meter, gram, or liter) corresponds to the *ones* place. All the places to the left are powers of the standard unit. That is, the value of *deka-* (some sources use *deca-*) is 10 times the standard unit; the value of *hecto-* is 100 times the standard unit; the value of *kilo-* is 1,000 times the standard unit; and so on. All the places to the right of the standard unit are subdivisions of the standard unit. That is, the value of *deci-* is $\frac{1}{10}$ of the standard unit; the value of *centi-* is $\frac{1}{100}$ of the standard unit; the value of *milli-* is $\frac{1}{1,000}$ of the standard unit; and so on.

EXAMPLE Give the value of the metric units using the standard unit (gram, liter, or meter).

(a) Kilogram (kg) = 1,000 times 1 gram or **1,000 g**
(b) Deciliter (dL) = $\frac{1}{10}$ of 1 liter or **0.1 L**
(c) Hectometer (hm) = 100 times 1 meter or **100 m**
(d) Dekaliter (dkL) = 10 times 1 liter or **10 L**
(e) Milliliter (mL) = $\frac{1}{1,000}$ of 1 liter or **0.001 L**
(f) Centigram (cg) = $\frac{1}{100}$ of 1 gram or **0.01 g**

There are other metric prefixes for very large and very small amounts. For measurements smaller than one-thousandth of a unit or larger than one thousand times a unit, prefixes that align with periods on a place-value chart are commonly used.

Metric Prefixes

Prefix	Relationship to Standard Unit*
atto-(a)	quintillionth part ($\times 10^{-18}$)
femto-(f)	quadrillionth part ($\times 10^{-15}$)
pico-(p)	trillionth part ($\times 10^{-12}$)
nano-(n)	billionth of ($\times 10^{-9}$)
micro-(µ)	millionth of ($\times 10^{-6}$)
milli-(m)	thousandth of ($\times 10^{-3}$)
centi-(c)	hundredth of ($\times 10^{-2}$)
deci-(d)	tenth of ($\times 10^{-1}$)

50. How many decimeters are in 4,389 cm?   **51.** How many grams are in 47 dg?
52. How many deciliters are in 2.25 cL?

Change each measure to the measure indicated. When using the metric-value chart, place the decimal immediately *after* the unit.

53. 2,743 mm = _____ m   **54.** 385 g = _____ kg
55. 15 dkm = _____ km   **56.** 8 cL = _____ L
57. 296,484 m = _____ hm   **58.** 29.83 dg = _____ dkg
59. 0.3 cm = _____ dkm   **60.** 40 dL = _____ kL
61. 2,857 mg = _____ kg   **62.** 15,285 m = _____ km
63. Change 297 cm to hectometers.   **64.** Change 0.03 mL to liters.

3 Add or subtract as indicated.

65. 3 m + 8 m   **66.** 7 hL + 5 hL   **67.** 15 cg − 9 cg   **68.** 2 dm + 4 cm
69. 5 cL + 9 mL   **70.** 4 m + 2 L   **71.** 14 kL − 39 hL   **72.** 1 g − 45 cg
73. 3 cg − 5 mL   **74.** 7 km + 2 m
75. A patient absorbs 175 mL of fluid through an IV. If   **76.** 653 dkL of orange juice concentrate is removed
the IV bag has 825 mL left, how much fluid was in   from a vat containing 8 kL of the concentrate.
the bag to begin with?   How much concentrate remains in the vat?

Multiply.

77. 43 m × 12   **78.** 3.4 m × 12   **79.** 50.32 dm × 3
80. A plot of ground is divided into seven plots,
each with road frontage of 138.5 m. What is the
total road frontage of the plot of ground?

Divide.

81. 99 m ÷ 9   **82.** $\frac{54\ cL}{6}$

83. A block of silver weighing 978 g is cut into six   **84.** Two pieces of steel each 12 m long are cut into a
equal pieces. How much does each piece weigh?   total of 60 equal pieces. How long is each piece?
85. 2.5 cg ÷ 0.5 cg   **86.** 3 m ÷ 10 cm
87. How many 250-mL prescriptions can be made   **88.** How many 500-g containers are needed to hold
from a container of 4 L of decongestant?   40 kg of grass seed?

Solve.

89. Add 4.6 cL + 5.28 dL of photographic developer.   **90.** Add 3 m + 2 dkm of fabric for draperies.
91. Subtract 19.8 km − 32.3 hm of paved highway.   **92.** Subtract 13 kL − 39 hL of stored liquid.
93. Multiply 0.25 cL of cologne by 5.   **94.** Multiply a 35-mm film size by 2.
95. A length of satin fabric 30 dm long is cut into 4 **96.** An IV bag holding 250 mL of an antibiotic is
equal pieces. How long is each piece?   calibrated (marked off) into five equal sections. How
                                                            many milliliters are represented by each section?
97. How many containers of jelly can be made from   **98.** How many 2-kg vials of hydrochloric acid (HC1)
8,500 L of jelly if each container holds 4 dL of   can be obtained from 38 kg of HC1?
jelly?

5–5 | *Metric–U.S. Customary Comparisons*

Learning Outcome   **1** Convert between U.S. customary measures and metric measures.

Converting a measure from the U.S. customary system to the metric system, and vice versa, is often necessary because both systems are used in the United States. Most industries use one system exclusively, making conversions from one system to another uncommon. However, an industry that uses the U.S. customary system in the United States often needs to convert to the metric system to market its products in other countries.

1 **Convert Between U.S. Customary Measures and Metric Measures.**

To convert measures from the U.S. customary system to the metric system, and vice versa, we need only one conversion relationship for each type of measure (length, weight, and capacity). The following equivalency relationships have been rounded to the nearest hundredth of a unit:

For length: 1 m = 1.09 yd
For weight: 1 kg = 2.20 lb
For capacity: 1 L = 0.91 dry qt (dry measure)
          1 L = 1.06 liquid qt (liquid measure)

Dry quarts are less common than liquid quarts; if the type of quart (dry or liquid) is not specified, use liquid quarts.

Additional conversion relationships, such as centimeters to inches and kilometers to miles, can be derived from these relationships using unity ratios.

The most common procedure for converting between the U.S. customary and metric systems is to use conversion factors. Conversion factors always involve multiplication, whereas unity ratios may involve multiplication or division. Many conversion factors are given in the table below. Calculators sometimes have conversion factors programmed into the calculator for greater accuracy.

U.S. Customary and Metric Conversion Factors

	From	To	Multiply By
Length			
1 meter = 39.37 inches	meters	inches	39.37
	inches	meters	0.0254
1 meter = 3.2808 feet	meters	feet	3.2808
	feet	meters	0.3048
1 meter = 1.0936 yards	meters	yards	1.0936
	yards	meters	0.9144
1 centimeter = 0.3937 inch	centimeters	inches	0.3937
	inches	centimeters	2.54
1 millimeter = 0.03937 inch	millimeters	inches	0.03937
	inches	millimeters	25.4
1 kilometer = 0.6214 miles	kilometers	miles	0.6214
	miles	kilometers	1.6093
Weight			
1 gram = 0.0353 ounce	grams	ounces	0.0353
	ounces	grams	28.3286
1 kilogram = 2.2046 pounds	kilograms	pounds	2.2046
	pounds	kilograms	0.4536
Liquid Capacity			
1 liter = 1.0567 quarts	liters	quarts	1.0567
	quarts	liters	0.9463

EXAMPLE   To change 50 ft to meters:

feet to meters: Conversion factor is 0.3048.

50 × 0.3048 = 15.24 m   Multiply.

Body mass index (BMI) is the standard unit for measuring a person's degree of obesity or emaciation. BMI is body weight in kilograms (kg) divided by height in meters squared, or BMI = w/h^2.

51. $12\frac{3}{8}$ in. and $23\frac{5}{8}$ in.

52. $14\frac{7}{8}$ in. and $25\frac{3}{4}$ in.

Find the midpoint between each pair of points.

53. 1.8 cm and 7 cm

54. 3.5 cm and 6.3 cm

55. 5.5 cm and 10.3 cm

56. 3 cm and 5.2 cm

57. $7\frac{1}{2}$ in. and $15\frac{1}{4}$ in.

58. 2 in. and $9\frac{3}{4}$ in.

5–8 | Temperature Formulas

Learning Outcomes

1 Make Celsius/Kelvin temperature conversions.

2 Make Rankine/Fahrenheit temperature conversions.

3 Convert Fahrenheit to Celsius temperatures.

4 Convert Celsius to Fahrenheit temperatures.

This section is concerned with temperature conversions involving the Celsius, Kelvin, Rankine, and Fahrenheit scales.

1 Make Celsius/Kelvin Temperature Conversions.

The *Kelvin* scale is one scale used to measure temperature in the metric system of measurement. Units on the scale are abbreviated with a capital K (without the symbol ° because these units are called *kelvins*) and are measured from absolute zero, the temperature at which *all* heat is said to be removed from matter. Another metric temperature scale is the *Celsius* scale (abbreviated °C), which has as its zero the freezing point of water. The Kelvin and Celsius scales are related such that absolute zero on the Kelvin scale is the same as −273°C on the Celsius scale. Each unit of change on the Kelvin scale is equal to 1 degree of change on the Celsius scale; that is, the size of a kelvin and a Celsius degree is the same on both scales.

To convert celsius degrees to kelvins:

Use the formula: K = °C + 273, where K = kelvins and °C = degrees Celsius.

Although temperature conversions can be made using unity ratios, we generally use formulas rather than unity ratios to make temperature conversions.

EXAMPLE A temperature of 21°C on the Celsius scale is the same as what temperature on the Kelvin scale?

K = °C + 273   Substitute 21 for °C.

K = 21 + 273   Add.

K = 294

21°C = 294 K.

To convert kelvins to celsius degrees:

Use the formula: °C = K − 273, where °C = degrees Celsius and K = kelvins.

EXAMPLE A temperature of 300 K on the Kelvin scale corresponds to what temperature on the Celsius scale?

$$°C = K - 273 \qquad \text{Substitute 300 for K.}$$
$$°C = 300 - 273 \qquad \text{Subtract.}$$
$$°C = 27$$

300 K = 27°C.

2 Make Rankine/Fahrenheit Temperature Conversions.

The U.S. customary system temperature scale that starts at absolute zero is called the *Rankine* scale. It is related to the more familiar *Fahrenheit* scale, which places the freezing point of water at 32°. One degree of change on the Rankine scale equals 1 degree of change on the Fahrenheit scale. Absolute zero (the zero for the Rankine scale) corresponds to 460 degrees *below* zero (−460°) on the Fahrenheit scale.

To convert fahrenheit degrees to rankine degrees:

Use the formula: °R = °F + 460, where °R = degrees Rankine and °F = degrees Farenheit

EXAMPLE What temperature on the Rankine scale corresponds to 40°F on the Fahrenheit scale?

$$°R = °F + 460 \qquad \text{Substitute 40 for °F.}$$
$$°R = 40 + 460 \qquad \text{Add.}$$
$$°R = 500$$

40°F = 500°R.

To convert rankine degrees to fahrenheit degrees:

Use the formula: °F = °R − 460, where °F = degrees Fahrenheit at °R = degrees Rankine

EXAMPLE What temperature on the Fahrenheit scale corresponds to 650°R on the Rankine scale?

$$°F = °R - 460 \qquad \text{Substitute 650 for °R.}$$
$$°F = 650 - 460 \qquad \text{Subtract.}$$
$$°F = 190$$

650°R = 190°F.

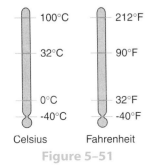

Figure 5–51

3 Convert Fahrenheit to Celsius Temperatures.

The Celsius and Fahrenheit scales are the most common temperature scales used for reporting air and body temperatures. The formulas for converting temperatures using these two scales are more complicated than the previous ones because 1 degree of change on the Celsius scale does *not* equal 1 degree of change on the Fahrenheit scale. (Fig. 5–51).

To convert fahrenheit degrees to celsius degrees:

Use the formula: $°C = \dfrac{5}{9}(°F - 32)$,   where $°C =$ degrees Celsius at $°F =$ degrees Fahrenheit.

EXAMPLE Change 212°F (the boiling point of water) to degrees Celsius.

$°C = \dfrac{5}{9}(°F - 32)$ Substitute 212 for °F in the formula.

$°C = \dfrac{5}{9}(212 - 32)$ Work within grouping; subtract $212 - 32$.

$°C = \dfrac{5}{9}(180)$ Multiply. $\dfrac{5}{\cancel{9}_1} \times \dfrac{\overset{20}{\cancel{180}}}{1} = 100.$

$°C = 100$

212°F = 100°C.

EXAMPLE According to Dr. Shotwell, an antifungal powder containing tolnaftate may be stored at a room temperature of 77°F. Change 77°F to degrees Celsius.

Estimation The Celsius temperature is a smaller value than the Fahrenheit temperature for values near 77°F. From Fig. 5–51 we see that 90°F is approximately 32°C. Then 77°F will be less than 32°C.

$°C = \dfrac{5}{9}(°F - 32)$ Substitute 77 for °F.

$°C = \dfrac{5}{9}(77 - 32)$ Work grouping. $77 - 32 = 45.$

$°C = \dfrac{5}{9}(45)$ Multiply. $\dfrac{5}{\cancel{9}_1} \times \dfrac{\overset{5}{\cancel{45}}}{1}$

$°C = 25$

Interpretation **77°F = 25°C.**   Based on the estimation the solution is reasonable.

---

4 Convert Celsius to Fahrenheit Temperatures.

To convert celsius degrees to fahrenheit degrees:

Use the formula: $°F = \dfrac{9}{5}°C + 32$, where $°F =$ degrees Fahrenheit and $°C =$ degrees Celsius.

SELF-STUDY EXERCISES 5–8

1 Make the temperature conversions.

1. 82°C = ____ K   **2.** 438 K = ____ °C   **3.** 17°C = ____ K   **4.** 273 K = ____ °C
5. 98 K = ____ °C   **6.** 71°C = ____ K   **7.** 60°C = ____ K   **8.** 192 K = ____ °C

2 Make the temperature conversions.

9. 460°F = ____ °R   **10.** 98°F = ____ °R   **11.** 710°R = ____ °F   **12.** 920°R = ____ °F   **13.** 180°F = ____ °R
14. 212°F = ____ °R **15.** 350°R = ____ °F **16.** 600°R = ____ °F   **17.** 0°R = ____ °F   **18.** 32°F = ____ °R

3 Change the Fahrenheit temperatures to Celsius.

19. 95°F   **20.** 32°F   **21.** 113°F   **22.** 41°F   **23.** 59°F
24. 50°F   **25.** 149°F   **26.** 122°F   **27.** 176°F   **28.** 248°F

4 Change the Celsius temperatures to Fahrenheit.

29. 70°C   **30.** 15°C   **31.** 45°C   **32.** 50°C   **33.** 20°C
34. 215°C   **35.** 310°C   **36.** 410°C   **37.** 185°C   **38.** 0°C

Section 5–1

Identify an appropriate U.S. customary measure for each item.
 1. Package of spaghetti
 2. Tank of gasoline
 3. Container of motor oil
 4. Distance from work to the hospital
 5. Package of taco shells
 6. Porterhouse steak
 7. Shipment of iron
 8. Bag of ammonium nitrate
 9. Size of an aluminum pot
 10. Height of a tree
 11. Sugar for a pie recipe
 12. Man's shirt size
 13. Fabric for a pair of kitchen curtains
 14. Hourly speed of an aircraft

Write two unity ratios that relate the given pair of measures.
 15. Cups and quarts
 16. Hours and days
 17. Pounds and tons
 18. Yards and miles

Using unity ratios or conversion factors, convert the given measures to the new units.

 19. 12 ft = ____ yd
 20. 11 yd = ____ ft
 21. $1\frac{1}{5}$ mi = ____ ft
 22. How many feet of wire are needed to put a fence along a property line $2\frac{1}{4}$ mi long?
 23. How many ounces are in 5 lb?
 24. An object weighing $57\frac{3}{5}$ lb weighs how many ounces?
 25. Find the number of pounds in 680 oz.
 26. A can of fruit weighs 22.4 oz. How many pounds is this?
 27. The net weight of a can of peas is 19 oz. If a case contains 16 cans, what is the net weight of a case in ounces? In pounds?
 28. How many quarts are in 8 pt?
 29. How many pints are in $7\frac{1}{2}$ qt?
 30. Find the number of gallons in 15 qt.
 31. Find the number of pints in 3 gal.
 32. How many gallons are in 36 pt?
 33. How many feet of wire are needed to fence a property line $1\frac{1}{4}$ mi long?
 34. A cook has 1 qt of vegetable oil. His recipe requires 2 c of oil. How many recipes can be made from the quart of oil?

Express the measures in standard notation.
 35. 6 ft 17 in.
 36. 1 mi 5,375 ft
 37. 12 lb $17\frac{1}{2}$ oz
 38. 2 gal 7 qt
 39. 1 gal 2 qt 5 pt
 40. 2 T 3,100 lb
 41. 3 yd 2 ft 16 in.
 42. 1 qt 3 c 12 oz
 43. $3\frac{1}{4}$ ft 10 in.
 44. 2 lb 21 oz
 45. 1 gal 3 qt 48 oz

Section 5–2

Add or subtract. Write answers in standard notation.
 46. 12 oz + 2 lb
 47. 8 ft − 49 in.
 48.   4 gal + 3 qt
 49.   2 ft 9 in.
      + 8 ft 2 in.
 50.   7 lb 8 oz
      + 5 lb 9 oz
 51.   5 gal   3 qt
      + 2 gal   3 qt
 52.   7 ft 9 in.
      − 4 ft 6 in.
 53.   4 lb 9 oz
      − 3 lb 11 oz
 54.   4 yd 1 ft 8 in.
      − 2 yd 2 ft 11 in.
 55. A rug 12 ft 6 in. long must fit in a room whose length is 10 ft 9 in. How much should be trimmed from the rug to make it fit the room?
 56. Two packages to be sent air express weigh 5 lb 4 oz each. What is the shipping weight of the two packages?
 57. A water hose purchased for an RV was 2 ft long. What was its length after 7 in. were cut off?
 58. A vinyl flooring installer cut 19 in. from a piece of vinyl 13 ft long. How long was the vinyl piece after it was cut?

Section 5–3

Multiply and write answers for mixed measures in standard notation.

 59. 42 ft
     12 ft
 60. 8 lb 3 oz
     9
 61. 9 in.
     7 in.
 62. 10 gal 3 qt
     7

Divide.
63. 20 yd 2 ft 6 in. ÷ 2
64. 5 gal 3 qt 2 pt ÷ 6
65. 65 ft ÷ 12 (Write answer in feet.)
66. 21 ft ÷ 4 (Write answer in feet and inches.)
67. If 18 lb of candy are divided equally into four boxes, express the weight of the contents of each box in pounds and ounces.
68. If 32 equal lengths of pipe are needed for a job and each length is to be 2 ft 8 in., how many feet of pipe are needed for the job?

69. 14 ft ÷ 4 ft
70. 2 mi 120 ft ÷ 15 ft
71. 400 lb ÷ 90 lb
72. 6 yd 2 ft ÷ 5 ft

73. $5\dfrac{\text{mi}}{\text{min}} = \underline{\quad}\dfrac{\text{mi}}{\text{hr}}$
74. $2{,}520\dfrac{\text{gal}}{\text{hr}} = \underline{\quad}\dfrac{\text{qt}}{\text{hr}}$
75. $88\dfrac{\text{ft}}{\text{sec}} = \underline{\quad}\dfrac{\text{mi}}{\text{hr}}$
76. $18\dfrac{\text{mi}}{\text{gal}} = \underline{\quad}\dfrac{\text{ft}}{\text{gal}}$

77. A pump that moves water at the rate of $75\frac{\text{gal}}{\text{hr}}$ can move how many gallons per minute?
78. A plane that travels at the rate of 240 mph is traveling how many feet per second?
79. How many quarts of milk are needed for a recipe that calls for 3 pt of milk?
80. How many $\frac{1}{2}$-oz servings of jelly can be made from a $1\frac{1}{2}$-lb container of jelly?

Section 5–4

Give the prefix that relates each number to the standard unit in the metric system.
81. 1,000 times
82. $\dfrac{1}{10}$ of
83. $\dfrac{1}{1{,}000}$ of
84. 10 times
85. $\dfrac{1}{100}$ of
86. 100 times

Give the value of the prefixes based on a standard measuring unit.
87. dekameter (dkm)
88. hectogram (hg)
89. milligram (mg)
90. centigram (cg)
91. kiloliter (kL)
92. deciliter (dL)

Choose the most reasonable answer.
93. Height of the Washington Monument
    (a) 200 m   (b) 200 cm
    (c) 200 mm   (d) 200 km
94. Height of Mt. Rushmore
    (a) 1.6 km   (b) 1.6 m
    (c) 1.6 cm   (d) 1.6 mm
95. Weight of an egg
    (a) 50 g   (b) 50 kg
    (c) 50 mg
96. Weight of a saccharin tablet
    (a) 50 kg   (b) 50 mg
    (c) 50 g
97. Weight of a man's shoe
    (a) 0.25 g   (b) 0.25 mg
    (c) 0.25 kg
98. Carton of milk
    (a) 4 L   (b) 4 mL
99. Bottle of medicine
    (a) 50 L   (b) 50 mL

Change to the unit indicated.
100. 0.4 dkm = ____ hm
101. 67.1 m = ____ dkm
102. 4 m = ____ dm
103. 2.3 m = ____ mm
104. 5 cm = ____ mm
105. 0.123 hm = ____ mm
106. How many millimeters are in 0.432 km?
107. 23 dkm = ____ mm
108. 42.7 cm = ____ dkm
109. 41,327 dkm = ____ km
110. A board is 1.82 m long. How many centimeters long is the board?
111. 394.5 g = ____ hg
112. 2.7 hg = ____ dg
113. 3,000,974 cg = ____ kg

Perform the operations indicated.
114. 25 mm − 14 mm
115. 12 g + 5 m
116. 17 mg − 8 mL
117. 8 g − 52 cg
118. 43 dkg × 7
119. 6.83 cg × 9
120. $\dfrac{18\text{ cm}}{9}$
121. 7.5 kg ÷ 0.5 kg
122. $\dfrac{8\text{ hL}}{20\text{ L}}$
123. 34 hL ÷ 4
124. 2.4 m ÷ 5 cm

125. Fabric must be purchased to make seven garments, each requiring 2.7 m of fabric. How much fabric must be purchased?
126. Candy weighing 526 g is mixed with candy weighing 342 g. What is the weight of the mixture?

127. A recipe calls for 5 mL of vanilla flavoring and 24 cL of milk. How much liquid is this?

128. 20 boxes, each weighing 42 kg, are to be moved. How much weight must be moved?

129. A metal rod 42 m long is cut into seven equal pieces. How long is each piece?

130. 32 kg of a chemical are distributed equally among 16 chemistry students. How many kilograms of chemical does each student receive?

131. A serving of punch is 25 cL. How many servings can be obtained from 25 L of punch?

132. How many containers of jelly can be made from 8,548 L of jelly if each container holds 4 dL of jelly?

133. A bolt of fabric contains 6.8 dkm. If a shirt requires 1.7 m, how many shirts can be made from the bolt?

Section 5–5

Make the conversions.

134. 7 m = ____ inches

135. 215 m = ____ yards

136. 69 km = ____ miles

137. 15 L = ____ liquid quarts

138. 12 qt = ____ liters

139. 32 kg = ____ pounds

140. 10 lb = ____ kilograms

141. 9 in. = ____ centimeters

142. 21 ft = ____ meters

143. 14.8 dkL = ____ quarts

144. $3\frac{1}{2}$ gal = ____ liters

145. How many meters long is 200 ft of pipe?

146. Concrete weighing 90 lb weighs how many kilograms?

147. Two cities 175 mi apart are how many kilometers apart?

148. A room 10 m wide is how many feet wide?

149. Wanda Williams is 5 ft 9 in. and weighs 142 lb. What is her body mass index?

150. Ravi Mehra weighs 168 lb and is 5 ft 7 in. What is his body mass index?

Section 5–6

Use unity ratios or conversion factors to convert the measures of time.

151. How many days are in 72 hr?

152. How many minutes are in 2.4 hr?

153. Convert 158 min to hr.

154. A doctor ordered a surgery patient not to drive for 3 weeks. How many days is this?

155. A heart surgery patient was held in intensive care for 96 hr. How many days is this?

156. A bone marrow patient remained hospitalized for 72 days. How many weeks is this?

157. If you have worked for your employer for 39 months, how many years have you worked?

158. There are 96 days remaining in a year. How many weeks remain?

Section 5–7

Determine the number of significant digits of each number.

159. 304,243

160. 2,401,000

161. 4.010

162. 0.023

Find the greatest possible error of each measurement.

163. $2\frac{1}{2}$ in.

164. $7\frac{3}{8}$ in.

165. $3\frac{5}{16}$ ft

166. $7\frac{9}{32}$ in.

167. 5.8 cm

168. 12.2 cm

169. 15.3 cm

170. 7.5 oz

Measure line segments 171–180 in Fig. 5–52 (tolerance $= \pm \frac{1}{32}$ in.).

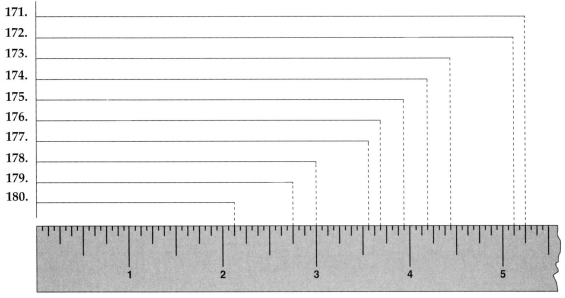

Figure 5–52

Measure line segments 181–190 in Fig. 5–53 to the nearest millimeter.

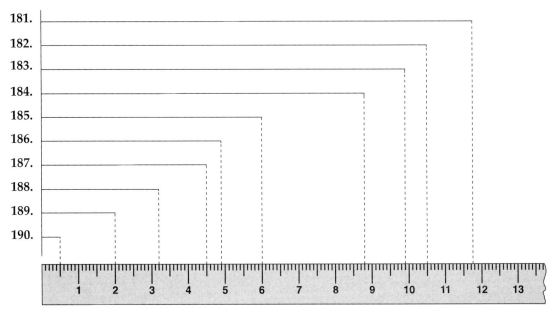

Figure 5–53

Find the distance between each pair of points.
191. $7\frac{5}{8}$ in. and $12\frac{7}{8}$ in.
192. $9\frac{5}{16}$ in. and $11\frac{3}{8}$ in.
193. $3\frac{5}{32}$ in. and 8 in.
194. 9 in. and $16\frac{1}{4}$ in.
195. 17.8 cm and 20.5 cm
196. 31.9 cm and 37 cm

Find the midpoint between each pair of points.
197. $15\frac{3}{8}$ in. and $25\frac{1}{8}$ in.
198. $12\frac{1}{2}$ in. and $15\frac{3}{4}$ in.
199. 2.3 cm and 8.9 cm
200. 1.2 cm and 7 cm
201. 5 cm and 9.8 cm
202. 8 cm and 8.3 cm

Make the temperature conversions.
203. $78°C =$ _____ K
204. $410 K =$ _____ °C
205. $12°F =$ _____ °R
206. $720°R =$ _____ °F
207. $95°C =$ _____ °F
208. $86°F =$ _____ °C
209. The label on a container of the acid-reducer famotidine states that storage above 40°C should be avoided. What is 40°C on the Fahrenheit scale?
210. The freezing point of benzene is 5°C. What is this temperature on the Fahrenheit scale?
211. Candy that should reach a cooking temperature of 365°F should reach what temperature using the Celsius scale?
212. Summer road surface temperatures of 122°F give what reading on the Celsius scale?

CHALLENGE PROBLEMS

213. Compare and contrast the U.S. customary and the metric systems of measurement.
214. Why do you think the metric system has not fully replaced the U.S. customary system in the United States?

CONCEPTS ANALYSIS

1. Give four items that are measured in pounds.
2. Give four items that are measured in feet.
3. If you were building a house, what units of linear measure would you be likely to use?
4. Explain how metric units for length, weight, and capacity are similar.
5. Which measure is longer, a yard or a meter?
6. Which metric measure would you use to dispense a liquid medicine?
7. If your medicine bottle reads 5 mg, is the medicine more likely to be in liquid or capsule form?

Identify the mistake, explain why it is wrong, and correct the mistake.

8. ⟨illegible⟩

9. $418 yd =$ _____ ft?

$$418 \text{ yd} \times \frac{3 \text{ ft}}{1 \text{ yd}} = 1{,}234 \text{ yd}$$

10. $24 pt =$ _____ gal?

$$24 \text{ pt} \times \frac{1 \text{ gal}}{4 \text{ pt}} = 6 \text{ gal}$$

11. A stack of plywood is 4 ft high. If each sheet of plywood is $\frac{3}{4}$ in. thick, how many pieces of plywood are in the stack?
4 ft = 48 in. The stack of plywood is 48 in. high. $48 \times \frac{3}{4} = 36$. There are 36 sheets of plywood in the stack.

CHAPTER SUMMARY

Learning Outcomes	What to Remember with Examples

Section 5–1

1 Identify uses of U.S. customary system measures of length, weight, and capacity (pp. 215–217).

Associate small, medium, and large objects with appropriate measuring units of comparable size.

Nose drops, oz; sofa, lb; coal, T; travel, mi; belt, in.; window, ft or in.; fabric, yd; oil, qt or gal; fuel, gal

2 Convert one U.S. customary unit of measure to another using unity ratios (pp. 217–220).

To write two equivalent measures as a unity ratio, write one measure in the numerator and its equivalent measure in the denominator. The ratio has a value of 1.

Write two unity ratios for the equivalent measures: 1 yard = 36 inches

$$\frac{1 \text{ yard}}{36 \text{ inches}} \quad \text{or} \quad \frac{36 \text{ inches}}{1 \text{ yard}}$$

To convert from one U.S. customary unit of measure to another using unity ratios:

1. Write the original measure in the numerator of a fraction with 1 in the denominator.
2. Multiply by a unity ratio with the original unit of measure in the denominator and the new unit in the numerator.

To estimate:

Multiplication can be used to change to smaller units. There will be more smaller units. Division can be used to change to larger units. There will be fewer larger units.

Change 5 ft to inches.

$$\frac{5 \text{ ft}}{1} \times \frac{12 \text{ in.}}{1 \text{ ft}} = 60 \text{ in.}$$

Change 285 ft to yards.

$$\frac{285 \text{ ft}}{1} \times \frac{1 \text{ yd}}{3 \text{ ft}} = 95 \text{ yd}$$

Change $3\frac{1}{2}$ qt to cups.

$$3\frac{1}{2} \text{ qt} = \frac{7}{2} \text{ qt}$$

$$\frac{7 \text{ qt}}{2} \times \frac{2 \text{ pt}}{1 \text{ qt}} \times \frac{2 \text{ c}}{1 \text{ pt}} = 14 \text{ c}$$

More than one unity ratio may be needed.

3 Convert from one U.S. customary unit of measure to another using conversion factors (pp. 220–221).

When we use conversion factors to convert units, we always multiply. Two equivalent measures have two conversion factors.

To develop a conversion factor:

1. Write a unity ratio that changes the given unit to the new unit. 2. Change the fraction (or ratio) to its decimal equivalent by dividing the numerator by the denominator.

Develop two conversion factors relating pints and quarts. 1 quart = 2 pints

Pints to Quarts

$$\frac{\text{pt}}{1}\left(\frac{\text{qt}}{\text{pt}}\right)$$ Write appropriate labels.

$$\frac{\text{pt}}{1}\left(\frac{1 \text{ qt}}{2 \text{ pt}}\right)$$ Insert numbers to form a unity ratio.

$$\frac{1}{2} = 0.5$$ Convert fraction to whole number, decimal, or mixed decimal equivalent.

$$\text{pt} \times 0.5 = \text{qt}$$

Quarts to Pints

$$\frac{\text{qt}}{1}\left(\frac{\text{pt}}{\text{qt}}\right)$$

$$\frac{\text{qt}}{1}\left(\frac{2 \text{ pt}}{1 \text{ qt}}\right)$$

$$\frac{2}{1} = 2$$

$$\text{qt} \times 2 = \text{pt}$$

To convert from one U.S. customary unit of measure to another using conversion factors:

1. Select the appropriate conversion factor.
2. Multiply the original measure by the conversion factor.

Change 9 pt to quarts.

$$9 \text{ pt} \times 0.5 = 4.5 \text{ qt}$$ Conversion factor is 0.5.

4 Write mixed U.S. customary measures in standard notation (p. 222).

Each measure is converted to the next larger unit of measure when possible.

$$2 \text{ ft } 16 \text{ in.} = 2 \text{ ft} + 1 \text{ ft } 4 \text{ in.} = 3 \text{ ft } 4 \text{ in.}$$ 16 in. = 1 ft 4 in.

Section 5–2

1 Add U.S. customary measures (pp. 224–225).

To add or subtract U.S. customary measures:

1. Convert inch measure to a measure with a common U.S. customary unit. 2. Add or subtract.

Chapter Summary

Add 5 lb and 7 oz.   Convert 5 lb to oz.

$$\frac{5 \text{ lb}}{1} \times \frac{16 \text{ oz}}{1 \text{ lb}} = 80 \text{ oz}$$ Multiply by unity ratio $\frac{16 \text{ oz}}{1 \text{ lb}}$.

80 oz + 7 oz = 87 oz   Combine like measures.

2 Subtract U.S. customary measures (pp. 225–226).

To add or subtract mixed U.S. customary measures:

1. Align like measures in columns. **2.** Add or subtract, carrying or borrowing as necessary.

Subtract 2 ft 8 in. from 7 ft.

$$
\begin{array}{r}
6 \text{ ft } 12 \text{ in.} \\
- \ 2 \text{ ft } \ \ 8 \text{ in.} \\
\hline
4 \text{ ft } \ \ 4 \text{ in.}
\end{array}
$$
   Rewrite 7 ft as 6 ft 12 in.
   Subtract.

Section 5–3

1 Multiply a U.S. customary measure by a number (p. 227).

To multiply a measure by a number:

1. Multiply the numbers associated with each unit of measure by the given number. **2.** Express the answer in standard notation.

Multiply 2 gal 3 qt by 5.

$$
\begin{array}{r}
2 \text{ gal } \ \ 3 \text{ qt} \\
\times \ \ \ \ \ \ \ \ \ 5 \\
\hline
10 \text{ gal } 15 \text{ qt} \\
13 \text{ gal } \ \ 3 \text{ qt}
\end{array}
$$
   15 qt = 3 gal 3 qt

2 Multiply a U.S. customary measure by a measure (pp. 227–228).

To multiply a length measure by a like length measure:

1. Multiply the numbers associated with each like unit. **2.** The product will be a square unit of measure.

Multiply 5 yd by 12 yd.

$$5 \text{ yd} \times 12 \text{ yd} = 60 \text{ yd}^2$$

3 Divide a U.S. customary measure by a number (pp. 228–229).

To divide a measure by a number:

1. Divide the largest measure by the number. **2.** Convert any remainder to the next smaller unit. **3.** Repeat Steps 1 and 2 until no other units are left. **4.** Express any remainder as a fraction of the smallest unit. **5.** The quotient will be a measure.

Divide 7 lb 5 oz by 5.

$$
\begin{array}{r}
1 \text{ lb} \quad \ \ 7\frac{2}{5} \text{ oz} \\
5\overline{)7 \text{ lb} \quad \ \ 5 \text{ oz}} \\
\underline{5 \text{ lb}} \quad \quad \quad \\
2 \text{ lb} = \underline{32} \text{ oz} \\
37 \text{ oz} \\
\underline{35} \text{ oz} \\
2
\end{array}
$$

Chapter 5 / Direct Measurement

4 Divide a U.S. customary measure by a measure (p. 229).	To divide a measure by a measure:

4 Divide a U.S. customary measure by a measure (p. 229).

To divide a measure by a measure:

1. Convert the measures to the same unit. **2.** Divide. The quotient is a number that indicates how many.

How many 5-oz glasses of juice can be poured from 1 gallon of juice?

$$\frac{1 \text{ gal}}{1} \times \frac{4 \text{ qt}}{1 \text{ gal}} \times \frac{2 \text{ pt}}{1 \text{ qt}} \times \frac{2 \text{ ¢}}{1 \text{ pt}} \times \frac{8 \text{ oz}}{1 \text{ ¢}} = 128 \text{ oz}$$

$$\frac{128 \text{ oz}}{5 \text{ oz}} = 25\frac{3}{5} \text{ glasses}$$

5 Change from one U.S. customary rate measure to another (pp. 230–231).

1. Compare the units of both numerators and both denominators to determine which units will change. **2.** Multiply by the unity ratio (or ratios) containing the new unit so that the unit to be changed will reduce.

Change $10\frac{\text{mi}}{\text{hr}}$ to $\frac{\text{ft}}{\text{hr}}$.

$$\frac{10 \text{ mi}}{1 \text{ hr}} \text{ to } \frac{\text{ft}}{\text{hr}} \qquad \text{Miles change to feet. Hours stay the same.}$$

$$\frac{10 \text{ mi}}{1 \text{ hr}} \left(\frac{5,280 \text{ ft}}{1 \text{ mi}} \right) = 52,800 \frac{\text{ft}}{\text{hr}}$$

Section 5–4

1 Identify uses of metric measures of length, weight, and capacity (pp. 232–236).

Associate small, medium, and large objects with appropriate measuring units of comparable size. Powers-of-10 prefixes are used, such as *kilo* for 1,000.

Perfume, mL; soda, L; travel, km; racetrack, m; vitamin, mg; potatoes, kg; eye drops, mL

2 Convert from one metric unit of measure to another (pp. 237–240).

Move the decimal point in the original measure to the left or right as many places as necessary to move from the original unit to the new unit on the metric chart of prefixes.

Change 5.04 cL to liters. Move the decimal two places to the left.

$$5.04 \text{ cL} = 0.0504 \text{ L}$$

3 Make calculations with metric measures (pp. 240–242).

To add or subtract metric measures:

1. Change measures to measures with like units if necessary. **2.** Add or subtract the numerical values. Give the answer in the common unit of measure.

To multiply or divide a metric measure by a number, multiply or divide the numbers and keep the same unit.

To divide a metric measure by a measure:

1. Change measures to measures with like units if necessary. **2.** Divide the numbers, canceling the units of measure. The answer will be a number.

7 km + 34 m =
7,000 m + 34 m = 7,034 m

or

7 km + 0.034 km = 7.034 km

7 mg × 3 = 21 mg
5 m ÷ 25 cm

or

$$\frac{500 \text{ cm}}{25 \text{ cm}} = 20$$

Section 5–5

1 Convert between U.S. customary measures and metric measures (pp. 245–246).

1. Set up the original amount as a fraction with the original unit in the numerator and 1 in the denominator. **2.** Multiply by a unity ratio with the original unit in the denominator and the new unit in the numerator.

How many feet are in 14 meters?

$$\frac{14 \text{ m}}{1} \times \frac{3.28 \text{ ft}}{1 \text{ m}} = 45.92 \text{ ft}$$ Use the conversion factor 1 m = 3.28 ft.

Section 5–6

1 Convert from one unit of time to another (p. 247).

Use unity ratios or conversion factors to convert one unit of time to another.

How many days are in 3,420 hours?

$$1 \text{ day} = 24 \text{ hours}$$

$$3,420 \text{ hr} \times \frac{1 \text{ day}}{24 \text{ hr}} = 142.5 \text{ days}$$

2 Make calculations with measures of time (pp. 247–248).

Measures of time can be added or subtracted if they are the same unit of measure by adding or subtracting the quantities and keeping the same unit of measure. A time measure can be multiplied or divided by a number by multiplying or dividing the measure by the number and keeping the same measuring unit.

A printer prints 1 page in 30 seconds. How long does it take to print 45 pages?

1 min = 60 sec; 1 hr = 60 min

30 sec × 45 = 1,350 sec

$$1,350 \text{ sec} \times \frac{1 \text{ min}}{60 \text{ sec}} = 22.5 \text{ min}$$ Minutes to complete job.

$$22.5 \text{ min} \times \frac{1 \text{ hr}}{60 \text{ min}} = 0.375 \text{ hr}$$ Portion of hour to complete job.

The printer can print 45 pages in 22.5 minutes or 0.375 hour.

3 Equate time of day among time zones (pp. 248–252).

To equate time for two locations in different time zones, subtract the GMT for each zone (GMT for unknown time – GMT for known time) to get the difference in time zones. Then adjust the known time by the difference in the time zones.

If it is 7 A.M. in Providence, Rhode Island (-5 GMT), what is the time in San Jose, California (-8 GMT)?

$-8 - (-5) = -8 + 5 = -3$. Thus, the time of day in San Jose is three hours earlier than the time of day in Providence. Therefore, subtract 3 from the time in Providence to get the local time in San Jose.

7 A.M. $- 3$ hours $= 4$ A.M. At 7:00 A.M. in Providence, Rhode Island, the time will be 4:00 A.M. in San Jose, California.

Chapter 5 / Direct Measurement

Section 5–7

1 Determine the significant digits of a number (pp. 253–254).

For whole numbers:

1. Start with the leftmost nonzero digit. **2.** Count each digit through the rightmost nonzero digit.

> 258 has three significant digits.
> 1,050 has three significant digits.

For decimal numbers:

1. Start with the leftmost nonzero digit. **2.** Count each digit through the last digit.

> 0.023 has two significant digits.
> 1.50 has three significant digits.

2 Find the greatest possible error of a measurement (pp. 254–255).

1. Determine the precision of the measurement of the smallest subdivision. **2.** Find half of the precision.

> The precision of $2\frac{5}{8}$ is $\frac{1}{8}$.
>
> $$\frac{1}{2} \times \frac{1}{8} = \frac{1}{16}.$$
>
> The greatest possible error is $\frac{1}{16}$.

3 Read the U.S. customary rule (pp. 255–256).

Align the rule along the object (Fig. 5–54). Count the number of whole and fractional inches ($\frac{1}{16}$'s, $\frac{1}{8}$'s, $\frac{1}{4}$'s, and so on) to determine the approximate length. Use eye judgment to estimate closeness of the object to a mark on the rule.

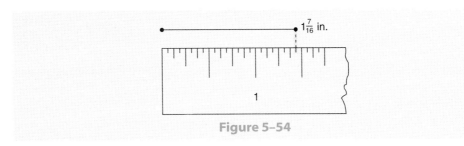

Figure 5–54

4 Read the metric rule (pp. 256–257).

See Fig. 5–55. Align the rule along the object. Count the number of millimeters and/or centimeters (10 mm = 1 cm) to determine the approximate length. Use eye judgment to estimate closeness of the object to a mark on the rule.

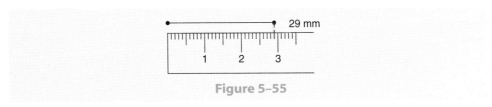

Figure 5–55

5 Find the distance and the midpoint of a line segment (pp. 257–258).

Distance: Subtract the coordinates of each point.
Distance = $P_2 - P_1$

> Find the distance between the points 3.5 cm and 6.3 cm on a metric rule.
>
> Distance = 6.3 − 3.5   Let P_1 = 3.5 cm and P_2 = 6.3 cm.
>   = 2.8 cm

Midpoint:

Average the coordinates of the end points (P_1 and P_2).

$$\text{Midpoint} = \frac{P_1 + P_2}{2}$$

Find the midpoint between the points 3.5 cm and 6.3 cm on a metric rule.

$$\text{Midpoint} = \frac{3.5 + 6.3}{2} \qquad \text{Let } P_1 = 3.5 \text{ cm and } P_2 = 6.3 \text{ cm.}$$
$$= \frac{9.8}{2}$$
$$= 4.9 \text{ cm}$$

Section 5–8

1 Make Celsius/Kelvin temperature conversions (pp. 260–261).

To change degrees Celsius to kelvins use the formula:
$K = °C + 273$

Change 20°C to K.

$K = °C + 273$ Substitute 20 for °C.
$K = 20 + 273$ Add.
$K = 293$

To change kelvins to degrees Celsius use the formula: $°C = K - 273$

Change 281 K to °C.

$°C = K - 273$ Substitute 281 for K.
$°C = 281 - 273$ Subtract.
$°C = 8$

2 Make Rankine/ Fahrenheit temperature conversions (p. 261).

To change degrees Fahrenheit to degrees Rankine, use the formula: $°R = °F + 460$

Change 50°F to °R.

$°R = °F + 460$ Substitute 50 for °F.
$°R = 50 + 460$ Add.
$°R = 510$

To change degrees Rankine to degrees Fahrenheit use the formula: $°F = °R - 460$

Change 750°R to °F.

$°F = °R - 460$ Substitute 750 for °R.
$°F = 750 - 460$ Subtract.
$°F = 290$

3 Convert Fahrenheit to Celsius temperatures (pp. 261–262).

To change degrees Fahrenheit to degrees Celsius use the formula: $°C = \dfrac{5}{9}(°F - 32)$

Change 14°F to °C.

$$°C = \frac{5}{9}(°F - 32)$$ Substitute 14 for °F.

$$°C = \frac{5}{9}(14 - 32)$$ Perform operation in parentheses.

$$°C = \frac{5}{9}(-18)$$ Multiply.

$$°C = -10$$

4 Convert Celsius to Fahrenheit temperatures (pp. 262–263).

To change degrees Celsius to degrees Fahrenheit use the formula: $°F = \frac{9}{5}°C + 32$

Change 28°C to °F.

$$°F = \frac{9}{5}°C + 32$$ Substitute 28 for °C.

$$°F = \frac{9}{5}(28) + 32$$ Multiply.

$$°F = \frac{252}{5} + 32$$ Divide.

$$°F = 50.4 + 32$$ Add.

$$°F = 82.4$$

CHAPTER TRIAL TEST

Change to the measure indicated (see the inside covers of the text for conversion factors).

1. 3 ft = _____ in.
2. 36 oz = _____ lb
3. 32 qt = _____ gal
4. Add 2 yd 6 ft 10 in. + 3 ft 7 in. Write your answer in standard notation.

5. $60 \dfrac{\text{gal}}{\text{min}} = $ _____ $\dfrac{\text{qt}}{\text{sec}}$

6. 21 in. ÷ 3 in.

7. A 495-ft section of highway is to be resurfaced. How many yards is this?

8. How many quarts are contained in a 55-gal drum?

9. If an automobile travels at 55 mph, how many feet does it travel per second?

10. A spark plug wire kit contains $18\frac{1}{2}$ ft of wire in a coil. The directions call for cutting off individual lengths of 28 in. each. How many 28-in. wires can be cut from the coil?

11. Find the measure of the line segment *AB* in Fig. 5–56 (tolerance = $\pm\frac{1}{32}$ in.).

12. Select the appropriate U.S. customary measure for the contents of a swimming pool.

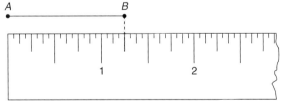

Figure 5–56

13. Write 1 gal 3 qt 6 c 20 oz in standard notation.
14. Which is the appropriate metric measure for eye drops: 28 mL, 28 dL, 28 L, or 28 kL?
15. If Lisbon, Portugal, is 0 GMT and Bern, Switzerland, is +1 GMT, how many hours time difference is there between the two cities?
16. Grayson Lee traveled to Wellington, New Zealand, to do research on geological formations. He arrived at 3:00 P.M. local time (NZST, GMT +12) and needed to contact his office located in Los Alamos, New Mexico (MST, GMT −7). What time was it in Los Alamos?

To find the area of a polygon:

1. Visualize the polygon.
2. Select the appropriate formula.
3. Substitute the known values into the formula.
4. Evaluate the formula.

EXAMPLE A carpet installer is carpeting a room measuring 16 ft by 20 ft. Projecting out from one wall is a fireplace whose hearth measures 3 ft by 6 ft (Fig. 6–11). How many square yards of carpet does the installer require for the job? How much is wasted?

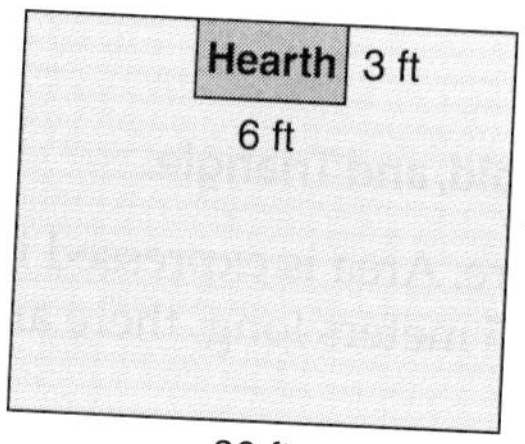

First, compute the area of the room without considering the area of the hearth. This is the amount of carpet needed. Second, compute the area of the fireplace hearth. This is the amount wasted. Finally, analyze the dimensions to see if the result is expressed in the desired unit. If not, convert to the desired unit.

Let A_{room} = the area of the room and A_{hearth} = the area of the hearth.

$A_{room} = lw$ Select the appropriate formula and substitute.

$A_{room} = 20 \times 16$ Multiply.

$A_{room} = 320 \text{ ft}^2$ Area is a square measure.

$A_{hearth} = lw$ Select the appropriate formula and substitute.

$A_{hearth} = 3 \times 6$ Multiply.

$A_{hearth} = 18 \text{ ft}^2$ Area is a square measure.

Because carpet is sold in square yards, we convert the square footage to square yards. Since $9 \text{ ft}^2 = 1 \text{ yd}^2$, we can use a *unity ratio*.

Recall from Chapter 5 that a unity ratio is a fraction with one unit of measure in the numerator and a different, but equivalent, unit of measure in the denominator. A unity ratio contains the original unit and the new unit. Use the unity ratio $\dfrac{1 \text{ yd}^2}{9 \text{ ft}^2}$.

$$\frac{320 \text{ ft}^2}{1} \times \frac{1 \text{ yd}^2}{9 \text{ ft}^2} = \frac{320 \text{ yd}^2}{9} = 35\frac{5}{9} \text{ yd}^2 \text{ or } 36 \text{ yd}^2 \text{ of carpet needed,}$$

$$\frac{\overset{2}{18} \text{ ft}^2}{1} \times \frac{1 \text{ yd}^2}{\underset{1}{9} \text{ ft}^2} = 2 \text{ yd}^2 \text{ carpet wasted on the hearth}$$

This job requires $35\frac{5}{9} \text{ yd}^2$ or 36 yd^2 of carpet. Of this, 2 yd^2 is waste from the hearth.

The preceding example brings up some interesting questions. Can a fraction of a yard of carpet be purchased? If carpet can only be purchased in 12-ft or 15-ft widths, will there be additional waste? Is it more practical to purchase extra carpet that will be wasted or to spend extra labor costs to install the carpet with several seams? You may want to investigate common practices in the industry.

EXAMPLE If a $\frac{1}{2}$-in. mortar joint is used in a wall of 2-in. × 4-in. × 8-in. bricks, six bricks cover 1 square foot. If a $\frac{1}{4}$-in. mortar joint is used, seven bricks cover 1 square foot. How many bricks are needed to build a 50-ft × 14-ft wall with $\frac{1}{2}$-in. joints? With $\frac{1}{4}$-in. joints?

Hearth 3 ft
6 ft
16 ft
20 ft
Figure 6–11

$$A_{\text{wall}} = 50 \times 14$$
$$A_{\text{wall}} = 700 \text{ ft}^2$$

$$\frac{700 \text{ ft}^2}{1} \times \frac{6 \text{ bricks}}{1 \text{ ft}^2} = 4{,}200 \text{ bricks}$$ Find the number of bricks at 6 per ft², using a unity ratio.

$$\frac{700 \text{ ft}^2}{1} \times \frac{7 \text{ bricks}}{1 \text{ ft}^2} = 4{,}900 \text{ bricks}$$ Find the number of bricks at 7 per ft², using a unity ratio.

Using $\frac{1}{2}$-in. mortar joints, 4,200 bricks are needed. Using $\frac{1}{4}$-in. mortar joints, 4,900 bricks are needed.

EXAMPLE Asphalt roofing shingles are sold in bundles. The number of bundles needed to cover a square (100 ft²) depends on the overlap when the shingles are installed. If the overlap allows 4 in. of each shingle to be exposed, then four bundles are needed per square. With a 5-in. exposure, 3.2 bundles are needed per square. Figure the number of bundles for a 4-in. exposure and for a 5-in. exposure for a roof measuring 30 ft × 20 ft.

$$A_{\text{roof}} = 30 \times 20$$
$$A_{\text{roof}} = 600 \text{ ft}^2$$

$$\frac{\overset{6}{600 \text{ ft}^2}}{1} \times \frac{1 \text{ square}}{\underset{1}{100 \text{ ft}^2}} = 6 \text{ squares}$$ Find the number of squares (100 ft²) using a unity ratio.

$$6 \times 4 = 24 \text{ bundles}$$ Find the number of bundles for a 4-in. exposure.
$$6 \times 3.2 = 19.2 \ or \ 20 \text{ bundles}$$ Find the number of bundles for a 5-in. exposure.

Therefore, 24 bundles of shingles are required for a 4-in. exposure, and 20 bundles (from 19.2 bundles) are required for a 5-in. exposure.

EXAMPLE Lynn Fly, a vinyl floor installer, discovers that one of the squares in the flooring pattern is damaged. He decides to cut out the damaged square and replace it. The damaged square measures 20 cm on each side. What is the area of the square to be replaced?

$$A_{\text{square}} = s^2$$ Select the appropriate formula.
$$A_{\text{square}} = (20 \text{ cm})^2$$ Substitute measurement of one side and square it.
$$A_{\text{square}} = 400 \text{ cm}^2$$

The vinyl square measures 400 cm².

TIP!

Dimension Analysis

When we evaluate formulas, the measuring units are sometimes omitted from the written steps. However, it is very important to use the correct unit in your calculations so that the unit in the solution is correct. Compare examples involving perimeter and area.

Example (molding for sink, p. 280)

$P_{square} = 4s$

$P_{square} = 4(40)$ The measuring unit is centimeters; the measure is multiplied by a number.

$P_{square} = \textbf{160 cm}$ The measuring unit in the solution is centimeters.

Perimeter is always a linear measure, which means the measuring unit should be to the first power.

Example (vinyl flooring in the preceding example)

$A_{square} = s^2$

$A_{square} = (20)^2$ The measuring unit is centimeters and a measure is squared or a measure is multiplied by a measure.

$A_{square} = \textbf{400 cm}^2$ The measuring unit in the solution is square centimeters.

Area is always a square measure, which means that measures of area are to the second power or have an exponent of 2.

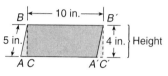

Figure 6–12

Since a rectangle is a parallelogram, the area of both types of polygons is related. The *height* of a parallelogram is the perpendicular distance between two parallel sides. Height is also called *altitude*.

If triangle ABC ($\triangle ABC$) in Fig. 6–12 were transposed to the right side of the parallelogram ($\triangle A'B'C'$), we would have a rectangle.

The base of the original parallelogram is the same as the length of the newly formed rectangle. Similarly, the height of the parallelogram is the same as the width of the rectangle. Thus, the area of the rectangle and parallelogram is 10×4 or 40 in².

EXAMPLE   Find the area of a parallelogram with a base of 16 in., an adjacent side of 8 in., and a height of 7 in. (Fig. 6–13).

Visualize the parallelogram.

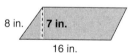

Figure 6–13

$A_{parallelogram} = bh$ Substitute.

$A_{parallelogram} = 16 \times 7$ Multiply.

$A_{parallelogram} = \textbf{112 in.}^2$

The areas of a trapezoid and a triangle are related to the area of a parallelogram.

TIP!

Where Does $\frac{1}{2}$ Come from in the Trapezoid Formula?

Placing two congruent trapezoids end to end forms a parallelogram. The area of one trapezoid is half the area of the larger parallelogram. Since the area of a parallelogram is the height times the base, the area of a trapezoid is *one-half* the height times the "base" of the parallelogram formed by joining two equal trapezoids end to end ($b_1 + b_2$).

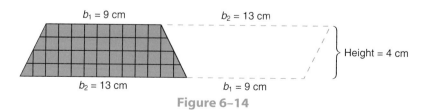

Figure 6–14

EXAMPLE Find the area of the shaded trapezoid in Figure 6–14.

$$A = \frac{1}{2}h\,(b_1 + b_2)$$ Select the appropriate formula and substitute the known values.

$$A = \frac{1}{2}(4)(9 + 13)$$ Add inside grouping.

$$A = \frac{1}{2}(4)\,(22)$$ Multiply.

$$A = 44 \text{ cm}^2$$ cm × cm = cm².

EXAMPLE A trapezoidal table top for special use in a reading classroom measures 60 cm on the shorter base and 90 cm on the longer base. If the height (perpendicular distance between the bases) is 50 cm, how much Formica is needed to re-cover the top?

Estimation A rectangle whose length equals the shorter base (60 cm) and height equals 50 cm has an area of (60)(50) or 3,000 cm². A rectangle whose length equals the longer base has an area of (90)(50) or 4,500 cm². The area of the trapezoid is between 3,000 cm² and 4,500 cm².

$$A = \frac{1}{2}h\,(b_1 + b_2)$$ Select the appropriate formula.
Sustitute in formula and evaluate. Simplify grouping.

$$A = \frac{1}{2}\,(50)\,(60 + 90)$$

$$A = \frac{1}{2}\,\overset{25}{(\cancel{50})}\,(150)\\ {\scriptstyle 1}$$ Reduce where possible.

$$A = 3{,}750 \text{ cm}^2$$ The solution is between the estimated areas.

Interpretation **3,750 cm² of Formica are needed.**

Like the trapezoid, the triangle forms a parallelogram if an identical triangle is placed beside the original triangle. A triangle, then, is half a parallelogram, and its area can be expressed as half the area of a parallelogram, that is, one-half the base times the height.

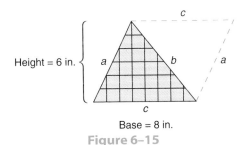

Height = 6 in.

Base = 8 in.

Figure 6–15

EXAMPLE Find the area of the triangle in Fig. 6–15.

$$A = \frac{1}{2}bh$$ Select the appropriate formula and substitute the known values.

$$A = \frac{1}{2}(8)(6)$$ Multiply.

$$A = 24 \text{ in}^2$$ in. × in. = in².

6–1 Perimeter and Area

EXAMPLE   A gable on a house has a rise of 7 ft 6 in. and a span of 30 ft 6 in. What is its area? The rise is the height of the gable, and the span is the base of the gable (Fig. 6–16).

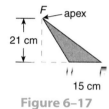

7 ft 6 in.

30 ft 6 in.

Figure 6–16

$A = \dfrac{1}{2}bh$

Select the appropriate formula.
Substitute in formula after converting to one common unit of measure.
7 ft 6 in. = 7.5 ft, 30 ft 6 in. = 30.5 ft because 1 ft = 12 in. and 6 in. = 0.5 ft.

$A = \dfrac{1}{2}(30.5)(7.5)$

$A = 0.5(30.5)(7.5)$

$A = 114.375 \text{ ft}^2$

The area of the gable is 114.375 ft².

In some triangles, the height is measured along an imaginary line outside the triangle from the base to the highest point of the triangle (the *apex*) (Fig. 6–17).

EXAMPLE   The base of $\triangle DEF$ in Fig. 6–17 is 15 cm and the height is 21 cm. Find the area.

F — apex

21 cm

15 cm

Figure 6–17

$A = \dfrac{1}{2}bh$ Substitute into the formula.

$A = \dfrac{1}{2}(15)(21)$ Multiply.

$A = \dfrac{1}{2}(315)$

$A = 157.5 \text{ cm}^2$

The area of $\triangle DEF$ is 157.5 cm².

When the three sides of a triangle are known and the height is not known, the area of a triangle can be found using *Heron's formula*.

To find the area of a triangle when the length of the three sides are known and the height is unknown:

1. Use the formula:

   $\text{Area} = \sqrt{s\,(s - a)\,(s - b)\,(s - c)}$ Heron's formula

   where $s = \frac{1}{2}(a + b + c)$ and a, b, and c are the lengths of the three sides.

2. Find s.
3. Substitute values for a, b, c, and s into the formula.
4. Evaluate the formula.

Chapter 6 / Perimeter, Area, and Volume

EXAMPLE   Find the area of the triangle by using Heron's formula.

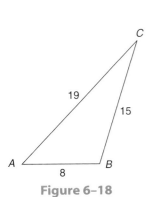

Figure 6–18

$$s = \frac{1}{2}(a + b + c)$$ Find s. Substitute length of sides.

$$s = \frac{1}{2}(19 + 15 + 8)$$ Add, then multiply.

$$s = 21$$

$$\text{Area} = \sqrt{s(s - a)(s - b)(s - c)}$$ Heron's formula.

$$\text{Area} = \sqrt{21(21 - 19)(21 - 15)(21 - 8)}$$ Substitute for s, a, b, and c, and perform each subtraction.

$$\text{Area} = \sqrt{21(2)(6)(13)}$$ Multiply.

$$\text{Area} = \sqrt{3,276}$$ Take the square root using a calculator.

$$\text{Area} = 57.23635209$$ Principal square root.

Area = 57.24 square units   Rounded to four significant digits.

SELF STUDY EXERCISES 6–1

1. Find the perimeter of the parallelogram in Fig. 6–19.

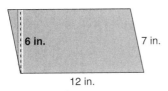

6 in.   7 in.

12 in.

Figure 6–19

2. Find the perimeter of the parallelogram in Fig. 6–20.

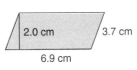

2.0 cm   3.7 cm

6.9 cm

Figure 6–20

Solve the problems involving perimeter.

3. An illuminated sign in the main entrance of a hospital is a parallelogram with a base of 48 in. and an adjacent side of 30 in. How many feet of aluminum molding are needed to frame the sign?

4. A customized van has a window cut in each side in the shape of a parallelogram with a base of 20 in. and an adjacent side of 11 in. How many inches of trim are needed to surround the two windows?

5. A contemporary building has a window in the shape of a parallelogram with a base of 50 in. and an adjacent side of 30 in. How many inches of trim are needed to surround the window?

6. A table for a reading lab has a top in the shape of a parallelogram with a base of 36 in. and an adjacent side of 18 in. How many inches of edge trim are needed to surround the tabletop?

7. Find the perimeter of the rectangle in Fig. 6–21.

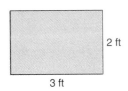

2 ft

3 ft

Figure 6–21

8. Find the perimeter of the rectangle in Fig. 6–22.

9.2 ft

12.7 ft

Figure 6–22

9. A rectangular parking lot is 340 ft by 125 ft. Find the perimeter of the parking lot.

11. How many feet of quarter-round molding are needed to finish around the baseboard after sheet vinyl flooring is installed if the room is 16 ft by 18 ft and there are three 3-ft-wide doorways?

10. A room is 15 ft by 12 ft. How many feet of chair rail are needed for the room? Disregard openings.

12. The swimming pool in Fig. 6–23 measures 32 ft by 18 ft. How much fencing is needed, including material for a gate, if the fence is to be built 7 ft from each side of the pool?

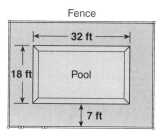

Figure 6–23

13. A Formica tabletop measures 40 in. by 62 in. How many feet of edge trim are needed (12 in. = 1 ft)?

14. A countertop requires rolled edging to be installed on all four sides. How much rolled edging material is needed if the countertop measures 25 in. by 40 in.?

15. Find the perimeter of Fig. 6–24.

16. Find the perimeter of Fig. 6–25.

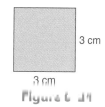

3 cm

3 cm

Figure 6–24

8.9 cm

8.9 cm

Figure 6–25

17. The square parking lot of a doctor's office is to have curbs built on all four sides. If the lot is 150 ft on each side, how many feet of curb are needed? Allow 10 ft for a driveway into the parking lot.

19. A 20-in. × 20-in. central heating and air conditioning return air vent is being installed in a wall. Find the perimeter of the wall opening.

18. A border of 4-in. × 4-in. wall tiles surrounds the floor of a shower stall that is 48 in. × 48 in. How many tiles are needed for this border? Disregard spaces for grout (connecting material between the tiles).

Find the perimeter of Figs. 6–26 and 6–27. Round to hundredths.

20.

21.

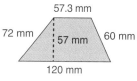

57.3 mm

72 mm   57 mm   60 mm

120 mm

Figure 6–26

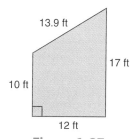

13.9 ft

17 ft

10 ft

12 ft

Figure 6–27

22. The six glass panes in a kitchen light fixture each measure $4\frac{1}{2}$ in. along the top and 10 in. along the bottom. The top and bottom are parallel. The two nonparallel sides of each pane are 10 in. What is the combined perimeter of the six trapezoidal panes?

23. A section of a hip roof is a trapezoid measuring 38 ft at the bottom, 14 ft at the top, 10 ft high, and 11 and 12 ft, respectively, on each side. Find the perimeter of this section of the roof.

24. A lot in an urban area has two nonparallel sides that are each 64 ft. The parallel sides are 120 ft and 154 ft. Find the perimeter of the trapezoidal property.

25. A swimming pool is fashioned in a trapezoidal design. The parallel sides are $18\frac{1}{2}$ ft and 31 ft. The other sides are $24\frac{1}{2}$ ft and 24 ft. What is the perimeter of the pool?

Find the perimeter of the triangles in Figs. 6–28 and 6–29. Round to hundredths.

26.

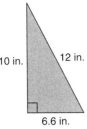

10 in.   12 in.

6.6 in.

Figure 6–28

27.

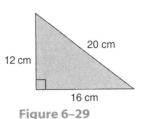

12 cm   20 cm

16 cm

Figure 6–29

28. Find the perimeter of the triangle in Fig. 6–30.

12.8 cm

6.5 cm

9 cm

Figure 6–30

29. Selena Henson is planning a patio that will adjoin the sides of her L-shaped home. One side of the patio is 24 ft and the other is 18 ft. The shape of the patio is triangular. Draw a representation of the patio and find the perimeter if the length of the third side is 30 ft.

30. Find the perimeter of a triangle that has sides measuring 2 ft 6 in., 1 yd 8 in., and 4 ft 6 in.

2

31. Find the area of the shape in Fig. 6–31.

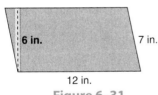

6 in.   7 in.

12 in.

Figure 6–31

32. Find the area of the parallelogram in Fig. 6–32.

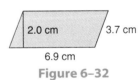

2.0 cm   3.7 cm

6.9 cm

Figure 6–32

33. Find the area of a parking lot that is a parallelogram 200 ft long if the perpendicular distance between the sides is 45 ft (Fig. 6–33).

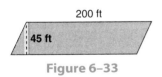

200 ft

45 ft

Figure 6–33

34. A table for a reading lab has a top in the shape of a parallelogram with a base of 36 in. and an adjacent side of 18 in. If the height is 14 in., find the area of the tabletop.

35. Find the area of the shape in Fig. 6–34.

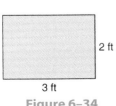

2 ft

3 ft

Figure 6–34

36. Find the area of the shape in Fig. 6–35.

9.2 ft

12.7 ft

Figure 6–35

37. A rectangular parking lot is 340 ft by 125 ft. Find the number of square feet in the parking lot.

38. A room is 15 ft by 12 ft. How many square feet of flooring are needed for the room?

39. A small kitchen 9 ft by 10 ft by 8-ft high is being wallpapered. How many square feet of paper are needed if there are 63 ft^2 of openings in the kitchen?

40. The dimensions of a sun porch are 9 ft 6 in. by 15 ft. How many board feet of 1-in.-thick tongue-and-groove flooring are needed to build the porch? (Board feet for 1-in.-thick lumber is the same number as ft^2.) Disregard waste.

41. Find the area of Fig. 6–36.

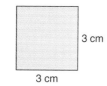

3 cm

3 cm

Figure 6–36

42. Find the area of the square in Fig. 6–37.

8.9 cm

Figure 6–37

43. Madison Duke is wallpapering a laundry room 8 ft by 8 ft by 8-ft high. How many square feet of paper will she need if there are 63 ft^2 of openings in the room?

44. Making no allowances for bases, the pitcher's mound, or the home plate area, how many square yards of artificial turf are needed to resurface an infield at an indoor baseball stadium? The infield is 90 ft on each side (9 ft^2 = 1 yd^2).

45. Ted Davis is a farmer who wants to apply fertilizer to a 40-acre field with dimensions $\frac{1}{4}$ mi $\times$ $\frac{1}{4}$ mi. Find the area in square miles.

46. A 36-in. $\times$ 36-in. ceramic tile shower stall is being installed. How many 4-in. $\times$ 4-in. tiles are needed to cover the floor? Disregard the drain opening and grout spaces.

47. Tiles that are 6 in. $\times$ 6 in. cover the floor of a shower. How many tiles are needed for the floor if the shower measures 4.5 ft by 6 ft?

48. A 20-in. $\times$ 20-in. central heating and air conditioning return air vent is being installed in a wall. Find the area of the wall opening.

49. A design in a marble floor is a larger rectangle with a length of 48 in. and a width of 24 in. If the design is made of 8-in. $\times$ 8-in. marble tiles, how many tiles are needed for each design, assuming no waste?

50. A machine cuts sheet metal into rectangles measuring 5 cm long and 4 cm wide. How many can be cut from a piece of sheet metal 100 cm $\times$ 200 cm?

51. Signs in the shape of rectangles with a length of 24 in. and a width of 10 in. will be cut from sheet metal. How many whole signs can be cut from a piece of sheet metal 48 in. $\times$ 48 in.? Illustrate your answer with a drawing.

Find the area of Figs. 6–38 and 6–39. Round to hundredths.

52.

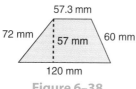

57.3 mm

72 mm   57 mm   60 mm

120 mm

Figure 6–38

53.

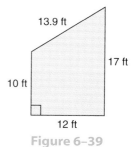

13.9 ft

17 ft

10 ft

12 ft

Figure 6–39

54. The six glass panes in a kitchen light fixture each measure $4\frac{1}{2}$ in. along the top and 10 in. along the bottom. The top and bottom are parallel. The height of each pane is 8 in. What is the combined area of the six trapezoidal panes?

55. A section of a hip roof is a trapezoid measuring 38 ft at the bottom, 14 ft at the top, and 10 ft high. Find the area of this section of the roof in square feet.

56. A trapezoidal lot in an urban area is 60 ft wide. The parallel sides are 120 ft and 154 ft, and they are perpendicular to the width of the lot. Find the area of the trapezoidal property.

Find the area of the triangles in Figs. 6–40 and 6–41.

57.

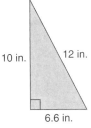

Figure 6–40

58.

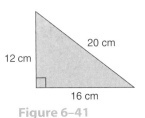

Figure 6–41

59. Find the area of the triangle in Fig. 6–42.

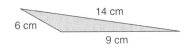

Figure 6–42

60. Carlee McAnally is planning a patio that will adjoin the sides of her L-shaped home. One side of the patio is 24 ft and the other is 18 ft. The shape of the patio is triangular. Draw a representation of the patio. Find the number of square feet of surface area to be covered with concrete.

61. A louver, or triangular vent, for a gable roof measures 6 ft 6 in. wide and stands 3 ft high. Find the area in square feet of the vented portion of the gable.

62. A mason lays tile to form a triangle. The height of the triangle is 12 ft and the base is 5 ft. What is the area?

63. A metal worker cuts a triangular plate with a base of 11 in. and a height of $4\frac{1}{2}$ in. from a piece of metal. What is the area of the plate?

64. If aluminum siding costs $6.75 a square yard installed, how much does it cost to put the siding on the two triangular gable ends of a roof under construction? Each gable has a span (base) of 30 ft 6 in. and a rise (height) of 7 ft 6 in. Any portion of a square yard is rounded to the next highest square yard.

6–2 | *Circles and Composite Figures*

Learning Outcomes

1 Find the circumference of a circle.

2 Find the area of a circle.

3 Find the perimeter and area of composite figures.

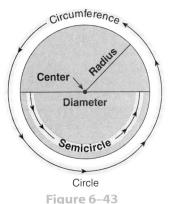

Figure 6–43

A *circle* is a closed curved line with points that lie in a plane and are the same distance from the *center* of the figure.

The *center* of a circle is the point that is the same distance from every point on the circumference of the circle.

The *radius* (plural: *radii*, pronounced "ray · dē · ī") is a straight line segment from the center of a circle to a point on the circle. It is half the diameter.

The *diameter* of a circle is a straight line segment from a point on the circle through the center to another point on the circle.

The *circumference* of a circle is the perimeter or length of the closed curved line that forms the circle.

A *semicircle* is half a circle and is created by drawing a diameter.

1 Find the Circumference of a Circle.

The circle is a geometric form with a special relationship between its circumference and its diameter. If we divide the circumference of any circle by its diameter, the quotient is always the same number.

$$\pi = \frac{\text{Circumference}}{\text{Diameter}} = \frac{C}{d}$$

This number is a nonrepeating, nonterminating decimal approximately equal to 3.1415927 to seven decimal places. The Greek letter π (pronounced "pie") represents this value. Convenient approximations often used in calculations involving π are $3\frac{1}{7}$ and 3.14. Many calculators have a π key.

TIP!

What Is π?

Here is a way to visualize π.

Find a circular object (Fig. 6–44). Use a tape measure to measure the circumference of the object. If a tape measure is not available, wrap a string around the object, then stretch the string along a rule to measure the object.

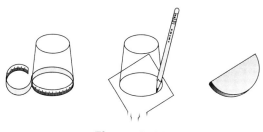

Figure 6–44

Next, trace the object on a sheet of paper and cut out the circle. Fold the circle in half and measure the straight edge (diameter).

Divide the circumference by the diameter $\frac{C}{d}$. Since measurements are approximate, the quotient will probably not be exactly 3.14, but it should be slightly over 3.

Repeat the activity again with different-sized circular objects. The larger the circle, the closer your approximation should be to π since the measured amount will have a smaller percent error from the actual amount.

The formulas for the circumference and area of a circle are:

Circumference (C)	Area (A)		
$C = \pi d$	$A = \pi r^2$	d is diameter ($d = 2r$)
$C = 2\pi r$		r is radius ($r = \frac{1}{2}d$)

To find the circumference or area of a circle:

1. Select the appropriate formula.
2. Substitute values for r or d as appropriate.
3. Evaluate the formula. Use 3.14 or the calculator value for π.

EXAMPLE   Find the circumference of a circle that has a diameter of 1.3 m.

$C = \pi d$ Select circumference formula with diameter d.
$C = 3.14(1.3)$ Use 3.14 for π.
$C = 4.082$ m   Evaluate.

Figure 6–45   **The circumference is 4.1 m (rounded).**   Circumference is a linear measure.

TIP!

Calculator Values of π

Calculations involving π are always approximations. Many calculators include a π function or menu option where π to seven or more decimal places is computed by pressing a single key. Other calculators require the use of more than one key to activate the ▣ key. However, for computations by hand or a calculator without the ▣ function, 3.14 is sometimes adequate. **We use the calculator value 3.141592654 for π in all examples and exercises unless stated otherwise**.

EXAMPLE   What is the circumference of a circle that has a diameter of 21 cm (Fig. 6–46)? Round to tenths.

$C = \pi d = \pi(21) = 65.97344573$ Use the ▣ key on your calculator for π.

The circumference is 66.0 cm (rounded).   Circumference is a linear measure.

Figure 6–46

TIP!

Fractions versus Decimals

When you use your calculator, you will find it easier if you convert mixed U.S. customary linear measurements to their decimal equivalents. For instance, if a diameter is 7 ft 6 in., convert it to 7.5 ft (from $7\frac{6}{12}$ ft, in which $\frac{6}{12} = 0.5$).

EXAMPLE   Find to the nearest hundredth the circumference of a circle with a radius of 1 ft 9 in. (Fig. 6–47).

$C = 2\pi r$

$C = 2\pi(1.75)$ Substitute for π and r: $1\frac{9}{12}$ ft = 1.75 ft

$C = 10.99557429$

Figure 6–47 $C = $ **11.00 ft (rounded)**   Circumference is a linear measure.

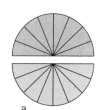

a

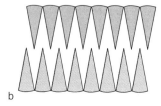

b

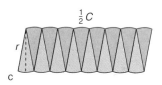

$\frac{1}{2}C$

r

c

Figure 6–48

2 Find the Area of a Circle.

The area of a circle, like the circumference, is obtained from relationships within the circle. If we divide a circle into two semicircles and then subdivide each semicircle into pie-shaped pieces, we get something like Fig. 6–48a. If we then spread the upper and lower pie-shaped pieces, we get Fig. 6–48b. Now if we push the upper and lower pieces together, the result approximates the rectangle in Fig. 6–48c, whose length is $\frac{1}{2}$ the circumference and whose width is the radius. Thus, the area of the circle is approximately the area of a rectangle, that is, length times width. Since the length of the rectangle is one-half the circumference and the width is the radius, the area of a circle equals one-half the circumference times the radius.

$$A = \frac{1}{2}C \times r \qquad \text{The formula for circumference is } C = 2\pi r, \text{ so we substitute } 2\pi r \text{ for } C.$$

$$A = \frac{1}{2}(2\pi r)(r) \qquad \text{Multiply. Reduce where possible.}$$

$$A = \frac{1}{\overset{1}{2}}(\overset{1}{2}\pi r)(r) \qquad \text{Area is a square measure: } r \cdot r = r^2$$

$$A = \pi r^2$$

EXAMPLE Find the area of a circle whose radius is 8.5 m (Fig. 6–49). Round to tenths.

Figure 6–49

$$A = \pi r^2 \qquad \text{Select appropriate formula and substitute for } r.$$
$$A = \pi (8.5)^2 \qquad \text{Square the radius.}$$
$$A = \pi(72.25) \qquad \text{Multiply by } \pi.$$
$$A = 226.9800692 \text{ m}^2 \qquad \text{Area is a square measure.}$$

The area of the circle is 227.0 m².

TIP!

Use a Continuous Calculator Sequence for Accuracy.

When mixed U.S. customary measures do not convert to convenient, terminating decimal equivalents, a continuous calculator sequence may be preferred for accuracy. For example,

$$3 \text{ ft } 4 \text{ in.} = 3 + \frac{4}{12} \text{ ft}$$

We can reduce and calculate with the mixed number or the improper fraction equivalent.

$$3 + \frac{4}{12} = 3 + \frac{1}{3} = 3\frac{1}{3} = \frac{10}{3}$$

Or we can make a continuous sequence using $3 + \frac{4}{12}$.

$$3 \boxed{+} 4 \boxed{\div} 12 \boxed{=} \Rightarrow 3.333333333$$

Many calculators apply the order of operations. Check with an appropriate example to verify your calculator's process.

Find the area to the nearest hundredth of the top of a circular tank with a diameter of 12 ft 8 in.

If we are using a calculator, we can calculate the area using continuous steps as shown in the tip following this example.

$A = \pi r^2$

$A = (\pi) \left(\dfrac{12 + \frac{8}{12}}{2} \right)^2$ Follow the order of operations. 8 in. $= \frac{8}{12}$ ft.

                               The diameter, $12 + \frac{8}{12}$, divided by 2 is the radius.

$A = 126.012772$ ft^2 Calculator result

$A = 126.01$ ft^2 (rounded)   Area is a square measure.

The area of the top of the tank is 126.01 ft^2 or 125.95 ft^2 (basic calculator using $\pi = 3.14$).

TIP!

Develop Your Calculator Proficiency.

It is important to know how your calculator performs continuous operations.

Let's try a problem we have already worked so we know what the correct answer should be. We will rework the preceding example.

Option 1 using the $\boxed{x^2}$ key and $\boxed{\pi}$ key:

$$12 \boxed{+} 8 \boxed{\div} 12 \boxed{=} \boxed{\div} 2 \boxed{=} \boxed{x^2} \boxed{\times} \boxed{\pi} \boxed{=} \Rightarrow 126.012772$$

Option 2 using the general power key $\boxed{\wedge}$ and the $\boxed{\pi}$ key:

$$12 \boxed{+} 8 \boxed{\div} 12 \boxed{=} \boxed{\div} 2 \boxed{=} \boxed{\wedge} 2 \boxed{=} \boxed{\times} \boxed{\pi} \boxed{=} \Rightarrow 126.0127721$$

An alternative sequence of keystrokes is required if the parentheses keys are used. Experiment with your calculator to find other sequences that work.

Option 3 using parentheses:

$$\boxed{\pi} \boxed{\times} \boxed{(} \boxed{(} 12 + 8 \div 12 \boxed{)} \boxed{\div} 2 \boxed{)} \boxed{x^2} \boxed{=} \Rightarrow 126.0127721$$

Work several problems both continuously and by recording and reentering intermediate calculations to learn how your calculator works and to develop your skills and confidence.

3   **Find the Perimeter and Area of Composite Figures.**

As we look about us, we see geometric shapes that are not squares, rectangles, parallelograms, trapezoids, triangles, or circles. Yet our work may require us to calculate the perimeter and area of these shapes. In such cases, we can use what we already know to find both perimeter and area. These shapes are called *composite* figures; that is, figures made up of two or more geometric figures.

To find a missing dimension of a composite shape:

1. Determine how the missing dimension is related to known dimensions.
2. Make a calculation of known dimensions according to the relationship found in Step 1.

Find the missing dimensions x and y on the slab foundation.

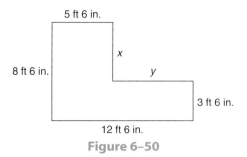

Figure 6–50

Separate the figure into parts.

In layout 1, the side of B opposite its 3'6" side is also 3'6" because opposite sides of a rectangle are equal. The side of A opposite its 8'6" side is, for the same reason, 8'6". Dimension x must therefore be the difference between 8'6" and 3'6".

$$x = 8'6'' - 3'6''$$

$$x = 5'$$

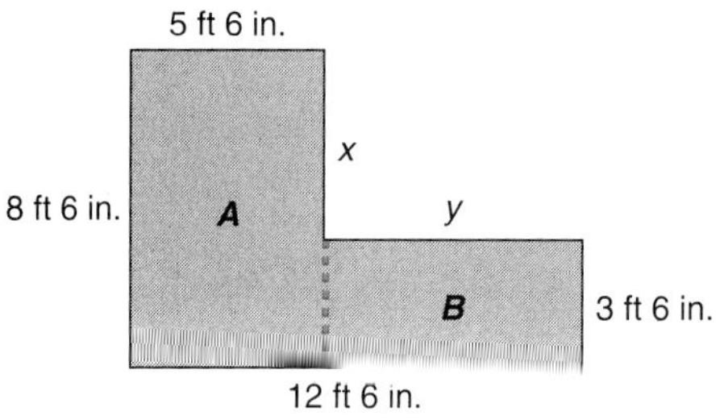

Layout I

Figure 6–51

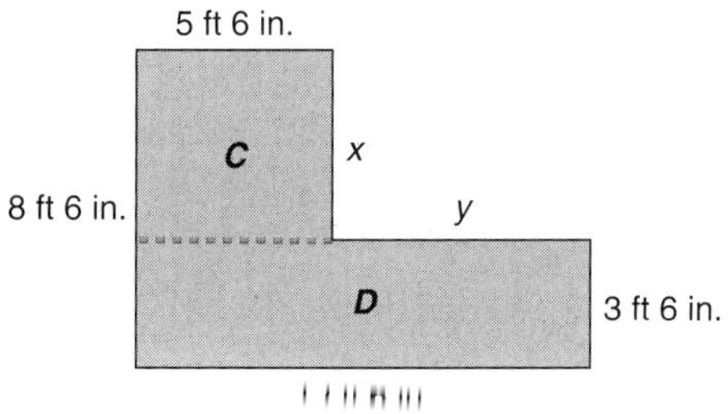

Layout II

Figure 6–52

If we think of layout II as two horizontal rectangles, we can find dimension y. The side opposite the 5'6" side of C must be 5'6". The side of D opposite the 12'6" side must also be 12'6". Dimension y must therefore be the difference between 12'6" and 5'6".

$$y = 12'6'' - 5'6''$$

$$y = 7''$$

The missing dimensions are $x = 5'$ and $y = 7'$.

Perimeter is the sum of the lengths of the sides of a figure. The number and the length of the sides vary from one composite figure to the next, so no specific formula covers the entire variety of composite shapes that exist. However, we can use a general formula.

Perimeter of a composite figure:

$$P = a + b + c + \cdots$$

where $a, b, c, \ldots$ are lengths of all the sides.

EXAMPLE Find the number of feet of 4-in. stock needed for the base plates of a room that has the layout shown in Fig. 6–53. Make no allowances for openings when calculating the linear footage of the base plates.

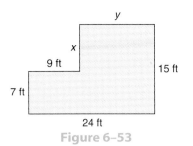

Figure 6–53

1. Find the missing dimensions.

$$x = 15 \text{ ft} - 7 \text{ ft} \qquad y = 24 \text{ ft} - 9 \text{ ft}$$
$$x = 8 \text{ ft} \qquad\qquad y = 15 \text{ ft}$$

2. Apply the general formula for the perimeter of a polygon.

$$P = a + b + c + \cdots$$

$$P = 7 \text{ ft} + 24 \text{ ft} + 15 \text{ ft} + 15 \text{ ft} + 8 \text{ ft} + 9 \text{ ft} = 78 \text{ ft}$$

3. Count the number of sides on the layout to make sure that each is substituted into the formula.

The room needs 78 ft of 4-in. stock for the base plates.

As with the perimeter, the area of a composite figure is found by finding the sum of the areas of the parts of the figure.

Area of a composite figure:

$$A = A_1 + A_2 + A_3 + \cdots$$

where $A_1, A_2, A_3, \ldots$ are the areas of all the parts of the composite figure.

EXAMPLE   Find the number of square yards of carpeting required for the room in the preceding example ($9 \text{ ft}^2 = 1 \text{ yd}^2$).

1. Divide the composite shape into two polygons with areas we can compute (Fig. 6–54). In this case, A is a rectangle and B is a square.

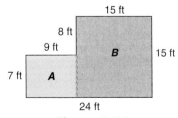

Figure 6–54

2. Find the areas of the smaller polygons and add them.

Rectangle A	**Square B**
$A_1 = lw$	$A_2 = s^2$
$A_1 = 9 \times 7$	$A_2 = 15^2$
$A_1 = 63 \text{ ft}^2$	$A_2 = 225 \text{ ft}^2$

$$A_1 + A_2 = \text{total area}$$

$$63 + 225 = 288 \text{ ft}^2$$

3. Convert square feet to square yards using a unity ratio.

$$\frac{\overset{32}{\cancel{288}} \text{ ft}^2}{1} \times \frac{1 \text{ yd}^2}{\underset{1}{\cancel{9}} \text{ ft}^2} = 32 \text{ yd}^2$$

The room requires 32 yd² of carpeting.

EXAMPLE A 15-in.-diameter wheel has a 3-in. hole in the center. Find the area of a side of the wheel to the nearest tenth (Fig. 6–55).

We are asked to find the area of the color portion of the wheel in Fig. 6–55. To do so, we find A_{outside}, the area of the larger circle (diameter 15 in.) and *subtract* the area of A_{inside}, the smaller circle (diameter 3 in.). The color portion is called a *ring*.

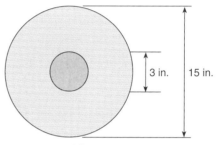

Figure 6–55

$A_{\text{outside}} = \pi r^2 \quad \left(r = \frac{15}{2} = 7.5 \right)$

$A_{\text{outside}} = \pi (7.5)^2$

$A_{\text{outside}} = \pi (56.25)$

$A_{\text{outside}} = 176.7145868 \text{ in}^2$

$A_{\text{inside}} = \pi r^2 \quad \left(r = \frac{3}{2} = 1.5 \right)$

$A_{\text{inside}} = \pi (1.5)^2$

$A_{\text{inside}} = \pi (2.25)$

$A_{\text{inside}} = 7.068583471 \text{ in}^2$

Area of wheel (ring) = $A_{\text{outside}} - A_{\text{inside}}$

$A_{\text{wheel}} = 176.7145868 - 7.068583471 = 169.6460033$ or 169.6 in^2

The area of the wheel (ring) is 169.6 in².

EXAMPLE A bandsaw has two 25-cm wheels spaced 90 cm between centers (Fig. 6–56). Find the length of the saw blade.

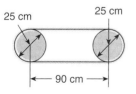

This layout is a composite figure which consists of a semicircle at each end and a rectangle in the middle. It is called a *semicircular-sided* figure. The two semicircles equal one whole circle, so we need to find the circumference of one circle (wheel) and add it to the lengths of the two sides of the rectangle.

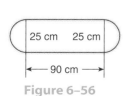

Figure 6–56

$C = \pi d$

$C = \pi (25)$

$C = 78.53981634 \text{ cm}$

Total length of blade = $C + 2l$

Total length of blade = $78.53981634 + 2(90)$

Total length of blade = $258.5 \text{ cm (rounded)}$

The bandsaw blade is 258.5 cm in length.

Specific applications for a particular industry or career often use the formulas for area, perimeter, or circumference.

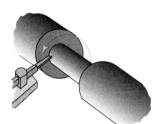

EXAMPLE Find the cutting speed of a lathe if a piece of work that has a 7 in. diameter turns on a lathe at 75 revolutions per minute (rpm) (Fig. 6–57).

The *cutting speed* is the speed of a tool that passes over the work, such as the speed of a lathe or a sander as it sands (passes over) a piece of wood. If the cutting speed is too fast or too slow, safety and quality are impaired. The formula for cutting speed is CS = C (in feet) × rpm.

CS = cutting speed

C = circumference or one revolution (in feet)

rpm = revolutions per minute

Cutting speed is measured in *feet per minute* (ft/min)

$C = \pi d$

$C = \pi(7)$

$C = 21.99114858$ in.   Convert 21.99114858 in. into feet using a unity ratio.

$C = 1.832595715$ ft $\dfrac{21.99114858 \text{ in.}}{1} \cdot \dfrac{1 \text{ ft}}{1 \text{ in.}} = 1.832595715$ ft.

Figure 6–57

CS = C × rpm = $1.832595715 \times 75 = 137$ ft/min   Rounded.

The cutting speed of the lathe is approximately 137 ft/min.

1 Find the circumference of circles with the following dimensions. Round to tenths.

1. Diameter = 8 cm
2. Diameter = 15 m
3. Radius = 3 in.
4. Radius = 16 yd
5. Radius = 1.5 ft
6. Radius = $8\frac{1}{2}$ ft
7. Diameter = 5.5 m
8. Diameter = $5\frac{1}{4}$ in.

2 Find the area of circles with the following dimensions. Round to tenths.

9. Diameter = 8 cm
10. Diameter = 15 m
11. Radius = 3 in.
12. Radius = 16 yd
13. Radius = 1.5 ft
14. Radius = $8\frac{1}{2}$ ft
15. Diameter = 5.5 m
16. Diameter = $5\frac{1}{4}$ in.

3 Find the missing dimensions x and y in Figs. 6–58 and 6–59.

17.

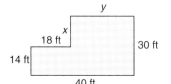

Figure 6–58

18.

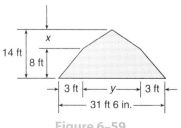

Figure 6–59

Find the color area of Figs. 6–60 through 6–63 to the nearest tenth.

19.

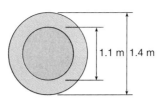

Figure 6–60

20.

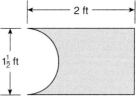

Figure 6–61

21.

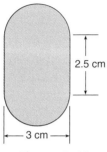

Figure 6–62

22.

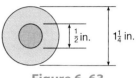

Figure 6–63

23. Find the perimeter of the layout in Fig. 6–64.

24. Find the area of the layout in Fig. 6–64.

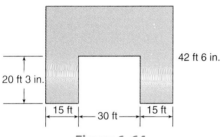

Figure 6–64

Solve the problems involving perimeter and area of composite figures.

25. A house features a family room with the layout shown in Fig. 6–65. Find the number of feet of 4 in. stock needed for the base plates around the perimeter. Make no allowances for doorways or other openings.

26. If the family room in Exercise 25 is to be covered with roll vinyl that costs $16.50 per yd^2 for materials and labor, how much would it cost to install the vinyl? (*Note*: Round to the next whole yd^2 before figuring cost.)

27. The metal piece in Fig. 6–66 has the dimensions indicated. If the 18-gauge steel weighs 2 lb per ft^2, how much does the piece weigh to the nearest pound? ($144\ in^2 = 1\ ft^2$)

28. The metal used in Exercise 27 costs $2.78 per pound. Any fraction of a pound is rounded to the next pound. How much will the material cost for making the metal piece?

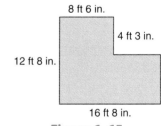

Figure 6–65

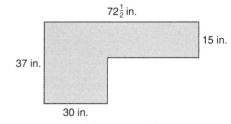

Figure 6–66

29. A swimming pool is in the form of a semicircular-sided figure. Its width is 20 ft and the parallel portions of the sides are each 20 ft (Fig. 6–67). What is the area of a 5-ft-wide walk surrounding the pool?

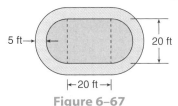

5 ft→   20 ft
|←20 ft→|

Figure 6–67

30. A belt connecting two 9-in.-diameter drums on a conveyor system needs replacing. How many inches must the new belt be if the centers of the drums are 10 ft apart (Fig. 6–68)?

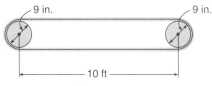

9 in.   9 in.
—— 10 ft ——

Figure 6–68

31. Find the area of the ring formed by the cross-cut section of Fig. 6–69.

35 mm
40 mm

Figure 6–69

32. The wall of a galvanized water pipe is 3.5 mm thick. If the outside circumference is 68 mm, what is the area (to nearest tenth) of the color inside cross-cut section (Fig. 6–70)?

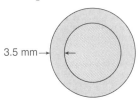

3.5 mm→

Figure 6–70

33. A 2-in.-inside-diameter pipe and a 4-in.-inside-diameter pipe empty into a third pipe whose inside diameter is 5 in. (Fig. 6–71). Is the third pipe large enough for the combined flow? (Justify your answer.)

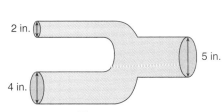

2 in.
4 in.
5 in.

Figure 6–71

34. A large pipe whose interior cross-sectional area is 20 in² empties into two smaller pipes that each have an interior diameter of 4 in. (Fig. 6–72). Are the smaller pipes together large enough to carry off the flow from the larger pipe? (Justify your answer.)

4 in.
20 in²
4 in.

Figure 6–72

35. Cutting speed, when applied to a grinding wheel, is called *surface speed*. What is the surface speed in ft/min of a 9-in.-diameter grinding wheel revolving at 1,200 rpm? (Surface speed = circumference in feet × rpm.)

36. A 12-in.-diameter polishing wheel revolves at 500 rpm. What is the surface speed? (See Exercise 35 for formula.)

37. A lamp lights up effectively an area 10 ft in diameter (Fig. 6–73). How many square feet does the lamp light effectively?

38. A steel sleeve with a 3-in. outside diameter measures $2\frac{1}{2}$ in. across the inside. The sleeve will be babbitted to fit a 1.604-in.-diameter motor shaft. What must the thickness of the babbitt be (in hundredths)? The babbitt is color shaded in Fig. 6–74. (A babbitt is a lining of babbitt metal, a soft antifriction alloy, that allows two metal parts to fit snugly together.)

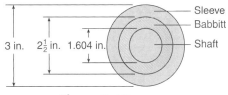

3 in. $2\frac{1}{2}$ in.   1.604 in.

Sleeve
Babbitt
Shaft

Figure 6–74

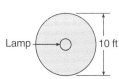

Lamp   10 ft

Figure 6–73

39. Two 12-cm-diameter drums are connected by a belt to form a conveyor system. The centers of the drums are 2 m apart (Fig. 6–75). How long must a replacement belt be (to the nearest tenth of a meter)?

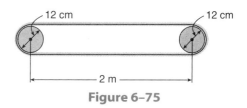

Figure 6–75

Learning Outcome   **1** Find the volume of a right prism and a right cylinder.

Common household items like cardboard storage boxes and toy building blocks are examples of three-dimensional geometric figures classified generally as *prisms*. Cans and pipes are examples of *cylinders*.

1 Find the Volume of a Right Prism and a Right Cylinder.

A *prism* is a three-dimensional figure with polygonal bases (ends) that are parallel and faces (sides) that are parallelograms, rectangles, or squares. In a *right prism*, the faces are perpendicular to the bases.

A *right circular cylinder* is a three-dimensional figure with a curved surface and two circular bases such that the height is perpendicular to the bases.

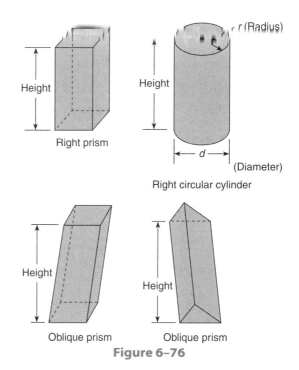

Figure 6–76

In an *oblique prism* or an *oblique cylinder* the faces are *not* perpendicular to the bases.

The *height* of a three-dimensional figure with two bases is the shortest distance between the two bases.

In right circular cylinders and in right prisms, the height is the same as the length of a side or face. However, in oblique prisms and cylinders the height is the perpendicular distance between the bases and is different from the length of a side or face.

The *volume* of an object, such as a container, is used to estimate how many containers can be loaded into a given-size storage area or shipped in a container of certain dimensions.

The *volume* of a three-dimensional geometric figure is the amount of space it occupies, measured in terms of three dimensions (length, width, and height).

If we have a rectangular box measuring 1 ft long, 1 ft wide, and 1 ft high (Fig. 6–77), it will be a cube representing 1 cubic foot (ft^3). Its volume is calculated by multiplying length × width × height, or 1 ft × 1 ft × 1 ft = 1 ft^3. We indicate a cubic measure with an exponent 3 after the unit of measure, meaning that the measure is "*cubed*."

From this concept of 1 ft^3 or 1 cubic foot comes the formula for the volume of a rectangular box as length × width × height or $V = lwh$. Note that $l × w$ is the formula for the area of the rectangle (or square) that forms the base of the rectangular box. If the base of the prism is a triangle, trapezoid, or other polygon, we use the appropriate formula for the area of the base. A general formula can be used for the volume of *any* right prism or right cylinder.

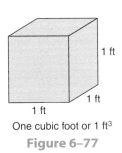

One cubic foot or 1 ft³

Figure 6–77

Formula for the volume of right prism or right cylinder:

$$V = Bh$$

where B is the area of the base and h is the height of the prism or cylinder.

EXAMPLE Find the volume of the triangular prism in Fig. 6–78 if the height is 15 cm and the bases are triangles 3 cm on a side and 2.6 cm in height.

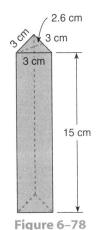

Figure 6–78

$V = Bh$ Substitute the formula for the area of the triangular base for *B*.

$V = \left(\dfrac{1}{2} bh_1\right)h_2$ $h_1 = 2.6$ cm (height of prism base), $h_2 = 15$ cm (height of prism).

$V = \left[\dfrac{1}{2}(3)(2.6)\right]15$ Substitute values and evaluate.

$V = 58.5 \text{ cm}^3$

The volume of the prism is 58.5 cm³.

EXAMPLE What is the cubic-inch displacement (space occupied) of a cylinder whose diameter is 5 in. and whose height is 4 in.?

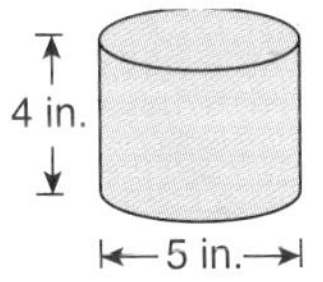

Figure 6–79

$V = Bh$ Substitute the formula for the area of the circular base for *B*.

$V = \pi r^2 h$ Substitute values; $r = \frac{1}{2}$ diameter, or 2.5.

$V = \pi(2.5)^2(4)$ Evaluate.

$V = 78.5 \text{ in}^3$ Rounded.

The cylinder displacement is 78.5 in³.

Learning Strategy *Be Realistic about Memorizing Formulas.*

Do you normally remember details or general information? Most of us remember the context of a particular conversation with a friend, but not the exact words that were spoken. The formulas you use often can generally be recalled, but formulas you use occasionally are easily forgotten.

Instead of memorizing formulas, focus on

- The purpose for which the formula is used
- The interpretation of each letter of that formula
- Terminology that is useful in locating a specific formula when needed
- Memorizing only formulas that are required for a specific objective (a test or exam, formulas used often on the job, and so on)
- Relating formulas to other formulas you have studied (area of parallelogram to area of rectangle, perimeter of square to perimeter of rectangle, and so on)

SELF-STUDY EXERCISES 6–3

1 Find the volume in Figs. 6–80 through 6–83. Round to the nearest hundredth if necessary.

1.

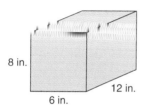

8 in.
6 in.
12 in.

Figure 6–80

2.

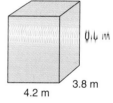

0.6 m
3.8 m
4.2 m

Figure 6–81

3.

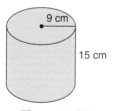

9 cm
15 cm

Figure 6–82

4.

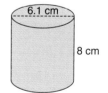

6.1 cm
8 cm

Figure 6–83

5. A right pentagonal (five-sided) prism is 10 cm high. If the area of each pentagonal base is 32 cm², what is the volume of the prism?

6. What is the volume of a triangular prism that has a height of 8 in., a triangular base that measures 4 in. on each side, and a height of 3.46 in.? Round to hundredths.

7. How many cubic inches are in an aluminum can with a $2\frac{1}{2}$ in. diameter and $4\frac{3}{4}$ in. height? Round to tenths.

8. What is the volume of a cylindrical oil storage tank that has a 40 ft diameter and 15 ft height? Round to the nearest whole number.

9. A cylindrical water well is 1,200 ft deep and 6 in. across. How much soil and other material were removed? Round to the nearest cubic foot. (*Hint*: Convert measures to a common unit.)

10. A cylindrical vegetable can is 10.7 cm tall and 7.3 cm across. Find the volume.

Section 6–1

Find the perimeter of Figs. 6–84 through 6–93.

1.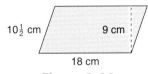
Figure 6–84

2.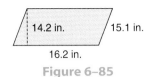
Figure 6–85

3.
Figure 6–86

4.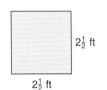
Figure 6–87

5.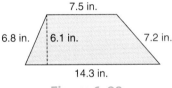
Figure 6–88

6.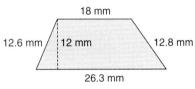
Figure 6–89

7.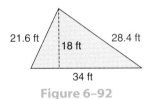
Figure 6–90

8.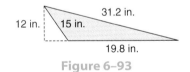
Figure 6–91

9.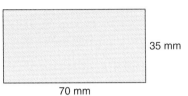
Figure 6–92

10.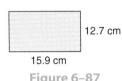
Figure 6–93

11. The Tennessee Highway Department has signs in the form of a parallelogram. One set of parallel sides each measures 15 ft and one set of parallel sides each measures 18 feet. Find the perimeter of the sign.

12. Antique tiles were often made in the form of a square that is 6 in. on each side. What is the perimeter of a tile?

13. A rectangular tablecloth measures 84 in. by 60 in. What length of lace is required to trim the edges of the cloth?

14. A classroom table in the form of a trapezoid has parallel sides that measure 26 in. and 48 in. The two nonparallel sides both measure 21.1 in. Find the length of trim needed to encase the edges of the table.

15. A triangular corner table has sides that measure 15 in., 20 in., and 25 in. What length of finish molding is needed to cover the edges of the table?

16. Find the number of feet of roll fencing needed to fence a square storage area measuring $15\frac{1}{2}$ ft on a side. A preassembled gate 4 ft wide will be installed.

Find the area of Figs. 6–94 through 6–103.

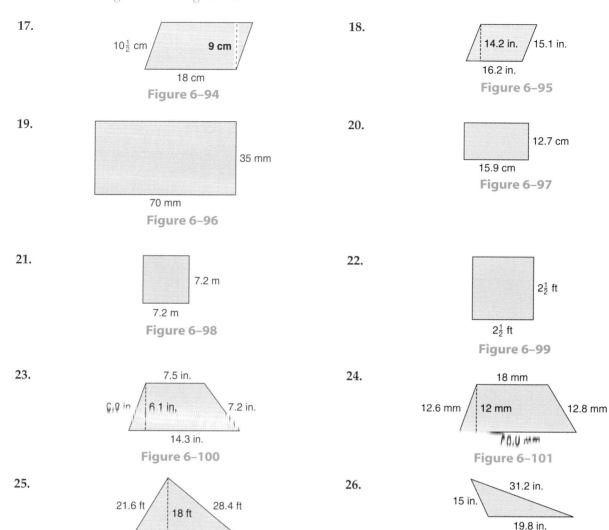

17.
$10\frac{1}{2}$ cm   **9 cm**
18 cm
Figure 6–94

18.
14.2 in.   15.1 in.
16.2 in.
Figure 6–95

19.
35 mm
70 mm
Figure 6–96

20.
12.7 cm
15.9 cm
Figure 6–97

21.
7.2 m
7.2 m
Figure 6–98

22.
$2\frac{1}{2}$ ft
$2\frac{1}{2}$ ft
Figure 6–99

23.
7.5 in.
6.9 in. 6.1 in. 7.2 in.
14.3 in.
Figure 6–100

24.
18 mm
12.6 mm 12 mm 12.8 mm
20.0 mm
Figure 6–101

25.
21.6 ft 18 ft 28.4 ft
34 ft
Figure 6–102

26.
31.2 in.
15 in.
19.8 in.
Figure 6–103

27. If a parking lot for a new hospital in the shape of a parallelogram measures 275 ft by 150 ft and has a height of 120 ft, how many square feet need to be paved?

28. A hall wall with no windows or doors measures 25 ft long by 8 ft high. Find the number of square feet to be covered if paneling is installed on the two walls.

29. A den 18 ft by $16\frac{1}{2}$ ft is to be carpeted. How many square yards of carpeting are needed?

30. Vincent Ores, a contractor, is to brick the storefront of a landscape service that has a doorway measuring 7 ft by 6 ft. How many bricks are needed if the storefront is 20 ft by 12 ft and the bricks cover at the rate of 6 per square foot using $\frac{1}{2}$-in. mortar joints?

31. A roof measuring 16 ft by 20 ft is to be covered with asphalt roofing cement. How much would the project cost if the asphalt roofing cement spreads at the rate of 150 square feet per gallon and costs $4.75 per gallon? The cement is purchased by the gallon only.

32. A square area of land is 10,000 m². If a baseball field must be at least 99.1 m along each foul line to the park fence, is the square adequate for regulation baseball?

33. Debbie Murphy is building a contemporary home with four front windows, each in the form of a parallelogram. If each window has a 5-ft base and a height of 2 ft, how many square feet of the 25 ft × 11 ft wall will require stain?

34. If the stain used on the wall in Exercise 33 is applied at a cost of $2.75 per square yard, find the cost to the nearest dollar of staining the front wall.

35. A dining room wall 12 ft long by 8 ft high will be wallpapered. How many single rolls (24 in. by 20 ft) of wallpaper will be needed? Assume there is no waste.

Find the area of Figs. 6–104 through 6–110. Round to hundredths.

36.

Figure 6–104

37.

Figure 6–105

38.

Figure 6–106

39.

Figure 6–107

40.

Figure 6–108

41.

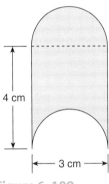

Figure 6–109

42.

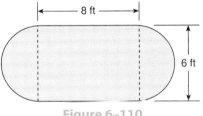

Figure 6–110

43. A circle has a diameter of 2.5 m. If its diameter were increased 1.2 m, what would be the difference in the areas (to the nearest tenth) of the two circles?

44. If a circular flower bed has a radius of 24 in., find its area.

45. Three water hoses, with interior cross-sectional areas of 1 in^2, $1\frac{1}{2}$ in^2, and 2 in^2, empty into a larger hose. What is the inside diameter of the larger hose if it carries the combined flow of the three smaller hoses? Round to tenths.

46. A 35-cm diameter circular window has what area? Round to tenths.

47. A $\frac{1}{4}$-in. electric drill with variable speed control turns as slowly as 25 rpm. If an abrasive disk with a 5-in. diameter is attached to the drill driveshaft, what is the disk's slowest cutting speed in ft/min? Round to the nearest whole number. (Cutting speed = circumference measured in feet × rpm.)

48. What is the cross-sectional area of the opening in a round flue tile whose inside diameter is 8 in.? Round any part of an inch to the next tenth of an inch.

49. Triple-wall galvanized chimney pipe for prefabricated sheet-metal fireplaces has a circumference of 47 in. What is the inside diameter of the fire-stop spacer through which the chimney pipe passes as it goes through the ceiling? Round to the nearest tenth.

50. Find the area of the cross-section of a wire whose diameter is $\frac{1}{16}$ in. (Express your answer to the nearest thousandth.)

51. Find the area of the color portion of the sidewalk in Fig. 6–111. The color portion is one-fourth of a circle. Express the answer in square yards rounded to tenths. ($144 \text{ in}^2 = 1 \text{ ft}^2$, $9 \text{ ft}^2 = 1 \text{ yd}^2$.)

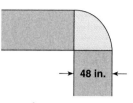

Figure 6–111

52. Find the area of the composite layout shown in Fig. 6–112. Round the answer to the nearest tenth.

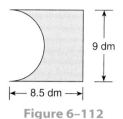

9 dm

8.5 dm

Figure 6–112

53. If a cogwheel makes one complete revolution and its radius is 9.4 in., how long is the path traveled by any one point on the cogwheel? Round to tenths.

54. Find the volume of the prism in Fig. 6–113.

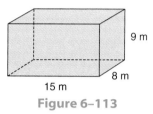

9 m

8 m

15 m

Figure 6–113

55. Find the volume of the triangular prism in Fig. 6–114.

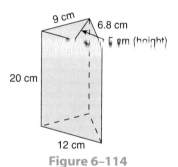

9 cm

6.8 cm

5 cm (height)

20 cm

12 cm

Figure 6–114

56. Find the volume of the cylinder of Fig. 6–115.

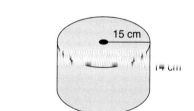

15 cm

14 cm

Figure 6–115

57. Find the volume of the cylinder in Fig 6–116.

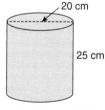

20 cm

25 cm

Figure 6–116

58. If concrete weighs 160 lb per cubic foot, what is the weight of a concrete circular slab 4 in. thick and 15 ft across?

59. How many cubic yards of topsoil are needed to cover an 85-ft by 65-ft area for landscaping if the topsoil is 6 in. deep? Round to the nearest whole number. ($27 \text{ ft}^3 = 1 \text{ yd}^3$)

60. An interstate highway is repaired in one section 48 ft across, 25 ft long, and 8 in. deep. If concrete costs $25.50 per cubic yard, what is the cost of the concrete needed to repair the highway rounded to the nearest dollar? ($27 \text{ ft}^3 = 1 \text{ yd}^3$)

61. A pipeline to carry oil between two towns 5 mi apart has an inside diameter of 18 in. If 1 mi = 5,280 ft and $1 \text{ ft}^3 = 7.48$ gal, how many gallons of oil will the pipeline hold (to the nearest gallon)?

62. Lou Ferrante plans to build a rectangular dog pen. He has 360 ft of fencing and would like to enclose as much area as possible for his dog. What length and width should he make the dog pen?

63. Tosha May, an interior designer, needs to order a tablecloth to cover a circular table that is 24 in. from the floor and 18 in. across the top (Fig. 6–117). The cloth must drape the table so that it just touches the floor. Determine the shape of the tablecloth. Find the number of square feet of fabric needed for the tablecloth.

64. Given that the formula for finding the area of a triangle is $A = \frac{1}{2}bh$, find the area of the triangle in Fig. 6–119 for Exercise 6 of the Concepts Analysis.

Figure 6–117

1. Describe five activities or jobs that require you to find the perimeter of a shape.

3. Draw five parallelograms of different sizes and cut out each parallelogram. Cut each parallelogram into two pieces by cutting across a diagonal (opposite corners). Describe similarities and/or differences between the two pieces of each parallelogram.

5. Draw a shape that has the same relationship to a parallelogram as a square has to a rectangle (Fig. 6–118).

2. Describe five activities or jobs that require you to find the area of a shape.

4. Each piece of the parallelogram in Exercise 3 forms a triangle. Use the comparisons from Exercise 3 to write a formula to find the area of a triangle as it relates to a parallelogram.

6. Explain how you could find the area of the composite figure (Fig. 6–119).

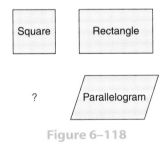

Figure 6–118

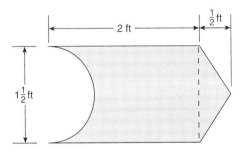

Figure 6–119

7. Discuss the similarities and differences between a square and a rectangle.

9. Drawing freely on the formulas for the circumference of a circle, devise your own formulas to find the radius and the diameter of a circle.

8. Discuss the similarities and differences between a rectangle and a parallelogram.

10. The new shape in Exercise 5 is called a *rhombus*. List the properties of a rhombus. Write a formula for finding the area and perimeter of a rhombus.

Learning Outcomes   *What to Remember with Examples*

Section 6–1

1 Find the perimeter of a parallelogram, rectangle, square, trapezoid, and triangle (pp. 278–281).

Formula for perimeter: $P_{parallelogram} = 2(b + s)$: b is the base, or the length of a side that is or can be rotated to a horizontal position; and s is the length of a side that joins with the base.

Find the perimeter of a parallelogram with a base of 1 m and a side of 0.5 m.

$$P_{parallelogram} = 2(b + s) \qquad \text{Substitute } b = 1 \text{ and } s = 0.5.$$
$$P_{parallelogram} = 2(1 + 0.5)$$
$$P_{parallelogram} = 2(1.5)$$
$$P_{parallelogram} = 3 \text{ m}$$

Formula for perimeter: $P_{rectangle} = 2(l + w)$: l is length of the long side; w is width of the short side.

A rectangular flower bed is 8 ft × 3 ft. How many feet of edging are needed to surround the bed?

$$P_{rectangle} = 2(l + w) \qquad \text{Substitute } l = 8 \text{ and } w = 3.$$
$$P_{rectangle} = 2(8 + 3)$$
$$P_{rectangle} = 2(11)$$
$$P_{rectangle} = 22 \text{ ft}$$

Formula for perimeter: P_{square} = 4s. s is the length of a side.

Find the perimeter of a square 6.5 cm on a side.

$$P_{square} = 4s \qquad \text{Substitute } s = 6.5.$$
$$P_{square} = 4(6.5)$$
$$P_{square} = 26 \text{ cm}$$

Formula for perimeter: $P_{trapezoid} = b_1 + b_2 + a + c$, where b_1 and b_2 are the lengths of the bases or parallel sides and a and c are the length of the sides adjacent to the bases.

Find the perimeter (P) of a trapezoid that has bases of 10 cm and 8 cm and has sides a and c of 7 cm each.

$$P_{trapezoid} = b_1 + b_2 + a + c \qquad \text{Substitute values.}$$
$$P_{trapezoid} = 10 + 8 + 7 + 7 \qquad \text{Add.}$$
$$P_{trapezoid} = 32 \text{ cm}$$

Formula for perimeter: $P_{triangle} = a + b + c$, where a, b, and c are the lengths of the three sides.

What is the perimeter (P) of a triangular roof vent that has sides a, b, and c of 6 ft, 4 ft, and 4 ft, respectively?

$$P_{triangle} = a + b + c \qquad \text{Substitute values.}$$
$$P_{triangle} = 6 + 4 + 4 \qquad \text{Add.}$$
$$P_{triangle} = 14 \text{ ft}$$

2 Find the area of a rectangle, square, parallelogram, trapezoid, and triangle (pp. 281–287).

Formula for area: $A_{\text{rectangle}} = lw$, where l is the length and w the width.

How much landscaping fabric is needed to cover a flower bed that is 8 ft × 3 ft?

$$A_{\text{rectangle}} = lw$$
$$A_{\text{rectangle}} = 8 \times 3$$
$$A_{\text{rectangle}} = 24 \text{ ft}^2$$

Formula for area: $A_{\text{square}} = s^2$, where s is the length of a side.

Find the area of a square 1.5 in. on a side.

$$A_{\text{square}} = s^2$$
$$A_{\text{square}} = (1.5 \text{ in.})^2$$
$$A_{\text{square}} = 2.25 \text{ in}^2$$

Formula for area: $A_{\text{parallelogram}} = bh$: b is the base; h is the height, or length of a perpendicular distance between the bases.

Find the area of a sign shaped like a parallelogram with a base of 36 in. and a height of 12 in.

$$A_{\text{parallelogram}} = bh$$
$$A_{\text{parallelogram}} = 36(12)$$
$$A_{\text{parallelogram}} = 432 \text{ in}^2$$

Formula for area: $A_{\text{trapezoid}} = \frac{1}{2}h(b_1 + b_2)$, where h is the height between the parallel bases and b_1 and b_2 are the lengths of the bases or two parallel sides.

Find the area of a trapezoid that has bases of 11 in. and 18 in. and height of 12 in.

$$A_{\text{trapezoid}} = \frac{1}{2}h(b_1 + b_2) \qquad \text{Substitute values.}$$

$$A_{\text{trapezoid}} = \frac{1}{2}(12)(11 + 18) \qquad \text{Perform operation in grouping.}$$

$$A_{\text{trapezoid}} = \frac{1}{2}(12)(29) \qquad \text{Multiply.}$$

$$A_{\text{trapezoid}} = 174 \text{ in}^2$$

Formula for area: $A_{\text{triangle}} = \frac{1}{2}(bh)$, where b is the base and h is the height.

A triangle has a base of 3 m and a height of 2 m. Find its area.

$$A_{\text{triangle}} = \frac{1}{2}(bh) \qquad \text{Substitute values.}$$

$$A_{\text{triangle}} = \frac{1}{2}(3)(2) \qquad \text{Multiply.}$$

$$A_{\text{triangle}} = 3 \text{ m}^2$$

Section 6–2

1 Find the circumference of a circle (pp. 292–293).

Formula for circumference: $C = \pi d$, or $C = 2\pi r$; d is the diameter or distance across the center of a circle; r is the radius, or half the diameter; π is approximated on a calculator as 3.141592654.

Find the circumference of a circle with a 3-cm diameter.

$$C = \pi d$$
$$C = \pi(3)$$
$$C = 9.4 \text{ cm (rounded)}$$

What is the distance around a circle with an 18-in. radius?

$$C = 2\pi r$$
$$C = 2\pi(18)$$
$$C = 113.1 \text{ in. (rounded)}$$

2 Find the area of a circle (pp. 294–295).

Formula for area: $A_{\text{circle}} = \pi r^2$; r is the radius.

Find the area of a circle whose diameter is 3 m.

First, find the radius: $r = \dfrac{d}{2}$ 3 m ÷ 2 = 1.5 m.

$$A_{\text{circle}} = \pi r^2$$
$$A_{\text{circle}} = \pi(1.5 \text{ m})^2$$
$$A_{\text{circle}} = 7.07 \text{ m}^2 \text{ (rounded)}$$

3 Find the perimeter and area of composite figures (pp. 295–299).

Some problems involve composite figures, which require finding several areas, circumferences, or perimeters.

A belt moving around two pulleys 5 in. in diameter must be replaced. How long must the belt be if the centers of the pulleys are 18 in. apart? This setup forms a semicircular-sided figure (Fig. 6–120). Both semicircles form one full circle. Find its circumference and add the upper and lower distances (2 × 18 = 36).

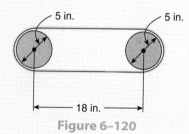

5 in.   5 in.

18 in.

Figure 6–120

$$C = \pi d$$
$$C = \pi(5)$$
$$C = 15.70796327$$

15.70796327 + 36 = 51.7 in. (rounded). The belt must be 51.7 in. long.

Section 6–3

1 Find the volume of a right prism and a right cylinder (pp. 302–304).

Use the formula for the volume of a prism and cylinder: $V = Bh$, where B is the area of the base and h is the height.

Find the volume of a cylinder that has a diameter of 20 mm and a height of 80 mm.

$$V = Bh$$
$$V = \pi r^2 h$$
$$V = \pi(10)^2(80)$$
$$V = \pi(100)(80)$$
$$V = 25{,}132.74123 \text{ mm}^3 = 25{,}132 \text{ mm}^3 \quad \text{Rounded}$$

$B = \pi r^2$

$r = \frac{1}{2}d = 10$ mm

CHAPTER TRIAL TEST

Find the perimeter and area of Figs. 6–121 through 6–128.

1.

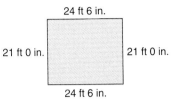

24 ft 6 in.
21 ft 0 in.
21 ft 0 in.
24 ft 6 in.

Figure 6–121

2.

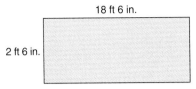

18 ft 6 in.
2 ft 6 in.

Figure 6–122

3.

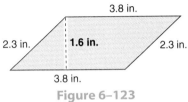

3.8 in.
2.3 in.
1.6 in.
2.3 in.
3.8 in.

Figure 6–123

4.

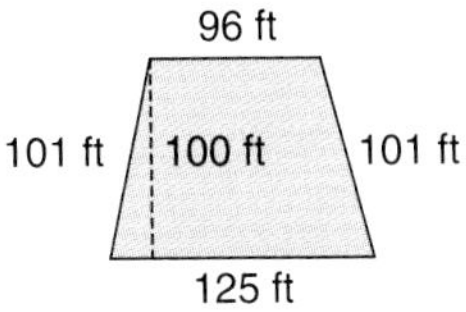

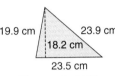

96 ft
101 ft
100 ft
101 ft
125 ft

Figure 6–124

5.

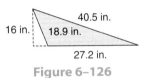

19.9 cm
23.9 cm
18.2 cm
23.5 cm

Figure 6–125

6.

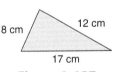

40.5 in.
16 in.
18.9 in.
27.2 in.

Figure 6–126

7.

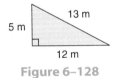

8 cm
12 cm
17 cm

Figure 6–127

8.

13 m
5 m
12 m

Figure 6–128

Find the circumference and area.

9.

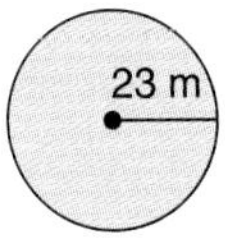

23 m

Figure 6–129

10.

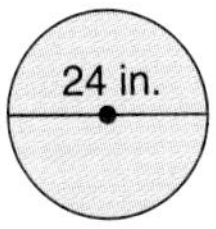

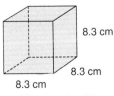

24 in.

Figure 6–130

Find the volume.

11.

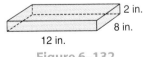

8.3 cm
8.3 cm
8.3 cm

Figure 6–131

12.

2 in.
8 in.
12 in.

Figure 6–132

13.

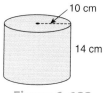

10 cm

14 cm

Figure 6-133

14.

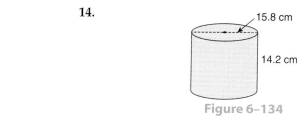

15.8 cm

14.2 cm

Figure 6-134

15. A company charges $3.00 labor per square yard to install carpet and padding. If a dining room is 12 ft × 13 ft, what is the labor cost to install the padding and carpet (9 ft² = 1 yd²)?

16. How many squares (100 ft²) of siding are needed for four sides of a garage 18 ft × 15 ft × 9 ft high if there is 125 ft² of openings? Allow one-fourth extra siding for overlap and waste.

17. How many square feet of floor space are there in the plan shown in Fig. 6-135? A full square yard must be puchased for any portion of a square yard.

18. A machine stamps 1-in. squares from sheet metal. How many can it cut from a 3 ft × 4 ft sheet of metal?

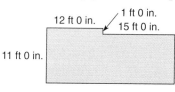

12 ft 0 in.

1 ft 0 in.

15 ft 0 in.

11 ft 0 in.

Figure 6-135

Solve these problems and when necessary, round to hundredths.

19. Find the circumference of a circle whose radius is 6 in.

20. Find the area of the circle in Exercise 19.

21. What is the perimeter of a roller-skating rink with the dimensions shown in Fig. 6-136?

22. Find the area to the nearest square foot of the skating rink in Exercise 21.

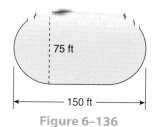

75 ft

150 ft

Figure 6-136

23. A water pipe with an inside diameter of $\frac{5}{8}$ in. and an outside diameter of $\frac{7}{8}$ in. is cut off at one end. What is the cross-sectional area of the ring formed by the wall of the pipe?

24. Find the area of the color portion of the tiled walk that surrounds a rectangular swimming pool (Fig. 6-137).

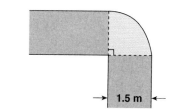

1.5 m

Figure 6-137

25. Find the area of a washer with an inside diameter of $\frac{1}{4}$ in. and an outside diameter of $\frac{1}{2}$ in.

26. If it costs $1.25 to sod each square foot of a circular lawn (Fig. 6-138), how much will it cost, to the nearest cent, if the lawn is $6\frac{1}{2}$ ft wide and surrounds a round goldfish pond with a diameter of 12 ft?

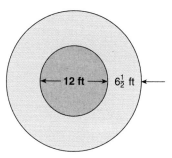

12 ft $6\frac{1}{2}$ ft

Figure 6-138

Interpreting and Analyzing Data

7

A *graph* shows information visually. Graphs may show how our tax dollars are divided among various government services, trace the fluctuations in a patient's temperature, or illustrate regional planting seasons. Other graphs may show equations, inequalities, and their solutions. Tables, on the other hand, usually list data. For example, income tax tables, list taxes due on different incomes.

7–1 | *Reading Circle, Bar, and Line Graphs*

Learning Outcomes

1 Read circle graphs.
2 Read bar graphs.
3 Read line graphs.

Figure 7–1 Distribution of wholesale price for a $425 color television.

Graphs give us useful information at a glance if we interpret them properly. Three common graphs used to represent data are the circle graph, the bar graph, and the line graph.

1 **Read Circle Graphs.**

A *circle graph* uses a divided circle to show pictorially how a whole quantity is divided into parts.

The complete circle represents one whole quantity. The circle is divided into parts so that the sum of all the parts equals the whole quantity. These parts can be expressed as fractions, decimals, or percents. Fig. 7–1 is a circle graph.

When we "read" a graph, we examine the information on the graph.

To read a circle, bar, or line graph:

1. Examine the title of the graph to find out what information is shown.
2. Examine the parts to see how they relate to one another and to the whole.
3. Examine the labels for each part of the graph and any explanatory remarks that may be given.
4. Use the given parts to calculate additional amounts.

EXAMPLE Use Fig. 7–1 to answer the questions.

(a) What percent of the wholesale price is the cost of labor?
(b) What percent of the wholesale price is the cost of materials?
(c) What would the wholesale price be if no tariff (tax) was paid on imported parts?

(a) $\dfrac{R}{100} = \dfrac{130}{425}$ *R* is the percent of the wholesale price ($425) that is attributed to labor cost. The labor cost is $130.

$R \times 425 = 13{,}000$

$R = \dfrac{13{,}000}{425}$

$R = 30.58823529$

$R = 30.6\%$ **(labor)**

(b) $\dfrac{R}{100} = \dfrac{125}{425}$ R is the percent of the wholesale price ($425) that is attributed to materials cost. The materials cost is $125.

$$R \times 425 = 12{,}500$$

$$R = \dfrac{12{,}500}{425}$$

$$R = 29.41176471$$

$$R = 29.4\% \text{ (materials)}$$

(c) Price − tariff = **425** − 40 = **$385 (cost without tariff)**

2 **Read Bar Graphs.**

Different types of graphs allow us to access different types of information. A *bar graph* uses two or more bars to compare two or more amounts.

The bar lengths represent the amounts being compared. Bars can be drawn either horizontally or vertically.

The *axis* or *reference line* that runs along the length of the bars is a scale of the amounts being compared; in Fig. 7–2 this line is horizontal. The other reference line (vertical in this case) labels the bars. Figure 7–2 is a horizontal bar graph.

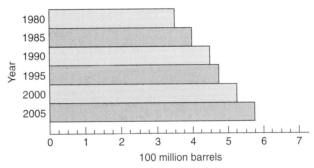

Figure 7–2 Company oil production.

EXAMPLE   Use Fig. 7–2 to answer the questions.

(a) How many more 100 million barrels of oil are indicated for the company in 2005 than in 1985?
(b) Judging from the graph, should company oil production in 2010 be more or less than in 2005?
(c) How many 100 million barrels of oil did the company produce in 1985?

(a) 2005 production − 1985 production = 5.75 − 4.00 = 1.75 hundred million barrels.
(b) More, because the trend has been toward greater production.
(c) Four hundred million barrels in 1985.

3 **Read Line Graphs.**

Line graphs are encountered in industrial reports, handbooks, and the like. A *line graph* uses one or more lines to show changes in data.

The horizontal axis on a line graph usually represents periods of time or specific times. The vertical axis represents numerical amounts. Line graphs show trends in data and high and low values at a glance. Fig. 7–3 is a line graph.

EXAMPLE   Use Fig. 7–3 to answer the questions regarding a patient's temperature.

(a) On what date and time of day did the patient's temperature first drop to within 0.2 degrees of normal (98.6° F)?

(b) On which post-op (post-operative) days did the patient's temperature remain within 0.2 degrees of normal?

(c) What was the highest temperature recorded for the patient?

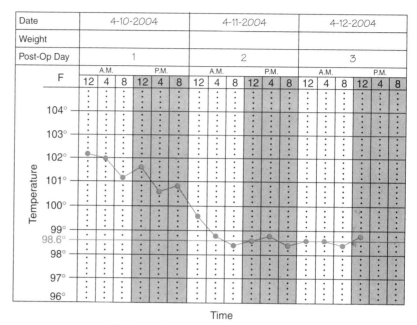

Figure 7–3 Graphic temperature chart

(a) **4-11-04 at 4 A.M.** (Each "dot" is 0.2 degrees, so the temperature was 98.8° F.)

(b) **Post-op days 2 and 3** (beginning at 4 A.M. on day 2)

(c) **102.2 degrees** (recorded at 12 A.M. on 4-10-04)

SELF-STUDY EXERCISES 7–1

1   Use Fig. 7–4 to answer Exercises 1–3.

1. What percent of the gross salary goes into retirement? Round to tenths.

2. What percent of the take-home pay is federal income tax? Round to tenths.

3. What percent of the gross pay is the take-home pay? Round to tenths.

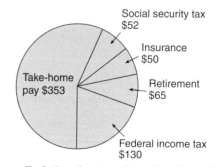

Figure 7–4 Distribution of weekly salary of $650.

Use Fig. 7–5 to answer Exercises 4–6. Round to tenths.

4. What percent of the day is spent working?
5. What percent of the day is spent sleeping?
6. The amount of time spent studying is what percent of the time spent in class?

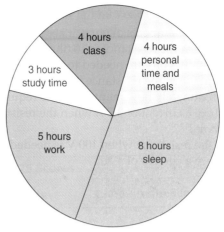

Figure 7–5 Distribution of a typical day.

2 Use Fig. 7–6 to answer Exercises 7–9.

7. What expenditure is expected to be the same next year as this year?
8. What two expenditures are expected to increase next year?
9. What two expenditures are expected to decrease next year?

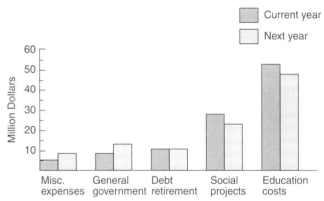

Figure 7–6 Distribution of local tax dollars.

Use Fig. 7–7 to answer Exercises 10–13.

10. What year had the largest daily number of barrels of oil produced on shore?
11. What was the total number of barrels of oil produced daily in 1990? In 2000?
12. By what percent did daily off shore oil production increase from 1960 to 2000?
13. In what year was the total daily oil production at its lowest level for the 5 decades reported?

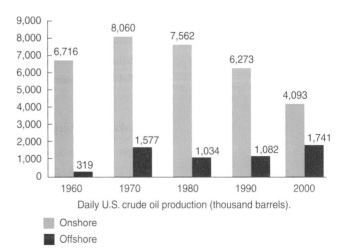

Figure 7–7 Daily U.S. crude oil production (thousand barrels).
Source: Energy Information Administration, U.S. Government.

Table 7–2	Frequency Distribution of Credit-Hour Loads		
Class Interval	Midpoint	Tally	Class Frequency
0–4	2	///	3
5–9	7	//// //	7
10–14	12	////	4
15–19	17	//	2

2 Make and Interpret Histograms.

A *histogram* is a bar graph that uses two scales, one for class intervals and one for class frequencies. The grouped frequency distribution in Table 7–2 can be made into a histogram. The frequencies in this histogram form the vertical scale. The class intervals form the horizontal scale.

To make a histogram for a data set:

1. Create a table (frequency distribution) with headings *Class Interval, Midpoint, Tally, Class Frequency*.
2. Use the data to complete the frequency distribution.
3. Show the class intervals on one axis of a bar graph with spacing to provide for the width of the bars. Increment the other axis based on the numbers shown in the class frequency column.
4. Use the class frequency data to draw a bar for each of the class intervals.
5. Label both axes to reflect the information being displayed. Provide a title for the histogram.

EXAMPLE Use the frequency distribution for credit-hour loads (Table 7–2) to make a histogram (Fig. 7–10)

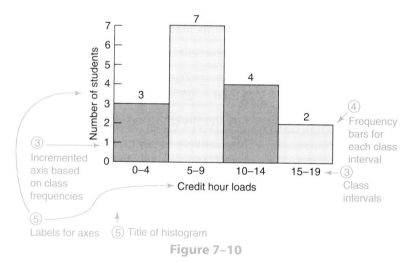

Figure 7–10

To use a histogram to provide desired information:

1. Determine the information that is desired.
2. Read from the histogram directly information regarding frequencies for particular class intervals, or vice versa.
3. Use mathematical operations and data from the histogram to calculate information not provided directly from the histogram.

EXAMPLE Use the histogram for credit-hour loads (Fig. 7–11) to find:

(a) How many students carried 5–9 credit hours?

The bar for 5–9 credit hours shows 7 students.

(b) How many students carried less than 15 hours?

$3 + 7 + 4 =$ **14 students**   Three class intervals show students that carried less than 15 hours, so we add.

(c) From the graph determine the number of students in the study.

$3 + 7 + 4 + 2 =$ **16 students**   Add the number of students in each category.

(d) Find the percent of students that carried 10–14 credit hours.

$\dfrac{4}{16} = \dfrac{1}{4} = 0.25 =$ **25%**   Since 4 students carried 10–14 credit hours and there are 16 students (scores) in the study, the percent is calculated using 16 as the base and 4 as the percentage.

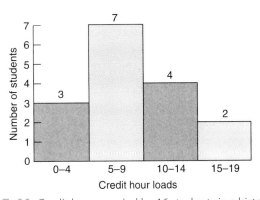

Figure 7–11 Credit hours carried by 16 students in a history class.

(e) How many students carried 10–14 hours?

4 carried 10–14 hours.   Read the value of the height of the bar from the vertical scale.

(f) How many students carried more than 14 hours?

2 carried more than 14 hours.

(g) What is the ratio of students carrying 0–4 hours to those carrying 5–9 hours?

$$\frac{3 \text{ students with } 0-4 \text{ hours}}{7 \text{ students with } 5-9 \text{ hours}} = \frac{3}{7}$$

The ratio is $\frac{3}{7}$.

EXAMPLE   A hospital department bases employee vacation leave on years of service. Employees fall into four categories: 0–2 years, eight employees; 3–5 years, six employees; 6–8 years, four employees; 9–11 years, two employees. Make a histogram showing this information.

The years of service are already arranged in class intervals and so the intervals may be used as given. The class frequency or number of employees in each interval is also provided and so the frequencies may also be used as given. Therefore, it is *not* necessary to make a frequency distribution. The class intervals form the horizontal scale and the frequencies form the vertical scale (see Fig. 7–12).

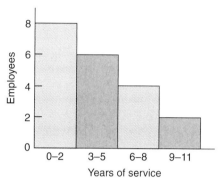

Figure 7–12 Years of service of 20 employees.

3 Make and Interpret Frequency Polygons.

A *frequency polygon* is a line graph that joins the midpoints of the bars of a histogram and begins with the beginning point of the first bar and ends with the ending point of the last bar.

To make a frequency polygon:

1. Begin with a histogram and identify each class midpoint with a dot (point) on the top of each bar.
2. Connect the points to make a line graph. Connect the end points of the line graph to the horizontal axis at the smallest value of the first interval and the largest value of the last interval on the right.
3. Remove the bars and label the horizontal scale with the class midpoints.

EXAMPLE   Make a frequency polygon using the data in the frequency distribution that illustrates the number of years of service for 20 hospital employees.

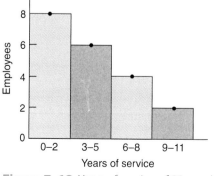

Identify each class midpoint with a point. Use the histogram from the previous example.

Figure 7–13 Years of service of 20 employees.

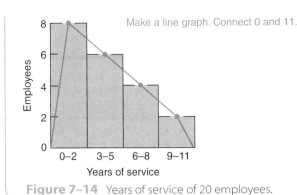

Make a line graph. Connect 0 and 11.

Figure 7–14 Years of service of 20 employees.

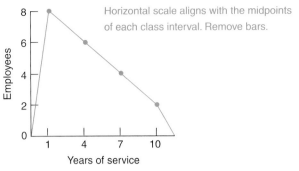

Horizontal scale aligns with the midpoints of each class interval. Remove bars.

Figure 7–15 Years of service of 20 employees.

EXAMPLE Use the frequency polygon to answer the questions.

(a) What can you say about the relationship between the years of service and the number of employees?

As the number of years of service increases the number of employees decreases.

(b) How many employees have an average of 4 years of service?

6 employees.

(c) What is the average number of years of service for employees in the 6 to 8 years of service bracket?

7 is the midpoint or average for the interval.

SELF-STUDY EXERCISES 7–2

1 Use Table 7–3 to answer the questions. The frequency distribution shows the ages of 25 college students in a landscaping class.

1. How many students are 22 or younger?
2. How many students are older than 34?
3. What is the ratio of the number of students 38–40 to the number of students 17–19?
4. What is the ratio of the smallest class frequency to the largest class frequency?
5. What percent of the total class are students age 17–19?

Table 7–3 Frequency Distribution of 25 Ages

Class Interval	Midpoint	Tally	Class Frequency
38–40	39	/	1
35–37	36	/	1
32–34	33	//	2
29–31	30	///	3
26–28	27	//	2
23–25	24	//// /	6
20–22	21	//// //	7
17–19	18	///	3

6. What percent of the total class are students age 20–22?
7. What two age groups make up the smallest number of students in the class?
8. What two age groups make up the largest number of students in the class?
9. How many students are over age 28?
10. How many students are under age 26?

Use the given hourly pay rates (rounded to the nearest whole dollar) for 33 support employees in a private college to complete a frequency distribution using the format shown in Table 7–4.

Table 7–4   Pay Rates of 33 Support Employees

	Class Interval	Midpoint	Tally	Class Frequency
11.	$14–16	_____	____	_____
12.	$11–13	_____	____	_____
13.	$8–10	_____	____	_____
14.	$5–7	_____	____	_____

$6 $6 $10 $7 $6 $6 $6
$6 $7 $7 $8 $8 $6 $6
$11 $10 $7 $11 $8 $16 $6
$6 $9 $6 $7 $9 $6 $6
$12 $13 $7 $15 $5

Use the given 40 test scores of two physics classes to complete a frequency distribution using the format in Table 7–5.

Table 7–5   Test Scores of 40 Physics Students

	Class Interval	Midpoint	Tally	Class Frequency
15.	91–95	_____	____	_____
16.	86–90	_____	____	_____
17.	81–85	_____	____	_____
18.	76–80	_____	____	_____
19.	71–75	_____	____	_____
20.	66–70	_____	____	_____
21.	61–65	_____	____	_____
22.	56–60	_____		

57 91 76 89 82 59 72 88
76 84 67 59 77 66 56 76
77 84 85 79 69 88 75 58
85 65 67 66 93 83 69 81
80 64 78 76 72 90 79 90

23. Students recorded the number of hours they studied each week as: 3, 15, 18, 0, 2, 9, 12, 16, 7, 8, 5, 10, 14, 9, 7, 3, 4, 14, 17, 16, 6, 4, 8, 11, 14, 13, 10, 15, 16, 5.

Create a grouped frequency distribution with four classes beginning with the class 0–4. Show the class interval, midpoint, tally, and class frequency for each class interval.

24. The number of animals at the city animal shelter varies daily. Use the data to make a frequency distribution with five classes beginning with the class 11–20. Show the class interval, midpoint, tally, and class frequency for each class interval: 22, 31, 32, 27, 29, 16, 12, 18, 21, 30, 46, 52, 43, 51, 42, 26, 42, 17, 19, 25.

2 Complete the following.

25. Use the information from Exercises 11–14 to make a histogram. Use the reference lines in Fig. 7–16 as guides.
26. Use the information from Exercises 15–22 to make a histogram.
27. Use the data in Table 7–3 to make a histogram.
28. Use the data in Exercise 23 to make a histogram.
29. Use the data in Exercise 24 to make a histogram.

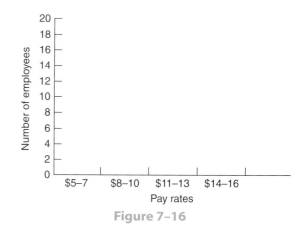

Figure 7–16

30. Use the information from Exercises 11–14 to make a frequency polygon. Use the reference lines in Fig. 7–16 as guides.
32. Use the data in Table 7–3 to make a frequency polygon.
34. Use the data from Exercises 24 and 29 to make a frequency polygon.

31. Use the information from Exercises 15–22 to make a frequency polygon.

33. Use the data from Exercises 23 and 28 to make a frequency polygon.

7–3 | *Finding Statistical Measures*

Learning Outcomes
1 Find the arithmetic mean or arithmetic average.
2 Find the median and the mode.
3 Find the range and the standard deviation.

In this age of information explosion we have massive amounts of data available to us. *Data* are facts or information from which conclusions may be drawn. To use these data effectively in the decision-making process, we need to examine the data and summarize key trends and characteristics. This summary is generally in the form of statistical measurements. A *statistical measurement* or *statistic* is a standardized, meaningful measure of a set of data that reveals a certain feature or characteristic of the data. Such statistics are called *descriptive statistics*.

1 Find the Arithmetic Mean or Arithmetic Average.

An *average* is an approximate number that is a central value of a set of data. The most common average is the arithmetic mean or arithmetic average. The *arithmetic mean* or *mean* is the sum of the quantities in the data set divided by the number of quantities.

Symbols can be used to write the procedures for statistical measures. The formula in symbols for the mean for a set of data is

$$\bar{x} = \frac{\Sigma x_i}{n}$$

The formula is read as "the mean $\bar{x}$ (read "x-bar") equals the sum Σ of each value of x (read "x sub i") divided by the number of values n. The Greek capital letter *sigma*, Σ, is a summation symbol and indicates the addition of a set of values. The notation x_i identifies each data value with a subscript: x_1 is the first value, x_2 is the second value, and x_n is the *n*th value.

To find the arithmetic mean or arithmetic average:

1. Add the quantities.
   Σx_i

2. Divide the sum by the number of quantities.

Symbolically, $\bar{x} = \dfrac{\Sigma x_i}{n}$, where $\bar{x}$ = arithmetic mean; Σ means sum;

x_i represents each x value;
n = number of values.

Find the mean of 22, 31, and 37.
$x_1 = 22, x_2 = 31, x_3 = 37$
$\Sigma x_i = 22 + 31 + 37$
$\Sigma x_i = 90$
$\bar{x} = \dfrac{\Sigma x_i}{n} = \dfrac{90}{3}$
$\bar{x} = 30$

EXAMPLE   Find the average or mean of each set of quantities.

(a) Pulse rates: 68, 84, 76, 72, 80

There are five pulse rates, so we find their sum and divide by 5:

$$\bar{x} = \frac{\Sigma x_i}{n} = \frac{68 + 84 + 76 + 72 + 80}{5} = \frac{380}{5} = 76$$

The mean pulse rate is 76.

(b) Pounds: 21, 33, 12.5, 35.2 (to the nearest hundredth)

There are four weights, so we find their sum and divide by 4.

$$\bar{x} = \frac{\Sigma x_i}{n} = \frac{21 + 33 + 12.5 + 35.2}{4} = \frac{101.7}{4} = 25.425$$

The mean weight is 25.43 lb (to the nearest hundredth).

EXAMPLE   An automobile used 41 gal of regular gasoline on a trip of 876 miles. What was the average miles per gallon (mpg) to the nearest tenth?

The number of miles traveled on the 41 gal of gasoline is 876 and represents the total miles. Divide the total miles by 41.

$$\frac{876 \text{ mi}}{41 \text{ gal}} = 21.4 \frac{\text{mi}}{\text{gal}} \qquad \text{To nearest tenth.}$$

The car averaged 21.4 mpg on the trip.

EXAMPLE   Table 7–6 lists the final grades of a horticulture student. Find the student's QPA (quality point average) to the nearest hundredth based on a 4-point system.

To find the QPA for a term, the quality points for the letter grade of each course are multiplied by the credit hours of each course to obtain the total quality points for each course. This total is divided by the total credit hours earned. The quality points awarded are A = 4, B = 3, and C = 2.

Table 7–6   Quality Points for Final Grades

Subject	Grade	Credit Hours		Quality Points per Hour		Total Points
Algebra 101	B	3	×	3	=	9
Spray chemicals 102	A	3	×	4	=	12
Landscape 301	A	3	×	4	=	12
English 101	C	4	×	2	=	8
		13 hours				41 points

$$\frac{41}{13} = 3.153846154 \text{ or } 3.15 \qquad \text{To nearest hundredth.}$$

Thus, the student's QPA for the term is 3.15.

EXAMPLE A community college student has the following grades in Physics 101: 73, 84, 80, 62, and 70. What grade is needed on the last test for the student to get a C on the final grade or a 75 average?

One way to find the needed grade is to assume that each grade is 75 for the 6 tests:

$$\frac{75 + 75 + 75 + 75 + 75 + 75}{6} = \frac{450}{6} = 75$$ The sum of the six grades is 450.

Then a total of 450 points are needed on six tests to have an average of 75.

$73 + 84 + 80 + 62 + 70 = 369$ Add the five test grades.

$450 - 369 = 81$ Find the difference between the sum of the five grades and 450.

The student must earn a score of 81 or higher on the last test to have an average of 75 or higher.

$$\text{Check:} \quad \frac{73 + 84 + 80 + 62 + 70 + 81}{6} = \frac{450}{6} = 75$$

2 **Find the Median and the Mode.**

Besides the mean or arithmetic average, we also use the *median* and the *mode* to describe groups of data. These are measures of *central tendency*. Measures of central tendency are descriptive statistics that determine the center of a data set from a different perspective.

The *median* is the middle value when the data values are arranged in order of size. The median is frequently used for data relating to annual incomes, housing price ranges, etc.

To find the median:

1. Arrange the values in order of size, either smaller to larger or larger to smaller.
2. If the number of values is odd, the median is the middle value.
3. If the number of values is even, the median is the average of the two middle values.

EXAMPLE A TPR chart shows a patient's *temperature, pulse rate,* and *respiration rate.* The following pulse rates were recorded on a TPR chart: 68, 88, 76, 64, 72. What is the patient's median pulse rate?

In order of size: 88
                      76
                      72   ← median or middle value in an odd number of values
                      68
                      64

The median pulse rate is 72.

EXAMPLE   The following temperatures were recorded: 56°, 48°, 66°, and 62°. What is the median temperature?

The number of temperatures is even, so we find the average of the two middle values.

$$\text{In order of size} \quad \left. \begin{matrix} 48 \\ 56 \\ 62 \\ 66 \end{matrix} \right\} \quad \frac{56 + 62}{2} = \frac{118}{2} = 59$$

Thus, the median temperature is 59°.

The *mode* is the most frequently occurring value in the data set.

To find the mode:

1. Identify the value or values that occur with the greatest frequency as the mode.
2. If no value occurs more than another value, there is *no mode* for the data set.
3. If more than one value occurs with the greatest frequency, the modes of the data set are the values that have the greatest frequency.

EXAMPLE   The hourly pay rates at a local fast-food restaurant are as follows: cooks, $7.50; servers, $7.15; bussers, $7.15; dishwashers, $9.25; managers, $10.50. Find the mode.

Identify the value or values that appear most often.

The hourly pay rate of $7.15 occurs more than any other rate. It is the mode.

EXAMPLE   The daily work shifts at a mall clothing store are 4, 6, and 8 hours. Find the mode.

No shift occurs more frequently than another, so there is no mode.

3   Find the Range and the Standard Deviation.

The mean, the median, and the mode are *measures of central tendency.* Another group of statistical measures are *measures of variation or dispersion.* The variation or dispersion of a set of data may also be referred to as the *spread.* One of these measures of dispersion is the *range.* The range is the difference between the highest value and the lowest value in a set of data.

To find the range:

1. Find the highest and lowest values.
2. Find the difference between the highest and lowest values.

EXAMPLE Find the range for the data described in the example for fast-food restaurant hourly pay rates.

The high value is $10.50. The low value is $7.15.

$$\text{Range} = \$10.50 - \$7.15 = \textbf{\$3.35}$$

TIP!

Use More Than One Statistical Measure.

A common mistake when making conclusions or inferences from statistical measures is to examine only one statistic, such as the range. To obtain a complete picture of the data requires looking at more than one statistic.

Although the range gives us some information about dispersion, it does not tell us whether the highest or lowest values are typical values or extreme *outliers*. We can get a clearer picture of the data set by examining how much each data point *differs* or *deviates* from the mean.

The *deviation from the mean* of a data value is the difference between the value and the mean.

To find the deviations from the mean:

Data set: 38, 43, 45, 44.

1. Find the mean of a set of data.

$$\bar{x} = \frac{\text{Sum of data values}}{\text{Number of values}} = \frac{\Sigma x_i}{n}$$

$$\frac{38 + 43 + 45 + 44}{4} = \frac{170}{4} = 42.5$$

2. Find the amount that each data value deviates or is different from the mean.

Deviation from the mean =
Data value − Mean = $x_i - \bar{x}$

$38 - 42.5 = -4.5$ (below the mean)
$43 - 42.5 = 0.5$ (above the mean)
$45 - 42.5 = 2.5$ (above the mean)
$44 - 42.5 = 1.5$ (above the mean)

When the value is smaller than the mean, the difference is represented by a *negative* number indicating the value is *below* or less than the mean. When the value is larger than the mean, the difference is represented by a positive number indicating the value is *above* or greater than the mean. *The absolute value of the sum of the deviations below the mean should equal the sum of the deviations above the mean.* In the example in the box, only one value is below the mean, and its deviation is −4.5. Three values are above the mean, and the sum of these deviations is $0.5 + 2.5 + 1.5 = 4.5$. We say that *the sum of all deviations from the mean is zero.* This is true for all sets of data.

We have not gained any statistical insight or new information by analyzing the sum of the deviations from the mean or even by analyzing the average of the deviations.

$$\text{Average deviation} = \frac{\text{Sum of deviations}}{\text{Number of values}} = \frac{0}{n} = 0$$

SE

1 Find

1. 12, 14
4. 85, 68
7. 32°F,
10. $32, $
13. Resp

7–3 Finding Statistical Measures

Solve the problems.

15. A baseball player batted 276 home runs over a 16-year period. What was the average number of home runs per year to the nearest tenth?

16. Noel Womack scored 83, 96, 86, 92, and 93 in English 101. What must he score on the last test to earn an average score of 92?

17. An automobile used 32 gal of regular gasoline on a 786-mi trip. What was the average miles per gallon to the nearest tenth?

18. A pickup truck used 25 gal of regular gasoline on a 256-mi trip. What was the average miles per gallon to the nearest tenth?

2 Find the median for each data set.

19. 32, 56, 21, 44, 87
20. 78, 23, 56, 43, 38
21. 12, 21, 14, 18, 15, 16
22. 21, 33, 18, 32, 19, 44
23. $22, $35, $45, $30, $29
24. $66, $54, $76, $55, $69

25. The following hourly pay rates are used at fast-food restaurants: cooks, $8.15; servers, $8.25; bussers, $8.15; dishwashers, $8.25; managers, $11.25. Find the median pay rate.

26. The following hourly pay rates are used at a locally owned store: clerks, $8.45; bookkeepers, $9.25; operators, $8.15; assistant managers, $10.95. Find the median pay rate.

Find the mode for each data set.

27. 2, 4, 6, 2, 8, 2
28. 5, 12, 5, 5, 20
29. 21, 32, 67, 34, 23, 22
30. 32, 45, 41, 23, 56, 77
31. $56, $67, $32, $78, $67, $20, $67, $56
32. $32, $87, $67, $32, $32, $87, $77, $22

33. These weekend work shifts are in effect at a mall clothing store: 4 hours in A.M., 6 hours in P.M., 4 hours in P.M. Find the mode for the number of hours.

34. These special prices are in effect at a fast-food restaurant: $1.75, hamburgers; $1.97, hot ham sandwiches; $2.38, chicken fillet sandwiches; $1.97, roast beef sandwiches. Find the mode.

3 Find the range for each data set.

35. 22, 36, 41, 41, 17
36. 28, 33, 36, 13, 28
37. 10, 23, 12, 17, 13, 16
38. 22, 23, 18, 32, 29, 14
39. $25, $15, $25, $40, $19
40. $36, $44, $20, $52, $10
41. 23°F, 37°F, 29°F, 54°F, 46°F, 71°F, 67°F

Find the standard deviation for each data set. Round to the nearest hundredth.

42. 12, 14, 16, 18, 20
43. 68, 54, 73, 69
44. 32°F, 41°F, 54°F
45. $27, $32, $65, $29, $21
46. Respiration rates: 16, 24, 20
47. Pulse rates: 68, 84, 76

7–4 | *Counting Techniques and Simple Probabilities*

Learning Outcomes

1 Count the number of ways objects in a set can be arranged.

2 Determine the probability of an event occurring if an activity is repeated over and over.

1 Count the Number of Ways Objects in a Set Can Be Arranged.

A *set* is a well-defined group of objects or *elements*. The numbers 2, 4, 6, 8, and 10 can be a set of even numbers between 1 and 12. Women, men, and children can be a set of people. A, B, and C can be a set of the first three capital letters in the alphabet.

Counting, in this section, means determining all the possible ways the elements in a set can be arranged. One way to count is to *list* all possible arrangements and then count the number of arrangements.

332

List and count the ways the elements in the set *A*, *B*, and *C* can be arranged.

A is first, 2 choices	*B* is first, 2 choices	*C* is first, 2 choices
A BC	*B AC*	*C AB*
A CB	*B CA*	*C BA*

Therefore, *A*, *B*, and *C* can be arranged in six ways. Each of these arrangements can also be called a set.

If more than three elements are in the set, the procedure becomes more challenging. It may be helpful to use a *tree diagram*, which allows each new set of possibilities to branch out from a previous possibility.

To make a tree diagram of possible arrangements of items in a set:

1. List the choices for putting an item in the first slot.
2. From each choice, list the remaining choices for the second slot (first branch).
3. From the end of each branch, make a branch to the remaining choices for the third slot.
4. Continue the process until there are no remaining choices.
5. To list an arrangement, start with the first slot and follow a branch to its end and record the choices.
6. To count the total number of possible arrangements, count the number of branch ends in the last slot.

EXAMPLE Make a tree diagram and count the number of ways the elements in the set containing letters *W*, *X*, *Y*, and *Z* can be arranged.

4 choices	**3 choices**	**2 choices**	**1 choice**
		Y	*Z* = WXYZ
	X	*Z*	*Y* = WXZY
		Z	*X* = WYZX
W	*Y*	*X*	*Z* = WYXZ
		X	*Y* = WZXY
	Z	*Y*	*X* = WZYX
		Y	*Z* = XWYZ
	W	*Z*	*Y* = XWZY
		W	*Z* = XYWZ
X	*Y*	*Z*	*W* = XYZW
		W	*Y* = XZWY
	Z	*Y*	*W* = XZYW
		X	*Z* = YWXZ
	W	*Z*	*X* = YWZX
		W	*Z* = YXWZ
Y	*X*	*Z*	*W* = YXZW
		X	*W* = YZXW
	Z	*W*	*X* = YZWX
		X	*Y* = ZWXY
	W	*Y*	*X* = ZWYX
		W	*Y* = ZXWY
Z	*X*	*Y*	*W* = ZXYW
		W	*X* = ZYWX
	Y	*X*	*W* = ZYXW

24 possibilities or sets

As evident in the previous example, the greater the number of elements in a set, the greater the complexity and time required to list all possible arrangements. We can use logic and common sense to obtain a count of the possible arrangements of a set of elements.

There are four possibilities for the first letter, *W*, *X*, *Y*, or *Z*. For each of the four possible first letters, we have three choices for second letters. Now, we have 4 × 3 or 12 possibilities. For each of the 12 possibilities two choices remain for the third letter: 12 × 2 = 24. Then, for each of the 24 three-letter combinations, only one choice is left: 24 × 1 = 24. So, there are a total of 24 possibilities.

Another way to visualize this is to think of drawing letters from a bag or bowl (Fig. 7–18).

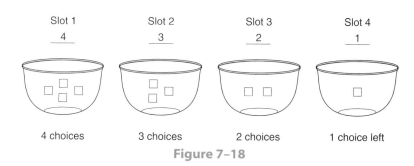

Figure 7–18

By multiplying the number of possible choices for each position, we can determine the total number of possibilities without listing them: 4 · 3 · 2 · 1 = 24.

To determine the number of choices for arranging a specified number of items:

1. Determine the number of slots to be filled.
2. Determine the number of choices for each slot.
3. Multiply the numbers from Step 2.

EXAMPLE   A coin is tossed three times. With each toss, the coin falls heads up or tails up. How many possible outcomes of heads and tails are there with three tosses of the coin?

There are three tosses. Each toss has only two possibilities, heads or tails; that is,

1st toss	2nd toss	3rd toss
2 possibilities	2 possibilities	2 possibilities

By multiplying the number of possible outcomes for each toss, we get 2 · 2 · 2 = 8. **So there are 8 possible outcomes.**

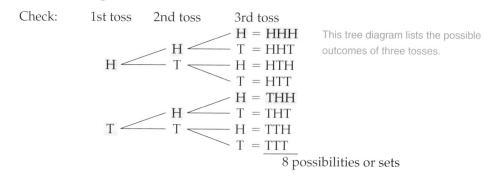

This tree diagram lists the possible outcomes of three tosses.

8 possibilities or sets

To Repeat or Not Repeat

In some situations, as in arranging W, X, Y, and Z, once a choice or selection is made, that choice cannot be repeated. In these situations the number of choices decreases with each selection.

$$\text{Number of possible arrangements of } W, X, Y, Z = 4 \cdot 3 \cdot 2 \cdot 1 = 24$$

In other situations, as in tossing a coin, every coin toss can result in a head or a tail. The results of any coin toss can repeat the result of a previous toss.

$$\text{Number of possible results of three coin tosses } = 2 \cdot 2 \cdot 2 = 2^3 = 8$$

When counting the number of possible outcomes, *first determine if repeats are allowed or not.*

EXAMPLE Henry has three ties: a red tie, a blue tie, and a green tie. He also has three shirts: a white shirt, a pink shirt, and a yellow shirt. How many sets of shirts and ties are possible?

If we start with the shirts, there are three possibilities (white, pink, and yellow). For each shirt, there are three possible ties (red, blue, and green). **So we have $3 \cdot 3 = 9$ possible outcomes.**

EXAMPLE Given the digits 1, 2, 3, 4, 5, and 6, how many three-digit numbers can be made without repeating a digit?

The numbers are to contain three digits, so there are three positions to fill. We have six digits to work with. The first digit can be one of six. For each of these six digits, there are five possible second digits. For each of the five second digits, there are four possible third digits.

Positions:	1st	2nd	3rd
Possibilities:	6	5	4

By multiplying the possibilities for each position, we get $6 \cdot 5 \cdot 4 = 120$ **possible outcomes or ways to make a three-digit number.**

2 Determine the Probability of an Event Occurring If an Activity Is Repeated Over and Over.

Probability means the chance of an event occurring if an activity is repeated over and over. The probability of an event occurring is expressed as a ratio or a percent.

Weather forecasters use percents when they forecast a 60% chance of rain or a 20% chance of snow. This text will use ratios like $\frac{3}{5}$ for 3 chances out of 5 or $\frac{2}{3}$ for 2 chances out of 3. The decimal equivalent of a ratio can also be used to express a probability: $\frac{3}{5} = 0.6$.

When a coin is tossed, two outcomes are possible, heads or tails. But only one side will be up. The probability of tossing heads is 1 out of 2, $\frac{1}{2}$, 0.5, or 50%.

8. How much are overhead and materials together?
9. What percent of the total cost is the profit?
10. What is the ratio of labor to total cost?
11. What is the ratio of the tariff to total cost?

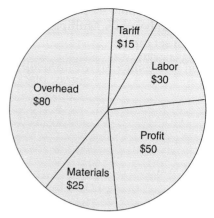

Figure 7–31 Manufacturing costs of a $200 electronic game.

Use the bar graph in Fig. 7–32 to answer the Questions 12–17 about the academic-year starting salaries of women and men college professors in various academic departments of a college.

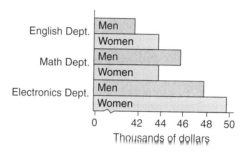

Figure 7–32 Salaries of women and men college professors.

12. In what departments do men make more than women?

13. In what departments do women make more than men?

14. What percent of women's salaries are men's salaries in the English Department (to the nearest tenth of a percent)?

15. What percent of men's salaries are women's salaries in the Electronics Department (to the nearest tenth of a percent)?

16. What is the ratio of men's salaries to women's salaries in the Math Department?

17. What is the ratio of men's salaries to women's salaries in the English Department?

18. Make a line graph to illustrate the following information about the average prices of homes in a subdivision from 1994 to 2004. Use the reference lines in Fig. 7–33 as guides.

   1994 average price: $80,000
   1999 average price: $90,000
   2004 average price: $95,000

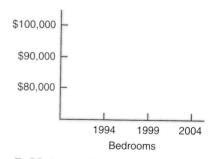

Figure 7–33 Prices of homes by number of bedrooms.

Use the frequency distribution shown in Table 7–11 to answer the Questions 19–22. The distribution shows the number of correct answers on a 30-question test in a science class.

19. How many students scored less than 16 correct?
20. How many students scored more than 25 correct?
21. What is the ratio of students scoring 6–10 correct to those scoring 21–25 correct?
22. What percent scored 21–30 correct?

Table 7–11	Frequency Distribution of Correct Answers		
Class Interval	Midpoint	Tally	Class Frequency
26–30	28	///	3
21–25	23	////	4
16–20	18	//// //	7
11–15	13	///	3
6–10	8	//	2
1–5	3	/	1

Use the format shown in Table 7–12 and the ages of 24 children in a day-care center to complete a frequency distribution.

Table 7–12	Ages of 24 Day-Care Children		
	Class Interval	Midpoint	Tally	Class Frequency
23.	4–6	_____	___	_____
24.	1–3	_____	___	_____

Ages of Children

1 1 1 1 1 1 1 2 2 2
3 3 3 3 3 4 4 4 4 5
5 5 5 6

Use the histogram in Fig. 7–34 to answer Questions 25–27 about a company's software programs and the number of employees trained to use them.

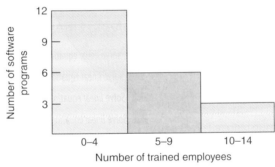

Figure 7–34 Company employees trained to use software.

25. How many employees at the most can use only three programs?
26. How many software programs does the company have?
27. How many employees does the company have?
28. Find the range, mean, median, and mode for these test scores: 77, 87, 77, 89, 70, 69, 82.
29. Find the range, mean, median, and mode for these test scores: 81, 78, 69, 75, 81, 93, 68.
30. Find the standard deviation of the data in Exercise 28.
31. Find the standard deviation of the data in Exercise 29.
32. List and count the ways the elements in the set L, M, N, and O can be arranged.
33. Rayford has two ties: a red tie and a blue tie. He also has three shirts: a white shirt, a green shirt, and a yellow shirt. How many combinations of shirts and ties are possible?
34. Millie has in her purse three green eye shadows, four white eye shadows, and two black eye shadows. What is the probability of her pulling out a black eye shadow?
35. An envelope contains the names of two men and three women to be interviewed for a promotion at Washington's Landscape Service. The interviewer draws names to determine the order of the interviews. What is the probability of drawing a woman's name first?
36. If a small boy has one red marble and three yellow marbles in his pocket, what is the probability of his pulling out the red one on the first try?

Unknown facts	The cost of each text.
Known facts	The notebook cost $3. The total cost was $198. The digital scanning text cost twice as much as the history text and notebook combined.
Relationships	Let x = the cost of the history text. The notebook cost $3. The digital scanning text cost 2 times the cost of the history text and the notebook, that is, $2(x + 3)$. The history text plus the digital scanning text plus the notebook equals $198. Therefore, the equation is $x + 2(x + 3) + 3 = 198$.
Estimation	If we ignore the notebook, the digital scanning text cost approximately twice the cost of the history text, so the ratio is 2 to 1. That is, the cost of the digital scanning text was about $\frac{2}{3}$ of the total bill and the history text was about $\frac{1}{3}$ ($\frac{1}{3}$ of $200 $\approx$ $70). So the digital scanning text was around $140 and the history text was around $70.

Calculations

$$x + 2(x + 3) + 3 = 198 \qquad \text{Distribute.}$$
$$x + 2x + 6 + 3 = 198 \qquad \text{Combine like terms.}$$
$$3x + 9 = 198 \qquad \text{Sort.}$$
$$3x = 198 - 9 \qquad \text{Combine.}$$
$$3x = 189 \qquad \text{Divide.}$$
$$\frac{3x}{3} = \frac{189}{3}$$
$$x = 63$$

Interpretation

The history text cost $63. The digital scanning text cost $2(x + 3)$. Thus, $2(63 + 3) = 126 + 6 = 132.

Check:

$$x + 2(x + 3) + 3 = 198 \qquad \text{Substitute.}$$
$$63 + 2(63 + 3) + 3 = 198 \qquad \text{Distribute.}$$
$$63 + 126 + 6 + 3 = 198 \qquad \text{Combine.}$$
$$198 \quad 198$$

EXAMPLE A horticulturist marks off a 32-m-wide rectangular nursery plot with 158 m of fencing. Because the plants must be properly spaced, she needs to know the length of the plot. Find the length in meters.

Unknown facts	The length of the nursery plot in meters.
Known facts	The nursery plot is rectangular. The width is 32 m. The fencing gives us the perimeter of 158 m.
Relationships	Using the formula for the perimeter of a rectangle, we know that the perimeter is twice the sum of the length and width, or $p = 2(l + w)$.
Estimation	The length must be more than the width of 32 m and less than half the perimeter, that is, less than 79 m.

Calculations

$$p = 2(l + w) \qquad \text{Substitute in the formula.}$$
$$158 = 2(l + 32) \qquad \text{Distribute.}$$
$$158 = 2(l) + 2(32)$$
$$158 = 2l + 64 \qquad \text{Sort.}$$
$$158 - 64 = 2l \qquad \text{Combine.}$$
$$94 = 2l \qquad \text{Divide.}$$
$$\frac{94}{2} = \frac{2l}{2}$$
$$47 = l$$

Interpretation

Thus, the length of the nursery plot is 47 m.

Chapter 8 / Linear Equations

Check:

$$p = 2(l + w) \qquad \text{Substitute.}$$
$$158 = 2(47 + 32) \qquad \text{Distribute.}$$
$$158 = 94 + 64 \qquad \text{Combine.}$$
$$158 = 158$$

EXAMPLE A machine part weighs 2.7 kg and is to be shipped in a carton weighing x kg. If the total weight of three packaged machine parts is 9.3 kg, how much does each carton weigh? (See Fig. 8–3.)

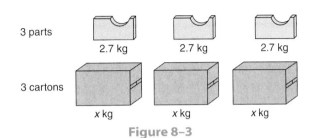

Figure 8–3

Unknown facts   The weight of each empty carton.

Known facts   There are three parts and three cartons. Each part weighs 2.7 kg. The three packaged parts (in cartons) weigh a total of 9.3 kg.

Relationships   Let x equal the weight of one empty carton. Each packaged part weighs the sum of the part (2.7 kg) plus the weight of a carton (x kg). The sum of the three packaged parts equals 9.3 kg; that is, $3(2.7 + x) = 9.3$ or $(2.7 + x) + (2.7 + x) + (2.7 + x) = 9.3$.

Estimation   Three packaged parts weigh 9.3 kg, and $\frac{9.3}{3} = 3.1$ kg per packaged carton. Since one part weighs 2.7 kg, then the carton weighs well under 1 kg.

Calculations

$$3(2.7 + x) = 9.3 \qquad \text{Distribute.}$$
$$3(2.7) + 3(x) = 9.3 \qquad \text{Make calculations.}$$
$$8.1 + 3x = 9.3 \qquad \text{Sort.}$$
$$3x = 9.3 - 8.1 \qquad \text{Combine.}$$
$$3x = 1.2 \qquad \text{Divide.}$$
$$\frac{3x}{3} = \frac{1.2}{3}$$
$$x = 0.4$$

Interpretation   Each carton weighs 0.4 kg.

Check:

$$3(2.7 + x) = 9.3 \qquad \text{Substitute.}$$
$$3(2.7 + 0.4) = 9.3 \qquad \text{Make calculations.}$$
$$3(3.1) = 9.3$$
$$9.3 = 9.3$$

Learning Strategy   *Additional Problem-Solving Strategies.*

- Read the problem carefully. Read it several times and phrase by phrase.
- Understand all the words in the problem.
- Analyze the problem:
  What are you asked to find?
  What facts are given?
  What facts are implied?

- Visualize the problem.
- State the conditions or relationships of the problem "symbolically."
- Examine the options.
- Develop a *plan* for solving the problem.
- Write your *plan* symbolically; that is, write an equation.
- Anticipate the characteristics of a reasonable solution.
- Solve the equation.
- Verify your answer with the conditions of the problem.

There are many different problem-solving plans. Additional strategies can be considered.

Spreadsheets.

Information is often displayed in a table of rows and columns called a *spreadsheet*. These rows and columns may show the results of calculations, such as totals or percents, in addition to the original data. Many software programs such as Excel help you build an *electronic spreadsheet* and draw appropriate graphs using formulas and equations. The calculations are made automatically once the formulas are defined in the spreadsheet.

In a spreadsheet key information is placed in a specific location called an *address*, and formulas are developed using these addresses as the variables. Spreadsheet programs are useful because key information can be changed while retaining the basic formulas of the spreadsheet. This process allows one to see quickly how various changes in the key data affect the results.

Let's examine a spreadsheet for the following situation. *The 7th Inning Sports Memorabilia Shop is developing an annual operating budget. The budget categories and the projected amount of expense for each category are shown in the spreadsheet in Fig. 8–4. The spreadsheet can be used to find the total operating budget for the year and the percent of total annual budget for each category in the projected annual budget. Formulas must be developed to calculate this information.*

The spreadsheet program labels columns with letters like A, B, and C, and the rows with numbers. We'll use row 1 for the title of the spreadsheet, row 3 to label the columns of data, and column A to label the rows of data. Each position on the spreadsheet is a *cell*, and the program identifies each cell by its column letter and row number. For example, the amount budgeted for taxes and insurance is in cell B9 and is $14,000.00.

TIP!

Spreadsheet Addresses.

A spreadsheet address is an individual cell that is the intersection of a column and a row. This address is generally given in two parts: column and row.

Address	Interpretation
C 8	Column C and Row 8
J 30	Column J and Row 30
AA 5	Column AA (follows Column Z) and Row 5

Now we develop formulas to make the calculations. Each spreadsheet program gives various shortcuts for writing formulas and formats for giving instructions unique to that program; however, we will write the basic concepts used, and add program-specific conventions as appropriate.

	A	B	C
1	The 7th Inning Budgeted Operating Expenses		
2			
3	Expense	Budget Amount	Percent of Total Budget
4			
5	Salaries	$42,000.00	
6	Rent	$36,000.00	
7	Depreciation	$9,000.00	
8	Utilities and Phone	$14,500.00	
9	Taxes and Insurance	$14,000.00	
10	Advertising	$3,500.00	
11	Purchases	$120,000.00	
12	Other	$500.00	
13			
14	Total		

Figure 8–4

EXAMPLE Write the formulas and make the calculations to complete the Fig. 8–4 spreadsheet.

To find the total budget to be placed in cell B14, we need to add the amounts in cells B5–B12. To do this, we give the addresses of the cells to be added in a formula: B14 = B5 + B6 + B7 + B8 + B9 + B10 + B11 + B12.

The percent of the total budget is calculated by dividing the specific amount by the total budget and then multiplying by 100. We will write a formula for each line of data. Most programs use an asterisk (*) to show multiplication, and a forward slash (/) to show division.

C5 = B5/B14*100	C6 = B6/B14*100	C7 = B7/B14*100
C8 = B8/B14*100	C9 = B9/B14*100	C10 = B10/B14*100
C11 = B11/B14*100	C12 = B12/B14*100	

There are two ways to determine the value for cell C14. If we use the percent method, the percentage and the base are the same amount, so the total percent is 100%. To build in a *check* against the spreadsheet formulas, however, it is advisable to find the total percent by adding the calculated percents. It is easy to make a typing error in the formulas or to place the formula in the wrong cell. The total should be 100% or extremely close. There may be a small discrepancy due to the effects of rounding. C14 = C5 + C6 + C7 + C8 + C9 + C10 + C11 + C12. The spreadsheet with the completed calculations is shown in Fig. 8–5.

	A	B	C
1	The 7th Inning Budgeted Operating Expenses		
2			
3	Expense	Budget Amount	Percent of Total Budget
4			
5	Salaries	$42,000.00	17.5
6	Rent	$36,000.00	15.0
7	Depreciation	$9,000.00	3.8
8	Utilities and Phone	$14,500.00	6.1
9	Taxes and Insurance	$14,000.00	5.8
10	Advertising	$3,500.00	1.5
11	Purchases	$120,000.00	50.1
12	Other	$500.00	0.2
13			
14	Total	$239,500.00	100.0

Figure 8–5

SELF-STUDY EXERCISES 8–4

1 Solve.

1. $4(x - 5) = 8$
2. $3(y - 7) = -15$
3. $6(x + 3) = -12$
4. $5(x + 7) = 25$
5. $-3(x + 8) = -39$
6. $-2(x - 7) = 26$
7. $3(2x - 1) = 3$
8. $4(x - 1) = 28$
9. $-4(-3x + 1) = 8$
10. $-9(2x - 3) = 27$
11. $7(x - 3) = 7x + 21$
12. $5x + 15 = 5(x + 3)$
13. $8(x - 5) = 8x + 12$
14. $6x + 7 = 3(2x + 4)$
15. $2(3x - 5) = 7x - 12$
16. $5(3x - 7) + 2 = 7x + 7$
17. $2x - 3 + 6x = 4(3x - 2) + 3$
18. $7(x - 1) = 2(x + 3) + 4(x - 7)$
19. $3(x - 4) - 2(3x - 1) = 17$
20. $6x - 4 + 2x = -(x - 5)$
21. $8x - (2x - 9) = -3$
22. $6 - 2(3x - 1) = 14$
23. $12 - 5(2x - 3) = -5$
24. $8 - 3(x - 2) = x - 6$

Write the statements as equations and solve.

25. The sum of x and 4 equals 12. Find x.

26. 4 less than 2 times a number is 6. Find the number.

27. Three parts totaling 27 lb are packaged for shipping. Two parts weigh the same. The third part weighs 3 lb less than each of the two equal parts. Find the weight of each part.

28. How many gallons of water must be added to 24 gal of pure cleaner to make 60 gal of diluted cleaner?

29. A wet casting weighing 4.03 kg weighs 3.97 kg after drying. Write and solve an algebraic equation to find the weight loss due to drying.

30. A plumber needs 3 times as much perforated pipe as solid pipe to lay a drain line 400 ft long. How much of each type of pipe is needed?

31. An engineering student purchased a new circuits text, a used Spanish text, and a graphing calculator. She remembered that the calculator cost twice as much as the Spanish book and that the circuits text cost $70.00. The total before tax was $235.00. What was the cost of the calculator and the Spanish text?

32. Mary Jefferson purchased a home on a square lot 150 feet on each side. She wants to enclose the entire lot with a cedar fence. One estimate for the job was for $14.00 per linear foot. How much will the fence cost?

33. Phil Chu, owner of Chu's Landscape Service, knows that one of his fertilizer tanks holds twice as many gallons of liquid fertilizer as a second tank. The two tanks together hold 325 gal. If both tanks are filled to capacity, how many gallons of fertilizer does each tank hold?

34. Carolina Villa hired a stone mason to build a triangular flower bed at one end of her patio. She needs enough mulch to cover 60 ft², the area of the flower bed. If the flower bed has an altitude of 10 ft, how long is the base?

35. Develop formulas to complete the spreadsheet in Fig. 8–6 to show data for the actual expenses for the 7th Inning Sports Memorabilia Shop.

36. Use the formulas to complete the spreadsheet for the 7th Inning Sports Memorabilia Shop.

	A	B	C	D	E	F
1	The 7th Inning Budgeted Operating Expenses And Actual Expenses					
2						
3	Expense	Budget Amount	Percent of Total Budget	Actual Expenses	% of Actual Total Expense	% Difference from Budget
4						
5	Salaries	$45,000.00		$42,000.00		
6	Rent	$37,000.00		$36,000.00		
7	Depreciation	$12,000.00		$14,000.00		
8	Utilities and Phone	$13,000.00		$10,862.56		
9	Taxes and Insurance	$15,000.00		$13,583.29		
10	Advertising	$2,000.00		$2,847.83		
11	Purchases	$125,000.00		$132,894.64		
12	Other	$2,000.00		$1,356.35		
13						
14	Total					

Figure 8–6

8–5 | *Numerical Procedures for Solving Equations*

Learning Outcome   **1** Solve an Equation Using Numerical Procedures.

The most traditional procedure for solving equations uses symbolic manipulation. This procedure is probably the most efficient procedure, but it requires the knowledge of many different techniques for solving various types of equations. Another procedure that can be used is called the *numerical procedure for solving equations*. This procedure applies a guess and check strategy for problem solving.

To solve an equation numerically:

1. Make a table of values by selecting various values to be substituted for the variable on each side of the equation.
2. Substitute a selected value for the variable and evaluate each side of the equation.
3. Repeat Step 2 with a different selected value.
4. Compare the results from Steps 2 and 3 for both sides of the equation. Were the results from your second selected value closer to being equal or farther apart? Make future selections based on your observations.
5. Repeat Steps 2–4 for additional selected values until a value is found that makes the results of the calculations for both sides of the equation equal.

EXAMPLE   Solve the equation $2x + 3 = 15$ numerically.

Variable x	Left Side $2x + 3$	Right Side 15	Observations
1	5	15	Left and right differ by 10.
2	7	15	Left and right differ by 8. Select larger value.
3	9	15	Left and right differ by 6. Examine the pattern developing. **When the value of x is increased by 1, the value of $2x + 3$ is increased by 2.**
4	11	15	Expand the table using the pattern. $x =$ previous value + 1 = 3 + 1 = 4; $2x + 3 =$ previous value + 2 = 9 + 2 = 11.
5	13	15	$x = 4 + 1 = 5$; $2x + 3 = 11 + 2 = 13$.
6	15	15	$x = 5 + 1 = 6$; $2x + 3 = 13 + 2 = 15$. This value of x solves the equation.

The solution of the equation $2x + 3 = 15$ is 6.   Check: $2(6) + 3 = 15$
$12 + 3 = 15$
$15 = 15$

EXAMPLE   Solve the equation $2x - 4 = 3x + 2$ numerically.

Variable x	Left Side $2x - 4$	Right Side $3x + 2$	Observations
1	−2	5	$2(1) - 4 = 2 - 4 = -2$. $3(1) + 2 = 3 + 2 = 5$. Expression values differ by 7.
2	0	8	$2(2) - 4 = 4 - 4 = 0$. $3(2) + 2 = 6 + 2 = 8$. Expression values differ by 8. **Difference is increasing so select smaller values of x (one unit less than original selection of 1).**
0	−4	2	$2(0) - 4 = 0 - 4 = -4$. $3(0) + 2 = 0 + 2 = 2$. Expression values differ by 6.
−1	−6	−1	$2(-1) - 4 = -2 - 4 = -6$. $3(-1) + 2 = -3 + 2 = -1$. Expression values differ by 5. Continue to select smaller values.

Variable x	Left Side $2x - 4$	Right Side $3x + 2$	Observations
-2	-8	-4	$2(-2) - 4 = -4 - 4 = -8$. $3(-2) + 2 = -6 + 2 = -4$. Expression values differ by 4. Observe pattern developing. **As x decreases by 1, $2x - 4$ decreases by 2 and $3x + 2$ decreases by 3**.
-3	-10	-7	Use the pattern to continue. $x =$ previous value $- 1 = -2 - 1 = -3$; $2x - 4 =$ previous value $- 2 = -8 - 2 = -10$; $3x + 2 =$ previous value $- 3 = -4 - 3 = -7$.
-4	-12	-10	$x = -3 - 1 = -4$; $2x - 4 = -10 - 2 = -12$; $3x + 2 = -7 - 3 = -10$.
-5	-14	-13	$x = -4 - 1 = -5$; $2x - 4 = -12 - 2 = -14$; $3x + 2 = -10 - 3 = -13$.
-6	-16	-16	$x = -5 - 1 = -6$; $2x - 4 = -14 - 2 = -16$; $3x + 2 = -13 - 3 = -16$ (left and right are equal).

The solution of the equation $2x - 4 = 3x + 2$ is -6.

$$\text{Check:} \quad 2(-6) - 4 = 3(-6) + 2$$
$$-12 - 4 = -18 + 2$$
$$-16 = -16$$

The numerical procedure for solving equations can also be used in solving applied problems. This procedure can often be used for problems when an algebraic procedure leads to skills that you have not yet developed.

EXAMPLE   Several students contributed a total of $120 to buy a small refrigerator to use in the dorm. Later, another student joined the group, and the cost to the original group dropped $10 per person. How many students were in the original group?

First, write variable expressions to match the facts of the problem.

$s =$ number of students in original group

$\dfrac{120}{s} =$ original cost per student

$s + 1 =$ number of students in new group

$\dfrac{120}{s + 1} =$ new cost per student

Next, make a table of values for selected values of s.

Number of Original Students	Cost per s Students	Cost per $s + 1$ Students	Observations
s	$\dfrac{\$120}{s}$	$\dfrac{\$120}{s + 1}$	**New cost is $10 less per student than the original cost per student.**
1	$\dfrac{\$120}{1} = \$120$	$\dfrac{\$120}{2} = \$60$	$\$120 - \$60 = \$60$ difference, too high.
2	$\dfrac{\$120}{2} = \$60$	$\dfrac{\$120}{3} = \$40$	$\$60 - \$40 = \$20$ difference, still too high.
3	$\dfrac{\$120}{3} = \$40$	$\dfrac{\$120}{4} = \$30$	$\$40 - \$30 = \$10$ difference, matches conditions of problem.

There were three students in the original group.

To solve the preceding example using symbolic manipulation, we will need to know how to solve rational equations (Chapter 14) and quadratic equations (Chapter 15). Using the numerical procedure to solve equations allows us to solve a wider variety of equations that require many different algebraic skills.

SELF-STUDY EXERCISES 8–5

1 Solve the equations numerically.

1. $4x - 3 = 9$

x	$4x - 3$	9	Observations

2. $7x - 1 = 4x + 17$

x	$7x - 1$	$4x + 17$	Observations

3. $7x = 8x + 4$

4. $4x - 13 = 2x + 15$

5. $7(x + 2) = -6 + 2x$

6. The shorter side of a carpenter's square in the shape of an L is 3 in. shorter than the longer side. If the total length of the L-shaped tool is 28 in., what is the measure of each side of the carpenter's square?

Longer Side	Shorter Side	Observations
t	$t - 3$	Total length = 28 in.

7. Nurse Vance can prepare dosages for her patients in 30 min. If she gets help from assistants who work at the same rate and together they complete the preparation in 5 min, how many helpers did she get?

Number of Helpers	Time Required to Complete Task	Observations
h	$\dfrac{30 \text{ min}}{h+1}$	Nurse Vance + helpers complete the task in 5 min.

8. A certain fabric sells at a rate of 3 yd for $8.00. How many yards can Kristin buy for $36?

Yards of Fabric	Total Cost	Observations
y	$\dfrac{\$ 8.00}{3} \times y$	Total cost should be $36.00.

9. A group of investors purchases land in Rhode Island for $120,000. When four more investors joined the group, the cost dropped $8,000 per person for the original group. How many investors were in the original group?

Number of Original Investors	Cost per n Investors	Cost per $n + 4$ Investors	Observations
n	$\dfrac{\$120,000}{n}$	$\dfrac{\$120,000}{n + 4}$	New cost is $8,000 less per investor than the original cost per investor.

10. What are the dimensions of a rectangular room if the area is 84.5 m^2 and the room is twice as long as it is wide?

Width	Length	Area	Observations
w	$2w$	$w (2w)$	Area is 84.5 m^2

8–6 | *Function Notation*

Learning Outcomes

1 Evaluate a function.

2 Find the range of a function for a given domain.

3 Graph independent and dependent values of functions.

In the study of mathematics, you will gain valuable insights into basic principles by examining patterns. Function notation helps us identify patterns.

1 Evaluate a Function.

Suppose you wish to rent a car and the salesperson quotes a rate of $25 per day plus $4 for each 100 miles (or any part of 100) that the car is driven. You want to calculate the total rental charge for several different mileage totals. You can best examine the pattern for various mileages by using function notation.

In *function notation*, we have two types of variables, an *independent variable* and a *dependent variable*. As the words imply, one variable is dependent on the other.

In our car rental example, the total cost of the rental (dependent variable) is dependent on the number of miles the car is driven (independent variable). We use different letters to represent the two variables, such as C for the total cost of the rental and m for the miles driven. Using C and m, a function that relates these two variables is $C = \$25 + \$4m$ or $C = 25 + 4m$. To reveal patterns that develop from different values of the independent variable, we commonly arrange the information in a table (see Table 8–1).

Table 8–1	Car Rental Costs		
Exact Miles	Hundred Miles, m	Calculations, $C = 25 + 4m$	Cost, C
278	3	$C = 25 + 4(3)$ $C = 25 + 12$ $C = 37$	$37
346	4	$C = 25 + 4(4)$ $C = 25 + 16$ $C = 41$	$41
412	5	$C = 25 + 4(5)$ $C = 25 + 20$ $C = 45$	$45

A common function notation expresses the independent and dependent variables as

$$f(x) = 25 + 4x \quad \text{or} \quad C(m) = 25 + 4m$$

In the first notation, $f(x) = 25 + 4x$, a standard notation called the "function of x" is used; x represents the independent variable and $f(x)$ represents the dependent variable. Think of the values of the independent variable as the *input values*. Calculations are performed. Then the values of the dependent variable are the results of the calculations, or the *output values* (Fig. 8–7).

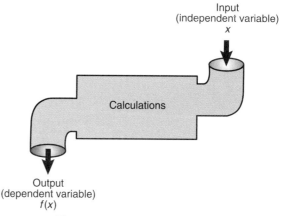

Input
(independent variable)
x

Calculations

Output
(dependent variable)
$f(x)$

Figure 8–7 Function Notation.

$$\text{Function of } x: f(x) = 25 + 4x \qquad\qquad \text{Cost function: } C(m) = 25 + 4m$$

dependent   independent   dependent   independent
variable   variable   variable   variable

In the second notation, $C(m) = 25 + 4m$, the letters serve as reminders for the meaning of the variables. That is, $C(m)$ is the dependent variable and represents the cost of the car rental based on the number of 100 miles driven. The independent variable m represents the number of hundred miles on which the cost is based. This type of function is often called a *cost function*.

Using function notation allows us to symbolically show various independent values that we want to examine.

To evaluate a function:

1. Identify or select the value of the independent variable to be used.
2. Substitute the value selected from Step 1 for the independent variable in each place that it appears in the function.
3. Perform the operations indicated by the function to find the corresponding dependent value.

EXAMPLE   In the car rental example, find $f(3)$, $f(4)$, and $f(5)$ for the function $f(x) = 25 + 4x$.

Table 8–2 Car Rental Costs

Exact Miles	Hundred Miles, x	Calculations, $f(x) = 25 + 4x$	Cost, $f(x)$	
278	3	$f(3) = 25 + 4(3)$ $f(3) = 25 + 12$ $f(3) = 37$	\$37	Substitute 3 for x.
346	4	$f(4) = 25 + 4(4)$ $f(4) = 25 + 16$ $f(4) = 41$	\$41	Substitute 4 for x.
412	5	$f(5) = 25 + 4(5)$ $f(5) = 25 + 20$ $f(5) = 45$	\$45	Substitute 5 for x.

EXAMPLE   Examine Table 8–2 and answer the questions.

(a) Without making any calculations, what would you expect the cost of the car rental to be for 194 miles?

(b) Are the rentals for 304 miles and 400 miles the same? Explain.

(a) The rate for 200 miles (from 194 miles) would be \$4 less than the rate for 300 miles, or \$8 more than the basic rate. The rate is \$37 − \$4 = \$33, or \$25 + \$8 = \$33.

(b) Both rates are calculated at 400 miles. The 4 miles over 300 miles count as a portion of the next 100 miles.

Visualize the meaning of function notation (Fig. 8–8).

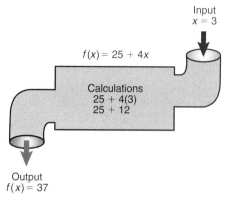

Figure 8–8 Input, calculations, output.

EXAMPLE   Evaluate the function $f(x) = x - 1$ for the given values.

(a) $f(4)$ (b) $f(8)$ (c) $f(10)$ (d) $f(-1)$

(a) $f(4) = 4 - 1$ $f(x)$ is one less than the value of x.
  $f(4) = 3$
(b) $f(8) = 8 - 1$ 1 less than 8 is 7.
  $f(8) = 7$
(c) $f(0) = 0 - 1$ 1 less than or to the left of 0 is -1.
  $f(0) = -1$
(d) $f(-1) = -1 - 1$ 1 less than or to the left of -1 is -2.
  $f(-1) = -2$

2   **Find the Range of a Function for a Given Domain.**

A function shows a relationship between independent and dependent variables. As values change for the input or independent variable, values of the output or dependent variable change accordingly.

Depending on the problem, there may be some values that cannot be considered. For example, in the car rental problem, an input value of 0 would mean that the car was driven no miles. Even if the car was driven 1 mile, the customer would be charged for 100 miles. Is this situation likely to occur? Also, can a car be driven a negative number of miles? The only appropriate values of this independent variable then are positive values.

Within the context of the car rental problem, only natural numbers, which represent the number of hundreds of miles, or fraction thereof, that the car was driven, are appropriate.

Is there a maximum number that is appropriate? Can the car be driven 2,000 miles in one day? Probably not. Thus, even though we may not know the exact maximum, we know that there is a reasonable amount for the maximum. These suitable values for the independent variable make up a set of values called the *domain* of the function. As you can see, the values that make up the domain are considered within the context of the situation (Fig. 8–9).

Once the values of the domain (input) are determined, then we can examine the possible values of the dependent variable (output). These output values make up a set of numbers called the *range*. In the car rental example, the range includes numbers

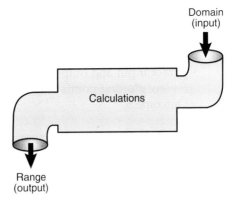

Domain
(input)

Calculations

Range
(output)

Figure 8–9 Domain—input, range—output.

like $29, $33, $37, $41, $45, $49, and so on. The upper limit of the range is determined by a reasonable maximum number of miles the car could be driven in one day.

To find the range of a function for a given domain:

1. Evaluate the function for three or more values of the domain.
2. Examine the corresponding value of the dependent variable for each value of the domain used in Step 1.
3. Identify the common characteristics of the values of the dependent variables found in Step 2.

EXAMPLE   Identify the characteristics of the range for various input values in the following.

(a) $f(x) = 3x$ for whole-number values in the domain.
(b) $f(x) = x - 5$ for any integer.

(a) $f(x) = 3x$ Substitute values of x.
$\quad f(1) = 3(1)$ $x = 1$
$\qquad = 3$
$\quad f(2) = 3(2)$ $x = 2$
$\qquad = 6$
$\quad f(3) = 3(3)$ $x = 3$
$\qquad = 9$

The output values (range) will also be whole numbers. The output values will be multiples of 3.

(b) $f(x) = x - 5$ Substitute values of x.
$\quad f(-1) = -1 - 5$ $x = -1$
$\qquad = -6$
$\quad f(0) = 0 - 5$ $x = 0$
$\qquad = -5$
$\quad f(5) = 5 - 5$ $x = 5$
$\qquad = 0$
$\quad f(6) = 6 - 5$ $x = 6$
$\qquad = 1$

The output values (range) will also be integers. Each output value will be a natural number if the input value for x is an integer greater than 5, the output value will be 0 if the input value is exactly 5, and the output value will be a negative integer if the input value is an integer less than 5.

Distributing -1 changes only the sign of each term inside the parentheses.

$$-(2x - 3) = -1(2x - 3) = -2x + 3$$

Section 8-3

1 Solve linear equations using the addition axiom (pp. 361–363).

1. Locate the variable in the equation.
2. Identify the constant that is associated with the variable by addition (or subtraction).
3. Add the opposite of the constant to both sides.

$x + 3 = 2$	Add -3 to both sides.
$x + 3 - 3 = 2 - 3$	Combine like terms.
$x + 0 = -1$	$x + 0 = x$
$x = -1$	
$9 = x - 6$	Add 6 to both sides.
$9 + 6 = x - 6 + 6$	Combine like terms.
$15 = x + 0$	$x + 0 = x$
$15 = x$	

2 Solve linear equations using the multiplication axiom (pp. 364–366).

Multiply both sides of the equation by the reciprocal of the coefficient of the letter term *or* divide both sides of the equation by the coefficient of the letter term.

Solve:

$$\frac{x}{5} = 8 \qquad\qquad -8x = 24 \qquad\qquad \frac{1}{3}x = \frac{4}{9}$$

$$\frac{5}{1}\left(\frac{x}{5}\right) = 8\left(\frac{5}{1}\right) \qquad \frac{-8x}{-8} = \frac{24}{-8} \qquad \left(\frac{3}{1}\right)\left(\frac{1}{3}x\right) = \left(\frac{4}{9}\right)\left(\frac{3}{1}\right)$$

$$x = 40 \qquad\qquad x = -3 \qquad\qquad$$

$$7.5 = 2.5x \qquad\qquad \frac{x}{0.6} = 2.9 \qquad\qquad x = \frac{4}{3}$$

$$\frac{7.5}{2.5} = \frac{2.5x}{2.5} \qquad 0.6\left(\frac{x}{0.6}\right) = 0.6(2.9)$$

$$3 = x \qquad\qquad x = 1.74$$

To check the solutions of equations, substitute the value of the variable in each place it appears in the equation. Perform operations on both sides of the equation. The two sides of the equation should be equal.

Verify that $x = 3$ is the solution for the equation $2x - 1 = 5$.

$$2(3) - 1 = 5$$
$$6 - 1 = 5$$
$$5 = 5$$

3 Solve linear equations with like terms on the same side of the equation (pp. 366–367).

Combine the like terms on the same side of the equation. Solve the remaining equation using the multiplication axiom.

Solve $3x - 5x = 12$ for x.

$$3x - 5x = 12$$
$$-2x = 12$$
$$\frac{-2x}{-2} = \frac{12}{-2}$$
$$x = -6$$

Chapter 8 / Linear Equations

| **4** Solve linear equations with like terms on opposite sides of the equation (pp. 367–371). | Use the addition axiom to move letter terms to one side of the equation and number terms to the other. Combine like terms. Solve the remaining basic equation. |

Solve $x - 5 = 7$ for x.

$$x - 5 = 7$$
$$x - 5 + 5 = 7 + 5$$
$$x = 12$$

Solve $7.2 = x - 3.5$ for x.

$$7.2 = x - 3.5$$
$$7.2 + 3.5 = x$$
$$10.7 = x$$

Solve $x - \frac{3}{8} = \frac{5}{8}$ for x.

$$x - \frac{3}{8} = \frac{5}{8}$$
$$x = \frac{5}{8} + \frac{3}{8}$$
$$x = \frac{8}{8}$$
$$x = 1$$

Solve $3x - 5 = 5x + 7$ for x.

$$3x - 5 = 5x + 7$$
$$3x - 5x = 7 + 5$$
$$-2x = 12$$
$$\frac{-2x}{-2} = \frac{12}{-2}$$
$$x = -6$$

Section 8–4

1 Solve linear equations that contain parentheses (pp. 372–380).

Remove parentheses using the distributive property. Continue solving the equation.

Solve $3(2x - 5) = 8x + 7$ for x.

$$3(2x - 5) = 8x + 7 \qquad \text{Distribute.}$$
$$6x - 15 = 8x + 7 \qquad \text{Sort terms.}$$
$$6x - 8x = 7 + 15 \qquad \text{Combine like terms.}$$
$$-2x = 22 \qquad \text{Divide by coefficient of } x.$$
$$\frac{-2x}{-2} = \frac{22}{-2}$$
$$x = -11$$

Six-Step Problem-Solving Plan

A shipment of college textbooks is sent in two boxes weighing 37 lb total. One box weighs 9 lb more than the other. What does each box weigh?

Unknown facts
Weight of each box.

Known facts
Total weight of two boxes is 37 lb.
One box weighs 9 lb more.

Relationships
x = weight of one box $x + x + 9 = 37$
$x + 9$ = weight of other box

Estimation
If the boxes were the same weight, each would weigh $18\frac{1}{2}$ lb.
Thus, one box will be less than and one box will be more than $18\frac{1}{2}$ lb.

Calculations

$$x + x + 9 = 37 \quad \text{Combine like terms.}$$
$$2x + 9 = 37 \quad \text{Sort terms.}$$
$$2x = 37 - 9 \quad \text{Combine like terms.}$$
$$2x = 28 \quad \text{Divide.}$$
$$\frac{2x}{2} = \frac{28}{2}$$
$$x = 14 \text{ lb} \quad \text{Weight of one box}$$
$$x + 9 = 23 \text{ lb} \quad \text{Weight of the other box}$$

Interpretation

The two boxes weigh 14 and 23 lb.

Section 8–5

1 Solve an equation using numerical procedures (pp. 381–384).

1. Make a table of values by selecting numbers to be substituted for the variable on each side of the equation.
2. Substitute a value, and evaluate both sides of the equation.
3. Repeat Step 1 with a different selected value.
4. Compare the results from Steps 2 and 3 for both sides of the equation. Were the results from your second selected value closer to being equal or farther apart? Make future selections based on your observations.
5. Repeat Steps 2–4 for additional selected values until a value is found that makes the results of the calculations for both sides of the equation equal.

Solve the equation $2x - 5 = x + 2$ numerically.

x	Left Side $2x - 5$	Right Side $x + 2$	Observations
1	$2(1) - 5 = -3$	$1 + 2 = 3$	Left and right differ by 6.
2	$2(2) - 5 = -1$	$2 + 2 = 4$	Left and right differ by 5. Two sides are getting closer. Continue to select larger values for x.
5	$2(5) - 5 = 5$	$5 + 2 = 7$	Left and right differ by 2.
6	$2(6) - 5 = 7$	$6 + 2 = 8$	Left and right differ by 1.
7	$2(7) - 5 = 9$	$7 + 2 = 9$	Both sides equal.

The solution is $x = 7$.

Section 8–6

1 Evaluate a function (pp. 386–388).

A function shows patterns that develop from using various values of the independent variable.

Universal Works charges $50 as a flat fee and $18 per hour (or any part of an hour) for leveling and clearing ground. Write a function that expresses the cost of service in relation to the number of hours of work. The total cost includes $50 in addition to the hourly cost of $18 times the number of hours. Thus, the function is $f(x) = 18x + 50$, where x represents the number of hours and $f(x)$ represents the total cost.

As values change for the input or independent variable, values of the output or dependent variable change accordingly. Evaluate a function by substituting a value for the independent variable. Then, use the order of operations to simplify.

Find the cost for 2 hours of work, 4 hours of work, and 10 hours of work by Universal Works. Use the function $f(x) = 18x + 50$.

Independent variable: x Dependent variable: $f(x)$

$f(x) = 18x + 50$	$f(x) = 18x + 50$	$f(x) = 18x + 50$
$f(2) = 18(2) + 50$	$f(4) = 18(4) + 50$	$f(10) = 18(10) + 50$
$f(2) = 36 + 50$	$f(4) = 72 + 50$	$f(10) = 180 + 50$
$f(2) = \$86$	$f(4) = \$122$	$f(10) = \$230$

2 Find the range of a function of a given domain (pp. 388–389).

The *domain* of a function is a set of numbers that includes all acceptable values for the independent variable. Acceptable values are determined by the context of the problem. The *range* of a function is a set of numbers that includes the results of the evaluation of the function for each value in the domain.

> The domain for the hours of work for Universal Works is the set of natural numbers (1, 2, 3, 4, . . .). Find the range.

> The number of hours worked is multiplied by $18 and $50 is added to the product. Therefore, the range will also be natural numbers. Only natural numbers that are 50 more than the multiples of 18 are included. Also, the range will be expressed in whole dollars.

3 Graph independent and dependent values of functions (pp. 390–391).

A function can be represented graphically on a coordinate system. The independent variable is represented on the horizontal axis and the dependent variable is represented on the vertical axis.

> Fig. 8–12 graphs the function used by Universal Works for leveling and clearing ground.

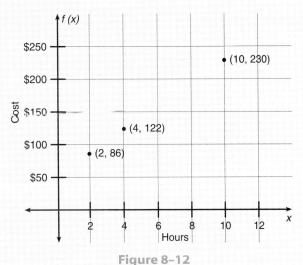

Figure 8–12

CHAPTER TRIAL TEST

Solve.

1. $x + 5 = 19$
2. $y - 8 = -15$
3. $5x = 30$
4. $-8y = 72$
5. $\dfrac{x}{2} = 5$
6. $-\dfrac{3}{5}m = -6$
7. $x = 5 - 8$
8. $5x + 2x = 49$

9. $y - 2y = 15$

10. $2x = 3x - 7$

11. $5 - 2x = 3x - 10$

12. $5x + 3 - 7x = 2x + 4x - 11$

13. $3(x + 4) = 18$

14. $2(x - 7) = -10$

15. $7 - (x - 3) = 4x$

16. $3x + 2 = 4(x - 1) - 1$

17. $4.5x - 3.4 = 2.1x - 0.4$

18. 5 added to x equals 16. Find x.

19. Twice the sum of a number and 3 is 10. Find the number.

20. One tank holds 4 times as many gallons as a smaller tank. If the tanks hold 2,580 gal together, how much does each tank hold?

21. The Ross Prairie Wildlife Management Area (WMA) is located in Marion County, Florida, and the Potts Wildlife Management Area is located in Citrus County, Florida. Together, the two WMAs have 11,034 acres. Potts WMA has 4,980 acres more than Ross Prairie. Write an equation to express the relationship. How many acres are in each WMA?

22. Acme Medical Corporation operates two hospitals in one small county. The chief executive officer reported for 1 month that an increased use of the outpatient center at one hospital was twice that of the other hospital's outpatient center. If the number of outpatients served by both facilities for the month was 2,550, how many outpatients used each facility for the month?

Evaluate the functions.

23. $f(x) = -6x - 5; f(-2); f(5)$

24. $g(x) = 6 - 7x; g(2); g(0)$

Equations with Fractions and Decimals

9

Thus far, we have used mostly integers in equations. However, real-world situations require us to deal with various fraction and decimal quantities when we solve equations.

9–1 | Solving Linear Equations with Fractions by Clearing the Denominators

Learning Outcomes
1 Solve fractional equations by clearing the denominators.
2 Solve applied problems involving rate, time, and work.

1 Solve Fractional Equations by Clearing the Denominators.

The techniques we used in Chapter 8 will also help us solve equations with fractions and decimals. Other techniques minimize the calculations with fractions and decimals and allow more steps to be performed mentally. One of these techniques involves *clearing* the equation of all denominators in the first step. The resulting equation, which contains no fractions, is then solved using the procedures we learned earlier.

This process, called *clearing the fractions*, is another application of the multiplication axiom.

To clear an equation of a single fraction:

Apply the multiplication axiom by multiplying the entire equation by the fraction's denominator.

EXAMPLE Solve $\dfrac{4d}{3} = 12$.

$$\dfrac{4d}{3} = 12 \qquad \text{Multiply both sides by the denominator 3. Reduce where possible.}$$

$$(\overset{1}{\cancel{3}})\dfrac{4d}{\underset{1}{\cancel{3}}} = (3)(12)$$

$$4d = 36 \qquad \text{Divide.}$$

$$\dfrac{4d}{4} = \dfrac{36}{4}$$

$$d = \mathbf{9}$$

Check: $\quad \dfrac{4(\overset{3}{\cancel{9}})}{\underset{1}{\cancel{3}}} = 12$

$$12 = 12$$

EXAMPLE Solve $2x + \dfrac{3}{4} = 1$ by clearing the fraction.

$$2x + \dfrac{3}{4} = 1 \qquad \text{Multiply each term by the denominator 4.}$$

$$4\,(2x) + 4\left(\dfrac{3}{4}\right) = 4\,(1) \qquad \text{Reduce and multiply.}$$

$$4(2x) + \overset{1}{4}\left(\frac{3}{\underset{1}{4}}\right) = 4(1)$$ Use procedures from Chapter 8 to finish the solution.

$$8x + 3 = 4$$ Sort.

$$8x = 4 - 3$$ Combine.

$$8x = 1$$ Divide.

$$\frac{8x}{8} = \frac{1}{8}$$

$$x = \frac{1}{8}$$

Check: $$2x + \frac{3}{4} = 1$$ Substitute $\frac{1}{8}$ in place of x.

$$\overset{1}{2}\left(\frac{1}{\underset{4}{8}}\right) + \frac{3}{4} = 1$$ Multiply.

$$\frac{1}{4} + \frac{3}{4} = 1$$ Add.

$$1 = 1$$

One advantage of clearing fractions is that it can be applied to equations in which the variable is in the *denominator* of the fraction. Let's use this procedure to solve $\frac{10}{x} = 2$. Note that the term $\frac{10}{x}$ is *not* the product of 10 and x. Therefore, the coefficient of x is *not* 10.

EXAMPLE   Solve $\frac{10}{x} = 2$.

$$\frac{10}{x} = 2$$ Multiply both sides by denominator x. Reduce where possible.

$$(\overset{1}{x})\frac{10}{\underset{1}{x}} = (x)2$$

$$10 = 2x$$ Divide.

$$\frac{10}{2} = \frac{2x}{2}$$

$$5 = x$$

Check: $$\frac{\overset{2}{10}}{\underset{1}{5}} = 2$$

$$2 = 2$$

TIP!

Check for Extraneous Roots.

When you solve equations with a variable in the fraction's denominator, you may get a "root" that does not make a true statement when substituted in the original equation. Solutions or roots that do not make a true statement in the original equation are called **extraneous roots. When solving an equation with the variable in the denominator, you *must* check the root to see if it makes a true statement in the original equation.**

One situation that produces an extraneous root is a value that causes the denominator of any fraction to be zero. Such values are called *excluded values*.

To identify an excluded value that causes a denominator of zero:

1. Write an equation for each fraction that has a variable in the denominator by setting the denominator equal to zero.
2. Solve each resulting equation.
3. The solution for each equation from Step 1 is an excluded value.

EXAMPLE Solve $\dfrac{0}{x} = 5$.

$$\dfrac{0}{x} = 5 \qquad \text{Multiply both sides by } x.$$

$$x\left(\dfrac{0}{x}\right) = 5(x)$$

$$\dfrac{0}{5} = \dfrac{5x}{5}$$

$$0 = x \qquad \text{Possible solution.}$$

Check: $\dfrac{0}{0} \neq 5 \qquad \text{Does not check. Zero is an excluded value.}$

The equation $\dfrac{0}{x} = 5$ has no solution.

Equations in which the fraction contains more than one term in its numerator or denominator can also be cleared of fractions.

EXAMPLE Solve $\dfrac{12}{Q + 6} = 1$. Identify any excluded values.

Excluded value:

$$Q + 6 = 0 \qquad \text{Set denominator equal to zero and solve for } Q.$$
$$Q = 0 - 6$$
$$Q = -6 \qquad \text{Excluded value.}$$

$$\dfrac{12}{Q + 6} = 1 \qquad \text{Multiply both sides of the equation by the denominator, the quantity } Q + 6. \text{ Reduce where possible.}$$

$$(Q + 6)\dfrac{12}{Q + 6} = (Q + 6)\, 1$$

$$12 = Q + 6 \qquad \text{Sort.}$$
$$12 - 6 = Q \qquad \text{Combine like terms.}$$
$$6 = Q \qquad \text{Because the numerical coefficient of } Q \text{ is already 1, the equation is solved.}$$

Chapter 9 / Equations with Fractions and Decimals

Check: $\dfrac{12}{6+6} = 1$ Substitute 6 for Q and evaluate.

$$\frac{12}{12} = 1$$

$$1 = 1$$

The solution of the equation is $Q = 6$.

When an equation has more than one fractional term, we expand our process.

To solve an equation by clearing all fractions:

1. Multiply each term of the *entire* equation by the least common multiple (LCM) of the denominators of the equation.
2. Apply the distributive property to remove parentheses.
3. Combine like terms on each side of the equation.
4. Sort terms to collect the variable or letter terms on one side and number terms on the other.
5. Combine like terms on each side of the equation.
6. Solve the basic equation by multiplying by the reciprocal of the coefficient of the variable term or dividing by the coefficient of the variable term.

EXAMPLE   Solve $-\dfrac{1}{4}x = 9 - \dfrac{2}{3}x$ by clearing all fractions first.

$$-\frac{1}{4}x = 9 - \frac{2}{3}x$$
Multiply each term by the LCM of the denominators. Reduce. LCM = 12 or 4(3).

$$(4)(3)\left(-\frac{1}{4}x\right) = (4)(3)(9) - (4)(3)\left(\frac{2}{3}x\right)$$

$$(\overset{1}{4})(3)\left(-\frac{1}{\underset{1}{4}}x\right) = (4)(3)(9) - (4)(\overset{1}{3})\left(\frac{2}{\underset{1}{3}}x\right)$$
Multiply the remaining factors.

$$-3x = 108 - 8x$$
This equation contains no fractions. Sort terms.

$$-3x + 8x = 108$$
Combine like terms.

$$5x = 108$$
Divide by the coefficient of x.

$$\frac{5x}{5} = \frac{108}{5}$$

$$x = \frac{108}{5}$$

Check using a calculator:

$$-\frac{1}{4}x = 9 - \frac{2}{3}x$$
Substitute $\dfrac{108}{5}$ for x.

$$-\frac{1}{4}\left(\frac{108}{5}\right) = 9 - \frac{2}{3}\left(\frac{108}{5}\right)$$

Left Side: $\boxed{(}\boxed{(}\boxed{(-)}\,1\,\boxed{\div}\,4\boxed{)}\,\boxed{\times}\,\boxed{(}108\,\boxed{\div}\,5\boxed{)}\,\boxed{=} \Rightarrow -5.4$

Right Side: $9\,\boxed{-}\,\boxed{(}2\,\boxed{\div}\,3\boxed{)}\,\boxed{\times}\,\boxed{(}108\,\boxed{\div}\,5\boxed{)} = \Rightarrow -5.4$

TIP!

Improper Fractions versus Mixed Numbers

The solutions of many equations are not whole numbers. The solution is usually written as a *proper* or *improper fraction* in lowest terms. You can also change the solution to its decimal or mixed-number equivalent. When we solve applied problems, the answer is interpreted in an appropriate form.

Clearing fractions creates larger numbers in the new equation, but the numbers are integers instead of fractions.

EXAMPLE Find the total resistance in a parallel dc circuit with three branches rated at 4 ohms, 10 ohms, and 20 ohms respectively. Solve $\frac{1}{R} = \frac{1}{4} + \frac{1}{10} + \frac{1}{20}$ by clearing fractions first.

$R = 0$ is the excluded value.

$$\frac{1}{R} = \frac{1}{4} + \frac{1}{10} + \frac{1}{20}$$

> The LCM is 20R because R, 4, 10, and 20 all divide evenly into **20R**.

$$20\overset{}{R}\left(\frac{1}{\cancel{R}}\right) = \overset{5}{\cancel{20}}R\left(\frac{1}{\underset{1}{\cancel{4}}}\right) + \overset{2}{\cancel{20}}R\left(\frac{1}{\underset{1}{\cancel{10}}}\right) + \overset{1}{\cancel{20}}R\left(\frac{1}{\underset{1}{\cancel{20}}}\right)$$

> Multiply each term in the *entire* equation by 20R and reduce.

$$20 = 5R + 2R + R$$

> Combine like terms.

$$20 = 8R$$

> Divide by the coefficient of R.

$$\frac{20}{8} = \frac{8R}{8}$$

> Reduce.

$$\frac{5}{2} = R$$

> $\frac{5}{2} = 2.5$

Interpretation The resistance is 2.5 ohms.

Check: $\dfrac{1}{R} = \dfrac{1}{4} + \dfrac{1}{10} + \dfrac{1}{20}$

> Substitute $\frac{5}{2}$ for R.

$$\frac{1}{\frac{5}{2}} = \frac{1}{4} + \frac{1}{10} + \frac{1}{20}$$

Left Side	**Right Side**
$\dfrac{1}{\frac{5}{2}} = 1\left(\dfrac{2}{5}\right) = \dfrac{2}{5}$	$\dfrac{1}{4} + \dfrac{1}{10} + \dfrac{1}{20} =$
	$\dfrac{5}{20} + \dfrac{2}{20} + \dfrac{1}{20} = \dfrac{8}{20} = \dfrac{2}{5}$

TIP!

Fractions versus Decimals

Sometimes applied problems that require fractions in their equations require that their solutions be expressed as decimal numbers. In these cases, you will perform the division indicated by the fraction.

The equation in the preceding example is derived from the formula for finding total resistance in a parallel dc circuit with three branches rated at 4, 10, and 20 ohms. Ohms are expressed in decimal numbers, so in an application the solution should be converted to a decimal equivalent.

$$R = \frac{5}{2} \quad \text{or} \quad 2.5 \text{ ohms}$$

2 **Solve Applied Problems Involving Rate, Time, and Work.**

A *rate measure* involves a unit of time. If car A travels 50 miles in 1 hour, then car A's *rate of work* (travel) is 50 miles per 1 hour, or $\dfrac{50 \text{ mi}}{1 \text{ hr}}$ expressed as a fraction. If car A travels for 3 hours, then the *amount of work* is $\dfrac{50 \text{ mi}}{\text{hr}} \times 3 \text{ hr} = 150$ miles.

To find the amount of work produced by one individual or machine:

1. Identify the rate of work and the time worked.
2. Use the formula for amount of work.

Formula for amount of work:

Amount of work completed = Rate of work × Time worked

EXAMPLE	A carpenter can install 1 door in 3 hr. Find the number of doors the carpenter can install in 30 hr.
Known facts	Rate of work = 1 door per 3 hr, $\dfrac{1}{3}$ door per hour, or $\dfrac{1 \text{ door}}{3 \text{ hr}}$
	Time worked = 30 hr
Unknown facts	W = Amount of work or number of doors installed
Relationships	Amount of work = Rate of work × Time worked
Estimation	Several doors can be installed in 30 hr., but less than 30 since 1 door would have to be installed every hour to install 30 doors.
Calculations	$W = \dfrac{1 \text{ door}}{\overset{}{\underset{1}{3} \text{ hr}}} \times \overset{10}{30} \text{ hr}$ Reduce and multiply.
	$W = 10$ doors
Interpretation	**Thus, 10 doors can be installed in 30 hr.**

If two workers or machines do a job together, we can find the amount of work done by each worker or machine. Combined, the amounts equal 1 total job.

To find the amount of work each individual or machine produces when working together:

1. Identify the rate of work for each individual or machine. If unknown, assign a letter to represent the unknown.
2. Identify the time worked for each individual or machine. If unknown, assign a letter to represent the unknown. *Note*: Only one letter should be used and other unknowns should be written in relationship to the one letter.
3. Use the formula for completing one job.

Formula for completing one job when A and B are working together:

$$\left(\begin{array}{c} \text{A's} \\ \text{amount of} \\ \text{work} \end{array}\right) + \left(\begin{array}{c} \text{B's} \\ \text{amount of} \\ \text{work} \end{array}\right) = 1 \text{ completed job}$$

or

$$\left(\begin{array}{c} \text{A's} \\ \text{rate of} \\ \text{work} \end{array} \times \begin{array}{c} \\ \text{time} \\ \text{worked} \end{array}\right) + \left(\begin{array}{c} \text{B's} \\ \text{rate of} \\ \text{work} \end{array} \times \begin{array}{c} \\ \text{time} \\ \text{worked} \end{array}\right) = 1 \text{ completed job}$$

EXAMPLE Pipe 1 fills a tank in 6 min and pipe 2 fills the same tank in 8 min (Fig. 9–1). How long does it take for both pipes together to fill the tank?

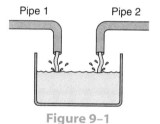

Figure 9–1

Known facts Pipe 1 fills the tank at a rate of 1 tank per 6 min, $\frac{1}{6}$ tank per minute, or $\frac{1 \text{ tank}}{6 \text{ min}}$.

Pipe 2 fills the tank at a rate of 1 tank per 8 min, $\frac{1}{8}$ tank per minute, or $\frac{1 \text{ tank}}{8 \text{ min}}$.

Unknown facts T = time (in minutes) for both pipes together to fill the tank.

Relationships Amount of work of pipe 1 = $\frac{1 \text{ tank}}{6 \text{ min}} (T)$. Amount of work of pipe 2 = $\frac{1 \text{ tank}}{8 \text{ min}} (T)$.

Amount of work together = Pipe 1's work + Pipe 2's work.

Estimation Both pipes together should fill the tank more quickly than the faster rate, or in less than 6 min.

Calculations

$$\frac{1 \text{ tank}}{6 \text{ min}}(T \text{ min}) + \frac{1 \text{ tank}}{8 \text{ min}}(T \text{ min}) = 1 \text{ tank}$$

The LCM is 24.

$$(24)\left(\frac{1}{6}T\right) + (24)\left(\frac{1}{8}T\right) = (24)(1)$$

Clear fractions.

$$(\overset{4}{24})\left(\frac{1}{6}T\right) + (\overset{3}{24})\left(\frac{1}{8}T\right) = (24)(1)$$

Reduce and multiply.

$$4T + 3T = 24$$

Combine.

$$7T = 24$$

Divide.

$$\frac{7T}{7} = \frac{24}{7}$$

$$T = \frac{24}{7}\left(\text{or } 3\frac{3}{7}\right) \text{ min}$$

Interpretation   Both pipes together fill the tank in $3\frac{3}{7}$ min.

TIP!

Improper Fractions versus Mixed Numbers or Decimal Equivalents

In solving equations, a solution that is an improper fraction, like $\frac{24}{7}$, is left as an improper fraction. However, when you solve applied problems, it is desirable to change improper fractions like $\frac{24}{7}$ into mixed numbers or decimal equivalents.

The problem statement generally dictates the interpretation. In the preceding example, $\frac{24}{7}$ min is more appropriately interpreted as $3\frac{3}{7}$ min or 3.4 min (rounded). In some instances, you may need to change $\frac{3}{7}$ min to seconds ($\frac{3}{7}$ min $\times \frac{60 \text{ sec}}{1 \text{ min}} = 25.8$ sec). Then, $3\frac{3}{7}$ min becomes 3 min 26 sec.

Two pipes, one a faucet and the other a drain, have opposite functions. Pipe 1 fills the tank. Pipe 2 is a drain and empties the tank. In this case, we subtract the work done by the drain from the work done by the faucet. This combined action, if the faucet fills at a faster rate than the drain empties, results in a full tank.

Formula for completing one job when A and B are working in opposition:

$$\begin{pmatrix} \text{A's} \\ \text{amount of} \\ \text{work} \end{pmatrix} - \begin{pmatrix} \text{B's} \\ \text{amount of} \\ \text{work} \end{pmatrix} = 1 \text{ completed job}$$

or

$$\begin{pmatrix} \text{A's} \\ \text{rate of} \times \text{time} \\ \text{work} \quad \text{worked} \end{pmatrix} - \begin{pmatrix} \text{B's} \\ \text{rate of} \times \text{time} \\ \text{work} \quad \text{worked} \end{pmatrix} = 1 \text{ completed job}$$

based on A's amount of work being greater than B's amount of work.

EXAMPLE A faucet fills a tank in 6 min. A drain empties the tank in 8 min (Fig. 9–2). If both the faucet and drain are open, in how many minutes will the tank start to overflow if the faucet is not turned off?

Known facts Faucet's rate of work = 1 tank filled per 6 min, $\frac{1}{6}$ tank per min, or $\dfrac{1 \text{ tank}}{6 \text{ min}}$.

Drain's rate of work = 1 tank emptied per 8 min, $\frac{1}{8}$ tank per min, or $\dfrac{1 \text{ tank}}{8 \text{ min}}$.

Unknown facts T = time (min) until tank overflows with faucet and drain both working.

Relationships Amount of work of faucet = $\dfrac{1 \text{ tank}}{6 \text{ min}}(T)$

Faucet

Amount of work of drain = $\dfrac{1 \text{ tank}}{8 \text{ min}}(T)$

Amount of work
when both faucet
and drain are open = Faucet's work − Drain's work

Estimation With both the faucet and drain open, it should take longer to fill the tank than if the faucet was open and the drain closed. It should take longer than 6 min.

Drain

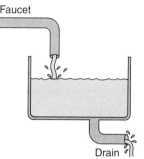

Figure 9–2

Calculations

$\dfrac{1 \text{ tank}}{6 \text{ min}}(T \text{ min}) - \dfrac{1 \text{ tank}}{8 \text{ min}}(T \text{ min}) = 1 \text{ tank}$ Analyze dimensions.

$\dfrac{1}{6}T - \dfrac{1}{8}T = 1$ Clear fractions. LCM = 24.

$(24)\left(\dfrac{1}{6}T\right) - (24)\left(\dfrac{1}{8}T\right) = (24)(1)$ Reduce.

$(\overset{4}{24})\left(\dfrac{1}{\underset{1}{6}}T\right) - (\overset{3}{24})\left(\dfrac{1}{\underset{1}{8}}T\right) = (24)(1)$ Multiply remaining factors.

$4T - 3T = 24$ Combine like terms.

$T = 24 \text{ min}$

Interpretation **With the faucet and drain both open, the tank will be full in 24 min.**

SELF-STUDY EXERCISES 9–1

1 Solve the equations. Identify excluded values as appropriate.

1. $\dfrac{2}{7}x = 8$

2. $-7 = \dfrac{21}{33}p$

3. $\dfrac{1}{3}r = \dfrac{6}{7}$

4. $0 = -\dfrac{2}{5}c$

5. $-\dfrac{5}{3}m = 9$

6. $-9m = \dfrac{5}{3}$

7. $\dfrac{5}{8}t = 1$

8. $-\dfrac{5}{7}p = -\dfrac{11}{21}$

9. $10 = -\dfrac{1}{35}t$

10. $\dfrac{5}{12}z = 20$

11. $\dfrac{2x}{3} = 18$

12. $\dfrac{7}{Q} = 21$

13. $\dfrac{y+1}{2} = 7$

14. $\dfrac{7}{p-4} = -8$

15. $0 = \dfrac{x}{4}$

16. $-\dfrac{8}{P} = -72$

17. $\dfrac{P}{-8} = -72$

18. $-8 = \dfrac{4B}{B-6}$

19. $\dfrac{3P}{7} = 12$

20. $\dfrac{8-R}{76} = 1$

21. $\dfrac{2}{9}c + \dfrac{1}{3}c = \dfrac{3}{7}$

22. $-\dfrac{1}{4}x = 9 - \dfrac{2}{3}x$

23. $\dfrac{2}{7}y + \dfrac{3}{8} = \dfrac{1}{7}y + \dfrac{5}{3}$

24. $\dfrac{1}{3}x + \dfrac{1}{2}x = \dfrac{20}{3}$

25. $\dfrac{7}{R} - \dfrac{2}{R} = -1$

26. $S = \dfrac{1}{15} + \dfrac{1}{5} + \dfrac{1}{30}$

27. $18 - \dfrac{1}{4}x = \dfrac{1}{2}$

28. $\dfrac{1}{7}H - \dfrac{1}{3}H = 0$

29. $\dfrac{7}{16}h + \dfrac{1}{9} = \dfrac{1}{3}$

30. $x + \dfrac{1}{4}x = 8$

31. $3y + 9 = \dfrac{1}{4}y$

32. $18 = \dfrac{4}{3x} - \dfrac{3}{2x}$

33. $\dfrac{2}{7}p + 1 = \dfrac{1}{3}p$

34. $S = \dfrac{1}{10} + \dfrac{1}{25} + \dfrac{1}{50}$

35. $-x + \dfrac{1}{7} = \dfrac{1}{2}x$

36. $0 = 1 + \dfrac{2}{9}c - c$

2 Set up an equation and solve.

37. A kiln fires 8 large vases in 2 hr. How many vases can be fired in 20 hr?

38. A printing press produces 1 day's newspaper in 4 hr. A higher-speed press does 1 day's newspaper in 2 hr. How much time does it take both presses to produce 1 day's newspaper?

39. One machine packs 1 day's salmon catch in 8 hr. A second machine packs 1 day's catch in 5 hr. How much time does it require for 1 day's catch to be packed if both machines are used?

40. A painter can paint a house in 6 days. Another painter takes 8 days to paint the same house. If they work together, how much time will it take them to paint the house?

41. One bottling machine can fill 400 bottles of water in 1 hr and another machine can fill 400 bottles of water in $1\frac{1}{2}$ hr. If both machines are working, how much time does it take to fill 400 bottles of water?

42. A tank has two pipes entering. Pipe 1 alone fills the tank in 4 min and pipe 2 takes 12 min to fill the tank. How much time does it take to fill the tank if both pipes are operating at the same time?

43. A tank has two pipes entering. Pipe 1 alone fills the tank in 30 min and together the pipes take 10 min to fill the tank. How much time does pipe 2 need to fill the tank alone?

44. A tank has two pipes entering it and one leaving it. Pipe 1 fills the tank in 3 min. Pipe 2 takes 7 min to fill the same tank. Pipe 3, however, empties the tank in 21 min. How much time does it take to fill the tank with all three pipes operating at the same time?

For exercises 45–48, see example on page 408.

45. Find the total resistance in a parallel dc circuit with two branches rated at 12 and 30 ohms, respectively.

46. A parallel dc circuit has three branches rated at 12 ohms, 15 ohms, and 20 ohms. Find the total resistance.

47. Three branches of a parallel dc circuit are rated at 16 ohms, 24 ohms, and 32 ohms. Find the total resistance.

48. A parallel dc circuit has a total resistance of 3 ohms. One branch alone produces 18 ohms resistance. What is the resistance of the other branch?

9–2 | *Solving Decimal Equations*

Learning Outcomes

1 Solve decimal equations.

2 Solve applied problems with decimal equations.

1 **Solve Decimal Equations.**

As in equations with fractions, we can also clear an equation of decimals before starting to solve the equation.

To solve a decimal equation by clearing decimals:

1. Multiply each term of the *entire* equation by the least common multiple (LCM) of the fractional amounts following the decimal points.
2. Follow the same steps used in solving a linear equation.

TIP!

LCM for Decimals

Digits to the right of the decimal point represent fractions whose denominators are determined by the place value. To find the LCM for all the decimal numbers in an equation, *look* at the decimal with the most digits after the decimal point. Use its denominator to clear the decimals.

This procedure allows you to avoid dividing by a decimal, which can be a common source of error when making calculations by hand.

EXAMPLE   Solve $0.38 + 1.1y = 0.6$ by first clearing the equation of decimals.

$$0.38 + 1.1y = 0.6 \qquad \text{The LCM is } 100.$$
$$100\,(0.38) + 100\,(1.1y) = 100\,(0.6) \qquad \text{Multiply by 100.}$$
$$38 + 110y = 60 \qquad \text{Sort.}$$
$$110y = 60 - 38 \qquad \text{Combine.}$$
$$\frac{110y}{110} = \frac{22}{110} \qquad \text{Divide.}$$
$$y = 0.2$$

2 Solve Applied Problems with Decimal Equations.

Many applied problems are solved with decimal equations. One common problem involves interest on a loan or on an investment.

The interest formula resembles the percentage formula, $P = RB$, but it includes the time period over which the money was borrowed or invested. If we know three of the four elements, we can find the fourth.

Formula for simple interest:

$$I = PRT$$

where $I =$ **interest**, $P =$ **principal**, $R =$ **rate** or percent in decimal form, and $T =$ **time.**

For comparison with the percentage formula, interest is the part or percentage and the principal is the base.

EXAMPLE A $1,000 investment is made for $2\frac{1}{2}$ years at 6.25%. Find the amount of interest.

When we use the interest formula, we change the percent to a decimal equivalent. When no time period is given for the rate, we assume the rate to be per year.

$I = PRT$ Substitute the values in the formula.

    principal   rate   time

$I = \$1{,}000 \times 0.0625 \times 2.5 \text{ years}$ 6.25% = 0.0625; $2\frac{1}{2}$ years = 2.5 years.

$I = \$156.25$

The interest for $2\frac{1}{2}$ years is \$156.25.

Hand-Clearing Decimals versus Using A Calculator

The preceding example illustrates that in some cases clearing decimals is not the most efficient way to solve an equation with decimals. The numbers produced may be extremely large and cumbersome, and the decimals may not all be cleared.

$I = \$1{,}000 \times 0.0625 \times 2.5$ LCM is 10,000.

$10{,}000I = \overset{1}{10{,}000}\left(\$1{,}\overset{100}{000} \times \dfrac{625}{\underset{1}{10{,}000}} \times \dfrac{25}{\underset{1}{10}}\right)$ Multiply by 10,000.

$10{,}000I = 100(625)(25)$ Multiply.

$10{,}000I = 1{,}562{,}500$ Divide.

$I = \dfrac{1{,}562{,}500}{10{,}000}$

$I = \$156.25$

A calculator gives the solution more quickly and more efficiently if we proceed using the decimals.

Let's use the interest formula to solve problems when facts other than the interest are missing.

EXAMPLE Aetna Photo Studio borrowed \$3,500 for some darkroom equipment and has to pay \$1,890 in interest over a 3-year period. What is the interest rate?

Known facts   Principal = \$3,500. Interest = \$1,890. Time = 3 years.

Unknown facts R = Rate.

Relationships $I = PRT$.

Estimation   One year's interest is about $\frac{1}{3}$ of the total interest, or about \$600.

$$10\% \text{ of } \$3{,}500 = \$350$$

$$20\% \text{ of } \$3{,}500 = \$350 \times 2 = \$700$$

Therefore, R is more than 10% and less than 20%.

9-2 Solving Decimal Equations   415

Calculations	$I = PRT$	Substitute known values.
	$1{,}890 = 3{,}500 \times R \times 3$	Multiply.
	$1{,}890 = 10{,}500R$	Solve for R.
	$\dfrac{1{,}890}{10{,}500} = \dfrac{10{,}500R}{10{,}500}$	
	$0.18 = R$	Translate as a percent. $0.18 = 18\%$.
Interpretation	**The interest rate is 18%.**	

EXAMPLE The distance formula is Distance = Rate × Time. If a car was driven 82.5 mi at an average of 55 mi per hr, how long did the trip take?

Known facts	Distance = 82.5 mi. Rate $= \dfrac{55 \text{ mi}}{\text{hr}}$.
Unknown facts	T = Time in hours.
Relationships	Distance = Rate × Time.
Estimation	At 55 mi per hr, a car would travel 110 mi in 2 hr. It would take less than 2 hr to travel 82.5 mi.
Calculations	$82.5 = 55 \times T$ Substitute known values.
	$82.5 = 55T$ Divide.
	$\dfrac{82.5}{55} \quad \dfrac{55T}{55}$
	$1.5 = T$
Interpretation	**The trip took 1.5 or $1\frac{1}{2}$ hr.**

EXAMPLE The formula for voltage V is wattage W divided by amperage A: $V = \frac{W}{A}$. Find the voltage to the nearest hundredth needed for a circuit of 1,280 W with a current of 12.23 A.

Estimation	$1{,}200 \div 12 = 100$
	$V = \dfrac{W}{A}$ Substitute values.
	$V = \dfrac{1{,}280}{12.23}$ Divide.
	$V = 104.6606705$ or $104.66 \; V$
Interpretation	**The voltage is 104.66 V.**

EXAMPLE Juan earned $41.85 working 6.75 hr at a fast-food restaurant. What was his hourly wage?

Known facts	Pay = $41.85. Hours = 6.75.
Unknown facts	Wage = W.
Relationships	Amount of pay = Hourly wage × Hours worked.

Chapter 9 / Equations with Fractions and Decimals

Estimation	7 hr at \$5 = \$35. And 7 hr at \$6 = \$42. Hourly wage is close to \$6.
Calculations	$41.85 = W \times 6.75$
	$41.85 = 6.75W$
	$\dfrac{41.85}{6.75} = \dfrac{6.75W}{6.75}$
	$6.2 = W$ or \$6.20 per hr
Interpretation	**Juan's hourly wage is \$6.20 per hour.**

SELF-STUDY EXERCISES 9–2

1 Solve the equations.

1. $2.3x = 4.6$

2. $0.8R = 0.6$ (round to nearest tenth)

3. $0.33x + 0.25x = 3.5$ (round to nearest hundredth)

4. $0.3a = 4.8$

5. $1.5p = 7$ (round to nearest tenth)

6. $0.04x = 0.08 - x$ (round to nearest hundredth)

7. $0.4p = 0.014$

8. $0.47 = R + 0.4R$ (round to nearest hundredth)

9. $2.3 = 5.6 + y$

10. $4.3 = 0.3x - 7.34$

11. $2x + 3.7 = 10.3$

12. $0.16 + 2.3x = -0.3$

13. $1.5x + 2.1 = 3$

14. $3.82 - 2.5y = 1$

15. $0.15p = 2.4$

2 Solve the problems using decimal equations.

16. If the formula for force is Force = Pressure × Area, how many pounds of force are produced by a pressure of 35 pounds per square inch on a piston whose surface area is 2.5 in²? Express the pounds of force as a decimal number.

17. The circumference of a circle equals π times the diameter. If a steel rod has a diameter of 1.5 in., what is the circumference of the rod to the nearest hundredth? (Circumference is the distance around a circle.)

18. The distance formula is Distance = Rate × Time. If a trucker drove 682.5 mi at 55 mi per hr (mi/hr), how long did she drive? (Answer to the nearest whole number.)

19. The distance formula is Distance = Rate × Time. If a tractor-trailer rig is driven 422.5 mi at 65 mi per hr (mph) on interstate highways, how long does the trip take?

20. Electrical resistance in ohms (Ω) is voltage V divided by amperage A. Find the resistance to the nearest tenth for a motor with a voltage of 12.4 V requiring 1.5 A.

21. Rita earned \$54.75 working 7.5 hr at a college bookstore. What was her hourly wage?

22. The formula for electrical power is $V = \frac{W}{A}$: voltage (V) equals wattage (W) divided by amperage (A). Find the voltage to the nearest hundredth needed for a circuit of 500 W with a current of 3.2 A.

23. Find the interest paid on a loan of \$2,400 for 1 year at an interest rate of 11%.

24. Find the interest paid on a loan of \$800 at $8\frac{1}{2}\%$ interest for 2 years.

25. Find the total amount of money (maturity value) that the borrower will pay back on a loan of \$1,400 at $12\frac{1}{2}\%$ simple interest for 3 years.

26. Find the rate of interest on an investment of \$2,500 made by Nurse Honda for a period of 2 years if she received \$612.50 in interest.

27. Maddy Brown needed start-up money for her landscape service. She borrowed \$12,000 for 30 months and paid \$360 interest on the loan. What interest rate did she pay?

28. Raul Fletes needs money to buy lawn equipment. He borrows $500 for 7 months and pays $53.96 in interest. What is his rate of interest?

29. Linda Davis agrees to lend money to Alex Luciano at a special interest rate of 9%, on the condition that he borrow enough that he will pay her $500 in interest over a 2-year period. What is the minimum amount Alex can borrow?

30. Rob Thweatt needs money for medical school. He borrows $6,000 at 12% interest. If he pays $360 interest, what is the duration of the loan?

9–3 | *Using Proportions to Solve Problems*

Learning Outcomes

1 Solve equations that are proportions.

2 Solve problems with direct variation.

3 Solve problems with inverse variation.

4 Solve problems that involve similar triangles.

1 **Solve Equations That Are Proportions.**

A common type of equation that contains fractions is a *proportion*. An example of a proportion is

$$\frac{N}{5} = \frac{1}{12}$$

In a proportion, *each side* of the equation is a fraction or ratio. (See Chapter 4, Section 2, Outcome 3 for an introduction to ratios and proportions.)

The relationship of the terms of the ratios or fractions in a proportion such as $\frac{6}{9} = \frac{2}{3}$ can be stated as "6 is to 9 as 2 is to 3."

Another way of representing a proportion is 6:9 = 2:3, which is also read "6 is to 9 as 2 is to 3."

As noted in Chapter 4, Section 2, Outcome 3, the *cross products* are equal in a proportion.

Property of proportions:

$$\text{If } \frac{a}{b} = \frac{c}{d}, \text{ then } ad = bc, \quad b, d \neq 0$$

Remember, properties do not apply when denominators are zero.

EXAMPLE Solve the proportions.

(a) $\dfrac{x}{4} = \dfrac{9}{6}$ (b) $\dfrac{4x}{5} = \dfrac{17}{20}$ (c) $\dfrac{x-2}{x+8} = \dfrac{3}{5}$ (d) $\dfrac{3}{x} = 7$

(a) $\dfrac{x}{4} = \dfrac{9}{6}$ Cross multiply. $x(6) = $ **6x**; $4(9) = $ **36**.

$6x = $ **36**   Solve for x.

$\dfrac{6x}{6} = \dfrac{36}{6}$

$x = 6$

(b) $\dfrac{4x}{5} = \dfrac{17}{20}$ Cross multiply. $4x(20) = \boxed{80x}$; $5(17) = \boxed{85}$.

$80x = \boxed{85}$ Solve for x.

$\dfrac{80x}{80} = \dfrac{85}{80}$

$x = \dfrac{85}{80}$ Reduce.

$x = \dfrac{17}{16}$

(c) $\dfrac{x-2}{x+8} = \dfrac{3}{5}$ Cross multiply.

$5(x-2) = \boxed{3(x+8)}$ Distribute.

$5x - 10 = 3x + 24$ Sort.

$5x - 3x = 24 + 10$ Combine terms.

$2x = 34$ Solve for x.

$\dfrac{2x}{2} = \dfrac{34}{2}$

$x = 17$

(d) $\dfrac{3}{x} = 7$ Write 7 as a fraction $7 = \dfrac{7}{1}$.

$\dfrac{3}{x} = \dfrac{7}{1}$ Cross multiply.

$3 = \boxed{7x}$ Solve for x.

$\dfrac{3}{7} = \dfrac{7x}{7}$

$\dfrac{3}{7} = x$

2 Solve Problems with Direct Variation.

Apples	Cost
4	$1
8	$2
12	$3
16	$4

Many problems in the workplace can be solved using proportions. These problems have a common characteristic: The details of the problem can be grouped into two pairs of data. These two pairs of data can be *directly* related.

A *direct variation* is one in which the quantities being compared are directly related, so that as one quantity increases (or decreases), the other quantity also increases (or decreases).

Data are often arranged in tables so these relationships can be examined. If 4 apples cost $1, set up a table to examine related costs of other amounts of apples.

We can also use this relationship to find the cost of any given number of apples or to determine how many apples can be purchased with any given amount of money.

To set up a direct variation:

1. Establish two pairs of related data.
2. Write one pair of data in the numerators of two ratios.

3. Write the other pair of data in the denominators of two ratios.
4. Form a proportion using the two ratios.

EXAMPLE   (a) Find the cost of 10 apples if 4 apples cost $1. (b) How many apples can be purchased for $10?

(a) Pair 1: 4 apples cost $1.

Pair 2: 10 apples cost c dollars.

Estimation   8 apples would cost $2. Therefore, 10 apples will cost more than $2.

$$\frac{4 \text{ apples}}{10 \text{ apples}} = \frac{\$1}{\$c}$$ Pair 1 is the numerator of each ratio.
Pair 2 is the denominator of each ratio. Cross multiply.

$$4c = 10$$ Divide.

$$\frac{4c}{4} = \frac{10}{4}$$

$$c = 2.50$$ To the nearest cent.

Interpretation   **10 apples cost $2.50.**

(a) Pair 1: 4 apples cost $1.

Pair 2: a apples cost $10 .

Estimation   If 10 apples cost $2.50, 4 times as many apples can be bought for $10. That is, 40 apples can be bought.

$$\frac{4 \text{ apples}}{a \text{ apples}} = \frac{\$1}{\$10}$$ Pair 1
Pair 2

$$40 = a$$ Cross multiply.

Interpretation   **40 apples can be bought for $10.**

Directly related data pairs can also be set up by making each pair a ratio. In the preceding example, we would write the ratio as

$$\frac{4 \text{ apples}}{\$1} = \frac{10 \text{ apples}}{\$c}$$ Pair 1
Pair 2

Cross multiply.

A *third way* to use directly related data is to identify a data pair in which both values are known then use that relationship to find a conversion factor. For example, if 4 apples cost $1, how much does 1 apple cost?

$$\$1 \div 4 \text{ apples} = \$0.25 \text{ per apple}$$

A conversion factor is used to multiply by the number of items. If each apple costs $0.25, then 10 apples cost $10 \times \$0.25$, or $2.50. A conversion factor is also called a *constant of direct variation*. The direct variation formula is $y = kx$, where x is the *independent variable*, y is the *dependent variable*, and k is the *constant of direct variation*. Another way to express this is $k = \frac{y}{x}$.

Chapter 9 / Equations with Fractions and Decimals

To find the constant of direct variation:

1. Identify a pair of related data in which both values are known.
2. Write the data pair as a ratio. The units in the denominator of the ratio should match the units of the in dependent variable. $k = \dfrac{y}{x}$
3. Leave the ratio as a fraction or change it to a decimal equivalent.

EXAMPLE A 15-oz box of cereal costs $2.29. Find the constant of direct variation for the cost per ounce. (This is also referred to as the *unit cost*.)

$$k = \frac{y}{x}$$
$$\begin{array}{c}\text{Unit cost}\\\text{or}\\\text{Constant of variation}\end{array} = \frac{\text{cost}}{\text{oz}} = \frac{\$}{\text{oz}}$$

$$k = \frac{\$2.29}{15 \text{ oz}}$$ Change to a decimal equivalent.

$$k = 0.1527$$ Rounded.

The unit cost or constant of direct variation is 0.1527 $\dfrac{\$}{\text{oz}}$.

EXAMPLE A 15-oz box of cereal costs $2.29 or a 36-oz box of cereal costs $4.89. Which is the better buy? Round to the nearest ten thousandth.

In the preceding example we found the cost per ounce (unit cost) of the 15-oz box of cereal to be $0.1527.

Find the unit cost of the 36-oz box of cereal.

$$k = \frac{\$4.89}{36} = \$0.1358 \qquad \text{Ratio of } \frac{\$}{\text{oz}}.$$

Compare the unit costs.

Unit cost of 15-oz box = $0.1527; Unit cost of 36-oz box = $0.1358

The 36-oz box of cereal costs less per ounce and is the better buy.

In the preceding example, we make a judgement solely on the basis of the mathematical facts. In reality, other factors are considered. Do you have the money to buy the larger box? Will you be able to use the larger amount before it gets stale?

Which method for direct variation is preferred? The proportion method is the most versatile for a variety of situations. Using the constant of direct variation is useful in computer programs and electronic spreadsheets.

EXAMPLE	A truck travels 102 mi on 6 gal of gasoline. How far will it travel on 30 gal of gasoline?
Known facts	Pair 1: 102 mi uses 6 gal of gasoline.
Unknown facts	Pair 2: m mi uses 30 gal of gasoline.
Estimation	30 gal ÷ 6 gal = 5. Then, approximately 5 times 100 miles can be driven. More than 102 miles can be traveled on 30 gal.

Dimension Analysis

Calculations	$\dfrac{102 \text{ mi}}{m \text{ mi}} = \dfrac{6 \text{ gal}}{30 \text{ gal}}$ $\dfrac{\text{distance}_1}{\text{distance}_2} = \dfrac{\text{gasoline}_1}{\text{gasoline}_2}$ Pair 1 / Pair 2
	$\dfrac{102}{m} = \dfrac{6}{30}$ $\dfrac{\text{mi}}{\text{mi}} = \dfrac{\text{gal}}{\text{gal}}.$
	$102(30) = 6m$ Cross multiply: mi(gal) = gal (mi).
	$\dfrac{3{,}060}{6} = \dfrac{6m}{6}$ Divide by gal. Reduce: $\dfrac{\text{mi}(\cancel{\text{gal}})}{\cancel{\text{gal}}} = \dfrac{\text{gal}(\text{mi})}{\cancel{\text{gal}}}.$
	$510 = m$ m is expressed in miles.
Interpretation	**The truck will travel 510 mi on 30 gal of gasoline.**

TIP!

Analyze Dimensions.

Even though we often remove the written dimensions from an equation, we should analyze the dimensions to be sure we use the correct units in the solution.

EXAMPLE	If a metal rod tapers 1 in. for every 24 in. of length, what is the amount of taper of a 30-in. piece of rod? (See Fig. 9–3.)

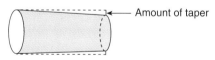

— Amount of taper

Figure 9–3

Known facts	Pair 1: 1-in taper for 24-in. length
Unknown facts	Pair 2: x-in taper for 30-in length
Estimation	A 30-in. rod will taper more than 1 in.
Calculations	$\dfrac{1\text{-in.taper}}{x\text{-in. taper}} = \dfrac{24\text{-in. length}}{30\text{-in.length}}$ Pair 1 / Pair 2
	$\dfrac{1}{x} = \dfrac{24}{30}$ Cross multiply.
	$1(30) = 24x$
	$30 = 24x$ Divide.

$$\frac{30}{24} = \frac{24x}{24}$$

$$\frac{5}{4} = x \quad \text{or} \quad 1\frac{1}{4} = x$$

Interpretation The amount of taper for a 30-in. length of rod is $1\frac{1}{4}$ in.

3 Solve Problems with Inverse Variation.

An *inverse variation* is one in which the quantities being compared are inversely related. That is, as one quantity increases, the other decreases, or as one quantity decreases, the other increases.

For example, as we *increase* pressure on a gas, the gas compresses and so *decreases* in volume. Or, as we *decrease* pressure on the gas, it *increases* in volume as it expands. This relationship is the opposite, or inverse, of direct variation.

If 3 workers frame a house in 2 weeks and the contractor *increases* the number of workers to 6, the framing time *decreases* to 1 week, assuming the workers work at the same rate. In other words, the framing time is *inversely proportional* to the number of workers on the job.

Unlike the directly related ratios in a proportion, inversely related ratios do not allow us the flexibility we had in setting up ratios of unlike measures.

To set up an inverse proportion:

1. Establish two pairs of related data.
2. Arrange one pair as the numerator of one ratio and the denominator of the other ratio.
3. Arrange the other pair so that each ratio contains like measures.
4. Form a proportion using the two ratios.

EXAMPLE If 5 machines take 12 days to complete a job, how long will it take for 8 machines to do the job?

As the number of machines *increases*, the amount of time required to do the job *decreases*. Thus, the quantities are *inversely proportional*.

Pair 1: 5 machines finish in 12 days.

Pair 2: 8 machines finish in x days.

Estimation We expect more machines to do the job in less than 12 days. Also, we did not double the number of machines, so it will take more than half of the time (6 days).

$$\frac{5 \text{ machines}}{8 \text{ machines}} = \frac{x \text{ days for 8 machines}}{12 \text{ days for 5 machines}}$$

Each ratio uses like measures. The pairs are arranged inversely.

Dimension Analysis

$$\frac{5}{8} = \frac{x}{12}$$

$$\frac{\text{machines}}{\text{machines}} = \frac{\text{days}}{\text{days}}$$

$$5(12) = 8x$$

Cross multiply: machines(days) = machines(days)

$$60 = 8x$$

Divide by machines.

$$\frac{60}{8} = x$$

$$\frac{\cancel{\text{machines}} (\text{days})}{\cancel{\text{machines}}} = \text{days}$$

$$x = \frac{15}{2} \quad \text{or} \quad 7\frac{1}{2} \text{ days}$$

Interpretation It will take $7\frac{1}{2}$ days for 8 machines to do the job.

Why Is Estimation So Important?

Suppose we arrange the data so that each pair forms a ratio of unlike measures and then we invert one of the ratios. Look at the value to see if the answer conforms to what we expect the answer to be.

Pair 1   Reciprocal of Pair 2

$$\frac{5 \text{ machines}}{12 \text{ days}} = \frac{x \text{ days}}{8 \text{ machines}}$$

$$12x = 40$$

$$x = 3.33 \text{ days}$$

How does this compare to our estimate? We expected our answer to be less than 12 days but more than 6 days.

$$\frac{\text{machines}}{\text{days}} = \frac{\text{days}}{\text{machines}}$$ Analyzing the dimensions we produce an incorrect statement.

$$\text{machines}(\text{machines}) = \text{days}(\text{days})$$

These measures are not equal. Estimation can identify proportions that are set up incorrectly!

Inverse variation has a conversion factor similar to the constant of direct variation. The inverse variation formula is $y = \frac{k}{x}$, where x is the independent variable, y is the dependent variable, and k is the *constant of inverse variation*. This formula can also be written as $k = xy$.

To find the constant of inverse variation:

1. Identify a pair of related data in which both values are known.
2. Write the related pair as a product. $k = xy$.
3. Multiply to find the constant of inverse variation.

The constant of inverse variation and the known data pair of 5 machines and 12 days can be used to solve the preceding example.

$$k = 12(5)$$ Using $k = xy$, find k for $y = 5$ and $x = 12$.

$$k = 60$$ Constant of inverse variation.

$$y = \frac{k}{x}$$ Substitute $k = 60$ and $y = 8$.

$$8 = \frac{60}{x}$$ Solve for x.

$$8x = 60$$ Divide.

$$x = \frac{60}{8}$$

$$x = 7\frac{1}{2} \qquad \text{Days for 8 machines to complete the job.}$$

Gears and pulleys involve *inverse* relationships. Suppose a large gear and a small gear are in mesh, or a large pulley is connected by a belt to a smaller pulley. The larger gear or pulley has the *slower* speed, and the smaller gear or pulley has the *faster* speed.

EXAMPLE   A 10-in.-diameter gear is in mesh with a 5-in.-diameter gear (Fig. 9–4). If the larger gear has a speed of 25 revolutions per minute (rpm), at how many rpm does the smaller gear turn?

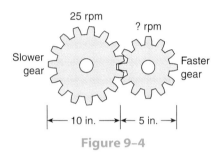

Figure 9–4

Because gears in mesh are *inversely* related, we set up an inverse proportion. Each ratio uses like measures and the ratios are in inverse order.

Pair 1: **25 rpm** of larger gear for **10-in**. size of larger gear.

Pair 2: **x rpm** of smaller gear for **5 in**. size of smaller gear.

Estimation   We expect the speed of the smaller gear to be faster than the 25-rpm speed of the larger gear.

$$\frac{25 \text{ rpm of larger gear}}{x \text{ rpm of smaller gear}} = \frac{5 \text{ in. of smaller gear}}{10 \text{ in. of larger gear}} \qquad \begin{array}{l} \text{Pair 2} \\ \text{Pair 1} \end{array} \qquad \text{Set up ratios in inverse order.}$$

$$\frac{25}{x} = \frac{5}{10} \qquad\qquad\qquad \text{Cross multiply.}$$

$$25(10) = 5x$$

$$250 = 5x \qquad\qquad\qquad \text{Divide.}$$

$$\frac{250}{5} = x$$

$$50 = x$$

Interpretation   Thus, the smaller gear turns at the faster speed of 50 rpm.

We can distinguish between direct and inverse variation by anticipating cause-and-effect situations.

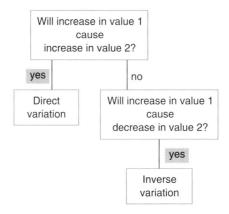

Similarly,

Decrease causes decrease = Direct variation

Decrease causes increase = Inverse variation

4 Solve Problems That Involve Similar Triangles.

Many applications involving proportions are based on the properties of similar triangles. Before we solve applied problems using similar triangles, let's examine the definitions of similar and congruent triangles. *Congruent triangles* have the same size and shape (Fig. 9–5). *Similar triangles* have the same shape but not the same size (Fig. 9–6).

Every triangle has six parts: three angles and three sides. Each angle or side of one similar or congruent triangle has a corresponding angle or side in the other similar or congruent triangle. The symbol for showing congruency is ≅ and is read "is congruent to." The symbol for a triangle is Δ.

Properties of congruent triangles:

Congruent triangles $\triangle ABC \cong \triangle DEF$

Figure 9–5

Corresponding angles of congruent triangles are the same size; that is, they are equal in measure.	Corresponding sides of congruent triangles are the same size; that is, they are equal in measure.
Angle A = angle D	Side a = side d
Angle B = angle E	Side b = side e
Angle C = angle F	Side c = side f

The symbol for showing similarity is ~ and is read "is similar to."

Properties of similar triangles:

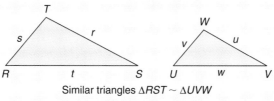

Similar triangles △RST ~ △UVW

Figure 9–6

Corresponding angles of similar triangles are equal in size.

Angle R = angle U
Angle S = angle V
Angle T = angle W

Corresponding sides of similar triangles are directly proportional.

Side r corresponds to side u
Side s corresponds to side v
Side t corresponds to side w

Examine the corresponding angles (Fig. 9–7). The symbol for angle is ∠.

∠ A corresponds to and is equal in measure to ∠ M.
∠ B corresponds to and is equal in measure to ∠ N.
∠ C corresponds to and is equal in measure to ∠ Q.

Examine the corresponding sides. Sides are named with the capital letters that identify the end points or with a lowercase letter that represents the angle opposite the side.

Side BC corresponds to side NQ a corresponds to m
Side AC corresponds to side MQ or b corresponds to n
Side AB corresponds to side MN c corresponds to q

The corresponding sides of similar triangles are proportional.

$$\frac{BC}{NQ} = \frac{AC}{MQ} = \frac{AB}{MN} \qquad \text{or} \qquad \frac{a}{m} = \frac{b}{n} = \frac{c}{q}$$

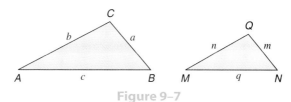

Figure 9–7

To find a missing side of a similar triangle:

1. Examine the given dimensions of the corresponding sides and find one pair for which both measures are known.
2. Pair the unknown side with its corresponding known side.
3. Set up a direct proportion with the two pairs.
4. Solve the proportion.

EXAMPLE   Find the missing side in Fig. 9–8 if $\triangle ABC \sim \triangle DEF$.

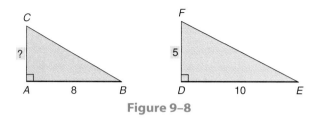

Figure 9–8

Write ratios of corresponding sides in a proportion. Use single lowercase letters to identify the sides. Side a is opposite $\angle A$, side b is opposite $\angle B$, and so on.

Pair 1: 8 from $\triangle ABC$ corresponds and is proportional to 10 from $\triangle DEF$.
Pair 2: b from $\triangle ABC$ corresponds and is proportional to 5 from $\triangle DEF$.

Estimation Side b should be less than 5 since 8 is less than 10.

$$\frac{\text{Pair 1}}{\text{Pair 2}} \quad \frac{8}{b} = \frac{10}{5}$$

Pair with both measures known.
Pair with one measure known.

$$10b = 8(5) \qquad \text{Cross multiply.}$$
$$10b = 40 \qquad \text{Divide.}$$
$$b = 4$$

Interpretation **Side b is 4 units long.**

EXAMPLE   A tree surgeon must know the height of a tree to determine which way to fell it so that it does not endanger lives, traffic, or property. A 6-ft pole casts a 4-ft shadow when the tree casts a 20-ft shadow (Fig. 9–9). What is the height of the tree?

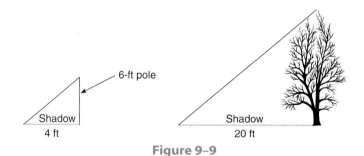

Figure 9–9

The triangles formed are similar. Let x = the height of the tree.

Pair 1: 6-ft pole casts a 4-ft shadow.   Both measures known.
Pair 2: x-ft tree casts a 20-ft shadow.   One measure known.

Estimation The tree is more than 20 ft tall.

Since the pole is taller than its shadow, the tree will be taller than its shadow of 20 ft.

$$\frac{6 \,(\text{height of pole})}{x \,(\text{height of tree})} = \frac{4 \,(\text{shadow of pole})}{20 \,(\text{shadow of tree})} \qquad \begin{array}{l} \text{Pair 1} \\ \text{Pair 2} \end{array}$$
$$6(20) = 4x \qquad \text{Cross multiply.}$$
$$120 = 4x \qquad \text{Divide.}$$
$$30 = x$$

Interpretation **The tree is 30 ft tall.**

A building lies between points A and B, so the distance between these points cannot be measured directly by a surveyor. Find the distance using the similar triangles shown in Fig. 9–10.

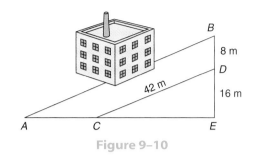

Figure 9–10

$\triangle ABE \sim \triangle CDE$. Note that CD must be made parallel to AB for the triangles to be similar. Parallel means the lines are the same distance apart from end to end.

Visualize the triangles as separate triangles (Fig. 9–11). Using the lowercase letter e for the missing AB measure is confusing because in $\triangle CDE$, side CD can also be thought of as side e.

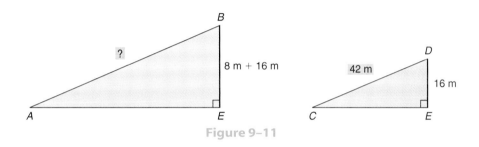

Figure 9–11

Pair 1: AB from $\triangle ABE$ corresponds to CD from $\triangle CDE$.
Pair 2: BE from $\triangle ABE$ corresponds to DE from $\triangle CDE$.

Estimation AB is more than 42 m.

$$\frac{AB}{BE} = \frac{CD}{DE}$$ Substitute values. $CD = 42$; $DE = 16$.

   $BE = BD + DE = 8 + 16 = 24$.

$$\frac{AB}{24} = \frac{42}{16}$$ Cross multiply.

$16AB = (24)42$ AB is interpreted as a single variable.
$16AB = 1,008$ Divide.
$AB = 63$ m

Interpretation   Thus, the distance from A to B is 63 m.

1 Solve the proportions.

1. $\dfrac{x}{5} = \dfrac{9}{15}$

2. $\dfrac{3x}{16} = \dfrac{3}{8}$

3. $\dfrac{x-1}{x+6} = \dfrac{4}{5}$

4. $\dfrac{5}{x} = 8$

5. $\dfrac{2}{7} = \dfrac{x-4}{x+3}$

6. $\dfrac{2x+1}{8} = \dfrac{3}{7}$

7. $\dfrac{3x-2}{3} = \dfrac{2x+1}{3}$

8. $\dfrac{5}{2x-2} = \dfrac{1}{8}$

9. $\dfrac{8}{3x+2} = \dfrac{8}{14}$

10. $\dfrac{2x}{8} = \dfrac{3x+1}{7}$

2 Solve using direct variation.

11. If 7 cans of dog food sell for $4.13, how much will 10 cans sell for?

12. If 6 cans of coffee sell for $22.24, how much will 20 cans sell for?

13. A mechanic took 7 hr to tune up 9 fuel-injected engines. At this rate, how many fuel-injected engines can be tuned up in 37.5 hr? Round to the nearest whole number.

14. A costume maker took 9 hr to make 4 headpieces for a Mardi Gras ball. At this rate, how many complete headpieces can be made in 35 hr?

15. How far can a family travel in 5 days if it travels at the rate of 855 mi in 3 days?

16. How far can a tractor-trailer rig travel in 8 days if it travels at the rate of 1,680 mi in 4 days?

17. How much crystallized insecticide does 275 acres of farmland need if the insecticide treats 50 acres per 100 lb?

18. How much fertilizer does 2,625 ft^2 of lawn need if the fertilizer treats 1,575 ft^2 per gallon? Express the answer to the nearest tenth of a gallon.

19. A can of tomatoes contains 793 g and costs $1.49. Find the constant of direct variation for the cost per gram.

20. A can of tomatoes contains 411 g and costs $0.89. Find the constant of direct variation for the cost per gram.

21. From Exercises 19 and 20 compare the unit cost of the small can of tomatoes with the unit cost of the large can to determine which size can is more economical.

22. A bottling machine can fill 800 bottles in 2.5 hr. Find the constant of direct variation for the number of bottles filled per hour.

3 Solve using inverse variation.

23. A 12-in. pulley turns at 30 rpm. Find the constant of inverse variation.

24. A 16-in. gear makes 60 rpm. Find the constant of inverse variation.

25. The fan pulley and alternator pulley are connected by a fan belt on an automobile engine. The fan pulley is 225 cm in diameter and the alternator pulley is 125 cm in diameter. If the fan pulley turns at 500 rpm, how many revolutions per minute does the faster alternator pulley turn?

26. The volume of a certain gas is inversely proportional to the pressure on it. If the gas has a volume of 160 in^3 under a pressure of 20 pounds per square inch (psi), what is the volume if the pressure is decreased to 16 psi?

27. A pulley that measures 15 in. across (diameter) turns at 1,600 rpm and drives a larger pulley at the rate of 1,200 rpm. What is the diameter of the larger pulley in this inverse relationship?

28. Six painters can trim the exterior of all the new brick homes in a subdivision in 9 weeks. The contractor wants to have the homes ready in just 3 weeks and so needs more painters. How many painters are needed for the job? Assume that the painters all work at the same rate.

29. Two groundskeepers take 25 hr to prepare a golf course for a tournament. How long would it take five groundskeepers to prepare the golf course?

30. A gear measures 5 in. across. It turns another gear 2.5 in. across. If the larger gear has a speed of 25 revolutions per minute (rpm), what is the rpm of the smaller gear?

31. A small pulley 3 in. in diameter turns 250 rpm and drives a larger pulley at 150 rpm. What is the diameter (distance across) of the larger pulley?

32. Two painters working at the same speed can paint 800 ft^2 of wall space in 6 hr. If a third painter paints at the same speed, how long will it take the three of them to paint the same wall space?

33. Three machines complete a printing project in 5 hr. How many machines are needed to finish the same project in 3 hr?

34. A gear measures 6 in. across. It is in mesh with another gear with a diameter of 3 in. If the larger gear has a speed of 60 revolutions per minute (rpm), what is the rpm of the smaller gear?

35. Nurse Lee prepares dosages for her patients in 30 min. If she gets help from assistants, who also work at her rate, and together they complete the preparation in 6 min, how many helpers did she get?

36. A 4.5-in. pulley turning at 1,000 revolutions per minute (rpm) is belted to a larger pulley turning at 500 rpm. What is the size of the larger pulley?

37. $AB = FE$
$BC = DE$
$AC = DF$

38. $\angle R$ and $\angle U = 90°$
$QR = TU$
$RP = SU$

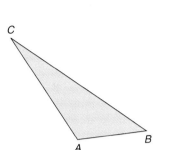
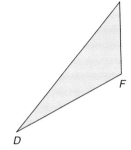

Figure 9–12

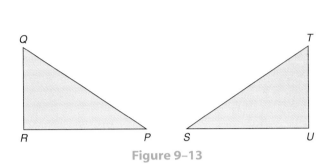

Figure 9–13

39. $JK = NM$
$\angle J = \angle M$
$\angle K = \angle N$

40. $\triangle ABC \sim \triangle EDF$. $\angle A = \angle E$, $\angle C = \angle F$. Find a and d.
Use Fig. 9–15.

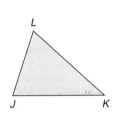
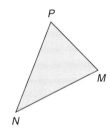

Figure 9–14

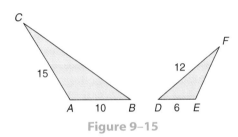

Figure 9–15

41. $\triangle ABC \sim \triangle DEC$. $\angle 1$ and $\angle 2$ have the same measure. Find DC and DE. (*Hint*: Let $DC = x$ and $AC = x + 3$. Use Fig. 9–16.)

42. Find the height of a tree that casts a 30-ft shadow when a 6-ft 6-in. pole casts a shadow of 3 ft.

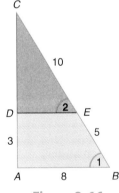

Figure 9–16

ASSIGNMENT EXERCISES

Section 9–1

Solve the equations.

1. $m + \dfrac{1}{4} = \dfrac{3}{4}$

2. $\dfrac{3}{5}y = 12$

3. $P = \dfrac{1}{2} + \dfrac{1}{3}$

4. $x + \dfrac{1}{7}x = 16$

5. $\dfrac{2}{5} - x = \dfrac{1}{2}x + \dfrac{4}{5}$

6. $\dfrac{R}{7} - 6 = -R$

7. $\frac{3}{7}m - \frac{1}{2} = \frac{2}{3}$

8. $0 = \frac{8}{9}c + \frac{1}{4}$

9. $\frac{2}{15}P - P = 4$

10. $\frac{1}{4}S = \frac{1}{4} + \frac{1}{10} + \frac{1}{20}$

11. $m = 2 + \frac{1}{4}m$

12. $\frac{7}{9} + 3 = \frac{1}{2}T$

13. $\frac{1}{12}x = \frac{1}{4} + \frac{1}{8}$

Solve the equations. Identify extraneous values.

14. $6x + \frac{1}{4} = 5$

15. $2x + 4 = \frac{1}{2}$

16. $\frac{2}{5}x + 6 = \frac{2}{3}x$

17. $\frac{3}{10}x = \frac{1}{8}x + \frac{2}{5}$

18. $\frac{2}{5} + 3x = \frac{1}{10} - x$

19. $\frac{1}{x} = \frac{2}{5} + \frac{3}{10}$

20. $\frac{1}{R} = \frac{1}{10} + \frac{1}{3} + \frac{1}{6}$

21. $\frac{2}{P} = \frac{1}{2} + \frac{1}{4} - \frac{5}{12}$

22. $\frac{3}{x} + 4 = \frac{1}{5} - 7$

23. $\frac{5}{12}x - \frac{3}{4} = \frac{1}{9} - \frac{2}{3}x$

Set up an equation with fractions and solve.

24. A plastic tube fills a container in 10 min. A drain in the container, however, empties the container in 30 min. If the drain is open, how long does it take to fill the container?

25. Melissa can complete a landscape project in 3 hr. Henry can complete the same project in 7 hr. How long would it take Melissa and Henry working together to complete the landscape project?

26. One optical scanner reads a stack of sheets in 20 min. A second scanner reads the same stack in 12 min. How long does it take for both scanners together to process the one stack of sheets?

27. An apprentice electrician can install 5 light fixtures in 2 hr. How many light fixtures can be installed in 10 hr?

28. A brick mason can erect a retaining wall in 6 hr. The brick mason's apprentice can do the same job in 10 hr. How much time does it take both of them working together to erect the retaining wall?

29. A parallel dc circuit has 3 branches rated at 2 ohms, 6 ohms, and 12 ohms. Find the total resistance. See example on p. 408.

30. Find the total resistance of a parallel circuit with 2 branches rated at 12 ohms and 16 ohms.

Section 9–2

Solve the equations. Round to tenths if necessary.

31. $3.4 = 1.5 + T$

32. $2y + 2.9 = 11.7$

33. $2.3x - 4.1 = 0.5$

34. $0.22 + 1.6x = -0.9$

35. $6.8 = 0.2y - 8.64$

36. $1.4x - 7.2 = 3.5x - 4.3$

37. $0.3x - 2.15 = 0.8x + 3.75$

38. $1.3x + 2 = 8.6x - 3.24$

39. $2.7 - x = 5 + 2x$

40. $4x - 3.2 + x = 3.3 - 2.4x$

41. $6.7y - y = 8.4$ (round to tenths)

42. $0.9R = 0.3$ (round to tenths)

43. $\frac{4x}{0.7} = \frac{3}{1.2}$ (round to hundredths)

44. $0.86 = R + 0.4R$ (round to hundredths)

45. $0.04y = 0.02 - y$ (round to hundredths)

46. $\frac{x}{6} = \frac{1.8}{3}$

47. $\frac{2.1}{x} = \frac{4.3}{7}$ (round to tenths)

48. The distance formula is Distance = Rate × Time. If a portable MRI unit traveled 350.8 mi to and from a rural hospital at 50 mph, how long to the nearest hour did the trip to and from the hospital take?

49. Electrical resistance in ohms (Ω) is voltage (V) divided by amperage (A). Find the resistance of a small motor with a voltage of 8.5 V requiring 0.5 A.

50. Lester earned $29.69 working $4\frac{3}{4}$ hr for a landscape service. What was his hourly wage?

51. The formula for voltage is $V = \frac{W}{A}$: voltage (V) equals wattage (W) divided by amperage (A). Find the voltage to the nearest hundredth needed for a circuit of 385 W with a current of 3.5 A.

52. If the formula for simple interest is Interest = Principal × Rate × Time, find the interest to the nearest cent on a principal of $1,000 at an 8.5% rate for 1.5 years.

53. If the formula for force is Force = Pressure × Area, how many pounds of force are produced by a pressure of 30 psi on a piston whose surface area is 12.5 in²?

Solve the proportions.

54. $\dfrac{x}{6} = \dfrac{5}{3}$

55. $\dfrac{3x}{8} = \dfrac{3}{4}$

56. $\dfrac{x-3}{x+6} = \dfrac{2}{5}$

57. $\dfrac{7}{x} = 6$

58. $\dfrac{2}{3} = \dfrac{x+3}{x-7}$

59. $\dfrac{4x+3}{15} = \dfrac{1}{3}$

60. $\dfrac{3x-2}{3} = \dfrac{2x+1}{4}$

61. $\dfrac{5}{4x-3} = \dfrac{3}{8}$

62. $\dfrac{8}{3x-2} = \dfrac{2}{3}$

63. $\dfrac{4x}{7} = \dfrac{2x+3}{3}$

64. $\dfrac{2x}{3x-2} = \dfrac{5}{8}$

65. $\dfrac{5x}{3} = \dfrac{2x+1}{4}$

66. $\dfrac{5}{9} = \dfrac{x}{2x-1}$

67. $\dfrac{7}{x} = \dfrac{5}{4x+3}$

68. $\dfrac{3}{5} = \dfrac{2x-3}{7x+4}$

Solve the problems using proportions.

69. If 15 machines can complete a job in 6 weeks, how many machines are needed to complete the job in 4 weeks?

70. In preparing a banquet for 30 people, Cedric Henderson uses 9 lb of potatoes. How many pounds of potatoes will be needed for a banquet for 175 people?

71. A car with a speed control device travels 100 mi at 50 mph. The trip takes 2 hr. If the car traveled at 40 mph, how much time would the driver need to reach the same destination?

72. A 6-ft landscape engineer casts a 5-ft shadow on the ground. How tall is a nearby tree that casts a 30 ft shadow?

73. There are 25 women in a class of 35 students. If this is typical of all classes in the college, how many women are enrolled in the college if it has 6,300 students altogether?

74. It takes five people 7 days to clear an acre of land of debris left by a tornado; inversely, more people can do the job in less time. How long would it take seven people all working at the same rate?

75. A gear with a diameter of 45 cm is in mesh with a gear that has a diameter of 30 cm. If the larger gear turns at 1,000 rpm, how many revolutions per minute does the smaller gear make?

76. A certain fabric sells at a rate of 3 yd for $7.00. How many yards can Clemetee Whaley buy for $35.00?

77. Three workers take 5 days to assemble a shipment of microwave ovens; inversely, more workers can do the job in less time. How long will it take five workers to do the same job?

78. An architect's drawing is scaled at $\frac{3}{4}$ in. = 6 ft. What is the actual height of a door that measures $\frac{7}{8}$ in. on the drawing?

79. A CAD program scales a blueprint so that $\frac{5}{8}$ in. = 2 ft. What is the actual measure of a wall that is shown as $1\frac{5}{16}$ in. on the blueprint?

80. A 10-in. pulley makes 900 revolutions every minute. It drives a larger pulley at 500 rpm. What is the diameter of the larger pulley in this inverse relationship?

81. On recent trips to rural health centers, a portable mammography unit used 81.2 gal of unleaded gasoline. If the travel involved a total of 845 mi, how many gallons of gasoline would be used for travel of 1,350 mi? Round to tenths.

82. A pulley whose diameter is 3.5 in. is belted to a pulley whose diameter is 8.5 in. In this inverse relationship, if the smaller, faster pulley turns at the rate of 1,200 rpm, what is the rpm of the slower pulley? Round to the nearest whole number.

83. A contractor estimates that a painter can paint 300 ft^2 of wall space in 3.5 hr. How many hours should the contractor estimate for the painter to paint 425 ft^2? Express your answer to the nearest tenth.

84. A coffee company mixes 1.6 lb of chicory with every 3.5 lb of coffee. At this ratio, how many pounds of chicory are needed to mix with 2,500 lb of coffee? Round to the nearest whole number.

85. A wire 825 ft long has a resistance of 1.983 ohms (Ω). How long is a wire of the same diameter if the resistance is 3.247 Ω? Round to the nearest whole foot.

86. The ceramic tile for a job weighs 510 lb. If the average weight of the ceramic tile is 4.25 lb per square foot, how many square feet are estimated for this job?

87. A gear is 4 in. across. It turns a smaller gear 2 in. across. If the larger gear has a speed of 30 rpm, what is the rpm of the smaller gear?

88. A small pulley with a 6-in. diameter turns 350 rpm and drives a larger pulley at 150 rpm. What is the diameter (distance across) of the larger pulley?

89. Two machines can complete a printing project in 6 hr. How many machines would be needed to finish the same project in 4 hr?

90. Waylon can install a hard drive in 10 PCs in 6 hr. He gets help from assistants who work at his rate and together they complete the installation in 2 hr. How many helpers did he get?

91. Write proportions for the similar triangles in Fig. 9–17.

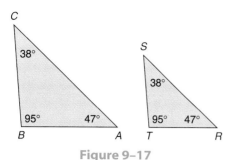

Figure 9–17

92. $\triangle LMN \sim \triangle XYZ$, $\angle L = \angle X$, $\angle M = \angle Y$. Find m and x (see Fig. 9–18).

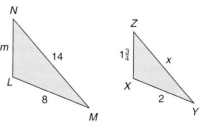

Figure 9–18

93. Find AB if $\triangle DEC \sim \triangle AEB$ (see Fig. 9–19).

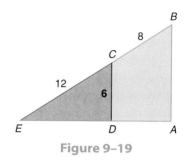

Figure 9–19

CHALLENGE PROBLEMS

94. Make up a word problem that can be solved with a direct proportion. Include a lawn mower, tanks of gasoline, and acres to be mowed.

96. Make up a word problem that can be solved with an inverse proportion. Include a belt, pulleys, rpm's, and diameters of the pulleys.

95. A gear turning at 130 rpm has 50 teeth. It is in mesh with another gear that turns at 65 rpm. How many teeth does the other gear have?

97. The ratio of water to antifreeze in a mixture of radiator solution is 2 to 5. If the radiator is filled with 10 gal of liquid, how much is water and how much is antifreeze?

CONCEPTS ANALYSIS

1. What is the rule for clearing an equation of fractions or denominators?

3. What is an extraneous root?

5. What is the rule for clearing an equation of decimals?

7. State the simple interest formula in words and symbols.

9. Explain the difference between a direct proportion and an inverse proportion.

11. Give some examples of situations that are directly proportional.

2. When the variable is in the denominator of a fraction, why should the solution always be checked?

4. How can the LCM for a decimal equation be identified easily?

6. Is it always wise to clear an equation of decimals? Explain.

8. Express in symbols the property of a proportion.

10. Explain how to set up a direct proportion and an inverse proportion.

12. Give some examples of situations that are inversely proportional.

Find a mistake in the examples, explain it, and correct it.

13.
$$\frac{2}{3}x = \frac{4}{5}$$
$$\left(\frac{3}{2}\right)\frac{2}{3}x = \left(\frac{4}{5}\right)\left(\frac{2}{3}\right)$$
$$x = \frac{8}{15}$$

14. $0.5x + x = 3.7$
$$0.6x = 3.7$$
$$\frac{0.6x}{0.6} = \frac{3.7}{0.6}$$
$$x = 6.167 \text{ (rounded)}$$

15. $3(0.2x - 3.1) = 4.7$
$$10[3(0.2x - 3.1)] = 10[4.7]$$
$$30(2x - 31) = 47$$
$$60x - 930 = 47$$
$$60x = 930 + 47$$
$$60x = 977$$
$$\frac{60x}{60} = \frac{977}{60}$$
$$x = 16.283 \text{ (rounded)}$$

Learning Outcomes   What to Remember with Examples

Section 9–1

1 Solve fractional equations by clearing the denominators (pp. 404–409).

To clear the equation of denominators: multiply the *entire* equation by the least common multiple (LCM) of the denominators of the equation.

$$\frac{3}{a} + \frac{1}{3} = \frac{2}{3a}$$ LCM = 3a.

$$(3a)\left(\frac{3}{a}\right) + (3a)\left(\frac{1}{3}\right) = (3a)\left(\frac{2}{3a}\right)$$ Multiply by 3a.

$$(\overset{3}{\cancel{3a}})\left(\frac{3}{\underset{1}{\cancel{a}}}\right) + (\overset{a}{\cancel{3a}})\left(\frac{1}{\underset{1}{\cancel{3}}}\right) = (\overset{1}{\cancel{3a}})\left(\frac{2}{\underset{1}{\cancel{3a}}}\right)$$

$$9 + a = 2$$
$$a = 2 - 9$$
$$a = -7$$

If a variable is in a denominator, check the root to see if it makes a true statement and is not an *extraneous root*.

Check:

$$\frac{3}{-7} + \frac{1}{3} = \frac{2}{3(-7)}$$ Substitute −7 for a.

$$\frac{3}{-7} + \frac{1}{3} = \frac{2}{-21}$$ LCM = 21.

$$(21)\left(\frac{3}{-7}\right) + (21)\left(\frac{1}{3}\right) = (21)\left(\frac{2}{-21}\right)$$

$$(\overset{-3}{\cancel{21}})\left(\frac{3}{\underset{1}{\cancel{-7}}}\right) + (\overset{7}{\cancel{21}})\left(\frac{1}{\underset{1}{\cancel{3}}}\right) = (\overset{-1}{\cancel{21}})\left(\frac{2}{\underset{1}{\cancel{-21}}}\right)$$

$$-9 + 7 = -2$$
$$-2 = -2$$ The root makes a true statement.

2 Solve applied problems involving rate, time, and work (pp. 409–412).

Amount of work: Rate × Time = Amount of work

A *rate* is a ratio such as 1 card per 15 min or $\dfrac{1\text{ card}}{15\text{ min}}$. Cancel units when solving.

Lashonda can install a PC video card in 15 min. How many cards can she install in $1\frac{1}{2}$ hours?

Amount of work (W) = Rate × Time

$$W = \frac{1\text{ card}}{15\text{ min}} \times 90\text{ min}$$ $1\frac{1}{2}$ hr = 90 min

$$W = 6\text{ cards}$$

Completing one job when working together:
(A's rate × time) + (B's rate × time) = 1 job

Let T = time. Let 1 = 1 completed job.

Galenda assembles a product in 10 min and Marcus assembles the product in 15 min. How long will it take them together to assemble one product? Galenda's rate × time: $\frac{1}{10}T$. Marcus's rate × time: $\frac{1}{15}T$.

$$\frac{1}{10}T + \frac{1}{15}T = 1$$

$$(30)\left(\frac{1}{10}T\right) + (30)\left(\frac{1}{15}T\right) = (30)(1)$$

$$(\overset{3}{30})\left(\frac{1}{10}T\right) + (\overset{2}{30})\left(\frac{1}{15}T\right) = (30)(1)$$

$$3T + 2T = 30$$

$$5T = 30$$

$$\frac{5T}{5} = \frac{30}{5}$$

$$T = 6 \text{ min}$$

Estimation: less than 10 min but more than half of 10, or 5 min.
Galenda's rate: $\frac{1}{10}$ per min.
Marcus's rate: $\frac{1}{15}$ per min.

Time in minutes: T.

LCM = 30.
Multiply by 30.

Combine like terms.

Divide.

Completing one job when working in opposition:
(A's rate × time) − (B's rate × time) = 1 job

Solve as above but *subtract* the two amounts of work.

An inlet valve fills a vat in 2 hr. A drain valve empties the vat in 5 hr. With both valves open, how long does it take for the vat to fill? Inlet valve's rate × time: $\frac{1}{2}T$. Drain valve's rate × time: $\frac{1}{5}T$.

$$\frac{1}{2}T - \frac{1}{5}T = 1$$

$$(10)\left(\frac{1}{2}T\right) - (10)\left(\frac{1}{5}T\right) = (10)1$$

$$(\overset{5}{10})\left(\frac{1}{2}T\right) - (\overset{2}{10})\left(\frac{1}{5}T\right) = (10)1$$

$$5T - 2T = 10$$

$$3T = 10$$

$$\frac{3T}{3} = \frac{10}{3}$$

$$T = 3\frac{1}{3} \text{ hr}$$

Estimation: more than 2 hr.

LCM = 10.
Multiply by 10.

Section 9–2

1 Solve decimal equations (pp. 413–414).

To clear the equation of decimals: multiply the *entire* equation by the least common multiple (LCM) of the fractional amounts following the decimal point. (The place value of the decimal amount with the most places after the decimal point will be the LCM.)

$$3.5x + 2.75 = 10$$

$$(100)(3.5x) + (100)(2.75) = (100)(10)$$

$$350x + 275 = 1{,}000$$

$$350x = 1{,}000 - 275$$

$$350x = 725$$

$$\frac{350x}{350} = \frac{725}{350}$$

$$x = 2.071428571$$

$$x = 2.07 \quad \text{(rounded)}$$

Place value of the LCM = 100.
Multiply by 100.
Sort.
Combine like terms.
Divide.

Chapter 9 / Equations with Fractions and Decimals

| 2 Solve applied problems with decimal equations (pp. 414–417). | In many decimal equations, it is just as convenient to work with the decimals as it is to clear the decimals. |

John worked 2.5 hr at $8.50 per hour. What was his pay? Let P = pay.

$P = 8.50(2.5)$ Pay = Rate per hr × hr. Estimation: 2.5 hr @ $10 per hr = $25. Pay < $25.

$P = \$21.25$

Interest formula: $I = PRT$, where I is interest, R is rate, P is principal, and T is time. If any three elements are given, the fourth one can be found. A calculator is useful in making the calculations.

Sequoia earned $300 on an investment of $1,500 for 2 years. What interest rate was she paid? Given: principal ($1,500), interest ($300), and time (2 years). The rate (R) is missing.

$$I = PRT$$
$$300 = 1,500 \times R \times 2$$
$$300 = 3,000R$$
$$\frac{300}{3,000} = \frac{3,000R}{3,000}$$
$$0.1 = R$$

Estimation: 10% of $1,500 for 1 yr = $150.
$150 × 2 yr = $300.

Change decimal to a percent: $0.1 = 10\%$ (move decimal 2 places to the right).

Section 9–3

| 1 Solve equations that are proportions (pp. 418–419). | Property of proportions: If $\dfrac{a}{b} = \dfrac{c}{d}$, then $ad = bc$ ($b, d \neq 0$). |

To solve: **1.** Find the cross products. **2.** Divide both sides by the coefficient.

$$\frac{4}{2} = \frac{x}{6}$$
$$(2)(x) = (4)(6) \quad \text{Cross products.}$$
$$2x = 24 \quad \text{Divide by coefficient.}$$
$$\frac{2x}{2} = \frac{24}{2}$$
$$x = 12$$

| 2 Solve problems with direct variation (pp. 419–423). | Verify that the problem involves direct proportion: As one amount increases (decreases), the other amount increases (decreases). |

Set up a direct proportion: **1.** Establish two pairs of related data. **2.** Write one pair of data in the numerators of two ratios. **3.** Write the other pair of data in the denominators of the two ratios. **4.** Form a proportion using the two ratios.

Kinta makes $72 for 8 hr of work in a hospital business office. How many hours must he work to make $150?

Proportion is direct: As hours increase, pay increases.

$$\frac{\$72}{\$150} = \frac{8 \text{ hr}}{x \text{ hr}} \quad \text{Pair 1. Estimation: hr > 8 hr.}$$
$$\qquad\qquad\qquad\qquad \text{Pair 2.}$$

$$\frac{72}{150} = \frac{8}{x} \quad \text{Cross multiply.}$$
$$(72)(x) = (8)(150)$$

Chapter Summary

$$72x = 1{,}200 \qquad \text{Divide.}$$
$$\frac{72x}{72} = \frac{1{,}200}{72}$$
$$x = 16\frac{2}{3}\ \text{hr}$$

3 Solve problems with inverse variation (pp. 423–426).

Verify that the problem involves inverse proportion: As one amount increases (decreases), the other amount decreases (increases).

Set up an inverse proportion: **1.** Establish two pairs of related data. **2.** Arrange one pair as the numerator of one ratio and the denominator of the other. **3.** Arrange the other pair so that each ratio contains like measures. **4.** Form a proportion using the two ratios.

Jane can paint a room in 4 hr. If she has two helpers who also can paint a room in 4 hr., how long will it take all three to paint the same room?

Proportion is inverse: As number of painters increases, time decreases.

$$\frac{1\ \text{painter}}{3\ \text{painters}} = \frac{x\ \text{hr}}{4\ \text{hr}} \qquad \text{Pair 1: 1 painter, 4 hr.} \qquad \text{Estimation: hr < 4 hr.}$$
$$\text{Pair 2: 3 painters, } x\ \text{hr.}$$
$$\frac{1}{3} = \frac{x}{4}$$
$$(3)(x) = (1)(4)$$
$$3x = 4$$
$$\frac{3x}{3} = \frac{4}{3}$$
$$x = 1\frac{1}{3}\ \text{hr for 3 painters to paint the room.}$$

4 Solve problems that involve similar triangles (pp. 426–429).

Each angle of one triangle is equal to its corresponding angle in the other similar triangle. Each side of one triangle is proportional to its corresponding side in the other triangle.

Find the height of a building that makes a shadow of 25 m when a meter stick makes a shadow of 1.5 m.

Direct proportion: As shadow increases, height increases.

$$\frac{x\text{-m building}}{1\text{-m stick}} = \frac{25\text{-m building shadow}}{1.5\text{-m stick shadow}} \qquad \begin{array}{l}\text{Pair 1.} \\ \text{Pair 2}\end{array} \quad \begin{array}{l}\text{Estimation: meter stick} \\ \text{< shadow, so building < 25 m.}\end{array}$$

$$\frac{x}{1} = \frac{25}{1.5} \qquad \text{Cross multiply.}$$

$$1.5x = 25 \qquad \text{Divide.}$$
$$x = \frac{25}{1.5}$$
$$x = 16.67\ \text{m}$$

CHAPTER TRIAL TEST

Solve the equations.

1. $\dfrac{3}{8}y = 6$

2. $4 = \dfrac{1}{3}x + 2$

3. $\dfrac{3a}{7} = 9$

4. $\dfrac{R}{7} = \dfrac{2}{5}$

5. $\dfrac{3 + Q}{1} = \dfrac{4}{5}$

6. $\dfrac{3}{y + 2} = \dfrac{2}{3}$

7. $\dfrac{8}{y+2} = -7$

8. $\dfrac{4}{5}z + z = 8$

9. $\dfrac{2}{7}x = \dfrac{1}{2}x + 4$

10. $-\dfrac{2}{3}x = \dfrac{1}{4}x - 11$

11. $5x + \dfrac{3}{5} = 2$

12. $3x + 2 = \dfrac{2}{3}$

13. $\dfrac{3}{5}x + \dfrac{1}{10}x = \dfrac{1}{3}$

14. $\dfrac{1}{x} = \dfrac{1}{3} + \dfrac{5}{6}$

Solve the equations. Round to hundredths when necessary.

15. $1.3x = 8.02$

16. $4.5y + 1.1 = 3.6$

17. $\dfrac{1.2}{x} = 4.05$

18. $\dfrac{3.8}{6} = \dfrac{0.05}{R}$

19. $0.18x = 300 - x$

20. $4.3 = 7.6 + x$

21. $3x + 1.4 = 8.9$

22. $7.9 = 0.5x - 8.35$

23. $0.23 + 7.1x = -0.8$

Solve the problems involving fractions, decimal numbers, and proportions.

24. A pipe fills 1 tank in 4 hr. If a second pipe empties 1 tank in 6 hr, how long does it take for the tank to fill with both pipes operating?

25. A 9-in. gear is in mesh with a 4-in. gear. If the larger gear makes 75 rpm, how many revolutions per minute does the smaller gear make in this inverse relationship?

26. One employee can wallpaper a room in 2 hr. Another employee can wallpaper the same room in 3 hr. How long will it take both employees to wallpaper the room when working together?

27. Using the formula Pressure $= \dfrac{\text{Force}}{\text{Area}}$, how much pressure does a force of 32.75 lb exert on a surface of 24.65 in^2 in a hydraulic system? Answer in pounds per square inch (psi) rounded to hundredths.

28. A 1-year loan for the purchase of an electronically controlled assembly machine in a factory cost the management $2,758 in simple interest. If the interest rate was 8%, how much did the machine cost (principal)? Round to the nearest dollar. Interest = Principal × Rate × Time.

29. If a compact car used 62.5 L of unleaded gasoline to travel 400 mi, how many liters of gasoline would the driver use to travel 350 mi? Round to tenths.

30. If three workers take 8 days to complete a job, how many workers would be needed to finish the same job in only 6 days if each worked at the same rate? (More workers take fewer days.)

31. Resistance in a parallel dc circuit equals voltage divided by amperage. What is the resistance (in ohms) if the voltage is 40 V and amperage is 3.5 A? Express the answer as a decimal rounded to thousandths.

32. The ratio of men to women in technical and trade occupations is estimated to be 3 to 1, that is, $\frac{3}{1}$. If 56,250 men are employed in such occupations in a certain city, how many employees are women?

33. If an ice maker produces 75 lb of ice in $3\frac{1}{2}$ hr, how many pounds of ice would it produce in 5 hr? Round to the nearest whole number.

34. Find HI if $\triangle ABC \sim \triangle GHI$ (Fig. 9–20).

35. Find DB if $\triangle ABE \sim \triangle CDE$, $CD = 9$, $AB = 12$, $DE = 15$ (Fig. 9–21).

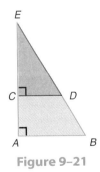

Figure 9–21

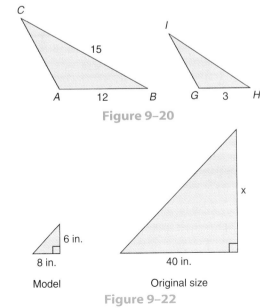

Figure 9–20

36. A model of a triangular part is shown in Fig. 9–22. Find side x if the part is to be similar to the model.

Figure 9–22

Powers and Polynomials

10

Learning Outcomes   **1** Multiply powers with like bases.
                      2 Divide powers with like bases.
                      3 Find a power of a power.

We have seen in various equations and formulas that a variable can represent all types of numbers: natural numbers, whole numbers, integers, fractions, decimals, and signed numbers. When raising a variable to a power, we must consider all the types of numbers the variable can be. As we study other types of numbers, we will look at their powers. Examine the chart of values for x^2, x^3, and x^4 in Table 10–1.

Table 10–1   Powers of *x*.

x	x^2	x^3	x^4
5	$(5)(5) = 25$	$(5)(5)(5) = 125$	$(5)(5)(5)(5) = 625$
-3	$(-3)(-3) = 9$	$(-3)(-3)(-3) = -27$	$(-3)(-3)(-3)(-3) = 81$
$\dfrac{1}{3}$	$\left(\dfrac{1}{3}\right)\left(\dfrac{1}{3}\right) = \dfrac{1}{9}$	$\left(\dfrac{1}{3}\right)\left(\dfrac{1}{3}\right)\left(\dfrac{1}{3}\right) = \dfrac{1}{27}$	$\left(\dfrac{1}{3}\right)\left(\dfrac{1}{3}\right)\left(\dfrac{1}{3}\right)\left(\dfrac{1}{3}\right) = \dfrac{1}{81}$
0.7	$(0.7)(0.7) = 0.49$	$(0.7)(0.7)(0.7) = 0.343$	$(0.7)(0.7)(0.7)(0.7) = 0.2401$
-2.1	$(-2.1)(-2.1) = 4.41$	$(-2.1)(-2.1)(-2.1) = -9.261$	$(-2.1)(-2.1)(-2.1)(-2.1) = 19.4481$

The *laws of exponents* have evolved from the patterns that formed when interpreting powers as repeated multiplication. These laws allow us to shorten our processes and to take advantage of the accessibility of technology. Be sure to notice for what circumstances a law is applicable. For example, many of the laws apply to factors with *like* bases.

1   **Multiply Powers with Like Bases.**

To multiply 2^3 by 2^2 using repeated multiplication, we have 2(2)(2) times 2(2) or 2(2)(2)(2)(2). Another way of writing this is $2^3 \times 2^2 = 2^5$. We can multiply x^4 and x^2. Even though the value of x is not known, $x^4(x^2)$ is $x(x)(x)(x)$ times $x(x)$ or $x(x)(x)(x)(x)(x)$.

$$x^4(x^2) = x^6$$

The product contains $4 + 2$ or 6 factors of x. The following law is a shortcut for using repeated multiplication.

To multiply powers that have like bases:

1. Verify that the bases are the same. Use this base as the base of the product.
2. Add the exponents for the exponent of the product.

This rule can be stated symbolically as

$$a^m(a^n) = a^{m+n} \qquad \text{where } a, m, \text{ and } n \text{ are real numbers.}$$

EXAMPLE   Write the products.

    (a) $y^4(y^3)$ (b) $a(a^2)$ (c) $b(b)$ (d) $x(x^3)(x^2)$ (e) $x^2(y^3)$ (f) $5x^3(2x^4)$

(a) $y^4(y^3) = y^{4+3} = \boldsymbol{y^7}$ The bases are the same, so add the exponents.
(b) $a(a^2) = a^{1+2} = \boldsymbol{a^3}$ When the exponent for a base is not written, it is 1.
(c) $b(b) = b^{1+1} = \boldsymbol{b^2}$

(d) $x(x^3)(x^2) = x^{1+3+2} = x^6$

(e) $x^2(y^3) = x^2y^3$ Bases are unlike.

(f) $5x^3(2x^4) = 10x^7$ Multiply coefficients. Add exponents of like bases.

TIP!

Do You Always Add Exponents When Multiplying?

The previous law applies *only* to expressions with *like* bases. Thus, $x^2(y^3)$ can only be written as x^2y^3. No other simplification can be made.

---

2 Divide Powers with Like Bases.

When reducing fractions, we can cancel or reduce to a factor of 1 any factors common to both the numerator and denominator. We use this concept when dividing powers that have like bases.

EXAMPLE Reduce or simplify the fractions; that is, perform the division indicated by each fraction.

(a) $\dfrac{x^5}{x^2}$ (b) $\dfrac{a^2}{a}$ (c) $\dfrac{b^7}{c^5}$ (d) $\dfrac{m^3}{m^3}$ (e) $\dfrac{y^3}{y^4}$

(a) $\dfrac{x^5}{x^2} = \dfrac{x(x)(x)(x)(x)}{x(x)} = x^3$ The quotient contains $5 - 2 = 3$ factors of x, or x^3.

(b) $\dfrac{a^2}{a} = \dfrac{a(a)}{a} = a$ **(or a^1)**   The quotient contains $2 - 1 = 1$ factor of a.

(c) $\dfrac{b^7}{c^5} = \dfrac{b^7}{c^5}$ Bases are unlike, so no simplification can be made.

(d) $\dfrac{m^3}{m^3} = \dfrac{(m)(m)(m)}{(m)(m)(m)} = 1$ All factors of m reduce to 1.

(e) $\dfrac{y^3}{y^4} = \dfrac{(y)(y)(y)}{(y)(y)(y)(y)} = \dfrac{1}{y}$ The denominator has more factors of y than the numerator.

In the example above examine the exponents in the division and the exponent of the quotient. In parts a and b, the exponent of the quotient is the difference of the exponents of the dividend and the divisor.

(a) $\dfrac{x^5}{x^2} = x^{5-2} = x^3$ (b) $\dfrac{a^2}{a} = a^{2-1} = a^1 = a$

The results in parts d and e illustrate that the interpretation of exponents goes beyond just repeated multiplication, a new notation for 1 and for reciprocals is needed.

(d) $\dfrac{m^3}{m^3} = m^{3-3} = y^{-3} = m^0 = 1$ Any nonzero base raised to the zero power equals 1.

(e) $\dfrac{y^3}{y^4} = y^{3-4} = y^{-1} = \dfrac{1}{y^1} = \dfrac{1}{y}$ An expression with a negative exponent equals the reciprocal of the expression written with a positive exponent having the same absolute value.

TIP!

Reciprocals and Negative Exponents

Any nonzero number raised to the zero power is equal to 1.

$$n^0 = 1 \quad \text{if} \quad n \neq 0.$$

An expression with a *negative exponent* can be written as an equivalent expression with a positive exponent.

$$n^{-1} = \frac{1}{n} \qquad \frac{1}{n^{-1}} = n \quad \text{if } n \neq 0$$

A nonzero number times its reciprocal equals 1.

$$n \cdot \frac{1}{n} = 1 \qquad n^1 \cdot n^{-1} = n^0 = 1$$

To divide powers that have like bases:

1. Verify that the bases are the same. Use this base as the base of the quotient.
2. Subtract the exponents for the exponent of the quotient.

This rule can be stated symbolically as

$$\frac{a^m}{a^n} = a^{m-n}$$

where a, m, and n are real numbers except that $a \neq 0$.

Because expressions with positive integral exponents are evaluated by using repeated multiplication, it is preferable to rewrite expressions with negative exponents as equivalent expressions with positive exponents. This manipulation is accomplished by applying the definition of negative exponents and exponents of zero.

EXAMPLE Write the quotients using positive exponents.

(a) $\dfrac{x^5}{x^8}$ (b) $\dfrac{a}{a^4}$ (c) $\dfrac{y^{-3}}{y^2}$ (d) $\dfrac{x^3}{x^{-5}}$ (e) $\dfrac{12\,x^7}{8\,x^5}$

(a) $\dfrac{x^5}{x^8} = x^{5-8} = x^{-3} = \dfrac{1}{x^3}$ (b) $\dfrac{a}{a^4} = a^{1-4} = a^{-3} = \dfrac{1}{a^3}$

(c) $\dfrac{y^{-3}}{y^2} = y^{-3-2} = y^{-5} = \dfrac{1}{y^5}$ (d) $\dfrac{x^3}{x^{-5}} = x^{3-(-5)} = x^{3+5} = x^8$

(e) $\dfrac{12\,x^7}{8\,x^5} = \dfrac{3\,x^{7-5}}{2} = \dfrac{3\,x^2}{2}$ or $\dfrac{3}{2}x^2$

TIP!

Manipulating Negative and Positive Exponents

The inverse relationship of multiplication and division allows us flexibility in applying the laws of exponents.

Look at part a of the previous example.

$$\frac{x^5}{x^8} \quad \text{is} \quad x^5 \div x^8 \quad \text{or} \quad x^5 \cdot \frac{1}{x^8} \quad \text{or} \quad x^5 \cdot x^{-8}$$

The multiplication law of exponents can be used.

$$x^5 \cdot x^{-8} = x^{5+(-8)} = x^{-3}$$

A factor can be moved from a numerator to a denominator (or vice versa) by changing the sign of its exponent.

$$x^{-2} = \frac{x^{-2}}{1} = \frac{1}{x^2}, \qquad \frac{1}{x^{-3}} = \frac{x^3}{1} = x^3$$

This property *does not apply* to a term that is part of a numerator or denominator that has two or more terms.

$$\frac{x^{-2}+1}{3} \qquad \textit{does not equal} \qquad \frac{1}{3x^2}.$$

In other words, *only factors* of the entire numerator or denominator can be moved. If more than one term is in the numerator, each term is divided by the denominator.

$$\frac{x^{-2}+1}{3} = \frac{x^{-2}}{3} + \frac{1}{3} = \frac{1}{3x^2} + \frac{1}{3}$$

EXAMPLE Simplify the expressions and make all exponents positive.

(a) $\dfrac{a^2 b^{-3}}{ab^{-1}}$ (b) $\dfrac{xy^{-1}}{xy^2}$ (c) $\dfrac{x^3 + y^2}{xy}$

There is more than one way to simplify the expressions.

(a)

Option 1

$$\frac{a^2 b^{-3}}{ab^{-1}} = a^{2-1} b^{-3-(-1)}$$ Apply the division law of exponents. $2 - 1 = 1; -3 - (-1) = -3 + 1 = -2$. Be careful with the signs.

$$= ab^{-2}$$ Make all exponents positive.

$$= \frac{a}{b^2}$$

Option 2

$$\frac{a^2 b^{-3}}{ab^{-1}} = \frac{a^2 a^{-1} b^{-3} b^1}{1}$$ Apply the property of negative exponents to write all factors in the numerator.

$$= ab^{-2} = \frac{a}{b^2}$$ Apply the multiplication law of exponents and make all exponents positive.

Chapter 10 / Powers and Polynomials

(b) $\dfrac{xy^{-1}}{xy^2} = x^{1-1}\,y^{-1-(2)}$ Apply the division law of exponents.

$ = y^{-3}$ Make exponent positive.

$ = \dfrac{1}{y^3}$

(c) $\dfrac{x^3 + y^2}{xy} = \dfrac{x^3}{xy} + \dfrac{y^2}{xy}$ Separate into two terms and apply division law to like bases.

$ = \dfrac{x^{3-1}}{y} + \dfrac{y^{2-1}}{x}$

$ = \dfrac{x^2}{y} + \dfrac{y}{x}$

3 **Find a Power of a Power.**

In the term $(2^3)^2$, we have a power raised to a power. 2^3 is the base of the expression and 2 is the exponent.

$$(2^3)^2 \quad \text{is} \quad (2^3)(2^3) = 2^{3+3} = 2^6 \quad \text{or} \quad 64$$

Let's look at some examples of raising numerical and variable powers to a power.

EXAMPLE Find the powers of powers by first expressing each expression as repeated multiplication and then applying the multiplication law of exponents.

(a) $(3^2)^3$ (b) $(x^3)^4$ (c) $(a^2)^2$ (d) $(n^3)^5$

(a) $(3^2)^3 = (3^2)(3^2)(3^2) = 3^{2+2+2} = 3^6 = \mathbf{729}$
(b) $(x^3)^4 = (x^3)(x^3)(x^3)(x^3) = x^{3+3+3+3} = \mathbf{x^{12}}$
(c) $(a^2)^2 = (a^2)(a^2) = a^{2+2} = \mathbf{a^4}$
(d) $(n^3)^5 = (n^3)(n^3)(n^3)(n^3)(n^3) = n^{3+3+3+3+3} = \mathbf{n^{15}}$

From the preceding example we can see a pattern developing. In each case, if we multiply the exponent, we get the exponent of the new power.

To raise a power to a power:

1. Multiply exponents.
2. Keep the same base.

$$(a^m)^n = a^{mn}$$

where a, m, and n are real numbers.

Applying this rule to the problems in the preceding example, we have

(a) $(3^2)^3 = 3^{2(3)} = 3^6 = 729$ (b) $(x^3)^4 = x^{3(4)} = x^{12}$

(c) $(a^2)^2 = a^{2(2)} = a^4$ (d) $(n^3)^5 = n^{3(5)} = n^{15}$

Other laws of exponents are extensions or combinations of the three laws we have already examined. Although these additional laws are not necessary, they are useful tools for simplifying expressions containing exponents.

To raise a fraction or quotient to a power:

1. Raise the numerator to the power.
2. Raise the denominator to the power.

$$\left(\frac{a}{b}\right)^n = \frac{a^n}{b^n} \qquad b \neq 0$$

EXAMPLE Raise the fractions to the indicated power.

(a) $\left(\frac{2}{3}\right)^2$ (b) $\left(\frac{-3}{4}\right)^2$ (c) $\left(\frac{-1}{3}\right)^3$ (d) $\left(\frac{x}{y^3}\right)^3$ (e) $\left(\frac{x^2}{y^3}\right)^4$

(a) $\left(\frac{2}{3}\right)^2 = \frac{2^2}{3^2} = \frac{4}{9}$

(b) $\left(\frac{-3}{4}\right)^2 = \frac{(-3)^2}{4^2} = \frac{9}{16}$ $(-3)(-3) = +9.$

(c) $\left(\frac{-1}{3}\right)^3 = \frac{(-1)^3}{3^3} = \frac{-1}{27}$ $(-1)(-1)(-1) = -1.$

(d) $\left(\frac{x}{y^3}\right)^3 = \frac{x^3}{(y^3)^3} = \frac{x^3}{y^9}$

(e) $\left(\frac{x^2}{y^3}\right)^4 = \frac{(x^2)^4}{(y^3)^4} = \frac{x^8}{y^{12}}$

To raise a product to a power:

Raise each factor to the indicated power.

$$(ab)^n = a^n b^n$$

EXAMPLE Raise the products to the indicated powers.

(a) $(ab)^2$ (b) $(a^2b)^3$ (c) $(xy^2)^2$ (d) $(3x)^2$ (e) $(2x^2y)^3$ (f) $(-5xy)^2$

(a) $(ab)^2 = a^{1(2)} b^{1(2)} = a^2 b^2$ Unwritten exponents are understood to be 1.

(b) $(a^2b)^3 = a^{2(3)} b^{1(3)} = a^6 b^3$

(c) $(xy^2)^2 = x^{1(2)} y^{2(2)} = x^2 y^4$

(d) $(3x)^2 = 3^{1(2)} x^{1(2)} = 3^2 x^2 = 9x^2$ Evaluate numerical factors.

(e) $(2x^2y)^3 = 2^{1(3)} x^{2(3)} y^{1(3)} = 2^3 x^6 y^3 = 8x^6 y^3$

(f) $(-5xy)^2 = (-5)^{1(2)} x^{1(2)} y^{1(2)} = (-5)^2 x^2 y^2 = 25x^2y^2$

TIP!

Limitations of the Laws of Exponents

It is very important to understand what the laws of exponents *do not* include.

- The product-raised-to-power law applies to factors, not terms.

$$(a + b)^3 \quad \textbf{does not equal} \quad a^3 + b^3$$

- The multiplication-of-powers law applies to like bases.

$$a^2(b^3) \quad \textbf{does not equal} \quad ab^5 \quad \text{or} \quad (ab)^5$$

- An exponent affects only the one factor or grouping immediately to the left.

$$3x^2 \quad \text{and} \quad (3x)^2 \quad \textbf{are not equal}$$

In the term $3x^2$, the numerical coefficient 3 is multiplied times the square of x. In the term $(3x)^2$, $3x$ is squared. Thus, $(3x)^2 = 3^2 x^2 = 9x^2$.

- A negative coefficient of a base is not affected by the exponent.

$$-x^3 \quad \textbf{means} \quad -(x)(x)(x)$$

If $x = 4$, $-x^3 = -(4)^3$ or $-(64) = -64$. If $x = -4$, $-x^3 = -(-4)^3$ or $-(-64) = 64$.

SELF-STUDY EXERCISES 10–1

1 Write the products.

1. $x^3(x^4)$
2. $m(m^3)$
3. $a(a)$
4. $x^2(x^3)(x^5)$
5. $y(y^2)(y^3)$
6. $a^2(b)$
7. $3a^5(4b^7)$
8. $2x^3(5x^8)$

2 Write the quotients with positive exponents.

9. $\dfrac{y^7}{y^2}$
10. $\dfrac{x^5}{x}$
11. $\dfrac{a^3}{a^4}$
12. $\dfrac{b^6}{b^5}$

13. $\dfrac{m^2}{m^2}$
14. $\dfrac{x}{x^3}$
15. $\dfrac{y^5}{y}$
16. $\dfrac{n^2}{n^{-5}}$

17. $\dfrac{x^{-4}}{x^6}$
18. $\dfrac{8n^2}{4n^3}$
19. $\dfrac{18x^7}{12x^0}$
20. $\dfrac{x^{-3}}{x^2}$

21. $\dfrac{x^2 y^{-2}}{xy^4}$
22. $\dfrac{ab^2}{a^{-2}b^4}$
23. $\dfrac{x^2 y^4}{xy^2}$
24. $\dfrac{a^3 b^2}{a^2 b^3}$

3 Simplify the expressions.

25. $(4^2)^3$
26. $(x^4)^2$
27. $(y^7)^0$
28. $(x^{10})^4$

29. $(a^7)^3$
30. $\left(-\dfrac{1}{2}\right)^3$
31. $\left(-\dfrac{2}{7}\right)^2$
32. $\left(\dfrac{a}{b}\right)^4$

33. $(2m^2 n)^3$
34. $\left(\dfrac{x^2}{y}\right)^3$
35. $(x^2 y^4)^3$
36. $(-2a)^2$

37. $(x^2 y)^3$
38. $(-3ab^2)^3$
39. $(-7x^2)^5$
40. $(4x^2 y^8)^2$

Learning Outcomes

1 Identify polynomials, monomials, binomials, and trinomials.
2 Identify the degree of terms and polynomials.
3 Arrange polynomials in descending order.

1 Identify Polynomials, Monomials, Binomials, and Trinomials.

A *polynomial* is an algebraic expression in which the exponents of the variables are nonnegative integers and there are no variables in a denominator.

EXAMPLE Identify which expressions are polynomials. If an expression is not a polynomial, explain why.

(a) $5x^2 + 3x + 2$ (b) $5x - \dfrac{3}{x}$ (c) 9 (d) $-\dfrac{1}{2}x + 3x^{-2}$

(a) $5x^2 + 3x + 2$ is a polynomial.

(b) $5x - \dfrac{3}{x}$ is not a polynomial because the term $-\dfrac{3}{x}$ is equivalent to $-3x^{-1}$. A polynomial cannot have a variable with a negative exponent.

(c) 9 is a polynomial because it is equivalent to $9x^0$ and the exponent, zero, is a nonnegative integer.

(d) $-\dfrac{1}{2}x + 3x^{-2}$ is not a polynomial because $3x^{-2}$ has a negative exponent.

Some polynomials have special names depending on the number of terms contained in the polynomial.

A *monomial* is a polynomial containing one term. A term may have more than one factor.

$$3, -2x, 5ab, 7xy^2, \dfrac{3a^2}{4} \text{ are monomials.}$$

A *binomial* is a polynomial containing two terms.

$$x + 3, 2x^2 - 5x, x + \dfrac{y}{4} \text{ are binomials.}$$

A *trinomial* is a polynomial containing three terms.

$$a + b + c, x^2 - 3x + 4, x + \dfrac{2a}{7} - 5 \text{ are trinomials.}$$

To identify polynomials, monomials, binomials, and trinomials:

1. Write all variables in the numerator if necessary.
2. Expressions that have a negative exponent in any term are not polynomials.
3. Identify the expression based on the number of terms it contains.

EXAMPLE Identify which of the expressions are polynomials, then state whether each polynomial is a monomial, binomial, or trinomial.

(a) $x^2y - 1$ (b) $4(x - 2)$

(c) $\dfrac{2x + 5}{2y}$ (d) $3x^2 - x + 1$

(a) $x^2y - 1$ — Binomial

(b) $4(x - 2)$ — Monomial

(c) $\dfrac{2x + 5}{2y}$ — This one-term expression is **not a monomial because it is not a polynomial** (the exponent of y would be -1 if it were in the numerator).

(d) $3x^2 - x + 1$ — Trinomial

2 Identify the Degree of Terms and Polynomials.

The *degree of a term* that has only one variable with a nonnegative exponent is the same as the exponent of the variable.

$$3x^4, \quad \text{fourth degree} \qquad -5x, \quad \text{first degree}$$

The degree of a *constant* is 0. A variable to the zero power is implied.

$$5 = 5x^0, \quad \text{degree zero}$$

If a term has more than one variable, the degree of the term is the *sum* of the exponents of all the variable factors.

$$2xy = 2x^1y^1, \quad \text{second degree} \qquad -2ab^2 = -2a^1b^2, \quad \text{third degree}$$

To identify the degree of a term:

1. Exponents of variable factors must be integers greater than zero.
2. For a term that is a constant, the degree is zero.
3. For a term that has only one variable factor, the degree is the same as the exponent of the variable.
4. For a term that has more than one variable factor, the degree is the sum of the exponents of the variables.

EXAMPLE Identify the degree of each term in the polynomial $5x^3 + 2x^2 - 3x + 3$.

$$5x^3, \quad \text{degree 3 (or third degree)}$$
$$2x^2, \quad \text{degree 2 (or second degree)}$$
$$-3x, \quad \text{degree 1 (or first degree)}$$
$$3, \quad \text{degree 0}$$

Special names are associated with terms of degree 0, 1, 2, and 3. A *constant term* has degree 0. A *linear term* has degree 1. A *quadratic term* has degree 2. A *cubic term* has degree 3.

The *degree of a polynomial* that has only one variable and only positive integral exponents is the degree of the term with the largest exponent.

A *linear polynomial* has degree 1. A *quadratic polynomial* has degree 2. A *cubic polynomial* has degree 3.

$$5x^3 - 2 \text{ has a degree of 3 and is a cubic polynomial.}$$

$$x + 7 \text{ has a degree of 1 and is a linear polynomial.}$$

$$7x^2 - 4x + 5 \text{ has a degree of 2 and is a quadratic polynomial.}$$

To identify the degree of a polynomial:

1. Identify the degree of each term of the polynomial.
2. Compare the degrees of each term of the polynomial and select the greatest degree as the degree of the polynomial.

EXAMPLE Identify the degree of the polynomials.

(a) $5x^4 + 2x - 1$ (b) $3x^3 - 4x^2 + x - 5$ (c) 7   (d) $x - \dfrac{1}{2}$

(a) $5x^4 + 2x - 1$ has a **degree of 4**.
(b) $3x^3 - 4x^2 + x - 5$ has a **degree of 3 and is a cubic polynomial**.
(c) 7 has a **degree of 0 and is a constant**.
(d) $x - \dfrac{1}{2}$ has a **degree of 1 and is a linear polynomial**.

3   **Arrange Polynomials in Descending Order.**

The terms of a polynomial are customarily arranged in order based on the degree of each term of the polynomial. The terms can be arranged beginning with the term with the highest degree (*descending order*) or beginning with the term with the lowest degree (*ascending order*).

Most Common Arrangement of Polynomials

Polynomials are most often arranged in **descending order** so that the degree of the polynomial is the degree of the first term.

The first term of a polynomial arranged in descending order is called the *leading term* of the polynomial. The coefficient of the leading term of a polynomial is called the *leading coefficient*.

To arrange polynomials in descending order of a variable:

1. Identify the variable on which the terms of the polynomial will be arranged if the polynomial has more than one variable.
2. Compare the degrees of the selected variable for each term.
3. List the term with highest degree of the specified variable first.
4. Continue to list the terms of the polynomial in descending order of the selected variable.

Arrange each polynomial in descending order and identify the degree, the leading term, and the leading coefficient of the polynomial.

(a) $5x + 3x^3 - 7 + 6x^2$ (b) $x^4 - 2x + 3$ (c) $4 + x$ (d) $x^2 + 5$

(a) $5x + 3x^3 - 7 + 6x^2 = 3x^3 + 6x^2 + 5x - 7$.
Third degree, leading term is $3x^3$, leading coefficient is 3.
(b) $x^4 - 2x + 3$ is already in descending order.
Fourth degree, leading term is x^4, leading coefficient is 1.
(c) $4 + x = x + 4$.
First degree or linear polynomial, leading term is x, leading coefficient is 1.
(d) $x^2 + 5$ is already in descending order.
Second degree or quadratic polynomial, leading term is x^2, leading coefficient is 1.

TIP!

Coefficients of Zero

A polynomial arranged in descending order can be written with every successive degree represented. Missing terms have a coefficient of 0.

$$x^4 - 2x + 3 \text{ is the same as } x^4 + 0x^3 + 0x^2 - 2x^1 + 3x^0.$$

Learning Strategy   *New Terminology*

Terminology sometimes seems overwhelming. To learn several new terms, look at the similarities and differences in terms.

- Examine prefixes and word stems.

Prefix	Stem
poly-	-nomial
mo- (one)	-nomial
bi- (two)	-nomial
tri- (three)	-nomial

In the mathematical context, polynomial is the global term, and monomials, binomials, and trinomials are specific types of polynomials.

- Group new words with similar mathematical concepts, and relate differences numerically if possible.

Constant	degree 0
Linear	degree 1
Quadratic	degree 2
Cubic	degree 3

SELF-STUDY EXERCISES 10–2

1 Identify each of the following expressions as a monomial, binomial, or trinomial.

1. $2x^3y - 7x$
2. $5xy^2 + 8y$
3. $3xy$
4. $7ab$
5. $5(3x - y)$
6. $(x - 5)(2x + 4)$
7. $5x^2 - 8x + 3$
8. $7y^2 + 5y - 1$
9. $\dfrac{4x - 1}{5}$
10. $\dfrac{x}{6} - 5$
11. $4(x^2 - 2) + x^3$
12. $3x^2 - 7(x - 8)$

2 Identify the degree of each term in each polynomial.

13. $6x$
14. $8x^2$
15. $6x^2 - 8x + 12$
16. $7x^3 - 8x + 12$
17. $x - 12$
18. $3x^2 - 8$
19. 15
20. 21
21. $2x - \dfrac{1}{4}$
22. $8x^2 + \dfrac{5}{6}$
23. $5x^2y^3 + 7xy$
24. $8r^4s - 5r^3s^5$

Identify the degree of each polynomial.

25. $5x^2 + 8x - 14$
26. $x^3 - 8x^2 + 5$
27. $9 - x^3 + x^6$
28. $12 - 15x^2 - 7x^5$
29. $2x - \dfrac{4}{5}x^2$
30. $\dfrac{7}{8} - x$
31. $5x^3y + 3x^2y^2 + 5xy^5$
32. $7x^2y - 4xy^5$

3 Arrange each polynomial in descending powers of x, and identify the degree, leading term, and leading coefficient of the polynomial.

33. $5x - 3x^2$
34. $7 - x^3$
35. $4x - 8 + 9x^2$
36. $5x^2 + 8 - 3x$
37. $7x^3 - x + 8x^2 - 12$
38. $7 - 15x^4 + 12x$
39. $-7x + 8x^6 - 7x^3$
40. $15 - 14x^8 + x$
41. $12x + 8x^4y^5 + 15x^3y^7$
42. $5r^3y - 7r^4y + 3ry^3 - 8y^6$

10–3 | Basic Operations with Polynomials

Learning Outcomes
1 Add or subtract polynomials.
2 Multiply polynomials.
3 Divide polynomials.

1 Add or Subtract Polynomials.

Now that variables with exponents have been introduced, we can broaden our concept of *like terms*. For letter terms to be like terms, the letters as well as the exponents of the letters must be exactly the same; that is, $2x^2$ and $-4x^2$ are like terms because the x's are both squared. But $2x^2$ and $4x$ are not like terms because the x's do not have the same exponents.

EXAMPLE Which pairs of terms are like terms?

(a) $3a^2b$ and $-\frac{2}{3}a^2b$
(b) $-8xy^2$ and $7x^2y$
(c) $10ab^2$ and $(-2ab)^2$
(d) $5x^4y^3$ and $-2y^3x^4$
(e) $2x^2$ and $3x^3$

(a) $3a^2b$ and $-\frac{2}{3}a^2b$ **are like terms**. All letters and their corresponding exponents are the same.

(b) $-8xy^2$ and $7x^2y$ are *not* like terms. In $-8xy^2$ the exponent of x is 1, and in $7x^2y$ the exponent of x is 2. Also, the exponents of y are not the same.

(c) $10ab^2$ and $(-2ab)^2$ are *not* like terms. In the term $10ab^2$, only the b is squared. In $(-2ab)^2$, the entire term is squared, and the result of the squaring is $4a^2b^2$.

(d) $5x^4y^3$ and $-2y^3x^4$ are like terms. Both x factors have an exponent of 4, and both y factors have an exponent of 3. Because multiplication is commutative, the order of factors does not matter.

(e) $2x^2$ and $3x^3$ are *not* like terms. The exponents of x are not the same.

Algebraic expressions containing several terms are simplified as much as possible by combining like terms.

To combine like terms:

1. Combine the coefficients using the rules for adding or subtracting signed numbers.
2. The letter factors and exponents do not change.

TIP!

Is Combining Like Terms the Same as Adding? And, What Does *Simplify* Mean?

- Combining versus adding

With the introduction of integers into our number system, we broaden our concept of addition to include both addition and subtraction. That is, when adding integers with like signs, we add absolute values, and when adding integers with unlike signs, we subtract absolute values. Also, we developed a strategy for interpreting any subtraction as an equivalent addition by changing the subtrahend to its opposite.

It is common to use the word *combine* when referring to addition or to subtraction of signed numbers.

- Simplifying

The instructions to "simplify the expression" are vague but often used in mathematics exercises. In general, to simplify an expression means to write the expression using fewer terms or reduced or with lower coefficients or exponents. When the instructions say "simplify," the intent is for you to examine the expression and see which laws or operations allow you to rewrite the expression in a simpler form.

EXAMPLE   Simplify the algebraic expressions by combining like terms.

(a) $5x^3 + 2x^3$ (b) $x^5 - 4x^5$ (c) $a^3 + 4a^2 + 3a^3 - 6a^2$
(d) $3m + 5n - m$ (e) $y + y - 5y^2 + y^3$

(a) $5x^3 + 2x^3 = (5 + 2)x^3 = 7x^3$ $5x^3$ and $2x^3$ are like terms; add the coefficients 5 and 2. The sum has the same letter factor and exponent as the like terms.

(b) $x^5 - 4x^5 = (1 - 4)x^5 = -3x^5$ x^5 and $-4x^5$ are like terms. $1 - 4 = -3$. The difference has the same letter factor and exponent as the like terms.

(c) $a^3 + 4a^2 + 3a^3 - 6a^2 = 4a^3 - 2a^2$ a^3 and $3a^3$ are like terms. $4a^2$ and $-6a^2$ are like terms. Combine coefficients mentally.

(d) $3m + 5n - m = 2m + 5n$ $3m$ and $-m$ are like terms.

(e) $y + y - 5y^2 + y^3 = 2y - 5y^2 + y^3$ y and y are like terms. Coefficients are 1.

When an algebraic expression contains a grouping preceded immediately by a negative sign, we subtract the entire grouping. Another interpretation is that the grouping is multiplied by -1, the implied coefficient of the grouping. Either interpretation causes *each* sign within the grouping to be changed to its opposite and at the same time removes the parentheses. If the grouping is preceded by a positive sign or an unexpressed positive sign, parentheses are removed without changing signs, as if each term in the grouping were multiplied by $+1$, the implied coefficient.

EXAMPLE Simplify the expressions.

(a) $y^2 + 2y - (3y^2 + 5y)$
(b) $(m^3 - 3m^2 - 5m + 4) - (4m^3 - 2m^2 - 5m + 2)$

(a) $y^2 + 2y - (3y^2 + 5y) =$

$y^2 + 2y \; -1 \; (3y^2 + 5y) =$ Distribute implied coefficient of -1.

$y^2 + 2y - 3y^2 - 5y =$ Combine like terms.
$-2y^2 - 3y$

(b) $(m^3 - 3m^2 - 5m + 4) \; - \; (4m^3 - 2m^2 - 5m + 2) =$

$(m^3 - 3m^2 - 5m + 4) \; -1 \; (4m^3 - 2m^2 - 5m + 2) =$ Distribute the implied coefficient of -1.

$m^3 - 3m^2 - 5m + 4 - 4m^3 + 2m^2 + 5m - 2 =$ Combine like terms.
$-3m^3 - m^2 + 2$

2   Multiply Polynomials.

Simplifying algebraic expressions may also involve multiplication.

To multiply by a monomial:

1. Multiply the coefficients using the rules for signed numbers.
2. Multiply the letter factors using the laws of exponents for factors with like bases.
3. Distribute if multiplying a polynomial by a monomial.

EXAMPLE Multiply.

(a) $(4x)(3x^2)$ (b) $(-6y^2)(2y^3)$ (c) $(-a)(-3a)$

(a) $(4x)(3x^2) = 4(3)\,(x^{1+2}) = 12\,x^3$ Multiply coefficients. Add exponents.

(b) $(-6y^2)(2y^3) = -6(2)\,(y^{2+3}) = -12\,y^5$ Multiply coefficients. Add exponents.

(c) $(-a)(-3a) = -1(-3)\,(a^{1+1}) = 3\,a^2$ Multiply coefficients. Add exponents. The coefficient of $-a$ is -1. The exponent of $-a$ is 1.

If factors are to be multiplied times more than one term, apply the distributive property.

EXAMPLE Perform the multiplications.

(a) $2x(x^2 - 4x)$ (b) $-2y^3(2y^2 + 5y - 6)$ (c) $4a(3a^3 - 2a^2 - a)$

(a) $2x(x^2 - 4x) = 2x^3 - 8x^2$ Distribute.

(b) $-2y^3(2y^2 + 5y - 6) = -4y^5 - 10y^4 + 12y^3$ Distribute.

(c) $4a(3a^3 - 2a^2 - a) = 12a^4 - 8a^3 - 4a^2$ Distribute.

3 **Divide Polynomials.**

Simplifying algebraic expressions may also involve division.

To divide algebraic expressions:

1. Divide the coefficients using the rules for signed numbers.
2. Divide the letter factors using the laws of exponents for factors with like bases.

EXAMPLE Divide. Express answers with positive exponents.

(a) $\dfrac{2x^4}{x}$ (b) $\dfrac{-6y^5}{2y^3}$ (c) $\dfrac{-4x}{4x^2}$ (d) $\dfrac{-5x^4}{15x^2}$ (e) $\dfrac{3x}{12x^3}$

(a) $\dfrac{2x^4}{x} = \dfrac{2}{1}(x^{4-1}) = 2\,x^3$ Divide coefficients. Subtract exponents. The coefficient of x in the denominator is 1.

(b) $\dfrac{-6y^5}{2y^3} = \dfrac{-6}{2}(y^{5-3}) = -3\,y^2$ Divide coefficients. Subtract exponents.

(c) $\dfrac{-4x}{4x^2} = \dfrac{-4}{4}(x^{1-2}) = -1\,x^{-1}$ or $-\dfrac{1}{x}$ Divide coefficients. Subtract exponents. Write factors with negative exponents as equivalent positive exponents.

This can also be written as $\dfrac{1}{-x}$ or $\dfrac{-1}{x}$.

(d) $\dfrac{-5x^4}{15x^2} = \dfrac{-5}{15}(x^{4-2}) = -\dfrac{1}{3}x^2$ or $-\dfrac{x^2}{3}$ Reduce coefficients. Subtract exponents.

$-\dfrac{1}{3}x^2$ is the same as $-\dfrac{1}{3}\left(\dfrac{x^2}{1}\right)$ or $-\dfrac{x^2}{3}$.

(e) $\dfrac{3x}{12x^3} = \dfrac{3}{12}(x^{1-3}) = \dfrac{1}{4}x^{-2} = \dfrac{1}{4x^2}$ Reduce coefficients. Subtract exponents. Make exponent positive.

Since $\dfrac{1}{4}x^{-2} = \dfrac{1}{4}\left(\dfrac{1}{x^2}\right)$, this can be written as $\dfrac{1}{4x^2}$.

When a polynomial is divided by a monomial, *each* term in the dividend (numerator) is divided by the divisor (denominator).

EXAMPLE   Perform the divisions.

(a) $\dfrac{18a^4 + 15a^3 - 9a^2 - 12a}{3a}$ (b) $\dfrac{3x^3 - x^2}{x^3}$ (c) $\dfrac{6x^3 + 2x^2}{2x^2}$

(a) $\dfrac{18a^4 + 15a^3 - 9a^2 - 12a}{3a} = \dfrac{18a^4}{3a} + \dfrac{15a^3}{3a} - \dfrac{9a^2}{3a} - \dfrac{12a}{3a}$ Distribute or write as separate terms.

$= 6a^3 + 5a^2 - 3a - 4$ Simplify each term.

(b) $\dfrac{3x^3 - x^2}{x^3} = \dfrac{3x^3}{x^3} - \dfrac{x^2}{x^3} = 3 - x^{-1}$ or $3 - \dfrac{1}{x}$ Write as separate terms and simplify each term.

(c) $\dfrac{6x^3 + 2x^2}{2x^2} = \dfrac{6x^3}{2x^2} + \dfrac{2x^2}{2x^2} = 3x + 1$ Write as separate terms and simplify each term.

TIP!

Why Can't We Cancel Terms?

You reduce or cancel *factors*, **but not** *terms*. Look at part c of the previous example, $\dfrac{6x^3 + 2x^2}{2x^2}$.

A common *mistake* is to cancel the terms.

$$\dfrac{\cancel{6x^3} + \cancel{2x^2}}{\cancel{2x^2}} = 6x^3$$ Incorrect!

Why is this not correct? To check a division, multiply the quotient by the divisor (denominator). The result will be the dividend (numerator).

Does $6x^3(2x^2) = 6x^3 + 2x^2$?   NO!

Expressions that have several operations can be simplified.

EXAMPLE   Simplify and write all exponents as positive exponents.

(a) $\dfrac{5x^5}{10x^2} + 3x(2x^2)$ (b) $\dfrac{3ab^2 - a^2b + 4}{ab}$ (c) $3x(4xy^2)^2$

(a) $\dfrac{5x^5}{10x^2} + 3x(2x^2) =$ Simplify each term.

$\dfrac{1}{2}x^3 + 6x^3 =$ Combine terms. $\dfrac{1}{2} + 6 = \dfrac{1}{2} + \dfrac{12}{2} = \dfrac{13}{2}$

$\dfrac{13}{2}x^3$ or $\dfrac{13x^3}{2}$

(b) $\dfrac{3ab^2 - a^2b + 4}{ab} =$ Write or mentally visualize as separate terms.

$\dfrac{3ab^2}{ab} - \dfrac{a^2b}{ab} + \dfrac{4}{ab} =$ Simplify each term.

$3b - a + \dfrac{4}{ab}$

(c) $3x(4xy^2)^2 =$ Follow the order of operations. Raise to a power first.

$3x(16x^2y^4) =$ Multiply.

$48x^3y^4$

SELF-STUDY EXERCISES 10–3

1 Simplify.

1. $3a^2 + 4a^2$
2. $5x^3 - 2x^3$
3. $b^2 + 3a^2 + 2b^2 - 5a^2$
4. $3a - 2b - a$
5. $x - 3x - 2x^2 - 3x^2$
6. $3a^2 - 2a^2 + 4a^2$
7. $x^2 + 3y - (2x^2 + 5y)$
8. $4m^2 - 2n^2 - (2m^2 - 3n^2)$
9. $7a + 3b + 8c + 2a - (b - 2c)$
10. $5x + 3y - (7x - 2z)$
11. $(4x^2 - 3) - (3x^2 + 2) - (x^2 - 1)$
12. $5x - (3x + 7) - (-x - 8)$

2 Multiply.

13. $7x(2x^2)$
14. $(-2m)(-m^2)$
15. $(-3m)(7m)$
16. $(-y^3)(2y^3)$
17. $4x^2(2x - 7)$
18. $-3ab(2a^2b - 5ab^2)$
19. $7x(5x^3 - 3x^2 - 7)$
20. $2xy(3x^2y - 5xy^2)$
21. $3x(x - 6)$
22. $4x(3x^2 - 7x + 8)$
23. $-4x(2x - 3)$
24. $2x^2(5 + 2x)$

3 Divide.

25. $\dfrac{6x^4}{3x^2}$
26. $\dfrac{-5a^2}{10a}$
27. $\dfrac{-7x}{-14x^3}$
28. $\dfrac{-9x^5}{12x^2}$
29. $\dfrac{6x^2 - 4x}{2x}$
30. $\dfrac{12x^5 - 6x^3 - 3x^2}{3x^2}$
31. $\dfrac{7x^4 - x^2}{x^3}$
32. $\dfrac{8x^4 + 6x^3}{2x^2}$
33. $\dfrac{6x^4 + 8x^2}{18x^2}$
34. $\dfrac{5a^2b^3 - 3ab^2 - 7}{ab}$
35. $\dfrac{15a^2b^3 - 3ab^2}{6ab}$
36. $\dfrac{18x^2 - 12y^2 - 6xy}{6xy}$
37. $\dfrac{15x^2}{3x} - 2x(7x^3)$
38. $\dfrac{6x(2x^3y^2)^3 + 8x}{2x}$
39. $\dfrac{-3x - (5x + 8x^3)}{2x}$
40. $\dfrac{3x^2 - 5y^2}{15xy} - \dfrac{8x^3}{4x}$

10–4 | *Powers of 10 and Scientific Notation*

Learning Outcomes

1 Multiply and divide by powers of 10.
2 Change a number from scientific notation to ordinary notation.
3 Change a number from ordinary notation to scientific notation.
4 Multiply and divide numbers in scientific notation.

1 Multiply and Divide by Powers of 10.

We learned that our number system is based on the number 10; that is, each place value is a *power of 10*. Look at the place-value chart in Fig. 10–1 to see how our number system relates to powers of 10.

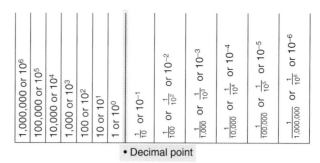

Figure 10–1 Base-10 place-value chart.

The ones place is 10^0. This is consistent with the definition of zero exponents. The tenths place is $\frac{1}{10}$ or 10^{-1}, which is consistent with the definition of negative exponents.

Whole-Number Part		Fractional or Decimal Part	
Millions	$1,000,000 = 10^6$	$\frac{1}{10} = 0.1 = 10^{-1}$	Tenths
Hundred-thousands	$100,000 = 10^5$	$\frac{1}{100} = 0.01 = 10^{-2}$	Hundredths
Ten-thousands	$10,000 = 10^4$	$\frac{1}{1,000} = 0.001 = 10^{-3}$	Thousandths
Thousands	$1,000 = 10^3$	$\frac{1}{10,000} = 0.0001 = 10^{-4}$	Ten-thousandths
Hundreds	$100 = 10^2$	$\frac{1}{100,000} = 0.00001 = 10^{-5}$	Hundred-thousandths
Tens	$10 = 10^1$	$\frac{1}{1,000,000} = 0.000001 = 10^{-6}$	Millionths
Ones	$1 = 10^0$		

TIP!

What Does the Exponent in a Power of 10 Tell Us?

- When the exponent of 10 is positive, the exponent is the same as the number of zeros in the equivalent whole number.
- When the exponent of 10 is negative, the exponent is the same as the number of zeros in the denominator of the equivalent fraction.
- When the exponent of 10 is negative, the exponent is the same as the number of decimal digits in the decimal equivalent.

To multiply by a power of 10:

1. If the exponent is positive, shift the decimal point to the *right* the number of places indicated by the *positive* exponent. Attach zeros as necessary.
2. If the exponent is negative, shift the decimal point to the *left* the number of places indicated by the *negative* exponent. Insert zeros as necessary.

EXAMPLE Perform the multiplications using powers of 10.

(a) 275×10 (b) 0.18×100 (c) $2.4 \times 1{,}000$ (d) 43×0.1

(a) $275 \times 10 = 275 \times 10^1 = \mathbf{2{,}750}$ The exponent is $+1$, so move the decimal one place to the right. Attach one zero.

(b) $0.18 \times 100 = 0.18 \times 10^2 = \mathbf{18}$ The exponent is $+2$, so move the decimal two places to the right.

(c) $2.4 \times 1{,}000 = 2.4 \times 10^3 = \mathbf{2{,}400}$ The exponent is $+3$, so move the decimal three places to the right. Attach two zeros.

(d) $43 \times 0.1 = 43 \times 10^{-1} = \mathbf{4.3}$ The exponent is -1, so move the decimal one place to the left.

To divide by a power of 10:

1. Change the division to an equivalent multiplication.
2. Use the rule for multiplying by a power of 10.

EXAMPLE Perform the divisions using powers of 10.

(a) $3.14 \div 10$ (b) $0.48 \div 100$ (c) $20.1 \div 1{,}000$

(a) $3.14 \div 10 = 3.14 \times \frac{1}{10}$ Change the division to an equivalent multiplication.

$= 3.14 \times 10^{-1}$ Express the fraction $\frac{1}{10}$ as a power of 10.

$= \mathbf{0.314}$ Because the exponent of 10 is -1, move the decimal one place to the *left*.

(b) $0.48 \div 100 = 0.48 \times \frac{1}{100} = 0.48 \times 10^{-2} = \mathbf{0.0048}$ Move the decimal two places to tho loft. Inoort two zoroo.

(c) $20.1 \div 1{,}000 = 20.1 \times \frac{1}{1{,}000} = 20.1 \times 10^{-3} = \mathbf{0.0201}$ Move decimal three places to the left. Insert one zero.

         A power of 10 can be multiplied or divided by another power of 10 by following the laws of exponents.

EXAMPLE Multiply or divide using the laws of exponents.

(a) $10^5(10^2)$ (b) $10^{-1}(10^2)$ (c) $10^0(10^3)$ (d) $\dfrac{10}{10^3}$

(e) $\dfrac{10^5}{10^4}$ (f) $\dfrac{10^2}{10^2}$ (g) $\dfrac{10^{-2}}{10^3}$

(a) $10^5(10^2) = 10^{5+2} = \mathbf{10^7}$ Like bases, add exponents.

(b) $10^{-1}(10^2) = 10^{(-1+2)} = \mathbf{10^1}$ or   **10**   Like bases, add exponents.

(c) $10^0(10^3) = 10^{0+3} = \mathbf{10^3}$ Like bases, add exponents.

(d) $\dfrac{10}{10^3} = 10^{1-3} = \mathbf{10^{-2}}$ or $\dfrac{1}{10^2}$ Like bases, subtract exponents.

(e) $\dfrac{10^5}{10^4} = 10^{5-4} = \mathbf{10^1}$ or   **10**   Like bases, subtract exponents.

(f) $\dfrac{10^2}{10^2} = 10^{2-2} = \mathbf{10^0}$ or   **1**   Like bases, subtract exponents.

(g) $\dfrac{10^{-2}}{10^3} = 10^{-2-3} = \mathbf{10^{-5}}$ or $\dfrac{1}{10^5}$ Like bases, subtract exponents.

2   Change a Number from Scientific Notation to Ordinary Notation.

A number is expressed in *scientific notation* if it is the product of two factors. The absolute value of the first factor is a number greater than or equal to 1 but less than 10. The second factor is a power of 10.

    A number that is written strictly according to place value is an *ordinary number*.

TIP!

Characteristics of Scientific Notation

- Numbers between 0 and 1 and between -1 and 0 require negative exponents when written in scientific notation.
- The first factor in scientific notation always has only one nonzero digit to the left of the decimal.
- The use of the times sign ($\times$) for multiplication is the most common representation for scientific notation.
- The second factor is a power of 10.

EXAMPLE   Which of the terms is expressed in scientific notation?

(a) 4.7×10^2 (b) 0.2×10^{-1} (c) -3.4×5^2 (d) 2.7×10^0

(e) 34×10^4 (f) 8×10^{-6} (g) $-2.8 \div 10^3$

(a) 4.7 is more than 1 but less than 10. 10^2 is a power of 10. Multiplication is indicated. **Thus, the term is in scientific notation.**

(b) Even though 10^{-1} is a power of 10, **this term is not in scientific notation** because the first factor (0.2) is less than 1.

(c) The absolute value of -3.4 is more than 1 and less than 10, but 5^2 is not a power of 10. **Thus, this term is not in scientific notation.**

(d) 2.7 is more than 1 and less than 10. 10^0 is a power of 10. **Thus, this term is in scientific notation.**

(e) 34 is greater than 10. Even though 10^4 is a power of 10, **this term is not in scientific notation** because the first factor (34) is 10 or more.

(f) 8 is more than 1 and less than 10. 10^{-6} is a power of 10. **Thus, this term is in scientific notation.**

(g) The absolute value of -2.8 is more than 1 and less than 10, but division rather than multiplication is the indicated operation. **This term is not in scientific notation.**

To change from a number written in scientific notation to an ordinary number:

1. Perform the indicated multiplication by moving the decimal point in the first factor the appropriate number of places. Affix or insert zeros as necessary.
2. Omit the power-of-10 factor.

When we multiply by a power of 10, the exponent of 10 tells us how many places and in which direction to move the decimal.

EXAMPLE   Change to ordinary numbers.

(a) 3.6×10^4 (b) 2.8×10^{-2} (c) 1.1×10^0 (d) -6.9×10^{-5}
(e) -9.7×10^6

(a) $3.6 \times 10^4 = 36000. = \mathbf{36{,}000}$ Move the decimal four places to the right.

(b) $2.8 \times 10^{-2} = .028 = \mathbf{0.028}$ Move the decimal two places to the left.

(c) $1.1 \times 10^0 = \mathbf{1.1}$ Move the decimal no (zero) places.

(d) $-6.9 \times 10^{-5} = -.000069 = \mathbf{-0.000069}$ Move the decimal five places to the left.

(e) $-9.7 \times 10^6 = -9700000. = \mathbf{-9{,}700{,}000}$ Move the decimal six places to the right.

3   **Change a Number from Ordinary Notation to Scientific Notation.**

To express an ordinary number in scientific notation, we reverse the procedures we used before. Shifting the decimal point changes the value of a number. The power-of-10 factor is used to offset or balance this change so the original value of the number is maintained.

To change from a number written in ordinary notation to scientific notation:

1. By inserting a caret (∧) in the proper place, indicate where the decimal should be positioned in the ordinary number so that the absolute value of the number is valued at 1 or between 1 and 10.
2. Determine the number of places and in which direction the decimal shifts *from* the new position (caret) *to* the old position (decimal point). This number is the exponent of the power of 10.

TIP!

A Balancing Act: Why Count from the New to the Old?

Moving the decimal in the ordinary number changes the value of the number unless you balance the effect of the move in the power-of-10 factor. When a decimal is moved in the first factor, the value is changed. To offset this change, an opposite change must be made in the power-of-10 factor. From the new to the old position indicates the proper number of places and the direction (positive or negative) for balancing with the power-of-10 factor.

Remember the word **NO**. Count from **N**ew to **O**ld.

$3{,}800 = 3.8 \times 10^3$ $3_\wedge 800. \times 10^3$ N → O = +3.
$0.0045 = 4.5 \times 10^{-3}$ $0.004_\wedge 5 \times 10^{-3}$ N → O = −3.

EXAMPLE Express in scientific notation.

(a) 285   (b) 0.007   (c) 9.1   (d) 85,000   (e) 0.00074

(a) $285 \rightarrow 2_\wedge 85 = 2.85 \times 10^2$
The unwritten decimal is after the 5. Place the caret between 2 and 8 so the number 2.85 is between 1 and 10. Count *from* the caret *to* the decimal to determine the exponent of 10. A move two places to the right represents the exponent +2.

(b) $0.007 \rightarrow 0.007_\wedge = 7 \times 10^{-3}$
7 is between 1 and 10. Count *from* the caret *to* the decimal. A move three places to the left represents the exponent −3.

(c) $9.1 = 9.1 \times 10^0$
9.1 is already between 1 and 10, so the decimal does not move; that is, the decimal moves zero places.

(d) $85,000 \rightarrow 8_\wedge 5000 = 8.5 \times 10^4$
From the caret *to* the decimal is four places to the right.

(e) $0.00074 \rightarrow 0.0007_\wedge 4 = 7.4 \times 10^{-4}$
From the caret *to* the decimal is four places to the left.

Occasionally a number is written in power-of-10 notation but not in scientific notation because the first factor is not equal to 1 or is not between 1 and 10. To write the number in scientific notation shift the decimal to the proper place and adjust the original power-of-10 factor appropriately.

TIP

Between 1 and 10: One Nonzero Digit to the Left of the Decimal

The expression "between 1 and 10" means any number that is more than 1 but less than 10. The first factor in scientific notation must be equal to 1 *or* between 1 and 10. That means **there will be one and only one nonzero digit to the left of the decimal point**.

EXAMPLE Express in scientific notation.

(a) 37×10^5 (b) 0.03×10^3

(a) $3_\wedge 7 \times 10^5 = 3.7 \times 10^1 \times 10^5$ N ▸ O = +1.
        $= 3.7 \times 10^6$

(b) $0.03_\wedge \times 10^3 = 3 \times 10^{-2} \times 10^3$ N ▸ O = −2.
        $= 3 \times 10^1$

4   **Multiply and Divide Numbers in Scientific Notation.**

We can multiply or divide numbers expressed in scientific notation without first having to convert them to ordinary numbers.

To multiply numbers in scientific notation:

1. Multiply the first factors using the rules of signed numbers.
2. Multiply the power-of-10 factors using the laws of exponents.
3. Examine the first factor of the product (Step 1) to see if its value is equal to 1 or between 1 and 10.
   (a) If so, write the results of Steps 1 and 2.
   (b) If not, shift the decimal so the first factor is equal to 1 or is between 1 and 10, and adjust the exponent of the power-of-10 factor accordingly.

EXAMPLE   Multiply.

(a) $(4 \times 10^2)(2 \times 10^3)$ (b) $(3.7 \times 10^3)(2.5 \times 10^{-1})$ (c) $(8.4 \times 10^{-2})(5.2 \times 10^{-3})$

(a) $(4 \times 10^2)(2 \times 10^3) = \mathbf{8 \times 10^5}$
Because 8 is between 1 and 10, we do not make any adjustments.

(b) $(3.7 \times 10^3)(2.5 \times 10^{-1}) = \mathbf{9.25 \times 10^2}$
Because 9.25 is between 1 and 10, we do not make any adjustments.

(c) $(8.4 \times 10^{-2})(5.2 \times 10^{-3}) = \mathbf{43.68 \times 10^{-5}}$
43.68 is not between 1 and 10. Adjustments are necessary.

$43.68 \rightarrow 4 {}_\wedge 3.68$ or 4.368×10^1 Adjust first factor.

$4.368 \times 10^1 \times 10^{-5} = 4.368 \times 10^{1-5}$ Multiply powers of 10.

$= \mathbf{4.368 \times 10^{-4}}$

Division involving numbers in scientific notation is similar to multiplication.

To divide numbers in scientific notation:

1. Divide the first factors using the rules of signed numbers.
2. Divide the power-of-10 factors using the laws of exponents.
3. Examine the first factor of the quotient (Step 1) to see if its value is equal to 1 or between 1 and 10.
   (a) If so, write the results of Steps 1 and 2.
   (b) If not, shift the decimal so that the first factor is equal to 1 or is between 1 and 10, and adjust the exponent of the power-of-10 factor accordingly.

EXAMPLE   Divide.

(a) $\dfrac{3 \times 10^5}{2 \times 10^2}$ (b) $\dfrac{1.44 \times 10^{-3}}{6 \times 10^{-5}}$ (c) $\dfrac{9.6 \times 10^{29}}{3.2 \times 10^{111}}$ (d) $\dfrac{1.25 \times 10^3}{5}$

(a) $\dfrac{3 \times 10^5}{2 \times 10^2} = \dfrac{3}{2} \times 10^{5-2} = \mathbf{1.5 \times 10^3}$
Because 1.5 is between 1 and 10, no adjustments are necessary. The first factor is usually written in decimal notation.

(b) $\dfrac{1.44 \times 10^{-3}}{6 \times 10^{-5}} = \dfrac{1.44}{6} \times 10^{-3-(-5)} = 0.24 \times 10^{-3+5} = 0.24 \times 10^2$

0.24 is less than 1, so adjustments are necessary.

$0.24 \rightarrow 0.2_\wedge 4 = 2.4 \times 10^{-1}$ Adjust first factor.

Then $2.4 \times 10^{-1} \times 10^2 = \mathbf{2.4 \times 10^1}$.   Multiply powers of 10.

(c) $\dfrac{9.6 \times 10^{29}}{3.2 \times 10^{111}} = \dfrac{9.6}{3.2} \times 10^{29-111} = \mathbf{3 \times 10^{-82}}$

(d) $\dfrac{1.25 \times 10^3}{5}$ 5 is the same as 5×10^0.

$\dfrac{1.25 \times 10^3}{5 \times 10^0} = \dfrac{1.25}{5} \times 10^{3-0} = 0.25 \times 10^3$ 0.25 is less than 1.

$0.25 \rightarrow 0.2_\wedge 5 = 2.5 \times 10^{-1}$ Adjust first factor.

Then, $2.5 \times 10^{-1} \times 10^3 = 2.5 \times 10^{-1+3} = \mathbf{2.5 \times 10^2}$.

TIP!

Negative Exponents and Mental Adjustment of Exponents

In scientific notation, we **do** leave exponents as negative numbers when appropriate.

Once we understand the concept of balancing the effect of moving a decimal by adjusting the exponent of the power of 10 factor, we can perform this adjustment mentally.

$43.68 \times 10^{-5} = 4_\wedge 3.68 \times 10^{-5}$ N ▸ O = +1.

$\quad\quad\quad\quad\; = 4.368 \times 10^{-5+1}$ Adjust mentally.

$\quad\quad\quad\quad\; = 4.368 \times 10^{-4}$

$0.25 \times 10^3 = 0.2_\wedge 5 \times 10^3$ N ▸ O = −1.

$\quad\quad\quad\quad\; = 2.5 \times 10^{3-1}$ Adjust mentally.

$\quad\quad\quad\quad\; = 2.5 \times 10^2$

TIP!

Scientific Notation and the Calculator

Power-of-10 Key

The power-of-10 key, labeled $\boxed{\text{EXP}}$ or $\boxed{\text{EE}}$ on most calculators, is a shortcut key for entering the following keys:

$$\boxed{\times}\; 10 \;\boxed{x^y}$$

The shortcut key is used only for power-of-10 factors, and only the *exponent* of 10 is entered. If you enter $\boxed{\times}\, 10$, then $\boxed{\text{EXP}}$, your answer will have one extra factor of 10.

Look at $\dfrac{(3 \times 10^5)}{(2 \times 10^2)}$ on the calculator.

$$3 \;\boxed{\text{EXP}}\; 5 \;\boxed{\div}\; 2 \;\boxed{\text{EXP}}\; 2 \;\boxed{=} \Rightarrow 1500$$

This result may be expressed in scientific notation if desired: $1{,}500 = 1.5 \times 10^3$.

The internal program of a calculator has predetermined how the output of a calculator is displayed. For example, even if you would like an answer displayed in scientific notation, it may fall within the guidelines for display as an ordinary number. You must make the conversion to scientific notation yourself. The reverse may also be true.

EXAMPLE A star is 4.2 light-years from Earth. If 1 light-year is 5.87×10^{12} miles, how many miles from Earth is the star?

To solve this problem, express the given information in a direct proportion of two pairs of information relating light-years to miles.

Pair 1: 1 light-year $= 5.87 \times 10^{12}$ mi
Pair 2: 4.2 light-years $= x$ mi

Estimation The star will be more than 4 times 5.87×10^{12} miles away since 4.2 light years is more than 4 times 1 light year.

$$\frac{1 \text{ light-year}}{4.2 \text{ light-years}} = \frac{5.87 \times 10^{12} \text{ mi}}{x \text{ mi}}$$

Pair 1 becomes the two numerators.
Pair 2 becomes the two denominators (direct proportion).

$$\frac{1}{4.2} = \frac{5.87 \times 10^{12}}{x}$$

Cross multiply.

$$x = (5.87 \times 10^{12})(4.2)$$

$4.2 = 4.2 \times 10^{0}$.

$$x = 24.654 \times 10^{12}$$

Perform scientific notation adjustment.

$$x = 2.4654 \times 10^{12+1}$$

$N \rightarrow O = +1$.

$$x = 2.4654 \times 10^{13}$$

or 2.5×10^{13} Round the first factor to tenths.

Interpretation The star is 2.5×10^{13} miles from Earth.

EXAMPLE An angstrom (Å) is 1×10^{-7} mm. What is the length in millimeters of 14.82 Å?

We will set up a direct proportion of fractions relating angstrom units to millimeters.

Pair 1: 1 Å $= 1 \times 10^{-7}$ mm
Pair 2: 14.82 Å $= x$ mm

Estimation 14.82 angstrom units are more than 10 times longer than 1×10^{-7} mm or more than 1×10^{-6}.

$$\frac{1 \text{ Å}}{14.82 \text{ Å}} = \frac{1 \times 10^{-7} \text{ mm}}{x \text{ mm}}$$

Pair 1 becomes the two numerators.

Pair 2 becomes the two denominators (direct proportion).

$$\frac{1}{14.82} = \frac{1 \times 10^{-7}}{x}$$

Cross multiply.

$$x = 14.82 \times 10^{-7}$$

Adjust the first factor and exponent.

$$x = 1.482 \times 10^{-7+1}$$

$N \rightarrow O = +1$.

$$x = 1.482 \times 10^{-6} \text{ mm}$$

Interpretation The length of 14.82 Å is 1.482×10^{-6} mm.

EXAMPLE One coulomb (C) is approximately 6.28×10^{18} electrons. How many coulombs do 2.512×10^{21} electrons represent?

Pair 1: 1 C = 6.28×10^{18} electrons
Pair 2: x C = 2.512×10^{21} electrons

Estimation Since 2.512×10^{21} is larger than 6.28×10^{18}, there is more than 1 coulomb.

$$\frac{1 \text{ C}}{x \text{ C}} = \frac{6.28 \times 10^{18} \text{ electrons}}{2.512 \times 10^{21} \text{ electrons}}$$

Pair 1 becomes the two numerators.

Pair 2 becomes the two denominators (direct proportion).

$$\frac{1}{x} = \frac{6.28 \times 10^{18}}{2.512 \times 10^{21}}$$

Cross multiply.

$$(6.28 \times 10^{18})\,(x) = 2.512 \times 10^{21}$$

Divide by the coefficient of x.

$$x = \frac{2.512 \times 10^{21}}{6.28 \times 10^{18}}$$

Divide coefficients, subtract exponents.

$$x = 0.4 \times 10^3$$

Adjust the first factor.

$$x = 4 \times 10^{3-1}$$

$N \rightarrow O = -1$.

$$x = 4 \times 10^2$$

Write as an ordinary number.

$$x = 400 \text{ C}$$

Interpretation 2.512×10^{21} electrons represent 400 C.

SELF-STUDY EXERCISES 10–4

1 Multiply or divide as indicated.

1. 0.37×10^2
2. 1.82×10^3
3. 5.6×10^{-1}
4. 142×10^{-2}
5. 78×10^4
6. 62×10^0
7. $4.6 \div 10^4$
8. $6.1 \div 10$
9. $7.2 \div 10^1$
10. $42 \div 10^0$
11. $10^4(10^6)$
12. $10^{\,3}(10^{\,4})$
13. $10^0(10^{-3})$
14. $10^{-3}(10^4)$
15. $10(10^2)$
16. $\dfrac{10^4}{10^2}$
17. $\dfrac{10}{10^4}$
18. $\dfrac{10^4}{10^4}$
19. $\dfrac{10^{-2}}{10^3}$
20. $\dfrac{10^0}{10^1}$

2 Write as ordinary numbers.

21. 4.3×10^2
22. 6.5×10^{-3}
23. 2.2×10^0
24. 7.3×10
25. 9.3×10^{-2}
26. 8.3×10^4
27. 5.8×10^{-3}
28. 8×10^4
29. 6.732×10^0
30. 5.89×10^{-3}

3 Express in scientific notation.

31. 392
32. 0.02
33. 7.03
34. 42,000
35. 0.081
36. 0.0021
37. 23.92
38. 0.101
39. 1.002
40. 721
41. 42×10^4
42. 32.6×10^3
43. 0.213×10^2
44. 0.0062×10^{-3}
45. $56,000 \times 10^{-3}$

4 Perform the indicated operations. Express answers in scientific notation.

46. $(6.7 \times 10^4)(3.2 \times 10^2)$
47. $(1.6 \times 10^{-1})(3.5 \times 10^4)$
48. $(5.0 \times 10^{-3})(4.72 \times 10^0)$
49. $(8.6 \times 10^{-3})(5.5 \times 10^{-1})$
50. $\dfrac{3.15 \times 10^5}{4.5 \times 10^2}$
51. $\dfrac{4.68 \times 10^3}{7.2 \times 10^7}$
52. $\dfrac{4.55 \times 10^{-1}}{6.5 \times 10^{-4}}$
53. $\dfrac{7.84 \times 10^{-2}}{9.8 \times 10^0}$

54. A star is 5.5 light years from Earth. If one light-year is 5.87×10^{12} miles, how many miles from Earth is the star?

55. An angstrom (Å) is 1×10^{-7} mm. How many angstroms are in 4.2×10^{-5} mm?

Section 10–1

Perform the indicated operations. Write the answers with positive exponents.

1. $x^5 \cdot x^5$

2. $x^2(x^4)$

3. $3x^4 \cdot 7x^5$

4. $5x^3 \cdot 8x$

5. $\dfrac{x^8}{x^5}$

6. $\dfrac{x^3}{x^5}$

7. $\dfrac{21x^4}{3x}$

8. $\dfrac{24y^7}{18y^{10}}$

9. $\dfrac{x^3 y^{-1}}{x^2 y^2}$

10. $\dfrac{x^3 + y^2}{x^2 y}$

11. $(x^3)^4$

12. $(-x^3)^3$

13. $(x^{-3})^{-5}$

14. $(x^2 y^5)^2$

15. $(-3x^2)^3$

16. $\dfrac{a^2 b^7}{ab^2}$

17. $\dfrac{xy^3}{xy^5}$

Section 10–2

Identify the degree of each polynomial.
18. $5a^4 - 7a^3 + 12a^2 - 38$

19. $2x^3 y^2 - 15xy^3 + 21y^4$

20. Is the expression $5x^3 - 3x^{-2}$ a polynomial? Why or why not?

Arrange the following polynomials in descending order and identify the degree of each polynomial, the leading term, and the leading coefficient.
21. $5x + 3x^3 - 8 + x^2$

22. $3y^5 - 7y - 8y^4 + 12$

23. $5x^2 - 12x + 2x^4 - 32$

Section 10–3

Simplify the following.
24. $4x^3 + 7x - 3x^3 - 5x$

25. $8x - 2x^4 - (3x^3 + 5x - x^3)$

26. $5x - 3x + (7x^2 - 8x)$

27. $4r^2 - (3y^2 + 7r^2 - 8y^2)$

28. $4r^3(-3r^4)$

29. $-7r^8(-3r^{-2})$

30. $5a(a^2 - 7)$

31. $2x(x^2 + 3x - 5)$

32. $-2y(3y^2 - 7y - 12)$

33. $-2x(x^3 - 7x^2 + 15)$

34. $\dfrac{12x^5}{6x^3}$

35. $\dfrac{12x^7}{-18x^4}$

36. $\dfrac{11x^4}{22x^7}$

37. $\dfrac{42x^3 y}{-15x^3 y^3}$

38. $\dfrac{-8x^3 y^5}{20x^4 y}$

39. $\dfrac{6x^3 - 12x^2 + 21x}{3x}$

40. $\dfrac{25y^5 - 85y^3 + 70y^2}{-5y}$

41. $\dfrac{4x^5}{8x^2} - 3x^2(2x^4)$

42. $2x(3x^2 y)^3 + \dfrac{7x^3}{21x^2}$

Section 10–4

Perform the indicated operations. Express as ordinary numbers.
43. $10^5 \cdot 10^7$

44. $10^{-2} \cdot 10^8$

45. $10^7 \cdot 10^{-10}$

46. 4.2×10^5

47. $8.73 \div 10^{-3}$

48. $5.6 \div 10^{-2}$

Write as ordinary numbers.
49. 3.75×10^5

50. 4.23×10^4

51. 3.87×10^{-5}

52. 7.37×10^{-9}

Write in scientific notation.

53. 52,000	**54.** 4,500	**55.** 670	**56.** 160
57. 0.00017	**58.** 3,800,000	**59.** 5,600	**60.** 78
61. 52,000	**62.** 80,000,000	**63.** 5,830,000	**64.** 1,736,500,000
65. 41,980,000	**66.** 0.78	**67.** 0.033	**68.** 0.0000011
69. 0.000000008	**70.** 0.0009832	**71.** 0.0000719	**72.** 0.0120307
73. 0.000675	**74.** 0.00000092	**75.** 0.004	**76.** 5,096,000
77. 0.0000008	**78.** 182,000	**79.** 1,600	**80.** 5,200,000
81. 97,000,000	**82.** 0.00049	**83.** 0.00052	**84.** 0.588
85. 10.53	**86.** 246.7	**87.** 5,082	**88.** 42,000
89. 7,800	**90.** 5,729	**91.** 25,000	**92.** 4,800
93. 4,000	**94.** 0.000001	**95.** 0.0047	

Perform the indicated operation and write the result in scientific notation.

96. $(4.2 \times 10^5)(3.9 \times 10^{-2})$ **97.** $(7.8 \times 10^{53})(5.6 \times 10^{72})$ **98.** $\dfrac{5.2 \times 10^8}{6.1 \times 10^5}$ **99.** $\dfrac{1.25 \times 10^3}{3.7 \times 10^{-8}}$

100. The national debt for the United States is approximately 6 trillion dollars. Write the number in scientific notation.

101. The United States population is approximately 250 million. Write the number in scientific notation.

102. If each individual in the United States paid an equal share of the national debt, how much would each person have to pay?

CHALLENGE PROBLEMS

103. The national debt for the United States is approximately 6 trillion dollars. A dollar bill is approximately 0.2 mm thick. Suppose the national debt is represented with dollar bills stacked on top of each other. How many kilometers high would the stack reach?

104. The distance from the Earth to the Moon is 380,000 km. How many times will the stack of bills represented by the national debt reach from the Earth to the Moon?

CONCEPTS ANALYSIS

Write the laws of exponents and properties in your own words and give an example illustrating each.

1. $a^m \cdot a^n = a^{m+n}$ **2.** $\dfrac{a^m}{a^n} = a^{m-n}, a \neq 0$ **3.** $a^0 = 1, a \neq 0$

4. $a^{-n} = \dfrac{1}{a^n}, a \neq 0$ **5.** $(a^m)^n = a^{mn}$ **6.** $\left(\dfrac{a}{b}\right)^n = \dfrac{a^n}{b^n}, b \neq 0$

7. $(ab)^n = a^n b^n$

8. What two conditions must be satisfied before a number is expressed in scientific notation?

Find the mistake in the examples. Explain the mistake and correct it.

9. $\dfrac{9x^3 - 12x^2 + 3x}{3x} = 3x^2 - 4x$ **10.** $x^5(x^3) = x^{15}$

11. $\dfrac{3.4 \times 10^5}{2 \times 10^{-2}} = 1.7 \times 10^3$ **12.** $4x^3 + 3x^3 = 7x^6$

Chapter 10 / Powers and Polynomials

Learning Outcomes	What to Remember with Examples

Section 10-1

1 Multiply powers with like bases (pp. 441–442).

To multiply powers with like bases, add the exponents and keep the common base as the base of the product. $a^m(a^n) = a^{m+n}$, where a, m, and n are real numbers.

Multiply. $x^5(x^{-7}) =$

$$x^{5+(-7)} = x^{-2} \text{ or } \frac{1}{x^2}$$

2 Divide powers with like bases (pp. 442–445).

To divide powers with like bases, subtract the exponents and keep the common base as the base of the quotient. $\dfrac{a^m}{a^n} = a^{m-n}$, where a, m, and n are real numbers except that $a \neq 0$.

Divide. $\dfrac{a^7}{a^4} =$

$$a^{7-4} = a^3$$

3 Find a power of a power (pp. 445–447).

To find a power of a power, multiply the exponents and keep the same base. $(a^m)^n = a^{mn}$, where a, m, and n are real numbers.

Simplify. $(y^5)^4 =$

$$y^{5(4)} = y^{20}$$

To raise a fraction or quotient to a power, raise both the numerator and denominator to the power. $\left(\dfrac{a}{b}\right)^n = \dfrac{a^n}{b^n}$ and $b \neq 0$.

$$\left(\frac{-2}{x^2}\right)^3 = \frac{(-2)^3}{(x^2)^3} = \frac{-8}{x^6}$$

To raise a product to a power, raise each factor to the power. $(ab)^n = a^n b^n$.

$$(5x^2y)^2 = 5^2x^4y^2 = 25x^4y^2$$

Section 10-2

1 Identify polynomials, monomials, binomials, and trinomials (pp. 448–449).

Polynomials are algebraic expressions in which the exponent of the variable is a non-negative integer and there are no variables in a denominator. A monomial is a polynomial with a single term. A binomial is a polynomial containing two terms. A trinomial is a polynomial containing three terms.

Give an example of a polynomial, a monomial, a binomial, and a trinomial.

Polynomial: $4x^3 + 6x^2 - x + 3$ Monomial: $8x$
Binomial: $6x + 3$ Trinomial: $8x^2 - 4x - 3$

2 Identify the degree of terms and polynomials (pp. 449–450).

The degree of a term containing one variable is the exponent of the variable. The degree of a term of more than one variable is the sum of the exponents of all the variable factors. The degree of a constant term is zero. The degree of a polynomial that has only one variable is the degree of the term that has the largest exponent.

Identify the degree of each term and the degree of the polynomial.

$4x^3$ $+$ $6x^2$ $-$ x $+$ 3 The polynomial has a degree of 3.
degree 3   degree 2   degree 1   degree 0

| **3** Arrange polynomials in descending order (pp. 450–451). | To arrange polynomials in descending order, list the term that has the highest degree first, the term that has the next highest degree second, and so on, until all terms have been listed. |

Arrange the polynomial in descending order.

$$4 - 2x + 7x^5 - 3x^2 + x^3$$

Descending order:

$$7x^5 + x^3 - 3x^2 - 2x + 4$$

Section 10–3

| **1** Add polynomials (pp. 452–454). | Like terms are terms that not only have the same letter factors, but also have the same exponent. To add or subtract like terms, add or subtract the coefficients and keep the variable factors and their exponents exactly the same. |

Simplify. $3x^4 + 8x^2 - 7x^2 + 2x^4 = 5x^4 + x^2$

| **2** Multiply polynomials (pp. 454–455). | To multiply expressions containing powers, multiply the coefficients; then add the exponents of like bases. Write with positive exponents. |

Simplify. $(3x^4y^5)(7x^2yz) = 21x^6y^6z$

| **3** Divide polynomials (pp. 455–457). | To divide expressions containing powers, divide (or reduce) the coefficients; then subtract the exponents of the like bases. Write with positive exponents. |

Simplify. $\dfrac{10x^7y^4}{5x^8y^2} = 2x^{-1}y^2 = \dfrac{2y^2}{x}$

Section 10–4

| **1** Multiply and divide by powers of 10 (pp. 457–460). | To multiply by powers of 10, add the exponents and keep the base of 10. To divide by powers of 10, subtract the exponents and keep the base of 10. |

Multiply. $10^6(10^7) = 10^{13}$ Divide. $10^3 \div 10^5 = 10^{-2}$

| **2** Change a number from scientific notation to ordinary notation (pp. 460–461). | To change a number from scientific notation to ordinary notation, perform the indicated multiplication by moving the decimal point in the first factor the appropriate number of places. Insert zeros as necessary. Omit the power-of-10 factor. |

Write 3.27×10^{-4} as an ordinary number.

00003.27 = 0.000327.   Shift the decimal 4 places to the *left*.

| **3** Change a number from ordinary notation to scientific notation (pp. 461–462). | To change a number written in ordinary notation to scientific notation, insert a caret in the proper place to indicate where the decimal should be positioned so that the absolute value of the number is valued at 1 or between 1 and 10. Then determine how many places and in which direction the decimal shifts from the new position (caret) to the old position (decimal point). This number is the exponent of the power of 10. |

Write 54,000 in scientific notation.

5ᴧ4000. = 5.4 × 10⁴   From New to Old is 4 places to the right so the exponent is +4.

4 Multiply and divide numbers in scientific notation (pp. 462–466).

To multiply numbers in scientific notation, multiply the first factors, then multiply the powers of 10 by using the laws of exponents. Next, examine the first factor of the product to see if its absolute value is equal to 1 or is between 1 and 10. If the absolute value of the factor is 1 or is between 1 and 10, the process is complete. If the absolute value of the factor is not 1 or not between 1 and 10, shift the decimal so that the factor is equal to 1 or is between 1 and 10, and adjust the exponent of the power of 10 accordingly.

Multiply.

$(4.5 \times 10^{89})(7.5 \times 10^{36}) =$ Multiply first factors then power-of-10 factors.
$33.75 \times 10^{125} =$ Write first factor in scientific notation.
$3_{\wedge}375 \times 10^{1} \times 10^{125} =$ Multiply powers-of-10.
3.375×10^{126}

To divide numbers in scientific notation, use steps similar to multiplication, but apply the rule for the division of signed numbers and the laws of exponents for division.

Divide.

$(3 \times 10^{-3}) \div (4 \times 10^{2}) =$ Divide first factors, then power-of-10 factors.
$0.75 \times 10^{-5} =$ Write first factor in scientific notation.
$07_{\wedge}5 \times 10^{-1} \times 10^{-5} =$ Multiply powers-of-10.
7.5×10^{-6}

CHAPTER TRIAL TEST

Perform the indicated operations. Write the answers with positive exponents.

1. $(x^4)(x)$

2. $\dfrac{x^0}{x^2}$

3. $\left(\dfrac{4}{7}\right)^2$

4. $(6a^2b)^2$

5. $\left(\dfrac{x^2}{y}\right)^2$

6. $\dfrac{12x^2}{4x^3}$

7. $4a(3a^2 - 2a + 5)$

8. $\dfrac{60x^3 - 45x^2 - 5x}{5x}$

9. $(10^3)^2$

10. $\dfrac{10^{-5}}{10^3}$

11. $5x^2 - 3x - (2x + 4x^2)$

12. $4x^3 + 2x - 7 + (3x^3 - 7x - 5)$

Write as ordinary numbers.

13. 42×10^3

14. 0.83×10^2

15. 5.9×10^{-2}

16. 7.8×10^{-7}

Write in scientific notation.

17. 240

18. 5.2301

19. 0.021

20. 52.3×10^2

21. 783×10^{-5}

Perform the indicated operations. Express the answers in scientific notation.

22. $(5.9 \times 10^5)(3.1 \times 10^4)$

23. $\dfrac{5.25 \times 10^4}{1.5 \times 10^2}$

24. A star is 3.4 light-years from Earth. If 1 light-year is 5.87×10^{12} mi, how many miles from Earth is the star?

25. The total resistance (in ohms) of a dc series circuit equals the total voltage divided by the total amperage. If the total voltage is 3×10^3 V and the total amperage is 2×10^{-3} A, find the total resistance (in ohms) expressed as an ordinary number.

Roots and Radicals

11

11-1 | *Irrational Numbers and Real Numbers*

Learning Outcomes

1 Write roots using radical and exponential notation.

2 Estimate an irrational number.

3 Position an irrational number on the number line.

4 Write powers and roots using rational exponents and radical notation.

In Chapter 1 we saw that the opposite or inverse operation for raising to powers was finding roots. The *square root* of a number is the number that is used as a factor 2 times to equal the perfect square. The *radical notation* for square root is $\sqrt{}$, $\sqrt{25} = 5$. The *index* of a square root is 2, but is not required in radical notation.

A *cube root* is the number that is used as a factor 3 times to give the cube. The index of a cube root is 3. For cube roots, the index 3 is written in the $\sqrt{}$ portion of the radical sign: $\sqrt[3]{8} = 2$. For *fourth roots*, the index 4 is written in the $\sqrt{}$ portion of the radical sign: $\sqrt[4]{81} = 3$.

The results when numbers are squared, cubed, raised to the fourth power, and so on are called *perfect powers*. The principal (or positive) root of a perfect power is a whole number.

Powers of 2	Powers of 3	Powers of 4	Powers of 5
$2^1 = 2$	$3^1 = 3$	$4^1 = 4$	$5^1 = 5$
$2^2 = 4$	$3^2 = 9$	$4^2 = 16$	$5^2 = 25$
$2^3 = 8$	$3^3 = 27$	$4^3 = 64$	$5^3 = 125$
$2^4 = 16$	$3^4 = 81$		
$4^4 = 256$	$5^4 = 625$		
$2^5 = 32$	$3^5 = 243$	$4^5 = 1,024$	$5^5 = 3,125$
$2^6 = 64$	$3^6 = 729$	$4^6 = 4,096$	$5^6 = 15,625$

Selected roots

$\sqrt[4]{16} = 2$ $\qquad$ $\sqrt[5]{243} = 3$ $\qquad$ $\sqrt[3]{64} = 4$ $\qquad$ $\sqrt[6]{15,625} = 5$

1 **Write Roots Using Radical and Exponential Notation.**

Another notation for indicating a root is a fractional exponent. To indicate a square root, use the exponent $\frac{1}{2}$. To indicate a cube root, use the exponent $\frac{1}{3}$. For a fourth root, use the exponent $\frac{1}{4}$. To generalize, the exponential notation for the root of a number is the number written with the fractional exponent. The index of the root is the denominator of the fractional exponent and 1 is the numerator.

$$\sqrt[n]{a} = a^{\frac{1}{n}}$$ where *n* is a natural number greater than 1 and *a* is a positive number unless otherwise indicated.

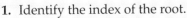

To write roots using radical and exponential notation:

1. Identify the index of the root.
2. For radical notation, place the index of the root in the $\sqrt{}$ portion of the radical sign with the radicand under the bar portion of the sign. The index 2 for square roots does not have to be written.
3. For exponential notation, write the radicand as the base and the index of the root as the denominator of a fractional exponent that has a numerator of 1.

Equivalent Notations for Roots

Just as $\frac{1}{2}$, 0.5, and 50% are three notations for writing equivalent values, radical notation and exponential notation are notations for writing equivalent values for roots.

In Words	Radical Notation	Exponential Notation
Square root of n	$\sqrt{n}$	$n^{1/2}$ or $n^{\frac{1}{2}}$ or $n^{0.5}$
Cube root of n	$\sqrt[3]{n}$	$n^{1/3}$ or $n^{\frac{1}{3}}$ or $n^{0.\overline{3}}$
Fourth root of n	$\sqrt[4]{n}$	$n^{1/4}$ or $n^{\frac{1}{4}}$ or $n^{0.25}$

EXAMPLE Write the roots using radical notation and exponential notation.

(a) Square root of 25 (b) Cube root of 125 (c) Fourth root of 625

(a) Square root of $25 = \sqrt{25} = 25^{1/2}$ or $25^{0.5}$

(b) Cube root of $125 = \sqrt[3]{125} = 125^{1/3}$ or $125^{0.\overline{3}}$

(c) Fourth root of $625 = \sqrt[4]{625} = 625^{1/4}$ or $625^{0.25}$

2 Estimate an Irrational Number.

Not all numbers are perfect powers. The root of a nonperfect power is an *irrational number*.

To find the two whole numbers that are closest to the value of an irrational number expressed as a radical:

1. Make a list of perfect powers that go beyond the given radicand.
2. Identify the two perfect powers that the radicand is between.
3. Find the principal roots of the two perfect powers from Step 2.
4. The root of the radicand is between the roots found in Step 3.

EXAMPLE Find two whole numbers that are closest to (a) $\sqrt{75}$ and (b) $\sqrt[3]{120}$.

(a) 1, 4, 9, 16, 25, 36, 49, 64, 81   Perfect squares. 75 is between 64 and 81.

$\quad\quad \sqrt{64} = 8 \quad\quad \sqrt{81} = 9$ Square roots of 64 and 81.

$\sqrt{75}$ **is between 8 and 9.**

(b) 1, 8, 27, 64, 125   Perfect cubes. 120 is between 64 and 125.

$\quad\quad \sqrt[3]{64} = 4 \quad\quad \sqrt[3]{125} = 5$ Cube roots of 64 and 125.

$\sqrt[3]{120}$ **is between 4 and 5.**

Real numbers include both rational and irrational numbers (Fig. 11–1).

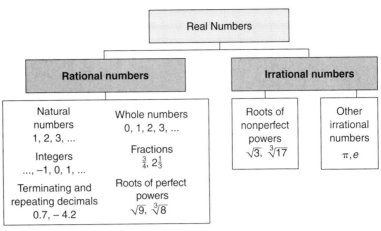

Figure 11–1

3 **Position an Irrational Number on the Number Line.**

Roots are ordered and have a position on the number line similar to the ordering of rational numbers.

To position an irrational number on the number line:

1. Find the perfect powers that the radicand is between.
2. Determine which perfect power the radicand is nearest.
3. Take the root of each perfect power found in Step 1.
4. Place the root of the radicand on the number line between the roots of the perfect powers and nearest the root of the perfect power identified in Step 2.

EXAMPLE   Position $\sqrt{11}$ on the number line.

1, 4, 9, 16   List the perfect squares until you have a number greater than 11.

11 is between 9 and 16.

11 is nearest 9. $11 - 9 = 2$; $16 - 11 = 5$.

$\sqrt{11}$ is between $\sqrt{9}$ and $\sqrt{16}$ or 3 and 4.   It is closer to 3. $\sqrt{9} = 3$; $\sqrt{16} = 4$.

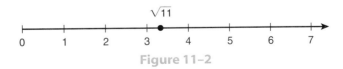

Figure 11–2

Because of the availability of scientific and graphing calculators, radical notation, which once was the most popular notation for roots, is now being replaced by the fractional exponent notation for roots.

There are numerous ways to find roots using scientific and graphing calculators.

Special root keys:

Some calculators have special function keys for the most common roots: square roots $\boxed{\sqrt{}}$ and cube roots $\boxed{\sqrt[3]{}}$. To use these, press the special root key, enter the base or radicand, and then press the $\boxed{=}$ or $\boxed{\text{ENTER}}$ key. Calculators may have a menu choice of special roots. From the menu, select the appropriate root, then enter the base or the radicand. To view the value of the root, press the $\boxed{=}$ or $\boxed{\text{ENTER}}$ key. To find roots in general, other calculator functions are necessary.

General root key:

Use the general root key, $\boxed{x^{1/y}}$, on your calculator, to find the fourth root of 16. Enter the base or radicand, press the general root key, enter only the index of the root (denominator of exponent), and then press $\boxed{=}$. Some calculators have a general root key in radical notation, $\boxed{\sqrt[y]{x}}$. Some typical sequences of keys are

$16 \boxed{x^{1/y}} 4 \boxed{=} \Rightarrow 2$

$16 \boxed{\sqrt[y]{x}} 4 \boxed{=} \Rightarrow 2$ Enter index second.

$4 \boxed{\sqrt[y]{x}} 16 \boxed{=} \Rightarrow 2$ Enter index first.

General power key:

Using the general power key, $\boxed{\wedge}$ or $\boxed{x^y}$, on a calculator, find the fourth root of 16.

Exponent as indicated division: $16 \boxed{\wedge} \boxed{(} \boxed{1} \boxed{\div} \boxed{4} \boxed{)} \boxed{=} \Rightarrow 2$
$$16^{1/4} = 2$$

Exponent as fraction: $16 \boxed{\wedge} 1 \boxed{a\,b/c} 4 \boxed{=} \Rightarrow 2$ If a fraction key is available.
$$16^{1/4} = 2$$

Exponent as decimal equivalent: $16 \boxed{\wedge} \boxed{\cdot} 25 \boxed{=} \Rightarrow 2$
$$16^{0.25} = 2$$

Test various sequences on your calculator to determine what options you have for finding roots.

EXAMPLE   Use calculator values to find the approximate positions of the square roots on the number line (Fig. 11–3).

     (a) $\sqrt{2}$ (b) $\sqrt{3}$ (c) $\sqrt{5}$ (d) $\sqrt{6}$

     (a) $\sqrt{2} \approx 1.414$ **(b)** $\sqrt{3} \approx 1.732$

     (c) $\sqrt{5} \approx 2.236$ **(d)** $\sqrt{6} \approx 2.449$

Figure 11–3

4 **Write Powers and Roots Using Rational Exponents and Radical Notation.**

Powers and roots are inverse operations for nonnegative values. In the order of operations, powers and roots (Exponents in "Please Excuse My Dear Aunt Sally") have the same priority. The order in which the operations are performed does not matter.

	Rational Exponent Notation	**Radical Notation**
For $x \geq 0$:	$(x^{1/n})^n = x$	$\left(\sqrt[n]{x}\right)^n = x$
	$(x^n)^{1/n} = x$	$\sqrt[n]{x^n} = x$

EXAMPLE Write both the rational exponent and radical notations.

(a) The fourth root of the square of x.
(b) The square of the fourth root of x.
(c) The square root of the cube of x.
(d) The cube of the square root of x.

Rational Exponent Notation **Radical Notation**

(a) $(x^2)^{1/4}$ $\sqrt[4]{x^2}$

(b) $(x^{1/4})^2$ $\left(\sqrt[4]{x}\right)^2$

(c) $(x^3)^{1/2}$ $\sqrt{x^3}$

(d) $(x^{1/2})^3$ $\left(\sqrt{x}\right)^3$

To write powers and roots using rational exponents and radical notation:

Rational exponent to radical:

1. Write the numerator of a rational exponent as the power.
2. Write the denominator of the rational exponent as the index of the root.

$$\text{Symbolically, } x^{\text{power}/\text{root}} = \sqrt[\text{root}]{x^{\text{power}}} \quad \text{or} \quad \left(\sqrt[\text{root}]{x}\right)^{\text{power}}$$

Radical to rational exponent:

1. Write the power as the numerator of the rational exponent.
2. Write the index of the root as the denominator of the rational exponent.

$$\text{Symbolically, } \sqrt[\text{root}]{x^{\text{power}}} = x^{\text{power}/\text{root}}$$

EXAMPLE Write $x^{2/3}$ in radical form.

$x^{2/3}$ Write the numerator of the rational exponent as the power.
Write the denominator of the rational exponent as the index of the root.

$\sqrt[3]{x^2}$

EXAMPLE Write $\sqrt[4]{x^3}$ in rational exponent notation.

$\sqrt[4]{x^3}$ Write the power of the radicand as the numerator of the fractional exponent.

Write the index as the denominator of the fractional exponent.

$x^{3/4}$

Interpreting an exponent depends on the type of number used as the exponent. Let's summarize the different types of numbers we have used as exponents.

Different Types of Numbers Used as Exponents Indicate Different Interpretations.

Exponent Type	Means
Natural number exponents greater than 1	Repeated multiplication
$4^3 = 4 \times 4 \times 4 = 64$	
Exponent of 1	Expression equals the base
$7^1 = 7$	
Exponent of 0	Expression equals 1
$5^0 = 1$	
Negative integral exponents	Reciprocals
$3^{-2} = \dfrac{1}{3^2} = \dfrac{1}{9}$	
Fractional or decimal exponents	Roots
$9^{1/2} = 9^{0.5} = \sqrt{9} = 3$	
$8^{1/3} = 8^{0.\overline{3}} = \sqrt[3]{8} = 2$	

SELF-STUDY EXERCISES 11–1

1 Write the roots in radical notation and exponential notation.

1. Square root of 36
2. Cube root of 27
3. Fourth root of 81
4. Fifth root of 32
5. Square root of 81
6. Cube root of 1,331

2 Find the two whole numbers that are closest to each root.

7. $\sqrt{40}$
8. $\sqrt{68}$
9. $\sqrt{12}$
10. $\sqrt{120}$
11. $\sqrt[3]{12}$
12. $\sqrt[3]{36}$
13. $\sqrt[3]{50}$
14. $\sqrt[3]{65}$

3 Position each root on a number line. Label the points. Estimate roots and check estimation with a calculator.

15. $\sqrt{25}$
16. $\sqrt{17}$
17. $\sqrt{20}$
18. $\sqrt{10}$
19. $\sqrt{18}$
20. $\sqrt{32}$
21. $\sqrt{7}$
22. $\sqrt{40}$

4 Write each expression in rational exponent notation and radical notation.

23. Square root of x^3
24. Cube root of the square of x
25. Fourth power of cube root of x
26. Square of cube root of x
27. Cube of fourth root of x
28. Fifth power of cube root of x

Write in radical form.

29. $x^{3/5}$
30. $x^{2/5}$
31. $x^{3/4}$
32. $x^{5/8}$
33. $y^{1/5}$
34. $y^{7/8}$
35. $y^{3/8}$
36. $y^{4/5}$

Write in rational exponent notation in simplest form.

37. $\sqrt[3]{x^2}$
38. $\sqrt[4]{x^2}$
39. $\sqrt[5]{x^2}$
40. $\sqrt[5]{x^3}$
41. $\sqrt[6]{x^5}$
42. $\sqrt[6]{x^3}$
43. $\sqrt[6]{x^4}$
44. $\sqrt[7]{x^4}$

Learning Outcomes

1 Find the square root of variables.

2 Simplify square-root radicals using rational exponents and the laws of exponents.

3 Simplify square-root radical expressions containing perfect-square factors.

1 **Find the Square Root of Variables.**

The square root of a variable is best understood by using the rational exponent notation and the laws of exponents. We will continue to consider only real numbers, and we will assume that the variables represent positive values.

$$\sqrt{x^2} = (x^2)^{1/2} = x^1 = x$$
$$\sqrt{x^4} = (x^4)^{1/2} = x^2$$
$$\sqrt{x^6} = (x^6)^{1/2} = x^3$$

TIP!

Perfect-Square Variable Factors

A positive variable with an even-number exponent is a perfect square. To find the square root of a variable factor, take $\frac{1}{2}$ of the exponent.

EXAMPLE Find the square root of the positive variables.

(a) $\sqrt{x^8}$ (b) $-\sqrt{x^{12}}$ (c) $\pm\sqrt{x^{18}}$ (d) $\sqrt{\dfrac{a^2}{b^4}}$

(a) $\sqrt{x^8} = (x^8)^{1/2} = x^4$ **(b)** $-\sqrt{x^{12}} = -(x^{12})^{1/2} = -x^6$

(c) $\pm\sqrt{x^{18}} = \pm(x^{18})^{1/2} = \pm x^9$ **(d)** $\sqrt{\dfrac{a^2}{b^4}} = \dfrac{(a^2)^{1/2}}{(b^4)^{1/2}} = \dfrac{a}{b^2}$

TIP!

Principal and Negative Square Roots

There are two square roots for a positive radicand. It is customary to indicate if the positive (principal) square root, the negative square root, or both square roots are to be considered by putting the appropriate sign in front of the radicand or its coefficient.

$$\sqrt{4} = 2 \qquad -\sqrt{4} = -2 \qquad \pm\sqrt{4} = \pm 2$$

2 **Simplify Square-Root Radicals Using Rational Exponents and the Laws of Exponents.**

Even though the calculator makes the pencil-and-paper method of taking roots outmoded, it is still helpful to perform some mental manipulations on expressions containing powers and roots before evaluating these expressions using the calculator. One common manipulation is to simplify rational exponents and radical expressions by applying the laws of exponents.

To simplify radicals using rational exponents and the laws of exponents:

1. Convert the radicals to equivalent expressions using rational exponents.
2. Apply the laws of exponents and the arithmetic of fractions.
3. Convert simplified expressions back to radical notation if desired.

EXAMPLE Convert the radical expressions to equivalent expressions using rational exponents and simplify if appropriate.

(a) $\sqrt[3]{x}$ (b) $\sqrt[5]{2y}$ (c) $(\sqrt{ab})^3$ (d) $\sqrt[4]{16b^8}$ (e) $(\sqrt[3]{27xy^5})^4$

(a) $\sqrt[3]{x} = x^{1/3}$ **(b)** $\sqrt[5]{2y} = (2y)^{1/5}$ **or** $2^{1/5}y^{1/5}$

(c) $(\sqrt{ab})^3 = (ab)^{3/2}$ **or** $a^{3/2}\,b^{3/2}$

(d) $\sqrt[4]{16b^8} =$ 16 is a perfect fourth root. $2^4 = 16$.

$(2^4 b^8)^{1/4} =$ Raise each factor to the $\frac{1}{4}$ power.

$(2^4)^{1/4}(b^8)^{1/4} =$ $(2^4)^{1/4} = 2^1 = 2; (b^8)^{1/4} = b^2$.

$\mathbf{2b^2}$

(e) $(\sqrt[3]{27xy^5})^4 =$ 27 is a perfect cube. $3^3 = 27$.

$(3^3xy^5)^{4/3} =$ Raise each factor to the $\frac{4}{3}$ power.

$(3^3)^{4/3}\,x^{4/3}\,(y^5)^{4/3} =$ $(3^3)^{4/3} = 3^4 = 81; (y^5)^{4/3} = y^{20/3}$.

$\mathbf{81\,x^{4/3}y^{20/3}}$

TIP!

Simplifying Coefficients

When coefficients of variable terms have rational exponents, the numerical equivalent can be determined if desired. Usually, if the coefficient is a perfect power of the indicated root, we will evaluate the coefficient. Otherwise, we may leave the coefficient with the rational exponent.

$$8^{2/3} = (8^{1/3})^2 = ((2^3)^{1/3})^2 = 2^2 = 4$$

Other laws of exponents are applied to rational exponents. The laws of exponents apply to factors having *like bases*. The following example illustrates other laws of exponents applied to rational exponents.

EXAMPLE Perform the following operations and simplify. Express answers with positive exponents in lowest terms.

(a) $(x^{3/2})(x^{1/2})$ (b) $(3a^{1/2}b^3)^2$ (c) $\dfrac{x^{1/2}}{x^{1/3}}$ (d) $\dfrac{10a^3}{2a^{1/2}}$

(a) $(x^{3/2})(x^{1/2}) = x^{3/2+1/2} = x^{4/2} = \mathbf{x^2}$ Add exponents.

(b) $(3a^{1/2}b^3)^2 = 3^2ab^6 = \mathbf{9ab^6}$ Multiply exponents: $\frac{1}{2} \cdot 2 = 1; 3 \cdot 2 = 6$.

(c) $\dfrac{x^{1/2}}{x^{1/3}} = x^{1/2-1/3} = \mathbf{x^{1/6}}$ Subtract exponents: $\frac{1}{2} - \frac{1}{3} = \frac{3}{6} - \frac{2}{6} = \frac{1}{6}$.

(d) $\dfrac{10a^3}{2a^{1/2}} = 5a^{3-1/2} = \mathbf{5a^{5/2}}$ Reduce coefficients. Subtract exponents: $3 - \frac{1}{2} = \frac{3}{1} - \frac{1}{2} = \frac{6}{2} - \frac{1}{2} = \frac{5}{2}$.

Chapter 11 / Roots and Radicals

EXAMPLE   Evaluate the expressions (a) and (d) from the preceding example for $x = 2$ and $a = 3$ both before and after simplifying.

(a) $(x^{3/2})(x^{1/2}) =$

Before simplifying: $(x^{3/2})(x^{1/2}) = (2^{3/2})(2^{1/2})$ Substitute $x = 2$.

$2\ \boxed{\land}\ \boxed{(}\ \boxed{(}\ 3\ \boxed{\div}\ 2\ \boxed{)}\ \boxed{\times}\ 2$ Enter into the calculator using the general power key $\boxed{\land}$ or $\boxed{x^y}$.

$\boxed{\land}\ \boxed{(}\ \boxed{(}\ 1\ \boxed{\div}\ 2\ \boxed{)}\ \boxed{=}\ \Rightarrow \mathbf{4}$

After symplifying: $(x^{3/2})(x^{1/2}) = x^{3/2+1/2}$ Add exponents.

$\qquad\qquad\qquad = x^2$ Substitute $x = 2$.

$\qquad\qquad\qquad 2^2 = \mathbf{4}$ Evaluate.

(d) $\dfrac{10\,a^3}{2a^{1/2}} =$

Before simplifying: $\dfrac{10\,a^3}{2a^{1/2}} = \dfrac{10 \cdot 3^3}{2 \cdot 3^{1/2}}$ Substitute $a = 3$.

$\boxed{(}\ 10\ \boxed{\times}\ 3\ \boxed{\land}\ 3\ \boxed{)}\ \boxed{\div}$ Both the numerator and denominators are groupings.

$\boxed{(}\ 2\ \boxed{\times}\ 3\ \boxed{\land}\ \boxed{(}\ 1\ \boxed{\div}\ 2\ \boxed{)}\ \boxed{)}\ \boxed{=}$

$\Rightarrow \mathbf{77.94228634}$

After simplifying: $\dfrac{10\,a^3}{2a^{1/2}} = \dfrac{10}{2} \cdot \dfrac{a^3}{a^{1/2}}$ Reduce coefficients. Subtract exponents.

$\qquad\qquad = 5a^{3-\frac{1}{2}}$ $3 - \dfrac{1}{2} = \dfrac{6}{2} - \dfrac{1}{2} = \dfrac{5}{2}.$

$\qquad\qquad = 5a^{5/2} \text{ or } 5a^{2.5}$ Substitute $a = 3$.

$\qquad\qquad = 5 \cdot 3^{2.5}$ Evaluate.

$5\ \boxed{\times}\ 3\ \boxed{\land}\ 2.5\ \boxed{=}\ \Rightarrow$

$\mathbf{77.94228634}$

Is it really worth the effort to simplify an expression before evaluating? Yes, in most cases. Long sequences of calculator steps are very tedious to enter.

As we saw in part (d) of the preceding example, *decimal exponents* can also be used to show roots.

EXAMPLE   Perform the operations. Express the answers with positive exponents.

(a) $a^{2.3}(a^3)$ (b) $(5a^{3.5}\,b^{0.5})^2$

(a) $a^{2.3}(a^3) = a^{2.3+3} = \mathbf{a^{5.3}}$

What does $a^{5.3}$ mean? If written as an improper fraction, $a^{5.3} = a^{5\frac{3}{10}} = a^{53/10}$.

This means we take the tenth root of a to the 53rd power or $\sqrt[10]{a^{53}}$.

(b) $(5a^{3.5}\,b^{0.5})^2 = 5^2 a^{3.5(2)}\,b^{0.5(2)}$

$\qquad\qquad\qquad = \mathbf{25a^7 b}$

3   Simplify Square-Root Radical Expressions Containing Perfect-Square Factors.

Some radicands are perfect squares, while others are not perfect squares. In the latter case, it may be useful to simplify the radicand. This is usually done by first writing the radicand as appropriate factors.

To *factor an algebraic expression* is to write it as the indicated product of two or more factors, that is, as a multiplication.

Some radicands that are not perfect squares can be *factored* into a perfect square times other factors. We simplify radicals by taking the square root of all the perfect-square factors. The factors that are not perfect squares are left under the radical sign, and the square roots of the perfect-square factors are outside the radical:

$$\sqrt{4x} = 2\sqrt{x};\ \sqrt{3x^2} = x\sqrt{3}.$$

A number such as 8 is not a perfect square; however, it does have a perfect-square factor: 4 is a factor of 8 and 4 is a perfect square. Thus, $\sqrt{8}$ can be written as $\sqrt{4 \cdot 2} = 2\sqrt{2}$.

Similarly, x^3 is not a perfect square because the exponent is not even. However, any power having an odd exponent greater than 1 has a perfect-square factor. Using the laws of exponents, $a^3 = a^2 a^1$. Then $\sqrt{a^3}$ can be simplified as $\sqrt{a^2 \cdot a^1} = a\sqrt{a}$. To find a^2 and a^1 as factors of the power a^3, we subtract the exponent 1 from the odd exponent 3.

To simplify square-root radicals containing perfect-square factors:

1. If the radicand is a perfect square, express it as a square root without the radical sign.
2. If the radicand is *not* a perfect square, factor the radicand into as many perfect-square factors as possible. The square roots of the perfect-square factors appear *outside* the radical and the other factors stay *inside* (under) the radical sign.

If the radicand is *not* a perfect square and *cannot* be factored into one or more perfect-square factors, it is in simplest radical form.

EXAMPLE Simplify the radicals.

(a) $\sqrt{32}$ (b) $\sqrt{y^7}$ (c) $\sqrt{18x^5}$ (d) $\sqrt{75xy^3z^5}$ (e) $\sqrt{7x}$

When factoring coefficients, some perfect squares that can be used are 4, 9, 16, 25, 36, 49, 64, and 81.

(a) $\sqrt{32} = \sqrt{16 \cdot 2}$

What is the *largest* perfect-square factor of 32? 4 is a factor of 32, but 16 is also a factor of 32. Use the *largest* perfect-square factor.

$$= \sqrt{16}\,(\sqrt{2}) = 4\sqrt{2}$$

(b) $\sqrt{y^7} = \sqrt{y^6 \cdot y^1} = \sqrt{y^6}\,(\sqrt{y})$

The largest perfect-square factor of y^7 is y^{7-1}, or y^6.

$$= y^3\sqrt{y}$$

(c) $\sqrt{18x^5}$

Factor radicand into as many perfect-square factors as you can.

$$= \sqrt{9 \cdot 2 \cdot x^4 \cdot x^1}$$

Write each factor as a separate radical.

$$= \sqrt{9}\,(\sqrt{2})\,(\sqrt{x^4})\,(\sqrt{x})$$

Take the square root of the two perfect squares.

$$= 3\,(\sqrt{2})\,(x^2)\sqrt{x}$$

Multiply coefficients. Multiply radicals.

$$= 3x^2\sqrt{2x}$$

In the previous example we showed each step in the simplifying process. However, we customarily do most of these steps mentally.

(d) $\sqrt{75xy^3z^5} = \sqrt{25 \cdot 3 \cdot x \cdot y^2 \cdot y \cdot z^4 \cdot z}$

Write the square roots of the perfect-square factors outside the radical sign. The remaining factors are written under the radical sign.

$$= 5yz^2\sqrt{3xyz}$$

(e) $\sqrt{7x} = \sqrt{7x}$

7 and x contain no perfect-square factors. The radical is in simplest form.

TIP!

Finding Perfect-Square Factors

- Whole-number perfect squares: 1, 4, 9, 16, 25, 36, 49, 64, 81, 100, 121, 144, ...
- 1 is a factor of any number: $8 = 8 \cdot 1$. To factor using the perfect square 1 does not simplify a radicand.
- Any variable with an exponent higher than 1 is a perfect square or has a perfect-square factor.
- Perfect-square variables:

$$x^2, x^4, x^6, x^8, x^{10}, \ldots$$

- Variables with a perfect-square factor:

$$x^3 = x^2 \cdot x^1, \qquad x^5 = x^4 \cdot x^1$$
$$x^7 = x^6 \cdot x^1, \qquad x^9 = x^8 \cdot x^1$$

- A convenient way to keep track of perfect-square factors is to circle them. The square roots of circled factors are written outside the radical sign. The uncircled factors stay in the radicand as is.

$$\sqrt{75ab^4c^3} = \sqrt{\textcircled{25} \cdot 3 \cdot a \cdot \textcircled{b^4} \cdot \textcircled{c^2} \cdot c^1}$$
$$= 5b^2c\sqrt{3ac}$$

EXAMPLE Simplify.

(a) $\sqrt{7x^2}$ (b) $\sqrt{9a}$ (c) $\sqrt{32m^5n^6}$

(a) $\sqrt{7\textcircled{x^2}} = x\sqrt{7}$ **(b)** $\sqrt{\textcircled{9}a} = 3\sqrt{a}$

(c) $\sqrt{32m^5n^6} = \sqrt{\textcircled{16} \cdot 2 \cdot \textcircled{m^4} \cdot m \cdot \textcircled{n^6}} = 4m^2n^3\sqrt{2m}$

SELF-STUDY EXERCISES 11–2

1 Find the square root of the positive variables.

1. $\sqrt{x^{10}}$ **2.** $\sqrt{x^6}$ **3.** $\pm\sqrt{x^{16}}$ **4.** $\sqrt{\dfrac{x^{14}}{y^{24}}}$

5. $\sqrt{a^6b^{10}}$ **6.** $-\sqrt{a^2b^4c^{12}}$ **7.** $\sqrt{\dfrac{x^2y^4}{z^{10}}}$ **8.** $\sqrt{\dfrac{a^4}{b^{10}c^{12}}}$

2 Convert the radical expressions to equivalent expressions using rational exponents, and simplify if appropriate.

9. $\sqrt[3]{y}$ **10.** $\sqrt[3]{5x}$ **11.** $\left(\sqrt{xy}\right)^5$ **12.** $\left(\sqrt{r}\right)^7$

13. $\sqrt[4]{81x^8}$ **14.** $\sqrt[4]{16r^{12}}$ **15.** $\left(\sqrt[3]{8x^3y^6}\right)^4$ **16.** $\left(\sqrt[4]{16xy^5}\right)^5$

17. $\sqrt[4]{x}$ **18.** $\sqrt[5]{7x}$ **19.** $\left(\sqrt[3]{xy}\right)^4$ **20.** $\sqrt[4]{81b^{20}}$

21. $\left(\sqrt[3]{8xy^7}\right)^2$

11-2 Simplifying Irrational Expressions

Perform the operations and simplify. Express answers with positive exponents in lowest terms.

22. $x^{5/2} \cdot x^{3/2}$ **23.** $y^{4/3} \cdot y^{5/3}$ **24.** $(4a^{1/3}b^2)^3$ **25.** $(2x^{2/3}y^{5/6})^3$

26. $\dfrac{x^{1/3}}{x^{2/3}}$ **27.** $\dfrac{x^{4/5}}{x^{1/5}}$ **28.** $\dfrac{y^{1/3}}{y^{1/2}}$ **29.** $\dfrac{b^{2/3}}{b^{1/2}}$

30. $\dfrac{12x^4}{3x^{1/2}}$ **31.** $\dfrac{20x^3}{4x^{2/3}}$

32. Evaluate $(x^{1/3})(x^{2/3})$ if $x = 2$. **33.** Evaluate $\dfrac{12x^2}{2x^{1/3}}$ if $x = 8$.

Perform the operations. Express the answers with positive exponents.

34. $a^{1.2}(a^4)$ **35.** $(7a^{1.7} \cdot a^{2.3})^2$ **36.** $\dfrac{x^{3.7}}{x^{1.7}}$

37. $\dfrac{12x^{2.9}}{4x^{0.9}}$ **38.** $2x^{2.1}(3x^{3.9})$

3 Simplify.

39. $\sqrt{24}$ **40.** $\sqrt{98}$ **41.** $\sqrt{48}$ **42.** $\sqrt{x^9}$ **43.** $\sqrt{y^{15}}$

44. $\sqrt{12x^3}$ **45.** $\sqrt{56a^5}$ **46.** $\sqrt{72a^3x^4}$ **47.** $\sqrt{44x^5y^2z^7}$

48. Create a square-root radical with the product of a constant factor that is not a perfect square but has a perfect-square factor and a variable factor with an odd exponent greater than 1. Then simplify.

11–3 | *Basic Operations with Square-Root Radicals*

Learning Outcomes   **1** Add or subtract square-root radicals.
2 Multiply square-root radicals.
3 Divide square-root radicals.
4 Rationalize a denominator.

1   **Add or Subtract Square-Root Radicals.**

As you recall, when adding or subtracting measures or algebraic terms, we only add or subtract like quantities. Similarly, only *like* radicals are added or subtracted. Square-root radicals are *like* radicals whenever the radicands are identical.

Like radicals are radical expressions with radicands that are identical and have the same order or index.

To add or subtract like square-root radicals:

1. Add or subtract the coefficients of like radicals.
2. Use the common radicand as a factor in the solution.

Symbolically, $a\sqrt{b} + c\sqrt{b} = (a + c)\sqrt{b}.$

EXAMPLE Add or subtract the radicals when possible.

(a) $3\sqrt{7} + 2\sqrt{7}$ 　　(b) $4\sqrt{2} + \sqrt{2}$ 　　(c) $5\sqrt{3} + 7\sqrt{5} + 2\sqrt{3} - 4\sqrt{5}$

(d) $3 + \sqrt{3}$ 　　(e) $3\sqrt{11} - 3\sqrt{11}$ 　　(f) $2\sqrt{2} - \sqrt{3}$

(a) $3\sqrt{7} + 2\sqrt{7} = 5\sqrt{7}$ 　　Radicals are like. Add coefficients.

(b) $4\sqrt{2} + \sqrt{2} = 5\sqrt{2}$ 　　When no coefficient is written in front of a radical, the coefficient is 1.

(c) $5\sqrt{3} + 7\sqrt{5} + 2\sqrt{3} - 4\sqrt{5} =$ 　　Combine like radicals.

$7\sqrt{3} + 3\sqrt{5}$

(d) $3 + \sqrt{3}$ 　　No addition can be performed. These are not like radicals.

(e) $3\sqrt{11} - 3\sqrt{11} = 0\sqrt{11} = 0$ 　　Zero times any number is zero.

(f) $2\sqrt{2} - \sqrt{3}$ 　　The terms cannot be combined. These are not like radicals.

We add or subtract square-root radical expressions only if they have *like* radicands. However, when radicals are not in simplest form, you may not be able to recognize like radicals. We sometimes can simplify the radical expressions and obtain like radicands. If we obtain like radicands, then we add or subtract.

EXAMPLE Add or subtract the radical expressions when possible.

(a) $12\sqrt{5} + 3\sqrt{20}$ 　　(b) $\sqrt{3} - \sqrt{27}$ 　　(c) $6\sqrt{3} + 2\sqrt{8}$

(a) $12\sqrt{5} + 3\sqrt{20}$ 　　$\sqrt{20}$ can be simplified. $\sqrt{20} = \sqrt{4 \cdot 5}$.

$12\sqrt{5} + 3\sqrt{4 \cdot 5}$ 　　$\sqrt{4} = 2$. The 2 becomes a coefficient. 3 and 2 are both coefficients of the same term.

$12\sqrt{5} + 3 \cdot 2\sqrt{5}$ 　　Multiply 3 and 2.

$12\sqrt{5} + 6\sqrt{5}$ 　　Now we have like radicands. Add coefficients.

$18\sqrt{5}$ 　　Keep like radicand.

(b) $\sqrt{3} - \sqrt{27}$ 　　$\sqrt{27}$ can be simplified. $\sqrt{27} = \sqrt{9 \cdot 3}$.

$\sqrt{3} - \sqrt{9 \cdot 3}$ 　　$\sqrt{9} = 3$. The 3 becomes a coefficient.

$\sqrt{3} - 3\sqrt{3}$ 　　Now we have like radicals. Add coefficients.

$-2\sqrt{3}$ 　　Keep like radicand.

(c) $6\sqrt{3} + 2\sqrt{8}$ 　　$\sqrt{8}$ can be simplified. $\sqrt{8} = \sqrt{4 \cdot 2}$.

$6\sqrt{3} + 2\sqrt{4 \cdot 2}$ 　　$\sqrt{4} = 2$. The 2 becomes a coefficient.

$6\sqrt{3} + 2 \cdot 2\sqrt{2}$ 　　$2 \cdot 2\sqrt{2} = 4\sqrt{2}$

$6\sqrt{3} + 4\sqrt{2}$ 　　Terms cannot be combined. Radicals are still unlike radicals.

2 **Multiply Square-Root Radicals.**

When multiplying two square-root radicals, the expressions under the radical signs (radicands) are multiplied together. Numbers in front of the radical signs are coefficients of the radicals and are multiplied separately.

11–3 Basic Operations with Square-Root Radicals 　　485

To multiply square-root radicals:

1. Multiply coefficients to give the coefficient of the product.
2. Multiply radicands to give the radicand of the product.
3. Simplify if possible.

$$\text{Symbolically, } a\sqrt{b} \cdot c\sqrt{d} = ac\sqrt{bd}.$$

Learning Strategy   *Express Procedures in Your Own Words.*

The precise details of a procedure can sometimes be overwhelming. The procedure written symbolically often guides you through the process. However, it is desirable to describe the procedure in your own words.

For example, the procedure for multiplying square-root radicals could be casually phrased in various ways.

$$\text{Outside times outside} \Rightarrow \text{Stays outside}$$
$$\text{Inside times inside} \Rightarrow \text{Stays inside}$$
$$(2\sqrt{3})(5\sqrt{2}) = 10\sqrt{6}$$

Another example might be used with the procedure for simplifying perfect-square radicands. The radical symbol defines a confined area. "Perfect" factors get out; "imperfect factors" stay in. As perfect factors come out, they are different (square root is taken).

$$\sqrt{9x^2y} = \qquad$$ 9 is "perfect." It comes out as 3.
    x^2 is "perfect." It comes out as x.
    y is "imperfect." It stays in as y.

$$3x\sqrt{y}$$

EXAMPLE   Multiply the following radicals.

(a) $\sqrt{3} \cdot \sqrt{5}$ (b) $\sqrt{\dfrac{7}{8}} \cdot \sqrt{\dfrac{2}{3}}$ (c) $3\sqrt{2} \cdot 4\sqrt{3}$ (d) $\sqrt{3} \cdot \sqrt{12}$

(a) $\sqrt{3} \cdot \sqrt{5} = \sqrt{15}$ $\sqrt{15}$ will not simplify.

(b) $\sqrt{\dfrac{7}{8}} \cdot \sqrt{\dfrac{2}{3}} = \sqrt{\dfrac{14}{24}} = \sqrt{\dfrac{7}{12}}$ Reduce.

(c) $3\sqrt{2} \cdot 4\sqrt{3} = 12\sqrt{6}$ Multiply coefficients 3 and 4. Then multiply radicands 2 and 3.

(d) $\sqrt{3} \cdot \sqrt{12} = \sqrt{36} = 6$ 36 is a perfect square. Take the square root of 36.

The distributive property can be used to multiply radical expressions.

EXAMPLE   Multiply by distributing.

(a) $5(\sqrt{7} - 2)$

(b) $\sqrt{3}(\sqrt{11} - 5)$

(c) $\sqrt{2}(\sqrt{5} - \sqrt{7})$

(a) $5(\sqrt{7} - 2) = 5\sqrt{7} - 5 \cdot 2$ Distribute. Multiply $5 \cdot 2$.

$\qquad\qquad = 5\sqrt{7} - 10$

(b) $\sqrt{3}(\sqrt{11} - 5) = \sqrt{3} \cdot \sqrt{11} - 5\sqrt{3}$ Distribute. Multiply $\sqrt{3} \cdot \sqrt{11}$.

$\qquad\qquad\qquad = \sqrt{33} - 5\sqrt{3}$

(c) $\sqrt{2}(\sqrt{5} - \sqrt{7}) = \sqrt{2} \cdot \sqrt{5} - \sqrt{2} \cdot \sqrt{7}$ Distribute. Multiply radicals.

$\qquad\qquad\qquad = \sqrt{10} - \sqrt{14}$

3   Divide Square-Root Radicals.

When dividing square-root radicals, we follow a similar procedure as with multiplication. Coefficients and radicands are divided (or reduced) separately.

To divide square-root radicals:

1. Divide coefficients to give the coefficient of the quotient.
2. Divide radicands to give the radicand of the quotient.
3. Simplify if possible.

$$\frac{a\sqrt{b}}{c\sqrt{d}} = \frac{a}{c}\sqrt{\frac{b}{d}} \qquad c \text{ and } d \neq 0$$

EXAMPLE   Divide the following radicals.

(a) $\dfrac{\sqrt{12}}{\sqrt{4}}$ (b) $\dfrac{\sqrt{\dfrac{2}{3}}}{\sqrt{\dfrac{7}{4}}}$ (c) $\dfrac{3\sqrt{6}}{6}$ (d) $\dfrac{5\sqrt{20}}{\sqrt{10}}$ (e) $\dfrac{8x^3\sqrt{9x^2}}{2x^2\sqrt{16x^3}}$

(a) $\dfrac{\sqrt{12}}{\sqrt{4}} = \sqrt{\dfrac{12}{4}} = \sqrt{3}$ Divide or reduce.

(b) $\dfrac{\sqrt{\dfrac{2}{3}}}{\sqrt{\dfrac{7}{4}}} = \sqrt{\dfrac{\dfrac{2}{3}}{\dfrac{7}{4}}} = \sqrt{\dfrac{2}{3}\left(\dfrac{4}{7}\right)} =$ Multiply $\frac{2}{3}$ by the reciprocal of $\frac{7}{4}$. $\left(\frac{2}{3} \div \frac{7}{4} = \frac{2}{3} \cdot \frac{4}{7}\right)$.

$\sqrt{\dfrac{8}{21}} = \sqrt{\dfrac{4 \cdot 2}{21}} = \dfrac{2\sqrt{2}}{\sqrt{21}}$ Factor and take the square root of perfect-square factors.

(c) $\dfrac{3\sqrt{6}}{6}$ The coefficient 3 and the denominator 6 can be divided (or reduced) because they are both outside the radical.

$\dfrac{\overset{1}{3}\sqrt{6}}{\underset{2}{6}} = \dfrac{\sqrt{6}}{2}$ A coefficient of 1 does not have to be written in front of the radical.

(d) $\dfrac{5\sqrt{20}}{\sqrt{10}}$ The 20 and 10 are divided because they are both square-root radicands.

$$\frac{5\sqrt{20}}{\sqrt{10}} = 5\sqrt{2}$$

(e) $\dfrac{8x^3\sqrt{9x^2}}{2x^2\sqrt{16x^3}} =$ Reduce the factors outside the radical.

$\dfrac{4x\sqrt{9x^2}}{\sqrt{16x^3}} =$ Reduce the factors inside the radical.

$\dfrac{4x\sqrt{9}}{\sqrt{16x}} =$ Remove perfect-square factors from all radicands.

$\dfrac{4x(3)}{4\sqrt{x}} =$ Reduce the resulting coefficients.

$\dfrac{3x}{\sqrt{x}} =$

4 Rationalize a Denominator.

Fractions with radical factors in the denominator are often rewritten in an equivalent form. Before the common use of calculators, this was a popular manipulation. Dividing by a rational number, and most often a whole number, was easier and more accurate than dividing by a rounded decimal approximation of the irrational number. Even now, finding common denominators and other procedures are easier if all denominators contain only rational numbers. Thus, $\frac{1}{\sqrt{3}}$ and $\sqrt{\frac{2}{5}}$ are generally rewritten so that the denominator is a rational number. This procedure is called *rationalizing the denominator*.

To rationalize a denominator:

1. If the denominator has an irrational factor, multiply the denominator by another irrational factor so that the resulting radicand is a perfect square.
2. To preserve the value of the fraction, multiply the numerator by the same radical factor used in step 1. Thus, in steps 1 and 2 together, we have multiplied by an equivalent of 1.
3. Simplify all radicals and reduce the resulting fraction, if possible.

EXAMPLE Rationalize all denominators. Simplify the answers if possible.

(a) $\dfrac{5}{\sqrt{7}}$ (b) $\dfrac{2}{\sqrt{x}}$ (c) $\dfrac{4}{\sqrt{8}}$ (d) $\dfrac{5\sqrt{2}}{x^2\sqrt{3x}}$ (e) $\sqrt{\dfrac{2}{3}}$

(a) $\dfrac{5}{\sqrt{7}} = \dfrac{5}{\sqrt{7}} \cdot \dfrac{\sqrt{7}}{\sqrt{7}} = \dfrac{5\sqrt{7}}{7}$ $\sqrt{7} \cdot \sqrt{7} = (\sqrt{7})^2 = 7$. Denominator now has no irrational factors.

(b) $\dfrac{2}{\sqrt{x}} = \dfrac{2}{\sqrt{x}} \cdot \dfrac{\sqrt{x}}{\sqrt{x}} = \dfrac{2\sqrt{x}}{x}$ $\sqrt{x} \cdot \sqrt{x} = (\sqrt{x})^2 = x$. Denominator now has no irrational factors.

(c) $\dfrac{4}{\sqrt{8}} = \dfrac{4}{\sqrt{4 \cdot 2}} = \dfrac{4}{2\sqrt{2}} = \dfrac{2}{\sqrt{2}}$ Simplify perfect-square factors in radicands and reduce.

$= \dfrac{2}{\sqrt{2}} \cdot \dfrac{\sqrt{2}}{\sqrt{2}} = \dfrac{2\sqrt{2}}{2} = \sqrt{2}$ Rationalize the denominator by multiplying by $\dfrac{\sqrt{2}}{\sqrt{2}}$.

Chapter 11 / Roots and Radicals

(d) $\dfrac{5\sqrt{2}}{x^2\sqrt{3x}} = \dfrac{5\sqrt{2}}{x^2\sqrt{3x}} \cdot \dfrac{\sqrt{3x}}{\sqrt{3x}} = \dfrac{5\sqrt{6x}}{x^2 \cdot 3x} = \dfrac{5\sqrt{6x}}{3x^3}$

(e) $\sqrt{\dfrac{2}{3}} = \dfrac{\sqrt{2}}{\sqrt{3}} \cdot \dfrac{\sqrt{3}}{\sqrt{3}} = \dfrac{\sqrt{6}}{3}$ Separate fractional radicand into a fraction with a radical factor in both the numerator and denominator.

When is rationalizing a denominator practical? Before calculators and computers, a radical in the denominator meant dividing by a decimal approximation of an irrational number. Calculations were more difficult and answers were generally less accurate. Are calculator values less accurate if the denominator is not rationalized? Examine the next example.

EXAMPLE Use a calculator to compare the approximate value of the expressions:

$$\dfrac{1}{\sqrt{2}} \quad \text{and} \quad \dfrac{\sqrt{2}}{2}$$

$$\dfrac{1}{\sqrt{2}} = 1 \div \boxed{\sqrt{}}\ 2\ \boxed{=} \Rightarrow 0.7071067812$$

$$\dfrac{\sqrt{2}}{2} = \boxed{\sqrt{}}\ 2 \div 2\ \boxed{=} \Rightarrow 0.7071067812$$

The radical expressions are equivalent.

TIP!

Why Do We Still Rationalize Denominators?

In reality, the importance of rationalizing radical expressions in technical applications is minimal. Approximate values and calculators and computers are most often used.

The most important use of rationalizing is to find alternative, but equivalent, representations of expressions. These alternative representations may make it easier to identify like radicals and other properties.

Traditionally, radical expressions are in simplest form if

1. There are no perfect-square factors in any radicand.
2. There are no fractional radicands.
3. The denominator of a radical expression is rational (contains no radicals).

A logical sequence for simplifying radical expressions follows.

To simplify a radical expression:

1. Reduce radicands and coefficients whenever possible.
2. Simplify expressions by removing perfect-square factors from all radicands.
3. Reduce radicands and coefficients whenever possible.
4. Rationalize denominators that contain radicals factors.
5. Reduce radicands and coefficients whenever possible.

EXAMPLE Perform the operation and simplify if possible.

(a) $\sqrt{y^3} \cdot \sqrt{8y^2}$ (b) $\sqrt{\dfrac{1}{3}} \cdot \sqrt{\dfrac{8}{3x}}$

(a) $\sqrt{y^3} \cdot \sqrt{8y^2} = \sqrt{8y^5} = \sqrt{4 \cdot 2 \cdot y^4 \cdot y} = \mathbf{2y^2\sqrt{2y}}$

(b) $\sqrt{\dfrac{1}{3}} \cdot \sqrt{\dfrac{8}{3x}} = \sqrt{\dfrac{8}{9x}} = \dfrac{\sqrt{8}}{\sqrt{9x}} = \dfrac{\sqrt{4 \cdot 2}}{\sqrt{9 \cdot x}} = \dfrac{2\sqrt{2}}{3\sqrt{x}} \cdot \dfrac{\sqrt{x}}{\sqrt{x}} = \dfrac{\mathbf{2\sqrt{2x}}}{\mathbf{3\,x}}$

SELF-STUDY EXERCISES 11–3

1 Add or subtract. Simplify radicals where necessary.

1. $5\sqrt{3} + 7\sqrt{3}$
2. $8\sqrt{5} - 12\sqrt{5}$
3. $4\sqrt{7} + 3\sqrt{7} - 5\sqrt{7}$
4. $2\sqrt{3} - 8\sqrt{5} + 7\sqrt{3}$
5. $9\sqrt{11} - 3\sqrt{6} + 4\sqrt{6} - 12\sqrt{11}$
6. $44\sqrt{2} + \sqrt{3} - \sqrt{2} + 5\sqrt{3}$
7. $2\sqrt{3} + 5\sqrt{12}$
8. $7\sqrt{5} + 2\sqrt{45}$
9. $4\sqrt{63} - \sqrt{7}$
10. $3\sqrt{6} - 2\sqrt{54}$
11. $8\sqrt{2} - 3\sqrt{28}$
12. $2\sqrt{3} + \sqrt{48}$
13. $3\sqrt{5} + 4\sqrt{180}$
14. $7\sqrt{98} - 2\sqrt{2}$
15. $6\sqrt{40} - 2\sqrt{90}$
16. $\sqrt{12} - \sqrt{27}$

2 Multiply and simplify if possible.

17. $\sqrt{6} \cdot \sqrt{7}$
18. $\sqrt{18} \cdot \sqrt{2}$
19. $3\sqrt{2} \cdot 5\sqrt{5}$
20. $8\sqrt{3} \cdot 5\sqrt{12}$
21. $5\sqrt{3x} \cdot 4\sqrt{5x^2}$
22. $2x\sqrt{3x^4} \cdot 7x^2\sqrt{8x}$
23. $7\left(\sqrt{2} + 5\right)$
24. $4\left(12 - \sqrt{3}\right)$
25. $\sqrt{5}\left(\sqrt{2} - 3\right)$
26. $\sqrt{7}\left(4 + \sqrt{3}\right)$
27. $\sqrt{3}\left(\sqrt{5} + \sqrt{2}\right)$
28. $\sqrt{11}\left(\sqrt{2} + \sqrt{3}\right)$

Perform the indicated operations and simplify if possible.

29. $\sqrt{x^2} \cdot \sqrt{3x}$
30. $\sqrt{\dfrac{2}{3}} \cdot \sqrt{\dfrac{4}{5y}}$
31. $\sqrt{\dfrac{1}{x}} \cdot \sqrt{\dfrac{8}{7x}}$
32. $\sqrt{\dfrac{4}{9y^2}} \cdot \sqrt{\dfrac{1}{2y}}$

33. $\sqrt{\dfrac{8x}{3}} \cdot \sqrt{\dfrac{2x^2}{3}}$
34. $\sqrt{7} \cdot \sqrt{x^2}$
35. $\sqrt{\dfrac{4x^2}{x^3}} \cdot \sqrt{12x}$
36. $\sqrt{\dfrac{1}{2}} \cdot \sqrt{\dfrac{x^2}{3}}$

37. $\sqrt{\dfrac{y^3}{2}} \cdot \sqrt{\dfrac{2}{7y}}$
38. $\sqrt{8x} \cdot \sqrt{x^2}$

3 Divide and simplify.

39. $\dfrac{\sqrt{18}}{\sqrt{2}}$
40. $\dfrac{5\sqrt{10}}{10}$
41. $\dfrac{4\sqrt{12}}{2\sqrt{6}}$

42. $\dfrac{15\sqrt{24}}{9\sqrt{2}}$
43. $\dfrac{12x^2\sqrt{8x}}{15x\sqrt{6x^3}}$
44. $\dfrac{6x^4\sqrt{25x^3}}{2x\sqrt{16x^2}}$

4 Rationalize the denominator and simplify.

45. $\dfrac{5}{\sqrt{3}}$
46. $\dfrac{6}{\sqrt{5}}$
47. $\dfrac{1}{\sqrt{8}}$
48. $\dfrac{\sqrt{5}}{\sqrt{11}}$

49. $\dfrac{\sqrt{8}}{\sqrt{12}}$
50. $\dfrac{5x}{\sqrt{3x}}$
51. $\dfrac{2x^2}{\sqrt{7x^2}}$
52. $\dfrac{4x^5}{\sqrt{12x^3}}$

Learning Outcomes

1. Write imaginary numbers using the letter i.
2. Raise imaginary numbers to powers.
3. Write real and imaginary numbers in complex form, $a + bi$.
4. Combine complex numbers.

1 Write Imaginary Numbers Using the Letter i.

Taking the square root of a negative number introduces a new type of number. This new type of number is called an *imaginary number*. An *imaginary number* has a factor of $\sqrt{-1}$, which is represented by i ($i = \sqrt{-1}$).

Imaginary numbers and real numbers combine to form the set of complex numbers (Fig. 11–4). A *complex number* is a number that can be written in the form $a + bi$, where a and b are real numbers and i is $\sqrt{-1}$.

The square root of a negative number, say, $\sqrt{-16}$, can be simplified as $\sqrt{-1 \cdot 16}$ or $4\sqrt{-1}$. Because the square root of -1 is an *imaginary* factor and 4 is a real factor, we will use the letter i to represent $\sqrt{-1}$ and rewrite $4\sqrt{-1}$ as $4i$. Similarly, $\sqrt{-4} = \sqrt{-1 \cdot 4} = 2\sqrt{-1} = 2i$.

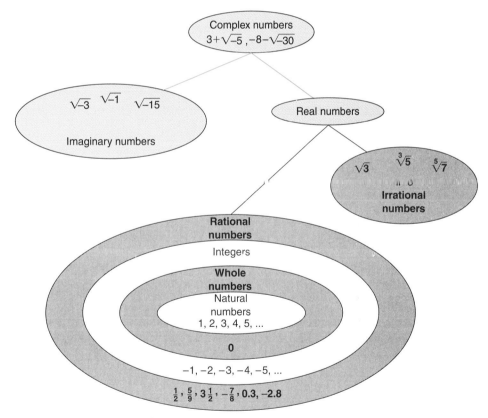

Figure 11–4 Complex-number system.

TIP!

Is i Different from j?

Electronics and other applications of imaginary numbers may use j rather than i to represent $\sqrt{-1}$. Thus, $3 + 4i$ and $3 + 4j$ represent the same amount.

EXAMPLE   Rewrite the imaginary numbers using the letter i for $\sqrt{-1}$.

(a) $\sqrt{-9}$ (b) $\sqrt{-25}$ (c) $\sqrt{-7}$

(a) $\sqrt{-9} = \sqrt{-1 \cdot 9} = 3\sqrt{-1} = 3\boldsymbol{i}$
(b) $\sqrt{-25} = \sqrt{-1 \cdot 25} = 5\sqrt{-1} = 5\boldsymbol{i}$
(c) $\sqrt{-7} = \sqrt{-1 \cdot 7} = \sqrt{7} \cdot \sqrt{-1} = \sqrt{7}\boldsymbol{i}$ or $\boldsymbol{i}\sqrt{7}$.

TIP!

Are Numerical Coefficients Always First?

$\sqrt{7}i$ and $\sqrt{7i}$ do not represent the same amount. In the first term, $\sqrt{7}i$, i is not under the radical symbol. In the second term, $\sqrt{7i}$, i is under the radical symbol. Because it is easy to confuse the two terms, when the coefficient of an imaginary number is an irrational number, we write the i factor first: $\sqrt{7}i = i\sqrt{7}$.

2 Raise Imaginary Numbers to Powers.

Some powers of imaginary numbers are real numbers. Examine the pattern that develops with powers of i.

$i = \sqrt{-1} = i$
$i^2 = (\sqrt{-1})^2 = -1$
$i^3 = i^2 \cdot i^1 = -1i = -i$
$i^4 = i^2 \cdot i^2 = -1(-1) = 1$

$i^5 = i^4 \cdot i^1 = 1(i) = i$
$i^6 = i^4 \cdot i^2 = 1(-1) = -1$
$i^7 = i^4 \cdot i^3 = 1(-i) = -i$
$i^8 = i^4 \cdot i^4 = 1(1) = 1$

$i^9 = i^4 \cdot i^4 \cdot i^1 = 1(1)(i) = i$
$i^{10} = i^4 \cdot i^4 \cdot i^2 = 1(1)(-1) = -1$
$i^{11} = i^4 \cdot i^4 \cdot i^3 = 1(1)(-i) = -i$
$i^{12} = i^4 \cdot i^4 \cdot i^4 = 1(1)(1) = 1$

- All powers of i simplify to either i, -1, $-i$, or 1.
- If the exponent of i is a multiple of 4, the result is 1.
- If the exponent of i is even but not a multiple of 4, the result is -1.
- All even powers of i are real numbers.

These and other patterns allow us to develop a shortcut process for simplifying powers of i.

To simplify a power of *i*:

1. Divide the exponent by 4 and examine the **remainder**.
2. The power of i simplifies as follows, based on the remainder in Step 1.

Remainder of $0 \Rightarrow i^0 = 1$ Remainder of $2 \Rightarrow i^2 = -1$
Remainder of $1 \Rightarrow i^1 = i$ Remainder of $3 \Rightarrow i^3 = -i$

EXAMPLE   Simplify the powers of i.

(a) i^{15} (b) i^{20} (c) i^{33} (d) i^{18}

(a) $i^{15} = -i$ $15 \div 4 = 3$ R3. Remainder of $3 \Rightarrow i^3 = -i$.
(b) $i^{20} = 1$ $20 \div 4 = 5$. Remainder $0 \Rightarrow i^0 = 1$.
(c) $i^{33} = i$ $33 \div 4 = 8$ R1. Remainder $1 \Rightarrow i^1 = i$.
(d) $i^{18} = -1$ $18 \div 4 = 4$ R2. Remainder $2 \Rightarrow i^2 = -1$.

3 Write Real and Imaginary Numbers in Complex Form, $a + bi$.

A complex number has two parts, a *real part* and an *imaginary part*. In $a + bi$, if $a = 0$, then $a + bi$ is the same as $0 + bi$ or bi, which is an imaginary number. If $b = 0$, then $a + bi$ is the same as $a + 0 \cdot i$ or a, which is a real number. Thus, real numbers and imaginary numbers are also complex numbers.

EXAMPLE   Rewrite in the form $a + bi$.

(a) 5   (b) $\sqrt{-36}$ (c) $-3i^2$ (d) $6 - \sqrt{-5}$

(a) $5 = 5 + 0i$ $a = 5, \ b = 0$.

(b) $\sqrt{-36} = 6i = 0 + 6i$ $a = 0, \ b = 6$.

(c) $-3i^2 = -3(-1) = 3 = 3 + 0i$ $a = 3, \ b = 0$.

(d) $6 - \sqrt{-5} = 6 - \sqrt{(5)(-1)} = 6 - i\sqrt{5}$ $a = 6, \ b = -\sqrt{5}$.

4 Combine Complex Numbers.

Complex numbers are combined by adding *like* parts. Real parts are added together and imaginary parts are added together.

To add or subtract complex numbers:

1. Add or subtract real parts for the real part of the answer.
2. Add or subtract imaginary parts for the imaginary part of the answer.

Symbolically, $(a + bi) + (c + di) = (a + c) + (b + d)i$

EXAMPLE   Combine the complex numbers.

(a) $(3 + 5i) + (8 - 2i)$ (b) $(-3 + i) - (7 - 4i)$ (c) $-2 + (5 + 6i)$

(a) $(\ 3 \ + \ 5i\ ) + (\ 8 \ - \ 2i\ ) = \ 11 \ + \ 3i$ $3 + 8 = 11 \ ; \ 5i - 2i = 3i$.

(b) $(-3 + i) - (7 - 4i) =$ Distribute -1.

$(-3 + i) + (-7 + 4i) = -10 + 5i$ $-3 - 7 = -10 \ ; \ i + 4i = 5i$.

(c) $-2 + (5 + 6i) = 3 + 6i$ $-2 + 5 = 3$.

1 Write the imaginary numbers using the letter i. Simplify if possible.

1. $\sqrt{-25}$ **2.** $\sqrt{-36}$ **3.** $\sqrt{-64x^2}$ **4.** $\sqrt{-32y^5}$

Simplify the powers of *i*.

5. i^{17} **6.** i^5 **7.** i^{20} **8.** i^{10}
9. i^{24} **10.** i^9 **11.** i^{32} **12.** i^{15}

3 Write the real and imaginary numbers in complex form.

13. 15   **14.** 17   **15.** $\sqrt{-49}$ **16.** $\sqrt{-81}$
17. $33i$ **18.** $-7i^3$ **19.** $4i^6$ **20.** $5 + \sqrt{-4}$
21. $8 + \sqrt{-32}$ **22.** $7 - \sqrt{-3}$

4 Combine the complex numbers.

23. $(4 + 3i) + (7 + 2i)$ **24.** $\sqrt{12} - 5\sqrt{-3} + \left(\sqrt{8} - \sqrt{-27}\right)$
25. $12 - 3i - (8 - 4i)$ **26.** $15 + 8i - (3 - 12i)$
27. $(4 + 7i) - (3 - 2i)$ **28.** $(8 - 5i) + (4 - 3i)$

11-5 | Equations with Squares and Square Roots

Learning Outcomes

  1   Solve equations with squared variables.
  2   Solve equations with square-root radical terms.
  3   Use the Pythagorean theorem to find the missing side of a right triangle.

  1   **Solve Equations with Squared Variables.**

In solving equations with squared variables, we use similar procedures for solving equations that we used in Chapter 8. Equations that we examine in this section are *quadratic equations* because they contain a quadratic or squared variable term. At this point, we will only examine quadratic equations with squared variable terms and number terms. In Chapter 15, we define quadratic equations and examine additional types of quadratic equations.

To solve an equation containing only squared variable terms and number terms:

1. Perform all normal steps to isolate the squared variable and to obtain a numerical coefficient of 1.
2. Take the square root of both sides.
3. Identify both roots of the equation. $\sqrt{x^2} = \pm x$, for $x \geq 0$.

EXAMPLE   Solve the equations.

    (a) $x^2 = 4$ (b) $x^2 + 9 = 25$ (c) $5x^2 + 10 = 30$
    (d) $\dfrac{3}{25} = \dfrac{x^2}{48}$ (e) $x^2 + 16 = 0$

  (a) $x^2 = 4$ The squared variable is isolated and has a coefficient of 1. Take the square root of both sides.

         $x = \pm 2$ Identify both roots. ± 2 indicates roots of $+2$ and -2.
  (b) $x^2 + 9 = 25$ Isolate the squared variable term. Apply the addition axiom to sort like terms.

         $x^2 = 25 - 9$ Combine like terms.

$$x^2 = 16$$ Take the square root of both sides.

$$x = \pm 4$$ Identify both roots. $x = 4$ or $x = -4$.

(c) $5x^2 + 10 = 30$ Isolate the squared variable term.

$$5x^2 = 30 - 10$$ Combine like terms.

$$5x^2 = 20$$ Divide by the coefficient of x^2.

$$\frac{5x^2}{5} = \frac{20}{5}$$

$$x^2 = 4$$ Take the square root of both sides.

$$x = \pm 2$$ Identify both roots. $x = 2$ or $x = -2$.

(d) $\dfrac{3}{25} = \dfrac{x^2}{48}$ To solve a proportion, cross multiply.

$$3(48) = 25x^2$$

$$144 = 25x^2$$ Divide by the coefficient of x^2.

$$\frac{144}{25} = \frac{25x^2}{25}$$

$$\frac{144}{25} = x^2$$ Take the square root of both sides.

$$\pm\frac{12}{5} = x$$ Identify both roots. $x = \dfrac{12}{5}$ or $x = -\dfrac{12}{5}$.

(e) $x^2 + 16 = 0$ Isolate the squared variable term.

$$x^2 = -16$$ Take the square root of both sides.

$$x = \pm\sqrt{-16}$$ Express the roots as imaginary numbers using i.

$$x = \pm 4i$$ Identify both roots. $x = 4i$ or $x = -4i$.

TIP!

Imaginary Roots versus No Real Roots

Sometimes within the context of an applied problem imaginary or complex roots are unacceptable. Often a problem will specify that only real roots are acceptable. If that is the case, part e in the preceding example would have *no real solutions*.

2 Solve Equations with Square-Root Radical Terms.

In this discussion we consider only equations containing a radical isolated on either or both sides of the equation. Whenever this is the case, we solve the equation by first squaring both sides of the equation. This clears the equation of radicals. Then, we use previously learned procedures to finish solving the equation.

To solve an equation containing square-root radicals that are isolated on either or both sides of the equation:

1. Square both sides to eliminate any radicals.
2. Perform all normal steps to isolate the variable and solve the equation.

EXAMPLE   Solve the following equations.

(a) $\sqrt{x} = \dfrac{3}{4}$ (b) $\sqrt{x-2} = 5$ (c) $\sqrt{5x^2 - 36} = 8$

(d) $\sqrt{1.7x^2} = \sqrt{6.8}$

(a) $\sqrt{x} = \dfrac{3}{4}$ Square both sides of the equation.

$$(\sqrt{x})^2 = \left(\dfrac{3}{4}\right)^2 \qquad (\sqrt{x})^2 = x;\ \left(\dfrac{3}{4}\right)^2 = \dfrac{9}{16}.$$

$$x = \dfrac{9}{16}$$

(b) $\sqrt{x - 2} = 5$ Square both sides of the equation.

$(\sqrt{x - 2})^2 = 5^2$ Any square-root radical squared equals the radicand.

$x - 2 = 25$ Isolate the variable.

$x = 25 + 2$ Combine like terms.

$\mathbf{x = 27}$

(c) $\sqrt{5x^2 - 36} = 8$ Square both sides of the equation.

$(\sqrt{5x^2 - 36})^2 = 8^2$ Do this step mentally.

$5x^2 - 36 = 64$ Isolate the variable term.

$5x^2 = 64 + 36$ Combine like terms.

$5x^2 = 100$ Divide by the coefficient of x^2.

$\dfrac{5x^2}{5} = \dfrac{100}{5}$ Do this step mentally.

$x^2 = 20$ Take the square root of both sides.

$x = \pm\sqrt{20}$ Identify both roots.

$\mathbf{x = \pm 2\sqrt{5}}$ Exact roots. $x = 2\sqrt{5}$ or $x = -2\sqrt{5}$.

$\mathbf{x \approx \pm 4.472}$ Approximate roots.

(d) $\sqrt{1.7\,x^2} = \sqrt{6.8}$ Square both sides to eliminate radicals.

$1.7x^2 = 6.8$ Divide by the coefficient of x^2.

$x' \dfrac{6.8}{1.7}$ Simplify.

$x^2 = 4$ Take the square root of both sides.

$\mathbf{x = \pm 2}$ Identify both roots. $x = 2$ or $x = -2$.

TIP!

When Do I Take the Square Root of Both Sides? When Do I Square Both Sides?

These are important questions.

- If an equation has a squared letter term, take the square root of both sides at the *end* of the solution. That is, when the squared letter has been isolated with a coefficient of $+1$, take the square root of both sides.
- If an equation has one or two radical terms but each radical is isolated on one side of the equal sign, square both sides of the equation at the *beginning* of the solution. Then, solve the remaining equation that contains no radicals.
- Some equations may have both radical terms and squared variables. Clear the radicals at the *beginning* of the procedure, and take the square root of both sides at the *end* of the procedure.

There are other situations in which an equation may have one or more radical terms that are not isolated. We will not discuss those situations at this time.

EXAMPLE   Shirley Riddle needs to prepare a flower bed for planting. She needs a square bed with an area of 128 ft². How long should each side of the bed be?

Estimation $11^2 = 121$

$12^2 = 144$

The bed should be between 11 and 12 ft on a side.

$$A = s^2 \qquad \text{Formula for the area of a square.}$$
$$128 = s^2 \qquad \text{Substitute 128 for } A.$$
$$\sqrt{128} = s \qquad \text{Take square root of both sides.}$$
$$8\sqrt{2} = s \qquad \text{Exact root}$$
$$11.3 \approx s \qquad \text{Approximate root}$$

Interpretation   Each side of the bed should be 11.3 ft.

TIP!

Examining Roots

Exact Roots Versus Approximate Roots

A root is an exact amount when no rounding has been done. If an exact root is desired, give the root in simplest form. This may include an irrational factor.

    If the root has been rounded, it becomes an approximate root. Approximate roots are often more meaningful than exact roots, especially in applied problems.

One Root Versus Two Roots

Look again at the four problems in the example starting at the bottom of page 495. Parts a and b have variables to the first power after the radical is removed. Therefore, there will be *at most* one solution. Parts c and d have variables that are squared when the radical is removed. These equations will have *at most* two roots or solutions.

Extraneous Roots

The techniques of squaring both sides of an equation is equivalent to multiplying by a variable, thus introducing the possibility of gaining an extraneous root. Be sure to check your roots.

3   **Use the Pythagorean Theorem to Find the Missing Side of a Right Triangle.**

One of the most famous and useful theorems in mathematics is the Pythagorean theorem. It is named for the Greek mathematician, Pythagoras. The theorem applies to a right triangle.

    A *right triangle* has two sides that form a right angle (square corner). These two sides are called the *legs* of the triangle. The side opposite the right angle, or the third side of the triangle, is called the *hypotenuse.* The *Pythagorean theorem* states that the square of the hypotenuse of a right triangle is equal to the sum of the squares of the two legs of the triangle.

Formula for Pythagorean theorem:

$$c^2 = a^2 + b^2$$

c is the hypotenuse of a right triangle; a and b are the legs (Fig. 11–5).

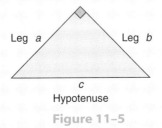

Leg a Leg b

c
Hypotenuse

Figure 11–5

To find a leg or the hypotenuse of a right triangle:

1. Identify the two known sides of the triangle and the missing side, and state the theorem symbolically.
2. Substitute known values in the Pythagorean theorem. $c^2 = a^2 + b^2$
3. Solve for the missing value.
4. The solution is the principal square root.

EXAMPLE   Find b when $a = 8$ mm and $c = 17$ mm.

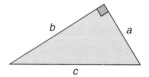

Figure 11–6

$$c^2 = a^2 + b^2$$ State the theorem symbolically and substitute known values.
$$17^2 = 8^2 + b^2$$
$$289 = 64 + b^2$$ Transpose to isolate.
$$289 - 64 = b^2$$ Combine like terms.
$$225 = b^2$$ Take the square root of both sides.
$$\sqrt{225} = b$$ Solution is the principal square root.
$$\mathbf{15\ mm} = b$$

Many times we are given problems that contain "hidden" triangles. In these cases, we need to visualize the triangle or triangles in the problems. Drawing one or more of the sides of the "hidden" triangle helps solve the problem.

EXAMPLE   Find the center-to-center distance between pulleys A and C (Fig. 11–7).

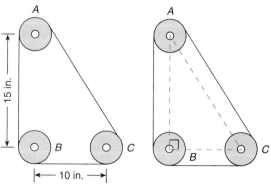

Figure 11–7

Connect the center points of the three pulleys to form a right triangle. The hypotenuse is the distance between the centers of pulleys A and C. Use the Pythagorean theorem and substitute the given values for the two known sides.

$$(AC)^2 = (AB)^2 + (BC)^2$$ State theorem symbolically and substitute values.
$$(AC)^2 = 15^2 + 10^2$$ Square both constants.

$$(AC)^2 = 225 + 100 \qquad \text{Combine like terms.}$$
$$(AC)^2 = 325 \qquad \text{Take the square root of both sides.}$$
$$AC = \sqrt{325} \qquad \text{Find the approximate principal square root.}$$
$$AC = 18.0277564$$

The distance from pulley A to pulley C is 18.0 in. (to the nearest tenth).

EXAMPLE   The head of a bolt is a square 0.5 in. on a side (distance across flats). What is the distance from corner to corner (distance across corners)? (See Fig. 11–8.)

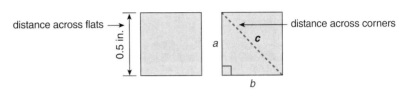

distance across flats

0.5 in.

distance across corners

a c

b

Figure 11–8

The sides of the bolt head form the legs of a triangle if a diagonal line is drawn from one corner to the opposite corner. This diagonal forms the hypotenuse of the right triangle. Because the legs of the triangle are known to be 0.5 in. each, we can substitute in the Pythagorean theorem to find the diagonal line, the hypotenuse, which is the distance across corners.

$$c^2 = a^2 + b^2 \qquad \text{State theorem symbolically.}$$
$$c^2 = 0.5^2 + 0.5^2 \qquad \text{Substitute values.}$$
$$c^2 = 0.25 + 0.25$$
$$c^2 = 0.5 \qquad \text{Take the square root of both sides.}$$
$$c = \sqrt{0.5} \qquad \text{Evaluate.}$$
$$c = 0.707106781$$

The distance across corners is 0.71 in. (to the nearest hundredth).

Sometimes right triangles are used to represent certain relationships, such as forces acting on an object at right angles or electrical and electronic phenomena related in the way the three sides of a right triangle are related. Let's look at an example.

EXAMPLE   Forces A and B come together at a right angle to produce force C (Fig. 11–9). If force A is 74.8 lb and the resulting force C is 91.5 lb, what is force B?

Since the forces are related in the way the sides of a right triangle are related, we may use the Pythagorean theorem to find the missing force B, a leg of the triangle.

Force $A = 74.8$ lb

Force $C = 91.5$ lb

Force B

Figure 11–9

$$A^2 + B^2 = C^2 \qquad \text{State theorem symbolically and substitute values.}$$
$$74.8^2 + B^2 = 91.5^2 \qquad \text{Square both constants.}$$
$$5{,}595.04 + B^2 = 8{,}372.25 \qquad \text{Transpose to isolate.}$$
$$B^2 = 8{,}372.25 - 5{,}595.04 \qquad \text{Combine constants.}$$
$$B^2 = 2{,}777.21 \qquad \text{Take the square root of both sides.}$$
$$B = \sqrt{2{,}777.21} \qquad \text{Find the approximate principal square root.}$$
$$B = 52.7 \qquad \text{To the nearest tenth.}$$

Force B is 52.7 lb.

SELF-STUDY EXERCISES 11–5

1 Solve the equations. Use your calculator to evaluate to the nearest thousandth when necessary.

1. $y^2 = 25$

2. $3x^2 = 75$

3. $3x^2 = 36$

4. $\dfrac{3}{81} = \dfrac{x^2}{12}$

5. $3 + R^2 = 12$

6. $\dfrac{1}{2}P^2 = \dfrac{3}{7}$

7. $\dfrac{4}{9} = \dfrac{x^2}{3}$

8. $16 - T^2 = 0$

9. $5x^2 - 10 = 20$

10. $x^2 + 25 = 0$

11. $\dfrac{x^2}{6} = \dfrac{12}{7}$

12. $-9 + 3x^2 = 18$

2 Solve the equations. Use your calculator if necessary. Evaluate to the nearest thousandth.

13. $\sqrt{x} = 9$

14. $\sqrt{\dfrac{y}{3}} = 8$

15. $8 = \sqrt{y - 2}$

16. $\sqrt{3x^2 - 3} = 9$

17. $\sqrt{\dfrac{3}{x + 2}} = 12$

18. $\sqrt{x^2} = 2$

19. $10 = \sqrt{2x}$

20. $\sqrt{\dfrac{2}{3}x^2} = 4$

21. $\sqrt{1.3y^2} = 2.4$

22. $\sqrt{y^2} = 2.9$

23. $\sqrt{2x^2 - 1} = 5$

24. $\sqrt{x - 5} = 2$

3 Find the missing side of the right triangle. Round the final answers to the nearest thousandth. Use Fig. 11–10 for Exercises 25–27.

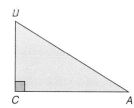

Figure 11–10

25. $AC = ?$
    $BC = 7$ cm
    $AB = 25$ cm

26. $AC = 24$ mm
    $BC = ?$
    $AB = 26$ mm

27. $AC = 15$ yd
    $BC = 8$ yd
    $AB = ?$

Use Fig. 11–11 for Exercises 28–30. Find the missing dimensions. Round to thousandths where appropriate.

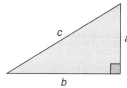

Figure 11–11

28. $a = 5$ cm
    $b = 4$ cm
    $c = ?$

29. $a = ?$
    $b = 9$ m
    $c = 11$ m

30. $a = 9$ ft
    $b = ?$
    $c = 15$ ft

Solve the problems. Round final answers to the nearest thousandth if necessary.

31. A light pole will be braced with a wire that is to be tied to a stake in the ground 18 ft from the base of the pole, which extends 26 ft above the ground. If the wire is attached to the pole 2 ft from the top, how much wire must be used to brace the pole? (See Fig. 11–12.)

32. Find the center-to-center distance between holes A and C in a sheet-metal plate if the distance between the centers of A and B is 16.5 cm and the distance between the centers of B and C is 36.2 cm (Fig. 11–13).

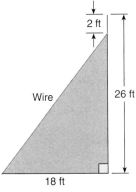

Wire 2 ft 26 ft 18 ft

Figure 11–12

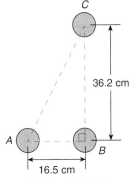

C 36.2 cm A 16.5 cm B

Figure 11–13

33. Find the length of a rafter that has a 10-in. overhang if the rise of the roof is 10 ft and the joists are 48 ft long (Fig. 11–14).

34. A stair stringer is 8 ft high and extends 10 ft from the wall (Fig. 11–15). How long will the stair stringer be?

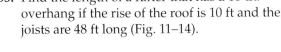
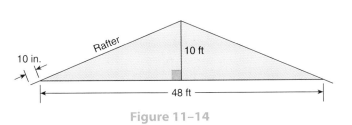

Rafter 10 in. 10 ft 48 ft

Figure 11–14

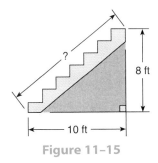

? 8 ft 10 ft

Figure 11–15

35. A machinist wishes to strengthen an L bracket that is 5 cm by 12 cm by welding a brace to each end of the bracket (Fig. 11–16). How much metal rod is needed for the brace?

36. To make a rectangular table more stable, a diagonal brace is attached to the underside of the table surface. If the table is 27 dm by 36 dm, how long is the brace?

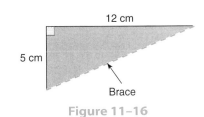

12 cm 5 cm Brace

Figure 11–16

37. Find the length of the side of the largest square nut that can be milled from a piece of round stock whose diameter is 15 mm (Fig. 11–17).

38. A rigid length of electrical conduit must be shaped as shown in Fig. 11–18 to clear an obstruction. What total length of the conduit is needed? (*Hint:* Don't forget to include *AB* and *CD* in the total length.)

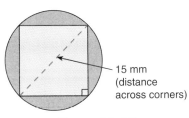

15 mm (distance across corners)

Figure 11–17

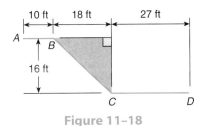

10 ft 18 ft 27 ft A B 16 ft C D

Figure 11–18

39. The vector diagram in Fig. 11–19 is used in electrical applications. Find the voltage of E_a.

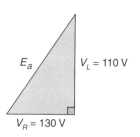

E_a $V_L = 110$ V

$V_R = 130$ V

Figure 11–19

ASSIGNMENT EXERCISES

Section 11–1

Write the roots in radical notation and exponential notation.
1. Square root of 49
2. Cube root of 125
3. Fourth root of 16
4. Fifth root of 3,125
5. Square root of 121
6. Cube root of 343

Find the two whole numbers that are closest to each root.
7. $\sqrt{38}$
8. $\sqrt{15}$
9. $\sqrt{135}$
10. $\sqrt{75}$
11. $\sqrt[3]{60}$
12. $\sqrt{135}$

Position each square root on a number line. Label the points. Estimate the roots and check estimation with a calculator.
13. $\sqrt{16}$
14. $\sqrt{11}$
15. $\sqrt{5}$
16. $\sqrt{38}$

Write in both rational exponent and radical notations.
17. The square root of the seventh power of x
18. The cube root of the fourth power of x
19. The square of the cube root of x
20. The fifth power of the fifth root of x

Write using rational exponents.
21. $\sqrt{x}$
22. $\sqrt[3]{x^5}$
23. $\sqrt[5]{x^4}$
24. $\sqrt[4]{9x}$
25. $\left(\sqrt[3]{xy}\right)^4$
26. $\sqrt[3]{64x^{10}}$
27. $\sqrt{7}$

Write in radical format.
28. $x^{5/8}$
29. $y^{3/5}$
30. $a^{1/4}$

Section 11–2

Find the square roots if all variables represent positive numbers.
31. $\sqrt{y^{12}}$
32. $\sqrt{a^{10}}$
33. $-\sqrt{b^{18}}$
34. $\pm\sqrt{\dfrac{x^4}{y^6}}$

Convert the radicals to equivalent expressions using rational exponents.
35. $\sqrt[3]{x}$
36. $\sqrt[4]{p}$
37. $\sqrt[5]{4y}$
38. $\left(\sqrt{ab}\right)^6$
39. $\sqrt[3]{8b^{12}}$
40. $\left(\sqrt{49\,x^2y^3}\right)^4$

Perform the operations. Express the answers with positive exponents in lowest terms.
41. $\left(a^{1/2}\right)\left(a^{3/2}\right)$
42. $\left(a^{4/3}\right)\left(a^{2/3}\right)$
43. $y^{3/4} \cdot y^{1/4}$
44. $y^{5/8} \cdot y^{1/8}$
45. $\left(3x^{1/4}y^2\right)^3$
46. $\left(2x^{3/4}\,y\right)^2$
47. $\left(4ax^{1/2}\right)^3$
48. $\left(x^{1/2}\right)^{1/3}$
49. $\dfrac{x^{3/4}}{x^{1/4}}$
50. $\dfrac{x^{1/6}}{x^{5/6}}$
51. $\dfrac{a^{5/6}}{a^{-1/3}}$
52. $\dfrac{a^{7/10}}{a^{2/5}}$
53. $\dfrac{x^{5/8}}{x^{3/4}}$
54. $\dfrac{y^{1/3}}{y^{5/6}}$
55. $\dfrac{a^3}{a^{1/3}}$

Evaluate the expressions in Exercises 56 to 62 for $a = 2$, $b = 1$, and $x = 3$ both before and after simplifying.

56. $\dfrac{a^2}{a^{3/5}}$ **57.** $\dfrac{12a^4}{6a^{1/2}}$ **58.** $\dfrac{27x^3}{9x^{2/3}}$ **59.** $\dfrac{15a^{3/5}}{10a^5}$

60. $\dfrac{14a^{5/6}}{24a^2}$ **61.** $a^{2.3}(a^4)$ **62.** $(3a^{1.2}b^2)^3$

Simplify the expressions.

63. $\left(\sqrt{5}\right)^2$ **64.** $\left(\sqrt{x^5}\right)^2$ **65.** $\sqrt{x^2}$ **66.** $\left(\sqrt{8}\right)^2$

67. $\sqrt{9P^3}$ **68.** $\sqrt{8^2}$ **69.** $\sqrt{18a^2b}$ **70.** $\sqrt{12x^2y^3}$

71. $\sqrt{32x^5y^2}$ **72.** $\sqrt{63x^4y^7}$ **73.** $\sqrt{75x^{10}y^9}$ **74.** $\sqrt{125xy^3}$

Section 11–3

Add or subtract the radicals.

75. $5\sqrt{3} - 7\sqrt{3}$ **76.** $4\sqrt{2} + 3\sqrt{5} - 8\sqrt{2} + 6\sqrt{5}$ **77.** $3\sqrt{7} - 2\sqrt{28}$

78. $\sqrt{2} - \sqrt{8}$ **79.** $2\sqrt{6} + 3\sqrt{54}$ **80.** $3\sqrt{5} - 2\sqrt{45}$

81. $4\sqrt{3} - 8\sqrt{48}$ **82.** $\sqrt{40} + \sqrt{90}$ **83.** $5\sqrt{8} - 3\sqrt{50}$

84. $5\sqrt{7} - 4\sqrt{63}$ **85.** $3\sqrt{2} - 5\sqrt{32}$

Multiply the radicals and simplify.

86. $\sqrt{6} \cdot \sqrt{3}$ **87.** $2\sqrt{8} \cdot 3\sqrt{6}$ **88.** $2\sqrt{a} \cdot \sqrt{b}$ **89.** $5\sqrt{3} \cdot 8\sqrt{7}$

90. $2\sqrt{3} \cdot 5\sqrt{18}$ **91.** $-8\sqrt{5} \cdot 4\sqrt{30}$ **92.** $5\left(\sqrt{3} - 2\right)$ **93.** $\sqrt{3}\left(\sqrt{12} - 5\right)$

94. $\sqrt{2}\left(\sqrt{6} - \sqrt{10}\right)$ **95.** $\sqrt{3}\left(\sqrt{6} - \sqrt{15}\right)$

Divide the radicals and simplify.

96. $\dfrac{4\sqrt{8}}{2}$ **97.** $\dfrac{3\sqrt{5}}{2\sqrt{20}}$ **98.** $\dfrac{2\sqrt{90}}{\sqrt{5}}$ **99.** $\dfrac{6\sqrt{18}}{8\sqrt{12}}$

100. $\dfrac{14\sqrt{56}}{7\sqrt{7}}$ **101.** $\dfrac{5\sqrt{48}}{20\sqrt{20}}$ **102.** $\dfrac{\sqrt{9x}}{\sqrt{3x}}$ **103.** $\dfrac{\sqrt{3y^3}}{\sqrt{y^3}}$

104. $\left(\sqrt{\dfrac{25}{36}}\right)^2$ **105.** $\left(\sqrt{\dfrac{9}{16}}\right)^2$ **106.** $\sqrt{\dfrac{9c^4}{25y^6}}$

Rationalize the denominator and simplify.

107. $\dfrac{5}{\sqrt{17}}$ **108.** $\dfrac{1}{\sqrt{8}}$ **109.** $\dfrac{\sqrt{7}}{\sqrt{12}}$ **110.** $\dfrac{\sqrt{3}}{\sqrt{7x}}$

111. $\dfrac{\sqrt{3}}{\sqrt{8}}$ **112.** $\dfrac{\sqrt{7}}{5\sqrt{18}}$ **113.** $\dfrac{5\sqrt{3}}{\sqrt{24}}$ **114.** $\dfrac{\sqrt{15}}{5\sqrt{7}}$

Section 11–4

Write the numbers using the letter i. Simplify if possible.

115. $\sqrt{-100}$ **116.** $-\sqrt{-16x^2}$ **117.** $\pm\sqrt{-24y^7}$

Simplify the powers of i.

118. i^5 **119.** i^{14} **120.** i^{98} **121.** i^{77}

Write as complex numbers in simplified form.

122. 5 **123.** $15i$ **124.** $3 + \sqrt{-9}$ **125.** $-12i^5$ **126.** $-6i^{11}$

Simplify.

127. $(5 + 3i) + (2 - 7i)$ **128.** $(4 - i) - (3 - 2i)$ **129.** $\left(7 - \sqrt{-9}\right) + \left(4 + \sqrt{-16}\right)$

Solve the equations containing squares and radicals. Use your calculator if needed. Evaluate to the nearest thousandth.

130. $q^2 = 81$

131. $x^2 - 36 = 0$

132. $3x^2 - 2 = 7$

133. $x^2 - 4 = 0$

134. $7 + P^2 = 107$

135. $x^2 + 4 = 0$

136. $x^2 + 81 = 0$

137. $18 = 2x^2$

138. $\sqrt{P + 2} = 12$

139. $\sqrt{\dfrac{27}{2}} = x$

140. $\sqrt{3 + x} = 14$

141. $\sqrt{q + 3} = 7$

142. $\sqrt{\dfrac{1}{4x}} = 2$

143. $\sqrt{1.3x^2} = 11.7$

144. $\sqrt{x^2 + 1} = 5$

145. $\sqrt{x^2 + 2} = 9$

146. $\sqrt{3 + y^2} = 10$

147. $\sqrt{Q^2 - 1} = 0$

148. $0 = \sqrt{z^2 - 4}$

149. $\sqrt{2 + y^2} = 8$

Use Fig. 11–20 to solve the following exercises. Round the final answers to the nearest thousandth if necessary.

150. $a = 3$ m
  $b = ?$
  $c = 5$ m

151. $a = 9$ in.
  $b = 12$ in.
  $c = ?$

152. $a = 8$ cm
  $b = 15$ cm
  $c = ?$

153. $a = 7$ ft
  $b = ?$
  $c = 10$ ft

154. $a = 8$ mm
  $b = ?$
  $c = 17$ mm

155. $a = ?$
  $b = 15$ yd
  $c = 17$ yd

156. $a = ?$
  $b = 12$ km
  $c = 15$ km

157. $a = 11$ mi
  $b = 17$ mi
  $c = ?$

158. $a = 10$ in.
  $b = 24$ in.
  $c = ?$

159. $a = ?$
  $b = 40$ cm
  $c = 50$ cm

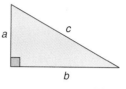

Figure 11–20

Solve the following problems. Round the final answers to the nearest thousandth if necessary.

160. If the base of a ladder is placed on the ground 4 ft from a house, how tall must the ladder be (to the nearest foot) to reach the chimney top that extends $18\frac{1}{2}$ ft above the ground?

161. In an automobile, three pulleys are connected by one belt. The center-to-center distance between the pulleys farthest apart cannot be measured conveniently. The other center-to-center distances are 12 in. and 18 in. (Fig. 11–21). Find the distance between the pulleys that are farthest apart.

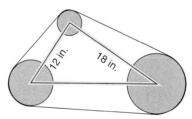

Figure 11–21

162. A central vacuum outlet is installed in one corner of a rectangular room that measures 9′ × 12′. How long must the nonelastic hose be to reach all parts of the room?

163. Find the diameter of a piece of round steel from which a 3-in. square nut can be milled.

164. Find the distance across the corners of a square nut that is 7.9 mm on a side.

CHALLENGE PROBLEMS

Perform the indicated operation and express in simplest form.

165. $5i(3 - 4i)$

166. $12(7 + 2i)$

Write the rules in words. Assume that all radicands represent positive values.

1. $a\sqrt{b} \cdot c\sqrt{d} = ac\sqrt{bd}$

2. $\dfrac{a\sqrt{b}}{c\sqrt{d}} = \dfrac{a}{c}\sqrt{\dfrac{b}{d}}$; $c, d \neq 0$

3. $a\sqrt{b} + c\sqrt{b} = (a + c)\sqrt{b}$

4. $\left(\sqrt{x}\right)^2 = x$ or $\sqrt{x^2} = x$, for positive values of x.

5. List the conditions for a radical expression to be in simplest form.

6. What does it mean to *rationalize* a denominator? What calculations or manipulations with fractions are easier if the denominator is a rational number?

7. What property of equality can be used to solve an equation that contains a squared variable term?

8. When should the procedure of taking the square root of both sides of an equation be used?

9. What property of equality can be used to solve an equation that contains a square-root radical?

10. When should the procedure of squaring both sides of an equation be used?

11. Write the property in words.

$$x^{1/n} = \sqrt[n]{x}$$

12. Write the following property in words:

$$x^{m/n} = \sqrt[n]{x^m} \text{ or } \left(\sqrt[n]{x}\right)^m$$

Illustrate with numerical examples the properties of radicals by using the laws of exponents and rational exponents. Assume that all radicands are positive values.

13. $\sqrt[n]{xy} = \sqrt[n]{x} \cdot \sqrt[n]{y}$

14. $\sqrt[m]{\sqrt[n]{x}} = \sqrt[n]{\sqrt[m]{x}} = \sqrt[mn]{x}$

15. $\sqrt[n]{\dfrac{x}{y}} = \dfrac{\sqrt[n]{x}}{\sqrt[n]{y}}$

16. $x^{-m/n} = \dfrac{1}{x^{m/n}}$

CHAPTER SUMMARY

Learning Outcomes	What to Remember with Examples

Section 11–1

1 Write roots using radical and exponential notation (pp. 473–474).

The root of a number is indicated by the denominator of a fractional exponent or the index of the radical.

The square root of 16
$$16^{1/2} = \sqrt{16} = 4;$$

The cube root of 125
$$125^{1/3} = \sqrt[3]{125} = 5;$$

The fifth root of 32
$$32^{1/5} = \sqrt[5]{32} = 2$$

2 Estimate an irrational number (pp. 474–475).

To estimate the square root of a number, identify the two perfect squares between which the number lies. The square root of the number will be between the square roots of the two perfect squares.

15 is between 9 and 16, so $15^{1/2}$ is between 3 and 4.
38 is between 36 and 49, so $38^{1/2}$ is between 6 and 7.

3 Position an irrational number on the number line (pp. 475–476).

Position irrational numbers on the number line (Fig. 11–22) by approximating the value of the irrational number mentally or with a calculator.

$6^{1/2} \approx 2.449489743$ $7^{1/2} \approx 2.645751311$ $8^{1/2} \approx 2.828427125$
$10^{1/2} \approx 3.16227766$

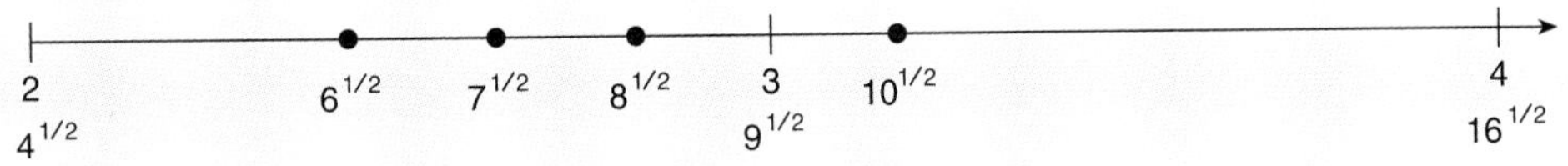

Figure 11–22

| 4 Write powers and roots using rational exponents and radical notation (pp. 477–478). | To write a radical expression as an expression with rational exponents, the radicand is the base of the expression. The exponent of the radicand is the numerator of the rational exponent, and the index of the root is the denominator of the rational exponent. |

Symbolically, $\sqrt[root]{x^{power}} = x^{power/root}$.

Write in rational exponent form.

$$\sqrt[5]{3^2} = 3^{2/5}$$

Write in radical form:

$$5^{1/2} = \sqrt{5}; \qquad 7^{3/5} = \sqrt[5]{7^3} \text{ or } (\sqrt[5]{7})^3$$

Section 11–2

| 1 Find the square root of variables (p. 479). | Variable factors are perfect squares if the exponent is divisible by 2. To find the square root of a perfect-square variable with an even-number exponent, take half of the exponent and keep the same base. |

Give the square root of the following: x^6, x^{10}, x^{24}.

$$\sqrt{x^6} = x^3, \qquad \sqrt{x^{10}} = x^5, \qquad \sqrt{x^{24}} = x^{12}$$

| 2 Simplify square-root radicals using rational exponents and the laws of exponents (pp. 479–481). | All the laws of exponents that were used for integral exponents apply to expressions with rational exponents. The arithmetic of fractions is also applied. |

Simplify.

$$x^{1/3} \cdot x^{2/3} = x^{1/3+2/3} = x^{3/3} = x; \qquad \frac{x^{7/8}}{x^{3/4}} = x^{7/8-3/4} = x^{7/8-6/8} = x^{1/8}$$

| 3 Simplify square-root radical expressions containing perfect-square factors (pp. 481–483). | To simplify square-root radical expressions containing perfect-square factors, factor constants and variables using the largest possible perfect-square factor of the constant and of each variable. The largest perfect square of a variable will be written with the largest possible even-numbered exponent. |

Simplify the following radical expressions: $\sqrt{98}, \sqrt{x^{13}}, \sqrt{72y^9}, 5\sqrt{12}$.

$$\sqrt{98} = \sqrt{49(2)} = 7\sqrt{2}; \qquad \sqrt{x^{13}} = \sqrt{x^{12}(x)} = x^6\sqrt{x}$$
$$\sqrt{72y^9} = \sqrt{36 \cdot 2 \cdot y^8 \cdot y} = 6y^4\sqrt{2y}$$
$$5\sqrt{12} = 5\sqrt{4 \cdot 3} = 5 \cdot 2\sqrt{3} = 10\sqrt{3}$$

Section 11–3

| 1 Add or subtract square-root radicals (pp. 484–485). | To add or subtract square-root radicals, first simplify all radical expressions; then add or subtract the coefficients of like radical terms. Like radical terms are terms that have exactly the same factors in the radicand and have the same index. |

Add or subtract: $3\sqrt{5x} + 7\sqrt{5x}; 2\sqrt{18} - 5\sqrt{8}$.

$$3\sqrt{5x} + 7\sqrt{5x} = 10\sqrt{5x}$$
$$2\sqrt{18} - 5\sqrt{8} = 2\sqrt{9 \cdot 2} - 5\sqrt{4 \cdot 2} = 2 \cdot 3\sqrt{2} - 5 \cdot 2\sqrt{2} =$$
$$6\sqrt{2} - 10\sqrt{2} = -4\sqrt{2}$$

| 2 Multiply square-root radicals (pp. 485–487). | To multiply radicals that have the same index, multiply the coefficients and write as the coefficient of the product, and multiply the radicands and write as the radicand of the product; then simplify the radical. |

Chapter 11 / Roots and Radicals

$$\text{Multiply. } 5\sqrt{7} \cdot 8\sqrt{14} = 40\sqrt{98} = 40\sqrt{49 \cdot 2} = 40 \cdot 7\sqrt{2} = 280\sqrt{2};$$
$$\sqrt{5}(\sqrt{7} - \sqrt{2}) = \sqrt{5} \cdot \sqrt{7} - \sqrt{5} \cdot \sqrt{2} = \sqrt{35} - \sqrt{10}$$

3 Divide square-root radicals (pp. 487–488).

To divide radicals that have the same index, divide the coefficients for the coefficient of the quotient; then divide the radicands for the radicand of the quotient. Simplify any remaining radical expressions that can be simplified; then simplify any resulting coefficients that can be simplified.

$$\text{Divide. } \quad \frac{12\sqrt{75}}{8\sqrt{6}} = \frac{3\sqrt{3 \cdot 25}}{2\sqrt{3 \cdot 2}} = \frac{3\sqrt{25}}{2\sqrt{2}} = \frac{3 \cdot 5}{2\sqrt{2}} = \frac{15}{2\sqrt{2}}$$

4 Rationalize a denominator (pp. 488–490).

To rationalize a denominator of a radical expression, remove perfect-square factors from all radicands. Multiply the denominator by another radical so that the resulting radicand is a perfect square. Then multiply the numerator by the same radical the denominator was multiplied by, simplify the resulting radicals, and reduce the resulting fraction, if possible.

$$\text{Rationalize the denominator. } \sqrt{\frac{3}{5}} = \frac{\sqrt{3}}{\sqrt{5}} \cdot \frac{\sqrt{5}}{\sqrt{5}} = \frac{\sqrt{15}}{5}$$

Section 11–4

1 Write imaginary numbers using the letter i (pp. 491–492).

The letter i is used to represent $\sqrt{-1}$. Thus, the square root of negative numbers can be expressed as imaginary numbers and simplified using the letter i. Be careful to distinguish when the negative is *outside* the radical and when it is *under* the radical.

$$\text{Simplify. } \sqrt{-48} = \sqrt{-1 \cdot 16 \cdot 3} = 4i\sqrt{3}$$

2 Raise imaginary numbers to powers (p. 492).

Imaginary numbers can be raised to powers by examining the remainder when the exponent is divided by 4. The remainder can be used as a simplified exponent. Then, $i^0 = 1$, $i^1 = i$, $i^2 = -1$, and $i^3 = -i$.

$$\text{Simplify. } i^{17} = i^1 = i; \; i^{42} = i^2 = -1$$

3 Write real and imaginary numbers in complex form, $a + bi$ (p. 493).

A complex number is a number that can be written in the form $a + bi$, where a and b are real numbers and i is $\sqrt{-1}$. Either a or b can be zero. If a is zero, the number is an imaginary number; if b is zero, the number is a real number.

Rewrite as complex numbers: $\sqrt{-81}, 38, \sqrt{9}, \sqrt{-19}$.

$$\sqrt{-81} = 9i = 0 + 9i; \qquad 38 = 38 + 0i; \qquad \sqrt{9} = 3 = 3 + 0i$$
$$\sqrt{-19} = i\sqrt{19} = 0 + i\sqrt{19}$$

4 Combine complex numbers (p. 493.)

To combine complex numbers, add the real parts for the real part of the result; then add the imaginary parts for the imaginary part of the result. Be careful with signs when subtracting complex numbers.

$$\text{Simplify. } (5 + 3i) + (7 - 8i) = (5 + 7) + (3 - 8)i = 12 - 5i;$$
$$(3 - 8i) - (4 + 6i) = (3 - 4) + (-8 - 6)i = -1 - 14i$$

Section 11–5

1 Solve equations with squared variables (pp. 494–495).

To solve an equation containing only squared variable terms and number terms, perform all normal steps to isolate the squared variable and to obtain a coefficient of 1 for the squared variable. Then take the square root of both sides. *Note:* When taking the square root of both sides, be sure to use both the positive and the negative square roots of the constant. Thus, two solutions are possible. Simplify radicals for exact solutions; use your calculator for approximate solutions.

Chapter Summary

507

Solve:

$$3x^2 + 5 = 29 \qquad \text{Sort and combine like terms.}$$
$$3x^2 = 24 \qquad \text{Divide.}$$
$$x^2 = 8 \qquad \text{Take square root of both sides.}$$
$$x = \pm\sqrt{8} \qquad \text{Simplify radicals.}$$
$$x = \pm2\sqrt{2} \qquad \text{Exact solution.}$$
$$x = \pm2.828 \qquad \text{Approximate solution.}$$

2 Solve equations with square-root radical terms (pp. 495–497).

To solve a basic equation containing square-root radicals, square both sides to eliminate any radical; then perform all normal steps to isolate the variable and solve the equation.

Solve:

$$\sqrt{2x + 1} = 5 \qquad \text{Square both sides.}$$
$$(\sqrt{2x + 1})^2 = 5^2$$
$$2x + 1 = 25 \qquad \text{Sort terms and combine terms.}$$
$$2x = 24 \qquad \text{Divide.}$$
$$x = 12$$

3 Use the Pythagorean theorem to find the missing side of a right triangle (pp. 497–499).

The square of the hypotenuse of a right triangle equals the sum of the squares of the other two sides. $c^2 = a^2 + b^2$

The hypotenuse (c) of a right triangle is 10 in. If side b is 6 in., find side a.

$$c^2 = a^2 + b^2 \qquad \text{Substitute values.}$$
$$10^2 = a^2 + 6^2 \qquad \text{Raise to powers.}$$
$$100 = a^2 + 36 \qquad \text{Sort.}$$
$$100 - 36 = a^2 \qquad \text{Combine.}$$
$$64 = a^2 \qquad \text{Take square root of both sides.}$$
$$\sqrt{64} = a \qquad \text{Find the principal square root.}$$
$$a = 8 \text{ in.}$$

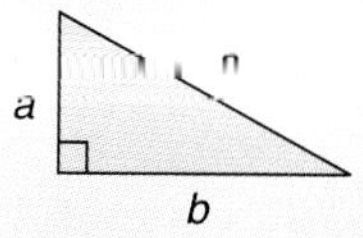

Figure 11–23

CHAPTER TRIAL TEST

Perform the indicated operations. Simplify if possible. Rationalize the denominators if needed.

1. $2\sqrt{7} \cdot 3\sqrt{2}$

2. $\dfrac{\sqrt{8}}{\sqrt{2}}$

3. $4\sqrt{3} + 2\sqrt{3}$

4. $3\sqrt{8} - 4\sqrt{8}$

5. $\dfrac{4\sqrt{2}}{\sqrt{3}}$

6. $\sqrt{3y^3} \cdot \sqrt{15y^2}$

7. $\dfrac{6\sqrt{8}}{2\sqrt{3}}$

8. $\dfrac{3\sqrt{a}}{\sqrt{b}}$

9. $\dfrac{3\sqrt{5}}{2} \cdot \dfrac{7}{\sqrt{3x}}$

Solve the equations containing square-root radicals or squared terms.
Evaluate to the nearest thousandth when appropriate.

10. $x^2 = 144$

11. $\sqrt{5 - x^2} = 2$

12. $8 = y^2 - 1$

13. $7 = \sqrt{Q}$

14. $\sqrt{\dfrac{x^2}{2}} = 6$

15. $\sqrt{\dfrac{6}{y}} = \sqrt{\dfrac{2}{3}}$

16. $\sqrt{4x} = 20$

17. $x^2 + 49 = 0$

Convert the radical expressions to equivalent expressions using rational exponents and simplify.

18. $\sqrt[6]{x}$

19. $\sqrt[3]{27x^{15}}$

20. $\left(\sqrt{5x^4}\right)^6$

21. $\sqrt[5]{x^{10}y^{15}z^{30}}$

Chapter 11 / Roots and Radicals

Perform the operations. Express answers with positive exponents in lowest terms.

22. $a^{4/5} \cdot a^{1/5}$　　**23.** $(125x^{1/2}y^6)^{1/3}$　　**24.** $\dfrac{b^{3/4}}{b^{1/4}}$　　**25.** $\dfrac{12x^{3/5}}{6x^{-2/5}}$　　**26.** $\dfrac{r^{-1/5}s^{1/3}}{r^{3/5}s^{-5/3}}$

Simplify.

27. i^{23}

28. i^{88}

Add or subtract.

29. $(5 + 3i) - (8 - 2i)$

30. $(7 - i) + (4 - 3i)$

Multiply and simplify.

31. $\sqrt{7}\left(\sqrt{5} - 4\right)$

32. $\sqrt{3}\left(\sqrt{6} - \sqrt{12}\right)$

Use Fig. 11–24 and the Pythagorean theorem to find the missing value.

33. $a = 5, b = 12$, find c.
34. $a = 3, b = 4$, find c.
35. $a = 8, c = 10$, find b.
36. $b = 10, c = 18$, find a.

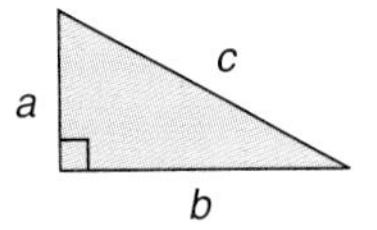

Figure 11–24

37. Three pulleys are designed so their centers form a right triangle when connecting lines are drawn. The distances between the centers forming the sides of the triangle are 8 in. and 15 in., respectively. What is the distance between the centers of the pulleys that form the hypotenuse?

38. Two forces A and B come together in a right angle to produce resulting force C. Force A is 42.7 kg and force B is 38.2 kg. Find the resulting force C.

Formulas and Applications

12

12–1 Formula Evaluation

1. Evaluate formulas.
2. Evaluate formulas in function notation.

12–2 Formula Rearrangement

1. Rearrange formulas to solve for a specified variable.

12–3 Geometric Formulas

1. Find the surface area of prisms and cylinders.
2. Find the volume of a pyramid and of a frustum of a pyramid.
3. Find the surface area and volume of a sphere.
4. Find the surface area and volume of a cone.

Formulas are procedures that have been used so frequently to solve certain types of problems that they have become the accepted means of solving these problems. Most formulas are expressed with one or more letter terms rather than words, and these procedures are written as symbolic equations, such as $A = \pi r^2$ and $P = 2(l + w)$, the formulas for the area of a circle and the perimeter of a rectangle, respectively. Electronic spreadsheets and calculator and computer programs are developed through formulas.

12–1 | Formula Evaluation

Learning Outcomes
1. Evaluate formulas.
2. Evaluate formulas in function notation.

The most common use of formulas is for finding missing values. If we know values for all but one variable of a formula, we can find the unknown value.

1   Evaluate Formulas.

To *evaluate* a formula is to substitute known values for some variables and perform the indicated operations to find the value of the variable in question. In so doing, we may need to use any or all of the steps and procedures for solving equations.

This section illustrates several formula evaluations to help develop a concept of formula evaluation.

To evaluate a formula:

1. Write the formula.
2. Rewrite the formula substituting known values for variables of the formula.
3. Solve the equation from Step 2 for the missing variable.
4. Interpret the solution within the context of the formula.

We can solve for a variable that is not isolated. In the next example the formula for the perimeter of a rectangle is used. The variable w is within parentheses and must be isolated.

EXAMPLE   Solve the formula $P = 2(l + w)$ for w if $P = 12$ ft and $l = 4$ ft.

$$P = 2(l + w)$$ Perimeter of a rectangle = 2 times the sum of the length and the width. Substitute values.

$$12 = 2(4 + w)$$ Apply the distributive property.

$$12 = 8 + 2w$$ Isolate the term with the variable.

$$12 - 8 = 2w$$ Combine like terms.

$$4 = 2w$$ Divide.

$$\frac{4}{2} = \frac{2w}{2}$$

$$2 = w$$ Interpret solution.

The width is 2 ft.

The formula for the area of a circle can be used to find the radius of a circle if we are given the area of the circle.

EXAMPLE   Evaluate the formula $A = \pi r^2$ for r if $A = 144$ in².

$$A = \pi r^2$$ Area of a circle = π times the radius squared. Substitute values.

$$144 = \pi r^2$$ Isolate the variable by dividing both sides by π.

$$\frac{144}{\pi} = \frac{\pi r^2}{\pi}$$

$$\frac{144}{\pi} = r^2$$ Take the square root of both sides.

$$r = \sqrt{\frac{144}{\pi}}$$ For convenience, rewrite with r on the left side. Use the calculator value for π.

$$r = \sqrt{45.83662361}$$ Only the principal square root is appropriate with measures.

$$r = 6.770275003$$ Interpret solution.

The radius is 6.77 in. (rounded).

The interest formula may be used to illustrate a variable that is one of several factors.

EXAMPLE Evaluate the formula $I = PRT$ for Principal (P) if Interest $(I) = \$94.50$, Rate $(R) = 21\%$, and Time $(T) = \frac{1}{2}$ year.

For convenience in using a calculator, convert $\frac{1}{2}$ year to 0.5 year. In this formula, the rate should be expressed as a decimal equivalent, $21\% = 0.21$.

$$I = PRT$$ Substitute values and solve for P.
$$94.50 = P(0.21)(0.5)$$ Multiply 0.21 and 0.5.
$$94.50 = 0.105\,P$$ Divide.
$$\frac{94.50}{0.105} = \frac{0.105\,P}{0.105}$$
$$900 = P$$ Interpret solution.

The principal is $900.

EXAMPLE Evaluate the formula $E = \dfrac{I - P}{I}$ to the nearest thousandth if $I = 24{,}000$ calories (cal) and $P = 8{,}600$ cal.

$$L = \frac{I - P}{I}$$ Engine efficiency = difference between heat input and output divided by heat input.

$$E = \frac{24{,}000 - 8{,}600}{24{,}000}$$ Substitute given values. Perform calculations in numerator grouping.

$$E = \frac{15{,}400}{24{,}000}$$ Divide.

$$E = 0.642$$ Dimension analysis: $\dfrac{\text{cal} - \text{cal}}{\text{cal}} = \dfrac{\cancel{\text{cal}}}{\cancel{\text{cal}}} = \text{no units}$

The engine efficiency is 0.642 (64.2% efficient).

EXAMPLE Evaluate the formula $R_T = \dfrac{R_1 R_2}{R_1 + R_2}$ if $R_1 = 10$ ohms and $R_2 = 6$ ohms.

$$R_T = \frac{R_1 R_2}{R_1 + R_2}$$ Total resistance = product of first resistance and second resistance divided by sum of first and second resistances.

$$R_T = \frac{10(6)}{10 + 6}$$ Substitute given values. Perform calculations in numerator and denominator groupings.

$$R_T = \frac{60}{16}$$ Divide.

$$R_T = 3.75$$ Dimension analysis: $\dfrac{\text{ohms}(\cancel{\text{ohms}})}{\cancel{\text{ohms}}} = \text{ohms}$

The total resistance in the circuit is 3.75 ohms.

Chapter 12 / Formulas and Applications

EXAMPLE Evaluate the formula $H = \dfrac{D^2N}{2.5}$ if $D = 4$ in. and $N = 8$.

$H = \dfrac{D^2N}{2.5}$ Horsepower = diameter in inches of the cylinder squared times the number of cylinders divided by 2.5.

$H = \dfrac{4^2(8)}{2.5}$ Perform power operation.

$H = \dfrac{16(8)}{2.5}$ Perform multiplication.

$H = \dfrac{128}{2.5}$ Divide.

$H = 51.2$

The engine is rated at 51.2 hp.

Sometimes we are asked to solve for a letter term that is not isolated. Let's look again at the horsepower formula and solve for a letter term other than horsepower.

EXAMPLE Solve the formula $H = \dfrac{D^2N}{2.5}$ for D (in inches) if $H = 45$ and $N = 6$.

$H = \dfrac{D^2N}{2.5}$ Substitute values.

$45 = \dfrac{D^2(6)}{2.5}$ Multiply to eliminate denominator.

$(2.5)45 = \dfrac{D^2(6)}{2.5}(2.5)$ Simplify.

$112.5 = 6D^2$ Divide by coefficient of D^2.

$\dfrac{112.5}{6} = D^2$

$18.75 = D^2$ Take the principal square root of both sides.

$D = 4.330$ in.   Rounded.

The diameter of the piston is 4.330 in.

TIP!

Interpret the Solution Within the Context of the Formula.

In applied problems where a negative square root is unrealistic, we use *only* the principal square root. In the previous example, for instance, we cannot have a negative diameter of a piston.

The next formula includes both powers and roots. Because the process for solving these equations involves multiplying and dividing by variables, the solution should be checked for extraneous roots.

12–1 Formula Evaluation

EXAMPLE In the formula $Z = \sqrt{R^2 + X^2}$, solve for R (in ohms) if $Z = 12.4$ ohms and $X = 12$ ohms.

$$Z = \sqrt{R^2 + X^2}$$ Impedance = square root of the sum of the resistance (R) squared plus the reactance (X) squared.

$$12.4 = \sqrt{R^2 + (12)^2}$$ Substitute values.

$$(12.4)^2 = (\sqrt{R^2 + 144})^2$$ Square both sides to eliminate radical.

$$153.76 = R^2 + 144$$ Isolate R^2 (sort terms).

$$153.76 - 144 = R^2$$ Subtract.

$$9.76 = R^2$$ Take the square root of both sides.

$$R = 3.12409987$$

The resistance is 3.1 ohms (rounded to tenths).

The next example has many steps and it is important to apply the order of operations. The formula is used to find the length of a belt connecting two pulleys.

EXAMPLE Evaluate $L = 2C + 1.57(D + d) + \dfrac{D + d}{4C}$ if $C = 24$ in., $D = 16$ in., and $d = 4$ in. See Fig. 12–1.

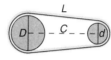

L = length of belt joining two pulleys
C = distance between centers of pulleys
D = diameter of large pulley
d = diameter of small pulley

Figure 12–1

$$L = 2C + 1.57(D + d) + \frac{D + d}{4C}$$ Substitute the given values.

$$L = 2(24) + 1.57(16 + 4) + \frac{16 + 4}{4(24)}$$ Work groupings in parentheses, numerator, and denominator.

$$L = 2(24) + 1.57(20) + \frac{20}{96}$$ Work multiplications and division.

$$L = 48 + 31.4 + 0.20833333$$ Add.

$$L = 79.60833333$$

The pulley belt is 79.61 in. long.

2 Evaluate Formulas in Function Notation.

A common business practice is to establish an equation that models a particular trend. This model can then be used to make projections and decisions. These equations are often written in function notation to facilitate evaluating the equation for various amounts. An example of a function that is commonly used is the *cost function*. A cost function in the manufacturing industry is a formula or equation written in function notation that is used to find the cost of producing various numbers of items. The function has two parts—a *fixed cost* and a *variable cost*. The fixed cost is determined by start-up costs, equipment and facility costs, and other related costs that are incurred no matter how many items are produced. The variable costs are costs for material, labor, packaging, shipping, and other costs that depend on the number of items produced. The basic form of a linear cost function in one variable is

Total cost = Variable cost (Number of items produced) + Fixed cost

or $$C(x) = mx + b$$

where m is the cost of producing one more item (marginal cost), x is the number of items produced, and b is the fixed cost.

EXAMPLE Find the cost of producing the following number of books if the cost function is $C(x) = 7.58x + 5,000$: (a) 100 (b) 1,000 (c) 10,000.

(a) $\quad C(x) = 7.58x + 5,000$ Substitute 100 for x.
$\quad\quad C(100) = 7.58(100) + 5,000$ Multiply.
$\quad\quad\quad\quad\quad = 758 + 5,000$ Add.
$\quad\quad\quad\quad\quad = \mathbf{\$5,758}$

(b) $\quad C(x) = 7.58x + 5,000$ Substitute 1,000 for x.
$\quad\quad C(1,000) = 7.58(1,000) + 5,000$ Multiply.
$\quad\quad\quad\quad\quad = 7,580 + 5,000$ Add.
$\quad\quad\quad\quad\quad = \mathbf{\$12,580}$

(c) $\quad C(x) = 7.58x + 5,000$ Substitute 10,000 for x.
$\quad\quad C(10,000) = 7.58(10,000) + 5,000$ Multiply.
$\quad\quad\quad\quad\quad = 75,800 + 5,000$ Add.
$\quad\quad\quad\quad\quad = \mathbf{\$80,800}$

Another type of function that relates to the cost function is a function that determines the *average* cost for producing a product. This function can be expressed as a *function within a function*.

EXAMPLE Use the function $A(x) = \dfrac{C(x)}{x}$, where $A(x)$ is the function for finding the average cost and $C(x)$ is the function for finding the total cost of producing x items, to find the average cost for producing (a) 100, (b) 1,000, and (c) 10,000 books from the preceding example.

(a) $\quad A(x) = \dfrac{C(x)}{x}$ Substitute 100 for x and C (100) for C(x).

$\quad\quad A(100) = \dfrac{C(100)}{100}$ Since $C(x)$ has already been evaluated at 100, substitute the result $5,758.

$\quad\quad\quad\quad = \dfrac{5,758}{100}$ Divide.

$\quad\quad\quad\quad = \mathbf{\$57.58}$

(b) $\quad A(x) = \dfrac{C(x)}{x}$ Substitute 1,000 for x and C (1,000) for C(x).

$\quad\quad A(1,000) = \dfrac{C(1,000)}{1,000}$ Since C(x) has already been evaluated at 1,000, substitute the result $12,580.

$\quad\quad\quad\quad = \dfrac{12,580}{1,000}$ Divide.

$\quad\quad\quad\quad = \mathbf{\$12.58}$

(c) $\quad A(x) = \dfrac{C(x)}{x}$ Substitute 10,000 for x and C(10,000) for C(x).

$\quad\quad A(10,000) = \dfrac{C(10,000)}{10,000}$ Since C(x) has already been evaluated at 10,000, substitute the result $80,800.

$\quad\quad\quad\quad = \dfrac{80,800}{10,000}$ Divide.

$\quad\quad\quad\quad = \mathbf{\$8.08}$

Functions for cost and average cost are used to determine an appropriate selling price of a product. Then, a *revenue function* can be used to determine the amount of revenue expected from the sale of a specified number of products.

EXAMPLE   Use the revenue function $R(x) = \$15x$, where x represents the number of books sold at $15, to find the amount of revenue generated by the sale of (a) 100, (b) 1,000, and (c) 10,000 books.

(a) $R(x) = \$15x$ Substitute 100 for x.
       $R(100) = 15(100)$ Multiply.
       $= \$1,500$

(b) $R(x) = \$15x$ Substitute 1,000 for x.
       $R(1,000) = 15(1,000)$ Multiply.
       $= \$15,000$

(c) $R(x) = \$15x$ Substitute 10,000 for x.
       $R(10,000) = 15(10,000)$ Multiply.
       $= \$150,000$

A *profit function* can be derived from the cost and revenue functions: $P(x) = R(x) - C(x)$, where $P(x)$ is the total profit for x products and $R(x)$ and $C(x)$ are the revenue and cost functions, respectively. When the total profit is a positive amount, we call this amount the *profit*. When the total profit is a negative amount, we call this amount the *loss*. When the total profit is zero, we call this value of x the *break-even point*.

EXAMPLE   Examine the profit for selling (a) 100, (b) 1,000, and (c) 10,000 books, using the cost and revenue functions from the preceding examples. Discuss your findings.

Compare the results of the revenue, cost, and profit functions for the three values. Calculate the average profit per book by using the function $A(x) = P(x)/x$.

Organize the data in a table.

x	$R(x)$	$C(x)$	$P(x) = R(x) - C(x)$	$A(x) = P(x)/x$
100	$1,500	$5,758	-$4,258	-$42.58
1,000	$15,000	$12,580	$2,420	$2.42
10,000	$150,000	$80,800	$69,200	$6.92

At a sale price of $15 per book, a significant loss ($42.48 per book) will be incurred if only 100 books are produced and sold. At 1,000 books, a modest profit of $2.42 per book will be realized. If 10,000 books are produced and sold, there will be a profit of $6.92 per book.

SELF-STUDY EXERCISES 12–1

1 Evaluate the formulas.

1. $D = RT$ if $R = 40$ mph and $T = 7$ hr
2. $S = C + M$ if $C = \$40$ and $M = \$80$
3. $A = 4\pi r^2$ if $\pi = 3.14$ and $r = 15$ in.
4. $c = b\sqrt{3}$ if $b = 12$ m
5. $A = \sqrt{s(s - a)(s - b)(s - c)}$ if $s = 69$ cm, $a = 50$ cm, $b = 43$ cm, and $c = 45$ cm

Evaluate the interest formula $I = PRT$ using the following values.

6. Find the interest if $P = \$800$, $R = 15.5\%$, and $T = 2\frac{1}{2}$ years.

7. Find the rate if $I = \$427.50$, $P = \$1,500$, and $T = 2$ years.

8. Find the time if $I = \$236.25$, $P = \$750$, and $R = 10.5\%$.

9. Find the principal if $I = \$838.50$, $R = 21\frac{1}{2}\%$, and $T = 1\frac{1}{2}$ years.

Evaluate using the percentage formula $P = \dfrac{RB}{100}$.

10. Find the percentage if $R = 15\%$ and $B = 600$ lb.

11. Find the rate if $P = 24$ kg and $B = 300$ kg.

12. Find the base if $P = \$250$ and $R = 7.4\%$. Round to hundredths.

13. Evaluate the rate formula $R = \dfrac{100P}{B}$ for P if $R = 16\%$ and $B = 85$.

14. Evaluate the base formula $B = \dfrac{100P}{R}$ for R if $B = \$2,200$ and $P = \$374$.

15. Find the length of a rectangular work area if the perimeter is 180 in. and the width is 24 in. Use the formula $P = 2(l + w)$.

16. Find the radius of a circle whose area is 254.5 cm² using the formula $r = \sqrt{\frac{A}{\pi}}$. Round to the nearest tenth.

17. Find the cost (C) if the markup (M) on an item is $5.25 and the selling price (S) is $15.75. Use the formula $M = S - C$.

18. Using the formula for the side of a square, $s = \sqrt{A}$, find the length of a side of a square field whose area is $\frac{1}{16}$ mi².

19. Evaluate the formula for the area of a circle, $A = \pi r^2$, if $r = 7$ in. Round to the nearest tenth.

20. Evaluate the formula for the area of a square, $A = s^2$, if $s = 2.5$ km.

21. Using the markup formula $M = S - C$, find the selling price (S) if the markup (M) is $12.75 and the cost (C) is $36.

22. Use the formula $R_T = \dfrac{R_1 R_2}{R_1 + R_2}$ to find the total resistance (R_T) if one resistance (R_1) is 12 ohms (Ω) and the second resistance (R_2) is 8 ohms (Ω).

23. What is the percent efficiency (E) of an engine if the input (I) is 25,000 calories and the output (P) is 9,600 calories? Use the formula $E = \dfrac{I - P}{I}$.

24. Distance is rate times time, or $D = RT$. Find the rate if the distance traveled is 140 mi and the time traveled is 4 hr.

25. The formula for voltage (Ohm's law) is $E = IR$. Find the amperes of current (I) if the voltage (E) is 120 V and the resistance (R) is 80 ohms (Ω).

26. According to Boyle's law, if temperature is constant, the volume of a gas is inversely proportional to the pressure on it. Find the final volume (V_2) of a gas using the formula $\dfrac{V_1}{V_2} = \dfrac{P_2}{P_1}$ if the original volume (V_1) is 15 ft³, the original pressure (P_1) is 60 lb per square inch (psi), and the final pressure (P_2) is 150 psi.

27. The formula for power (P) in watts (W) is $P = I^2R$. Find the current (I) in amperes if a device draws 63 W and the resistance (R) is 7 ohms (Ω).

28. Use the formula $H = \dfrac{D^2 N}{2.5}$ to find the number of cylinders (N) required in an engine of 3.2 hp (H) if the cylinder diameter (D) is 2 in.

29. The formula for the speed (s) of a driven pulley in revolutions per minute (rpm) is $s = \dfrac{DS}{d}$. Find the speed of a driven pulley with a diameter (d) of 5 in. if the diameter (D) of the driving pulley is 10 in. and its speed (S) is 800 rpm.

30. If the distance (C) between the centers of the pulleys in Exercise 29 is 24 in., find the length (L) to the nearest hundredth of the belt connecting them using the formula
$$L = 2C + 1.57(D + d) + \dfrac{D + d}{4C}$$

31. Find the reactance (X) in ohms using the formula $Z = \sqrt{R^2 + X^2}$ if the impedance (Z) is 10 ohms and the resistance (R) is 9 ohms. Round to tenths.

2

32. Use the cost function $C(x) = 12.98x + 20,000$ to find the cost of producing 1,000 blankets if x is the number of blankets produced.

33. Use the cost function $C(x) = 12.98x + 20,000$ to find the cost of producing 10,000 blankets.

12–1 Formula Evaluation

34. Find the average cost $A(x)$ of producing the 1,000 blankets in Exercise 32 if $A(x) = \dfrac{C(x)}{x}$.

35. Find the average cost of producing 10,000 blankets.

36. Use the revenue function $R(x) = 40x$, where x represents the number of blankets sold at \$40, to find the revenue generated by 1,000 blankets; 10,000 blankets.

37. Use the profit function $P(x) = R(x) - C(x)$ and the results of previous exercises to find the profit for producing (a) 1,000 and (b) 10,000 blankets.

38. Find the average profit per blanket for (a) 1,000 blankets and (b) 10,000 blankets if the function $A(x) = \dfrac{P(x)}{x}$ is used to find average profit.

12–2 │ *Formula Rearrangement*

Learning Outcome   **1** Rearrange formulas to solve for a specified variable.

We saw in Section 12–1 that we can solve a formula for any missing variable wherever the missing variable is located in the formula. In developing formulas for spreadsheets or in programming calculators or computers, the missing values must be indicated by a variable that is isolated on the left side of the equation. This may require that the formula be rearranged.

1 Rearrange Formulas to Solve for a Specified Variable.

Formula rearrangement generally refers to isolating a letter term other than the one already isolated in the formula. Solving formulas in this manner shortens our work when doing repeated formula evaluations. After we solve the formula for the desired variable, we rewrite the formula with the variable on the left side for convenience.

The following formulas require applying the addition axiom (transposing or sorting) and careful handling of signs.

> **EXAMPLE**   Solve the markup formula $M = S - C$, for S (selling price).
>
> $$M = S - C \qquad \text{Markup} = \text{Selling Price} - \text{Cost. Transpose to isolate } S.$$
> $$M + C = S \qquad \text{The coefficient of } S \text{ is positive, so the formula is solved.}$$
> $$S = M + C \qquad \text{Rewrite } S \text{ on the left for convenience.}$$

TIP!

Where Is the Variable or Unknown in a Formula?

It may help to think of the one letter we are solving for as the unknown or variable, and to think of the other letters *as if* they were coefficients or constants in an ordinary equation.

> **EXAMPLE**   Solve the formula $M = S - C$ for C (cost).
>
> $$M = S - C \qquad \text{Markup} = \text{Selling Price} - \text{Cost. Transpose to isolate } C.$$
> $$M - S = -C \qquad \text{The coefficient of } C \text{ is negative, so divide both sides by } -1.$$
> $$\frac{M - S}{-1} = \frac{-C}{-1} \qquad \text{Note effect on signs after division by } -1.$$

$$-M + S = C$$ Write the positive term first.

$$S - M = C$$ Rewrite with C on the left for convenience.

$$C = S - M$$

To rearrange a formula:

1. Determine which variable of the formula will be isolated (solved for).
2. Highlight or mentally locate all instances of the variable to be isolated.
3. Treat all other variables of the formula as you would treat numbers in an equation, and perform normal steps for solving an equation.
4. If the isolated variable is on the right side of the equation, interchange the sides so that it appears on the left side.

When a formula contains a term of several factors and we need to solve for one of those factors, we treat the factor being solved for as the variable and the other factors as its coefficient.

EXAMPLE Solve for R in the formula $I = PRT$.

$$I = P \, R \, T$$ Interest = Principal × Rate × Time. Since we are solving for R, PT is its coefficient. Divide both sides by the coefficient of the variable.

$$\frac{I}{PT} = \frac{P \, R \, T}{PT}$$

$$\frac{I}{PT} = R$$ Rewrite with R on the left.

$$R = \frac{I}{PT}$$

Sometimes formulas contain addition (or subtraction) and multiplication in which we must use the distributive property to solve for a particular letter. In the formula $A = P(1 + ni)$, we must use the distributive property to solve for either n or i because each appears inside the grouping.

EXAMPLE Solve for n in the formula for compound amount, $A = P(1 + ni)$.

$$A = P(1 + n \, i)$$ Identify the variable to be isolated. Use the distributive property to remove n from parentheses.

$$A = P + P n i$$ Transpose P to isolate the term with n.

$$A - P = P n i$$ Divide both sides by the coefficient of n.

$$\frac{A - P}{Pi} = \frac{P \, n \, i}{Pi}$$

$$\frac{A - P}{Pi} = n$$ Rewrite with n on the left.

$$i = \frac{A - P}{Pi}$$

When the formula contains division, we cleared the denominator before taking further steps.

EXAMPLE The formula $V = \dfrac{P}{I}$ represents the relationship among the voltage drop (V), electrical power (P), and the current (I). Rearrange the formula to solve for P.

$$V = \dfrac{P}{I}$$ Identify the variable to be isolated. Multiply both sides by the denominator I to clear it.

$$(I)V = \dfrac{P}{\cancel{I}}(\cancel{I})$$

$$IV = P$$ Rewrite with P on the left.

$$P = IV$$

TIP!

Subscripted Variables

Many formulas express relationships between two similar measurements by noting the different measurements with *subscripts*.

The average temperature for a 3-hour period is found by adding the temperatures for each of the periods and dividing by 3.

$$\text{Average temp.} = \frac{\text{Temp. 1st hour} + \text{Temp. 2nd hour} + \text{Temp. 3rd hour}}{3}$$

Using an abbreviated form of the formula, we have

$$t_{av} = \frac{t_1 + t_2 + t_3}{3}$$

Since we are referring to temperatures throughout the formula, we use subscripts to distinguish the various temperatures. To read this formula with subscripts, we say "t sub av equals the sum of t sub 1 plus t sub 2 plus t sub 3, all divided by 3."

When we wish to rearrange formulas written in the form of a proportion, we use the property of proportions that allows cross multiplication. Once we have cross multiplied, we use methods previously discussed to solve for the appropriate letter.

EXAMPLE The formula $\dfrac{T_1}{T_2} = \dfrac{V_1}{V_2}$ represents the proportional relationship between the temperature and volume of a gas. Solve the formula for T_1.

$$\dfrac{T_1}{T_2} = \dfrac{V_1}{V_2}$$ Identify the variable to be isolated. Cross multiply.

$$T_1 V_2 = T_2 V_1$$ Divide both sides by V_2 which is the coefficient of T_1.

$$\dfrac{T_1 V_2}{V_2} = \dfrac{T_2 V_1}{V_2}$$ Simplify.

$$T_1 = \dfrac{T_2 V_1}{V_2}$$

To solve a formula for a letter that is squared, we first solve for the squared letter and then take the square root of both sides of the formula.

EXAMPLE Solve for a in the formula $c^2 = a^2 + b^2$ (Pythagorean theorem).

$$c^2 = a^2 + b^2$$ Identify the variable to be isolated. Transpose b^2 to isolate the letter term being solved for.

$$c^2 - b^2 = a^2$$ Take the square root of both sides. Simplify.

$$\sqrt{c^2 - b^2} = \sqrt{a^2}$$ Rewrite with a on the left.

$$\sqrt{c^2 - b^2} = a$$

$$a = \sqrt{c^2 - b^2}$$ The context of the formula determines if both square roots are appropriate.

To rearrange a formula with a single term on each side, when one side is a square root, we *begin* by squaring both sides of the formula. Then, we solve for the desired letter by using the same techniques used earlier to rearrange formulas.

EXAMPLE Solve for A in the formula for the side of a square, $s = \sqrt{A}$.

$$s = \sqrt{A}$$ Identify the variable to be isolated. Square both sides to eliminate the $\sqrt{\ }$.

$$s^2 = \left(\sqrt{A}\right)^2$$ Notice that $\left(\sqrt{A}\right)^2 = A$, for $A > 0$.

$$s^2 = A$$ Rewrite with A on the left.

$$A = s^2$$

SELF-STUDY EXERCISES 12–2

1 Rearrange the formulas.

1. Solve $S_n = S + R$ for R, where S_n = new salary, S = current salary, and R = raise.
4. Solve $y = mx + b$ for b.

2. Solve $I_n = I - S$ for S, where I_n = new inventory, I = current inventory, and S = sales.
5. Solve $y = mx + b$ for m.

3. Solve $S = 2\pi rh$ for r.

6. Solve $\dfrac{T_1}{T_2} = \dfrac{V_1}{V_2}$ for T_2.

7. Solve $V = \pi r^2 h$ for r.

8. Solve $c = \sqrt{a^2 + b^2}$ for a.

9. Solve $\dfrac{R}{100} = \dfrac{P}{B}$ for R.

10. Solve $P = 2(b + s)$ for b.
13. Solve $R = AC - BC$ for C.
16. Solve $S = P - D$ for D.

11. Solve $C = 2\pi r$ for r.
14. Solve $A = s^2$ for s.
17. Solve $A = \pi r^2$ for r.

12. Solve $A = lw$ for l.
15. Solve $D = RT$ for R.
18. Solve $a^2 + b^2 = c^2$ for c.

19. The formula for finding the amount of a repayment on a loan is $A = I + P$, where A is the amount of the repayment, I is the interest, and P is the principal. Solve the formula for interest.

20. The formula for finding interest is $I = PRT$, where I represents interest, P represents principal, R represents rate, and T represents time. Rearrange the formula to find the time.

Learning Outcomes

1 Find the surface area of prisms and cylinders.

2 Find the volume of a pyramid and of a frustum of a pyramid.

3 Find the surface area and volume of a sphere.

4 Find the surface area and volume of a cone.

Geometry involves the study and measurement of shapes. We have already used formulas to find perimeters, areas, and volume of several geometric figures. We will now examine additional formulas for three-dimensional geometric figures.

1 **Find the Surface Area of Prisms and Cylinders.**

Some jobs require us to find the surface area of a three-dimensional figure. The surface area of a three-dimensional figure can refer to just the area of the *sides* of the figure. Or surface area can refer to the overall area, including the bases along with the sides.

The *lateral surface area* (LSA) of a three-dimensional figure is the area of its sides only.

The *total surface area* (TSA) of a three-dimensional figure is the area of the sides plus the area of its base or bases.

To find the lateral surface area, we find the sum of the area of each of the sides using the formula for the area of the appropriate plane surface. We can also find the lateral surface area (LSA) of a right prism or right circular cylinder (Fig. 12–2) by multiplying the perimeter of the base times the height of the figure.

Lateral surface area of a right prism or a right circular cylinder:

$$\text{LSA} = ph$$

where p is the perimeter of the base and h is the height of the three-dimensional figure.

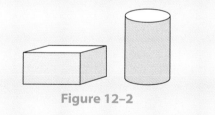

Figure 12–2

To get the total surface area (TSA) (Fig. 12–3), add the areas of the two bases to the lateral surface area.

Total surface area of a right prism or a right circular cylinder:

$$\text{TSA} = ph + 2B$$

where p is the perimeter of the base, h is the height of the three-dimensional figure, and B is the area of the base.

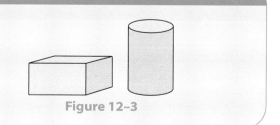

Figure 12–3

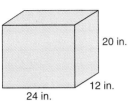

EXAMPLE Find the lateral surface area of a rectangular shipping carton measuring 24 in. in length, 12 in. in width, and 20 in. in height (Fig. 12–4).

20 in.

12 in.

24 in.

Figure 12–4

LSA = ph	Substitute the formula for the perimeter of a rectangle, $p = 2l + 2w$.
LSA = $(2l + 2w)h$	Substitute numerical values.
LSA = $[2(24) + 2(12)]20$	Perform calculations inside grouping.
LSA = $[72]20$	Multiply.
LSA = $1{,}440$ in.^{2}	Area requires square units.

The lateral surface area of the carton is 1,440 in.^{2}

EXAMPLE How many square centimeters of sheet metal are required to manufacture a can that has a radius of 4.5 cm and height of 9 cm? Assume no waste or overlap.

4.5 cm

9 cm

Figure 12–5

TSA = $ph + 2B$	Total surface area is needed. Substitute formulas $p = 2\pi r$, $B = \pi r^2$.
TSA = $2\pi rh + 2\pi r^2$	Substitute values.
TSA = $2(\pi)(4.5)(9) + 2(\pi)(4.5)^2$	Square 4.5 and perform multiplications.
TSA = $254.4690049 + 127.2345025$	Add.
TSA = 381.70 cm^2	Rounded. Area requires square units.

The can requires 381.70 cm^2 of sheet metal.

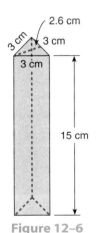

EXAMPLE Find the total surface area of the triangular prism shown in Fig. 12–6.

2.6 cm

3 cm

3 cm

3 cm

15 cm

Figure 12–6

TSA = $ph + 2B$	Perimeter of triangular base is $a + b + c$. Area of triangular base B is $\frac{1}{2}bh_1$, where h_1 is the height of the triangular base.
TSA = $(a + b + c)h + 2\left(\frac{1}{2}bh_1\right)$	Substitute values.
TSA = $(3 + 3 + 3)(15) + 2\left(\frac{1}{2}\right)(3)(2.6)$	Add inside grouping.
TSA = $9(15) + (2)\left(\frac{1}{2}\right)(3)(2.6)$	Perform multiplications.
TSA = $135 + 7.8$	Add.
TSA = 142.8 cm^2	Area requires square units.

The total surface area of the triangular prism is 142.8 cm^2.

2 Find the Volume of a Pyramid and of a Frustum of a Pyramid.

A *pyramid* is a three-dimensional geometric shape that has a polygon for a base and lateral faces that are triangles with a common vertex (apex). The *height or altitude* of the pyramid is the perpendicular distance from the vertex to the base. If the base of the pyramid is a polygon that has all sides equal, such as a square, the height meets the base at the center of the base and is a right or regular pyramid. Figure 12–7 shows right pyramids with 3-, 4-, and 5-sided bases.

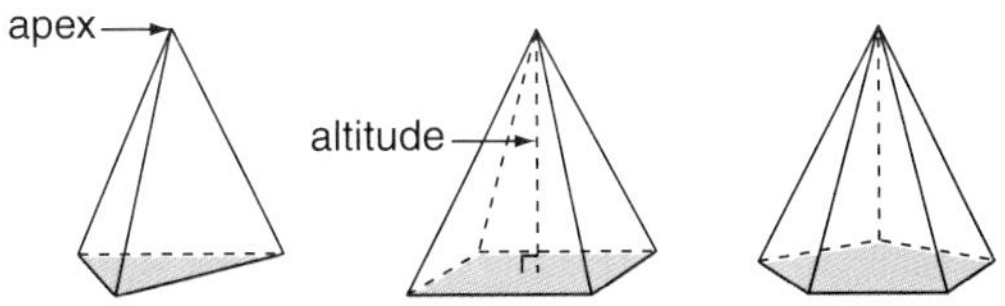

apex

altitude

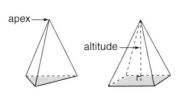

Figure 12–7

12–3 Geometric Formulas

Volume of any pyramid:

$$V = \frac{1}{3} Bh$$

where V is the volume, B is the area of the base, and h is the height of the pyramid.

EXAMPLE Find the volume of a pyramid that has a square base that is 15 cm on a side and a height of 28 cm (Fig. 12–8).

$$V = \frac{1}{3} Bh$$

Find the area of the base and substitute values for B and h.
$B = s^2 = 15^2 = 225$ cm^2

$$V = \frac{1}{3}(225)(28)$$ Multiply.

$$V = \textbf{2,100 cm}^3$$ Volume is cubic units.

Figure 12–8

A *frustum of a pyramid* is a part of a pyramid between the base and a plane passing through the pyramid that is parallel to the base.

Volume of the frustum of a pyramid:

$$V = \frac{1}{3} h \left(B_1 + B_2 + \sqrt{B_1 B_2} \right)$$

where V is the volume, B_1 is the area of the base, B_2 is the area of the top, and h is the height of the frustum.

EXAMPLE Find the volume of the frustum of a pyramid with a square base that is 4 in. on each side, a square top that is 2 in. on each side, and a height of 3 in. (Fig. 12–9).

$$V = \frac{1}{3} h \left(B_1 + B_2 + \sqrt{B_1 B_2} \right)$$

Substitute the values for B_1, B_2, and h.
$B_1 = 4^2 = 16$ in^2; $B_2 = 2^2 = 4$ in^2

$$V = \frac{1}{3}(3)\left(16 + 4 + \sqrt{16 \cdot 4}\right)$$ Simplify the radicand.

$$= \frac{1}{3}(3)\left(16 + 4 + \sqrt{64}\right)$$ Take the square root.

$$= \frac{1}{3}(3)(16 + 4 + 8)$$ Simplify the grouping.

$$= \frac{1}{3}(3)(28)$$ Multiply.

$$V = \textbf{28 in}^3$$ Volume is cubic units.

Figure 12–9

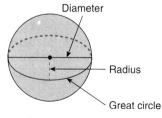

Figure 12–10 Sphere.

3 **Find the Surface Area and Volume of a Sphere.**

Soccer balls, golf balls, tennis balls, baseballs, and ball bearings are *spheres*. Spheres are also used as tanks to store gas and water because spheres hold the greatest volume for a specified amount of surface area. A *sphere* is a three-dimensional figure formed by a curved surface with points that are all equidistant from a point inside called the center (Fig. 12–10). A *great circle* divides the sphere in half and is formed by a plane passing through the center of the sphere.

Chapter 12 / Formulas and Applications

A sphere does not have bases like prisms and cylinders. The surface area of a sphere includes *all* the surface so there is only one formula. Because of the relationship of the sphere to the circle, the formula includes elements of the formula for the area of a circle. The total surface area of the sphere is 4 times the area of a circle with the same radius.

Total surface area of a sphere:

$$TSA = 4\pi r^2$$

where r = radius.

The formula for the volume of a sphere also contains elements found in formulas for a circle.

Volume of a sphere:

$$V = \frac{4\pi r^3}{3}$$

where r = radius.

Note that the radius is *cubed*, or raised to the power of 3, indicating volume.

EXAMPLE Find the surface area and volume of a sphere that has a diameter of 90 cm.

$TSA = 4\pi r^2$ Substitute values. $\frac{1}{2}d = r$, so $r = 45$ cm.

$TSA = 4(\pi)(45)^2$ Square 45 and multiply.

TSA = 25,447 cm²   Rounded.

$V = \frac{4\pi r^3}{3}$ Substitute values.

$V = \frac{4(\pi)(45)^3}{3}$ Cube 45, multiply, and divide.

V = 381,704 cm³   Rounded.

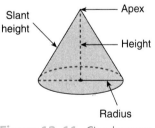

Slant height

Apex

Height

Radius

Figure 12–11 Circular cone.

4 **Find the Surface Area and Volume of a Cone.**

A right *cone* is a three-dimensional figure whose base is a circle and whose side surface tapers to a point, called the *vertex* or *apex*, and whose height is a perpendicular line between the base and apex (Fig. 12–11). One example of a *cone* is the funnel or the circular rain cap placed on top of stove vent pipes extending through the roof of some homes.

The *slant height* of a cone is the distance along the side from the base to the apex. Figure 12–11 shows the perpendicular height and the *slant height* of a cone.

The lateral surface area of a cone equals the circumference of the base times $\frac{1}{2}$ the slant height, or LSA = $2\pi r \left(\frac{1}{2}s\right)$, which simplifies as πrs.

Lateral surface area of a cone:

$$LSA = \pi r s$$

where r is the radius and s is the slant height.

The total surface area, then, is the lateral surface area plus the area of the base.

Total surface area of a cone:

$$TSA = \pi rs + \pi r^2$$

where r is the radius of the circular base, s is the slant height, and πr^2 is the area of the base.

The volume of a cone equals one-third the area of the base times the *height of the cone* (*not* the slant height).

Volume of a cone:

$$V = \frac{\pi r^2 h}{3}$$

where πr^2 is the area of the circular base and h is the height of the cone.

EXAMPLE Find the lateral surface area, total surface area, and volume of a cone that has a diameter of 8 cm, height of 6 cm, and slant height of 7 cm. Round to hundredths.

$LSA = \pi rs$	Substitute values: $r = \frac{1}{2}d$, or 4.
$LSA = (\pi)(4)(7)$	Multiply.
$\mathbf{LSA = 87.96\ cm^2}$	Rounded from 87.9645943.
$TSA = \pi rs + \pi r^2$	Substitute values. Use full calculator value for πrs.
$TSA = 87.9645943 + (\pi)(4)^2$	Perform operations using the proper order of operations.
$\mathbf{TSA = 138.23\ cm^2}$	Rounded.
$V = \frac{\pi r^2 h}{3}$	Substitute values.
$V = \frac{(\pi)(4)^2(6)}{3}$	Perform operations using the proper order of operations.
$\mathbf{V = 100.53\ cm^3}$	Rounded.

Learning Strategy *Be Realistic About Memorizing Formulas.*

Do you normally remember details or general information? Most of us remember the context of a particular conversation with a friend, but not the exact words that were spoken. The formulas you use often are generally easy to memorize, but formulas you use occasionally are easily forgotten.

Instead of memorizing formulas, focus on

- The purpose for which the formula is used
- The interpretation of each letter of that formula
- Terminology that is useful in locating a specific formula when needed
- Memorizing only formulas that are required for a specific objective (a test or exam, formulas used often on the job, and so on)
- Relating formulas to other formulas you have studied (area of parallelogram to area of rectangle, perimeter of square to perimeter of rectangle, and so on)

Formulas that are memorized but not used regularly are generally only in your short-term memory.

Find the weight of the cast-iron object shown in Fig. 12–12 if cast iron weighs 0.26 lb per cubic inch. Round to the nearest whole pound.

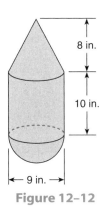

8 in.

10 in.

← 9 in. →

Figure 12–12

The solution requires finding the volume of the cone that forms the top of the object, the volume of the cylinder that forms the middle portion of the object, and the volume of the *hemisphere* (half sphere) that forms the bottom of the object.

$$V_{\text{cone}} = \frac{\pi r^2 h}{3} \qquad V_{\text{cylinder}} = \pi r^2 h \qquad V_{\text{hemisphere}} = \frac{1}{2}\left[\frac{4\pi r^3}{3}\right]$$

$$V_{\text{cone}} = \frac{(\pi)(4.5)^2(8)}{3} \qquad V_{\text{cylinder}} = (\pi)(4.5)^2(10) \qquad V_{\text{hemisphere}} = \frac{1}{2}\left[\frac{4(\pi)(4.5)^3}{3}\right]$$

$$V_{\text{cone}} = 169.6460033 \text{ in}^3 \qquad V_{\text{cylinder}} = 636.1725124 \text{ in}^3 \qquad V_{\text{hemisphere}} = 190.8517537 \text{ in}^3$$

$$\text{Total volume} = V_{\text{cone}} + V_{\text{cylinder}} + V_{\text{hemisphere}}$$

$$\text{Total volume} = 996.6702694 \text{ in}^3$$

Convert to pounds:

$$\frac{996.6702694 \text{ in}^3}{1} \times \frac{0.26 \text{ lb}}{1 \text{ in}^3} = 259 \text{ lb} \qquad \text{Rounded.}$$

To the nearest whole pound, the cast-iron object weighs 259 lb.

SELF-STUDY EXERCISES 12–3

1 Solve.

1. How many square inches are in the lateral surface area and total surface area of an aluminum can with a $2\frac{1}{2}$ in. diameter and $4\frac{3}{4}$ in. height? Round to tenths.

2. What is the lateral surface area and total surface area of a cylindrical oil storage tank that has a 40 ft diameter and 15-ft height? Round to the nearest whole number.

3. A right pentagonal prism is 10 cm high. If the area of each pentagonal base is 32 cm², and the perimeter is 20 cm, what is the lateral and total surface area of the prism?

4. What is the lateral and total surface area of a triangular prism that has a height of 8 in. and a triangular base that measures 4 in. on each side with an altitude of 3.46 in.? Round to hundredths.

5. A cylindrical water well is 1,200 ft deep and 6 in. across. Find the lateral surface area of the well. Round to the nearest square foot. (*Hint:* Convert measures to a common unit.)

2

6. Find the volume of a pyramid that has a square base of 30 cm on a side and a height of 42 cm.

7. Find the volume of a pyramid that has a square base of 48 m on a side and a height of 100 m.

8. Find the volume of a pyramid that has an equilateral triangular base with altitude 12 m and a side of 10.39 m. The height is 20 m. Round to tenths.

9. A frustum of a pyramid has a square base that is 18 in. on each side, a square top that is 10 in. on each side, and a height of 13 in. Find the volume of the frustum.

10. A frustum of a pyramid has a triangular base that has an area of 32 cm² and a triangular top that has a surface area of 28 cm². The height of the frustum is 81 cm. Find the volume of the frustum. Round to hundredths.

11. Cap blocks for a fence are molded in the shape of a frustum of a pyramid that has a square base and top. The base of the frustum is 30 in. on each side and the top is 24 in. on each side. The cap block is 5 in. thick (height of frustum). What is the volume of the frustum?

12. One component of a fountain is an inverted frustum that has a square base and opening at the top. The base is 18 in. on each side and the top is 30 in. on each side. If the height of the frustum is 7 in., what is the capacity (volume) of the fountain?

3 Solve. Round to tenths.

13. Find the surface area of a sphere with a radius of 5 cm.

14. Find the volume of a sphere with a radius of 6 in.

15. How many square feet of steel are needed to manufacture a spherical water tank with a diameter of 45 ft? What is the capacity?

16. If 1 ft^3 = 7.48 gal, how many gallons can the water tank in Exercise 15 hold?

17. A spherical propane tank has a diameter of 4 ft. How many square feet of surface area need to be painted? What is the capacity of the tank?

18. If a propane tank is filled to 90% of its capacity, how many gallons of propane does the tank in Exercise 17 hold? (1 ft^3 = 7.48 gal.)

4 Solve. Round to tenths.

19. Find the lateral surface area, total surface area, and volume of the cone in Fig. 12–13.

20. How many cubic feet are in a conical pile of sand that is 30 ft in diameter and is 20 ft high?

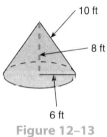

10 ft

8 ft

6 ft

Figure 12–13

21. How many square centimeters of sheet metal are needed to form a conical rain cap 25 cm in diameter if the slant height is 15 cm?

22. A cone-shaped storage container holds a photographic chemical. If the container is 80 cm wide and 30 cm high, how many liters of the chemical does it hold if 1 L = 1,000 cm^3?

23. Find the total surface area of a conical tank that has a radius of 15 ft and a slant height of 20 ft.

24. Find the height of a conical tank with a volume of 261.67 ft^3 and a radius of 5 ft. (*Hint:* Rearrange the volume formula to find the height.)

25. A cylindrical water tower with a conical top and hemispheric bottom (see Fig. 12–14) needs to be painted. If the cost is $2.19 per square foot, how much does it cost (to the nearest dollar) to paint the tank?

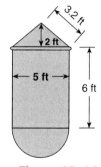

3.2 ft

2 ft

5 ft

6 ft

Figure 12–14

Section 12–1

Evaluate the percentage formula $P = \dfrac{RB}{100}$ using the given values.

1. Find the percentage if $R = 30\%$ and $B = \$2.70$.
2. Find the percentage if $R = 10\%$ and $B = 300$ lb.
3. Find the rate if $P = 12$ kg and $B = 125$ kg.
4. Find the base if $P = \$28.05$ and $R = 8.5\%$.
5. Evaluate the rate formula $R = \dfrac{100P}{B}$ for P if $R = 12\%$ and $B = 90$.
6. Evaluate the base formula $B = \dfrac{100P}{R}$ for R if $B = \$5,000$ and $P = \$450$.

Evaluate the interest formula $I = PRT$ using the given values.

7. Find the interest if $P = \$440$, $R = 16\%$, and $T = 2\frac{3}{4}$ years.
8. Find the rate if $I = \$2,484$, $P = \$4,600$, and $T = 3$ years.
9. Find the time if $I = \$387.50$, $P = \$1,550$, and $R = 12.5\%$.
10. Find the principal if $I = \$1,665$, $R = 18\frac{1}{2}\%$ per yr, and $T = 1\frac{1}{2}$ yr.
11. Find the length of a rectangular work area if the perimeter is 160 in. and the width is 30 in. Use the formula $P = 2(l + w)$.
12. Find the radius of a circle whose area is 132.7 mm^2 using the formula $r = \sqrt{\dfrac{A}{\pi}}$. Round to the nearest tenth.
13. Find the cost (C) if the markup (M) on an item is $\$25.75$ and the selling price (S) is $\$115.25$. Use the formula $M = S - C$.
14. Using the formula for the side of a square, $s^2 = A$, find the length of a side of a square field that has an area of $\frac{1}{4}$ mi^2.
15. Evaluate the formula for the area of a circle, $A = \pi r^2$, if $r = 5.5$ in. Round to the nearest tenth.
16. Evaluate the formula for the area of a square, $A = s^2$, if $s = 3.25$ km.
17. Ohm's law is $E = IR$. Find the amperes of current (I) if the voltage (E) is 220 V and the resistance (R) is 80 ohms (Ω).
18. Use the formula $R_t = \dfrac{R_1 R_2}{R_1 + R_2}$ to find the total resistance (R_t) if one resistance (R_1) is 10 ohms (Ω) and the second resistance (R_2) is 9 ohms (Ω). Round to tenths.
19. According to Boyle's law, if temperature is constant, the volume of a gas is inversely proportional to the pressure on it. Find the final volume (V_2) of a gas using the formula $\dfrac{V_1}{V_2} = \dfrac{P_2}{P_1}$ if the original volume (V_1) is 30 ft^3, the original pressure (P_1) is 75 psi (pounds) per square inch), and the final pressure (P_2) is 225 psi.
20. Distance is rate times time, or $D = RT$. Find the rate if the distance traveled is 260 mi and the time traveled is 4 hr.
21. The formula for power (P) in watts (W) is $P = I^2 R$. Find the current (I) in amperes if a device draws 80 W and the resistance (R) is 8 Ω. Round to tenths.
22. Use the formula $E = \dfrac{I - P}{I}$ to find the percent efficiency (E) of an engine if the input (I) is 22,600 calories and the output (P) is 5,600 calories. Round to the nearest tenth of a percent.
23. The formula for the speed (s) of a driven pulley in revolutions per minute (rpm) is $s = \dfrac{DS}{d}$. Find the speed of a driven pulley with diameter (d) of 3 in. if the diameter (D) of the driving pulley is 7 in. and its speed (S) is 600 rpm.
24. Find the impedance (Z) in ohms using the formula $Z = \sqrt{R^2 + X^2}$ if the resistance (R) is 4 Ω and the reactance (X) is 7 Ω. Round to tenths.
25. Use the formula $H = \dfrac{D^2 N}{2.5}$ to find the number of cylinders (N) to the nearest hundredth required in an engine of 5 hp (H) if the cylinder diameter (D) is 2.5 in.
26. If the distance (C) between the centers of the pulleys in Exercise 23 is 30 in., find the length (L) of the belt connecting them using the formula
$$L = 2C + 1.57(D + d) + \dfrac{D + d}{4C}$$
27. Use the cost function $C(x) = 3.48x + 10,000$ to find the cost of producing (a) 100, (b) 1,000, and (c) 10,000 toy trains if x is the number of trains produced.
28. Use the average cost function $A(x) = \dfrac{C(x)}{x}$ to find the average cost of producing (a) 100, (b) 1,000, and (c) 10,000 trains.
29. Use the revenue function $R(x) = 40x$, where x represents the number of trains sold, to find the revenue generated by (a) 100, (b) 1,000, and (c) 10,000 trains.
30. Use the profit function $P(x) = R(x) - C(x)$ and the results of previous exercises to find the profit for producing (a) 100, (b) 1,000, and (c) 10,000 trains.
31. Find the average profit per train for (a) 100 trains, (b) 1,000 trains, and (c) 10,000 trains. $A(x) = \dfrac{P(x)}{x}$.

Solve the formulas for the indicated variable.

32. $M = S - C$ for C **33.** $V = lwh$ for w **34.** $I = \dfrac{PRD}{365}$ for D

35. $B = cr^2x$ for r **36.** $r = \sqrt{s^2 - t^2}$ for t **37.** $PB = A$ for B

38. $V = lwh$ for h **39.** $I = Prt$ for r **40.** $s = c + m$ for c

41. $s = r - d$ for r **42.** $s = r - d$ for d **43.** $v = v_0 - 32t$ for t

44. $P = 2(l + w)$ for w **45.** $A = P(1 + rt)$ for t

46. $V = \dfrac{1}{3}Bh$ for h

47. The formula for finding the sale price on an item is $S = P - D$, where S is the sale price, P is the original price, and D is the discount. Solve the formula for the original price.

48. The formula for finding tax is $T = RM$, where T represents tax, R represents the tax rate, and M represents the marked price. Rearrange the formula to find the marked price.

49. Find the total surface area of the triangular prism in Fig. 12–15.

50. Find the lateral surface area of the cylinder in Fig. 12–16.

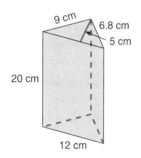

Figure 12–15

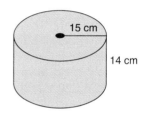

Figure 12–16

51. Find the total surface area of the cylinder in Fig. 12–16.

52. Find the total surface area of a cylinder that has a radius of 10 cm and a height of 30 cm.

53. What is the lateral surface area of a hexagonal column 20 ft high that measures 6 in. on a side?

54. If concrete weighs 160 lb per cubic foot, what is the lateral surface area of a concrete circular slab 4 in. thick and 15 ft across? Round to whole number ft².

55. A pipeline to carry oil between two towns 5 mi apart has an inside diameter of 18 in. If 1 mi = 5,280 ft, find the lateral surface area to the nearest ft².

56. How many barrels of oil will a conical tank hold if its height is $65\frac{1}{2}$ ft and its radius is 20 ft? Round to the nearest whole barrel. (31.5 gal = 1 barrel and 1 ft³ = 7.48 gal.)

57. Find the volume of a pyramid that has a square base of 12 m on a side and a height of 36 m.

58. Find the volume of a pyramid that has a square base of 7.2 ft on a side and a height of 24 ft.

59. Find the volume of a frustum of a pyramid that has a square base and top. The base is $5\frac{3}{8}$ ft on a side and the top is $3\frac{5}{8}$ ft on a side. The height is 6 ft.

60. Find the volume of a frustum of a pyramid that has a square base of 15 yd on a side and a top of 12 yd on a side. The height is 12 yd.

Solve. Round the final answer to the nearest tenth unless otherwise specified.

61. Find the total surface area of a sphere with a radius of 9 m.

62. Find the total surface area of a sphere if its diameter is 20 cm.

63. Find the volume of a sphere that has a radius of 12 ft.

64. Find the volume of a sphere that has a diameter of 30 cm.

65. Find the lateral surface area of a cone with a radius of 6 cm and a slant height of 9 cm.

66. Find the total surface area of a cone that has a radius of 4 m and a slant height of 8 m.

67. Find the volume of a cone with a radius of 6 in. and a height of 10 in.

68. If 1 gal = 231 in³, find the number of gallons that a conical oil container 18 in. high and 23 in. across holds.

69. The entire exterior surface of a conical tank with a slant height of 12 ft and a diameter of 18 ft is being painted. If the paint covers at a rate of 350 ft² per gallon, how many gallons of paint are needed for the job? Round any fraction of a gallon to the next whole gallon.

70. A hopper deposits sand in a cone-shaped pile with a diameter of 9′6″ and a height of 8′3″. To the nearest cubic foot, how much sand is deposited?

71. How many square feet of plastic material are needed to devise a wind-tunnel cone that has a base of 6 ft across and height of 12 ft (Fig. 12–17). (*Hint:* To find the slant height, consider it to be the hypotenuse of a right triangle.)

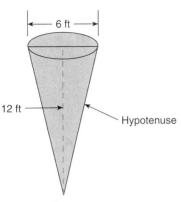

6 ft

12 ft

Hypotenuse

Figure 12–17

72. To the nearest square foot, how much nylon material is needed to construct a conical tentlike pavilion 30 ft in diameter and 5 ft from the apex to the base? (*Hint:* See Exercise 71.)

73. A No. 5 soccer ball has a diameter of 8.8 in. How much leather is needed to cover the surface?

74. How much does a 4-in. lead ball weigh if lead weighs 1 lb per 2.4 in^3?

75. A spherical tank is anchored halfway in the ground. How many cubic feet of packed earth is excavated to install the tank? The tank measures 20 ft across. Round to the nearest whole cubic foot.

CHALLENGE PROBLEMS

76. Devise your own formulas for the following relationships.
  (a) An electrical power company computes the monthly charges by multiplying the kilowatts of power used times the cost per kilowatt and adds to that a fixed monthly fee.
  (b) A store calculates the ending balance on a charge account by multiplying the interest rate times the previous unpaid balance and then adding the previous balance and purchases and subtracting payments.
  (c) Profit on the sale of a particular item is the product of the number of items sold and the difference between the selling price of the item and its cost to the seller.

77. Explain the usefulness of formula rearrangement in solving applied problems. Use at least one formula to illustrate.

CONCEPTS ANALYSIS

1. Briefly describe the procedure for evaluating a formula when numerical values are given for all but one variable.

3. Describe the similarities and differences in the formulas for the volume of a cylinder and a prism.

2. Describe at least two occasions when it is desirable to rearrange a formula.

Find the mistake in each problem. Explain the mistake and rework the problem correctly.

4. In the formula $a = 3b - c$, find b if $a = 5$ and $c = 1$.

$$5 = 3b - 1$$
$$5 + 1 = 3b$$
$$6 = 3b$$
$$3 = b$$

5. Solve the formula for y: $3x + 2y = 6$.

$$3x = -2y + 6$$
$$x = \frac{-2y + 6}{3}$$
$$x = -\frac{2}{3}y + 2$$

6. A calculator manufacturer estimates that it will cost \$40,000 to set up a production line to mass-produce calculators. If the cost to produce each calculator is expected to be \$8, write a cost function to represent the production cost.

7. Describe the similarities and differences in the lateral surface area and total surface area of a pyramid.

8. Write explanatory comments for each step in rearranging the formula

$$Z = \sqrt{R^2 + X^2} \text{ for } R.$$
$$(Z)^2 = \left(\sqrt{R^2 + X^2}\right)^2$$
$$Z^2 = R^2 + X^2$$
$$Z^2 - X^2 = R^2$$
$$\sqrt{Z^2 - X^2} = \sqrt{R^2}$$
$$\sqrt{Z^2 - X^2} = R$$
$$R = \sqrt{Z^2 - X^2}$$

CHAPTER SUMMARY

Learning Outcomes

What to Remember with Examples

Section 12–1

1 Evaluate formulas (pp. 511–514).

Substitute values. Solve using rules for solving equations and/or the order of operations.

Find the area (A) if the length (l) is 4 ft and the width (w) is 2 ft.

$$A = lw \qquad \text{Substitute values.}$$
$$A = (4)(2) \qquad \text{Multiply.}$$
$$A = 8 \text{ ft}^2$$

Sometimes the letter that is to be solved for must be isolated.

If the perimeter (P) of a square is 12 in., find the length of a side.

$$P = 4s \qquad \text{Substitute value for } P.$$
$$12 = 4s \qquad \text{Divide.}$$
$$3 = s \qquad \text{Interchange sides.}$$
$$s = 3 \text{ in.}$$

Apply appropriate techniques for solving equations when evaluating formulas.

The formula for the distance (d) an object falls is $d = \frac{1}{2}gt^2$. Find the distance if gravity (g) is 32 ft per second squared and seconds (t) is 3. Feet per second squared is $\frac{\text{ft}}{\text{sec}^2}$.

$$d = \frac{1}{2}gt^2 \qquad \text{Dimension analysis: } \frac{\text{ft}}{\text{sec}^2}(\text{sec}^2) = \text{ft}$$

$$d = \frac{1}{2}(32)(3)^2 \qquad \text{Raise to power.}$$

$$d = \frac{1}{2}(32)(9) \qquad \text{Multiply.}$$

$$d = 144 \text{ ft}^2$$

2 Evaluate formulas in function notation (pp. 514–516).

To evaluate a formula in function notation substitute the given value for the independent variable in every occurrence of the variable, and use the order of operations to simplify.

Find the cost $C(x)$ of producing 1,000 calendars using the cost function $C(x) = 2x + 2,000$.

$$C(1,000) = 2(1,000) + 2,000$$
$$C(1,000) = 2,000 + 2,000$$
$$C(1,000) = \$4,000$$

Section 12–2

1 Rearrange formulas to solve for a specified variable (pp. 518–521).

Isolate the desired variable so that it appears on the left. This can make evaluation simpler. Apply appropriate rules for solving equations.

Solve the formula $S = \dfrac{R + P}{2}$ for R.

$$(2)S = \frac{R + P}{2}(2) \qquad \text{Clear the denominator.}$$

$$2S = R + P \qquad \text{Isolate } R \text{ (sort terms).}$$

$$2S - P = R \qquad \text{Interchange sides so the variable appears on the left.}$$

$$R = 2S - P$$

Section 12–3

1 Find the surface area of prisms and cylinders (pp. 522–523).

Use the formula for lateral surface area (area of sides) of a right prism or cylinder: $LSA = ph$, where p is the perimeter of the base and h is the height.

Find the lateral surface area of a triangular prism that has a base of 4 in. on each side and a height of 10 in.

$$LSA = ph \qquad \text{Substitute values: } p = 4 + 4 + 4, h = 10.$$
$$LSA = (4 + 4 + 4)(10)$$
$$LSA = (12)(10)$$
$$LSA = 120 \text{ in}^2 \qquad \text{Area is square units.}$$

Use the formula for total surface area (sides plus bases) of a right prism or cylinder: $TSA = ph + 2B$, where p is the perimeter of the base, h is the height, and B is the area of a base.

Find the total surface area of the preceding triangular prism if the height of each base is 3.464 in.

$$TSA = ph + 2B \qquad LSA = ph = 120 \text{ in}^2, B = \frac{1}{2}bh.$$

$$TSA = 120 + 2\left(\frac{1}{2}\right)(4)(3.464) \qquad \text{Multiply.}$$

$$TSA = 120 + 13.856 \qquad \text{Add.}$$

$$TSA = 133.856 \text{ in}^2$$

Chapter Summary

2 Find the volume of a pyramid and of a frustum of a pyramid (pp. 523–524).

Use the formula for volume of a pyramid: $V = \dfrac{1}{3} Bh$, where B is the area of the base and h is the height of the pyramid.

Find the volume of a pyramid that has a square base that is 12 cm on a side and a height of 20 cm.

$$V = \frac{1}{3} Bh$$

Find the area of the base. $B = s^2 = 12^2 = 144 \text{ cm}^2$

$$V = \frac{1}{3} (144 \text{ cm}^2)20 \text{ cm}$$

$$V = (48 \text{ cm}^2)(20 \text{ cm})$$

$$V = 960 \text{ cm}^3$$

Volume is cubic units.

Use the formula for area of a frustum of a pyramid: $V = \frac{1}{3} h (B_1 + B_2 + \sqrt{B_1 B_2})$, where h is the height of the frustum, B_1 is the area of the base, and B_2 is the area of the top.

Find the volume of a frustum of a pyramid with a square base that is 7 in. on each side, a square top that is 3 in. on each side, and a height of 6 in.

$$V = \frac{1}{3} h \left(B_1 + B_2 + \sqrt{B_1 B_2}\right)$$

Find B_1 and B_2. $B_1 = 7^2 = 49$; $B_2 = 3^2 = 9$

$$V = \frac{1}{3} (6)\left(49 + 9 + \sqrt{49(9)}\right)$$

$$V = \frac{6}{3} \left(58 + \sqrt{441}\right)$$

$$V = 2 (58 + 21)$$

$$V = 2 (79)$$

$$V = 158 \text{ in}^3$$

3 Find the surface area and volume of a sphere (pp. 524–525).

Use the formula for total surface area of a sphere: $\text{TSA} = 4\pi r^2$, where r is the radius.

What is the surface area of a sphere 6 in. in diameter?

$$\text{TSA} = 4\pi r^2 \qquad r = \frac{1}{2}d = 3 \text{ in.}$$

$$\text{TSA} = 4\pi(3)^2$$
$$\text{TSA} = 4\pi(9)$$
$$\text{TSA} = 113.10 \text{ in}^2 \quad \text{Rounded.}$$

Use the formula for volume of a sphere: $V = \dfrac{4\pi r^3}{3}$, where r is the radius.

A spherical gas tank is 10 ft in diameter. Find its volume.

$$V = \frac{4\pi r^3}{3} \qquad r = \frac{1}{2}d = 5 \text{ ft.}$$

$$V = \frac{4\pi(5)^3}{3}$$

$$V = \frac{4\pi(125)}{3}$$

$$V = 523.60 \text{ ft}^3 \quad \text{Rounded.}$$

Chapter 12 / Formulas and Applications

4 Find the surface area and volume of a cone (pp. 525–527).

Use the formula for lateral surface area of a cone: $\text{LSA} = \pi r s$, where r is the radius and s is the slant height.

A conical pile of gravel has a diameter of 30 ft and a slant height of 40 ft. What is the lateral surface area?

$$\text{LSA} = \pi r s$$
$$\text{LSA} = \pi(15)(40) \qquad r = \frac{1}{2}d = \frac{1}{2}(30) = 15.$$
$$\text{LSA} = 1{,}884.96 \text{ ft}^2 \qquad \text{Rounded from } 1{,}884.955592.$$

Use the formula for total surface area of a cone: $\text{TSA} = \pi r s + \pi r^2$, where r is the radius and s is the slant height. πr^2 is the area of the base.

Find the total surface area of the preceding cone.

$$\text{TSA} = \pi r s + \pi r^2 \qquad \text{Substitute full calculator value for } \pi rs.$$
$$\text{TSA} = 1{,}884.955592 + \pi(15)^2$$
$$\text{TSA} = 1{,}884.955592 + \pi(225)$$
$$\text{TSA} = 1{,}884.955592 + 706.8583471$$
$$\text{TSA} = 2{,}591.81 \text{ ft}^2 \qquad \text{Rounded.}$$

Use the formula for volume of a cone: $V = \dfrac{\pi r^2 h}{3}$, where r is the radius and h is the height.

Find the volume of a cone with a radius of 6 in. and a height of 12 in.

$$V = \frac{\pi r^2 h}{3} \qquad \text{Substitute.}$$
$$V = \frac{\pi(6)^2(12)}{3} \qquad \text{Raise to power.}$$
$$V = \frac{\pi(36)(12)}{3} \qquad \text{Multiply and divide.}$$
$$V = 452.39 \text{ in}^3 \qquad \text{Rounded.}$$

CHAPTER TRIAL TEST

1. The electrical resistance of a wire is found from the formula $R = \dfrac{PL}{A}$. Rearrange the formula to find the length L of the wire.

2. The formula for the volume (V) of a solid rectangular figure is $V = lwh$ (length $\times$ width $\times$ height). If the volume of a mailing container is 1.5 m³, its length is 1.5 m, and its width is 0.5 m, what is its height?

3. Engine displacement d is found using the formula $d = \pi r^2 sn$. Solve to find r (the radius of the bore).

4. Using $d = 351$ in³, $s = 3.5$-in. stroke, and $n = 8$ cylinders, calculate to the nearest tenth the radius of the bore with the rearranged formula in Exercise 3.

5. A pentagonal prism (five sides) measures 1 in. on each side of its base and has a height of 10 in. What is its lateral surface area?

6. Find the total surface area of the pentagonal prism in Exercise 5.

7. A cylinder that has a height of 15 ft has two circular bases that each base has a diameter of 12 ft. What is the total surface area?

8. Find the lateral surface area of the cylinder described in Exercise 7.

9. A spherical tank 12 ft in diameter can hold how many gallons of fluid if 1 ft³ = 7.48 gal? Answer to the nearest whole gallon.

10. The base of a brass pyramid is an equilateral triangle with sides of 3 in. and altitude of 2.6 in. If the pyramid's height is 8 in., what is the volume of the pyramid?

11. Hard coal broken into small pieces is dumped into a cone-shaped pile. The base is 35 ft across and the pile stands 12 ft tall. How many cubic feet of coal are in the pile? Round to tenths.

12. A spherical gas storage tank 2.5 m wide needs to be sandblasted, primed, and refinished. The owner received an estimate of $8.50 per square meter. How much should the job cost to the nearest dollar?

13. A sheet-metal worker wants to make a tin cone. If the base of the cone is 30 cm across and its height is 25 cm, what is the total surface area required to the nearest tenth? (*Hint:* Use your knowledge of a right triangle to find the slant height.)

14. A cold water pipe with an outside diameter of $\frac{7}{8}$ in. is 23 ft long. How many whole rolls of insulating wrap are needed if one roll covers $5\frac{1}{2}$ ft^2 and no allowance is made for overlap? (*Hint:* $\frac{7}{8}$ in. is $\frac{7}{8} \times \frac{1}{12} = 0.0729$ ft.)

15. A steel rod has a diameter of 20 in. Find the lateral surface area of a 5-ft length of the rod measured in square feet to the nearest tenth.

16. Find the total surface area of a cylindrical storage tank if it is 30 ft tall and has a diameter of 12 ft. Round to hundredths.

17. Using the formula $P = \dfrac{1.27F}{D^2}$, calculate force F (in pounds) if the pressure P is 180 psi and the piston diameter is 3.25 in. Round to the nearest hundredth.

18. Use the formula $R_t = \dfrac{R_1 R_2}{R_1 + R_2}$ to find the total resistance (R_t) if one resistance (R_1) is 9 Ω and the second resistance (R_2) is 8 Ω. Round to tenths.

19. If the efficiency (E) of an engine is 70% and the input (I) is 40,000 calories, find the output (P) in calories. Use the formula $E = \dfrac{I - P}{I}$.

20. Find the volume of a pyramid that has a square base measuring 16 ft on each side and a height of 24 ft.

21. Find the volume of a frustum of a pyramid if both bases are squares. The top measures 8 m on each side and the bottom measures 12 m on each side. The height is 9 m.

22. Find the volume of a steel ball that has a radius of $1\frac{3}{5}$ in.

Products and Factors

13

Throughout our study of mathematics, we have examined products and factors. To reduce fractions, we looked for factors common to both the numerator and denominator. In examining perfect squares, we looked for two identical factors to find the square root of a value. We looked for perfect-square factors of a radicand so that the radical could be written in a simpler form. In this chapter, we again find it useful to examine products and factors.

13–1 | The Distributive Property and Common Factors

Learning Outcome **1** Factor an expression containing a common factor.

We applied the distributive property and finding common factors earlier in the text and applied them in different contexts. In this section, rather than use the distributive property to multiply and obtain a product, we will start with a product and regenerate the factors that produce the product. In other words, we want to undo the multiplication. Factoring resembles division, which is the inverse operation of multiplication.

1 Factor an Expression Containing a Common Factor.

The multiplication problem $7a(3a + 2)$ is the indicated product of $7a$ and the grouped quantity $3a + 2$. This is the *factored* form of the expression. After the expression is multiplied, we have two terms written as the indicated sum $21a^2 + 14a$. This is the *expanded* form. To rewrite the expression $21a^2 + 14a$ as the indicated product $7a(3a + 2)$ is to *factor* it.

Let's look at a general example of the distributive property:

$$a(x + y) = ax + ay$$

factored form   expanded form

Notice that a appears as a factor in both terms in the expanded form. When a factor appears in each of several terms, it is called a *common factor* of the terms. The distributive property in reverse can be used to write the addition as a multiplication. In other words, we can *factor* the expression.

$$ax + ay = a(x + y)$$

To factor an expression containing a common factor:

1. Find the *greatest* factor common to *each* term of the expression.
2. Divide each term by the common factor. Divide mentally if practical.
3. Rewrite the expression as the indicated product of the greatest common factor (GCF) and the quotients in Step 2.

EXAMPLE Write $3a + 3b$ in factored form.

We can use the distributive property to factor the expression.

$$3a + 3b = \qquad \text{Write 3 as a factor and divide each term by 3.}$$

$$3\left(\frac{3a}{3} + \frac{3b}{3}\right) = \qquad \text{Simplify.}$$

$$\mathbf{3(a + b)} \qquad \text{Factored form.}$$

The distributive property also applies if we have more than two terms.

EXAMPLE Write $3ab + 9a + 12b$ in factored form.

$$3\ ab\quad + 9a\quad +\quad 12b \qquad\qquad \text{3 is the common factor. Divide.}$$

$$3 \cdot 3 \qquad 3 \cdot 4$$

$$3\left(\frac{3ab}{3} + \frac{3 \cdot 3a}{3} + \frac{3 \cdot 4b}{3}\right) \qquad \text{Simplify.}$$

$$\mathbf{3\,(ab + 3a + 4b)} \qquad\qquad \text{Factored form}$$

When looking for a common factor, we always look for *all* common factors.

EXAMPLE Factor $10a^2 + 6a$ completely.

$$10a^2 + 6a = \qquad \text{The GCF is 2a. Write 2a as a factor and divide each term by 2a.}$$

$$2a\left(\frac{10a^2}{2a} + \frac{6a}{2a}\right) = \qquad \text{Simplify.}$$

$$\mathbf{2a(5a + 3)} \qquad\qquad \text{Factored form}$$

EXAMPLE Factor $2x^2 + 4x^3$ completely.

$$2x^2 + 4x^3 = \qquad \text{The GCF is } 2x^2 \text{. Write } 2x^2 \text{ as a factor and divide each term by } 2x^2.$$

$$2x^2\left(\frac{2x^2}{2x^2} + \frac{4x^3}{2x^2}\right) = \qquad \text{Simplify.}$$

$$\mathbf{2x^2(1 + 2x)} \qquad\qquad \text{Term of 1 must be written.}$$

EXAMPLE Write $2x + 3y$ in factored form.

$$2x + 3y = \qquad \text{The GCF is 1. The expression can only be written in factored form as } 1(2x + 3y).$$

$$1(2x + 3y) \qquad \text{Factored form}$$

TIP!

When an expression can only be written in factored form as 1 times the entire expression, the expression is a *prime polynomial*.

When Is It Necessary to Write a 1?

We have found that it is not always necessary to write the number 1. When is this the case?

When 1 is a term, it must be written.

$$2x^2 - x = x\left(\frac{2x^2}{x} - \frac{x}{x}\right) = x(2x - 1)$$

When 1 is a factor, writing the 1 is optional: $1 \cdot n = n$.

$$2a + 2b = 2\left(\frac{2a}{2} + \frac{2b}{2}\right) = 2(1a + 1b) \qquad \text{or} \qquad 2(a + b)$$

When **1** is an exponent, writing the 1 is optional: $n^1 = n$.

$$2x^3 - 5x^2 = x^2\left(\frac{2x^3}{x^2} - \frac{5x^2}{x^2}\right) = x^2(2x^1 - 5x^0) = x^2(2x - 5)$$

Also, recall that $n^0 = 1$.

Sometimes a binomial factor is the common factor.

EXAMPLE   Factor $7y(2y - 5) + 3(2y - 5)$.

$$7y(2y - 5) + 3(2y - 5) = \qquad \text{Common factor is } (2y - 5).$$

$$(2y - 5)\left[\frac{7y\,\cancel{(2y - 5)}}{\cancel{(2y - 5)}} + \frac{3\,\cancel{(2y - 5)}}{\cancel{(2y - 5)}}\right] =$$

$$\mathbf{(2y - 5)\,(7y + 3)}$$

If the leading coefficient of a polynomial is negative, it is often helpful to factor a common factor of -1.

EXAMPLE   Factor $-3x^2 + 2x - 5$.

$$-3x^2 + 2x - 5 = \qquad \text{Common factor is } -1.$$

$$-1\left(\frac{-3x^2}{-1} + \frac{2x}{-1} - \frac{5}{-1}\right) =$$

$$\mathbf{-1(3x^2 - 2x + 5)}$$

SELF-STUDY EXERCISES 13–1

1 Factor completely. Check.

1. $7a + 7b$
2. $12x + 12y$
3. $m^2 + 2m$
4. $5y^3 + 8y^2$
5. $6x^2 + 3x$
6. $12y^3 + 18y^4$
7. $12x^5 - 6x^4$
8. $5x - 15xy$
9. $5y + 3z$
10. $8x - 7y$
11. $5ab + 10a + 20b$
12. $4ax^2 + 6a^2x + 10a^2x^2$
13. $5a - 7ab + 35b$
14. $12a^2 - 15a + 6$
15. $3x^3 - 9x^2 - 6x$
16. $8a^2b + 14ab^3 + 28a^3b^3$
17. $3m^2 - 6m^3 + 12m^4$

Write in factored form so the leading coefficient of the polynomial factor is positive.

18. $-x - 7$
19. $-3x - 8$
20. $-5x + 2$
21. $-12x + 7$
22. $-x^2 + 3x - 8$
23. $-2x^2 - 7x - 11$
24. $-2x^2 + 6x - 8$
25. $-3x^2 - 9x + 15$
26. $-7x^2 - 21x + 14$
27. $-12x^2 + 18x + 6$
28. $5x(x + 3) + 8y(x + 3)$
29. $3x(2x - 1) + 5(2x - 1)$
30. $4y(3y - 5) + 7(3y - 5)$
31. $7a(a - b) + 2b(a - b)$
32. $5m(2m - 3n) - 7n(2m - 3n)$
33. $y(y - 2) - 3(y - 2)$
34. $3x(2x - 7) - 8(2x - 7)$
35. $7y(9y - 2) - 5(9y - 2)$
36. $5\sqrt{7} + 10$
37. $8\sqrt{3} - 12$
38. $3\sqrt{2} - 9\sqrt{3}$

Learning Outcomes
1   Multiply polynomials.
2   Use the FOIL method to multiply two binomials.
3   Multiply polynomials that result in special products.
4   Divide polynomials using long division.

1 Multiply Polynomials.

As we expand our experiences with multiplication, we need to use the appropriate terminology associated with polynomials. Recall:

- A *polynomial* is an expression with constants and variables with whole-number exponents and contains one or more terms with at least one variable term.

  Monomials, binomials, and trinomials are polynomials.

- A *monomial* contains one term, such as $5x^2$.
- A *binomial* contains two terms, such as $3a + 4$.
- A *trinomial* contains three terms, such as $4x^2 + x - 2$.

We multiply two binomials by using the distributive property. According to the distributive property, each term of the first factor is multiplied by each term of the second factor. This means we are required to use the distributive property more than once.

EXAMPLE   Multiply $(x + 4)(x + 2)$.

$$(x + 4)(x + 2) = x(x + 2) + 4(x + 2) \qquad \text{Apply the distributive property.}$$
$$= x^2 + 2x + 4x + 8 \qquad \text{Combine like terms.}$$
$$= x^2 + 6x + 8$$

To multiply two polynomials:

1. Use the distributive property to multiply each term of the first polynomial times the entire second polynomial.
2. Combine like terms.

Symbolically,

$$(a + b)(c + d) = a(c + d) + b(c + d) = ac + ad + bc + bd$$
$$(a + b + c)(d + e + f) = a(d + e + f) + b(d + e + f) + c(d + e + f)$$
$$= ad + ae + af + bd + be + bf + cd + ce + cf$$

EXAMPLE   Multiply $(2x^2 + 3x - 2)(3x^2 - 5x + 6)$.

$$(2x^2 + 3x - 2)(3x^2 - 5x + 6) =$$
$$2x^2(3x^2 - 5x + 6) + 3x(3x^2 - 5x + 6) - 2(3x^2 - 5x + 6) = \qquad \text{Distribute.}$$
$$6x^4 - 10x^3 + 12x^2 + 9x^3 - 15x^2 + 18x - 6x^2 + 10x - 12 = \qquad \text{Combine like terms.}$$
$$6x^4 - x^3 - 9x^2 + 28x - 12$$

Mentally multiply $(m + p)(m^2 - mp + p^2)$.

$$(a + b)(a^2 - ab + b^2) = a^3 + b^3$$ Use the pattern and substitute m for a and p for b.

$$(m + p)(m^2 - mp + p^2) = \mathbf{m^3 + p^3}$$

Mentally multiply $(4c - d)(16 + 4d + d^2)$.

$$(a - b)(a^2 + ab + b^2) = a^3 - b^3$$ Substitute 4 for a and d for b.

$$(4c - d)(16c^2 + 4cd + d^2) = \mathbf{64c^3 - d^3}$$ $4^3 = 64$

4 Divide Polynomials Using Long Division.

A polynomial can be divided by a polynomial by using a long division procedure. If the remainder is 0, both the divisor and the quotient are factors of the dividend.

To divide a polynomial by a polynomial using long division:

1. Divide the first term of the dividend by the first term of the divisor. The partial quotient is placed above the first term of the dividend.
2. Multiply the partial quotient times the divisor and align the product under like terms of the dividend.
3. Subtract (change subtrahend to opposite and use addition rules).
4. Bring down the next term of the dividend and repeat Steps 1–3.
5. Repeat Steps 1–4 until all terms of the dividend have been brought down. The result of the last subtraction is the remainder.
6. Write the remainder as a fraction with the remainder as the numerator and the divisor as the denominator.

Is $x - 4$ a factor of $2x^3 - 9x^2 + 7x - 12$?

$$
\begin{array}{r}
2x^2 - x + 3 \\
x - 4 \overline{) 2x^3 - 9x^2 + 7x - 12} \\
2x^3 - 8x^2 \\
\hline
-x^2 + 7x \\
-x^2 + 4x \\
\hline
3x - 12 \\
3x - 12 \\
\hline
0
\end{array}
$$

Divide: $\frac{2x^3}{x} = 2x^2$.
Multiply: $2x^2(x - 4) = 2x^3 - 8x^2$.
Subtract: $2x^3 - 2x^3 = 0$; $-9x^2 - (-8x^2) = -x^2$.
Divide: $\frac{-x^2}{x}$. Multiply: $-x(x - 4)$.
Subtract.
Divide: $\frac{3x}{x}$. Multiply: $3(x - 4)$.
Subtract.
Remainder $= 0$

Yes, $x - 4$ is a factor of $2x^3 - 9x^2 + 7x - 12$, and $(x - 4)(2x^2 - x + 3)$ is the factored form of $2x^3 - 9x^2 + 7x - 12$.

Is $x - 1$ a factor of $x^3 + 1$?

$$
\begin{array}{r}
x^2 + x + 1 + \frac{2}{x - 1} \\
x - 1 \overline{) x^3 + 0x^2 + 0x + 1} \\
x^3 - x^2 \\
\hline
x^2 + 0x \\
x^2 - x \\
\hline
x + 1 \\
x - 1 \\
\hline
2
\end{array}
$$

Represent missing powers of x with terms having a coefficient of 0.
Subtract: $0x^2 - (-x^2) = x^2$.

Subtract: $0x - (-x) = x$.

Subtract: $1 - (-1) = 2$.

No, $x - 1$ is not a factor of $x^3 + 1$ since the remainder is not zero.

1 Multiply.

1. $(x + 7)(x + 3)$
2. $(x + 8)(x + 5)$
3. $(2x - 1)(x + 2)$
4. $(3x + 7)(x - 5)$
5. $(x + 5)(x^2 + 2x - 1)$
6. $(x + 7)(x^2 - 3x + 2)$
7. $(x - 7)(x^2 - 5x + 2)$
8. $(x - 3)(x^2 - 8x + 1)$
9. $(2x - 5)(x^2 - x + 1)$
10. $(3x - 2)(x^2 + x - 3)$
11. $(2x + 7)(3x^2 - 5x + 2)$
12. $(5x - 3)(4x^2 - 2x - 3)$
13. $(x - 2)(x^2 + 2x + 4)$
14. $(2x - 3)(4x^2 + 6x + 9)$
15. $(x + 3)(x^2 - 3x + 9)$
16. $(3x + 2)(9x^2 - 6x + 4)$
17. $(x + 5)(x^2 - 5x + 25)$
18. $(11x - 3)(121x^2 + 33x + 9)$

2 Use the FOIL method to find the products. Practice combining the outer and inner products mentally.

19. $(a + 3)(a + 8)$
20. $(x - 4)(x + 5)$
21. $(y - 2)(y - 9)$
22. $(y - 7)(y - 3)$
23. $(2a + 3)(a + 4)$
24. $(3a - 5)(a + 1)$
25. $(3a - 2b)(a - 2b)$
26. $(5x - y)(x - 5y)$
27. $(3x - 4)(2x - 3)$
28. $(a - b)(2a - 5b)$
29. $(7 - m)(3 - 7m)$
30. $(5 - 2x)(8 - x)$
31. $(x + 7)(x + 4)$
32. $(y - 7)(y - 5)$
33. $(m + 3)(m - 7)$
34. $(3b - 2)(x + 6)$
35. $(4r - 5)(3r + 2)$
36. $(5 - x)(7 - 3x)$
37. $(4 - 2m)(1 - 3m)$
38. $(2 + 3x)(3 + 2x)$
39. $(x + 3)(2x - 5)$
40. $(5x - 7y)(4x + 3y)$
41. $(2a + 3b)(7a - b)$
42. $(5a + 2b)(6a - 5b)$
43. $(9x - 2y)(3x + 4y)$
44. $(5x - 8y)(4x - 3y)$
45. $(7m - 2n)(3m + 5n)$
46. $\left(\sqrt{3} - 2\right)\left(\sqrt{3} - 1\right)$
47. $\left(2\sqrt{3} + 5\right)\left(\sqrt{5} - 5\right)$

3 Find the special products using patterns.

48. $(a + 3)(a - 3)$
49. $(2x + 3)(2x - 3)$
50. $(a - y)(a + y)$
51. $(4r + 5)(4r - 5)$
52. $(5x + 2)(5x - 2)$
53. $(7 + m)(7 - m)$
54. $(3y - 5)(3y + 5)$
55. $(8y + 3)(8y - 3)$
56. $(3a - 11b)(3a + 11b)$
57. $(5y - 3)(5y + 3)$
58. $(x - 7)(x + 7)$
59. $(x - 11)(x + 11)$
60. $(2 - 3x)^2$
61. $(3x + 4)^2$
62. $(Q + L)^2$
63. $(a^2 + 1)^2$
64. $(2d - 5)^2$
65. $(3a + 2x)^2$
66. $(3x - 7)^2$
67. $(6 + Q)^2$
68. $(y + 5x)^2$
69. $(4 - 3j)^2$
70. $(3m - 2p)^2$
71. $(m^2 + p^2)^2$
72. $(2a - 7c)^2$
73. $(9 - 13a)^2$
74. $\left(\sqrt{3} - 2\right)\left(\sqrt{3} + 2\right)$
75. $\left(\sqrt{7} - \sqrt{2}\right)\left(\sqrt{7} + \sqrt{2}\right)$
76. $\left(5\sqrt{2} - 7\right)\left(5\sqrt{2} + 7\right)$
77. $(8 - i)(8 + i)$
78. $(5 - i)(5 + i)$
79. $(i - 1)(i + 1)$
80. $(4 - i)(4 + i)$

Use the extension of the FOIL method or the special products patterns to find the products.

81. $(x + p)(x^2 - xp + p^2)$
82. $(Q + L)(Q^2 - QL + L^2)$
83. $(3 + a)(9 - 3a + a^2)$
84. $(2x + 4p)(4x^2 - 8xp + 16p^2)$
85. $(3m + 2)(9m^2 - 6m + 4)$
86. $(6 - p)(36 + 6p + p^2)$
87. $(5y - p)(25y^2 + 5yp + p^2)$
88. $(x - 2y)(x^2 + 2xy + 4y^2)$
89. $(Q - 6)(Q^2 + 6Q + 36)$
90. $(3t - 2)(9t^2 + 6t + 4)$

4 Perform the indicated division and determine if the binomial is a factor of the dividend.

91. $x - 5 \overline{)x^2 - x - 20}$
92. $x - 3 \overline{)x^2 + 3x - 18}$
93. $x + 1 \overline{)x^3 - 2x^2 - x + 2}$
94. $x + 2 \overline{)x^3 + 3x^2 - 3x - 10}$
95. $x + 3 \overline{)3x^3 + 7x^2 - 3x + 8}$
96. $x - 5 \overline{)2x^3 - 3x^2 - 33x + 8}$
97. $x + 1 \overline{)x^3 + 2x^2 - 5x - 6}$
98. $x + 3 \overline{)x^3 + 2x^2 - 5x - 6}$
99. $x - 3 \overline{)x^3 - 27}$
100. $x + 5 \overline{)x^3 + 125}$

Learning Outcomes

1. Recognize and factor the difference of two perfect squares.
2. Recognize and factor a perfect-square trinomial.
3. Recognize and factor the sum or difference of two perfect cubes.

To factor any of the special products of the preceding section, we apply the inverse of the process. That is, we start with the product and "work back" to the factors that produce these special products.

To rewrite a special product in factored form, we must recognize the product as a pattern. Once we identify the special product, then we must know the pattern of the product in factored form.

1 **Recognize and Factor the Difference of Two Perfect Squares.**

Before we can factor such a special product, we must be able to recognize an expression as a special product.

To identify a binomial as the difference of two perfect squares:

1. Verify that the expression is a binomial.
2. Verify that the absolute value of each term is a perfect square.
3. Verify that the second term is subtracted from the first term.

EXAMPLE Identify the special products that are the difference of two perfect squares.

(a) $x^2 - 9$ (b) $a^2 + 49$ (c) $m^2 - 27$ (d) $3y^2 - 25$
(e) $9x^2 - 4$ (f) $4x^2 - 4x + 1$

(a) Difference of two perfect squares.
(b) Not the difference of two perfect squares; this is a *sum*, not a difference.
(c) Not the difference of two perfect squares; 27 is not a perfect square.
(d) Not the difference of two perfect squares; in $3y^2$, 3 is not a perfect square.
(e) Difference of two perfect squares.
(f) Not the difference of two perfect squares; this is a trinomial, not a binomial.

To factor the difference of two perfect squares:

1. Take the square root of the first term.
2. Take the square root of the second term.
3. Write one factor as the *sum* of the square roots found in Steps 1 and 2, and write the other factor as the *difference* of the square roots from Steps 1 and 2.

Symbolically,

$$a^2 - b^2 = (a + b)(a - b)$$

EXAMPLE Factor the special products, which are the differences of two perfect squares.

(a) $a^2 - 9$ (b) $x^2 - 36$ (c) $4x^2 - 1$ (d) $-49 + 16m^2$

(a) $a^2 - 9 = (a + 3)(a - 3)$ (b) $x^2 - 36 = (x + 6)(x - 6)$
(c) $4x^2 - 1 = (2x + 1)(2x - 1)$ (d) $-49 + 16m^2 = 16m^2 - 49 = (4m + 7)(4m - 7)$

Order of Factors

Because multiplication is commutative, the factors given as answers in the preceding example may be expressed in any order, such as $(a + 3)(a - 3)$ or $(a - 3)(a + 3)$, $(x + 6)(x - 6)$ or $(x - 6)(x + 6)$, and so on.

2 **Recognize and Factor a Perfect-Square Trinomial.**

A trinomial is a *perfect-square trinomial* if the first and last terms are positive perfect squares and the absolute value of the middle term is *twice* the product of the square roots of the first and last terms. Again, we need to be able to distinguish these special products from other expressions before we factor them.

To identify a perfect-square trinomial:

1. Verify that the expression is a trinomial.
2. Verify that the first and last terms are positive and perfect squares.
3. Mentally take the square root of the first and last terms and multiply the results. Two times this product should equal the absolute value of the middle term of the original trinomial.

EXAMPLE Verify that the trinomials are perfect-square trinomials.

(a) $x^2 + 14x + 49$ (b) $4m^2 - 12m + 9$ (c) $9x^2 + 24xy + 16y^2$

(a) The first and last terms, x^2 and 49, are positive perfect squares. The middle term, $14x$, has an absolute value that is twice the product of the square roots of x^2 and 49. And, $2(7x) = 14x$.

(b) The first and last terms, $4m^2$ and 9, are positive perfect squares. The middle term, $-12m$, has an absolute value that is twice the product of the square roots of $4m^2$ and 9. And, $2(2m \cdot 3) = 12m$.

(c) The first and last terms, $9x^2$ and $16y^2$, are positive perfect squares. The middle term, $24xy$, has an absolute value that is twice the product of the square roots of $9x^2$ and $16y^2$. And, $2(3x \cdot 4y) = 24xy$.

EXAMPLE Tell why the trinomials are *not* perfect-square trinomials.

(a) $x^2 + 2x - 1$ (b) $4x^2 + 6x + 9$
(c) $x^2 - 5x + 4$ (d) $-4x^2 - 4x + 1$

(a) The last term, -1, is negative. This term must be positive in a perfect-square trinomial.
(b) The middle term, $6x$, is not *twice* the product of $2x$ and 3.
(c) The middle term, $-5x$, is not *twice* the product of x and 2.
(d) The first term, $-4x^2$, is negative. It should be positive.

1. Write the square root of the first term.
2. Write the sign of the middle term.
3. Write the square root of the last term.
4. Indicate the square of this binomial quantity.

EXAMPLE Factor the perfect-square trinomials.

(a) $x^2 + 14x + 49$ (b) $4m^2 - 12m + 9$ (c) $9x^2 + 24xy + 16y^2$

	Square root of *first* term	Sign of *middle* term	Square root of *last* term	*Square* the quantity
(a) $x^2 + 14x + 49$	x	$+$	7	$(x + 7)^2$
(b) $4m^2 - 12m + 9$	$2m$	$-$	3	$(2m - 3)^2$
(c) $9x^2 + 24xy + 16y^2$	$3x$	$+$	$4y$	$(3x + 4y)^2$

---

3 **Recognize and Factor the Sum or Difference of Two Perfect Cubes.**

Before we factor the sum or difference of two perfect cubes, we should be able to recognize an expression that matches the pattern.

1. Verify that the expression is a binomial.
2. Verify that each term is a perfect cube.

EXAMPLE Identify the expressions that are the sum or difference of two perfect cubes.

(a) $a^3 - 27$ (b) $8x^3 + 9$ (c) $8y^3 - 125$ (d) $64 - 8a^3$
(e) $a^3 + 16b^3$ (f) $c^3 + y^3 - 27$ (g) $27x^3 + 8$ (h) $3a^3 - 64$

(a) Difference of two perfect cubes.
(b) Not the sum or difference of two perfect cubes; 9 is not a perfect cube.
(c) Difference of two perfect cubes.
(d) Difference of two perfect cubes.
(e) Not the sum or difference of two perfect cubes; 16 is not a perfect cube.
(f) Not the sum or difference of two perfect cubes; this is a trinomial.
(g) Sum of two perfect cubes.
(h) Not the sum or difference of two perfect cubes; 3 is not a perfect cube.

To factor the sum of two perfect cubes:

1. Write the binomial factor as the *sum* of the cube roots of the two terms.
2. Write the trinomial factor as the square of the first term from Step 1, *minus* the product of the two terms from Step 1, plus the square of the second term from Step 1.

Symbolically,

$$a^3 + b^3 = (a + b)(a^2 - ab + b^2)$$

To factor the difference of two perfect cubes:

1. Write the binomial factor as the *difference* of the cube roots of the two terms.
2. Write the trinomial factor as the square of the first term from Step 1, *plus* the product of the two terms from Step 1, plus the square of the second term from Step 1.

Symbolically,

$$a^3 - b^3 = (a - b)(a^2 + ab + b^2)$$

EXAMPLE Factor the expressions that are the sum or difference of two perfect cubes.

(a) $27a^3 - 8$ (b) $m^3 + 125n^3$

(a) $27a^3 - 8 =$

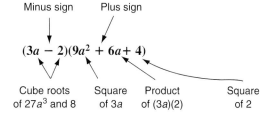

(b) $m^3 + 125n^3 =$

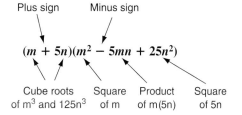

1 Identify the special products that are the difference of two perfect squares.

1. $r^2 - s^2$
2. $d^2 - 4d + 10$
3. $4y^2 - 16$
4. $36m^2 - 9n^2$
5. $9 + 4a^2$
6. $64p^2 - q^2$

Factor the following special products.

7. $y^2 - 49$
8. $16x^2 - 1$
9. $9a^2 - 100$
10. $4m^2 - 81n^2$
11. $9x^2 - 64y^2$
12. $25x^2 - 64$
13. $100 - 49x^2$
14. $4x^2 - 49y^2$
15. $121m^2 - 49n^2$
16. $81x^2 - 169$
17. $-9 + 4a^2$
18. $-16 + 25r^2$
19. $36x^2 - 49y^2$
20. $49 - 144x^2$
21. $16x^2 - 81y^2$

2 Verify which of the trinomials are perfect-square trinomials.

22. $4y^2 + 2y + 16$
23. $9m^2 - 24mn + 16n^2$
24. $16a^2 + 8a - 1$
25. $-9r^2 + 12r + 4$
26. $y^2 - 14y + 49$
27. $p^2 + 10p + 25$

Factor the special products.

28. $x^2 + 6x + 9$
29. $x^2 + 14x + 49$
30. $x^2 - 12x + 36$
31. $x^2 - 16x + 64$
32. $4a^2 + 4a + 1$
33. $25x^2 - 10x + 1$
34. $9m^2 - 48m + 64$
35. $4x^2 - 36x + 81$
36. $x^2 - 12xy + 36y^2$
37. $4a^2 - 20ab + 25b^2$
38. $y^2 - 10y + 25$
39. $9x^2 + 60xy + 100y^2$
40. $-x^2 - 12x - 36$
41. $-9x^2 + 6x - 1$
42. $-x^2 - 8x - 16$

3 Identify the special products that are the sum or difference of two perfect cubes.

43. $8b^3 - 125$
44. $y^2 - 14y + 49$
45. $T^9 + 27$
46. $c^3 + 16$
47. $125x^3 - 8y^3$
48. $64 + a^3$

Factor the special products, which are the sum or difference of two perfect cubes.

49. $m^3 - 8$
50. $y^3 - 125$
51. $Q^3 + 27$
52. $c^3 + 1$
53. $125d^3 - 8p^3$
54. $a^3 + 64$
55. $216a^3 - b^3$
56. $x^3 + Q^3$
57. $8p^3 - 125$
58. $27 - 8y^3$
59. $-a^3 - 8$
60. $-x^3 - 27$

13–4 | *Factoring General Trinomials*

Learning Outcomes

1 Factor general trinomials whose squared term has a coefficient of 1.
2 Remove common factors after grouping an expression.
3 Factor a general trinomial by grouping.
4 Factor any binomial or trinomial that is not prime.

Many trinomials do not have a common factor and do not match the pattern of a special product. Some will still factor as the product of two binomials. First, let's examine a trinomial whose squared term has a coefficient of 1.

1 **Factor General Trinomials Whose Squared Term Has a Coefficient of 1.**

Trinomials that are not perfect-square trinomials are *general trinomials*.

To factor a trinomial with a squared term that has a coefficient of 1:

1. Ensure the trinomial is arranged in descending powers of one variable.
2. Determine the signs of the second term of each binomial by examining the sign of the third term of the trinomial.

   $+ \Rightarrow$ like signs (both $+$ or both $-$, matching the middle sign)

   $- \Rightarrow$ unlike signs (one $+$ and one $-$, with the sign of the larger absolute value of the factor pair matching the sign of the middle term of the trinomial)

3. Write all factor pairs of the coefficient of the third term.
4. Select the factor pair that adds (algebraically) to the coefficient of the middle term of the trinomial. Include appropriate signs.
5. Write the binomial factors of the trinomial with the square root of the variable as the first term of each factor and the two factors from Step 3 as the second terms of the binomials.

EXAMPLE   Factor $x^2 + 6x + 5$.

$x^2 + 6x + 5$ Terms are already in descending order.

$(\ +\ )(\ +\ )$ Last term $+$ and middle term $+$ means signs are alike and both plus.

                          Only factor pair of 5 is 1(5). The algebraic sum is $+1 + (+5) = +6$.

$(x + 1)(x + 5)$ First terms are x. Second terms are $+1$ and $+5$.

Check by multiplying.

EXAMPLE   Factor $x^2 + 4x - 12$.

$x^2 + 4x - 12$ Terms are already in descending order. Last term $-$ means signs are unlike.

$(\ +\ )(\ -\ )$ Middle term $+$ means the factor with the larger absolute value in the selected factor pair will be $+$.

                          Factor pairs of 12:

                          $1 \cdot 12$

                          $\boxed{2 \cdot 6}$ Algebraic sum (difference when signs are unlike) of -2 and $+6$ is $+4$.

                          $3 \cdot 4$

$(x - 2)(x + 6)$ First terms are x. Second terms are -2 and 6.

EXAMPLE   Factor $x^2 - 20 - x$.

$x^2 - x - 20$ Arrange in descending order of x. Last sign $-$ means signs are unlike.

                          Middle sign $-$ means the factor with the larger absolute value in the selected factor pair will be $-$.

                          Factor pairs of 20:

                          $1 \cdot 20$

                          $2 \cdot 10$

                          $\boxed{4 \cdot 5}$ Algebraic sum (difference when signs are unlike) of $+4$ and -5 is -1.

$(x + 4)(x - 5)$ First terms are x. Second terms are $+4$ and -5.

2 **Remove Common Factors After Grouping an Expression.**

Algebraic expressions that have more than three terms and have no factors common to every term in the expression may have common factors for some groups of terms. In these cases, the expression may be written as groupings.

To remove common factors after grouping an expression:

1. Identify groupings.
2. Factor all common factors from each grouping.
3. Examine each term to see if there are common groupings. If so, factor out the common grouping.

EXAMPLE Write the expression $2x^2 - 2xb + ax - ay$ as an alternate expression by grouping pairs of terms. Factor common factors from each grouping.

$2x^2 - 2xb + ax - ay$	Group two terms in each grouping.
$(2x^2 - 2xb) + (ax - ay)$	Factor common factors from each grouping.
$2x(x - b) + a(x - y)$	There is no common grouping in the terms.

TIP!

Groupings Are Not Necessarily Factors.

In the preceding example the groupings $(2x^2 - 2xb)$ and $(ax - ay)$ are not factors. Notice they are groupings that are added. To rewrite the expression as $2x(x - b) + a(x - y)$ is not factored form. The expression has two terms and each term has two or more factors. An expression in factored form is only one term.

EXAMPLE Write the expressions in factored form by using grouping.

(a) $mx + 2m - 4x - 8$ (b) $y^2 + 2xy + 3y + 6x$
(c) $3x^2 - 9x - 7x + 21$ (d) $2x^2 + 8x + 5y - 15$

(a) $\quad mx + 2m - 4x - 8 =$	Group the four-termed expression into two terms.
$(mx + 2m) + (-4x - 8) =$	Factor any common factor in each of the two terms.
$m(x + 2) + -4(x + 2) =$	Convert double signs to an equivalent single sign.
$m(x + 2) - 4(x + 2) =$	Factor into one term by factoring out the common binomial factor $(x + 2)$.
$(x + 2)(m - 4)$ or $(m - 4)(x + 2)$	Check the result by using the FOIL method.
$(x + 2)(m - 4) = xm - 4x + 2m - 8$	Rearrange terms and factors (commutative properties of multiplication and addition).
$\quad\quad\quad\quad = mx + 2m - 4x - 8$	The factoring checks.
(b) $\quad y^2 + 2xy + 3y + 6x =$	Group into two terms.
$(y^2 + 2xy) + (3y + 6x) =$	Factor the common factor in each term.
$y(y + 2x) + 3(y + 2x) =$	Factor into one term by factoring the common binomial factor.
$(y + 2x)(y + 3)$ or $(y + 3)(y + 2x)$	

(c)

$$3x^2 - 9x - 7x + 21 =$$ Group into two terms.

$$(3x^2 - 9x) + (-7x + 21) =$$ Factor each term.

$$3x(x - 3) + -7(x - 3) =$$ Convert double signs to a single sign.

$$3x\,(x - 3) - 7\,(x - 3) =$$ Factor into one term by factoring the common binomial factor.

$$(x - 3)(3x - 7) \text{ or } (3x - 7)(x - 3)$$

(d)

$$2x^2 + 8x + 5y - 15 =$$ Group into two terms.

$$(2x^2 + 8x) + (5y - 15) =$$ Factor each term.

$$2x(x + 4) + 5(y - 3)$$

In this example, the two terms do not have a common factor. Thus, the expression cannot be factored into one term. Even if we rearrange the terms, we will not be able to write the expression as a single term.

$$2x^2 + 8x + 5y - 15 \text{ is prime.}$$

TIP!

Make Leading Coefficient of Binomial Positive.

When factoring by grouping, manipulate the signs of the common factor so that the leading coefficient of the binomial is positive.

$$-3x + 9$$

Factor as $-3(x - 3)$, *not* $3(-x + 3)$. Parts a and c of the preceding example illustrate this tip.

3   **Factor a General Trinomial by Grouping.**

We can now use a systematic method to factor general trinomials with leading integral coefficients that are positive and not equal to 1.

To factor a general trinomial of the form $ax^2 + bx + c$ by grouping:

1. Verify that a is a positive integer or rewrite with -1 as the common factor.
2. Multiply the coefficient of the first term of the trinomial by the coefficient of the last term.
3. Factor the product from Step 2 into a pair of factors:
   (a) whose *sum* is the coefficient of the middle term if the sign of the last term is positive, or
   (b) whose *difference* is the coefficient of the middle term if the sign of the last term is negative.
4. Rewrite the trinomial as a polynomial with four terms by replacing the middle term with two terms that have the coefficients identified in Step 3.
5. Group the polynomial with four terms from Step 4 into two groups with a common factor in each.
6. Factor the common factors from each of the two groups.
7. Factor the common binomial.

EXAMPLE   Factor $6x^2 + 19x + 10$ by grouping.

$$6(10) = 60$$ Multiply the coefficients of the first and third terms.

$$60 = (1)60$$ List all factor pairs of 60.

$$(2)30$$

$$(3)20$$

13–4 Factoring General Trinomials   557

$$\text{(4)}15$$
$$\text{(5)}12$$
$$\text{(6)}10$$

Identify the pair that *adds* to 19, since the sign of the last term is positive.

$$6x^2 + 19x + 10 =$$

Separate $+19x$ into two terms using the coefficients 4 and 15.

$$6x^2 + 4x + 15x + 10 =$$

Group into two terms.

$$(6x^2 + 4x) + (15x + 10) =$$

Factor common factors in each term.

$$2x(3x + 2) + 5(3x + 2) =$$

Factor common binomial.

$$(3x + 2)(2x + 5) \text{ or } (2x + 5)(3x + 2)$$

Check using the FOIL method.

EXAMPLE Factor $20x^2 - 23x + 6$ by grouping.

$$20(6) = 120$$ Find the product of 20 and 6.

List the factor pairs of 120 and select the pair that has a sum of 23.

$$120 = \text{(1)}120$$ List all factor pairs of 120.
$$\text{(2)}60$$
$$\text{(3)}40$$
$$\text{(4)}30$$
$$\text{(5)}24$$
$$\text{(6)}20$$
$$\text{(8)}15$$ Identify the pair that adds to 23, since the sign of the last term is positive.
$$\text{(10)}12$$

$$20x^2 - 23x + 6 =$$

Separate $-23x$ into two terms using the coefficients -8 and -15.

$$20x^2 - 8x - 15x + 6 =$$

Group into two terms. Be sure the second grouping keeps the negative sign with 15x.

$$(20x^2 - 8x) + (-15x + 6) =$$

Factor common factors in each term.

$$4x(5x - 2) - 3(5x - 2) =$$

Factor the common binomial.

$$(5x - 2)(4x - 3)$$

Check using the FOIL method.

If the third term of a trinomial has a negative sign, we use the same procedure but look for two factors whose *difference* is the coefficient of the middle term.

EXAMPLE Factor $10x^2 + 19x - 15$ by grouping.

$$10(15) = 150$$ Find the product of 10 and 15.

List the factor pairs of 150 and select the pair that has a difference of 19.

$$150 = \text{(1)}150$$ List all factor pairs of 150.
$$\text{(2)}75$$
$$\text{(3)}50$$
$$\text{(5)}30$$
$$\text{(6)}25$$ Identify the pair that subtracts to 19, since the last term is negative.
$$\text{(10)}15$$

The factors 25 and 6 have a difference of 19. When we rewrite the trinomial, we write $19x$ as $25x - 6x$.

$$10x^2 + 19x - 15 =$$

Separate $+19x$ into two terms using the coefficients 25 and -6. The signs will be different and the larger coefficient will be positive.

$$10x^2 + 25x - 6x - 15 =$$

Group into two terms.

$$(10x^2 + 25x) + (-6x - 15) =$$

Factor common factors in each term.

$$5x\,(2x + 5) - 3\,(2x + 5) =$$

Factor the common binomial.

$$(2x + 5)(5x - 3)$$

Check using the FOIL method.

TIP!

Shortening the Process for Factoring by Grouping

The following is a variation of the factor-by-grouping method. The variation is mathematically sound and employs a strategy using the property of 1 that is often overlooked.

Factor $6x^2 + 7x - 20$.

$6x^2 + 7x - 20$

Multiply the coefficients of the first and third terms: $6(-20) = -120$.

$120 =$	1	120
	2	60
	3	40
	4	30
	5	24
	6	20
	8	15
	10	12

A negative product means the factor pair has unlike signs. List all factor pairs of 120.

Identify the factor pair with a *difference* of $+7$.

The larger factor will be positive. $-8 + 15 = 7$

Variation in procedure begins here.

$$\frac{(6x\quad)(6x\quad)}{6}$$

Use the coefficient of the first term as the coefficient of the first term in *each* binomial.

This gives us an extra factor of 6, and we compensate by *dividing the expression by 6*.

$$\frac{(6x - 8)(6x + 15)}{6} =$$

Use the factors of -120 that have an algebraic sum of 7 as the second term of each binomial. Signs will be unlike.

$$\frac{2(3x - 4)(3)(2x + 5)}{6} =$$

Factor the common factors from each binomial and simplify the numerical factors.

$$\frac{\cancel{6}(3x - 4)(2x + 5)}{\cancel{6}} =$$

The factors of 6 in the numerator and denominator reduce.

$$(3x - 4)(2x + 5)$$

Check using the FOIL method.

Now, let's shorten the process again.

Once we have selected the pair of factors whose sum or difference matches the middle term, *make two fractions using the selected factor pair as the denominators. The leading coefficient of the trinomial will be the numerator of each fraction.*

$$6\,x^2 + 7x - 20 \qquad 6(-20) = -120$$

The factor pair of -120 that has a sum of $+7$ is -8 and $+15$. Make fractions and reduce each fraction.

$$\frac{6}{-8} = \frac{3}{-4} \qquad \frac{6}{+15} = \frac{2}{+5}$$ The leading coefficient of the trinomial is the numerator of both fractions.

Each reduced fraction gives the coefficients of one of the binomial factors. 3, -4 and 2, $+5$.

$$(3x - 4)(2x + 5)$$

Let's try this process with another example.

If the trinomial has a common factor, we must factor the common factor before finding the factor pair and writing the fractions. The common factor will be part of the final answer.

Factor $30x^2 + 8 - 32x$.

$30x^2 - 32x + 8$	Arrange in descending powers of x.
$\boxed{2}(15x^2 - 16x + 4)$	Factor any common factors.
$15(4) = 60$	Write the factor pairs of 60.

1	60
2	30
3	20
4	15
5	12
6	10

Identify the pair that has a *sum* of -16. Signs will be alike. **$-6 + (-10) = -16$**

Make fractions with the factor pair -6 and -10 as the denominators and the leading coefficient of the trinomial, 15, as the numerators.

$$\frac{15}{-6} = \frac{5}{-2} \qquad \frac{15}{-10} = \frac{3}{-2}$$ Reduce each fraction.

Write the factors using the coefficients 5, -2 and 3, -2. Don't forget the common factor.

$$\mathbf{2(5x - 2)(3x - 2)}$$

Learning Strategy *Practice Moves You from Systematic Processes to Intuitive Processes.*

Systematic processes help you understand the concepts and build confidence. They also help you develop your mathematical senses like your number sense, your spatial sense, and so on. The more you practice, the more you develop your intuitive senses.

Many students instinctively and automatically move to shortened and mental processes. Many times you can test a few combinations of coefficients and find the correct factors without going through the entire systematic process. With practice, you can write expressions in factored form more readily.

4   **Factor Any Binomial or Trinomial That Is Not Prime.**

Now let's develop a strategy for factoring any binomial or trinomial that can be factored. This strategy allows us to say with confidence that a particular binomial or trinomial will not factor or is prime. When the word *prime* refers to algebraic expressions, it describes algebraic expressions that have only themselves and 1 as factors.

To factor any binomial or trinomial:

Perform the following steps in order.

1. Factor out the greatest common factor (if any).
2. Check the binomial or trinomial to see if it is a special product.
   (a) If it is the difference of two perfect squares, use the pattern.
   (b) If it is a perfect-square trinomial, use the pattern.
   (c) If it is the sum or difference of two perfect cubes, use the appropriate pattern.
3. If there is no special product, factor by grouping or by any appropriate process.
4. Examine each factor to see if it can be factored further.
5. Check factoring by multiplying.

The steps for factoring any binomial or trinomial are presented visually in the flowchart (Fig. 13–1).

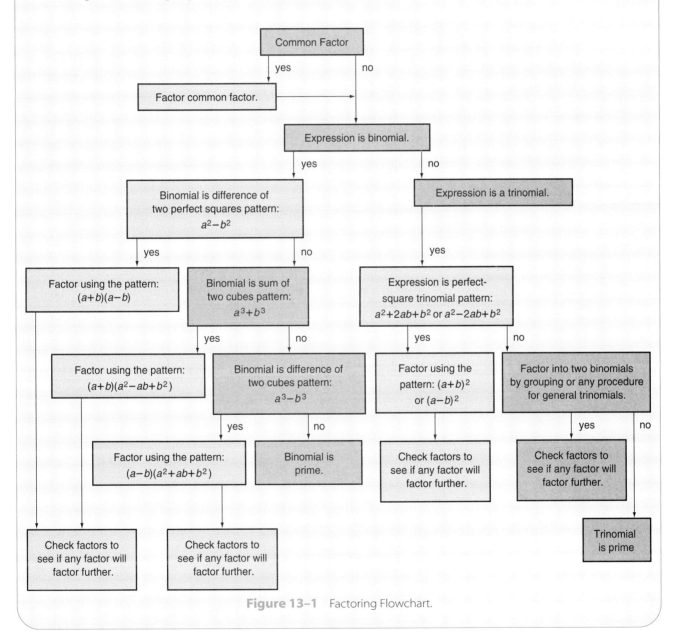

Figure 13–1 Factoring Flowchart.

EXAMPLE   Completely factor $4x^3 - 2x^2 - 6x$.

$2x(2x^2 - x - 3)$ $2x$ is the common factor. The trinomial is not a special product.

$2x(2x - 3)(x + 1)$ Factor the trinomial using any appropriate process but *keep* the $2x$ factor. Check by multiplying.

EXAMPLE   Completely factor $12x^2 - 27$.

$3(4x^2 - 9)$ 3 is the common factor. The binomial is the difference of two perfect squares.

$3(2x - 3)(2x + 3)$ Factor the difference of two perfect squares. Keep the common factor of 3. Check.

EXAMPLE   Completely factor $-18x^3 + 24x^2 - 8x$.

$-2x(9x^2 - 12x + 4)$ Look for a common factor: $-2x$. We factor a negative when the leading coefficient is negative. It is preferred that the leading coefficient in the binomial or trinomial be positive. The resulting trinomial is a perfect-square trinomial.

$-2x(3x - 2)^2$ Factor the perfect-square trinomial.

TIP!

Why Do We Factor?

There is discussion among mathematics professionals on the importance of factoring polynomials. Examine the arguments for and against using factoring techniques.

For factoring:

- Many problem-solving techniques that use factoring can be done more quickly than with the formula-based or systematic technique.
- Properties and patterns are generally easier to recognize in factored form.
- Computers (and even humans) can often evaluate factored expressions more quickly.

Evaluate the expression $8x^3 - 12x^2y - 8xy^2$ when $x = 4$ and $y = 3$.

Factoring Technique:	**Formula-based Technique:**
$4x(2x + y)(x - 2y)$	$8x^3 - 12x^2y - 8xy^2$
$4 \cdot 4(2 \cdot 4 + 3)(4 - 2 \cdot 3)$	$8 \cdot 4^3 - 12 \cdot 4^2 \cdot 3 - 8 \cdot 4 \cdot 3^2$
$16(11)(-2)$	$8 \cdot 64 - 12 \cdot 16 \cdot 3 - 8 \cdot 4 \cdot 9$
-352	$512 - 576 - 288$
	-352

Against factoring:

- Real-world applications rarely have numbers and expressions that factor.
- Factoring techniques only apply to special cases. Formula-based techniques apply in general.
- With the accessibility of calculators and inexpensive computer software, formula-based techniques are more practical than they were in the past.

1 Factor.

1. $x^2 + 7x + 6$
2. $x^2 - 7x + 6$
3. $x^2 - 5x + 6$
4. $x^2 + 5x + 6$
5. $x^2 - 11x + 28$
6. $x^2 + 8x + 12$
7. $x^2 - 8x + 12$
8. $x^2 + 13x + 12$
9. $x^2 - 13x + 12$
10. $x^2 + 7x + 12$
11. $x^2 - 7x + 12$
12. $x^2 - 4x + 3$
13. $x^2 + 8x + 7$
14. $x^2 + 7x + 10$
15. $x^2 - x - 6$
16. $x^2 + x - 6$
17. $x^2 - 5x - 6$
18. $x^2 + 5x - 6$
19. $x^2 - x - 12$
20. $x^2 + x - 12$
21. $x^2 + 4x - 12$
22. $x^2 - 4x - 12$
23. $x^2 - 11x - 12$
24. $x^2 + 11x - 12$
25. $y^2 - 3y - 10$
26. $y^2 - y - 6$
27. $b^2 + 2b - 3$
28. $-14 - 5b + b^2$
29. $12 - 7x + x^2$
30. $-30 - x + x^2$
31. $11x + 18 + x^2$
32. $-9x + x^2 + 18$
33. $-7x - 18 + x^2$
34. $-18 + x^2 + 17x$
35. $x^2 + 9x + 20$
36. $x^2 - 12x + 20$
37. $x^2 - 10x + 16$
38. $x^2 - 17x + 16$
39. $x^2 - 13x - 14$
40. $x^2 - 5x - 14$

2 Factor the polynomials by removing the common factors after grouping.

41. $x^2 + xy + 4x + 4y$
42. $6x^2 + 4x - 3xy - 2y$
43. $3mx + 5m - 6nx - 10n$
44. $30xy - 35y - 36x + 42$
45. $x^2 - 2x + 8x - 16$
46. $6x^2 - 2x - 21x + 7$
47. $x^2 - 4x + x - 4$
48. $8x^2 - 4x + 6x - 3$
49. $x^2 - 5x + 4x - 20$
50. $3x^2 - 6x + 5x - 10$
51. $4x^2 + 8x - 3x - 6$
52. $4x^2 - 8x + 3x - 6$
53. $4x^2 + 8x + 3x + 6$
54. $4x^2 - 8x - 3x + 6$
55. $8x^2 + 4x - 6x - 3$

3 Factor the trinomials by grouping.

56. $3x^2 + 7x + 2$
57. $3x^2 + 14x + 8$
58. $6x^2 + 13x + 6$
59. $8x^2 + 2x - 3$
60. $6x^2 - 17x + 12$
61. $2x^2 - 9x + 10$
62. $6x^2 - 13x + 5$
63. $8x^2 + 10x + 3$
64. $6x^2 - 11x + 5$
65. $8x^2 + 26x + 15$
66. $15x^2 - 22x - 5$
67. $8x^2 - 10x + 3$
68. $2x^2 - 5x - 7$
69. $12x^2 + 8x - 15$
70. $10x^2 + x - 3$
71. $12x^2 + 5x - 2$
72. $12x^2 - 5x - 2$
73. $12x^2 + 11x + 2$
74. $12x^2 - 11x + 2$
75. $24x^2 + 5x - 1$
76. $24x^2 - 11x + 1$
77. $6x^2 + 7xy - 10y^2$
78. $6a^2 - 17ab - 14b^2$
79. $18x^2 - 3x - 10$
80. $20x^2 - xy - 12y^2$
81. $7x^2 + x - 8$
82. $2a^2 + 21a + 19$

4 Factor the polynomials completely. Identify common factors first, then identify special cases.

83. $4x - 4$
84. $x^2 + x - 6$
85. $2x^2 + x - 3$
86. $x^2 - 9$
87. $4x^2 - 16$
88. $m^2 + 2m - 15$
89. $2a^2 + 6a + 4$
90. $b^2 + 6b + 9$
91. $16m^2 - 8m + 1$
92. $x^2 + 8x + 7$
93. $2m^2 + 5m + 2$
94. $2m^2 - 5m - 3$
95. $2a^2 - 3a - 5$
96. $3x^2 + 10x - 8$
97. $6x^2 + x - 15$
98. $8x^2 + 10x - 3$
99. $-2x^2 + 6x - 4$
100. $x^4 - 16$
101. $x^6 - 81$
102. $x^6 - 27$
103. $3x^4 - 48$
104. $6x^3 - 48$
105. $3x^2 - 6x - 24$

Section 13–1

Factor by removing the greatest common factor.

1. $5x + 5y$
2. $2x + 5x^2$
3. $12m^2 - 8n^2$
4. $25x^2y - 10xy^3 + 5xy$
5. $2a^3 - 14a^2 - 2a$
6. $30a^3 - 18a^2 - 12a$
7. $15x^3 - 5x^2 - 20x$
8. $9a^2b^3 - 6a^3b^2$
9. $18a^3 + 12a^2$
10. $a^2b + ab^2$
11. $-6x^2 - 10x$
12. $-12n^2 - 18$

Section 13–2

Use the FOIL method to find the products. Combine the outer and inner products mentally.

13. $(x + 7)(x + 4)$
14. $(y - 7)(y - 5)$
15. $(m + 3)(m - 7)$
16. $(3b - 2)(x + 6)$
17. $(4r - 5)(3r + 2)$
18. $(5 - x)(7 - 3x)$

19. $(4 - 2m)(1 - 3m)$

20. $(2 + 3x)(3 + 2x)$

21. $(x + 3)(2x - 5)$

22. $(5x - 7y)(4x + 3y)$

23. $(2a + 3b)(7a - b)$

24. $(5a + 2b)(6a - 5b)$

25. $(9x - 2y)(3x + 4y)$

26. $(5x - 8y)(4x - 3y)$

27. $(7m - 2n)(3m + 5n)$

Find the special products mentally.

28. $(y - 4)(y + 4)$

29. $(6x - 5)(6x + 5)$

30. $(3m + 4)(3m - 4)$

31. $(7y + 11)(7y - 11)$

32. $(5x - 2y)(5x + 2y)$

33. $(8a - 5b)(8a + 5b)$

34. $(12r - 7s)(12r + 7s)$

35. $(\sqrt{8} + 2)(\sqrt{8} - 2)$

36. $(\sqrt{5} + \sqrt{2})(\sqrt{5} - \sqrt{2})$

37. $(3 + i)(3 - i)$

38. $(7 - i)(7 + i)$

39. $(x + 9)^2$

40. $(x - 7)^2$

41. $(x - 3)^2$

42. $(2x - 3)^2$

43. $(4x - 15)^2$

44. $(5 + 3m)^2$

45. $(8 + 7m)^2$

46. $(5x - 13)^2$

47. $(4x - 11)^2$

48. $(K + L)(K^2 - KL + L^2)$

49. $(g - h)(g^2 + gh + h^2)$

50. $(4 + a)(16 - 4a + a^2)$

51. $(2H - 3T)(4H^2 + 6HT + 9T^2)$

52. $(3a - 5)(9a^2 + 15a + 25)$

53. $(6 + i)(36 - 6i + i^2)$

54. $(9y - p)(81y^2 + 9yp + p^2)$

55. $(z + 2t)(z^2 - 2zt + 4t^2)$

56. $(g - 2)(g^2 + 2g + 4)$

57. $(7T + 2)(49T^2 - 14T + 4)$

Find the quotients.

58. $x + 4 \overline{)\, x^2 - 3x - 28}$

59. $x - 9 \overline{)\, x^2 - 11x + 18}$

60. $2x - 3 \overline{)\, 6x^2 - 13x + 21}$

61. $5x + 2 \overline{)\, 15x^2 - 4x - 4}$

62. $x + 2 \overline{)\, x^3 - x^2 - 4x + 4}$

63. $x - 5 \overline{)\, x^3 - 4x^2 - 10x + 25}$

64. $2x - 1 \overline{)\, 6x^3 - 7x^2 + 4x - 3}$

65. $2x^2 - x - 1 \overline{)\, 6x^3 - x^2 - 4x - 1}$

Section 13–3

Identify the special products that are the difference of two perfect squares.

66. $25m^2 - 4n^2$

67. $64 + 4a^2$

68. $4f^2 - 9g^2$

69. $H^2 - G^2$

70. $s^2 - 4s + 10$

71. $64b^2 - 49$

Verify which of the following trinomials are perfect-square trinomials

72. $16c^2 + 8c - 1$

73. $-9x^2 + 12x + 4$

74. $4d^2 + 2d + 16$

75. $9t^2 - 24tp + 16p^2$

76. $a^2 - 14a + 49$

77. $j^2 + 10j + 25$

Identify the special products that are the sum or difference of two perfect cubes.

78. $R^3 + 81$

79. $125a^3 - 8b^3$

80. $64 + m^3$

81. $8z^3 - 125$

82. $d^3 - 12d + 36$

83. $64W^3 + 27$

Factor the expressions that are special products. Explain why the other expressions are not special products.

84. $x^2 - 81$

85. $25y^2 - 4$

86. $100a^2 - 8ab^2$

87. $a^2b^2 + 49$

88. $121 - 9m^2$

89. $a^2 + 2a + 1$

90. $4x^2 + 12x + 9$

91. $16c^2 - 24bc + 9b^2$

92. $y^2 - 12y - 4$

93. $n^2 + 169 - 26n$

94. $16d^2 - 20d + 25$

95. $36a^2 + 84ab + 49b^2$

96. $4x^2 - 25y^2$

97. $49 - 14x + x^2$

98. $9x^2y^2 - 49z^2$

99. $64 + 25x^2$

100. $4x^2 + 12xy + 9y^2$

101. $16x^2 + 24x + 9y^2$

102. $36 - x^2$

103. $49 - 81y^2$

104. $64x^2 - 25y^2$

105. $9x^2 - 100y^2$

106. $a^2 - 10a + 25$

107. $9x^2 - 6xy + y^2$

108. $25 - 16a^2b^2$

109. $9x^2y^2 - z^2$

110. $x^2 - y^2$

111. $x^2 + 4x + 4$

112. $\frac{1}{4}x^2 - \frac{1}{9}y^2$

113. $\frac{4}{25}x^2 - \frac{1}{16}y^2$

114. $27v^3 - 8$

115. $T^3 - 8$

116. $8r^3 + 27$

117. $d^3 + 729$

118. $125c^3 - 216d^3$

119. $27K^3 + 64$

Section 13–4

Factor the trinomials.

120. $x^2 + 10x + 21$

121. $x^2 + 11x + 24$

122. $x^2 + 29x + 28$

123. $x^2 + 13x + 30$

124. $x^2 - 13x + 40$

125. $x^2 - 9x + 8$

126. $x^2 - 17x - 18$

127. $x^2 - 11x - 26$

128. $x^2 + x - 30$

129. $x^2 + 5x - 24$

130. $6x^2 + 25x + 14$

131. $6x^2 + 25x + 4$

132. $4x^2 - 23x + 15$
133. $5x^2 - 34x + 24$
134. $3x^2 - x - 14$
135. $6x^2 - x - 35$
136. $3x^2 + 11x - 4$
137. $7x^2 - 13x - 24$

Factor the polynomials. Look for common factors and special cases.

138. $5mn - 25m$
139. $9a^2 - 100$
140. $a^2 - b^2$
141. $2x^2 - 3x - 2$
142. $b^3 - 8b^2 - b$
143. $a^2 - 81$
144. $x^2 - 14x + 13$
145. $y^2 - 14y + 49$
146. $m^2 - 3m + 2$
147. $b^2 + 8b + 15$
148. $x^2 - 13x + 30$
149. $169 - m^2$
150. $5x^2 + 13x + 6$
151. $x^2 - 4x - 32$
152. $x^2 + 9x + 14$
153. $x^2 + 19x - 20$
154. $x^2 + 8x - 20$
155. $2x^2 - 4x - 16$
156. $5x^2 - 20$
157. $2x^3 - 10x^2 - 12x$

CHALLENGE PROBLEMS

158. Write the pattern for factoring the difference of two cubes and illustrate it with an example.

159. What is the value of being able to recognize the special products when we factor algebraic expressions?

A standard-sized rectangular swimming pool is 25 ft long and 15 ft wide. This gives a water-surface area of 375 ft². A customer wants to examine some options for varying the size of the pool. If x is the amount of adjustment to the length and y is the amount of adjustment to the width, find the following.

160. Write an expression in both factored and expanded form for finding the water-surface area of a pool that is adjusted x feet in length and y feet in width.

161. Write an expression using one variable in both factored and expanded form for the water-surface area of a pool when the length and width are increased the same amount.

162. Will the expressions in Exercises 160 and 161 work for both increasing and decreasing the size of the pool?

163. Illustrate your answer in Exercise 162 with numerical examples.

CONCEPTS ANALYSIS

1. Explain how the FOIL process for multiplying two binomials is an application of the distributive property.

2. (a) How is the distributive property used to multiply a binomial and a trinomial?
(b) How is the distributive property used to multiply two trinomials?

3. Give an example of the product of a binomial and a trinomial to illustrate your answer to Question 2a.

4. Give an example of the product of two trinomials to illustrate your answer to Question 2b.

5. List the properties of a binomial that is the difference of two perfect squares.

6. List the properties of a perfect-square trinomial.

7. Explain in sentence form how the sign of the third term of a general trinomial affects the signs between the terms of its binomial factors.

8. What do we mean if we say that a polynomial is prime?

9. Write a brief comment explaining each lettered step of the following example.

10. Find a mistake in each of the following. Briefly explain the mistake, then work the problem correctly.
(a) $(3x - 2y)^2 = 9x^2 + 4y^2$
(b) $5xy^2 - 45x = 5x(y^2 - 9)$
(c) $2x^2 - 28x + 96 = (2x - 12)(x - 8)$
(d) $x^2 - 5x - 6 = (x - 2)(x - 3)$

Factor $10x^2 - 19x - 12$ using grouping.

(a)
$$\frac{120}{}$$
1, 120
2, 60
3, 40
4, 30
5, 24*   Factors wtih difference of -19.
6, 20
8, 15
10, 12

(b) $\dfrac{10x^2 - 19x - 12}{10x^2 + 5x - 24x - 12}$

(c) $5x(2x + 1) - 12(2x + 1)$

(d) $(2x + 1)(5x - 12)$

Learning Objectives   What to Remember with Examples

Section 13–1

1 Factor an expression containing a common factor (pp. 538–540).

To factor an expression containing a common factor:
1. Find the greatest factor common to each term of the expression.   **2.** Divide each term by the common factor.   **3.** Rewrite the expression as the indicated product of the greatest common factor and the remaining quantity.

Factor $3ab + 9b^2$.

$$3b\left(\frac{3ab}{3b} + \frac{9b^2}{3b}\right) = 3b(a + 3b)$$

Section 13–2

1 Multiply polynomials (pp. 541–542).

To multiply two polynomials distribute by multiplying each term of the first factor times each term of the second factor. Combine like terms.

Multiply $(x^2 + 3x + 1)(x^2 - 3x + 2)$.

$$x^2(x^2 - 3x + 2) + 3x(x^2 - 3x + 2) + 1(x^2 - 3x + 2) =$$
$$x^4 - 3x^3 + 2x^2 + 3x^3 - 9x^2 + 6x + x^2 - 3x + 2 =$$
$$x^4 - 6x^2 + 3x + 2$$

n Use the FOIL method to multiply two binomials (pp. 542–544).

To multiply two binomials by the FOIL method: Multiply First terms, Outer terms, Inner terms, and Last terms. Combine like terms.

Multiply $(3a - 2)(a + 3)$.

$$\begin{array}{cccc} \text{F} & \text{O} & \text{I} & \text{L} \end{array}$$
$$3a^2 + 9a - 2a - 6 \qquad \text{Combine like terms.}$$
$$3a^2 + 7a - 6$$

3 Multiply polynomials that result in special products (pp. 544–548).

To multiply the sum and difference of the same two terms:
1. Square the first term.   **2.** Insert minus sign.   **3.** Square the second term.

Multiply $(3b - 2)(3b + 2)$.

$$9b^2 - 4 \qquad (3b)^2 = 9b^2; (2)^2 = 4$$

To square a binomial:
1. Square the first term.   **2.** Double the product of the two terms.   **3.** Square the second term.

Multiply $(2a - 3)^2$.

$$4a^2 - 12a + 9 \qquad (2a)^2 = 4a^2; 2(2a)(-3) = -12a; 3^2 = 9$$

To multiply a binomial and a trinomial of the types $(a + b)(a^2 - ab + b^2)$ and $(a - b)(a^2 + ab + b^2)$:
1. Cube the first term of the binomial.   **2.** If the binomial is a sum, insert a plus sign; if the binomial is a difference, insert a minus sign.   **3.** Cube the second term of the binomial.

Multiply $(8b - 2)(64b^2 + 16b + 4)$.

$512b^3 - 8$ $(8b)^3 = 512b^3; 2^3 = 8$

Multiply $(2x + 5)(4x^2 - 10x + 25)$.

$8x^3 + 125$ $(2x)^3 = 8x^3; 5^3 = 125$

4 Divide polynomials using long division (p. 548).

To divide a polynomial by a polynomial using long division:

1. Divide the first term of the dividend by the first term of the divisor. The partial quotient is placed above the first term of the dividend. **2.** Multiply the partial quotient times the divisor and align the product under like terms of the dividend. **3.** Subtract (change subtrahend to opposite and use addition rules). **4.** Bring down the next term of the dividend and repeat Steps 1–3. **5.** Repeat Steps 1–4 until all terms of the dividend have been brought down. The result of the last subtraction is the remainder. **6.** Write the remainder as a fraction with the remainder as the numerator and the divisor as the denominator.

Divide $(3x^2 + 13x - 10) \div (x + 5)$

$$
\begin{array}{r}
3x - 2 \\
x + 5 \overline{\smash{\big)}\ 3x^2 + 13x - 10} \\
\underline{3x^2 + 15x} \\
-2x - 10 \\
\underline{-2x - 10} \\
0
\end{array}
$$

Section 13–3

1 Recognize and factor the difference of two perfect squares (pp. 550–551).

The difference of two perfect squares is a binomial in the form $a^2 - b^2$.

Identify which expression is the special product, the difference of two squares.

(a) $36 - 27A^2$; no, 27 is not a perfect square.
(b) $9c^2 - 4$; yes, both terms are perfect squares. The terms are subtracted.

To factor the difference of two perfect squares:
1. Take the square root of the first term. **2.** Take the square root of the second term. **3.** Write one factor as the sum and one factor as the difference of the square roots from Steps 1 and 2.
Symbolically, $a^2 - b^2 = (a + b)(a - b)$.

Factor $9c^2 - 4$.

$$(3c + 2)(3c - 2)$$

2 Recognize and factor a perfect-square trinomial (pp. 551–552).

A perfect-square trinomial is a trinomial in the form $a^2 + 2ab + b^2$. The first and last terms are positive perfect squares and the absolute value of the middle term is twice the product of the square roots of the first and last terms.

Identify which expression is a perfect-square trinomial.

(a) $4x^2 + 18x + 9$; not a perfect-square trinomial. The middle term is not twice the product of the square roots of the first and last terms. $(2)(2x)(3) = 12x$
(b) $9a^2 + 6a + 4$; the first and last terms are perfect squares. The middle term is twice the product of the square roots of the first and last terms, so it is a perfect-square trinomial.

To factor a perfect-square trinomial:
1. Write the square root of the first term. **2.** Write the sign of the middle term. **3.** Write the square root of the last term. **4.** Indicate the square of the quantity.
Symbolically, $a^2 + 2ab + b^2 = (a + b)^2$ and $a^2 - 2ab + b^2 = (a - b)^2$.

Factor $9a^2 + 6a + 4$.

$$(3a + 2)^2$$

3 Recognize and factor the sum or difference of two perfect cubes (pp. 552–553).

The sum of two perfect cubes is a binomial in the form $a^3 + b^3$. The difference of two perfect cubes is a binomial in the form $a^3 - b^3$.

Identify the sum or difference of two perfect cubes:

(a) $b^3 - 27$; yes, both terms are perfect cubes; difference of two cubes.
(b) $x^3 - 6$; no, 6 is not a perfect cube.
(c) $8 + y^3$; yes, both terms are perfect cubes; sum of two cubes.

To factor the sum (difference) of two perfect cubes:
1. Write the binomial factor as the sum (difference) of the cube roots of the two terms.
2. Write the trinomial factor as the square of the first term from Step 1, insert a minus sign if factoring a sum (a plus sign if factoring a difference). Write the product of the two terms from Step 1, plus the square of the second term from Step 1.

Symbolically, $a^3 + b^3 = (a + b)(a^2 - ab + b^2)$ and $a^3 - b^3 = (a - b)(a^2 + ab + b^2)$.

Factor $8 + y^3$.

$$(2 + y)(4 - 2y + y^2)$$

Factor $b^3 - 27$.

$$(b - 3)(b^2 + 3b + 9)$$

Section 13–4

1 Factor general trinomials whose squared term has a coefficient of 1 (pp. 554–555).

To factor a general trinomial whose squared term has a coefficient of 1: Factor using the FOIL method in reverse. Use the factors of the third term that will give the desired middle term and sign.

Factor $a^2 - 5a + 6$.

( )( )	Factors of 6 that add to 5 are 3 and 2.
$(a \quad 3)(a \quad 2)$	Terms with like signs that are negative add to -5.
$(a - 3)(a - 2)$	These factors of 6, -3 and -2, give $-5a$ as the middle term.

$$(a - 3)(a - 2)$$

2 Remove common factors after grouping an expression (pp. 556–557).

Arrange terms of expression into groups of two terms. Factor common factors from each group. Factor the common binomial factor if there is one.

Factor $2x^2 - 4x + xy - 2y$ completely.

$(2x^2 - 4x) + (xy - 2y)$	Arrange into groups of two terms. Factor common factors.
$2x(x - 2) + y(x - 2)$	Factor common binomial.
$(x - 2)(2x + y)$	

3 Factor a general trinomial by grouping (pp. 557–560).

To factor a general trinomial by grouping:
1. Verify that the first term is positive or rewrite with -1 as the common factor.
2. Multiply the coefficients of the first term and last term.
3. Factor the product from Step 2 into two factors a) whose sum is the coefficient of the middle term if the last term is positive, or b) whose difference is the coefficient of the middle term if the last term is negative.
4. Rewrite the trinomial so the middle term is the sum or difference from Step 3.
5. Group the polynomial from Step 4 in two groups of two terms.
6. Factor the common factor.
7. Factor the common binomial factor.

Factor $6x^2 - 17x + 12$.

$6 \cdot 12 = 72$ Factor pairs of 72.
$1 \cdot 72$ Find the factor pair that has a sum of -17.
$2 \cdot 36$
$3 \cdot 24$
$4 \cdot 18$
$6 \cdot 12$
$8 \cdot 9$ Selected pair: $-8 + (-9) = -17$.

$6x^2 - 8x - 9x + 12$ Rewrite the middle term. $-17x = -8x + (-9x)$.
$(6x^2 - 8x) + (-9x + 12)$ Factor common factor from each binomial.
$2x(3x - 4) - 3(3x - 4)$ Factor the common binomial.
$(3x - 4)(2x - 3)$

4 Factor any binomial or trinomial that is not prime (pp. 560–562).

To factor any binomial or trinomial:
1. Factor the greatest common factor (if any).
2. Check for any special products.
3. If there is no special product, factor by grouping.
4. Examine each factor to see if it can be factored further.
5. Check factoring by multiplying.

Factor $12x^3 + 6x^2 - 18x$ completely.

$$6x(2x^2 + x - 3) =$$
$$6x[(2x^2 + 3x) + (-2x - 3)] =$$
$$6x[x(2x + 3) - (2x + 3)] =$$
$$6x(2x + 3)(x - 1)$$

(After removing the common factor, procedures for factoring general trinomials can be used to factor the remaining trinomial.)

CHAPTER TRIAL TEST

Find the products.
1. $(m - 7)(m + 7)$
2. $(3x - 2)(3x + 2)$
3. $(a + 3)^2$
4. $(2x - 7)^2$
5. $(x - 3)(2x - 5)$
6. $(7x - 3)(2x + 1)$
7. $(x - 2)(x^2 + 2x + 4)$
8. $(3y + 4)(9y^2 - 12y + 16)$
9. $(5a - 3)(25a^2 + 15a + 9)$
10. $(a + 6)(a^2 - 6a + 36)$

Divide using long division.
11. $x - 3 \overline{) 2x^2 + x - 21}$
12. $x + 3 \overline{) 2x^3 - 3x^2 + 7x - 6}$

Factor by removing the greatest common factor.
13. $7x^2 + 8x$
14. $6ax + 15bx$
15. $7a^2b - 14ab$

Factor into special products.
16. $a^2 - 25$
17. $9x^2 - 25$
18. $x^3 - 216$
19. $8y^3 + 27x^3$
20. $x^2 + 4x + 4$
21. $x^2 - 18x + 81$
22. $x^2 + 6x + 9$
23. $x^2 - 12x + 36$

Factor into two binomials.

24. $x^2 + 5x - 24$
25. $6x^2 - 5x - 6$
26. $y^2 - y - 6$
27. $a^2 + 16ab + 64b^2$
28. $x^2 - 7x + 10$
29. $b^2 - 3b - 10$
30. $a^2 - 4a + 3$
31. $3x^2 + 23x + 30$
32. $3x^2 + x - 4$
33. $3m^2 - 5m + 2$
34. $4c^2 - 28c + 49$
35. $3x^2 - 11x + 6$
36. $3x^2 + 11xy + 3y^2$

Factor completely.

37. $3x^3 - 24$
38. $3x^2 - 3x - 18$
39. $5x^2 - 20$
40. $27a^3 - 8$
41. $3x^2 + 12x + 12$
42. $3x^2 - 15x - 18$

Rational Expressions and Equations

14

Once we become proficient in factoring algebraic expressions, we can use this skill to simplify *rational expressions*, also called *algebraic fractions*.

14–1 | *Simplifying Rational Expressions*

Learning Outcome **1** Simplify or reduce rational expressions.

1 **Simplify or Reduce Rational Expressions.**

In arithmetic, we learned to reduce fractions by reducing or dividing *factors* that are common to both the numerator and denominator. In our study of the laws of exponents, we also *reduced* or *simplified* algebraic fractions by reducing common factors. This process is accomplished whenever the numerator and denominator of the fraction are written in factored form. In the following example, we review the arithmetic process with fractions and the division laws of exponents. To focus on key aspects of the process, let's write steps that we often do mentally.

EXAMPLE Simplify or reduce the following fractions.

(a) $\dfrac{12}{15}$ (b) $\dfrac{24}{36}$ (c) $\dfrac{2x^2yz^3}{4xy^2z}$ (d) $\dfrac{9a^2b^3}{3ab}$

(a) $\dfrac{12}{15} = \dfrac{(?)(?)(3)}{(3)(5)} =$ Write in factored form using prime factorization.

$\dfrac{(2)(2)(\cancel{3})}{(\cancel{3})(5)} = \dfrac{4}{5}$ Reduce the common factors and multiply the remaining factors.

(b) $\dfrac{24}{36} = \dfrac{(2)(2)(2)(3)}{(2)(2)(3)(3)} =$ Write in factored form using prime factorization.

$\dfrac{(\cancel{2})(\cancel{2})(2)(\cancel{3})}{(\cancel{2})(\cancel{2})(\cancel{3})(3)} = \dfrac{2}{3}$ Reduce the common factors.

(c) $\dfrac{2x^2yz^3}{4xy^2z} = \dfrac{2x^{2-1}y^{1-2}z^{3-1}}{(2)(2)} =$ Write coefficients in factored form and reduce. Apply the laws of exponents to letter factors with like bases.

$\dfrac{xy^{-1}z^2}{2} = \dfrac{xz^2}{2y}$ Express all letter factors with positive exponents.

(d) $\dfrac{9a^2b^3}{3ab} = \dfrac{(3)(\cancel{3})a^{2-1}b^{3-1}}{\cancel{3}}$ Write coefficients in factored form and reduce. Apply the laws of exponents to letter factors with like bases.

$= 3ab^2$

Before continuing, we must understand that we are always reducing common *factors*. This procedure *will not* apply to *addends* or *terms*.

Can Common Addends Be Reduced?

Does $\frac{5}{10}$ reduce to $\frac{3}{8}$? We know from previous experience that $\frac{5}{10}$ reduces to $\frac{1}{2}$ or 0.5. Then what is wrong with the following argument?

Both the numerator and the denominator of the fraction $\frac{5}{10}$ can be rewritten as addends or terms:

Does $\dfrac{5}{10} = \dfrac{2 + 3}{2 + 8} = \dfrac{3}{8}$?   *Can common addends be reduced? No!*

We can verify with our calculator that this is an incorrect statement. Common *addends* cannot be reduced. To correct the process, we reduce *only factors*. Thus, we rewrite $\frac{5}{10}$ as factors:

$$\frac{5}{10} = \frac{(1)(\cancel{5})}{(2)(\cancel{5})} = \frac{1}{2}$$

If common factors are reduced,

$$\frac{5}{10} = \frac{1}{2}$$

Now let's use our knowledge of factoring *polynomials* to simplify rational expressions. A *rational expression* is an algebraic fraction in which the numerator or denominator or both are polynomials.

First, let's look at some examples that are written in factored form.

EXAMPLE   Simplify the rational expressions that are already in factored form.

(a) $\dfrac{x(x + 2)}{(x + 2)(x + 3)}$ (b) $\dfrac{3ab(a + 4)}{6ab(a + 3)}$ (c) $\dfrac{2x^2(2x - 1)}{x^3(x - 1)}$ (d) $\dfrac{(x + 3)(x - 4)}{(x - 3)(x + 4)}$

(a) $\dfrac{x(x + 2)}{(x + 2)(x + 3)} = \dfrac{x(\cancel{x + 2})}{(\cancel{x + 2})(x + 3)}$ Reduce common binomial factors.

$= \dfrac{x}{x + 3}$

The remaining x and $x + 3$ cannot be reduced. The x in the numerator is a factor ($x = 1 \cdot x$); however, the x in the denominator is an *addend* or *term*.

(b) $\dfrac{3ab(a + 4)}{6ab(a + 3)} = \dfrac{\overset{1}{\cancel{3ab}}(a + 4)}{\underset{2}{\cancel{6ab}}(a + 3)} =$ Reduce common factors: $1(a + 4) = a + 4$.

$\dfrac{a + 4}{2(a + 3)}$ or $\dfrac{a + 4}{2a + 6}$ Reduced rational expressions can be written in *either* factored or expanded form.

(c) $\dfrac{2x^2(2x - 1)}{x^3(x - 1)} =$ Reduce common factor: $\dfrac{x^2}{x^3} = x^{2-3} = x^{-1} = \dfrac{1}{x}$

$\dfrac{2(2x - 1)}{x(x - 1)}$ or $\dfrac{4x - 2}{x^2 - x}$ Reduced rational expressions can be written in *either* factored or expanded form.

(d) $\dfrac{(x + 3)(x - 4)}{(x - 3)(x + 4)}$ or $\dfrac{x^2 - x - 12}{x^2 + x - 12}$ There are no common factors; therefore, the fraction is already in lowest terms. Reduced rational expressions can be written in *either* factored or expanded form.

To simplify rational expressions:

1. Factor *completely* both the numerator and denominator.
2. Reduce factors common to both the numerator and denominator.
3. Write the simplified expression in either factored or expanded form.

EXAMPLE Reduce each rational expression to its simplest form.

(a) $\dfrac{x + y}{4x^2 + 4xy}$ (b) $\dfrac{a^2 + b^2}{a^2 - b^2}$ (c) $\dfrac{x^2 - 6x + 9}{x^2 - 9}$

(d) $\dfrac{3x^2 - 12}{6x + 12}$ (e) $\dfrac{a - b}{b - a}$

(a) $\dfrac{x + y}{4x^2 + 4xy} = \dfrac{(x + y)}{4x(x + y)} =$

Factor the common factor in the denominator and recall that the numerator is a grouping. Reduce common binomial factor.

$\dfrac{(x + y)}{4x(x + y)} = \dfrac{1}{4x}$

Numerator of 1 is necessary: $x + y = 1(x + y)$.

(b) $\dfrac{a^2 + b^2}{a^2 - b^2} \quad \dfrac{a^2 + b^2}{(a + b)(a - b)}$

Factor the difference of the squares in the denominator.

The numerator, which is the *sum* of the squares, will not factor. There are no factors common to both the numerator and denominator. Thus, the fraction is in simplest form:

$$\dfrac{a^2 + b^2}{(a + b)(a - b)} \quad \text{or} \quad \dfrac{a^2 + b^2}{a^2 - b^2}$$

(c) $\dfrac{x^2 - 6x + 9}{x^2 - 9} = \dfrac{(x - 3)(x - 3)}{(x + 3)(x - 3)}$

Write both the numerator and the denominator in factored form.

$= \dfrac{(x - 3)(x - 3)}{(x + 3)(x - 3)} = \dfrac{x - 3}{x + 3}$

Reduce common binomial factors.

(d) $\dfrac{3x^2 - 12}{6x + 12} = \dfrac{3(x^2 - 4)}{6(x + 2)} = \dfrac{3(x + 2)(x - 2)}{(3)(2)(x + 2)}$

Factor the numerator and the denominator completely.

$= \dfrac{3(x + 2)(x - 2)}{(3)(2)(x + 2)} = \dfrac{x - 2}{2}$

Reduce common numerical and binomial factors.

(e) $\dfrac{a - b}{b - a} =$

Write terms in numerator and denominator in same order.

$\dfrac{a - b}{-a + b} =$

Factor the denominator so that the leading coefficient of the binomial is positive.

$\dfrac{a - b}{-1(a - b)} =$

Reduce the common binomial factor.

$\dfrac{1}{-1} = -1$

Don't Forget 1 or −1.

In the preceding example, parts (a) and (e), we have some very important observations to make.

- When all other factors of a numerator or denominator reduce, a factor of 1 remains.
- Polynomial factors are opposites when every term of one grouping has an opposite in the other grouping.

$$(a - b) \text{ and } (b - a) \text{ are opposites.}$$
$$(2m - 5) \text{ and } (5 - 2m) \text{ are opposites.}$$

- Opposites differ by a factor of −1.

$$b - a = -1(-b + a) = -1(a - b)$$
$$5 - 2m = -1(-5 + 2m) = -1(2m - 5)$$

- When a numerator and denominator are opposites, the fraction reduces to −1.

$$\frac{a - b}{b - a} = \frac{1(a - b)}{-1(a - b)} = -1 \qquad \frac{2m - 5}{5 - 2m} = \frac{1(2m - 5)}{-1(2m - 5)} = -1$$

- It is helpful in recognizing common factors to rearrange or factor with −1 as a common factor so that leading coefficients within a grouping are positive: $(-x + 5) = -1(x - 5)$ or $(5 - x)$.
- It is helpful in recognizing common factors for the terms in both the numerator and denominator to be arranged in the same order.

$$\frac{x + 5}{5 + x} = \frac{x + 5}{x + 5} \qquad \frac{x - 2}{2 - x} = \frac{x - 2}{-1(x - 2)}$$

SELF-STUDY EXERCISES 14–1

1 Simplify the rational expressions.

1. $\dfrac{8}{18}$

2. $\dfrac{9}{24}$

3. $\dfrac{4a^2b^3}{2ab}$

4. $\dfrac{27a^3bc^2}{18a^2b^4c^2}$

5. $\dfrac{x(x + 3)}{(x + 3)(x + 2)}$

6. $\dfrac{3x^2(3x + 2)}{x^3(2x - 1)}$

7. $\dfrac{(x + 2)(x - 5)}{(x + 5)(x - 2)}$

8. $\dfrac{x + y}{2x + 2y}$

9. $\dfrac{a + b}{a^2 - b^2}$

10. $\dfrac{x^2 - 4x + 4}{x^2 - 4}$

11. $\dfrac{4x^2 - 16}{6x + 12}$

12. $\dfrac{3m - 11}{11 - 3m}$

13. $\dfrac{x^2 - x - 6}{3 - x}$

14. $\dfrac{b - a}{2a - 2b}$

15. $\dfrac{3x - 6}{2 - x}$

16. $\dfrac{2x - 3y}{6y - 4x}$

17. $\dfrac{5m - 10n}{2n - m}$

18. $\dfrac{2x - 6}{-x^2 + 5x - 6}$

19. $\dfrac{x^2 - 3x - 10}{-x^2 - 3x - 2}$.

20. $\dfrac{-x^2 - 9x - 14}{x^2 + 9x + 14}$

21. $\dfrac{x^2 - 4x - 21}{x^2 - 5x - 14}$

22. $\dfrac{2x^2 - 7x + 3}{2x^2 + 5x - 3}$

23. $\dfrac{3x^2 - 4x - 7}{3x^2 + 5x - 28}$

24. $\dfrac{6x^2 + x - 2}{9x^2 - 4}$

Learning Outcomes

1 Multiply and divide rational expressions.

2 Simplify complex rational expressions.

3 Use multiplication and conjugates to rationalize a numerator or denominator of a fraction that has a binomial with an irrational term.

1 **Multiply and Divide Rational Expressions.**

We can connect our knowledge of arithmetic of fractions and the laws of exponents to expand our mathematical experience. We express results in lowest terms or in simplified form.

EXAMPLE Multiply or divide as indicated. Express answers in simplest form.

(a) $\dfrac{5}{8} \cdot \dfrac{16}{25}$ (b) $\dfrac{7}{16} \div 2$ (c) $\dfrac{15xy^2}{xy^3} \cdot \dfrac{(xy)^3}{5x}$

(a) $\dfrac{5}{8} \cdot \dfrac{16}{25} = \dfrac{\overset{1}{\cancel{5}}}{\underset{1}{\cancel{8}}} \cdot \dfrac{\overset{2}{\cancel{16}}}{\underset{5}{\cancel{25}}}$ Reduce factors common to a numerator and denominator. Multiply remaining factors.

$= \dfrac{2}{5}$

If all factors that are common to a numerator and denominator are reduced or canceled before multiplying, the product will be in lowest terms.

(b) $\dfrac{7}{16} \div 2$ Rewrite 2 as a fraction.

$\dfrac{7}{16} \div 2 = \dfrac{7}{16} \div \dfrac{2}{1}$ Rewrite division as multiplication and multiply.

$= \dfrac{7}{16} \cdot \dfrac{1}{2} = \dfrac{7}{32}$

(c) $\dfrac{15xy^2}{xy^3} \cdot \dfrac{(xy)^3}{5x}$ Raise grouping to power.

$= \dfrac{15xy^2}{xy^3} \cdot \dfrac{x^3y^3}{5x} = \dfrac{15x^4y^5}{5x^2y^3}$ Apply the laws of exponents.

$= \dfrac{\overset{3}{\cancel{15}}\overset{x^2}{\cancel{x^4}}\overset{y^2}{\cancel{y^5}}}{\underset{1}{\cancel{5}}\underset{1}{\cancel{x^2}}\underset{1}{\cancel{y^3}}}$ Reduce. (Reducing could have occurred before multiplication.)

$= 3x^2y^2$

When applying the process of multiplying or dividing fractions to rational expressions, the key fact to remember is that the numerators and denominators should be written in *factored* form whenever possible.

To multiply or divide rational expressions:

1. Convert any division to an equivalent multiplication.
2. Factor completely every numerator and denominator.
3. Reduce factors that are common to a numerator and denominator.
4. Multiply remaining factors.
5. The result can be written in factored or expanded form.

EXAMPLE Perform the operation and simplify.

(a) $\dfrac{x^2 - 4x - 12}{2x - 12} \cdot \dfrac{x - 4}{x^2 + 4x + 4}$

(b) $\dfrac{4y^2 - 9}{2y^2} \div \dfrac{y^2 - 2y - 15}{4y^2 + 12y}$

(c) $\dfrac{2x - 6}{1 - x} \div \dfrac{x^2 - 2x - 3}{x^2 - 1}$

(a) $\dfrac{x^2 - 4x - 12}{2x - 12} \cdot \dfrac{x - 4}{x^2 + 4x + 4}$ Factor each numerator and denominator.

$= \dfrac{(x + 2)(x - 6)}{2(x - 6)} \cdot \dfrac{x - 4}{(x + 2)(x + 2)}$ Reduce factors common to a numerator and denominator.

$= \dfrac{\cancel{(x + 2)}\cancel{(x - 6)}}{2\cancel{(x - 6)}} \cdot \dfrac{x - 4}{\cancel{(x + 2)}(x + 2)}$ Multiply remaining factors.

$= \dfrac{x - 4}{2(x + 2)}$ or $\dfrac{x - 4}{2x + 4}$ Write in factored or expanded form.

(b) $\dfrac{4y^2 - 9}{2y^2} \div \dfrac{y^2 - 2y - 15}{4y^2 + 12y}$ Convert to multiplication.

$= \dfrac{4y^2 - 9}{2y^2} \cdot \dfrac{4y^2 + 12y}{y^2 - 2y - 15}$ Factor each numerator and denominator.

$= \dfrac{(2y + 3)(2y - 3)}{2y^2} \cdot \dfrac{4y(y + 3)}{(y + 3)(y - 5)}$ Reduce factors common to a numerator and denominator.

$= \dfrac{(2y + 3)(2y - 3)}{\underset{y}{\cancel{2y^2}}} \cdot \dfrac{\overset{2}{\cancel{4y}}\cancel{(y + 3)}}{\cancel{(y + 3)}(y - 5)}$ Multiply remaining factors.

$= \dfrac{2(2y + 3)(2y - 3)}{y(y - 5)}$ or $\dfrac{2(4y^2 - 9)}{y^2 - 5y}$ or $\dfrac{8y^2 - 18}{y^2 - 5y}$

(c) $\dfrac{2x - 6}{1 - x} \div \dfrac{x^2 - 2x - 3}{x^2 - 1}$ Convert to multiplication.

$= \dfrac{2x - 6}{1 - x} \cdot \dfrac{x^2 - 1}{x^2 - 2x - 3}$ Factor each numerator and denominator. Note the effect of factoring -1 in the denominator of the first fraction so the $x - 1$ will reduce with $x - 1$ in the numerator of the second fraction.

$= \dfrac{2(x - 3)}{-1(x - 1)} \cdot \dfrac{(x + 1)(x - 1)}{(x + 1)(x - 3)}$ Reduce factors common to a numerator and denominator.

$= \dfrac{2\cancel{(x - 3)}}{-1\cancel{(x - 1)}} \cdot \dfrac{\cancel{(x + 1)}\cancel{(x - 1)}}{\cancel{(x + 1)}\cancel{(x - 3)}}$ Multiply remaining factors.

$= \dfrac{2}{-1} = -2$

2 Simplify Complex Rational Expressions.

A *complex rational expression* is a rational expression that has a rational expression in its numerator or its denominator or both. Some examples of complex rational expressions are

$$\dfrac{4xy}{\dfrac{2x}{5}} \qquad \dfrac{\dfrac{x^2 - y^2}{2x}}{\dfrac{x - y}{3x^2}} \qquad \dfrac{1 + \dfrac{1}{x}}{\dfrac{2}{3}} \qquad \dfrac{\dfrac{2}{x} - \dfrac{5}{2x}}{\dfrac{3}{4x} + \dfrac{3}{x}}$$

To simplify a complex rational expression:

1. Rewrite the complex rational expression as a division then convert to an equivalent multiplication of rational expressions.
2. Multiply the rational expressions.

Before we look at examples involving complex rational expressions, the following Tip will eliminate some of our written steps.

TIP!

Mentally Converting from Complex Form to Division and Then to Multiplication

Examine the symbolic representation for converting a complex expression to multiplication. The numerator is multiplied by the reciprocal of the denominator.

$$\frac{\dfrac{a}{b}}{\dfrac{c}{d}} = \frac{a}{b} \div \frac{c}{d} = \frac{a}{b} \cdot \frac{d}{c} \quad b, c, \text{ and } d \neq 0$$

When simplifying a complex rational expression, we can eliminate some written steps by making this conversion mentally; for example,

$$\frac{\dfrac{4xy}{2x}}{5} \quad \text{becomes} \quad \frac{4xy}{1} \cdot \frac{5}{2x}$$

EXAMPLE Simplify.

(a) $\dfrac{\dfrac{4xy}{2x}}{5}$ (b) $\dfrac{\dfrac{x^2 - y^2}{2x}}{\dfrac{x - y}{3x^2}}$

(a) $\dfrac{\dfrac{4xy}{2x}}{5} = \dfrac{4xy}{1} \cdot \dfrac{5}{2x}$ Multiply the numerator by the reciprocal of the denominator.

$= \dfrac{4\overset{2}{x}y}{1} \cdot \dfrac{5}{2x}$ Reduce. Multiply remaining factors.

$= \dfrac{10y}{1} = 10y$ Simplify.

(b) $\dfrac{\dfrac{x^2 - y^2}{2x}}{\dfrac{x - y}{3x^2}} = \dfrac{x^2 - y^2}{2x} \cdot \dfrac{3x^2}{x - y}$ Multiply the numerator by the reciprocal of the denominator.

$= \dfrac{(x + y)(x - y)}{2x} \cdot \dfrac{3x^2}{x - y}$ Factor the numerator of the first fraction.

$$= \frac{(x + y)(x \cancel{- y})}{2\cancel{x}} \cdot \frac{3\overset{x}{\cancel{x^2}}}{\cancel{x - y}}$$ Reduce. Multiply remaining factors.

$$= \frac{3x(x + y)}{2} \quad \text{or} \quad \frac{3x^2 + 3xy}{2}$$ Factored or expanded form.

3 Use Multiplication and Conjugates to Rationalize a Numerator or Denominator of a Fraction That has a Binomial with an Irrational Term.

Our understanding of fractions and rational expressions can be used to manipulate fractional expressions that have irrational terms. In an advanced study of mathematics, there are occasions when you may want a numerator or denominator clear of irrational terms.

To use multiplication and conjugates to rationalize a numerator or denominator of a fraction that has a binomial with an irrational term:

1. Determine which part of the fraction (numerator or denominator) is to be cleared of a binomial with an irrational term.
2. Multiply by 1 in the form of $\frac{n}{n}$, where n is the conjugate of the binomial with the irrational term.
3. Simplify the resulting expression.

EXAMPLE Rationalize the denominator of the fraction $\dfrac{1 + 2\sqrt{3}}{2 - 5\sqrt{2}}$.

$$\frac{1 + 2\sqrt{3}}{2 - 5\sqrt{2}}$$ The conjugate of $2 - 5\sqrt{2}$ is $2 + 5\sqrt{2}$. Multiply by 1 in the form of $\frac{2 + 5\sqrt{2}}{2 + 5\sqrt{2}}$.

$$\frac{1 + 2\sqrt{3}}{2 - 5\sqrt{2}} \cdot \frac{2 + 5\sqrt{2}}{2 + 5\sqrt{2}}$$ FOIL the numerators and use the special product for the sum and difference of two terms for the denominator.

$$\frac{2 + 1(5\sqrt{2}) + 2(2\sqrt{3}) + 2\sqrt{3}(5\sqrt{2})}{2^2 - (5\sqrt{2})^2}$$ Simplify by multiplying coefficients.

$$\frac{2 + 5\sqrt{2} + 4\sqrt{3} + 10\sqrt{6}}{4 - 25(2)} =$$ Simplify the denominator.

$$\frac{2 + 5\sqrt{2} + 4\sqrt{3} + 10\sqrt{6}}{4 - 50} =$$ Combine like terms in the denominator.

$$\frac{2 + 5\sqrt{2} + 4\sqrt{3} + 10\sqrt{6}}{-46}$$ Manipulate the signs of the fraction to make the denominator positive.

or

$$-\frac{2 + 5\sqrt{2} + 4\sqrt{3} + 10\sqrt{6}}{46}$$

EXAMPLE   Rationalize the numerator of the fraction $\dfrac{1 + 2\sqrt{3}}{2 - 5\sqrt{2}}$.

$$\frac{1 + 2\sqrt{3}}{2 - 5\sqrt{2}} =$$

The conjugate of $1 + 2\sqrt{3}$ is $1 - 2\sqrt{3}$. Multiply by 1 in the form of $\dfrac{1 - 2\sqrt{3}}{1 - 2\sqrt{3}}$.

$$\frac{1 + 2\sqrt{3}}{2 - 5\sqrt{2}} \cdot \frac{1 - 2\sqrt{3}}{1 - 2\sqrt{3}} =$$

Use the special product for the sum and difference of two terms for the numerator and FOIL the denominator.

$$\frac{1^2 - \left(2\sqrt{3}\right)^2}{2 + 2\left(-2\sqrt{3}\right) + 1\left(-5\sqrt{2}\right) - 5\sqrt{2}\left(-2\sqrt{3}\right)} =$$

Simplify.

$$\frac{1 - 4(3)}{2 - 4\sqrt{3} - 5\sqrt{2} + 10\sqrt{6}} =$$

$$\frac{-11}{2 - 4\sqrt{3} - 5\sqrt{2} + 10\sqrt{6}}$$

TIP!

Does Rationalizing a Numerator or Denominator of a Fraction Make the Expression Easier to Evaluate?

Probably not. Rationalizing is done whenever you need to manipulate a fraction so that one term has only rational amounts. In more advanced studies of mathematics, there are specific situations when it is desirable to perform this manipulation.

SELF-STUDY EXERCISES 14–2

1   Multiply or divide the fractions. Reduce to simplest form.

1. $\dfrac{7}{12} \cdot \dfrac{18}{21}$

2. $\dfrac{3}{8} \cdot \dfrac{4}{15}$

3. $\dfrac{5x^2}{3y} \cdot \dfrac{2x}{3y}$

4. $\dfrac{8a^2}{15ab} \cdot \dfrac{21ab}{24a^2}$

5. $\dfrac{(x + 2)(x - 7)}{(x + 1)(x - 2)} \cdot \dfrac{3(x + 1)}{(x - 7)(x + 2)}$

6. $\dfrac{-4(x - 6)}{(x - 3)(x + 5)} \cdot \dfrac{(3 - x)(x + 5)}{(x - 6)(x - 2)}$

7. $\dfrac{a^2 - b^2}{4} \cdot \dfrac{12}{a + b}$

8. $\dfrac{2u + 2v}{5} \cdot \dfrac{10}{u + v}$

9. $\dfrac{a^2 - 49}{b^2 - 25} \cdot \dfrac{b - 5}{a + 7}$

10. $\dfrac{x^2 + 2x + 1}{5x - 5} \cdot \dfrac{15}{x + 1}$

11. $\dfrac{13r^2}{20a^2} \div \dfrac{39r^2}{5a}$

12. $\dfrac{4}{5} \div \dfrac{8}{15}$

13. $\dfrac{3}{8} \div \dfrac{9}{16}$

14. $\dfrac{5x^2y}{3y} \div \dfrac{10x^3}{9y^3}$

15. $\dfrac{7a^2b}{15b} \div \dfrac{14a}{9b}$

16. $\dfrac{a - b}{4} \div \dfrac{a - b}{2}$

17. $\dfrac{b - a}{7} \div \dfrac{a - b}{14}$

18. $\dfrac{2x - y}{x + y} \div \dfrac{y - 2x}{-x - y}$

19. $\dfrac{x}{x^2 - 4x + 4} \div \dfrac{1}{x - 2}$

20. $\dfrac{5a^2 - 5b^2}{a^2b^2} \div \dfrac{a + b}{10ab}$

21. $\dfrac{x^2 + 4x + 3}{x^2 - 4x - 5} \div \dfrac{x + 3}{x - 5}$

22. $\dfrac{\dfrac{5}{9}}{\dfrac{3}{5}}$

23. $\dfrac{\dfrac{7}{8}}{\dfrac{5}{6}}$

24. $\dfrac{\dfrac{x-5}{4}}{3x}$

25. $\dfrac{\dfrac{6}{x-2}}{9x}$

26. $\dfrac{\dfrac{4x}{8x}}{x+3}$

27. $\dfrac{\dfrac{7x}{21x}}{x-3}$

28. $\dfrac{\dfrac{x^2-y^2}{3x+y}}{\dfrac{y-x}{-3x-y}}$

29. $\dfrac{\dfrac{x^2-5x+4}{x^2+4x+3}}{\dfrac{4-x}{x+3}}$

30. $\dfrac{\dfrac{x^2-36}{5x}}{\dfrac{x+6}{15x}}$

31. $\dfrac{\dfrac{x^2-5x}{8}}{\dfrac{2x-10}{12x}}$

3 Rationalize the denominator of each fraction.

32. $\dfrac{3}{2-\sqrt{5}}$

33. $\dfrac{13}{5-2\sqrt{3}}$

34. $\dfrac{5-\sqrt{2}}{4+\sqrt{2}}$

35. $\dfrac{7+\sqrt{7}}{5+\sqrt{7}}$

36. $\dfrac{2+3\sqrt{5}}{7-2\sqrt{3}}$

37. $\dfrac{5-\sqrt{6}}{5-2\sqrt{6}}$

Rationalize the numerator of each fraction.

38. $\dfrac{2-\sqrt{5}}{7}$

39. $\dfrac{5-\sqrt{11}}{15}$

40. $\dfrac{4+\sqrt{6}}{20}$

41. $\dfrac{3+\sqrt{2}}{7}$

42. $\dfrac{4+3\sqrt{2}}{7}$

43. $\dfrac{5+2\sqrt{3}}{8}$

14–3 | Adding and Subtracting Rational Expressions

Learning Outcomes

1 Add and subtract rational expressions.
2 Use addition and subtraction of rational expressions to simplify complex fractions.

1 Add and Subtract Rational Expressions.

In adding and subtracting fractions, we can add or subtract only fractions with like denominators. Whenever we have unlike fractions, we must first find a common denominator for the fractions. We then convert each fraction to an equivalent fraction with the common denominator and add or subtract the numerators. Let's refresh our memory of the procedure for adding and subtracting fractions.

EXAMPLE Add or subtract as indicated.

(a) $\dfrac{3}{8}+\dfrac{1}{8}$ (b) $\dfrac{7}{12}-\dfrac{1}{3}$ (c) $5-\dfrac{2}{3}$ (d) $\dfrac{5}{2x}+\dfrac{3}{x}+\dfrac{9}{4}$

(a) $\dfrac{3}{8}+\dfrac{1}{8}=\dfrac{4}{8}=\dfrac{1}{2}$ Add the numerators and *keep* the *like* denominator. Reduce.

(b) $\dfrac{7}{12} - \dfrac{1}{3} =$

Select a common denominator and change to equivalent fractions with the common denominator of 12: $\frac{1}{3} = \frac{1}{3}\left(\frac{4}{4}\right) = \frac{4}{12}$.

$\dfrac{7}{12} - \dfrac{4}{12} = \dfrac{3}{12} = \dfrac{1}{4}$

Subtract and reduce.

(c) $5 - \dfrac{2}{3} =$

Convert 5 to a fraction with a denominator of 3: $\frac{5}{1}\left(\frac{3}{3}\right) = \frac{15}{3}$.

$\dfrac{15}{3} - \dfrac{2}{3} = \dfrac{13}{3}$ or $4\dfrac{1}{3}$

Subtract. Write improper fraction as a mixed number if desired.

(d) $\dfrac{5}{2x} + \dfrac{3}{x} + \dfrac{9}{4} =$

Convert to equivalent fractions with a common denominator of $4x$.

$\dfrac{5(2)}{2x(2)} + \dfrac{3(4)}{x(4)} + \dfrac{9(x)}{4(x)} =$

Multiply by 1 in the form of $\frac{n}{n}$ to get an equivalent fraction with a denominator of $4x$.

$\dfrac{10}{4x} + \dfrac{12}{4x} + \dfrac{9x}{4x} =$

Add numerators and *keep* the like (common) denominator.

$\dfrac{22 + 9x}{4x}$

TIP!

Writing Improper Fractions Versus Mixed Numbers

In arithmetic, we often write an improper fraction as a mixed number. In algebra, we use the mixed-number form only when the final result contains only numbers and when we are interpreting the result within the context of an applied problem. A rational expression like the solution in the preceding example, part (d), can be written in an alternative form.

$$\dfrac{22 + 9x}{4x} = \dfrac{22}{4x} + \dfrac{9x}{4x} = \dfrac{11}{2x} + \dfrac{9}{4}$$

Unless the context of the problem requires the step, there's usually no reason to do the extra work.

To add or subtract rational expressions with unlike denominators:

1. Find the *least common denominator (LCD)*.
2. Change *each* fraction to an equivalent fraction with the least common denominator.
3. Add or subtract numerators.
4. Keep the same (common) denominator.
5. Reduce (or simplify) if possible.

EXAMPLE Add $\dfrac{4}{x + 3} + \dfrac{3}{x - 3}$.

First, convert to equivalent expressions with a common denominator. The LCD is the product $(x + 3)(x - 3)$.

$$\dfrac{4}{x + 3} = \dfrac{4(x - 3)}{(x + 3)(x - 3)} = \dfrac{4x - 12}{(x + 3)(x - 3)}$$

Multiply the numerator and denominator by $(x - 3)$.

$$\frac{3}{x-3} = \frac{3(x+3)}{(x-3)(x+3)} = \frac{3x+9}{(x+3)(x-3)}$$

Multiply the numerator and denominator by $(x + 3)$. Order does not matter in multiplication.

Next, use the equivalent expressions to proceed.

$$\frac{4x-12}{(x+3)(x-3)} + \frac{3x+9}{(x+3)(x-3)} =$$

Add numerators.

$$\frac{4x-12+3x+9}{(x+3)(x-3)} =$$

Combine like terms in the numerator.

$$\frac{7x-3}{(x+3)(x-3)} \quad \text{or} \quad \frac{7x-3}{x^2-9}$$

Factored or expanded form.

Some of the steps in the preceding example can be combined into one step, thus requiring fewer *written* steps, but the example illustrates each step that must be performed, either mentally or on paper, to add the rational expressions. Look at the next example.

EXAMPLE Subtract $\dfrac{x}{x-2} - \dfrac{5}{x+4}$.

Change to equivalent expressions with a common denominator. The LCD is the product $(x - 2)(x + 4)$.

$$\frac{x}{x-2} = \frac{x(x+4)}{(x-2)(x+4)} = \frac{x^2+4x}{(x-2)(x+4)}$$

Multiply the numerator and denominator by $(x + 4)$.

$$\frac{5}{x+4} = \frac{5(x-2)}{(x+4)(x-2)} = \frac{5x-10}{(x-2)(x+4)}$$

Multiply the numerator and denominator by $(x - 2)$.

Use equivalent expressions and proceed.

$$\frac{x^2+4x}{(x-2)(x+4)} - \frac{5x-10}{(x-2)(x+4)} =$$

Subtract numerators.

$$\frac{x^2+4x-(5x-10)}{(x-2)(x+4)} =$$

Note: the *entire* numerator following the subtraction sign is subtracted.

$$\frac{x^2+4x-5x+10}{(x-2)(x+4)} =$$

Be careful with the signs when the parentheses are removed.

$$\frac{x^2-x+10}{(x-2)(x+4)} \quad \text{or} \quad \frac{x^2-x+10}{x^2+2x-8}$$

$x^2 - x + 10$ will not factor, so the fraction cannot be reduced.

2 **Use Addition and Subtraction of Rational Expressions to Simplify Complex Fractions.**

Recall that a complex rational expression may have rational expressions in the numerator or denominator or both. Thus, there may be complex expressions that require the addition or subtraction of rational expressions before conversion to an equivalent multiplication.

To simplify a complex fraction that requires adding or subtracting rational terms:

1. Use addition or subtraction to combine terms in the numerator or denominator or both.
2. Rewrite the complex expression as a division then multiplication of rational expressions.
3. Proceed using the procedure for multiplying rational expressions.

EXAMPLE Simplify.

(a) $\dfrac{1 + \dfrac{1}{x}}{\dfrac{2}{3}}$ (b) $\dfrac{\dfrac{2}{x} - \dfrac{5}{2x}}{\dfrac{3}{4x} + \dfrac{3}{x}}$

(a) $\dfrac{1 + \dfrac{1}{x}}{\dfrac{2}{3}} = \dfrac{\dfrac{x}{x} + \dfrac{1}{x}}{\dfrac{2}{3}} =$ Add the terms in the numerator. The common denominator for the numerator is x: $1 = \dfrac{x}{x}$.

$\dfrac{\dfrac{x+1}{x}}{\dfrac{2}{3}} =$ Multiply the numerator by the reciprocal of the denominator.

$\dfrac{x+1}{x} \cdot \dfrac{3}{2} =$

$\dfrac{3(x+1)}{2x}$ or $\dfrac{3x+3}{2x}$ Factored or expanded form.

(b) $\dfrac{\dfrac{2}{x} - \dfrac{5}{2x}}{\dfrac{3}{4x} + \dfrac{3}{x}} = \dfrac{\dfrac{4}{2x} - \dfrac{5}{2x}}{\dfrac{3}{4x} + \dfrac{12}{4x}} =$ Add or subtract fractions in the numerator and in the denominator. Find the common denominator for the numerator, $2x$, and the denominator, $4x$, and convert to equivalent fractions.

$\dfrac{\dfrac{-1}{2x}}{\dfrac{15}{4x}} =$ Multiply the numerator by the reciprocal of the denominator.

$\dfrac{-1}{2x} \cdot \dfrac{4x}{15} =$ Reduce and multiply.

$\dfrac{-1}{\underset{1}{2x}} \cdot \dfrac{\overset{2}{4x}}{15} =$

$\dfrac{-2}{15} = -\dfrac{2}{15}$

SELF-STUDY EXERCISES 14–3

1 Add or subtract the rational expressions. Reduce to simplest terms.

1. $\dfrac{3}{7} + \dfrac{2}{7}$

2. $\dfrac{3}{8} + \dfrac{7}{16}$

3. $\dfrac{2x}{3} + \dfrac{5x}{6}$

4. $\dfrac{3x}{8} + \dfrac{7x}{12}$

5. $\dfrac{2}{x} + \dfrac{3}{2}$

6. $\dfrac{5}{2x} + \dfrac{6}{x} + \dfrac{2}{3x}$

7. $\dfrac{5}{x+1} + \dfrac{4}{x-1}$

8. $\dfrac{6}{x+3} + \dfrac{4}{x-1}$

9. $\dfrac{5}{2x+1} - \dfrac{2}{x-1}$

10. $\dfrac{7}{x-6} - \dfrac{6}{x-5}$

2 Simplify.

11. $\dfrac{2 + \dfrac{1}{x}}{\dfrac{3}{5}}$

12. $\dfrac{3 + \dfrac{2}{x}}{\dfrac{5}{8}}$

13. $\dfrac{\dfrac{3}{x} + \dfrac{5}{3x}}{\dfrac{3}{6x} - \dfrac{2}{x}}$

14. $\dfrac{\dfrac{x}{6} - \dfrac{3x}{9}}{\dfrac{5x}{12} + \dfrac{3x}{4}}$

15. $\dfrac{\dfrac{4}{x} + \dfrac{3}{2x}}{\dfrac{7x}{5} + \dfrac{x}{10}}$

16. $\dfrac{\dfrac{2}{3x} + \dfrac{3}{5}}{\dfrac{1}{5x} + \dfrac{2}{3}}$

17. $\dfrac{\dfrac{7}{x+2} - \dfrac{5}{x+2}}{\dfrac{3}{2-x} + \dfrac{2}{x-2}}$

18. $\dfrac{\dfrac{5}{x-3} + \dfrac{2}{3-x}}{\dfrac{4}{2x-6} - \dfrac{5}{3x-9}}$

14–4 | *Solving Equations with Rational Expressions*

Learning Outcomes

1 Exclude certain values as solutions of rational equations.

2 Solve rational equations with variable denominators.

1 **Exclude Certain Values as Solutions of Rational Equations.**

A rational equation is an equation that contains one or more rational expressions. Some examples of rational equations are

$$\frac{1}{2} + \frac{1}{x} = \frac{1}{3}, \qquad x - \frac{3}{x} = 4, \qquad \frac{x}{x-2} = \frac{1}{x+3}$$

We solved some types of rational equations in Chapter 9 by first clearing the equation of all fractions. We cleared the fractions by multiplying both sides of the equation by the *least common multiple* (*LCM*) of all the denominators in the equation. Then we solved the equation as any other equation. However, when the variable is in the denominator of the equation, an important caution must be observed.

TIP!

Excluded Values

Because division by zero is impossible, any value of the variable that makes the value of any denominator zero is excluded as a possible solution or root of an equation. Therefore, it is important (1) to determine which values are *excluded values* and cannot be used as possible solutions, and (2) to check each possible solution.

To find excluded values of rational equations:

1. Set each denominator containing a variable equal to zero and solve for the variable.
2. Each equation in Step 1 produces an excluded value; however, some values may be repeats.

EXAMPLE Determine the value or values that must be excluded as possible solutions.

(a) $\dfrac{1}{x} + \dfrac{1}{2} = 5$

(b) $\dfrac{1}{x} = \dfrac{1}{x-3}$

(c) $\dfrac{1}{x+2} + \dfrac{1}{x-2} = \dfrac{1}{x^2-4}$

(d) $\dfrac{3x}{x+2} = 3 - \dfrac{5}{2x}$

(a) $\dfrac{1}{x} + \dfrac{1}{2} = 5$　　　　Set each denominator containing a variable equal to 0.

$x = 0$

Excluded value is 0.

(b) $\dfrac{1}{x} = \dfrac{1}{x-3}$　　　　Set each denominator containing a variable equal to 0.

$x = 0$,　　$x - 3 = 0$　　Solve each equation.

$x = 3$

Excluded values are 0 and 3 .

(c) $\dfrac{1}{x+2} + \dfrac{1}{x-2} = \dfrac{1}{x^2-4}$　　　　Set each denominator containing a variable equal to 0.

$x + 2 = 0$　　$x - 2 = 0$　　　　　$x^2 - 4 = 0$　Solve each equation.

$x = -2$　　$x = 2$　　　　$(x+2)(x-2) = 0$

$x + 2 = 0$　　$x - 2 = 0$

$x = -2$　　$x = 2$

Excluded values are −2 and 2 .

(d) $\dfrac{3x}{x+2} = 3 - \dfrac{5}{2x}$　　　　Set each denominator containing a variable equal to 0.

$x + 2 = 0$　　$2x = 0$　　Solve each equation.

$x = -2$　　$x = 0$

Excluded values are −2 and 0.

7 **Solve Rational Equations with Variable Denominators.**

Finding excluded values is another way to check whether roots are *extraneous roots* (Chapter 9). Even though we have found the excluded values, it is a good idea to check the solution to make sure it makes a true statement in the original equation.

To solve rational equations:

1. Determine the excluded values.
2. Clear the equation of all denominators by multiplying the entire equation by the least common multiple (LCM) of the denominators.
3. Complete the solution using previously learned strategies.
4. Eliminate any excluded values as solutions.
5. Check the solutions.

EXAMPLE　Solve the rational equations. Check.

(a) $\dfrac{2}{y} + \dfrac{1}{3} = 1$　　(b) $\dfrac{2}{y-2} = \dfrac{4}{y+1}$　　(c) $\dfrac{5}{x-2} - \dfrac{3x}{x-2} = -\dfrac{1}{x-2}$

(a) $\dfrac{2}{y} + \dfrac{1}{3} = 1$　　　　Excluded value: $y = 0$.

$3y\left(\dfrac{2}{y}\right) + 3y\left(\dfrac{1}{3}\right) = 3y(1)$　　Multiply by LCM, $3y$, and reduce.

$(3\cancel{y})\left(\dfrac{2}{\cancel{y}}\right) + (\cancel{3}y)\left(\dfrac{1}{\cancel{3}}\right) = 3y(1)$　　Multiply remaining factors.

$6 + y = 3y$　　Sort.

$6 = 3y - y$　　Combine.

$6 = 2y$　　Divide.

$$\frac{6}{2} = \frac{2y}{2}$$ Divide.

$$3 = y$$ 3 is not an excluded value.

Check: $\dfrac{2}{y} + \dfrac{1}{3} = 1$ Substitute 3 for y.

$$\frac{2}{3} + \frac{1}{3} = 1$$ Combine like terms.

$$\frac{3}{3} = 1$$ Simplify.

$$1 = 1$$ Solution checks.

(b) $\dfrac{2}{y-2} = \dfrac{4}{y+1}$ Excluded values: $y = 2$, $y = -1$.

$$(y-2)(y+1)\left(\frac{2}{y-2}\right) = (y-2)(y+1)\left(\frac{4}{y+1}\right)$$ Multiply by LCM, $(y-2)(y+1)$, and reduce.

$$(\cancel{y-2})(y+1)\frac{2}{\cancel{y-2}} = (y-2)(\cancel{y+1})\frac{4}{\cancel{y+1}}$$ Multiply remaining factors.

$$2(y+1) = 4(y-2)$$ Distribute.

$$2y + 2 = 4y - 8$$ Sort.

$$2y - 4y = -8 - 2$$ Combine like terms.

$$-2y = -10$$ Divide.

$$\frac{-2y}{-2} = \frac{-10}{-2}$$

$$y = 5$$ 5 is not an excluded value.

Check: $\dfrac{2}{y-2} = \dfrac{4}{y+1}$ Substitute 5 for y.

$$\frac{2}{5-2} = \frac{4}{5+1}$$ Simplify.

$$\frac{2}{3} = \frac{4}{6}$$ Reduce $\frac{4}{6}$.

$$\frac{2}{3} = \frac{2}{3}$$ Solution checks.

(c) $\dfrac{5}{x-2} - \dfrac{3x}{x-2} = -\dfrac{1}{x-2}$ Excluded value: $x - 2 = 0$, $x = 2$.

$$(x-2)\frac{5}{x-2} - (x-2)\frac{3x}{x-2} = (x-2)\left(-\frac{1}{(x-2)}\right)$$ Multiply by LCM, $x - 2$, and reduce.

$$(\cancel{x-2})\frac{5}{\cancel{x-2}} - (\cancel{x-2})\frac{3x}{\cancel{x-2}} = (\cancel{x-2})\left(-\frac{1}{(\cancel{x-2})}\right)$$

$$5 - 3x = -1$$ Sort.

$$-3x = -1 - 5$$ Combine.

$$-3x = -6$$ Divide.

$$\frac{-3x}{-3} = \frac{-6}{-3}$$

$$x = 2$$ This is an excluded value. The root will not check.

There is no solution.

Check: $\dfrac{5}{x-2} - \dfrac{3x}{x-2} = -\dfrac{1}{x-2}$ Substitute 2 for x.

$$\frac{5}{2-2} - \frac{3(2)}{2-2} = -\frac{1}{2-2}$$

$$\frac{5}{0} - \frac{6}{0} = -\frac{1}{0}$$ Division by zero is impossible. The equation has no solution.

A number of application problems can be solved by using rational equations. Let's look at an example.

EXAMPLE At the January office products sale, Isom Tibbs purchased a package of 3.5-in. computer disks for $16 and a package of writing pens for $6. He paid 40¢ more per pen than he paid for each computer disk, and the computer disk package contained 4 times as many disks as the number of pens. How many disks and how many pens did he purchase? Give the price of each disk and each pen.

Unknown facts Let n = the number of pens. The number of pens, the number of disks, the cost per pen, and the cost per disk are all unknown.

Known facts Disks cost $16 per package. Pens cost $6 per package. There are 4 times as many disks as pens. Each pen costs 40¢ more than each disk.

Relationships Number of disks = $4n$

Cost ÷ Number of items = Cost per item

Cost per pen $= \dfrac{6}{n}$

Cost per disk $= \dfrac{16}{4n}$

Cost per pen = Cost per disk plus 40¢.   40¢ = $\frac{40}{100}$ dollars.

$$\frac{6}{n} = \frac{16}{4n} + \frac{40}{100}$$

Estimation Pens cost more than disks and only $6 was spent for pens, so there will be only a small number of pens.

Calculations $\dfrac{6}{n} = \dfrac{16}{4n} + \dfrac{40}{100}$ Reduce. $\dfrac{40}{100} = \dfrac{2}{5}$

$\dfrac{6}{n} = \dfrac{16}{4n} + \dfrac{2}{5}$

Interpretation Excluded value: $n = 0$. This excluded value would also be inappropriate within the context of the problem since we know that some pens and disks were purchased.

$$20n\left(\frac{6}{n}\right) = 20n\left(\frac{16}{4n}\right) + 20n\left(\frac{2}{5}\right)$$ Multiply by LCM, $20n$ and reduce.

$$20\overset{}{n}\left(\frac{6}{\overset{}{n}}\right) = \overset{5}{20}n\left(\frac{16}{\overset{}{4n}}\right) + \overset{4}{20}n\left(\frac{2}{\overset{}{5}}\right)$$ Multiply remaining factors.

$120 = 80 + 8n$ Sort.

$120 - 80 = 8n$ Combine.

$40 = 8n$ Divide.

$5 = n$

$n = 5$ pens

There are 5 pens, and that fact can be used to find the other missing facts.

$$4n = 4(5) = \textbf{20 disks}$$

$$\frac{6}{n} = \frac{\$6}{5} = \textbf{\$1.20 per pen}$$

$$\frac{16}{4n} = \frac{\$16}{20} = \textbf{\$0.80 per disk}$$

Check: $\dfrac{6}{n} = \dfrac{16}{4n} + \dfrac{2}{5}$ Substitute 5 for n.

$\dfrac{6}{5} = \dfrac{16}{4(5)} + \dfrac{2}{5}$ Simplify grouping (denominator).

$\dfrac{6}{5} = \dfrac{16}{20} + \dfrac{2}{5}$ Combine fractions on the right.

$\dfrac{6}{5} = \dfrac{16}{20} + \dfrac{8}{20}$

$\dfrac{6}{5} = \dfrac{24}{20}$ Reduce: $\dfrac{24}{20} = \dfrac{6}{5}$.

$\dfrac{6}{5} = \dfrac{6}{5}$ The solution checks.

SELF-STUDY EXERCISES 14–4

1 Determine the value or values that must be excluded as possible solutions.

1. $\dfrac{5}{x} + \dfrac{3}{5} = 7$

2. $\dfrac{6}{x} + 5 = \dfrac{5}{12}$

3. $\dfrac{9}{x - 5} = \dfrac{7}{x}$

4. $\dfrac{15}{2x} = \dfrac{6}{x + 9}$

5. $\dfrac{7}{x + 8} + \dfrac{2}{x - 8} = \dfrac{5}{x^2 - 64}$

6. $\dfrac{9}{x - 2} + \dfrac{5}{x + 2} = \dfrac{1}{x^2 - 4}$

7. $\dfrac{2x}{4x - 3} - 4 = \dfrac{5}{6x}$

8. $\dfrac{1}{9x} + \dfrac{8x}{5x + 15} = 11$

2 Solve the equations. Check for extraneous roots.

9. $\dfrac{6}{7} + \dfrac{5}{x} = 1$

10. $\dfrac{4}{x} - \dfrac{3}{4} = \dfrac{1}{20}$

11. $\dfrac{2}{x} + \dfrac{5}{2x} - \dfrac{1}{2} = 4$

12. $\dfrac{5}{2x} - \dfrac{1}{x} = 6$

13. $\dfrac{2}{x - 3} + \dfrac{7}{x + 3} = \dfrac{21}{x^2 - 9}$

14. $\dfrac{1}{x^2 - 25} = \dfrac{3}{x - 5} + \dfrac{2}{x + 5}$

Use rational equations to solve the problems. Check for extraneous roots.

15. Shawna can complete an electronic lab project in 3 days. Tom can complete the same project in 4 days. How many days will it take to complete the project if Shawna and Tom work together? (*Hint:* Rate × Time = Amount of work.)

16. A group of investors purchased land in Maine for $12,000. More investors joined the group and the cost dropped $500 per person in the original group. How many investors were in the original group if the new group had $1\frac{1}{2}$ times as many investors as the original group?

17. Frank commutes 50 miles on a motorbike to a college, but on the return trip he travels 40 miles per hour and makes the trip in $\frac{3}{4}$ hour less time. Find Frank's speed going to school. (*Hint:* Time = Distance ÷ Rate.)

18. Pipe 1 fills a tank in 2 min and pipe 2 empties the tank in 6 min. How long does it take to fill the tank if both pipes are open? (*Hint:* Rate × Time = Amount of work.)

19. Kim Denley bought $150 worth of medium-roast coffee and, at the same cost per pound, $100 worth of dark-roast coffee. If she bought 25 pounds more of the medium-roast coffee than the dark-roast coffee, how many pounds of each did she buy and what was the cost per pound?

20. Find resistance$_2$ if resistance$_1$ is 10 ohms and resistance$_t$ is 3.75 ohms, using the formula

$$R_t = \frac{R_1 R_2}{R_1 + R_2}.$$

ASSIGNMENT EXERCISES

Section 14–1

Simplify the fractions.

1. $\dfrac{18}{24}$

2. $\dfrac{24}{42}$

3. $\dfrac{5a^2 b^3 c}{10 a^3 bc^2}$

4. $\dfrac{3x^2 y}{9x^3 y^3}$

5. $\dfrac{4xy(x-3)}{8xy(x+3)}$

6. $\dfrac{(x+7)(3-x)}{(x-3)(x-7)}$

7. $\dfrac{(x-4)(x+2)}{(x+2)(4-x)}$

8. $\dfrac{x+1}{3x+3}$

9. $\dfrac{m^2 - n^2}{m^2 + n^2}$

10. $\dfrac{3x^2 + 8x - 3}{2x^2 + 5x - 3}$

11. $\dfrac{x}{x+xy}$

12. $\dfrac{2x}{4x+6}$

13. $\dfrac{5x+15}{x+3}$

14. $\dfrac{x^3 + 2x^2 - 3x}{3x}$

15. $\dfrac{y^2 + 2y + 1}{y+1}$

16. $\dfrac{y-1}{y^2 - 2y + 1}$

17. $\dfrac{2x-6}{x^2 + 3x - 18}$

18. $\dfrac{x^2 - 4x + 4}{x^2 - 2x}$

19. $\dfrac{3x-9}{x-3}$

20. $\dfrac{4x-12}{x-3}$

Section 14–2

Multiply or divide the fractions. Reduce to simplest terms.

21. $\dfrac{3x^2}{2y} \cdot \dfrac{5x}{6y}$

22. $\dfrac{x^2 - y^2}{6} \cdot \dfrac{18}{x-y}$

23. $\dfrac{9}{x+b} \cdot \dfrac{5x + 5b}{3}$

24. $\dfrac{81 - x^2}{16 - f^2} \cdot \dfrac{4-f}{9+x}$

25. $\dfrac{4y^2 - 4y + 1}{6y - 6} \cdot \dfrac{24}{2y - 1}$

26. $\dfrac{x-3}{x+5} \cdot \dfrac{2x^2 + 10x}{2x - 6}$

27. $\dfrac{5-x}{x-5} \cdot \dfrac{x-1}{1-x}$

28. $\dfrac{3-x}{x-2} \cdot \dfrac{2x-4}{x-3}$

29. $\dfrac{x^2 + 6x + 9}{x^2 - 4} \cdot \dfrac{x-2}{x+3}$

30. $\dfrac{x^2}{x^2 - 9} \cdot \dfrac{x^2 - 5x + 6}{x^2 - 2x}$

31. $\dfrac{2a+b}{8} \div \dfrac{2a+b}{2}$

32. $\dfrac{17a^2}{21y^2} \div \dfrac{34a^2}{68}$

33. $\dfrac{y^2 - 2y + 1}{y} \div \dfrac{1}{y-1}$

34. $\dfrac{x^2 y^2}{3x^2 - 3y^2} \div \dfrac{8xy}{x-y}$

35. $\dfrac{y^2 + 6y + 9}{y^2 + 4y + 4} \div \dfrac{y+3}{y+2}$

36. $\dfrac{2x+2y}{3} \div \dfrac{x^2 - y^2}{y-x}$

37. $\dfrac{3x^2 + 6x}{x} \div \dfrac{2x+4}{x^2}$

38. $\dfrac{x^2 - 7x}{x^2 - 3x - 28} \div \dfrac{1}{-x-4}$

39. $\dfrac{y^2 - 16}{y+3} \div \dfrac{y-4}{y^2 - 9}$

40. $\dfrac{12x + 24}{36x - 36} \div \dfrac{6x + 12}{8x - 8}$

Simplify.

41. $\dfrac{\frac{5}{x-3}}{4}$

42. $\dfrac{6ab}{\frac{3a}{4}}$

43. $\dfrac{\frac{x^2 - 4x}{6x}}{\frac{x-4}{8x^2}}$

44. $\dfrac{\frac{2}{7}}{\frac{3}{4}}$

Rationalize the denominator of each fraction.

45. $\dfrac{12}{6 - \sqrt{5}}$

46. $\dfrac{8}{3 - \sqrt{2}}$

47. $\dfrac{7 + \sqrt{3}}{7 - \sqrt{3}}$

Chapter 14 / Rational Expressions and Equations

48. $\dfrac{4 - \sqrt{7}}{3 + \sqrt{5}}$

49. $\dfrac{5 + \sqrt{2}}{7 - 3\sqrt{5}}$

50. $\dfrac{4 - \sqrt{7}}{8 + 2\sqrt{11}}$

Rationalize the numerator of each fraction.

51. $\dfrac{8 + 2\sqrt{3}}{5}$

52. $\dfrac{7 - \sqrt{5}}{11}$

53. $\dfrac{4 - \sqrt{13}}{12}$

54. $\dfrac{1 - 3\sqrt{7}}{8}$

55. $\dfrac{5 - \sqrt{7}}{16}$

56. $\dfrac{2 + \sqrt{15}}{15}$

Section 14–3

Add or subtract the fractions. Reduce to simplest terms.

57. $\dfrac{2}{9} + \dfrac{4}{9}$

58. $\dfrac{2}{3} + \dfrac{5}{12}$

59. $\dfrac{3x}{7} + \dfrac{2x}{14}$

60. $\dfrac{5x}{3} - \dfrac{2x}{4}$

61. $\dfrac{3x}{4} + \dfrac{5x}{6}$

62. $\dfrac{3}{4} + \dfrac{7}{x}$

63. $\dfrac{5}{x} - \dfrac{7}{3}$

64. $\dfrac{3}{x} + \dfrac{5}{7}$

65. $\dfrac{3}{4x} + \dfrac{2}{x} + \dfrac{3}{6x}$

66. $\dfrac{6}{2x + 3} + \dfrac{2}{2x - 3}$

67. $\dfrac{7}{x - 3} + \dfrac{3}{x + 2}$

68. $\dfrac{4}{3x + 2} - \dfrac{3}{x - 2}$

69. $\dfrac{8}{x + 3} - \dfrac{2}{x - 4}$

70. $\dfrac{3}{x - 2} + \dfrac{5}{2 - x}$

71. $\dfrac{x}{x - 5} - \dfrac{3}{5 - x}$

Simplify.

72. $\dfrac{2 + \dfrac{3}{x}}{\dfrac{2}{x}}$

73. $\dfrac{\dfrac{5}{x} - \dfrac{3}{4x}}{\dfrac{1}{3x} + \dfrac{2}{x}}$

74. $\dfrac{5 - \dfrac{x - 2}{x}}{\dfrac{x - 4}{2x} - 2}$

75. $\dfrac{\dfrac{3x}{6} - \dfrac{5}{x}}{\dfrac{x}{3} + \dfrac{4}{2x}}$

Section 14–4

Determine the value or values that must be excluded as possible solutions of the following.

76. $\dfrac{3}{x} - \dfrac{4}{5} = 2$

77. $\dfrac{4}{x} = \dfrac{3}{x - 2}$

78. $\dfrac{3}{x - 5} - \dfrac{4}{x + 5} = \dfrac{1}{x^2 - 25}$

79. $\dfrac{5}{2x - 1} - \dfrac{6}{x} = \dfrac{4}{3x}$

Solve the equations. Check for extraneous roots.

80. $\dfrac{3}{x} + \dfrac{2}{3} = 1$

81. $\dfrac{4}{x} = \dfrac{1}{x + 5}$

82. $\dfrac{3}{x - 4} + \dfrac{1}{x + 4} = \dfrac{1}{x^2 - 16}$

83. $-\dfrac{4x}{x + 1} = 3 - \dfrac{4}{x + 1}$

Use rational equations to solve the following problems. Check for extraneous or inappropriate roots.

84. Fugita can complete the assembly of 50 widgets in 6 days. Ohn can complete the same job in 8 days. How many days will it take to complete the project if Fugita and Ohn work together? (*Hint:* Rate × Time = Amount of work.)

85. Several students chipped in a total of $120 to buy a small refrigerator to use in the dorm. If the group was increased to four students, the cost to the original group dropped $10 per person. How many students were in the original group?

CHALLENGE PROBLEMS

86. Pipe 1 fills a tank in 4 min and pipe 2 fills the same tank in 3 min. What proportion of the tank will be filled if both pipes are open for 1 min? (*Hint:* Rate × Time = Amount of work.)

87. One machine can do a job in 5 hr alone. How many hours would it take a second machine to complete the job alone if both machines together can do the job in 3 hr?

1. How are the properties $\frac{n}{n} \times 1$ and $1 \times n = n$ applied in the following example: $\frac{4}{6} = \frac{2}{3}$?

2. Why is the following problem incorrect?
$$\frac{5}{10} = \frac{2+3}{2+8} = \frac{3}{8}$$

3. Write a brief comment for each step of the following example.

$$\frac{x^2 + 3x + 2}{x^2 - 9} \div \frac{2x^2 + 3x - 2}{2x^2 - 7x + 3} =$$

$$\frac{x^2 + 3x + 2}{x^2 - 9} \cdot \frac{2x^2 - 7x + 3}{2x^2 + 3x - 2} =$$

$$\frac{(x+1)(x+2)}{(x+3)(x-3)} \cdot \frac{(2x-1)(x-3)}{(2x-1)(x+2)} = \frac{x+1}{x+3}$$

Find the mistake or mistakes in each problem and briefly explain each mistake. Then, rework the problem correctly.

4. $\frac{x}{x+3} = \frac{\cancel{x}}{\cancel{x}+3} = \frac{1}{3}$

5. $\frac{x^2 - 4}{x^2 + 4x + 4} \div x + 2 =$

$$\frac{(x+2)(x-2)}{(x+2)(x+2)} \cdot \frac{x+2}{1} = x - 2$$

6. $\frac{5}{x+2} + \frac{3}{x-3} = \frac{8}{(x+2)(x-3)}$

7. $\frac{3}{x-1} - \frac{5}{x+1} =$

$$\frac{3(x+1)}{(x-1)(x+1)} - \frac{5(x-1)}{(x-1)(x+1)} =$$

$$\frac{3x + 3 - 5x - 5}{(x+1)(x-1)} =$$

$$\frac{-2x - 2}{(x+1)(x-1)} =$$

$$\frac{-2(x+1)}{(x+1)(x-1)} = \frac{-2}{x-1}$$

8. $\frac{14}{3x} = \frac{12}{7x}$

$$\frac{\overset{2}{\cancel{14}}}{\underset{1}{\cancel{3x}}} = \frac{\overset{4}{\cancel{12}}}{\underset{1}{\cancel{7x}}}$$

$$8 = x^2$$
$$\pm\sqrt{8} = x$$
$$\pm 2\sqrt{2} = x$$

9. How is a division of rational expressions related to a multiplication of rational expressions?

10. In your own words, write a rule for multiplying rational expressions.

Learning Outcomes | **What to Remember with Examples**

Section 14–1
1 Simplify or reduce rational expressions (pp. 572–575).

If the rational expression is not in factored form, factor completely the numerator and the denominator. Then, reduce factors common to both the numerator and denominator.

Simplify $\frac{a+b}{16a^2 + 16ab}$. Factor common factor of $16a$ in the denominator.

$$\frac{a+b}{16a(a+b)}$$ Reduce the common binomial. Factor.

$$\frac{\cancel{a+b}}{16a(\cancel{a+b})} = \frac{1}{16a}$$

Section 14–2

1 Multiply and divide rational expressions (pp. 576–577).

Write the numerator and denominator in factored form whenever possible. For multiplication, reduce factors common to both the numerator and denominator. Then, multiply remaining factors.

$$\text{Multiply } \frac{x}{2x+8} \cdot \frac{2}{x+1}.$$

Write each numerator and denominator in factored form.

$$\frac{x}{2(x+4)} \cdot \frac{2}{x+1} =$$

Reduce.

$$\frac{(\cancel{2})(x)}{\cancel{2}(x+4)(x+1)} =$$

Multiply.

$$\frac{x}{(x+4)(x+1)} \quad \text{or} \quad \frac{x}{x^2+5x+4}$$

Factored or expanded form.

For division, multiply by the reciprocal of the second fraction. Factor, reduce factors common to both a numerator and denominator, then multiply remaining factors.

$$\text{Divide } \frac{x}{x+2} \div \frac{2}{x+2}.$$

Multiply by the reciprocal of the divisor. Reduce.

$$\frac{x}{\cancel{x+2}} \cdot \frac{\cancel{x+2}}{2} = \frac{x}{2}$$

Multiply remaining factors.

2 Simplify complex rational expressions (pp. 577–579).

To simplify complex rational expressions, rewrite as the division of two rational expressions. Next, rewrite the division as a multiplication of the first fraction by the reciprocal of the second fraction. Factor and perform the indicated multiplication.

Simplify.

$$\frac{\dfrac{x-3}{5x}}{\dfrac{4x-12}{15x^2}}$$

Write as division.

$$\frac{x-3}{5x} \div \frac{4x-12}{15x^2} =$$

Rewrite as multiplication, factor, and reduce.

$$\frac{x\cancel{-3}}{\cancel{5x}} \cdot \frac{\overset{3x}{\cancel{15x^2}}}{4(x\cancel{-3})} =$$

Multiply remaining factors.

$$\frac{3x}{4}$$

3 Use multiplication and conjugates to rationalize a numerator or denominator of a fraction that has a binomial with an irrational term (pp. 579–580).

1. Determine which term of the fraction (numerator or denominator) is to be cleared of a binomial with an irrational term.
2. Multiply by 1 in the form of $\frac{n}{n}$, where n is the conjugate of the binomial with the irrational term.
3. Simplify the resulting expression.

Rationalize the denominator in the expression $\dfrac{5x}{3+2\sqrt{5}}$.

$$\frac{5x}{3+2\sqrt{5}} \cdot \frac{3-2\sqrt{5}}{3-2\sqrt{5}} =$$

Multiply by 1 in the form of $\dfrac{3-2\sqrt{5}}{3-2\sqrt{5}}$.

$$\frac{5x(3-2\sqrt{5})}{3^2-(2\sqrt{5})^2} =$$

Distribute and simplify.

$$\frac{15x-10x\sqrt{5}}{9-4(5)} = \frac{15x-10x\sqrt{5}}{9-20}$$

$$\frac{15x-10x\sqrt{5}}{-11} \quad \text{or} \quad -\frac{15x-10x\sqrt{5}}{11}$$

Chapter Summary

Section 14–3

1 Add and subtract rational expressions (pp. 581–583).

Add (or subtract) only rational expressions with the same denominator. For rational expressions with unlike denominators:

1. Find the LCD.
2. Change each fraction to an equivalent fraction having the LCD.
3. Add (or subtract) the numerators.
4. Keep same LCD.
5. Reduce if possible.

Perform the following operations:

(a) $\dfrac{x}{x + 4} + \dfrac{2x}{x + 4} = \dfrac{3x}{x + 4}$ *Add numerators. Keep like denominator.*

(b) $\dfrac{3}{x + 2} - \dfrac{2}{x + 1}$ *LCD $= (x + 2)(x + 1)$. Change each fraction to an equivalent fraction with the LCD.*

$$\dfrac{3}{x + 2} = \dfrac{3(x + 1)}{(x + 2)(x + 1)}$$ *First fraction.*

$$= \dfrac{3x + 3}{(x + 2)(x + 1)}$$

$$\dfrac{2}{x + 1} = \dfrac{2(x + 2)}{(x + 1)(x + 2)}$$ *Second fraction.*

$$= \dfrac{2x + 4}{(x + 1)(x + 2)}$$

$$\dfrac{3x + 3}{(x + 2)(x + 1)} - \dfrac{2x + 4}{(x + 2)(x + 1)} = \dfrac{3x + 3 - (2x + 4)}{(x + 1)(x + 2)}$$ *Subtract numerators. Keep common denominator.*

$$\dfrac{3x + 3 - 2x - 4}{(x + 1)(x + 2)} = \dfrac{x - 1}{(x + 1)(x + 2)}$$

2 Use addition and subtraction of rational expressions to simplify complex fractions (pp. 583–584).

1. Use addition or subtraction to combine terms in the numerator or denominator or both.
2. Rewrite the complex fraction as a division then convert to an equivalent multiplication of rational expressions.
3. Use the procedure for multiplying rational expressions.

Simplify $\dfrac{\dfrac{3}{x} + 5}{\dfrac{2}{3x} - \dfrac{1}{x}}$.

$$\dfrac{\dfrac{3}{x} + \dfrac{5}{1}\left(\dfrac{x}{x}\right)}{\dfrac{2}{3x} - \dfrac{1}{x}\left(\dfrac{3}{3}\right)}$$ *Change to equivalent fractions in numerator with LCD of x.*

Change to equivalent fractions in denominator with LCD of $3x$.

$$\dfrac{\dfrac{3}{x} + \dfrac{5x}{x}}{\dfrac{2}{3x} - \dfrac{3}{3x}}$$ *Add fractions in the numerator.*

Subtract fractions in the denominator.

$$\dfrac{\dfrac{3 + 5x}{x}}{\dfrac{-1}{3x}}$$ *Rewrite as multiplication of reciprocal.*

Chapter 14 / Rational Expressions and Equations

$$\frac{3 + 5x}{\cancel{x}} \cdot \frac{3\cancel{x}}{-1} = \quad \text{Reduce and multiply.}$$

$$\frac{(3 + 5x)3}{-1} = \quad \text{Distribute.}$$

$$\frac{9 + 15x}{-1} = -(9 + 15x) \quad \text{or} \quad -9 - 15x$$

Section 14–4

1 Exclude certain values as solutions of rational equations (pp. 585–586).

To find excluded values, set each denominator with a variable equal to zero. The excluded values obtained cannot be accepted as solutions of the rational equation.

Find the excluded values for the equation:

$$4 = \frac{2}{3 + y} \quad \text{Set denominator equal to zero.}$$
$$3 + y = 0$$
$$y = -3 \quad \text{Excluded value.}$$

2 Solve rational equations with variable denominators (pp. 586–589).

To solve rational equations with variable denominators, clear the equation of fractions by multiplying each term of the equation by the LCM of all denominators in the equation. Solve the resulting equation. Check all solutions to determine if any are extraneous roots.

Solve $\dfrac{2}{x - 3} - \dfrac{1}{x + 3} = \dfrac{1}{x^2 - 9}$.

Excluded values are 3 and −3.
LCM is $x^2 - 9$ or $(x - 3)(x + 3)$.

$$\cancel{(x - 3)}(x + 3)\frac{2}{\cancel{x - 3}} - (x - 3)\cancel{(x + 3)}\frac{1}{\cancel{x + 3}} = \cancel{(x - 3)}\cancel{(x + 3)}\frac{1}{\cancel{(x - 3)}\cancel{(x + 3)}}$$
$$2(x + 3) - 1(x - 3) = 1$$
$$2x + 6 - x + 3 = 1$$
$$x + 9 = 1$$
$$x = -8$$

Check:

$$\frac{2}{-8 - 3} - \frac{1}{-8 + 3} = \frac{1}{(-8)^2 - 9}$$
$$\frac{2}{-11} - \frac{1}{-5} = \frac{1}{64 - 9}$$
$$\frac{-2}{11} + \frac{1}{5} = \frac{1}{55}$$
$$\frac{-10}{55} + \frac{11}{55} = \frac{1}{55}$$
$$\frac{1}{55} = \frac{1}{55}$$

CHAPTER TRIAL TEST

Perform the indicated operations.

1. $\dfrac{x - 3}{2x - 6}$

2. $\dfrac{x^2 - 16}{x - 4}$

3. $\dfrac{6x^2 - 11x + 4}{2x^2 + 5x - 3}$

4. $\dfrac{x^2 - 6x + 8}{2 - x}$

5. $\dfrac{(x - 2)(x - 4)}{(4 - x)(x + 2)}$

6. $\dfrac{y^2 + x^2}{y^2 - x^2}$

7. $\dfrac{6xy}{ab} \cdot \dfrac{a^2b}{2xy^2}$

8. $\dfrac{x^2 - y^2}{2x + y} \cdot \dfrac{4x^2 + 2xy}{y - x}$

9. $\dfrac{x - 2y}{x^3 - 3x^2y} \div \dfrac{x^2 - 4y^2}{x - 3y}$

10. $\dfrac{2a^2 - ab - b^2}{6x^2 + x - 1} \div \dfrac{a^2 - b^2}{8x + 4}$

11. $\dfrac{2x^2 + 3x + 1}{x} \div \dfrac{x + 1}{1}$

12. $\dfrac{4y^2}{2x} \cdot \dfrac{x}{8x}$

13. $\dfrac{1}{x + 2} - \dfrac{1}{x - 3}$

14. $\dfrac{2}{x + 2} + \dfrac{3}{x - 1}$

15. $\dfrac{3}{x} + \dfrac{1}{4}$

16. $\dfrac{2}{3y} - \dfrac{7}{y}$

17. $\dfrac{5}{3x - 2} + \dfrac{7}{2 - 3x}$

18. $\dfrac{2x}{3} - \dfrac{5x}{2}$

19. $\dfrac{x - 2y}{x^2 - 4y^2}$

20. $\dfrac{3}{2y} + \dfrac{2}{y} + \dfrac{1}{5y}$

21. $\dfrac{2x}{1 - \dfrac{3}{x}}$

22. $\dfrac{\dfrac{1}{a} + \dfrac{1}{b}}{ab}$

Determine the excluded values of x.

23. $\dfrac{5}{x} = \dfrac{2}{x + 3}$

24. $\dfrac{3}{x - 4} + \dfrac{5}{x + 4} = 6$

Solve the rational equations. Check for extraneous roots.

25. $\dfrac{3x}{x - 2} + 4 = \dfrac{3}{x - 2}$

26. $\dfrac{x}{x + 3} = 5$

27. $\dfrac{5}{x - 2} = \dfrac{-4}{x + 1}$

Use rational equations to solve the problems. Check for extraneous roots.

28. Henry can assemble four computers in 3 hr. Lester can assemble four computers in 4 hr. How long will it take to assemble four computers if both work together? (*Hint:* Rate × Time = Amount of work.)

29. A medical group purchased land in Colorado for $100,000. When the size of the group doubled, the cost dropped $10,000 per person in the original group. How many persons were in the original group?

30. Cedric Partee drove 300 mi in one day while on a vacation to the mountains, but on the return trip by the same route he drove 10 mi per hour less and the return trip took 1 hr longer. Find Partee's speed to and from the mountains. (*Hint:* Time = Distance ÷ Rate.)

Quadratic and Higher-Degree Equations

15

Most of the equations we have solved so far are called linear equations. In such equations, the variable appears only to the first power, such as x or y. In Chapter 11, we looked briefly at one type of quadratic equation. In this chapter, we continue to examine quadratic equations.

15–1 | Solving Quadratic Equations by the Square-Root Method

Learning Outcomes

1. Write quadratic equations in standard form.
2. Identify the coefficients of the quadratic, linear, and constant terms of a quadratic equation.
3. Solve pure quadratic equations by the square-root method ($ax^2 + c = 0$).

1 Write Quadratic Equations in Standard Form.

There are several methods for solving quadratic equations. Each method has strengths and weaknesses depending on the characteristics of the equation being solved. First, let's examine the basic characteristics of quadratic equations, starting with quadratic equations having only one variable. We shall then examine quadratic equations written as functions, which involve both independent and dependent variables.

A *quadratic equation* is an equation in which at least one letter term is raised to the second power and no letter term has a power more than 2 or less than 0. The *standard form* for a quadratic equation is $ax^2 + bx + c = 0$, where a, b, and c are real numbers and $a > 0$.

EXAMPLE   Arrange the quadratic equations in standard form.

(a) $3x^2 - 3 + 5x = 0$ (b) $7x - 2x^2 - 8 = 0$
(c) $12 = 7x^2 - 4x$ (d) $7x^2 = 3$

(a) $3x^2 - 3 + 5x = 0$ Arrange the terms in descending powers of x.
$\quad 3x^2 + 5x - 3 = 0$ Standard form.
(b) $7x - 2x^2 - 8 = 0$ Arrange the terms in descending powers of x.
$\quad -2x^2 + 7x - 8 = 0$ Multiply both sides by -1 to make the coefficient of the
x^2 term positive.

$\quad -1(-2x^2 + 7x - 8) = -1(0)$
$\quad\quad 2x^2 - 7x + 8 = 0$ Standard form.
(c) $12 = 7x^2 - 4x$ Rearrange the terms so that one side of the equation
equals zero.

$\quad 0 = 7x^2 - 4x - 12$ Interchange sides of equation.
$\quad 7x^2 - 4x - 12 = 0$
(d) $\quad\quad 7x^2 = 3$ Rearrange so that one side equals zero.

$\quad 7x^2 - 3 = 0$

A quadratic equation is a second-degree equation, which means that it can have as many as two real solutions. As we examine the process for solving quadratic equations, we shall see quadratic equations that have two, one, or no solutions.

2 Identify the Coefficients of the Quadratic, Linear, and Constant Terms of a Quadratic Equation.

Some methods for solving quadratic equations can be used for any type of quadratic equation but are very time-consuming. Other methods are quicker but apply only to certain types of quadratic equations. In choosing an appropriate method it is helpful to be able to recognize similar and different characteristics of quadratic equations.

The standard form of quadratic equations, $ax^2 + bx + c = 0$, has three types of terms.

1. ax^2 is a *quadratic term*; that is, the degree of the term is 2. In standard form, this term is the *leading term* and a is the *leading coefficient*.
2. bx is a *linear term*; that is, the degree of the term is 1 and the coefficient is b.
3. c is a *number* term or *constant term*; that is, the degree of the term is 0. The coefficient is c ($c = cx^0$).

A quadratic equation that has all three types of terms is sometimes referred to as a *complete quadratic equation*.

$ax^2 + bx + c = 0$ Complete quadratic equation.

A quadratic equation that has a quadratic term (ax^2) and a linear term (bx) but no constant term (c) is sometimes referred to as an *incomplete quadratic equation*.

$ax^2 + bx = 0$ Incomplete quadratic equation.

A quadratic equation that has a quadratic term (ax^2) and a constant (c) but no linear term (bx) is sometimes referred to as a *pure quadratic equation*, the type that we worked in Chapter 11.

$ax^2 + c = 0$ Pure quadratic equation.

It is helpful in planning our strategy for solving a quadratic equation to be able to identify the types of terms in a quadratic equation and to identify the coefficients a, b, and c.

EXAMPLE Write each equation in standard form and identify a, b, and c.

(a) $3x^2 + 5x - 2 = 0$ (b) $5x^2 = 2x - 3$
(c) $3x = 5x^2$ (d) $-6x^2 + 2x = 0$
(e) $5x^2 - 4 = 0$ (f) $x^2 = 9$
(g) $x^2 + 3x = 5x$ (h) $7 - x^2 = 3 + x$

(a) $3x^2 + 5x - 2 = 0$ In standard form.
$a = 3, \quad b = 5, \quad c = -2$

(b) $5x^2 = 2x - 3$ Write in standard form.
$5x^2 - 2x + 3 = 0$
$a = 5, \quad b = -2, \quad c = 3$

(c) $3x = 5x^2$ Write in standard form.
$0 = 5x^2 - 3x$ Interchange sides of equation.
$5x^2 - 3x = 0$
$a = 5, \quad b = -3, \quad c = 0$ No constant term means $c = 0$.

(d) $-6x^2 + 2x = 0$ Multiply by -1.
$-1(-6x^2 + 2x = 0)$
$6x^2 - 2x = 0$ Standard form.
$a = 6, \quad b = -2, \quad c = 0$ No constant term.

(e) $5x^2 - 4 = 0$ Standard form.
$a = 5, \quad b = 0, \quad c = -4$ No linear term means $b = 0$.

(f) $x^2 = 9$ Write in standard form.
$x^2 - 9 = 0$
$a = 1, \quad b = 0, \quad c = -9$ No linear term.

(g) $x^2 + 3x = 5x$ Write in standard form.
$x^2 + 3x - 5x = 0$ Combine like terms.
$x^2 - 2x = 0$
$a = 1, \quad b = -2, \quad c = 0$ No constant term.

15–1 Solving Quadratic Equations by the Square-Root Method

599

(h) $7 - x^2 = 3 + x$ Rearrange.
$-x^2 - x + 7 - 3 = 0$ Combine like terms.
$-x^2 - x + 4 = 0$ Multiply by -1.
$-1(-x^2 - x + 4 = 0)$
$x^2 + x - 4 = 0$ Standard form.
$a = 1, \quad b = 1, \quad c = -4$

3   **Solve Pure Quadratic Equations by the Square-Root Method ($ax^2 + c = 0$).**

To solve pure quadratic equations, solve for the squared letter, then take the square root of both sides of the equation. This procedure is the same one we used in Chapter 11. All roots or solutions should continue to be checked.

To solve a pure quadratic equation ($ax^2 + c = 0$):

1. Rearrange the equation, if necessary, so that quadratic or squared-letter terms are on one side and constants or number terms are on the other side of the equation.
2. Combine like terms, if appropriate.
3. Solve for the squared letter so that the coefficient is $+1$.
4. Apply the square-root principle of equality by taking the square root of both sides.
5. Check each solution in the original equation.

EXAMPLE   Solve $3y^2 = 27$

$3y^2 = 27$ Divide both sides by the coefficient of the quadratic term.
$y^2 = 9$ Take the square root of both sides.
$y = \pm 0$ Solutions are $y = 2$ or $y = -2$

Check $y = +3$	Check $y = -3$
$3y^2 = 27$	$3y^2 = 27$
$3(3)^2 = 27$	$3(-3)^2 = 27$
$3(9) = 27$	$3(9) = 27$
$27 = 27$	$27 = 27$ Both solutions check.

EXAMPLE   Solve $4x^2 - 9 = 0$.

$4x^2 - 9 = 0$ Transpose or sort terms.
$4x^2 = 9$ Divide by 4.
$x^2 = \dfrac{9}{4}$ Take the square root of the numerator and denominator if they are perfect-square whole numbers.
$x = \pm \dfrac{3}{2}$ Divide if a decimal answer is desired.
$x = \pm 1.5$ Solutions are $x = 1.5$ or $x = -1.5$.

Check $x = +1.5$	Check $x = -1.5$
$4x^2 - 9 = 0$	$4x^2 - 9 = 0$
$4(1.5)^2 - 9 = 0$	$4(-1.5)^2 - 9 = 0$
$4(2.25) - 9 = 0$	$4(2.25) - 9 = 0$
$9 - 9 = 0$	$9 - 9 = 0$
$0 = 0$	$0 = 0$

In some applications, the fractional answer may be more convenient. In other applications, the decimal answer may be more convenient.

Not all pure quadratic equations have *real solutions*. The next example illustrates such an equation.

EXAMPLE Solve $3m^2 + 48 = 0$.

$$3m^2 + 48 = 0$$
$$3m^2 = -48 \qquad \text{Isolate the variable.}$$
$$m^2 = \frac{-48}{3} \qquad \text{Solve for } m^2.$$
$$m^2 = -16 \qquad \text{Apply the square-root principle.}$$
$$m = \pm\sqrt{-16}$$

There are *no real solutions* to this equation. **If imaginary roots are acceptable, the roots are $\pm 4i$.**

TIP!

Roots of Pure Quadratic Equations

In a pure quadratic equation, the value of b is zero: $ax^2 + 0x + c = 0$.

• The equation has real solutions only if c is negative when the equation is in standard form. Remember, standard form requires that a be positive.
• The two solutions have the same absolute value.

SELF-STUDY EXERCISES 15–1

1 Arrange the quadratic equations in standard form.

1. $5 - 4x + 7x^2 = 0$
2. $8x^2 - 3 = 6x$
3. $7x^2 = 5$
4. $x^2 = 6x - 8$
5. $x^2 - 3x = 6x - 8$
6. $8 - x^2 + 6x = 2x$
7. $3x^2 + 5 - 6x = 0$
8. $5 = x^2 - 6x$
9. $x^2 - 16 = 0$
10. $8x^2 - 7x = 8$
11. $8x^2 + 8x - 2 = 8$
12. $0.3x^2 - 0.4x = 3$

2 Write each equation in standard form and identify a, b, and c.

13. $5x = x^2$
14. $7x - 3x^2 = 5$
15. $7x^2 - 4x = 0$
16. $8 + 3x^2 = 5x$
17. $5x - 6 = x^2$
18. $11x^2 = 8x$
19. $x = x^2$
20. $9x^2 - 7x = 12$
21. $5 = x^2$
22. $-x^2 - 6x + 3 = 0$
23. $0.2x - 5x^2 = 1.4$
24. $\frac{2}{3}x^2 - \frac{5}{6}x = \frac{1}{2}$
25. $1.3x^2 - 8 = 0$
26. $\sqrt{3}x^2 + \sqrt{5}x - 2 = 0$

Write equations in standard form.

27. The coefficient of the quadratic term is 8, the coefficient of the linear term is -2, the constant term is -3.

28. The constant is 0, the coefficient of the linear term is 3, and the coefficient of the quadratic term is 1.

29. $a = 5$, $b = 2$, $c = -7$

30. $a = 2.5$, $c = -0.8$

Solve the equations. Round to thousandths when necessary.

31. $x^2 = 9$ **32.** $x^2 - 49 = 0$ **33.** $9x^2 = 64$
34. $16x^2 - 49 = 0$ **35.** $0.09y^2 = 0.81$ **36.** $0.04x^2 = 0.81$
37. $2x^2 = 10$ **38.** $3x^2 + 4 = 7$ **39.** $5x^2 - 8 = 12$
40. $7x^2 - 2 = 19$

41. The label on a new tarpaulin shows it is a square and contains 529 ft². What is the length of each side of the tarpaulin?

42. A farmer has a field designed in the shape of a circle for efficiency in irrigation. The area of the field is known to be 21,000 yd². What length of irrigation line is required to provide water to the entire field? (*Note:* The line moves from the center of the field in a circle around the field.)

43. Describe the process for solving a pure quadratic equation.

44. What is the relationship between the roots of a pure quadratic equation?

15–2 | *Solving Quadratic Equations by Factoring*

Learning Outcomes

1 Solve incomplete quadratic equations by factoring ($ax^2 + bx = 0$).
2 Solve complete quadratic equations by factoring ($ax^2 + bx + c = 0$).

1 **Solve Incomplete Quadratic Equations by Factoring ($ax^2 + bx = 0$).**

An incomplete quadratic equation always has a variable common factor, so we can factor the expression $x^2 + 2x = 0$ to $x(x + 2) = 0$. We have two factors that have a product of 0. The *zero-product property* shown below can be used to solve the equation.

Zero-product property

If $ab = 0$, then a or b or both equal zero.

EXAMPLE Solve $x^2 + 2x = 0$ by factoring.

$x(x + 2) = 0$ — Factor. Set each factor equal to zero using the zero-product property.
$x = 0 \quad x + 2 = 0$ — Solve each equation for x.
$x = 0 \quad x = -2$

If $x = 0$, we can check by substituting 0 for x wherever x occurs in the original equation.

For $x = 0$, $x^2 + 2x = 0$
$0^2 + 2(0) = 0$
$0 + 0 = 0$
$0 = 0$

If $x = -2$, we can check by substituting -2 for x wherever x occurs in the original equation.

$x^2 + 2x = 0$
$(-2)^2 + 2(-2) = 0$
$4 + (-4) = 0$
$0 = 0$

1. Sort or transpose so that all terms are on one side of the equation in standard form, with zero on the other side.
2. Factor all common factors and set each variable factor equal to zero.
3. Solve for the variable in each equation formed in Step 2.
4. Check each solution in the original equation.

EXAMPLE   Solve $2x^2 = 5x$ for x.

$$2x^2 = 5x \qquad \text{Put in standard form.}$$
$$2x^2 - 5x = 0 \qquad \text{Factor common factor.}$$
$$x(2x - 5) = 0 \qquad \text{Set each factor equal to zero.}$$
$$x = 0 \quad 2x - 5 = 0 \qquad \text{Solve each equation.}$$
$$x = 0 \qquad 2x = 5$$
$$x = \frac{5}{2}$$

$$\text{Check } x = 0 \qquad \text{Check } x = \frac{5}{2}$$
$$2x^2 = 5x \qquad\qquad 2x^2 = 5x$$
$$2(0)^2 = 5(0) \qquad 2\left(\frac{5}{2}\right)^2 = 5\left(\frac{5}{2}\right)$$
$$2(0) = 5(0) \qquad \overset{1}{2}\left(\frac{25}{4}\right) = \frac{25}{2}$$
$$0 = 0 \qquad\qquad \frac{25}{2} = \frac{25}{2}$$

EXAMPLE   Solve $4x^2 - 8x = 0$ for x.

$$4x^2 - 8x = 0 \qquad \text{Standard form. Factor common factors.}$$
$$4x(x - 2) = 0 \qquad \text{Set each variable factor equal to zero.}$$
$$4x = 0 \qquad x - 2 = 0 \qquad \text{Solve each equation.}$$
$$x = \frac{0}{4} \qquad\qquad x = 2$$
$$x = 0$$

$$\text{Check } x = 0 \qquad\qquad \text{Check } x = 2$$
$$4(0)^2 - 8(0) = 0 \qquad 4(2)^2 - 8(2) = 0$$
$$4(0) - 8(0) = 0 \qquad 4(4) - 8(2) = 0$$
$$0 - 0 = 0 \qquad\qquad 16 - 16 = 0$$
$$0 = 0 \qquad\qquad 0 = 0$$

Roots of Incomplete Quadratic Equations

In an incomplete quadratic equation, the value of $c = 0$: $ax^2 + bx + 0 = 0$.

$$x(ax + b) = 0$$
$$x = 0 \qquad ax + b = 0$$
$$\frac{ax}{a} = \frac{-b}{a}$$
$$x = \frac{-b}{a}$$

- One root is always zero.
- The other root is $\dfrac{-b}{a}$.

2 **Solve Complete Quadratic Equations by Factoring ($ax^2 + bx + c = 0$).**

Complete quadratic equations have all three types of terms. If the expression on the left will factor into two binomials, we can apply the zero-product property. If the trinomial will not factor, we must use another method for solving the quadratic equation.

To solve a complete quadratic equation by factoring:

1. Arrange the equation in standard form.
2. Factor the trinomial into the product of two binomials, and set each factor equal to zero.
3. Solve for the variable in each equation formed in Step 2.

EXAMPLE   Solve $x^2 + 6x + 5 = 0$ for x.

$$x^2 + 6x + 5 = 0 \qquad \text{Standard form. Factor into the product of two binomials.}$$
$$(x + 5)(x + 1) = 0 \qquad \text{Set each factor equal to 0.}$$
$$x + 5 = 0 \qquad x + 1 = 0 \qquad \text{Solve for } x \text{ in each equation.}$$
$$x = -5 \qquad x = -1$$

Check $x = -5$	Check $x = -1$
$x^2 + 6x + 5 = 0$	$x^2 + 6x + 5 = 0$
$(-5)^2 + 6(-5) + 5 = 0$	$(-1)^2 + 6(-1) + 5 = 0$
$25 + (-30) + 5 = 0$	$1 + (-6) + 5 = 0$
$-5 + 5 = 0$	$-5 + 5 = 0$
$0 = 0$	$0 = 0$

EXAMPLE   Solve $6x^2 + 4 = 11x$ for x.

$$6x^2 + 4 = 11x$$ Transpose into standard form.

$$6x^2 - 11x + 4 = 0$$ Factor trinomial by grouping. Write $-11x$ as $-3x - 8x$.

$$6x^2 - 3x - 8x + 4 = 0$$ Group.

$$(6x^2 - 3x) + (-8x + 4) = 0$$ Factor common factors.

$$3x(2x - 1) - 4(2x - 1) = 0$$ Factor the common grouping.

$$(3x - 4)(2x - 1) = 0$$ Set each factor equal to 0.

$$3x - 4 = 0 \qquad 2x - 1 = 0$$ Solve each equation.

$$3x = 4 \qquad\qquad 2x = 1$$

$$x = \frac{4}{3} \qquad\qquad x = \frac{1}{2}$$

Check $x = \dfrac{4}{3}$

$$6x^2 + 4 = 11x$$

$$6\left(\frac{4}{3}\right)^2 + 4 = 11\left(\frac{4}{3}\right)$$

$$\overset{2}{6}\left(\frac{16}{\underset{3}{9}}\right) + 4 = \frac{44}{3}$$

$$\frac{32}{3} + \frac{12}{3} = \frac{44}{3} \qquad \left(4 = \frac{12}{3}\right)$$

$$\frac{44}{3} = \frac{44}{3}$$

Check $x = \dfrac{1}{2}$

$$6x^2 + 4 = 11x$$

$$6\left(\frac{1}{2}\right)^2 + 4 = 11\left(\frac{1}{2}\right)$$

$$\overset{3}{6}\left(\frac{1}{\underset{2}{4}}\right) + 4 = \frac{11}{2}$$

$$\frac{3}{2} + \frac{8}{2} = \frac{11}{2} \qquad \left(4 = \frac{8}{2}\right)$$

$$\frac{11}{2} = \frac{11}{2}$$

TIP!

Roots of Complete Quadratic Equations

Because all three types of terms are included in complete quadratic equations, the roots do *not* have the same characteristics as the roots of pure or incomplete quadratic equations.

- Zero is not a root.
- The two roots are not opposites.

SELF-STUDY EXERCISES 15–2

1 Solve.

1. $x^2 - 3x = 0$
2. $x^2 - 6x = 0$
3. $5x^2 - 10x = 0$
4. $2x^2 + x = 0$
5. $8x^2 - 4x = 0$
6. $5x^2 - 15x = 0$
7. $3x^2 - 7x = 0$
8. $y^2 + 4y = 0$
9. $3x^2 + 2x = 0$
10. $9x^2 = 12x$
11. $6x^2 = 18x$
12. $9x^2 = 6x$
13. The square of a number is 8 times the number. Find the number.
14. A square rug has an area 15 times the length of one of the sides. What is the length of a side?
15. How does an incomplete quadratic equation differ from a pure quadratic equation?
16. Will one of the two roots of an incomplete quadratic equation always be zero? Justify your answer.

Solve the equations by factoring.

17. $x^2 + 5x + 6 = 0$ **18.** $x^2 - 6x + 9 = 0$ **19.** $x^2 - 5x - 14 = 0$
20. $x^2 + 3x - 18 = 0$ **21.** $x^2 + 7x + 12 = 0$ **22.** $y^2 - 8y = -15$
23. $a^2 - 13a = 14$ **24.** $b^2 - 9b = -18$ **25.** $2x^2 - 7x + 3 = 0$
26. $3x^2 + 13x + 4 = 0$ **27.** $10x^2 - x - 3 = 0$ **28.** $6x^2 + 11x + 3 = 0$
29. $2x^2 + 13x + 15 = 0$ **30.** $3x^2 - 10x + 8 = 0$ **31.** $6x^2 + 17x - 3 = 0$
32. $3x^2 + 14x + 8 = 0$ **33.** $2x^2 - 13x + 15 = 0$ **34.** $6x^2 - 7x + 2 = 0$
35. $8x^2 + 3 = 10x$ **36.** $5x^2 + 3x = 2$ **37.** $6x^2 = x + 15$
38. $9x^2 + 18x = -5$ **39.** $6x^2 + 3 = 11x$ **40.** $5x^2 + 13x = 6$

41. A rectangular hallway is 6 ft longer than its width. The area is 55 ft². What are the length and width of the hallway?

42. A rectangular metal plate covering a spare tire well that has an opening of 378 in² is broken and must be reconstructed. The width is 3 in. less than the length. What dimensions should the metalsmith use when making the replacement part?

15–3 | Solving Quadratic Equations by Completing the Square

Learning Outcome **1** Solve quadratic equations by completing the square.

1 **Solve Quadratic Equations by Completing the Square.**

Not all complete quadratic equations can be solved by factoring. One procedure for solving these equations is called *completing the square*.

This method incorporates manipulations that result in a perfect-square trinomial that factors into the square of a binomial. Look again at the relationship between a perfect-square trinomial and the square of a binomial.

$$a^2 + 2ab + b^2 = (a + b)^2$$

In the expression $x^2 + 6x + c$, the value of c that would make the expression a perfect-square trinomial is $\left(\frac{6}{2}\right)^2 = 3^2 = 9$. Then, $x^2 + 6x + 9 = (x + 3)^2$. In a quadratic equation such as $x^2 + 6x = 2$, we can complete the square on the left (make the left side of the equation a perfect-square trinomal) by adding 9 to both sides of the equation.

$$x^2 + 6x + 9 = 2 + 9$$

Then,

$$(x + 3)^2 = 11$$

and take the square root of both sides.

$$x + 3 = \pm\sqrt{11}$$

To find the exact solutions, solve for x.

$$x = -3 \pm \sqrt{11}$$

or

$$x = -3 + \sqrt{11} \quad \text{and} \quad x = -3 - \sqrt{11}.$$

This method assumes that the coefficient of the quadratic term is 1.

To solve a quadratic equation by completing the square:

1. Write the equation in the form $ax^2 + bx = c$.
2. Divide the equation by a so that the coefficient of x^2 is 1.

$$x^2 + \frac{b}{a}x = \frac{c}{a}$$

Chapter 15 / Quadratic and Higher-Degree Equations

3. Form a perfect-square trinomial on the left by adding $\left(\dfrac{b}{2a}\right)^2$ or $\dfrac{b^2}{4a^2}$ to both sides of the equation.

$$x^2 + \dfrac{b}{a}x + \dfrac{b^2}{4a^2} = \dfrac{c}{a} + \dfrac{b^2}{4a^2}$$

4. Factor the perfect-square trinomial into the square of a binomial.

$$\left(x + \dfrac{b}{2a}\right)^2 = \dfrac{c}{a} + \dfrac{b^2}{4a^2}$$

5. Take the square root of both sides of the equation.

$$x + \dfrac{b}{2a} = \pm\sqrt{\dfrac{c}{a} + \dfrac{b^2}{4a^2}}$$

6. Solve for x.

$$x = -\dfrac{b}{2a} \pm \sqrt{\dfrac{c}{a} + \dfrac{b^2}{4a^2}}$$

EXAMPLE   Solve $x^2 - 5x - 2 = 0$ by completing the square.

$$x^2 - 5x - 2 = 0$$ The coefficient of x^2 is 1. Isolate the terms with variable factors.

$$x^2 - 5x = 2$$ Add $\left(\dfrac{-5}{2(1)}\right)^2$ to both sides ($a = 1$, $b = -5$). $\left(\dfrac{b}{2a}\right)^2$

$$x^2 - 5x + \left(\dfrac{-5}{2}\right)^2 = 2 + \left(\dfrac{-5}{2}\right)^2$$ Simplify.

$$x^2 - 5x + \dfrac{25}{4} = 2 + \dfrac{25}{4}$$

$$x^2 - 5x + \dfrac{25}{4} = \dfrac{8}{4} + \dfrac{25}{4}$$

$$x^2 - 5x + \dfrac{25}{4} = \dfrac{33}{4}$$ Write the perfect-square trinomial as the square of a binomial. $\left(x + \dfrac{b}{2a}\right)^2$

$$\left(x - \dfrac{5}{2}\right)^2 = \dfrac{33}{4}$$ Take the square root of both sides.

$$x - \dfrac{5}{2} = \pm\sqrt{\dfrac{33}{4}}$$ Simplify and solve for x.

$$x = \dfrac{5}{2} \pm \dfrac{\sqrt{33}}{2}$$ Identify each root.

$x = \dfrac{5}{2} + \dfrac{\sqrt{33}}{2}$ or $x = \dfrac{5}{2} - \dfrac{\sqrt{33}}{2}$ Exact roots.

$x = 2.5 + 2.872281323$ $x = 2.5 - 2.872281323$

$x = 5.37$ or $x = -0.37$ Approximate roots.

EXAMPLE   Solve $3x^2 - 3x - 7 = 0$ by completing the square.

$3x^2 - 3x - 7 = 0$ Make the leading coefficient 1 by dividing the equation by 3.

$\dfrac{3x^2}{3} - \dfrac{3x}{3} - \dfrac{7}{3} = 0$

$x^2 - x - \dfrac{7}{3} = 0$ Isolate the terms with variable factors.

$$x^2 - x = \frac{7}{3}$$

Add $\left(\frac{-1}{2(1)}\right)^2$ to both sides.

$$x^2 - x + \left(\frac{-1}{2}\right)^2 = \frac{7}{3} + \left(\frac{-1}{2}\right)^2$$

Simplify.

$$x^2 - x + \frac{1}{4} = \frac{7}{3} + \frac{1}{4}$$

$\frac{7}{3} + \frac{1}{4} = \frac{28}{12} + \frac{3}{12} = \frac{31}{12}$. Write the perfect-square trinomial as the square of the binomial.

$$x^2 - x + \frac{1}{4} = \frac{31}{12}$$

$$\left(x - \frac{1}{2}\right)^2 = \frac{31}{12}$$

Take the square root of both sides.

$$x - \frac{1}{2} = \pm\sqrt{\frac{31}{12}}$$

Simplify and solve for x,

$\sqrt{\frac{31}{12}} = \frac{\sqrt{31}}{2\sqrt{3}} \cdot \frac{\sqrt{3}}{\sqrt{3}} = \frac{\sqrt{93}}{6}$.

$$x = \frac{1}{2} \pm \frac{\sqrt{93}}{6}$$

Identify each root.

$$x = \frac{1}{2} + \frac{\sqrt{93}}{6} \qquad \text{or} \qquad x = \frac{1}{2} - \frac{\sqrt{93}}{6}$$

Exact roots.

$$x = 0.5 + 1.607275127 \qquad\qquad x = 0.5 - 1.607275127$$

$$\mathbf{x = 2.11} \qquad \text{or} \qquad \mathbf{x = -1.11}$$

Approximate solutions.

Completing-the-Square Method Works for All Types of Quadratic Equations

The completing-the-square method works for all types of quadratic equations. However, factoring is more efficient for incomplete quadratic equations and complete quadratic equations that factor.

Solve $x^2 - 7x = 0$.

By Factoring

$$x^2 - 7x = 0$$

$$x(x - 7) = 0$$

$$x = 0 \qquad x - 7 = 0$$

$$x = 7$$

By Completing the Square

$$x^2 - 7x + \left(\frac{-7}{2}\right)^2 = 0 + \left(\frac{-7}{2}\right)^2$$

$$x^2 - 7x + \frac{49}{4} = \frac{49}{4}$$

$$\left(x - \frac{7}{2}\right)^2 = \frac{49}{4}$$

$$x - \frac{7}{2} = \pm\frac{7}{2}$$

$$x = \frac{7}{2} \pm \frac{7}{2}$$

$$x = \frac{7}{2} + \frac{7}{2} \qquad x = \frac{7}{2} - \frac{7}{2}$$

$$x = \frac{14}{2} \qquad\qquad x = 0$$

$$x = 7$$

Chapter 15 / Quadratic and Higher-Degree Equations

Solve $x^2 + 5x + 6 = 0$.

By Factoring

$$x^2 + 5x + 6 = 0$$
$$(x + 2)(x + 3) = 0$$

$$x + 2 = 0 \qquad x + 3 = 0$$

$$x = -2 \qquad x = -3$$

By Completing the Square

$$x^2 + 5x + 6 = 0$$
$$x^2 + 5x = -6$$
$$x^2 + 5x + \left(\frac{5}{2}\right)^2 = -6 + \left(\frac{5}{2}\right)^2$$
$$x^2 + 5x + \frac{25}{4} = -6 + \frac{25}{4}$$
$$\left(x + \frac{5}{2}\right)^2 = \frac{-24}{4} + \frac{25}{4}$$
$$\left(x + \frac{5}{2}\right)^2 = \frac{1}{4}$$
$$x + \frac{5}{2} = \pm\sqrt{\frac{1}{4}}$$
$$x = -\frac{5}{2} \pm \frac{1}{2}$$
$$x = \frac{-5}{2} + \frac{1}{2} \qquad x = \frac{-5}{2} - \frac{1}{2}$$
$$x = \frac{-4}{2} \qquad x = \frac{-6}{2}$$
$$x = -2 \qquad x = -3$$

SELF-STUDY EXERCISES 15–3

1 Solve by completing the square. Simplify radicals and reduce.

1. $x^2 + 2x - 48 = 0$
2. $x^2 - 8x - 9 = 0$
3. $x^2 - 10x + 9 = 0$
4. $x^2 - 10x + 24 = 0$
5. $x^2 + 2x = 24$
6. $x^2 + 16x = -28$
7. $x^2 - 3x - 5 = 0$
8. $x^2 - 5x - 2 = 0$
9. $x^2 - 5x - 3 = 0$
10. $x^2 - 3x - 1 = 0$
11. $2x^2 - 4x - 3 = 0$
12. $2x^2 - 2x - 1 = 0$
13. $3x^2 - 6x + 2 = 0$
14. $3x^2 + 12x + 8 = 0$
15. $4x^2 - 12 = 8x$
16. $7x^2 = 3x + 7$

15–4 | *Solving Quadratic Equations Using the Quadratic Formula*

Learning Outcomes

1 Solve quadratic equations using the quadratic formula.

2 Determine the nature of the roots of a quadratic equation by examining the discriminant.

1 **Solve Quadratic Equations Using the Quadratic Formula.**

Another method for solving quadratic equations is to use the *quadratic formula*. This formula results from solving the standard quadratic equation, $ax^2 + bx + c = 0$, by completing the square.

Completing the Square and the Quadratic Formula

The completing-the-square method for solving quadratic equations works for all types of quadratic equations. We can use this method to show how the quadratic formula was derived.

$$ax^2 + bx + c = 0$$

a, b, and c are real numbers and $a > 0$.

$$ax^2 + bx = -c$$

Isolate the terms containing variables.

$$x^2 + \frac{bx}{a} = -\frac{c}{a}$$

Divide the entire equation by a so that the coefficient of x^2 is 1.

$$x^2 + \frac{bx}{a} + \frac{b^2}{4a^2} = \frac{b^2}{4a^2} - \frac{c}{a}$$

Add a constant to each side so that the left side becomes a perfect-square trinomial. The appropriate constant will be $\left(\dfrac{b}{2a}\right)^2$ or $\dfrac{b^2}{4a^2}$

$$\left(x + \frac{b}{2a}\right)^2 = \frac{b^2}{4a^2} - \frac{c}{a}$$

Factor the left side.

$$\sqrt{\left(x + \frac{b}{2a}\right)^2} = \sqrt{\frac{b^2}{4a^2} - \frac{c}{a}}$$

Take the square root of both sides.

$$x + \frac{b}{2a} = \pm\sqrt{\frac{b^2}{4a^2} - \frac{c}{a}}$$

Get a common denominator for the terms under the radical. $-\dfrac{c}{a} \cdot \dfrac{4a}{4a} = -\dfrac{4ac}{4a^2}$.

$$x + \frac{b}{2a} = \pm\sqrt{\frac{b^2}{4a^2} - \frac{4ac}{4a^2}}$$

Simplify the radicand. $\pm\sqrt{\dfrac{b^2 - 4ac}{4a^2}} = \dfrac{\pm\sqrt{b^2 - 4ac}}{2a}$.

$$x + \frac{b}{2a} = \pm\frac{\sqrt{b^2 - 4ac}}{2a}$$

Solve for x.

$$x = -\frac{b}{2a} \pm \frac{\sqrt{b^2 - 4ac}}{2a}$$

Combine like fractions.

$$x = \frac{-b \pm \sqrt{b^2 - 4ac}}{2a}$$

Quadratic formula.

Quadratic formula:

$$x = \frac{-b \pm \sqrt{b^2 - 4ac}}{2a}$$

where a, b, and c are coefficients of a quadratic equation in the form $ax^2 + bx + c = 0$.

To evaluate the quadratic formula:

1. Write the equation in standard form.
2. Identify a, b, and c.
3. Substitute numbers for a, b, and c in the formula.
4. Use the order of operations to simplify the expression under the radical.
5. Simplify the radical.
6. Factor any common factors in the numerator.
7. Simplify by reducing.
8. Write as two distinct solutions.
9. Write as exact solution or approximate solution.

Common Cause for Error

When you use the quadratic formula to solve problems, begin by writing the formula to help you remember it. When you write the formula, be sure to extend the fraction bar beneath the *entire* numerator. This omission is a common cause for errors.

EXAMPLE   Use the quadratic formula to solve $x^2 + 5x + 6 = 0$ for x.

$a = 1,$ $b = 5,$ $c = 6$ Identify a, b, and c.

$x = \dfrac{-b \pm \sqrt{b^2 - 4ac}}{2a}$ Quadratic formula. Substitute for a, b, and c.

$x = \dfrac{-5 \pm \sqrt{5^2 - 4 \cdot 1 \cdot 6}}{2 \cdot 1}$ Raise to power and multiply in grouping.

$x = \dfrac{-5 \pm \sqrt{25 - 24}}{2}$ Combine terms under radical.

$x = \dfrac{-5 \pm \sqrt{1}}{2}$ Evaluate radical.

$x = \dfrac{-5 \pm 1}{2}$

At this point, we separate the formula into two parts, one using the $+1$ and the other using the -1.

$x = \dfrac{-5 + 1}{2}$ $x = \dfrac{-5 - 1}{2}$

$x = \dfrac{-4}{2}$ $x = \dfrac{-6}{2}$

$x = -2$ $x = -3$

Check $x = -2$ Check $x = -3$

$x^2 + 5x + 6 = 0$ $x^2 + 5x + 6 = 0$

$(-2)^2 + 5(-2) + 6 = 0$ $(-3)^2 + 5(-3) + 6 = 0$

$4 - 10 + 6 = 0$ $9 - 15 + 6 = 0$

$-6 + 6 = 0$ $-6 + 6 = 0$

$0 = 0$ $0 = 0$

EXAMPLE   Use the quadratic formula to solve $3x^2 - 3x - 7 = 0$ for x. Round answers to the nearest hundredth.

$a = 3,$ $b = -3,$ $c = -7$ Identify a, b, and c.

$x = \dfrac{-b \pm \sqrt{b^2 - 4ac}}{2a}$ Quadratic formula. Substitute.
                                            $-(b) = -(-3) = +3$.

$x = \dfrac{+3 \pm \sqrt{(-3)^2 - 4(3)(-7)}}{2 \cdot 3}$ Perform calculations under the radical.

$x = \dfrac{+3 \pm \sqrt{9 + 84}}{6}$

$$x = \frac{+3 \pm \sqrt{93}}{6}$$

Evaluate the radical.

$$x = \frac{+3 \pm 9.643650761}{6}$$

Separate two solutions.

$$x = \frac{3 + 9.643650761}{6} \qquad x = \frac{3 - 9.643650761}{6}$$

$$x = \frac{12.643650761}{6} \qquad x = \frac{-6.643650761}{6}$$

$$x = 2.11 \quad \text{Rounded.} \qquad x = -1.11 \quad \text{Rounded.}$$

Notice that we round to hundredths *after* making the final calculation.

Check $x = 2.11$	Check $x = -1.11$
$3x^2 - 3x - 7 = 0$	$3x^2 - 3x - 7 = 0$
$3(2.11)^2 - 3(2.11) - 7 = 0$	$3(-1.11)^2 - 3(-1.11) - 7 = 0$
$3(4.4521) - 6.33 - 7 = 0$	$3(1.2321) + 3.33 - 7 = 0$
$13.3563 - 6.33 - 7 = 0$	$3.6963 + 3.33 - 7 = 0$
$0.0263 \approx 0$	$0.0263 \approx 0$

TIP!

Rounding Discrepancies

The symbol $\doteq$ means "is approximately equal to." Another symbol for approximations is $\approx$. If we had checked *before* rounding, the checks would have been "closer" to zero.

TIP!

Check with Calculator.

With a formula as complex as the quadratic formula, it is important to check your solutions.
    Use the full calculator value to check. As we mentioned in the previous tip, the full calculator value improves the accuracy of the check. Calculate the first root in the previous example.

$$3 \boxed{+} \boxed{\sqrt{}} \ 93 \boxed{=} \boxed{\div} \ 6 \boxed{=} \Rightarrow 2.107275127$$

Check:
    If your calculator allows you to insert the answer of your last calculation, use this function in checking. Otherwise, you can store an answer in memory and recall as appropriate.

$$3 \boxed{\text{ANS}} \boxed{x^2} \boxed{-} 3 \boxed{\text{ANS}} \boxed{-} 7 \boxed{=} \Rightarrow 0$$

Calculate the second root.

$$3 \boxed{-} \boxed{\sqrt{}} \ 93 \boxed{=} \boxed{\div} \ 6 \boxed{=} \Rightarrow -1.107275127$$

Check:

$$3 \boxed{\text{ANS}} \boxed{x^2} \boxed{-} 3 \boxed{\text{ANS}} \boxed{-} 7 \boxed{=} \Rightarrow 0.$$

In the next example, the equation is a rational equation and contains no squared term at first. The squared term appears *after* the multiplication to eliminate the denominator.

EXAMPLE Solve $\dfrac{3}{5 - x} = 2x$ for x. Round answers to the nearest hundredth.

$$\dfrac{3}{5 - x} = 2x$$

Eliminate the denominator and write the equation in standard form.

$$(5 - x)\dfrac{3}{5 - x} = 2x(5 - x)$$

$$3 = 2x(5 - x)$$

Distribute.

$$3 = 10x - 2x^2$$

Write the equation in standard form.

$$2x^2 - 10x + 3 = 0$$

$$a = 2, \qquad b = -10, \qquad c = 3$$

Identify a, b, and c.

$$x = \dfrac{-b \pm \sqrt{b^2 - 4ac}}{2a}$$

Write the formula. Substitute.

$$x = \dfrac{+10 \pm \sqrt{(-10)^2 - 4(2)(3)}}{2(2)}$$

Simplify radicand.

$$x = \dfrac{10 \pm \sqrt{100 - 24}}{4}$$

$$x = \dfrac{10 \pm \sqrt{76}}{4}$$

$$x = \dfrac{10 \pm 8.717797887}{4}$$

Separate into two solutions.

$$x = \dfrac{10 + 8.717797887}{4} \qquad x = \dfrac{10 - 8.717797887}{4}$$

$$x = \dfrac{18.717797887}{4} \qquad x = \dfrac{1.282202113}{4}$$

$$x = \mathbf{4.68} \qquad x = \mathbf{0.32}$$

Round after final calculation.

$$\text{Check } x = 4.68 \qquad\qquad \text{Check } x = 0.32$$

$$\dfrac{3}{5 - x} = 2x \qquad\qquad \dfrac{3}{5 - x} = 2x$$

$$\dfrac{3}{5 - (4.68)} = 2(4.68) \qquad \dfrac{3}{5 - (0.32)} = 2(0.32)$$

$$\dfrac{3}{0.32} = 9.36 \qquad\qquad \dfrac{3}{4.68} = 0.64$$

$$9.375 \doteq 9.36 \qquad 0.641025641 \doteq 0.64$$

TIP!

Extraneous Roots

When an equation is multiplied by a factor that contains a variable, as in the preceding example, an *extraneous root* can be introduced. An extraneous root is a root that will *not* check when substituted into the original equation. Thus, it is very important to check each root.

Even though quadratic equations have two roots, a root may be disregarded in an applied problem because it is not appropriate within the context of the problem.

EXAMPLE Find the length and width of a rectangular table if the length is 8 in. more than the width and the area is 260 in².

Unknown facts Length and width of rectangle

Known facts Area = 260 in², length = 8 in. more than width

Relationships Let x = number of inches in the width

$x + 8$ = number of inches in the length

Area = length times width, or $A = lw$

Estimation The square root of 260 is between 16 and 17; the width should be less than 16 and the length more than 16.

Calculations

$260 = (x + 8)(x)$ — Substitute into area formula and distribute.

$260 = x^2 + 8x$ — Write in standard form.

$x^2 + 8x - 260 = 0$

$a = 1, \qquad b = 8, \qquad c = -260$

$x = \dfrac{-b \pm \sqrt{b^2 - 4ac}}{2a}$ — Quadratic formula. Substitute.

$x = \dfrac{-8 \pm \sqrt{(8)^2 - 4(1)(-260)}}{2(1)}$ — Simplify the radicand.

$x = \dfrac{-8 \pm \sqrt{64 + 1{,}040}}{2}$ — Combine terms in radicand.

$x = \dfrac{-8 \pm \sqrt{1{,}104}}{2}$

$x = \dfrac{-8 \pm 33.22649545}{2}$ — Separate into two solutions.

$x = \dfrac{-8 + 33.22649545}{2} \qquad\qquad x = \dfrac{-8 - 33.22649545}{2}$

$x = \dfrac{25.22649545}{2} \qquad\qquad x = \dfrac{-41.22649545}{2}$

$x = 12.6 \qquad\qquad\qquad x = -20.6$ — Disregard.

$x + 8 = 20.6$

Interpretation Measurements are positive, so disregard the negative solution. **The width is 12.6 in. and the length is 20.6 in.**

---

2 **Determine the Nature of the Roots of a Quadratic Equation by Examining the Discriminant.**

In the previous sections, we made various observations about the roots of the different types of quadratic equations. Now, we make additional observations about the roots of quadratic equations.

Which method for solving quadratic equations is best? Since any method that is mathematically sound and produces consistently correct solutions is good, a "best" method may be evaluated differently. Factoring methods are generally quicker, especially when a calculator or computer is unavailable.

In general, follow these suggestions for solving quadratic equations.

1. Write the equation in standard form: $ax^2 + bx + c = 0$.
2. Identify the type of quadratic equation.

3. Solve pure quadratic equations by the square-root method.
4. Solve incomplete quadratic equations by finding common factors.
5. Solve complete quadratic equations by factoring into two binomials, if possible.
6. If factoring is not possible or is difficult, use the completing-the-square method or the quadratic formula to solve complete quadratic equations.

All types of quadratic equations can be solved using the completing-the-square method or the quadratic formula. Only certain types of quadratic equations can be solved by factoring or applying the square-root property. The radicand of the radical portion of the quadratic formula, $b^2 - 4ac$, is called the *discriminant*. The general characteristics of the roots of a quadratic equation can be determined by examining the discriminant.

Properties of the discriminant, $b^2 - 4ac$:

1. If $b^2 - 4ac \geq 0$, the equation has real-number roots.
   a. If $b^2 - 4ac$ is a perfect square, there are two rational roots.
   b. If $b^2 - 4ac = 0$, there is one rational root (sometimes called a *double root*).
   c. If $b^2 - 4ac$ is not a perfect square, there are two irrational roots.
2. If $b^2 - 4ac < 0$, the equation has no real-number roots. The roots are imaginary or complex.

EXAMPLE Examine the discriminant of each equation and determine the nature of the roots. Then solve the equation.

(a) $5x^2 + 3x - 1 = 0$ (b) $3x^2 + 5x = 2$ (c) $4x^2 + 2x + 3 = 0$

(a) $5x^2 + 3x - 1 = 0$ $a = 5, b = 3, c = -1$.

$\quad b^2 - 4ac = 3^2 - 4(5)(-1)$ Examine the discriminant.

$\quad\quad\quad = 9 + 20$ **There will be two irrational roots.**

$\quad\quad\quad = \boxed{29}$ Use the quadratic formula.

$\quad x = \dfrac{-b \pm \sqrt{b^2 - 4ac}}{2a}$ Substitute 29 for the discriminant.

$\quad x = \dfrac{-3 \pm \sqrt{29}}{2(5)}$ Identify each root.

$\quad x = \dfrac{-3 + \sqrt{29}}{10}$ or $x = \dfrac{-3 - \sqrt{29}}{10}$ Exact irrational roots.

(b) $\quad 3x^2 + 5x = 2$

$\quad 3x^2 + 5x - 2 = 0$ Standard form: $a = 3, b = 5, c = -2$.

$\quad b^2 - 4ac = 5^2 - 4(3)(-2)$ Examine the discriminant.

$\quad\quad\quad = 25 + 24$

$\quad\quad\quad = 49$ The discriminant is a perfect square.
    There are two rational roots and the
    trinomial will factor.

$\quad (3x - 1)(x + 2) = 0$ Factor. Set each factor equal to zero.

$\quad 3x - 1 = 0 \qquad x + 2 = 0$

$\quad\quad 3x = 1$

$\quad\quad x = \dfrac{1}{3} \qquad\qquad x = -2$ Exact rational roots.

(c) $4x^2 + 2x + 3 = 0$

$b^2 - 4ac = 2^2 - 4(4)(3)$

$= 4 - 48$

$= -44$

$x = \dfrac{-2 \pm \sqrt{-44}}{2 \cdot 4}$

$x = \dfrac{-2 \pm 2i\sqrt{11}}{8}$

$x = \dfrac{2(-1 \pm i\sqrt{11})}{8}$

$x = \dfrac{-1 \pm i\sqrt{11}}{4}$

$a = 4, b = 2, c = 3$.
Examine the discriminant.

The discriminant is negative. There are no real roots. **The roots are imaginary or complex.**

Substitute -44 for the discriminant in the quadratic formula.

Simplify the radical and factor.

Reduce.

The roots are complex.

Learning Strategy   *Understand the Concepts*

When calculators and computers can so easily find solutions to all types of equations, why do we spend so much time with paper-and-pencil techniques? For our knowledge of algebra and mathematics to be useful in real-world situations, we must understand the concepts and know when to use them, and we must understand what the solutions represent.

In developing this understanding of the concepts, we must examine a wide range of situations that might be encountered with the concept. In practice, technological tools can be used to solve equations once you have established appropriate equations through a critical examination of the situation.

SELF-STUDY EXERCISES 15–4

1 Solve the quadratic equations by using the quadratic formula.

1. $3x^2 - 7x - 6 = 0$

2. $x^2 + x - 12 = 0$

3. $5x^2 - 6x = 11$

4. $x^2 - 6x + 9 = 0$

5. $x^2 - x - 6 = 0$

6. $8x^2 - 2x = 3$

Solve the quadratic equations by using the quadratic formula. Round each final answer to the nearest hundredth.

7. $x^2 = -9x + 20$

8. $3x^2 + 6x + 1 = 0$

9. $\dfrac{2}{x} + \dfrac{5}{2} = x$

10. $\dfrac{2}{x} + 3x = 8$

11. $2x^2 + 3x + 3 = 0$

12. $x^2 - x + 2 = 0$

Solve the applied problems.

13. A rectangular tabletop is 3 cm longer than it is wide. Find the length and width to the nearest hundredth if the area is 47.5 cm². (Area = length × width, or $A = lw$.)

14. A rectangular instrument case has an area of 40 in². If the length is 6 in. more than the width, find the dimensions (length and width) of the instrument case.

15. A bricklayer plans to build an arch with a span (s) of 8 m and a radius (r) of 4 m. How high (h) is the arch? (Use the formula $h^2 - 2hr + \dfrac{s^2}{4} = 0$.)

16. A rectangular patio slab is 180 ft². If the length is 1.5 times the width, find the width and length of the slab to the nearest whole number.

17. A farmer normally plants a field 80 ft by 120 ft. This year the government requires the planting area to be decreased by 20%, and the farmer chooses to decrease the width and length of the field by an equal amount. Draw both the original field and the reduced-size field. If x is the amount by which the length and width are decreased, find the length and width of the new field.

18. The recommended dosage of a certain type of medicine is determined by the patient's weight. The formula to determine the dosage is given by $D = 0.1w^2 + 5w$, where D is the dosage in milligrams (mg) and w is the patient's body weight in kilograms. A doctor ordered a dosage of 1,800 mg. This dosage is to be administered to what weight patient?

2 Examine the discriminant of each equation and determine the nature of the roots. Solve each equation. Round to hundredths if necessary.

19. $3x^2 + x - 2 = 0$
20. $x^2 - 3x = -1$
21. $2x^2 + x = 2$
22. $3x^2 - 2x + 1 = 0$
23. $x^2 - 3x - 7 = 0$
24. $3x^2 + 5x - 6 = 0$
25. Describe the discriminant of a quadratic equation that has real roots.
26. Describe the discriminant of a quadratic equation that has rational and unequal roots.

15–5 | *Solving Higher-Degree Equations by Factoring*

Learning Outcomes

1 Identify the degree of an equation.
2 Solve higher-degree equations by factoring.

Some equations contain terms that have a higher degree than 2. Equations of higher degree that can be written in factored form are examined.

1 Identify the Degree of an Equation.

In Section 15–1, we defined a *quadratic* equation as an equation that has at least one letter term raised to the second power and no letter terms having a power higher than 2. Similarly, a *cubic equation* is an equation that has at least one letter term raised to the third power and no letter terms having a power higher than 3. A quadratic equation can also be referred to as a *second-degree* equation, and a cubic equation as a *third-degree* equation. The *degree of an equation* in one variable is the highest power of any letter term that appears in the equation.

When we solved linear or first-degree equations, we obtained at most one solution for the equation. Quadratic or second-degree equations have at most two solutions. Cubic or third-degree equations have at most three solutions, and fourth-degree equations have at most four solutions.

EXAMPLE State the degree of each of the equations.

(a) $x^2 + 3x + 4 = 0$
(b) $x^3 = 27$
(c) $x^4 + 3x^3 + 2x^2 + x + 4 = 0$
(d) $x^8 = 256$
(e) $3x + 7 = 5x - 3$

(a) $x^2 + 3x + 4 = 0$
Quadratic or second-degree equation
(b) $x^3 = 27$
Cubic or third-degree equation
(c) $x^4 + 3x^3 + 2x^2 + x + 4 = 0$
Fourth-degree equation
(d) $x^8 = 256$
Eighth-degree equation
(e) $3x + 7 = 5x - 3$
Linear or first-degree equation

2 Solve Higher-Degree Equations by Factoring.

Higher-degree equations that are written in factored form or can be written in factored form are presented here. We solve these equations by using the zero-product property; that is, if the product of the factors equals 0, then all factors containing a variable may be equal to 0. (See Section 15–2.)

To solve a higher-degree equation by factoring:

1. Write the equation in standard form.
2. Factor the polynomial side of the equation.
3. Set each factor containing a variable equal to zero.
4. Solve each equation from Step 3.

EXAMPLE Solve the equations by using the zero-product property.

(a) $x(x + 4)(x - 3) = 0$ (b) $(x + 5)(x - 7)(2x - 1) = 0$
(c) $2x^3 - 14x^2 + 20x = 0$ (d) $3x^3 - 15x = 0$

(a) $x(x + 4)(x - 3) = 0$ Already factored. Set each factor equal to zero.

$x = 0$ $x + 4 = 0$ $x - 3 = 0$ Solve each equation.

$x = 0$ $x = -4$ $x = 3$

(b) $(x + 5)(x - 7)(2x - 1) = 0$ Already factored. Set each factor equal to zero.

$x + 5 = 0$ $x - 7 = 0$ $2x - 1 = 0$ Solve each equation.

       $x = -5$ $x = 7$ $2x = 1$

$$\frac{2x}{2} = \frac{1}{2}$$

$$x = \frac{1}{2}$$

(c) $2x^3 - 14x^2 + 20x = 0$ Factor the common factors.

$2x(x^2 - 7x + 10) = 0$ Factor the trinomial.

$2x(x - 5)(x - 2) = 0$ Set each factor equal to zero.

$2x = 0$ $x - 5 = 0$ $x - 2 = 0$ Solve each equation.

$\dfrac{2x}{2} = \dfrac{0}{2}$ $x = 5$ $x = 2$

$x = 0$

(d) $3x^3 - 15x = 0$ Factor the common factors.

$3x(x^2 - 5) = 0$ Factor the difference of two squares, if possible.

$3x(x^2 - 5) = 0$ Set each factor equal to zero.

$3x = 0$ $x^2 - 5 = 0$

$\dfrac{3x}{3} = \dfrac{0}{3}$ $x^2 = 5$ Solve each equation.

$x = 0$ $x = \pm\sqrt{5}$ The roots are $x = 0$, $x = \sqrt{5}$, or $x = -\sqrt{5}$.

SELF-STUDY EXERCISES 15–5

1 State the degree of the equations.

1. $x^2 + 2x - 3 = 0$ **2.** $3x + 2x = 5$ **3.** $x^4 = 42$
4. $2x - 7 - 4x = 3$ **5.** $3x^3 + 2x^4 + 3 = x^2$ **6.** $x^7 = 128$

7. $x(x - 2)(x + 3) = 0$ 8. $2x(2x - 1)(x + 3) = 0$ 9. $3x(2x - 5)(3x - 2) = 0$
10. $x^3 - 7x^2 + 10x = 0$ 11. $3x^3 - 3x^2 = 18x$ 12. $4x^3 + 10x^2 + 4x = 0$
13. $2x^3 - 18x = 0$ 14. $12x^3 = 3x$ 15. $16x^3 = 9x$

16. 32 times a number is subtracted from twice the cube of the number and the result is zero. How many numbers meet these conditions? What are they?

17. Find all the numbers that satisfy the following conditions: A number cubed is increased by 5 times the number and the result is zero.

18. A shipping container is a large cardboard box. The volume is found by multiplying the length times the width times the height. The length of the box is 3 ft more than the width, and the height is 7 ft more than the width. The volume is 421 ft³. Write an equation to find the length, width, and height. Can this problem be solved by the methods found in Section 15–5? If so, solve it. If not, explain why it can't be solved.

ASSIGNMENT EXERCISES

Section 15–1

Identify the quadratic equations as pure, incomplete, or complete.
1. $x^2 = 49$ 2. $x^2 - 5x = 0$ 3. $5x^2 - 45 = 0$
4. $3x^2 + 2x - 1 = 0$ 5. $8x^2 + 6x = 0$ 6. $5x^2 + 2x + 1 = 0$
7. $x^2 - 32 = 0$ 8. $x^2 + x = 0$ 9. $3x^2 + 6x + 1 = 0$

Write the equations in standard form.
10. $x^2 - 7x = 8$ 11. $2x^2 - 5 = 8x$ 12. $7x = 2x^2 - 3$
13. $5 + x^2 - 7x = 0$ 14. $x - 4 = 3x^2$ 15. $3x = 1 - 4x^2$

Solve the equations by the square-root method. Round to thousandths when necessary.
16. $x^2 = 121$ 17. $x^2 = 100$ 18. $x^2 - 64 = 0$
19. $4x^2 = 9$ 20. $64x^2 - 49 = 0$ 21. $0.36y^2 = 1.09$
22. $0.16x^2 = 0.64$ 23. $5x^2 = 40$ 24. $2x^2 - 5 = 3$
25. $6x^2 + 4 = 34$ 26. $5x^2 - 6 = 19$ 27. $3x^2 = 12$
28. $3x^2 - 4 = 8$ 29. $2x^2 = 34$ 30. $5x^2 - 9 = 30$
31. $3y^2 - 36 = -8$ 32. $2x^2 + 3 = 51$ 33. $\frac{1}{2}x^2 = 8$
34. $\frac{2}{3}x^2 = 24$ 35. $\frac{1}{4}x^2 - 1 = 15$ 36. $\frac{2}{5}x^2 + 2 = 8$

37. A circle has an area of 845 cm². What is the radius of the circle?

38. The area of an oil painting that is a square is known to be 9,072 cm². What are the inside dimensions of its picture frame?

Section 15–2

Solve the equations by factoring.
39. $x^2 - 5x = 0$ 40. $4x^2 = 8x$ 41. $6x^2 - 12x = 0$
42. $3x^2 + x = 0$ 43. $10x^2 + 5x = 0$ 44. $3y^2 = 12y$
45. $y^2 - 5y = 0$ 46. $x^2 = 16x$ 47. $12x^2 + 8x = 0$
48. $8x^2 - 12x = 0$ 49. $x^2 + 3x = 0$ 50. $4x^2 - 28x = 0$
51. $5x^2 = 45x$ 52. $7x^2 = 28x$ 53. $y^2 + 8y = 0$
54. $z^2 - 6z = 0$ 55. $3m^2 - 5m = 0$ 56. $4n^2 - 3n = 0$
57. $2x^2 = x$ 58. $5y^2 = y$

59. 3 times the square of a number is the same as 12 times the number. What is the number?

60. Describe the steps for solving an incomplete quadratic equation. Include a clear description of the nature of the roots.

Solve the equations by factoring.

61. $x^2 - 4x + 3 = 0$
62. $x^2 + 7x + 12 = 0$
63. $x^2 + 3x = 10$
64. $x^2 - 7x + 12 = 0$
65. $x^2 + 7x = -6$
66. $x^2 + 3 = -4x$
67. $x^2 - 6x + 8 = 0$
68. $6y + 7 = y^2$
69. $6y^2 - 5y - 6 = 0$
70. $5y^2 + 23y = 10$
71. $10y^2 - 21y - 10 = 0$
72. $6x^2 - 16x + 8 = 0$
73. $4x^2 + 7x + 3 = 0$
74. $3x^2 = -7x + 6$
75. $12y^2 - 5y - 3 = 0$
76. $x^2 - 3x = 18$
77. $x^2 + 19x = 42$
78. $3x^2 + x - 2 = 0$
79. $3y^2 + y - 2 = 0$
80. $2x^2 - 4x - 6 = 0$
81. $2x^2 - 10x + 12 = 0$
82. $y^2 + 18y + 45 = 0$
83. $x^2 - 3x - 18 = 0$
84. $3x^2 - 9x - 30 = 0$
85. $2y^2 + 22y + 60 = 0$
86. $x^2 + 13x + 12 = 0$
87. $x^2 + 7x - 18 = 0$

88. An office building in the shape of a rectangle is known to have 47,500 ft^2 of space on the ground floor. The tenant wants to landscape the two longer sides of the building. The tenant also knows that the building is about 60 ft longer than it is wide. How many feet of land along the building need to be landscaped?

89. Jerri Amour is an architect who is designing a hospital. She knows that a kidney dialysis machine needs a space that is 7 ft longer than it is wide. The total area needed is 228 ft^2 of space. Advise Jerri about the number of feet of space she should include in the length and width of the planned space.

Section 15–3

Solve each equation by completing the square.

90. $x^2 + 2x - 3 = 0$
91. $x^2 - 4x + 4 = 0$
92. $x^2 - 6x + 8 = 0$
93. $x^2 - 8x + 12 = 0$
94. $x^2 - 6x + 7 = 0$
95. $x^2 - 8x + 14 = 0$
96. $x^2 - 6x + 4 = 0$
97. $x^2 - 6x + 12 = 0$
98. $x^2 - 2x + 6 = 0$
99. $x^2 - 5x + 4 = 0$
100. $x^2 - 3x - 18 = 0$
101. $x^2 - 3x = 7$

Section 15–4

Indicate the values for a, b, and c in the quadratic equations.

102. $5x^2 + x + 6 = 0$
103. $x^2 - 2x = 8$
104. $x^2 - 7x + 12 = 0$
105. $x^2 + 3x + 9 = 1$
106. $3x^2 - 2x + 7$
107. $x^2 - 3x = -2$

Solve the quadratic equations by using the quadratic formula.

108. $x^2 - 9x + 20 = 0$
109. $x^2 - 8x - 9 = 0$
110. $x^2 - 5x = -6$
111. $x^2 + 2x = 8$
112. $x^2 - x - 12 = 0$
113. $2x^2 - 3x - 2 = 0$

Solve the quadratic equations by using the quadratic formula. Round each final answer to the nearest hundredth.

114. $3x^2 + 6x + 2 = 0$
115. $2x^2 - 3x - 1 = 0$
116. $5x^2 + 4x - 8 = 0$
117. $3x^2 + 5x + 1 = 0$

118. A bricklayer plans to build an arch with a span (s) of 10 m and a radius (r) of 5 m. How high (h) is the arch? (Use the formula $h^2 - 2hr + \frac{s^2}{4} = 0$.)

119. A rectangular kitchen contains 240 ft^2. If the length is 2 times the width, find the length and width of the room to the nearest whole number. (Area = length × width, or $A = lw$.)

120. What are the dimensions of a rectangular tool storage room if the area is 45.5 m^2 and the room is 0.5 m longer than it is wide? Round to the nearest tenth.

121. Find the length and width of a piece of fiberglass if its length is 3 times the width and the area is 591 in^2. Round to the nearest inch.

Use the discriminant of the quadratic formula to determine the nature of the roots of the equations.

122. $x^2 - 3x + 2 = 0$
123. $x^2 + 8x + 16 = 0$
124. $2x^2 - 3x - 5 = 0$
125. $5x^2 - 100 = 0$
126. $3x^2 - 2x + 4 = 0$
127. $2x = 5x^2 - 3$

Section 15–5

State the degree of each equation.

128. $2x + 5x = 15$
129. $3x - 2x^3 + 8 = 0$
130. $16 = x^4$
131. $6 - 3x - 3 = 2x + 4$
132. $y^6 = 729$
133. $5y^8 + 2y^3 - 6 = y^2$

Find the roots of the following equations. Factor if necessary.

134. $x(x + 2)(x - 3) = 0$
135. $2x(3x - 2)(x - 2) = 0$
136. $3x(2x + 1)(x + 4) = 0$
137. $2x^3 + 10x^2 + 12x = 0$
138. $x^3 = 2x^2$
139. $2x^3 + 9x^2 = 5x$
140. $6x^3 + 3x^2 - 18x = 0$
141. $3x^3 - 6x^2 = 0$
142. $3x^3 - x^2 - 2x = 0$
143. $x^3 + 6x^2 + 8x = 0$
144. $x^3 - 8x^2 + 15x = 0$
145. $x^3 - x^2 - 20x = 0$
146. $x^3 + x^2 - 20x = 0$
147. $y^3 - 6y^2 + 7y = 0$
148. $y^3 + 2y^2 + 5y = 0$
149. $x^3 - 3x^2 - 4x = 0$
150. $y^3 + 7y^2 + 12y = 0$
151. $2y^3 + 6y^2 + 4y = 0$

Chapter 15 / Quadratic and Higher-Degree Equations

152. A number raised to the third power is the same as 121 times the number. What is the number? Give all possibilities.

153. A number multiplied by 51 is the same as 3 times the cube of the number. What is the number?

CHALLENGE PROBLEM

154. Many objects are designed with dimensions according to the ratio of the *Golden Ratio*. Objects that have measurements according to this ratio are said to be most pleasing to the eye. The *Golden Rectangle* has dimensions of length (*l*) and height (*h*) that satisfy the formula

$$\frac{l+h}{l} = \frac{l}{h}$$

You can cross multiply to obtain a quadratic equation for the Golden Rectangle.

You have been commissioned to construct a wall hanging for the lobby of a new office building. The wall is 32 ft long and the ceiling is 20 ft high. The owners want the wall hanging to be at least 2 ft from the ceiling, the floor, and each of the side corners. They also want the wall hanging to have dimensions according to the Golden Ratio.

Using the given information, determine the largest-size wall hanging that can be placed in the lobby that also has the dimensions of the Golden Rectangle.

CONCEPTS ANALYSIS

1. State the zero-product property and explain how it applies to solving quadratic equations.

2. What is an extraneous root?

3. What technique for solving equations can produce an extraneous root?

4. When will a quadratic equation have no real solutions?

5. When will a quadratic equation have irrational roots?

6. When will a quadratic equation have rational roots?

Find a mistake in each of the following. Correct and briefly explain the mistake.

7.
$$2x^2 - 2x - 12 = 0$$
$$2(x^2 - x - 6) = 0$$
$$2(x + 2)(x - 3) = 0$$
$$2 = 0 \quad x + 2 = 0 \quad x - 3 = 0$$
$$\qquad\qquad x = -2 \qquad x = 3$$
The roots are 0, −2, and 3.

8.
$$2x^3 + 5x^2 + 2x = 0$$
$$\frac{x(2x^2 + 5x + 2)}{x} = \frac{0}{x}$$
$$2x^2 + 5x + 2 = 0$$
$$(2x + 1)(x + 2) = 0$$
$$2x + 1 = 0 \qquad x + 2 = 0$$
$$2x = -1 \qquad x = -2$$
$$x = -\frac{1}{2}$$
The roots are $-\frac{1}{2}$ and −2.

9. Why is it incorrect to divide both sides of an equation by a variable?

10. What is the maximum number of roots the equation $x^4 = 16$ *could* have? How many real roots does the equation have?

CHAPTER SUMMARY

Learning Outcomes   **What to Remember with Examples**

Section 15–1

1 Write quadratic equations in standard form (p. 598).

Write the equation with all terms on the left side of the equation and zero on the right and arrange with the terms in descending powers. The leading coefficient should be positive.

Write the equation $5x + 3 = 4x^2$ in standard form.

$$5x + 3 = 4x^2 \qquad \text{Transpose and arrange in descending order.}$$
$$-4x^2 + 5x + 3 = 0 \qquad \text{Make leading coefficient positive.}$$
$$-1(-4x^2 + 5x + 3) = -1(0)$$
$$4x^2 - 5x - 3 = 0$$

2 Identify the coefficients of the quadratic, linear, and constant terms of a quadratic equation (pp. 598–600).

The coefficient of the quadratic term is the coefficient of the squared variable, the coefficient of the linear term is the coefficient of the first-power variable, and the constant term is its own coefficient.

Identify the coefficient of the quadratic and linear terms and identify the constant term in the following:

$$5x = 4x^2 - 3 \qquad \text{Write the equation in standard form. } a = 4, b = -5, c = -3.$$
$$4x^2 - 5x - 3 = 0$$

The coefficient of the quadratic term is 4. The coefficient of the linear term is -5 and the constant term is -3.

3 Solve pure quadratic equations by the square-root method ($ax^2 + c = 0$) (pp. 600–601).

To solve pure quadratic equations, solve for the squared variable; then take the square root of both sides of the equation. The two roots have the same absolute value.

Solve $2x^2 - 72 = 0$.

$$\begin{aligned} 2x^2 - 72 &= 0 && \text{Sort or transpose.} \\ 2x^2 &= 72 && \text{Divide by 2.} \\ x^2 &= 36 && \text{Take the square root of both sides.} \\ x &= \pm 6 \end{aligned}$$

Section 15–2

1 Solve incomplete quadratic equations by factoring ($ax^2 + bx = 0$) (pp. 602–604).

Arrange the equation in standard form. Factor the common factor. Then set each of the two factors equal to zero and solve for the variable in each equation. Both roots are rational and one root is always zero.

Solve $5x^2 - 15x = 0$.

$$\begin{aligned} 5x^2 - 15x &= 0 && \text{Factor common factors.} \\ 5x(x - 3) &= 0 && \text{Set each factor equal to 0.} \\ 5x = 0 \quad & x - 3 = 0 && \text{Solve each equation.} \\ x = 0 \quad & x = 3 \end{aligned}$$

2 Solve complete quadratic equations by factoring ($ax^2 + bx + c = 0$) (pp. 604–605).

Arrange the quadratic equation in standard form and factor the trinomial. Then set each factor equal to zero and solve for the variable. If the expression factors, the roots are rational.

Solve $2x^2 - 5x - 3 = 0$.

$$\begin{aligned} 2x^2 - 5x - 3 &= 0 && \text{Factor the trinomial.} \\ (x - 3)(2x + 1) &= 0 && \text{Set each factor equal to 0.} \\ x - 3 = 0 \quad & 2x + 1 = 0 && \text{Solve each equation.} \\ x = 3 \quad & 2x = -1 \\ & x = -\frac{1}{2} \end{aligned}$$

Section 15–3

1 Solve quadratic equations by completing the square (pp. 606–609).

1. Write the equation in the form $ax^2 + bx = c$.
2. Divide each term of the equation by a.
3. Form a perfect-square trinomial on the left by adding $\left(\frac{b}{2a}\right)^2$ or $\frac{b^2}{4a^2}$ to both sides.
4. Factor the perfect square trinomial into the square of a binomial. $\left(x + \frac{b}{2a}\right)^2$
5. Take the square root of both sides of the equation.
6. Solve for x.

Solve $2x^2 - x + 4 = 0$ by completing the square.

$$2x^2 - x + 4 = 0 \qquad \text{Isolate the variable terms.}$$

$$2x^2 - x = -4 \qquad \text{Divide equation by 2.}$$

$$x^2 - \frac{x}{2} = \frac{-4}{2} \qquad \text{Simplify and add } \left(\frac{-1}{2}\right)^2 \text{ to both sides.}$$

$$x^2 - \frac{x}{2} + \left(\frac{-1}{2}\right)^2 = -2 + \left(\frac{-1}{2}\right)^2 \qquad \text{Simplify.}$$

$$x^2 - \frac{x}{2} + \frac{1}{4} = -2 + \frac{1}{4} \qquad \text{Write the left side of the equation as the square of the binomial.}$$

$$\left(x - \frac{1}{2}\right)^2 = \frac{-8}{4} + \frac{1}{4}$$

$$\left(x - \frac{1}{2}\right)^2 = -\frac{7}{4} \qquad \text{Take the square root of both sides.}$$

$$x - \frac{1}{2} = \pm\sqrt{-\frac{7}{4}} \qquad \text{Solve for } x.$$

$$x = \frac{1}{2} \pm \frac{i\sqrt{7}}{2} \qquad \text{Exact roots.}$$

Section 15–4

1 Solve quadratic equations using the quadratic formula (pp. 609–614).

In the quadratic formula $x = \dfrac{-b \pm \sqrt{b^2 - 4ac}}{2a}$

a is the coefficient of the quadratic term, b is the coefficient of the linear term, and c is the constant when the equation is written in standard form. The variable is x.

Solve using the quadratic formula:

$$5x^2 + 7x - 6 = 0 \qquad \text{Identify } a, b, \text{ and } c.$$

$$a = 5, \qquad b = 7, \qquad c = -6 \qquad \text{Substitute values.}$$

$$x = \frac{-b \pm \sqrt{b^2 - 4ac}}{2a}$$

$$x = \frac{-7 \pm \sqrt{7^2 - 4(5)(-6)}}{2(5)}$$

$$x = \frac{-7 \pm \sqrt{49 + 120}}{10} \qquad \text{Add in the radicand.}$$

$$x = \frac{-7 \pm \sqrt{169}}{10} \qquad \text{Evaluate the square root.}$$

$$x = \frac{-7 \pm 13}{10} \qquad \text{Separate into two cases.}$$

$$x = \frac{-7 + 13}{10} \qquad x = \frac{-7 - 13}{10} \qquad \text{Solve each equation.}$$

$$x = \frac{6}{10} \qquad x = -\frac{20}{10}$$

$$x = \frac{3}{5} \qquad x = -2 \qquad \text{Exact roots.}$$

Solving applied problems requires knowledge of other mathematical formulas, for Example, $A = lw$ (area of a rectangle).

Chapter Summary

Find the length and width of a rectangular parking lot if the length is to be 12 m longer than the width and the area is to be 6,205 m².

Let x = number of meters in width
$x + 12$ = number of meters in length

Estimate: If the parking lot was a square, one side would be $s = \sqrt{A}$ or $s = 78.8$ m. In a rectangle where the length and width are different, the length is more than 78.8 m and the width is less than 78.8 m.

$$lw = A$$
$$x(x + 12) = 6{,}205$$
$$x^2 + 12x = 6{,}205$$
$$x^2 + 12x - 6{,}205 = 0$$

Use a calculator with the formula.

$$x = \frac{-12 \pm \sqrt{12^2 - 4(1)(-6{,}205)}}{2(1)}$$ Substitute values for a, b, and c.

$$x = \frac{-12 \pm \sqrt{144 + 24{,}820}}{2}$$ Simplify radicand.

$$x = \frac{-12 \pm \sqrt{24{,}964}}{2}$$ Evaluate square root.

$$x = \frac{-12 \pm 158}{2}$$ Separate into two cases.

$$x = \frac{-12 + 158}{2} \qquad x = \frac{-12 - 158}{2}$$ Solve each case.

$$x = \frac{146}{2} = 73 \qquad x = \frac{-170}{2} = -85$$ Disregard the negative root for a distance measure.

$x = 73$ width

$x + 12 = 85$ length

2 Determine the nature of the roots of a quadratic equation by examining the discriminant (pp. 614–616).

The radicand of the quadratic formula, $b^2 - 4ac$, is the discriminant of the quadratic equation.

1. If $b^2 - 4ac \geq 0$, the equation has real-number solutions.
   a. If $b^2 - 4ac$ is a perfect square, the two solutions are rational.
   b. If $b^2 - 4ac = 0$, there is one rational solution.
   c. If $b^2 - 4ac$ is not a perfect square, the two roots are irrational.
2. If $b^2 - 4ac < 0$, the equation has no real-number solutions or the solutions are complex.

Use the discriminant to determine the characteristics of the roots of the equation $3x^2 - 5x + 7 = 0$.

$a = 3, \qquad b = -5, \qquad c = 7$ Identify a, b, and c.
$(-5)^2 - 4(3)(7) = 25 - 84$ Substitute values into $b^2 - 4ac$ and simplify.
$\qquad\qquad\qquad = -59$

Since -59 is less than zero, the roots are not real or they are complex.   Interpret result.

Section 15–5

1 Identify the degree of an equation (p. 617).

The degree of an equation in one variable is the highest power of any term that appears in the equation.

State the degree: $x^3 + 4x = 0$ is a cubic or third-degree equation.
$x^4 = 81$ is a fourth-degree equation.

2 Solve higher-degree equations by factoring (pp. 617–618).

The higher-degree equations discussed in this section have a common variable factor and can be solved by factoring.

Solve $x^3 + 2x^2 - 3x = 0$.

$$x^3 + 2x^2 - 3x = 0 \qquad \text{Factor common factor.}$$
$$x(x^2 + 2x - 3) = 0 \qquad \text{Factor trinomial.}$$
$$x(x + 3)(x - 1) = 0 \qquad \text{Set each factor equal to 0.}$$
$$x = 0 \qquad x + 3 = 0 \qquad x - 1 = 0 \qquad \text{Solve each equation.}$$
$$x = -3 \qquad x = 1$$

CHAPTER TRIAL TEST

Identify the quadratic equations as pure, incomplete, or complete.

1. $3x^2 = 42$

2. $7x^2 - 3x + 2 = 0$

3. $5x^2 = 7x$

4. $4x^2 - 1 = 0$

Solve the quadratic equations. Find the exact solutions. Also, give approximate solutions to the nearest hundredth when appropriate.

5. $x^2 = 81$

6. $x^2 - 32 = 0$

7. $9x^2 = 16$

8. $81x^2 - 64 = 0$

9. $0.09x^2 = 0.49$

10. $2x^2 + 3x + 1 = 0$

11. $3x^2 - 6x = 0$

12. $3x^2 - 6x - 1 = 0$

13. $x - 5x + 6 = 0$

14. $x^2 - 3x - 4 = 0$

15. $2x^2 + 12 = 11x$

16. $3x^2 - 5x + 4 = 0$

17. Find the diameter (d) in mils to the nearest hundredth of a copper wire conductor whose resistance (R) is 1.314 Ω and whose length (L) is 3,642.5 ft. (Formula: $R = \dfrac{KL}{d^2}$, where K is 10.4 for copper wire.)

18. Find the radius (r) of a circle whose area (A) is 35.15 cm². Round the answer to the nearest hundredth centimeter. (Formula: $A = \pi r^2$.)

19. What is the current (I) in amps to the nearest hundredth if the resistance (R) of the circuit is 52.29 Ω and the watts (W) used are 205? $\left(\text{Formula: } R = \dfrac{W}{I^2}. \right)$

20. A square parcel of land is 156.25 m² in area. What is the length of a side to the nearest hundredth? (Use the formula $A = s^2$, where A is the area and s is the length of a side.)

21. In the formula $E = 0.5\,mv^2$, solve for v if $E = 180$ and $m = 10$ and $v > 0$.

Solve each equation.

22. $x(2x - 5)(x - 3) = 0$

23. $6x^3 + 21x^2 = 45x$

24. $2x^3 - x^2 - 6x = 0$

25. $6x^3 - 18x^2 = 0$

Exponential and Logarithmic Equations

16

16–1 Exponential Expressions, Equations, and Formulas

1. Evaluate formulas with at least one exponential term.
2. Evaluate formulas that contain a power of the natural exponential, e.
3. Solve exponential equations in the form $b^x = b^y$, where $b > 0$ and $b \neq 1$.

16–2 Logarithmic Expressions

1. Write exponential expressions as equivalent logarithmic expressions.
2. Write logarithmic expressions as equivalent exponential expressions.
3. Evaluate common and natural logarithmic expressions using a calculator.
4. Evaluate logarithms with a base other than 10 or e.
5. Evaluate formulas containing at least one logarithmic term.
6. Simplify logarithmic expressions by using the properties of logarithms.

16–1 | Exponential Expressions, Equations, and Formulas

Learning Outcomes

1. Evaluate formulas with at least one exponential term.
2. Evaluate formulas that contain a power of the natural exponential, e.
3. Solve exponential equations in the form $b^x = b^y$, where $b > 0$ and $b \neq 1$.

Many scientific, technical, and business phenomena have the property of exponential growth; that is, the growth rate does not remain constant as certain physical properties increase. Instead, the growth rate increases exponentially. For example, notice the difference between $2x$ and 2^x when x increases. If we write the expressions in function notation, we have two different functions of x.

$$f(x) = 2x$$
$$g(x) = 2^x$$

Examine the values in Table 16–1 and the graphical representation of these two functions (Fig. 16–1). 2^x is said to increase *exponentially*. $g(x) = 2^x$ is an *exponential function*. On the other hand, $2x$ increases at a constant rate. $f(x) = 2x$ is a *linear function* since $2x$ is a linear polynomial.

Table 16–1   Values of f(x) and g(x)

x	$f(x) = 2x$	$g(x) = 2^x$
1	$f(1) = 2(1) = 2$	$g(1) = 2^1 = 2$
2	$f(2) = 2(2) = 4$	$g(2) = 2^2 = 4$
3	$f(3) = 2(3) = 6$	$g(3) = 2^3 = 8$
4	$f(4) = 2(4) = 8$	$g(4) = 2^4 = 16$
5	$f(5) = 2(5) = 10$	$g(5) = 2^5 = 32$
6	$f(6) = 2(6) = 12$	$g(6) = 2^6 = 64$

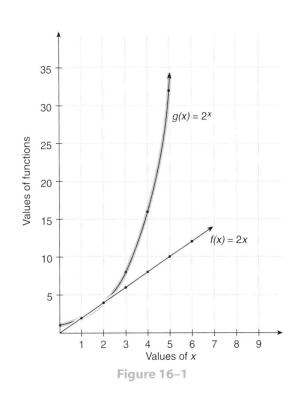

Figure 16–1

Before the easy availability of scientific and graphing calculators, examining and evaluating exponential functions was very tedious and time-consuming. Exponential expressions are found in many formulas. Even with the use of a scientific or graphing calculator, an understanding of exponential expressions is necessary.

1 Evaluate Formulas with at Least One Exponential Term.

Many formulas have terms that contain exponents. An *exponential expression* is an expression that contains at least one term that has a *variable exponent*. A variable exponent is an exponent that has at least one letter factor. Exponential expressions can be evaluated on a calculator by using the general power key ^ or other power keys as appropriate. The base of the term is entered first, then ^, followed by the value of the exponent. An *exponential equation* or formula contains at least one term that has a variable exponent.

A commonly used formula that contains a variable exponent is the formula for calculating the compound amount for compound interest. *Compound interest* is the interest calculated at the end of each period and then added to the principal for the next period. The *compound amount* or the accumulated amount is the combined principal and interest accumulated over a period of time.

Compound amount (accumulated amount):

$$A = P\left(1 + \frac{r}{n}\right)^{nt}$$

where A = accumulated amount
   P = original principal
   t = time in years
   r = rate per year expressed as a decimal equivalent
   n = compounding periods per year

EXAMPLE   Using the formula $A = P\left(1 + \frac{r}{n}\right)^{nt}$ and a calculator, find the accumulated amount on an investment of $1,500, invested at an interest rate of 9% for 3 years, if the interest is compounded quarterly.

Estimation   We expect to have more than $1,500.

$A = P\left(1 + \dfrac{r}{n}\right)^{nt}$ $P = \$1,500$; $r = 9\%$ or 0.09; $n =$ quarterly or 4 times a year; $t = 3$ years.

$A = 1,500\left(1 + \dfrac{0.09}{4}\right)^{(4)(3)}$ Simplify exponent and division term in grouping.

$A = 1,500(1 + 0.0225)^{12}$ Combine terms in grouping.

$A = 1,500(1.0225)^{12}$ 1.0225 ☐ 12 ☐ ⟹ 1.30604999.

$A = 1,500(1.30604999)$ Multiply.

$A = 1,959.07$ Rounded.

Interpretation   The accumulated amount of the $1,500 investment after 3 years is $1,959.07 to the nearest cent.

TIP!

Your Mind Is Often Quicker (and More Accurate) Than Your Fingers.

Should we do every step on the calculator? The following sequence of keystrokes is one option for performing all the calculations in the preceding example in a continuous series.

1500 ⊠ ☐(1 ⊞ .09 ÷ 4 ☐) ∧ ☐(4 ⊠ 3 ☐) ⊟

Whenever possible, it is advisable to do some calculations mentally. This can greatly decrease the complexity of the calculator sequences. 4(3) = 12.

1500 ⊠ ☐(1 ⊞ .09 ÷ 4 ☐) ∧ 12 ⊟ ⟹ 1959.074985

Try each sequence on your calculator.

TIP!

The accumulated amount is also the *future value* or *maturity value*. Many business calculators or computer software programs have a future value function (FV).

Interpreting a Formula

In working with formulas it is important to understand what is represented by each letter of the formula. The compound interest formula can also be given in terms of the number of *compounding periods* and the *interest rate per period*.

$$\text{Total compounding periods } (N) = \frac{\text{Compounding periods per year } (n)}{\text{times}} \\ \text{Number of years } (t)$$

$$N = nt$$

$$\text{Interest rate per period } (R) = \frac{\text{Interest rate per year } (r)}{\text{Compounding periods per year } (n)}$$

$$R = \frac{r}{n}$$

The compound amount or future value formula can also be written as

$$A = P(1 + R)^N \qquad \text{or} \qquad FV = P(1 + R)^N$$

where A or FV = accumulated amount or future value
R = interest rate per compounding period
N = total number of compounding periods

To find the compound interest:

1. Find the accumulated amount or future value using the formula
$$A = P(1 + R)^N$$

2. Subtract the original principal from the accumulated amount.
$$I = A - P$$

Combining both formulas:

$$I = P(1 + R)^N - P$$

or

$$I = P\left((1 + R)^N - 1\right)$$

EXAMPLE Find the compound interest on an investment of $10,000 at 8% annual interest compounded semiannually for 3 years.

$$P = 10{,}000, R = \frac{0.08}{2} = 0.04 \text{ per period}, N = 2 \times 3 = 6 \text{ periods}$$

$I = P(1 + R)^N - P$	Substitute values.
$I = 10{,}000\,(1 + 0.04)^6 - 10{,}000$	Combine terms inside grouping.
$I = 10{,}000(1.04)^6 - 10{,}000$	Raise to the power.
$I = 10{,}000(1.265319018) - 10{,}000$	Multiply.
$I = 12{,}653.19 - 10{,}000$	Subtract.
$I = \$2{,}653.19$	

The interest is $2,653.19.

The *present value* of an investment is the *lump sum* amount that should be invested now at a given interest rate for a specific period of time to yield a specific accumulated amount in the future.

$$PV = \frac{FV}{(1 + R)^N}$$

where PV = present value
FV = future value
R = interest rate per period
N = total number of periods

EXAMPLE The 7th Inning Sports Shop needs $20,000 in 10 years to replace engraving equipment. Find the amount the firm must invest at the present if it receives 10% interest compounded annually.

$$FV = 20,000, R = \frac{0.10}{1} = 0.1, N = 10(1) = 10$$

$PV = \dfrac{20,000}{(1 + 0.1)^{10}}$ Combine terms inside grouping.

$PV = \dfrac{20,000}{(1.1)^{10}}$ Raise denominator to the power.

$PV = \dfrac{20,000}{2.59374246}$ Divide.

$PV = \$7,710.87$

$7,710.87 should be invested now.

In advertising or in stating the terms of an investment or loan it is common to equate the compound interest rate to a comparable simple interest rate. This rate is referred to as the *effective rate, annual percentage rate* (*APR*) and *annual percentage yield* (*APY*).

To find the effective rate:

$$E = \left(1 + \frac{r}{n}\right)^n - 1$$

where E = effective rate
r = interest rate per year
n = number of compounding periods per year

EXAMPLE Find the effective interest rate for a loan of $600 at 10% compounded semiannually.

$E = \left(1 + \dfrac{r}{n}\right)^n - 1$ Substitute values. $r = 0.1$; $n = 2$.

$E = \left(1 + \dfrac{0.1}{2}\right)^2 - 1$ Simplify inside grouping.

$E = (1 + 0.05)^2 - 1$
$E = (1.05)^2 - 1$ Raise to the power.
$E = 1.1025 - 1$ Subtract.
$E = 0.1025$

Effective interest rate is 10.25%.

An *annuity* is a fund that accumulates compound interest as periodic payments add to the principal. An *ordinary annuity* has periodic payments that are made at the end of each payment period.

To find the future value of an ordinary annuity:

Apply the formula

$$FV = P\left[\frac{(1 + R)^N - 1}{R}\right]$$

where FV = future value of an ordinary annuity
P = amount of the periodic payment
R = interest rate per period
N = total number of periods

EXAMPLE Find the future value of an ordinary annuity of $6,000 for 5 years at 6% annual interest compounded semiannually.

$$FV = P\left[\frac{(1 + R)^N - 1}{R}\right], \qquad P = \$6,000, \qquad R = \frac{0.06}{2} = 0.03, \qquad N = 5(2) = 10$$

$$FV = 6,000\left[\frac{(1 + 0.03)^{10} - 1}{0.03}\right] \qquad \text{Simplify innermost grouping.}$$

$$FV = 6,000\left[\frac{(1.03)^{10} - 1}{0.03}\right] \qquad \text{Raise to the power.}$$

$$FV = 6,000\left[\frac{1.343916379 - 1}{0.03}\right] \qquad \text{Subtract in numerator.}$$

$$FV = 6,000\left[\frac{0.343916379}{0.03}\right] \qquad \text{Divide in grouping.}$$

$$FV = 6,000(11.46387931) \qquad \text{Multiply.}$$

$$FV = \mathbf{\$68,783.28}$$

When you have a specific future goal or target amount that you want to accumulate, a *sinking fund payment* is the amount you would invest in periodic payments to reach this goal. To determine the sinking fund payment, the *known values* are the future goal or amount, the amount of time, and the expected or guaranteed interest rate.

To find the sinking fund payment to produce a specified future value:

Apply the formula

$$P = FV\left[\frac{R}{(1 + R)^N - 1}\right]$$

where P = sinking fund payment
FV = future value or goal
R = interest rate per period
N = total number of periods

EXAMPLE A municipality has established a sinking fund to retire a bond issue of $500,000, which is due in 10 years. The account pays 8% quarterly interest. Find the amount of the quarterly sinking fund payment.

$$P = FV\left[\frac{R}{(1 + R)^N - 1}\right]$$ Substitute known values.

$$FV = \$500{,}000, \qquad R = \frac{0.08}{4} = 0.02, \qquad N = 10(4) = 40$$

$$P = 500{,}000\left[\frac{0.02}{(1 + 0.02)^{40} - 1}\right]$$ Simplify grouping in denominator.

$$P = 500{,}000\left[\frac{0.02}{(1.02)^{40} - 1}\right]$$

$$P = 500{,}000\left[\frac{0.02}{2.208039664 - 1}\right]$$

$$P = 500{,}000\left[\frac{0.02}{1.208039664}\right]$$ Divide.

$$P = 500{,}000[0.0165557478]$$ Multiply.

$$P = \$8{,}277.87$$

Finding the monthly payment to repay a loan is similar to the process for finding the sinking fund payment. The repayment of the loan in equal installments that are ~~applied to the principal and interest over a specified amount of time~~ is called the *amortization of a loan.*

To find the monthly payment for an amortized loan:

Apply the formula

$$M = P\left[\frac{R}{1 - (1 + R)^{-N}}\right]$$

where M = monthly payment
P = principal or initial amount of the loan
R = interest rate per month
N = total number of months

EXAMPLE Find the monthly payment on a 25-year home mortgage of $135,900 at 8%.

$$P = \$135{,}900, \qquad R = \frac{0.08}{12} = 0.0066666667, \qquad N = 25(12) = 300$$

$$M = P\left[\frac{R}{1 - (1 + R)^{-N}}\right]$$ Substitute values.

$$M = 135{,}900\left[\frac{0.0066666667}{1 - (1 + 0.0066666667)^{-300}}\right]$$ Simplify denominator.

$$M = 135{,}900\left[\frac{0.0066666667}{1 - (1.0066666667)^{-300}}\right]$$

$$M = 135{,}900 \left[\frac{0.0066666667}{1 - 0.1362365146} \right]$$

$$M = 135{,}900 \left[\frac{0.0066666667}{0.8637634854} \right]$$

Divide in grouping and multiply.

$$M = \$1{,}048.89$$

2 **Evaluate Formulas That Contain a Power of the Natural Exponential, e.**

In many applications involving circles, the irrational number π (approximately equal to 3.141592654) is used. Another irrational number, e, arises in the discussion of many physical phenomena. Many formulas contain a power of the natural exponential, e.

Exponential change is an interesting phenomenon. Let's look at the value of the expression $\left(1 + \dfrac{1}{n}\right)^n$ as n gets larger and larger. See Table 16–2.

Table 16–2 **Values of $\left(1 + \frac{1}{n}\right)^n$**

n	$\left(1 + \frac{1}{n}\right)^n$	Result
1	$\left(1 + \frac{1}{1}\right)^1$	2
2	$\left(1 + \frac{1}{2}\right)^2$	2.25
3	$\left(1 + \frac{1}{3}\right)^3$	2.37037037
10	$\left(1 + \frac{1}{10}\right)^{10}$	2.59374246
100	$\left(1 + \frac{1}{100}\right)^{100}$	2.704813829
1,000	$\left(1 + \frac{1}{1{,}000}\right)^{1{,}000}$	2.716923932
10,000	$\left(1 + \frac{1}{10{,}000}\right)^{10{,}000}$	2.718145927
100,000	$\left(1 + \frac{1}{100{,}000}\right)^{100{,}000}$	2.718268237
1,000,000	$\left(1 + \frac{1}{1{,}000{,}000}\right)^{1{,}000{,}000}$	2.718280469

The value of the expression changes very little as the value of n gets larger. We can say that the value approaches a given number. We call the number e, the *natural exponential*. The natural exponential, e, like π, is an irrational number and will never terminate nor repeat as more decimal places are examined.

The natural exponential, e, like π, is a constant. That is, the value is always the same; it does not vary. The natural exponential, e, is the limit that the value of the expression $\left(1 + \dfrac{1}{n}\right)^n$ approaches as n gets larger and larger without bound. The value of e to nine decimal places is 2.718281828.

To evaluate formulas containing the natural exponential, e, we can use a calculator or computer.

TIP!

The Natural Exponential, e, on Your Calculator

The natural exponential key is generally labeled $\boxed{e^x}$. On most calculators, the exponent is entered after pressing the $\boxed{e^x}$ key. To find $e^{2.3}$, enter the following sequence:

$$\boxed{e^x} \ 2.3 \ \boxed{=} \ \Rightarrow 9.974182455$$

Always test your calculator. A good test for the $\boxed{e^x}$ key is to find e^0.

$e^0 = 1$ Also, $e^1 = 2.718281828$

EXAMPLE The formula for the atmospheric pressure (in millimeters of mercury) is $P = 760e^{-0.00013h}$, where h is the height above sea level in meters. Find the atmospheric pressure at 100 m above sea level ($h = 100$).

Estimation Developing your number sense with powers of e comes after much examination of the power of e. For now you will minimize errors by making the calculations in the calculator two times and preferably two different ways. A negative exponent means you will multiply by a number less than 1 and the product will be smaller than the original value.

$P = 760e^{-0.00013h}$ $h = 100$.

$P = 760e^{-0.00013(100)}$ Simplify exponent. Multiply -0.00013 times 100 mentally.

$P = 760e^{-0.013}$ $e^{-0.013} = 0.987084135$ (leave in calculator)

$P = 760(0.987084135)$ Multiply by 760.

One continuous calculator sequence: $760 \times e^x (-) .013 =$

$P = 750.1839426$

Interpretation **The atmospheric pressure at 100 m above sea level is 750.18 mm.**

As the number of compounding periods per year increases, the effect of compounding levels off or reaches a limit. Therefore the natural exponential e can be substituted into the compound interest formula to accomplish *continuous compounding*.

To find the accumulated amount (future value) for continuous compounding:

1. Determine the principal (P), rate per year (r), and the number of years (t).
2. Evaluate the formula

$$A = Pe^{rt}$$

where A = accumulated amount or future value
   P = principal
   r = rate per year
   t = time in years

EXAMPLE Find the compound amount of $5,000 invested at an annual rate of 4% compounded continuously for 5 years.

$P = 5,000, r = 0.04$ (from 4%), $t = 5$

$A = Pe^{rt}$ Substitute values.

$A = 5,000e^{(0.04)(5)}$ Simplify exponents.

$A = 5,000e^{0.2}$ Raise to the power.

$A = 5,000(1.221402758)$ Multiply.

$A = 6,107.01$

The accumulated amount is $6,107.01.

3 Solve Exponential Equations in the Form $b^x = b^y$, where $b > 0$ and $b \neq 1$.

Some applications of exponents may require us to find the value of an unknown exponent. If we can write an equation in a specific form, $b^x = b^y$, where $b > 0$ and $b \neq 1$, we will be able to solve the equation using many of our previously learned skills for solving linear equations. Otherwise, we will need to learn some new strategies involving logarithms, which are introduced in the next section.

To illustrate a property of exponential equations, we look at the equation $2^x = 32$. In the equation, the value of x is the power of 2 that gives a result of 32. We can rewrite 32 as 2^5. Thus, $2^x = 2^5$. When an equation can be written in this form, a special property applies.

To solve an exponential equation in the form of $b^x = b^y$:

Apply the following property and solve for x. If $b^x = b^y$, and $b > 0$ and $b \neq 1$, then $x = y$.
Note: This property only applies when bases are *like*.

Learning Strategy *Using Conditional Properties (if . . . then)*

Many mathematical properties are phrased using an "if . . . then" format. That means, before the property can be used, the conditions or restrictions must be examined. It also means that the property is not appropriate under other conditions.

What are the conditions of the previous property?

- Bases must be like.
- Bases must be positive.
- Bases cannot equal 1.

To illustrate why the property does not work unless these conditions are met, look at this equation, $1^5 = 1^8$. This is a true statement since $1^5 = 1$ and $1^8 = 1$. But, applying the property "if $b^x = b^y$, then $x = y$," does $5 = 8$? No!

EXAMPLE Solve the equation $2^x = 32$.

$2^x = 32$ Rewrite 32 as a power of 2.
$2^x = 2^5$ $b = 2$, so $b > 0$, $b \neq 1$.
If $2^x = 2^5$, then $\mathbf{x = 5}$.   Apply property.

Check to see if the solution is appropriate.   Does $2^5 = 32$? Yes, so 5 is the correct solution.

EXAMPLE Solve the equation, $3^{x+1} = 27$.

$3^{x+1} = 27$ Rewrite 27 as a power of 3.
$3^{x+1} = 3^3$ $b = 3$, so $b > 0$, $b \neq 1$.
If $3^{x+1} = 3^3$, then $x + 1 = 3$.   Apply property.
$x + 1 = 3$ Solve for x.
$\quad x = 3 - 1$
$\quad \mathbf{x = 2}$

Check to see if the solution, 2, is correct.   Does $3^{2+1} = 27$?
$$3^3 = 27$$

Thus, the solution, $x = 2$, is correct.

It will not always be possible to rewrite an exponential equation as an equation with like bases. In such cases, other methods for solving the exponential equation are used.

SELF-STUDY EXERCISES 16–1

1 Using the compound amount formula, $A = P\left(1 + \dfrac{r}{n}\right)^{nt}$, find the accumulated amount.

1. Principal = $1,500, rate = 10%, compounded annually, time = 5 years.

2. Principal = $1,750, rate = 8%, compounded quarterly, time = 2 years.

3. The number of grams of a chemical that will dissolve in a solution is given by the formula $C = 100e^{0.02t}$, where t = temperature in degrees Celsius. Evaluate when
  (a) $t = 10$ **(c)** $t = 25$
  (b) $t = 20$ **(d)** $t = 30$

4. The compound amount when an investment is compounded continually (every instant) is expressed by the formula $A = Pe^{rt}$, where A = compounded amount, P = principal, t = time in years, and r = rate per year. Find the compound amount when
  (a) Principal = $1,000, interest = 9%, for 2 years
  (b) Principal = $1,500, interest = 10%, for 6 months

Evaluate using a scientific or graphing calculator.

5. 4^3　　　　**6.** 3^{-5}　　　　**7.** 5^{10}　　　　**8.** 8^{-3}　　　　**9.** $9^{2.5}$　　　　**10.** $10^{-\frac{5}{2}}$

2 A formula for electric current is $I = 1.50e^{-200t}$, where t is the time in seconds. Calculate the current for each time. Express answers in scientific notation.

11. 1 sec　　　　　　　**12.** 1.1 sec　　　　　　　**13.** 0.5 sec

Evaluate using a scientific or graphing calculator. Round to hundredths.

14. e^2　　　　　**15.** e^{-3}　　　　　**16.** $e^{0.21}$　　　　　**17.** $e^{-3.5}$

Use the compound interest formula for exercises 18–27.

18. First State Bank loaned Doug Morgan $2,000 for 4 years compounded annually at 8%. How much interest was Doug required to pay on the loan?

19. A loan of $8,000 for two acres of woodland is compounded quarterly at an annual rate of 12% for 5 years. Find the compound amount and the compound interest.

20. Compute the compound amount and the interest on a loan of $10,500 compounded annually for 4 years at 10%.

21. Find the future value of an investment of $10,500 if it is invested for 4 years and compounded quarterly at an annual rate of 8%.

22. You have $8,000 that you plan to invest in a compound-interest-bearing instrument. Your investment agent advises you that you can invest the $8,000 at 8% compounded quarterly for 3 years or you can invest the $8,000 at $8\frac{1}{4}\%$ compounded annually for 3 years. Which investment should you choose to receive the most interest?

23. Find the future value of $50,000 at 6%, compounded semiannually for 10 years.

24. Find the compound interest on $2,500 at $6\frac{3}{4}\%$, compounded daily by Leader Financial Bank for 20 days.

25. How much compound interest is earned on a deposit of $1,500 at 6.25%, compounded daily for 30 days?

26. Ezell Allen has found a short-term investment opportunity. He can invest $8,000 at 8.5% interest for 15 days. How much interest will he earn on this investment if the interest is compounded daily?

27. What is the compound interest on $8,000 invested at 8% for 180 days if it is compounded daily?

28. Find the effective interest rate for the investment described in exercise 21. Use the formula for the effective rate.
30. Betty Veteto has a loan of $8,500, compounded quarterly for 4 years at 6%. What is the effective interest rate for the loan? Use the formula.

Use the present value formula for exercises 32–39.
32. Compute the amount of money to be set aside today to ensure a future value of $2,500 in 1 year if the interest rate is 11% annually, compounded annually.
34. Ronnie Cox has just inherited $27,000. How much of this money should he set aside today to have $21,000 to pay cash for a Ventura Van, which he plans to purchase in 1 year? He can invest at 7.9% annually, compounded annually.

36. Joe Brozovich needs $2,000 in 3 years to make the down payment on a new car. How much must he invest today if he receives 8% interest annually, compounded annually?
38. Kristen Bieda plans to open a business in 4 years when she retires. How much must she invest today to have $10,000 when she retires if the bank pays 10% annually, compounded quarterly.

Use the future value formula for exercises 40–51.
40. Find the future value of an ordinary annuity of $3,000 annually for 2 years at 9% annual interest. Find the total interest earned.

42. Bert Simmons plans to pay an ordinary annuity of $5,000 annually for 3 years so he can take a year's sabbatical to study for a master's degree in business. The annual rate of interest is 8%. How much will Harry have at the end of 3 years?

44. Find the future value of an ordinary annuity of $6,500 semiannually for 7 years at 10% annual interest, compounded semiannually. How much was invested?

46. Rosa Kavanaugh established an ordinary annuity of $1,000 annually at 7% annual interest. What is the future value of the annuity after 15 years?
48. You invest in an ordinary annuity of $1,000 annually at 8% annual interest. What is the future value of the annuity at the end of 5 years?

50. Find the monthly payment on a home mortgage of $128,600 if the home is financed for 25 years at 7% interest.

29. What is the effective interest rate for a loan of $5,000 at 10%, compounded semiannually for 3 years? Use the effective rate formula.
31. What is the effective interest rate for a loan of $20,000 for 3 years if the interest is compounded quarterly at a rate of 12%?

33. How much should Marylynne Wessell set aside now to buy equipment that costs $8,500 in 1 year? The current interest rate is 7.5% annually, compounded annually.
35. Shirley Riddle received a $10,000 gift from her mother and plans a minor renovation to her home and an investment for 1 year, at which time she plans to take a trip projected to cost $6,999. The current interest rate is 8.3% annually, compounded annually. How much should be set aside today for her trip?

37. Calculate the amount of money that must be invested now at 6% annually, compounded quarterly, to obtain $1,500 in 3 years.

39. Charlie Bryant has a child who will be college age in 5 years. How much must he set aside today to have $20,000 for college tuition in 5 years if he gets 8% annually, compounded annually?

41. Len and Sharron Smith are saving money for their daughter Heather to attend college. They set aside an ordinary annuity of $4,000 annually for 2 years at 7% annual interest. How much will Heather have for college when she graduates from high school in 2 years? Find the total interest earned.
43. Joe Freeman is planning to establish a small business to provide consulting services in computer networking. He is committed to an ordinary annuity of $3,000 annually at 8.5% annual interest. How much will Joe have to establish the business after 3 years?
45. Jimmie Van Alphen pays an ordinary annuity of $2,500 quarterly at 8% annual interest, compounded quarterly, to establish supplemental income for retirement. How much will Jimmie have available at the end of 5 years?
47. You invest in an ordinary annuity of $500 annually at 8% annual interest. Find the future value of the annuity at the end of 10 years.
49. Make a chart comparing your results for exercises 47 and 48. Use these headings: Years, Total Investment, Total Interest. What general conclusion might you draw about effective investment strategy?
51. Find the monthly payment on a home mortgage of $128,600 if the home is financed for 15 years at 7% interest.

3 Solve for x.

52. $3^x = 3^7$
53. $5^x = 5^{-3}$
54. $3^{x+4} = 3^6$
55. $2^{x-3} = 2^7$

56. $6^x = 6^7$
57. $3^x = 27$
58. $2^x = 64$
59. $3^x = \frac{1}{81}$

60. $2^x = \frac{1}{64}$
61. $4^{3x} = 128$
62. $3^{2x} = 243$
63. $4^{3-x} = \frac{1}{16}$

64. $2^{4-x} = \frac{1}{16}$

16–2 | Logarithmic Expressions

Learning Outcomes

1 Write exponential expressions as equivalent logarithmic expressions.
2 Write logarithmic expressions as equivalent exponential expressions.
3 Evaluate common and natural logarithmic expressions using a calculator.
4 Evaluate logarithms with a base other than 10 or e.
5 Evaluate formulas containing at least one logarithmic term.
6 Simplify logarithmic expressions by using the properties of logarithms.

Logarithms were first introduced as a relatively fast way of carrying out lengthy calculations. The calculator and computer have diminished the importance of logarithms as a computational device; however, the importance of logarithms in advanced mathematics, electronics, and theoretical work is more evident than ever. Many formulas use logarithms to express the relationships of physical properties. Logarithms are also used to solve many exponential equations.

1 WRITE EXPONENTIAL EXPRESSIONS AS EQUIVALENT LOGARITHMIC EXPRESSIONS

A *logarithmic expression* is an expression that contains at least one term with a logarithm. The three basic components of a power are the base, exponent, and result of exponentiation or power. The word *power* has more than one meaning, so we use the *result of exponentiation* terminology.

$$3^4 = 81 \leftarrow \text{result of exponentiation (power)}$$

exponent ↗ base

When finding a power or the result of exponentiation, you are given the base and exponent. When finding a root, you know the base (radicand) and the index of the root (exponent). A third type of calculation involves finding the exponent when the base and the result of exponentiation are given. The exponent in this process is called the *logarithm*. In the logarithmic form $\log_b x = y$, b is the *base*, y is the *exponent or logarithm*, and x is the *result of exponentiation*. This equation is read, "The log of x to the base b is y." Logarithm is written in abbreviated form as "log." In the exponential form $x = b^y$, b is also the base, y is the exponent, and x is the result of the exponentiation.

Algebraic expressions written in logarithmic or exponential form have the same three components: base, exponent, result of exponentiation. Let's examine the mapping from one form to the other.

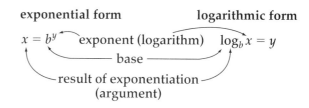

exponential form   **logarithmic form**

$x = b^y$ exponent (logarithm) $\log_b x = y$
base
result of exponentiation
(argument)

To convert an exponential expression to a logarithmic equation:

If $x = b^y$, then $\log_b x = y$, provided that $b > 0$ and $b \neq 1$.

1. The exponent in the exponential form is the variable solved for in the logarithmic form.
2. The base in the exponential form is the base in the logarithmic form.
3. The variable solved for in the exponential form is the result of exponentiation in logarithmic form. This result is sometimes referred to as the *argument*.

EXAMPLE Write in logarithmic form.

(a) $2^4 = 16$ (b) $3^2 = 9$ (c) $2^{-2} = \dfrac{1}{4}$

(a) $2^4 = 16$ converts to $\log_2 \mathbf{16 = 4}$
  Base = 2, exponent = 4, result of exponentiation = 16.

(b) $3^2 = 9$ converts to $\log_3 \mathbf{9 = 2}$
  Base = 3, exponent = 2, result of exponentiation = 9.

(c) $2^{-2} = \dfrac{1}{4}$ converts to $\log_2 \dfrac{\mathbf{1}}{\mathbf{4}} = \mathbf{-2}$
  Base = 2, exponent = −2, result of exponentiation = $\frac{1}{4}$.

2 **Write Logarithmic Expressions as Equivalent Exponential Expressions.**

The inverse of the process in the preceding learning outcome changes a logarithmic expression to an exponential expression.

To convert a logarithmic equation to an exponential equation:

$\log_b x = y$ converts to $x = b^y$, provided that $b > 0$ and $b \neq 1$.

1. The dependent variable (variable solved for) in the logarithmic form is the exponent in the exponential form.
2. The base in the logarithmic form is the base in the exponential form.
3. The result of the exponentiation in logarithmic form is the dependent variable (variable solved for) in exponential form.

EXAMPLE Write in exponential form.

(a) $\log_2 32 = 5$ (b) $\log_3 81 = 4$ (c) $\log_5 \dfrac{1}{25} = -2$ (d) $\log_{10} 0.001 = -3$

(a) $\log_2 32 = 5$ converts to $\mathbf{2^5 = 32}$
  Base = 2, exponent = 5, result of exponentiation = 32.

(b) $\log_3 81 = 4$ converts to $\mathbf{3^4 = 81}$
  Base = 3, exponent = 4, result of exponentiation = 81.

(c) $\log_5 \dfrac{1}{25} = -2$ converts to $\mathbf{5^{-2} = \dfrac{1}{25}}$
  Base = 5, exponent = −2, result of exponentiation = $\frac{1}{25}$.

(d) $\log_{10} 0.001 = -3$ converts to $\mathbf{10^{-3} = 0.001}$
  Base = 10, exponent = −3, result of exponentiation = 0.001.

When the base of a logarithm is 10, the logarithm is referred to as a *common logarithm.* If the base is omitted in a logarithmic expression, the base is assumed to be 10. Thus, $\log_{10} 1,000 = 3$ is normally written as $\log 1,000 = 3$. On a calculator, expressions containing common logarithms can be evaluated using the $\boxed{\log}$ key.

A logarithm with a base of e is a *natural logarithm* and is abbreviated as ln. Calculators normally have a $\boxed{\ln}$ key. Thus, $\log_e 1 = 0$ is normally written as $\ln 1 = 0$. Expressions containing natural logarithms can be evaluated using the $\boxed{\ln}$ key.

TIP!

Finding Logarithms Using Your Calculator

Calculators generally have two logarithm keys, $\boxed{\log}$ for *common logarithms* and $\boxed{\ln}$ for *natural logarithms.*

Evaluate log 2 and ln 2.

Most Calculators:

To find the common log of 2, press the $\boxed{\log}$ key, followed by the result of exponentiation, 2.

$$\boxed{\log}\ 2\ \boxed{=}\ \Rightarrow 0.3010299957$$

We can interpret this as $10^{0.3010299957} = 2$.

To find the natural log of 2, press the $\boxed{\ln}$ key, followed by the result of exponentiation, 2.

$$\boxed{\ln}\ 2\ \boxed{=}\ \Rightarrow 0.6931471806$$

That is, $e^{0.6931471806} = 2$.

Experiment with your calculator to determine the necessary keystrokes. A good test is to verify that $\log 1 = 0$ and $\ln 1 = 0$.

EXAMPLE   Evaluate using a calculator or exponential equations.

(a) $\log 10,000$ (b) $\log 0.000001$ (c) $\log_4 256$ (d) $\log_3 \dfrac{1}{27}$

(a) $\log 10,000 = \mathbf{4}$ $\boxed{\log} 10000 \boxed{=}$. Check: $10^4 = 10,000$.

(b) $\log 0.000001 = \mathbf{-6}$ Enter $\boxed{\log}.000001\boxed{=}$. Check: $10^{-6} = 0.000001$.

(c) Calculators generally do not have a key for logarithms with a base different from 10 or e. Rewrite the logarithmic equation as an exponential equation with like bases and solve.

$\log_4 256 = x$ is $4^x = 256$ Write 256 as a power of 4. We have $256 = 4^4$.

$\qquad\quad 4^x = 4^4$ Like bases.

$\qquad\qquad \mathbf{\mathit{x} = 4}$ Check: $4^4 = 256$.

(d) Rewrite the logarithmic equation as an exponential equation with like bases and solve.

$\log_3 \dfrac{1}{27} = x$ is $3^x = \dfrac{1}{27}$ Write $\frac{1}{27}$ as a power of 3. That is, $27 = 3^3$; $\frac{1}{27} = 3^{-3}$.

$\qquad\quad 3^x = 3^{-3}$

$\qquad\qquad \mathbf{\mathit{x} = -3}$ Check: $3^{-3} = \frac{1}{27}$.

EXAMPLE Evaluate using a calculator. Express the answers to the nearest ten-thousandth.

(a) ln 5   (b) ln 4.5   (c) ln 948

(a) ln 5 = **1.6094**   **(b)** ln 4.5 = **1.5041**   **(c)** ln 948 = **6.8544**

4 Evaluate Logarithms with a Base Other Than 10 or *e*.

Many situations require us to find a logarithm with a base other than 10 or *e*. A conversion formula is used to find the logarithm for a base other than 10 or *e* using a calculator.

To evaluate a logarithm with a base *b* other than 10 or *e*:

$$\log_b a = \frac{\log a}{\log b}$$

A similar process can be used with natural logarithms.

$$\log_b a = \frac{\ln a}{\ln b}$$

To check, evaluate b^x where b is the base of the logarithm and x is the logarithm. The result should equal the argument.

EXAMPLE Find $\log_7 343$.

$$\log_7 343 = \frac{\log 343}{\log 7} = \mathbf{3}$$ $\boxed{\text{log}}$ 343 $\boxed{\div}$ $\boxed{\text{log}}$ 7 $\boxed{-}$ $\Rightarrow$ 3; Check: $7^3 = 343$.

5 Evaluate Formulas Containing at Least One Logarithmic Term.

Many formulas use common and natural logarithms.

EXAMPLE The loudness of sound is measured by a unit called a *decibel* (dB). A very faint sound, called the *threshold sound*, is assigned an intensity of I_0. Other sounds have an intensity of I, which is a specified number times the threshold sound ($I = nI_0$).

Then the decibel rate is given by the formula dB = 10 log $\frac{I}{I_0}$. Find the decibel rating for sounds having the following intensities (I).

(a) A whisper, $110I_0$
(b) A speaking voice, $230I_0$
(c) A busy street, $9,000,000I_0$
(d) Loud music, $875,000,000,000I_0$
(e) A jet plane at takeoff, $109,000,000,000,000I_0$

(a) A whisper, $I = 110I_0$ $n = 110$

$$dB = 10 \log \frac{110I_0}{I_0}$$ $\frac{I_0}{I_0} = 1$

$$dB = 10 \log 110$$ $\log 110 = 2.041392685$

dB = 20 (rounded)

(b) A speaking voice, $I = 230I_0$

$$dB = 10 \log \frac{230I_0}{I_0}$$

$$dB = 10 \log 230$$

dB = 24 (rounded)

$n = 230$

$\frac{I_0}{I_0} = 1$

$\log 230 = 2.361727836$

(c) A busy street, $I = 9{,}000{,}000I_0$

$$dB = 10 \log \frac{9{,}000{,}000I_0}{I_0}$$

$$dB = 10 \log 9{,}000{,}000$$

dB = 70 (rounded)

$n = 9{,}000{,}000$

$\frac{I_0}{I_0} = 1$

$\log 9{,}000{,}000 = 6.954242509$

(d) Loud music, $I = 875{,}000{,}000{,}000I_0$

$$dB = 10 \log \frac{875{,}000{,}000{,}000I_0}{I_0}$$

$$dB = 10 \log 875{,}000{,}000{,}000$$

$$dB = 10 \log (8.75 \times 10^{11})$$

dB = 119 (rounded)

$n = 875{,}000{,}000{,}000$

$\frac{I_0}{I_0} = 1$

$875{,}000{,}000{,}000 = 8.75 \times 10^{11}$

$10 \boxed{\log} 8.75 \boxed{\text{EXP}} 11 \boxed{=} \Rightarrow 119.4200805$

(e) A jet plane at takeoff, $I = 109{,}000{,}000{,}000{,}000I_0 = (1.09 \times 10^{14})I_0$

$$dB = 10 \log \frac{(1.09 \times 10^{14})I_0}{I_0}$$

$$dB = 10 \log (1.09 \times 10^{14})$$

$$dB = 10(14.0374265)$$

dB = 140 (rounded)

$n = 1.09 \times 10^{14}$

$\frac{I_0}{I_0} = 1$

To find the time for an investment to reach a specified amount if compounded continuously:

$$t = \frac{(\ln A - \ln P)}{r}$$

t = time (in years) of investment
A = accumulated amount
P = initial invested amount
r = annual rate of interest

EXAMPLE Find the length of time for \$32,750 to reach \$47,650.97 when invested at 5% compounded continuously.

$A = \$47{,}650.97, \qquad P = \$32{,}750, \qquad r = 0.05$

$$t = \frac{\ln A - \ln P}{r}$$

Substitute known values.

$$t = \frac{\ln 47{,}650.97 - \ln 32{,}750}{0.05}$$

Evaluate numerator.

$$t = \frac{10.77165827 - 10.39665824}{0.05}$$

$$t = \frac{0.3750000246}{0.05}$$

Divide.

t = 7.5 years

Chapter 16 / Exponential and Logarithmic Equations

EXAMPLE How long does it take an investment of $10,000 to double if it is invested at 5% annual interest compounded continuously?

$$A = \$20,000, \qquad P = 10,000, \qquad r = 0.05$$

$$t = \frac{\ln A - \ln P}{r} \qquad\qquad \text{Substitute known values.}$$

$$t = \frac{\ln 20,000 - \ln 10,000}{0.05} \qquad\qquad \text{Evaluate numerator.}$$

$$t = \frac{9.903487553 - 9.210340372}{0.05}$$

$$t = \frac{0.6931471806}{0.05} \qquad\qquad \text{Divide.}$$

$$t = \textbf{13.9 years}$$

6 **Simplify Logarithmic Expressions by Using the Properties of Logarithms.**

Many applications of logarithms and exponential expressions require the understanding and use of the properties of logarithms. The laws of exponents are similar to the properties of logarithms because of the relationship between logarithms and exponents. Similar laws are appropriate for natural logarithms.

Properties of logarithms:

Logarithm of product	$\log_b mn = \log_b m + \log_b n$	The log of a product is the sum of the logs of the factors.
Logarithm of quotient	$\log_b \dfrac{m}{n} = \log_b m - \log_b n$	The log of a quotient is the difference of the logs of the numerator and denominator.
Logarithm of power	$\log_b m^n = n \log_b m$	The log of a quantity raised to an exponent is the exponent times the log of the quantity.
Logarithm of quantity with same base	$\log_b b = 1$	The log of a quantity with the same base as the quantity equals 1.

These laws can be illustrated with examples using a calculator.

EXAMPLE Show that the statements are true by using a calculator.

(a) $\log 6 = \log 2 + \log 3$ (b) $\log 2 = \log 6 - \log 3$
(c) $\log 2^3 = 3 \log 2$ (d) $\log 20 = 1 + \log 2$

(a) $\log 6 = \log 2(3) = \log 2 + \log 3$

$$\boxed{\log}\ 6\ \boxed{=}\ \Rightarrow \textbf{0.77815125,} \qquad \boxed{\log}\ 2\ \boxed{+}\ \boxed{\log}\ 3\ \boxed{=}\ \Rightarrow \textbf{0.77815125}$$

(b) $\log 2 = \log \dfrac{6}{3} = \log 6 - \log 3$

$$\boxed{\log}\ 2\ \boxed{=}\ \Rightarrow \textbf{0.301029996,} \qquad \boxed{\log}\ 6\ \boxed{-}\ \boxed{\log}\ 3\ \boxed{=}\ \Rightarrow \textbf{0.301029996}$$

(c) $\log 2^3 = 3 \log 2$

$$\boxed{\log}\ \boxed{(}\ 2\ \boxed{\wedge}\ 3\ \boxed{)}\ \boxed{=}\ \Rightarrow \textbf{0.903089987,} \qquad 3\ \boxed{\times}\ \boxed{\log}\ 2\ \boxed{=}\ \Rightarrow \textbf{0.903089987}$$

(d) $\log 20 = \log 10(2) = \log 10 + \log 2 = 1 + \log 2$

$$\boxed{\log}\ 20\ \boxed{=}\ \Rightarrow \textbf{1.301029996,} \qquad 1 + \boxed{\log}\ 2\ \boxed{=}\ \Rightarrow \textbf{1.301029996}$$

SELF-STUDY EXERCISES 16–2

1 Write as logarithmic equations.

1. $3^2 = 9$

2. $2^5 = 32$

3. $9^{\frac{1}{2}} = 3$

4. $16^{\frac{1}{4}} = 2$

5. $4^{-2} = \dfrac{1}{16}$

6. $3^{-4} = \dfrac{1}{81}$

2 Write as exponential equations.

7. $\log_3 81 = 4$

8. $\log_{12} 144 = 2$

9. $\log_2 \dfrac{1}{8} = -3$

10. $\log_5 \dfrac{1}{25} = -2$

11. $\log_{25} \dfrac{1}{5} = -0.5$

12. $\log_4 \dfrac{1}{2} = -0.5$

Solve for x by using an equivalent exponential expression.

13. $\log_4 64 = x$

14. $\log_3 x = -4$

15. $\log_6 36 = x$

16. $\log_7 \dfrac{1}{49} = x$

17. $\log_5 x = 4$

18. $\log_4 \dfrac{1}{256} = x$

3 Evaluate with a calculator. Express the answer to the nearest ten-thousandth.

19. $\log 3$

20. $\log 6$

21. $\log 2.4$

22. $\log 4.2$

23. $\log 150$

24. $\log 0.0012$

25. $\ln 4$

26. $\ln 2.5$

27. $\ln 0.15$

28. $\ln 275$

29. $\ln 100$

4 Evaluate with a calculator.

30. $\log_5 125$

31. $\log_3 729$

32. $\log_{\frac{1}{2}} 0.03125$

33. $\log_7 49$

34. $\log_8 56$

5 The intensity of an earthquake is measured on the *Richter scale* by the formula

$$\text{Richter scale rating} = \log \frac{I}{I_0}$$

where I_0 is the measure of the intensity of a very small (faint) earthquake. Find the Richter scale rating of earthquakes having the following intensities:

35. $1{,}000 I_0$

36. $100{,}000 I_0$

37. $100{,}000{,}000 I_0$

38. How long does it take an investment of $20,000 to double if it is invested at 5% annual interest, compounded continuously? Round to tenths.

39. How long does it take for $48,000 to reach $52,000 when invested at 4%, compounded continuously? Round to tenths.

6 Solve for x by using an equivalent logarithmic expression and the facts that $\log_3 2 = 0.631$ and $\log_3 5 = 1.465$.

40. $\log_3 10 = x$

41. $\log_3 8 = x$

42. $\log_3 6 = x$

ASSIGNMENT EXERCISES

Section 16–1

Solve for x.

1. $5^x = 5^8$

2. $2^x = 2^6$

3. $3^x = 3^{-2}$

4. $2^{x+3} = 2^7$

5. $4^{x-2} = 4^2$

6. $5^{2x-1} = 5^2$

7. $6^{3x+2} = 6^{-3}$

8. $2^x = 16$

9. $3^x = 81$

10. $3^x = \dfrac{1}{9}$

11. $2^x = \dfrac{1}{32}$

12. $4^{2x} = 64$

13. $5^{3x} = 125$

14. $3^{4-x} = \dfrac{1}{27}$

15. $6^{2-x} = \dfrac{1}{36}$

Evaluate.

16. e^3

17. e^{-4}

18. $e^{-0.12}$

19. e^{-10}

20. Calculate the compound interest on a loan of $200 at 6%, compounded annually for 4 years.

21. Calculate the compound interest on a 13% loan of $1,600 for 3 years if the interest is compounded annually.

22. Calculate the compound interest on a loan of $6,150 at $11\frac{1}{2}$% annual interest, compounded annually for 3 years.

23. Maria Sanchez invested $2,000 for 2 years at 12% annual interest, compounded semiannually. Calculate the interest she earned on her investment.

24. EZ Loan Company loaned $500 at 8% annual interest, compounded quarterly for 1 year. Calculate the amount the loan company will earn in interest.

25. Find the future value on an investment of $3,000 made by Ling Lee for 5 years at 12% annual interest, compounded semiannually.

26. Find the interest on $2,500 invested for 3 years at 3.75%, compounded quarterly.

27. Find the interest on a certificate of deposit (CD) of $10,000 for 5 years at 4%, compounded semiannually.

28. Find the interest on $10,000 for 2 years compounded monthly at a 6% annual interest rate.

29. Find the future value on an investment of $8,000, compounded quarterly for 7 years at 8%.

30. Find the compound interest on $5,000 for 2 years if the interest is compounded continuously at 12%.

31. Find the compound interest on $5,000 for 2 years if the interest is compounded continuously at 12%.

32. Find the effective interest rate for the loan described in Exercise 28.

33. Find the effective interest rate for the loan described in Exercise 29.

34. Find the accumulated amount on an investment of $8,000 invested for 5 years compounded quarterly at 5%.

35. Find the amount of interest on $1,500 invested for 4 years at a 3% annual rate compounded monthly.

36. Tommye Adams wishes to have $8,000 1 year from now to make a down payment on a lake house. How much should she invest at 9.5% annual interest to have her payment in 1 year?

37. Billy Hill wished to have $4,000 in 4 years to tour Europe. How much must he invest today at 8% annual interest, compounded quarterly, to have the $4,000 in 4 years?

38. Kristin Ammons was offered $15,000 cash or $19,500 to be paid after 2 years for a resort cabin. If money can be invested in today's market for 3% annual interest, compounded quarterly, which offer should Kristin accept?

39. An art dealer offered a collector $8,000 for a painting. The collector could sell the painting to an individual for $11,000 to be paid in 18 months. Currently investments yield 12% annual interest, compounded monthly. Which is the better deal for the collector?

40. If you were offered $700 today or $800 in 2 years, which would you accept if money can be invested at 12% annual interest, compounded monthly?

41. How much should a family invest now at 10% compounded annually to have a $7,000 house down payment in 4 years.

42. Dennis and Deb Walker had a baby in 1998. At the end of that year they began putting away $900 per year at 10% compounded annual interest for a college fund. When their child is 18 years old in 2016, the cost for 4 years of college is estimated to be about $20,000 per year.
   (a) How much money will be in the account when the child is 18 years old?
   (b) Will the Walkers have enough saved to send their child to college for 4 years?

43. Skip Quinn plans to deposit $2,000 at the end of every 6 months for the next 5 years to save up for a boat. If the interest rate is 12% annually, compounded semiannually, how much money will Skip have in his boat fund after 5 years?

44. A business deposits $4,500 at the end of each quarter in an account that earns 8% annual interest, compounded quarterly. What is the value of the annuity in 5 years?

45. How much must be set aside at the end of each six months in a sinking fund by the Fabulous Toy Company to replace a $155,000 piece of equipment at the end of 8 years if the account pays 8% annual interest, compounded semiannually?

46. Tasty Food Manufacturers, Inc., has a bond issue of $1,400,000 due in 30 years. If it wants to establish a sinking fund to meet this obligation, how much must be set aside at the end of each year if the annual interest rate is 6%?

47. Lausanne Private School System needs to set aside funds for a new computer system. What monthly sinking fund payment would be required to amount to $45,000, the approximate cost of the computer, in $1\frac{1}{2}$ years at 12% annual interest, compounded monthly?

48. Zachary Alexander owns a limousine that will need to be replaced in 4 years at a cost of $65,000. How much must he put aside each year in a sinking fund at 8% annual interest to be able to afford the new limousine?

49. Braddy's Department Store has a fleet of delivery trucks that needs to be replaced at a cost of $75,000. How much must it set aside every 3 months for 3 years in a sinking fund at 8% annual interest, compounded quarterly, to have enough money to replace the trucks?

50. Danny Lawrence Properties, Inc., has a bond issue that matures in 25 years for $1 million. How much must the company set aside each year in a sinking fund at 12% annual interest to meet this future obligation?

51. Brenda Pearson wants to save $25,000 for a new boat in 6 years. How much must be put aside in equal payments each year in an account earning 8% annual interest for Brenda to be able to purchase the boat?

52. How much money needs to be set aside today at 10% annual interest, compounded semiannually, to have the same amount as a semiannual ordinary annuity of $500 for 5 years at 10% annual interest, compounded semiannually?

53. Find the monthly payment on a mortgage of $238,000 if it is financed for 20 years at 7.5% annual interest.

54. Erma Thornton Braddy purchased a property with a mortgage of $528,260 at 6.8% annual interest. Find the monthly payment if the mortgage is amortized for 30 years.

A formula for electric current is $I = 1.50e^{-200t}$, where t is time in seconds. Calculate the current for each time. Express the answers in scientific notation.

55. 0.07 sec

56. 0.2 sec

57. 0.4 sec

58. The number of grams of a chemical that will dissolve in a solution is given by the formula $C = 100e^{0.05t}$, where t = temperature in degrees Celsius. Evaluate when

(a) $t = 10$ (b) $t = 20$
(c) $t = 45$ (d) $t = 50$

Section 16–2

Rewrite the following as logarithmic equations.

59. $2^3 = 8$

60. $5^2 = 25$

61. $3^4 = 81$

62. $81^{\frac{1}{2}} = 9$

63. $27^{\frac{1}{3}} = 3$

64. $5^{-3} = \frac{1}{125}$

65. $4^{-3} = \frac{1}{64}$

66. $8^{-\frac{1}{3}} = \frac{1}{2}$

67. $9^{-\frac{1}{2}} = \frac{1}{3}$

68. $121^{\frac{1}{2}} = 11$

69. $12^{-2} = \frac{1}{144}$

70. Write a true exponential expression; convert it to logarithmic form.

Rewrite the following as exponential equations. Verify if the equation is true.

71. $\log_{11} 121 = 2$

72. $\log_3 81 = 4$

73. $\log_{15} 1 = 0$

74. $\log_{25} 5 = \frac{1}{2}$

75. $\log_7 7 = 1$

76. $\log_3 3 = 1$

77. $\log_4 \frac{1}{16} = -2$

78. $\log_2 \frac{1}{16} = -4$

79. $\log_9 \frac{1}{3} = -0.5$

80. $\log_{16} \frac{1}{4} = -0.5$

81. $\log_{10} 1,000 = 3$

82. $\log_{10} 100 = 2$

Evaluate the following with a calculator. Express the answers to the nearest ten-thousandth.

83. log 5

84. log 3.8

85. log 180

86. log 0.0015

87. log 0.4

88. ln 12

89. ln 270

90. ln 0.134

91. ln 0.8

92. ln 80

93. $\log_5 30$

94. $\log_7 120$

Solve for x by using an equivalent exponential expression.

95. $\log_4 16 = x$

96. $\log_7 49 = x$

97. $\log_7 x = 3$

98. $\log_5 x = -2$

99. $\log_6 \frac{1}{36} = x$

100. $\log_4 \frac{1}{64} = x$

101. The intensity of an earthquake is measured on the Richter scale by the formula

$$\text{Richter scale rating} = \log \frac{I}{I_0}$$

where I_0 is the measure of the intensity of a very small (faint) earthquake. Find the Richter scale rating of the earthquakes having the following intensities:

(a) $100 I_0$ (b) $10,000 I_0$ (c) $150,000,000 I_0$

Solve for x by using an equivalent logarithmic expression and the facts that $\log_2 3 = 1.585$ and $\log_2 7 = 2.807$.

102. $\log_2 21$

103. $\log_2 9$

104. $\log_2 6$

105. An investment of $100,000 is invested at 4.1%, compounded continuously. How long will it take for the investment to accumulate to $150,000?

106. Quenesha McGee has $10,000 invested at 5.9%, compounded continuously. She needs to know how long it will take for the investment to accumulate to $18,000.

107. The Environmental Protection Agency (EPA) monitors atmosphere and soil contamination by dangerous chemicals. When possible, the chemical contamination is decomposed by using microorganisms that change the chemicals so they are no longer harmful. A particular microorganism can reduce the contamination level to about 65% of the existing level every 30 days. A soil test for a contaminated site shows 72,000,000 units per cubic meter of soil.

  (a) Write a formula for determining the contamination level after x 30-day periods: the initial contamination times the percent reduction (65% or 0.65) raised to the xth power.

  (b) What is the level of contamination after 60 days? After 150 days?

  (c) A "safe level" is 60,000 units of contamination per cubic meter of soil. Estimate how long it would take for the soil to reach this safe level. Discuss your method of arriving at the estimate.

108. Complete the spreadsheet showing the Accumulated Amount and Interest for each entry (Fig. 16–2). The formula written for the spreadsheet is: $= P*(1+r)\wedge t$, where P is principal, r is rate per period, and t is number of periods.

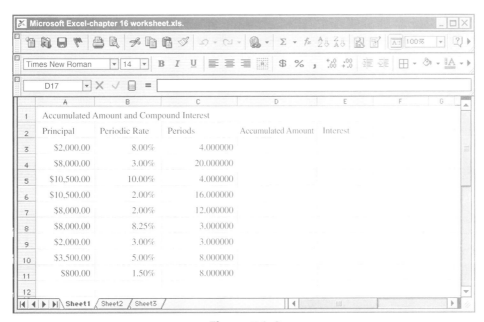

Figure 16–2

1. Give an example of an application that has a linear relationship and an application that has an exponential relationship.

3. Describe how to recognize an equation as an exponential equation.

5. Show symbolically the relationship between an exponential equation and a logarithmic equation. Give a numerical example illustrating this relationship.

2. Briefly describe how a linear relationship is different from an exponential relationship.

4. Explain why an equation of the form $b^x = b^y$ is only appropriate when $b > 0$ and $b \neq 1$.

6. Explain the difference between a common logarithm and a natural logarithm.

Find the mistakes in the examples. Explain the mistake and correct it.

7. $2^{x-3} = 4^2$
   $x - 3 = 2$
   $x = 2 + 3$
   $x = 5$

8. Change $\log_5 125 = 3$ to an equivalent exponential equation.
   $3^5 = 125$

Learning Outcomes	What to Remember with Examples

Section 16–1

1 Evaluate formulas with at least one exponential term (pp. 627–631).

A scientific or graphing calculator can be used along with the order of operations to evaluate formulas with at least one exponential term.

> Use the formula for compound interest to find the compound amount for a loan of $5,000 for 3 years at an annual interest rate of 6% if the principal is compounded semiannually.
>
> $A = P\left(1 + \dfrac{r}{n}\right)^{nt}$ $P = \$5{,}000; r = 0.06; n = 2; t = 3.$
>
> $A = 5{,}000\left(1 + \dfrac{0.06}{2}\right)^{2(3)}$ Perform operations inside grouping.
> Perform operation in exponent.
>
> $A = 5{,}000\,(1.03)^6$ Raise to power.
> $A = 5{,}000(1.194052297)$ Multiply.
> $A = \$5{,}970.26$ Compound amount (rounded).

2 Evaluate formulas that contain a power of the natural exponential, e (pp. 631–634).

Use a scientific or graphing calculator to evaluate formulas containing the natural exponential, e.

> Use the formula $P = 760\,e^{-0.00013h}$ for atmospheric pressure to find P at 50 m above sea level (h).
>
> $P = 760e^{-0.00013(50)}$
> $P = 760e^{-0.0065}$
> $P = 760(0.9935210793)$
> $P = 755.08$ Rounded.

3 Solve exponential equations in the form $b^x = b^y$, where $b > 0$ and $b \neq 1$ (pp. 635–636).

To solve an exponential equation in the form $b^x = b^y$, apply the property: When the bases are equal, the exponents are equal. If the bases are not equal, rewrite the bases as powers so that the bases are equal. The exponents will then be equal. This property applies only when b is positive and not equal to 1.

> Solve the equation $4^{3x+1} = 8$. Rewrite the bases: $4 = 2^2$ and $8 = 2^3$.
>
> $2^{2(3x+1)} = 2^3$ The bases are equal, so exponents are equal.
> $2(3x+1) = 3$ Distribute.
> $6x + 2 = 3$ Sort terms.
> $6x = 1$ Divide.
> $x = \dfrac{1}{6}$

Section 16–2

1 Write exponential expressions as equivalent logarithmic expressions (pp. 638–639).

To write exponential expressions as equivalent logarithmic expressions, use the format: $x = b^y$ converts to $\log_b x = y$.

> Write $16 = 2^4$ in logarithmic form. $\log_2 16 = 4$

2 Write logarithmic expressions as equivalent exponential expressions (p. 639).

To write logarithmic expressions as equivalent exponential expressions, use the format: $\log_b x = y$ converts to $x = b^y$.

> Write $\log_5 125 = 3$ in exponential form. $5^3 = 125$

3 Evaluate common and natural logarithmic expressions using a calculator (pp. 640–641).

Use the $\boxed{\log}$ key to find common logarithms and the $\boxed{\ln}$ key to find natural logarithms. Some calculators require the number to be entered before the $\boxed{\log}$ or $\boxed{\ln}$ key. Most calculators require the $\boxed{\log}$ or $\boxed{\ln}$ key to be entered, followed by the number.

Use a calculator to find log 25. Scientific calculator steps:

$$\log\ 25\ \boxed{=}\qquad \text{or}\qquad 25\ \boxed{\log}\ \Rightarrow 1.397940009$$

4 Evaluate logarithms with a base other than 10 or e (p. 641).

To evaluate logarithms with bases other than 10 or e, divide the common or natural log of the number by the common or natural log of the base.

Evaluate $\log_4 64$:

$$\log_4 64 = \frac{\log 64}{\log 4} = 3 \qquad \boxed{\log}\ 64\ \div\ \boxed{\log}\ 4\ \boxed{=}\ \Rightarrow 3$$

5 Evaluate formulas containing at least one logarithmic term (pp. 641–643).

To evaluate formulas containing at least one logarithmic term, use a calculator and follow the order of operations.

Use the formula $dB = 10\log\left(\dfrac{I}{I_0}\right)$ to find the decibel rating for a sound that is 350 times the threshold sound ($350I_0$).

$$dB = 10\log\left(\frac{350I_0}{I_0}\right)$$
$$dB = 10\log 350 \qquad\qquad 10\ \boxed{\times}\ \boxed{\log}\ 350\ \boxed{=}$$
$$dB = 25.44 \qquad\qquad\qquad \text{Rounded.}$$

6 Simplify logarithmic expressions by using the properties of logarithms (p. 643).

The laws of exponents also apply to expressions with logarithms.

$\log_b mn = \log_b m + \log_b n$

$\log_b \dfrac{m}{n} = \log_b m - \log_b n$

$\log_b m^n = n\log_b m$

$\log_b b = 1$

Write log (3)(8) in another way. Use a calculator to verify that the equation is true.

$$\log (3)(8) = \log 3 + \log 8$$
$$\boxed{\log}\ \boxed{(}\ 3\ \boxed{\times}\ 8\ \boxed{)}\ \boxed{=}\ \Rightarrow 1.380211242$$
$$\boxed{\log}\ 3\ \boxed{+}\ \boxed{\log}\ 8\ \boxed{=}\ \Rightarrow 1.380211242$$

CHAPTER TRIAL TEST

Evaluate using a calculator.
1. 1.2^{45}
2. 5^{-8}
3. $15^{\frac{3}{2}}$
4. $e^{-0.25}$
5. 12^5

Solve for x.
6. $4^x = 4^{-3}$
7. $2^{x-4} = 2^5$
8. $3^x = \dfrac{1}{9}$
9. $2^{2x-1} = 8$

Write as logarithmic equations.
10. $2^8 = 256$
11. $4^{-\frac{1}{2}} = \dfrac{1}{2}$

Write as exponential equations.
12. $\log_5 625 = 4$
13. $\log_3 \dfrac{1}{27} = -3$

Evaluate with a calculator. Express the answer to the nearest ten-thousandth.

14. $\log 4.8$

15. $\ln 32$

Solve for x by using an equivalent exponential expression and a calculator.

16. $\log_4 x = -2$

17. $\log_6 216 = x$

Evaluate the following with a calculator. Round to ten-thousandths.

18. $\log_7 2$

19. $\log_8 21$

20. The formula for the population growth of a certain species of insect in a controlled research environment is
$$P = 1,000,000\, e^{0.05t}$$
where t = time in weeks. Find the projected population after **(a)** 2 weeks and **(b)** 3 weeks.

21. The revenue in thousands of dollars from sales of a product is approximated by the formula
$$S = 125 + 83 \log (5t + 1)$$
where t is the number of years after a product was marketed. Find the projected revenue from sales of the product after 3 years.

22. The height in meters of the male members of a certain group is approximated by the formula
$$h = 0.4 + \log t$$
where t represents age in years for the first 20 years of life ($1 \le t \le 20$). Find the projected height of a 10-year-old male.

23. Find the accumulated amount on an investment of $5,000 at a rate of 5.8% per year, compounded annually for 2 years. Use the formula
$$A = P\left(1 + \frac{r}{n}\right)^{nt}$$

24. What is the future value of an ordinary annuity of $300 every 3 months for 4 years at 8% annual interest, compounded quarterly?

25. Find the future value of an ordinary annuity of $9,000 per year for 2 years at 15% annual interest.

26. What is the future value of an ordinary annuity of $985 every 6 months for 8 years at 8% annual interest, compounded semiannually?

27. What is the sinking fund payment required at the end of each year to accumulate to $125,000 in 16 years at 4% annual interest?

28. Find the compound interest on a loan of $3,000 for 1 year at 12% annual interest if the interest is compounded quarterly.

29. Find the effective interest rate for the loan described in exercise 28.

30. Find the compound interest on an investment of $2,000 invested at 5.75% for 28 years compounded quarterly.

31. If you were offered $600 today or $680 in 1 year, which would you accept if money can be invested at 12% annual interest, compounded monthly?

32. Mabel Langston needs $12,000 in 10 years for her daughter's college education. How much must be invested today at 8% annual interest, compounded semiannually, to have the needed funds?

33. Which of the two options yields the greatest return on your investment of $2,000?
Option 1: 8% annual interest compounded quarterly for 4 years
Option 2: $8\frac{1}{4}$% annual interest compounded annually for 4 years

34. Harvey Barton plans to buy a house in 4 years. He will make an $8,000 down payment on the property. How much should he invest today at 6% annual interest, compounded quarterly, to have the required amount in 4 years?

35. If you invest $1,000 today at 4% annual interest compounded continuously, how much will you have after 20 years?

36. If you invest $2,000 today at 8% annual interest, compounded quarterly, how much will you have after 3 years?

37. How much money should Bryan Trailer Sales set aside today to have $15,000 in 1 year to purchase a forklift if the interest rate is 10.4%, compounded annually?

38. Ginger Canoy has purchased a home for $122,000. She plans to finance $100,000 for 15 years at $5\frac{1}{2}$% interest. Calculate the monthly payment and the total interest.

Inequalities and Absolute Values

17

Learning Outcomes

1 Use set terminology.

2 Show inequalities on a number line and write inequalities in interval notation.

An *inequality* is a mathematical statement showing quantities that *are not equal*. The symbol $\neq$ is read "is not equal to."

$$5 \neq 7 \qquad 5 \text{ is not equal to } 7.$$

In most cases, showing that quantities are not equal to each other does not give enough information. We may need to know which quantity is larger (or smaller). When we want specific information, such as 5 *is less than* 7, or 7 *is greater than* 5, we use specific inequality symbols, $<$ (is less than) and $>$ (is greater than). The symbol $\leq$ indicates *is less than or equal to*, and the symbol $\geq$ indicates *is greater than or equal to*.

$5 < 7$	5 *is less than* 7.	$5 \leq 7$	5 *is less than or equal to* 7.
$7 > 5$	7 *is greater than* 5.	$7 \geq 5$	7 *is greater than or equal to* 5.

As in equations, we represent missing amounts in inequalities by letters. To *solve* an inequality, we find the value or set of values of the unknown quantity that makes the statement true.

1 **Use Set Terminology.**

A *set* is a group or collection of items. For example, a set of days of the week that begin with the letter T includes Tuesday and Thursday. In this chapter, we examine sets of numbers. Numbers that belong to a set are called *members* or *elements* of a set. The description of a set clearly distinguishes between the elements that belong to the set and those that do not belong. This description can be given in words or by using *set notation*.

To illustrate set notation, we examine the set of whole numbers between 1 and 8. One notation is to make a list or *roster* of the elements of a set. These elements are enclosed in braces and separated with commas.

$$\text{Set of whole numbers between 1 and 8} = \{2, 3, 4, 5, 6, 7\}$$

TIP!

Common Symbols for Sets

The following capital letters are used to denote the indicated set of numbers:

N = natural numbers	W = whole numbers
Z = integers	Q = rational numbers
I = irrational numbers	R = real numbers
M = imaginary numbers	C = complex numbers

Symbols that substitute for phrases that are often used in describing sets are

$\mid$ is read "such that"

$\in$ is read "is an element of"

Another method of illustrating a set is *set-builder notation*. The elements of the set are written in the form of an inequality using a variable to represent all the elements of the set.

$$\text{Set of whole numbers between 1 and 8} = \{x \mid x \in W \text{ and } 1 < x < 8\}$$

This statement is read "the set of values of x *such that* each x *is an element of* the set of whole numbers and x is between 1 and 8."

A special set is the empty set. The *empty set* is a set containing no elements. Symbolically, the empty set is identified as { } or φ. The symbol φ is the Greek letter phi, (pronounced "fee"). An example of an empty set is the set of whole numbers between 1 and 2. The set of rational numbers between 1 and 2 includes numbers like $1\frac{1}{2}$ and 1.3, but there are no whole numbers between 1 and 2. Thus, the set of whole numbers between 1 and 2 is the empty set.

EXAMPLE Answer the statements as true or false.

(a) 5 is an element of the set of whole numbers.
(b) $\frac{3}{4}$ is an element of the set of Z.
(c) $-8 \in$ the set of real numbers with the property $\{x \mid x < -5\}$.
(d) 3.7 is an element of the set of rational numbers with the property $\{x \mid x > 3.7\}$.
(e) The set of prime numbers that are evenly divisible by 2 is an empty set.

(a) 5 is an element of the set of whole numbers. **True.**
(b) $\frac{3}{4}$ is an element of the set of integers. **False.** Integers include only whole numbers and their opposites.
(c) -8 is an element of the set of real numbers with the property $\{x \mid x < -5\}$. **True.** -8 is a real number and it is less than -5. Both conditions are satisfied.
(d) 3.7 is an element of the set of rational numbers with the property $\{x \mid x > 3.7\}$. **False.** 3.7 is a rational number. A number can equal itself, but a number cannot be greater than itself.
(e) The set of prime numbers evenly divisible by 2 is the empty set. **False.** 2 is a prime number, and 2 is divisible by 2. Therefore, the set contains the element 2.

2 **Show Inequalities on a Number Line and Write Inequalities in Interval Notation.**

The set of numbers that is represented by an inequality in one variable can be shown visually on a number line.

To graph an inequality in one variable:

1. Determine the boundaries of the inequality, if they exist.
2. Locate the boundaries on a number line.
   (a) If the inequality is < or >, the boundary is *not* included. Represent this with a parenthesis or an open circle.
   (b) If the inequality is ≤ or ≥, the boundary *is* included. Represent this with a bracket.
3. Shade the portion of the number line between the boundaries. If there is no boundary in one or either direction, the graph continues indefinitely in that direction.

Another type of notation used to represent inequalities is *interval notation*. The two *boundaries* are separated by a comma and enclosed with a symbol that indicates whether the boundary is included or not. If there is no boundary in one or both directions, an infinity symbol, ∞, is used. We can say the set is *unbounded*.

To write an inequality in interval notation:

1. Determine the boundaries of the interval, if they exist.
2. Write the left boundary or $-\infty$ first and write the right boundary or $+\infty$ second. Separate the two with a comma.
3. Enclose the boundaries with the appropriate grouping symbols. A parenthesis "(" or ")" indicates that the boundary is *not* included. A bracket "[" or "]" indicates that a boundary *is* included.

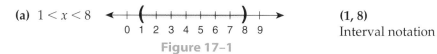

EXAMPLE Represent the sets of numbers on the number line and by using interval notation.

(a) $1 < x < 8$ (b) $1 \le x \le 8$ (c) $x < 3$
(d) $x \ge 3$ (e) all real numbers

(a) $1 < x < 8$ (1, 8)
Interval notation

Figure 17–1

Parentheses are used for both boundaries to indicate that 1 and 8 are *not* included in the solution set.

(b) $1 \le x \le 8$ [1, 8]
Interval notation

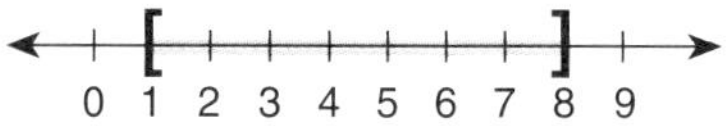

Figure 17–2

Brackets are used for both boundaries to indicate that 1 and 8 *are* included in the solution set.

(c) $x < 3$ $(-\infty, 3)$
Interval notation

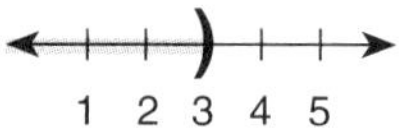

Figure 17–3

The left boundary is $-\infty$, and the right boundary is 3 but is not in the solution set so a parenthesis is used.

(d) $x \ge 3$ $[3, \infty)$
Interval notation

Figure 17–4

The left boundary is 3 and since 3 *is* included in the solution set, a bracket is used. The right boundary is ∞.

(e) All real numbers $(-\infty, \infty)$
Interval notation

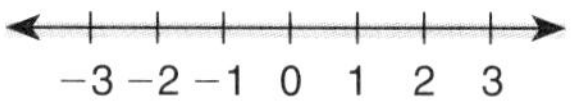

Figure 17–5

All real numbers is an unbounded set of numbers, thus parentheses in the interval notation are used. The number line extends indefinitely in both directions.

TIP!

Alternate Representation of Inequality on a Number Line

Boundaries on a number line can also be indicated with open or shaded circles (Fig. 17–6).

$-1 \le x < 2$ $[-1, 2)$

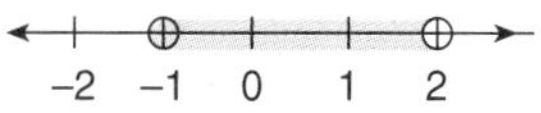

Figure 17–6

SELF-STUDY EXERCISES 17–1

1 Answer the statements as true or false.

1. 0 is an element of the natural numbers.
2. $\frac{5}{8} \in N$
3. $-6 \in Z$
4. 5.3 is an element of the set of rational numbers with the property $\{x \mid x \le 5.3\}$.

Write the sets as a roster.

5. The whole numbers between 5 and 12.
6. The even numbers between 3 and 8.
7. The whole numbers that are multiples of 5 and between 10 and 40.
8. The negative integers greater than -5.

Write the sets in set-builder notation.

9. $\{1, 3, 5, 7, 9\}$
10. $\{-7, -5, -3, -1\}$

2 Represent the sets on the number line and by using interval notation.

11. $5 < x < 9$
12. $-7 < x < -3$
13. $-5 \le x \le -3$
14. $8 \le x \le 12$
15. $x < 7$
16. $x < -2$
17. $x \ge 2$
18. $x \le 3$
19. $x > 5$
20. $x \ge -3$
21. $x \le -2$

22. All real numbers greater than -6.

23. University Trailer Company had sales of $843,000 for the previous year. The projected sales for the current year are more than the previous year, but less than $1,000,000, which is projected for the next year. Express sales for the current year using interval notation.

24. All real numbers with a minimum of 4 and a maximum of 18.

25. College classes generally must have a minimum number of students to avoid cancellation. If that number is 12, write and graph an inequality to represent the number of students that are in a cancelled class.

17–2 *Solving Simple Linear Inequalities*

Learning Outcome   **1**   Solve a simple linear inequality.

1   **Solve a Simple Linear Inequality.**

The procedures for solving simple inequalities are similar to the procedures for solving equations. The solution to an inequality is a *set* of numbers that satisfies the conditions of the statement.

The statement $x \le 9$ means the *solution set* is 9 or any number less than 9.

Compare Figs. 17–7 and 17–8. In Fig. 17–7, 9 is *not* part of the solution set; thus, 9 is represented on the number line with a parenthesis. In Fig. 17–8, 9 *is* part of the solution set; thus, 9 is represented on the number line with a bracket.

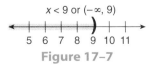

Figure 17–7

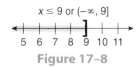

Figure 17–8

Look at the simalarities in solving a linear equation and a linear inequality.

EXAMPLE Solve the equation $4x + 6 = 2x - 2$ and the inequality $4x + 6 > 2x - 2$.

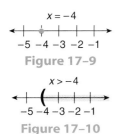

$x = -4$

Figure 17–9

$x > -4$

Figure 17–10

$4x + 6 = 2x - 2$	$4x + 6 > 2x - 2$	Sort.
$4x - 2x = -6 - 2$	$4x - 2x > -6 - 2$	Combine.
$2x = -8$	$2x > -8$	Divide.
$\dfrac{2x}{2} = \dfrac{-8}{2}$	$\dfrac{2x}{2} > \dfrac{-8}{2}$	
$x = -4$	$x > -4$ or $(-4, \infty)$	Interval notation.

The solution is -4 (Fig. 17–9). The solution set is any number greater than, but not including, -4 (Fig. 17–10).

TIP!

Three Ways to Represent the Solution Set of Inequalities: Symbolically, Graphically, and Interval Notation

In the preceding example, the solution to the inequality $4x + 6 > 2x - 2$ was written three ways.

- Symbolic representation: $x > -4$

- Graphical representation:

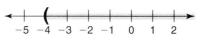

Figure 17–11

- Interval notation: $(-4, \infty)$

The symbolic representation evolves from the process of solving the inequality using the properties of inequalities and techniques for isolating the variable. Both the graphical representation and the interval notation help us interpret and visualize the solution.

EXAMPLE Solve the equation $3x + 5 = 7x + 13$ and the inequality $3x + 5 < 7x + 13$.

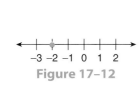

Figure 17–12

Figure 17–13

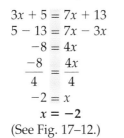
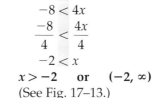

$3x + 5 = 7x + 13$	$3x + 5 < 7x + 13$	Sort.
$5 - 13 = 7x - 3x$	$5 - 13 < 7x - 3x$	Combine.
$-8 = 4x$	$-8 < 4x$	Divide.
$\dfrac{-8}{4} = \dfrac{4x}{4}$	$\dfrac{-8}{4} < \dfrac{4x}{4}$	
$-2 = x$	$-2 < x$	Interpret solution.
$x = -2$	$x > -2$ or $(-2, \infty)$	We prefer writing the variable on the left.
(See Fig. 17–12.)	(See Fig. 17–13.)	Saying -2 *is less than x* is the same as saying *x is greater than* -2.

The preceding example illustrates a very important difference in solving equations and solving inequalities. The sides of an equation can be interchanged without making any changes in the equal sign. This interchangeability is due to the *symmetric* property of equality; that is, if $a = b$, then $b = a$. The symmetric property does *not* apply to inequalities.

Chapter 17 / Inequalities and Absolute Values

When the sides of an inequality are interchanged, the sense of the inequality is reversed.

$$\text{If } a < b, \text{ then } b > a. \qquad \text{If } a > b, \text{ then } b < a.$$
$$\text{If } a \le b, \text{ then } b \ge a. \qquad \text{If } a \ge b, \text{ then } b \le a.$$

The *sense of an inequality* is the appropriate comparison symbol: less than, greater than, less than or equal to, and greater than or equal to.

To determine the effect of multiplying or dividing an inequality by a negative number, look at specific numbers and the number line (Fig. 17–14.). We start with a true statement.

$$2 < 5 \qquad \text{2 is to the left of 5 on the number line.}$$

Then, multiply each side of the inequality by -1.

$$-1(2) \overset{?}{<} -1(5)$$
$$-2 \overset{?}{<} -5 \qquad \text{This is a false statement: } -2 \text{ is to the right of } -5 \text{ and is greater than } -5.$$

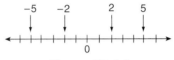

Figure 17–14

To make the statement true, the sense of the inequality must be reversed.
$$-2 > -5$$

If both sides of an inequality are multiplied or divided by a negative number, the sense of the inequality is reversed.

$$\text{If } a < b, \text{ then } -a > -b. \qquad \text{If } a > b, \text{ then } -a < -b.$$
$$\text{If } a \le b, \text{ then } -a \ge -b. \qquad \text{If } a \ge b, \text{ then } -a \le -b.$$

1. Follow the same sequence of steps that is normally used to solve a similar linear equation.
2. The sense of the inequality remains the same unless one of the following situations occurs:
   (a) The sides of the inequality are interchanged.
   (b) The steps used in solving the inequality require that the entire inequality (both sides) be multiplied or divided by a negative number.
3. If either situation (a) or (b) in Step 2 occurs in solving an inequality, *reverse* the sense of the inequality; that is, less than ($<$) becomes greater than ($>$), and vice versa.

EXAMPLE   Solve the inequality, $4x - 2 \leq 3(25 - x)$.

$$4x - 2 \leq 3(25 - x)$$ Distribute.
$$4x - 2 \leq 75 - 3x$$ Collect letter terms on the left and number terms on the right (sort).
$$4x + 3x \leq 75 + 2$$ Combine like terms.
$$7x \leq 77$$ Divide by the coefficient of the letter.
$$\frac{7x}{7} \leq \frac{77}{7}$$
$$x \leq 11 \qquad \text{or} \qquad (-\infty, 11]$$ Solution set. (See Fig. 17–15.)

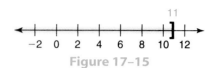

Figure 17–15

EXAMPLE   Solve the inequality, $2x - 3(x + 2) > 5x - (x - 5)$.

Figure 17–16

$$2x - 3(x + 2) > 5x - (x - 5)$$ Distribute.
$$2x - 3x - 6 > 5x - x + 5$$ Combine like terms.
$$-x - 6 > 4x + 5$$ Sort.
$$-x - 4x > 5 + 6$$ Combine like terms.
$$-5x > 11$$ Divide by the coefficient of the letter.
$$\frac{-5x}{-5} < \frac{11}{-5}$$ Both sides were divided by a negative number, so reverse the sense of the inequality.
$$x < -\frac{11}{5} \qquad \text{or} \qquad \left(-\infty, -\frac{11}{5}\right)$$

EXAMPLE   An electrically controlled thermostat is set so the heating unit automatically comes on and continues to run when the temperature is equal to or below 72°F. At what Celsius temperatures will the heating unit come on? One formula relating Celsius and Fahrenheit temperatures is $°F = \frac{9}{5}°C + 32$.

Using the formula $°F = \frac{9}{5}°C + 32$, the heating unit will operate when the expression $\frac{9}{5}°C + 32$ is less than or equal to 72.

$$\frac{9}{5}°C + 32 \leq 72$$

Estimation   The Celsius value will be less than the Fahrenheit value.

$$\frac{9}{5}°C \leq 72 - 32$$ Combine like terms.

$$\frac{9}{5}°C \leq 40$$ Multiply by the reciprocal of the coefficient of the letter.

$$\frac{5}{9}\left(\frac{9}{5}\right)°C \leq \left(\frac{5}{9}\right)(40)$$

$$°C \leq \frac{200}{9}$$ Interpret solution.

$$°C \leq 22.22 \qquad \text{or} \qquad (-\infty, 22.22]$$ To nearest hundredth.

Interpretation   **The heating unit comes on and continues to run if the temperature is less than or equal to 22.22°C.**

Chapter 17 / Inequalities and Absolute Values

1 Solve the inequalities. Show the solution set on a number line and by using interval notation.

1. $y - 3 > 5$
2. $x + 7 < 8$
3. $x - 9 \leq -12$
4. $3x + 7x < 60$
5. $5a - 6a \leq 3$
6. $y + 3y \leq 32$
7. $b + 6 > 5$
8. $5t - 18 < 12$
9. $2y + 7 < 17$
10. $3a - 8 \geq 7a$
11. $4x + 7 \leq 8$
12. $2t + 6 \leq t + 13$
13. $4y - 8 > 2y + 14$
14. $3(7 + x) \geq 30$
15. $6(3x - 1) < 12$
16. $2x > 7 - (x + 6)$
17. $4a + 5 \leq 3(2 + a) - 4$
18. $15 - 3(2x + 2) > 6$

19. Kevin Presley sold $196 more than twice as much merchandise as Robyn Presley. If Kevin sold at least $52,800, how much did Robyn sell? Write an inequality to represent the facts and solve.

20. A supplier prices one brand of recordable CD at $3.60 and another brand of recordable CD at more than twice this price. Write an inequality to represent the facts and express the cost of the more expensive CD. Show the solution set on a number line.

17–3 | *Compound Inequalities*

Learning Outcomes

1 Identify subsets of sets and perform set operations.

2 Solve compound inequalities with the conjunction condition.

3 Solve compound inequalities with the disjunction condition.

In many applications of inequalities, more than one condition is placed on the solution. As an example, when measurements are made, a range of acceptable values is specified. All measurements are approximations, and the range of acceptable values is generally stated as a tolerance. If an acceptable measurement is specified to be within ±0.005 in., then the range of acceptable values can be from 0.005 in. *less than* the ideal measurement to 0.005 in. *more than* the ideal measurement. This range can be stated symbolically as a *compound inequality*. A compound inequality is a mathematical statement that combines two statements of inequality.

The conditions placed on a compound inequality may use the connective *and* to indicate both conditions must be met simultaneously. Such compound inequalities may be written as a continuous statement. If an ideal measurement is 3 in. with a tolerance of ±0.005 in., the range of acceptable values would be

$$3 - 0.005 \leq x \leq 3 + 0.005$$

$$2.995 \leq x \leq 3.005$$

The conditions placed on a compound inequality may use the connective *or* to indicate that either condition may be met. Such compound inequalities **must** be written as two separate statements using the connective *or*.

1 **Identify Subsets of Sets and Perform Set Operations.**

Before solving compound inequalities, let's look at additional properties of sets. A concept related to sets is the concept of subsets. The symbol $\subset$ is read "is a subset of." A set is a *subset* of a second set if every element of the first set is also an element of the second set. If $A = \{1, 2, 3\}$ and $B = \{2\}$, then B is a subset of A. This is written in symbols as $B \subset A$ and is illustrated in Fig. 17–17.

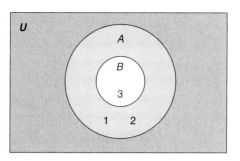

$B \subset A$

Figure 17–17

Two special sets are the universal set and the empty set. The *universal set* includes all the elements for a given description. The universal set is sometimes identified as U. The *empty set* is a set containing no elements. The empty set is identified as { } or ϕ.

EXAMPLE Using the given set definitions, answer the statements (a)–(h) as true or false.

$$U = \{0, 1, 2, 3, 4, 5, 6, 7, 8, 9\}; A = \{1, 3, 5, 7, 9\};$$

$$B = \{0, 2, 4, 6, 8\}; C = \{1, 2, 3\}; D = \{1\}; E = \{ \ \}$$

(a) $A \subset U$ (b) $B \subset U$ (c) $A \subset B$ (d) $C \subset A$
(e) $C \subset U$ (f) $D \subset A$ (g) $E \subset U$ (h) $E \subset B$

(a) $A \subset U$ **True;** every element of A is also an element of U.
(b) $B \subset U$ **True;** every element of B is also an element of U.
(c) $A \subset B$ **False;** 1, 3, 5, 7, and 9 are not elements of B.
(d) $C \subset A$ **False;** 2 is not an element of A.
(e) $C \subset U$ **True;** every element of C is also an element of U.
(f) $D \subset A$ **True;** 1 is an element of A.
(g) $E \subset U$ **True;** the empty set is a subset of every set.
(h) $E \subset B$ **True;** the empty set is a subset of every set.

Two common set operations are union and intersection. The *union* of two sets is a set that includes all elements that appear in *either* of the two sets. Union is generally associated with the condition "or." The symbol for union is $\cup$. The colored portion in Fig. 17–18 represents $A \cup B$. The *intersection* of two sets is a set that includes all elements that appear in *both* of the two sets. Intersection is generally associated with the condition "and." The symbol for intersection is $\cap$. The colored portion of Fig. 17–19 represents $A \cap B$.

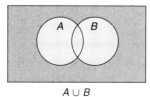

$A \cup B$

Figure 17–18

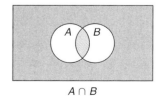

$A \cap B$

Figure 17–19

EXAMPLE If $A = \{1, 2, 3, 4, 5\}$, $B = \{1, 3, 5, 7, 9\}$, and $C = \{2, 4, 6, 8, 10\}$, list the elements in the following sets.

(a) $A \cup B$ (b) $A \cup C$ (c) $B \cup C$ (d) $A \cap B$ (e) $A \cap C$ (f) $B \cap C$

(a) $A \cup B = \{1, 2, 3, 4, 5, 7, 9\}$ All elements in either A or B.
(b) $A \cup C = \{1, 2, 3, 4, 5, 6, 8, 10\}$ All elements in either A or C.
(c) $B \cup C = \{1, 2, 3, 4, 5, 6, 7, 8, 9, 10\}$ All elements in either B or C.
(d) $A \cap B = \{1, 3, 5\}$ Elements common to both A and B.
(e) $A \cap C = \{2, 4\}$ Elements common to both A and C.
(f) $B \cap C = \{\ \ \}$ or ϕ Elements common to both B and C.

Another set operation is complement. The *complement* of a set is a set that includes every element of the universal set that is *not* an element of the given set. The symbol for the complement of a set is ′ and is read "prime." A' is represented by the colored portion of Fig. 17–20.

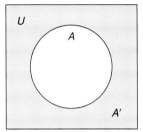

Figure 17–20

EXAMPLE If $U = \{0, 1, 2, 3, 4, 5\}$, $A = \{1, 3, 5\}$, $B = \{0, 2, 4\}$, and $C = \{1, 2, 3\}$, list the elements in the following sets.

(a) A' (b) B' (c) C' (d) $(A \cup B)'$ (e) $(A \cap C)'$

(a) $A' = \{0, 2, 4\}$ Elements of U not in A.
(b) $B' = \{1, 3, 5\}$ Elements of U not in B.
(c) $C' = \{0, 4, 5\}$ Elements of U not in C.
(d) $(A \cup B)'$
   $A \cup B = \{0, 1, 2, 3, 4, 5\}$ Elements in either A or B.
   $(A \cup B)' = \{\ \ \}$ or ϕ Elements of U not in either A or B.
(e) $(A \cap C)'$
   $A \cap C = \{1, 3\}$ Elements common to both A and C.
   $(A \cap C)' = \{0, 2, 4, 5\}$ Elements of U not common to both A and C.

2 Solve Compound Inequalities with the Conjunction Condition.

As we said before, a compound inequality is a statement that places more than one condition on the variable of the inequality. The solution set is the set of values for the variable that meets all conditions of the problem. One type of compound inequality is *conjunction*. A *conjunction* is an intersection or "and" set relationship. Both conditions must be met simultaneously in a conjunction. A conjunction can be written as a continuous statement.

Property for conjunctions:

If $a < x$ and $x < b$, then $a < x < b$.

If $a > x$ and $x > b$, then $a > x > b$.

Similar compound inequalities may also use $\leq$ and $\geq$.

If $a \leq x$ and $x \leq b$, then $a \leq x \leq b$.

If $a \geq x$ and $x \geq b$, then $a \geq x \geq b$.

To solve a compound inequality that is a conjunction:

1. Separate the compound inequality into two simple inequalities using the conditions of the conjunction.
2. Solve each simple inequality.
3. Determine the solution set that includes the *intersection* of the solution sets of the two simple inequalities.

EXAMPLE Find the solution set for each compound inequality. Graph the solution set on a number line and write the solution set in interval notation.

(a) $5 < x + 3 < 12$ (b) $-7 \le x - 7 \le 1$
(c) $-6 < 3x < -3$ (d) $17 \le 5x + 7 \le 32$

(a) $5 < x + 3 < 12$ — Separate into two simple inequalities.

$5 < x + 3$ and $x + 3 < 12$ — Solve each inequality.

$5 - 3 < x$ $x < 12 - 3$

$2 < x$ or $x > 2$ $x < 9$ — Find the overlap.

Figure 17–21

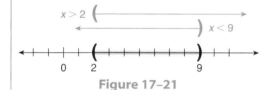

Figure 17–21 shows the solution set graphically. The solution set as a continuous statement is **$2 < x < 9$ and in interval notation, (2, 9).**

(b) $-7 \le x - 7 \le 1$ — Separate into two simple inequalities.

$-7 \le x - 7$ and $x - 7 \le 1$ — Solve each inequality.

$-7 + 7 \le x$ $x \le 1 + 7$

$0 \le x$ or $x \ge 0$ $x \le 8$ — Find the overlap.

Figure 17–22

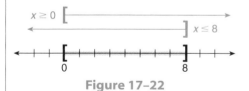

Figure 17–22 shows the solution set graphically. The solution set as a continuous statement is **$0 \le x \le 8$ and in interval notation, [0, 8].**

(c) $-6 < 3x < -3$ — Separate into two simple inequalities.

$-6 < 3x$ and $3x < -3$ — Solve each inequality.

$\frac{-6}{3} < \frac{3x}{3}$ $\frac{3x}{3} < \frac{-3}{3}$

$-2 < x$ or $x > -2$ $x < -1$ — Find the overlap.

Figure 17–23

Figure 17–23 shows the solution set graphically. The solution set is **$-2 < x < -1$ and in interval notation is (−2, −1).**

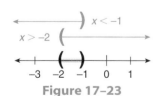

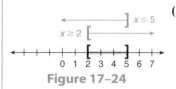

(d) $17 \le 5x + 7 \le 32$ — Separate into two simple inequalities.

$17 \le 5x + 7$ and $5x + 7 \le 32$ — Solve each inequality.

$17 - 7 \le 5x$ $5x \le 32 - 7$

$10 \le 5x$ $5x \le 25$

$2 \le x$ or $x \ge 2$ $x \le 5$ — Find the overlap.

Figure 17–24

Figure 17–24 shows the solution set graphically. The solution set as a continuous statement is **$2 \le x \le 5$ and in interval notation, [2, 5].**

TIP!

Greater Than Versus Less Than

Even though continuous compound inequalities can be written using *greater than* or *less than* symbols, using *less than* symbols follow the natural positions of the boundaries on the number line. If $a > b > c$, then $c < b < a$. For instance, if $3 > 0 > -2$, then $-2 < 0 < 3$. Compound inequalities using the "less than" symbols are used most often. Do not mix symbols in a compound inequality.

As with equations and simple inequalities, a compound inequality may have no solution.

EXAMPLE Find the solution set for the compound inequality, $7 < 2x - 1 < 5$.

Separate into two simple inequalities.

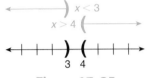

Figure 17–25

$$7 < 2x - 1 \quad \text{and} \quad 2x - 1 < 5$$
$$8 < 2x \qquad\qquad\qquad 2x < 6$$
$$4 < x \qquad\qquad\qquad x < 3$$
$$x > 4 \quad \text{and} \quad x < 3 \qquad\qquad \text{Find the overlap.}$$

As shown in Fig. 17–25, there is no overlap, so it is impossible for both conditions to be met at the same time. **The compound inequality has no solution.**

TIP!

The Solution of Conjunction Is the Overlap

The importance of a graphical representation of the solution set of a conjunction is that you can visualize the overlap. Or, in the case of no solution to a conjunction, you can see there is no overlap.

A good way to emphasize the overlapped portion on a graph is to use two different colors of highlighters. Yellow and blue highlighters are good choices. The overlapped portion of the graph will be green.

3 **Solve Compound Inequalities with the Disjunction Condition.**

Disjunction is another type of compound inequality. A *disjunction* is a union or "or" set relationship. Either condition is met in a disjunction. A disjunction uses the union symbol to indicate the intervals in the solution set.

To solve a compound inequality that is a disjunction:

1. Solve each simple inequality.
2. Determine the solution set that includes the union of the solution sets of the two simple inequalities.

EXAMPLE   Find the solution set for each compound inequality. Graph the solution set on a number line and write the solution in symbolic and interval notation.

(a) $x + 3 < -2$ or $x + 3 > 2$ (b) $x - 5 \le -3$ or $x - 5 \ge 3$

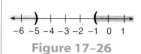

Figure 17–26

(a) $x + 3 < -2$ or $x + 3 > 2$ Solve each simple inequality.

$x < -2 - 3$ $x > 2 - 3$

$x < -5$ or $x > -1$ Solution set. See Fig. 17–26.

$(-\infty, -5)$ $\cup$ $(-1, \infty)$

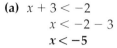

Figure 17–27

(b) $x - 5 \le -3$ or $x - 5 \ge 3$ Solve each simple inequality.

$x \le -3 + 5$ $x \ge 3 + 5$

$x \le 2$ or $x \ge 8$ Solution set. See Fig. 17–27.

$(-\infty, 2]$ $\cup$ $[8, \infty)$

TIP!

Let Your Graph Be Your Guide

The graphical representation of your solution set can be a great guide for writing the symbolic solution or the solution in interval notation.

Overlap on graph (Fig. 17–28)
Inequality is one continuous statement.

Two intervals on graph (Fig. 17–29)
Inequality is two separate statements.

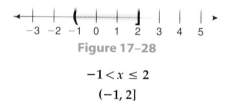

Figure 17–28

$-1 < x \le 2$

$(-1, 2]$

Figure 17–29

$x < 21$ or $x \ge 2$

$(-\infty, -1)$ $\cup$ $[2, \infty)$

SELF-STUDY EXERCISES 17–3

1 Use the sets for Exercises 1–10.

$U = \{-5, -4, -3, -2, -1, 0, 1, 2, 3, 4, 5\}$, $A = \{-4, -2, 0\}$, $B = \{1, 2, 3, 4, 5\}$,
$C = \{2, -2\}$, $D = \{0\}$, $E = \{\ \}$

Answer each statement as true or false and justify your answer.
1. $B \subset U$ 2. $C \subset A$ 3. $E \subset B$ 4. $A \subset B$

List the elements in each set.
5. $A \cap B$ 6. $C \cup D$ 7. $A \cup B$
8. A' 9. $(A \cap D)'$ 10. $(B \cup C)'$

2 Show the solution set of each inequality on a number line and by using symbolic and interval notation.

11. $6 \le 2x \le 8$ 12. $15 \le 3x \le 21$ 13. $-3 \le 3x \le 9$ 14. $-3 \le 2x - 1 \le 3$
15. $-5 \le -x + 1 \le -7$ 16. $1 < x - 2 < 5$ 17. $0 < 5x < 15$ 18. $4 \le 6 - x \le 8$
19. $6 < -2x < 12$ 20. $1 < 6 - x < 3$ 21. $-8 \le -3x - 2 \le 4$ 22. $5 \le 5 + 7x \le 9$

Chapter 17 / Inequalities and Absolute Values

Show the solution set of each inequality on a number line and by using symbolic and interval notation.

23. $x + 3 < 5$ or $x - 7 > 2$

24. $x - 1 \leq 2$ or $x \geq 7$

25. $2x - 1 \leq 7$ or $3x \geq 15$

26. $5 - x \leq 2$ or $x + 1 \leq 2$

27. $5x - 2 \leq 3x + 1$ or $2x \geq 7$

28. $-2x > 8$ or $3x \geq 27$

29. The blueprint specifications for a part show it has a measure of 5.27 cm with a tolerance of ±0.05 cm. Express the limit dimensions of the part with a compound inequality.

30. An automobile parts manufacturer makes a part that, according to the blueprint, measures 15.2 cm and has a tolerance of ±0.5 cm. Write an inequality that expresses the range of measures an acceptable part must have.

17–4 | Solving Quadratic and Rational Inequalities in One Variable

Learning Outcomes

1 Solve quadratic inequalities in one variable.

2 Solve rational inequalities in one variable.

Solving *quadratic* and *rational inequalities* is similar to solving quadratic and rational equations. The solution to these inequalities, like the solution to linear inequalities, is a *set* of numbers with specific boundaries.

1 **Solve Quadratic Inequalities in One Variable.**

Quadratic inequalities can be solved by all of the same methods we used to solve quadratic equations. You first treat the inequality as an equation to find the *critical values* or *boundaries*.

To solve quadratic inequalities by factoring:

1. Rearrange the inequality in standard form so the right side of the inequality is zero.
2. Write the left side of the inequality in factored form.
3. Determine the critical values by finding the values that make each factor equal to zero.
4. Plot the critical values on a number line to form the three regions.
5. Test each region of values by selecting any point within the region, substituting that value into the inequality, solving the inequality, and deciding if the resulting inequality is a *true* statement.
6. The solution set for the quadratic inequality is the region or regions that produce a *true* statement in Step 5.
7. The critical values or boundary points are included in the solution set if the inequality is "inclusive" ($\leq$ or $\geq$). The critical values or boundary points are not included in the solution set if the inequality is "exclusive" ($<$ or $>$).

EXAMPLE   Solve the inequality $x^2 + 5x + 6 \leq 0$.

$(x + 3)(x + 2) \leq 0$ Write in factored form. Set each factor equal to 0.

$x + 3 = 0$ $x + 2 = 0$ Determine the critical values by solving each equation.

$x = -3$ $x = -2$

Plot the critical values and label the corresponding regions (Fig. 17–30).

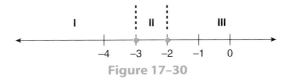

Figure 17–30

Region I: $x \leq -3$ To the left of the smaller critical value, -3.

Region II: $-3 \leq x \leq -2$ Between the critical values, -3 and -2.

Region III: $x \geq -2$ To the right of the larger critical value, -2.

Test one point in each region.

Region I: $x \leq -3$	Region II: $-3 \leq x \leq -2$
Region I test point: $x = -4$	**Region II test point**: $x = -2.5$
$(x + 3)(x + 2) \leq 0$	$(-2.5 + 3)(-2.5 + 2) \leq 0$
$(-4 + 3)(-4 + 2) \leq 0$	$(0.5)(-0.5) \leq 0$
$(-1)(-2) \leq 0$	$-0.25 \leq 0$
$2 \leq 0$	

The inequality is **false,** so Region I is *not* in the solution set.

The inequality is **true,** so Region II *is* in the solution set.

Region III: $x \geq -2$

Region III test point: $x = -1$
$$(-1 + 3)(-1 + 2) \leq 0$$
$$(2)(1) \leq 0$$
$$2 \leq 0$$

The inequality is **false** , so Region III is *not* in the solution set.

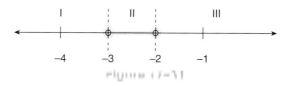

The solution set is $-3 \leq x \leq -2$ or $[-3, -2]$ (Fig. 17–31).

Selecting Test Points

Boundary points should not be used as test points. When possible, select integers as test points, and if zero is in a region, it makes an excellent test point.

EXAMPLE   Solve the inequality $2x^2 + x - 6 > 0$.

$(2x - 3)(x + 2) > 0$ Write in factored form. Set each factor equal to 0.

$2x - 3 = 0$ $x + 2 = 0$ Determine the critical values by solving each equation.

$2x = 3$ $x = -2$

$x = \dfrac{3}{2}$

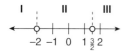

Figure 17–32

Plot the critical values and label the corresponding regions (see Fig. 17–32).

Region I: $x < -2$ Region II: $-2 < x < \frac{3}{2}$ Region III: $x > \frac{3}{2}$

Test each region.

                Chapter 17 / Inequalities and Absolute Values

Region I: $x < -2$

Region II: $-2 < x < \dfrac{3}{2}$

Region I test point: $x = -3$

$$(2x - 3)(x + 2) > 0$$
$$[2(-3) - 3][-3 + 2] > 0$$
$$(-6 - 3)(-1) > 0$$
$$(-9)(-1) > 0$$
$$9 > 0$$

The inequality is true, so Region I *is* in the solution set.

Region II test point: $x = 0$

$$(2x - 3)(x + 2) > 0$$
$$[2(0) - 3][0 + 2] > 0$$
$$(0 - 3)(2) > 0$$
$$(-3)(2) > 0$$
$$-6 > 0$$

The inequality is false, so Region II is *not* in the solution set.

Region III: $x > \dfrac{3}{2}$

Region III test point: $x = 2$

$$(2x - 3)(x + 2) > 0$$
$$[2(2) - 3][2 + 2] > 0$$
$$(4 - 3)(4) > 0$$
$$(1)(4) > 0$$
$$4 > 0$$

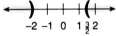

Figure 17–33

The inequality is true, so Region III *is* in the solution set.

The solution set is $x < -2$ or $x > \frac{3}{2}$. In interval notation the solution is $(-\infty, -2) \cup \left(\frac{3}{2}, \infty\right)$. See Fig. 17–33.

There are other ways to determine the solution set of a quadratic inequality. One way that is similar to finding a test point in each region is to determine the sign of each factor both to the left and right of each critical value. Then, using the rules for signed numbers, you can determine which region or regions match the conditions of the inequality. Let's examine the inequalities in the two previous examples again.

EXAMPLE   Determine the solution set of the inequalities in the two previous examples by examining the signs of each factor in each region.

(a) $x^2 + 5x + 6 \le 0$ (b) $2x^2 + x - 6 > 0$

(a) $x^2 + 5x + 6 \le 0$ Original inequality.
       $(x + 3)(x + 2) \le 0$ Factored form of inequality.

Critical values: $x = -3$ and $x = -2$

The critical value of a factor is the point on the number line where the factor equals zero. *Negative values* of the factor are to the *left* of the critical value and *positive values* are to the *right* (see Fig. 17–34.)

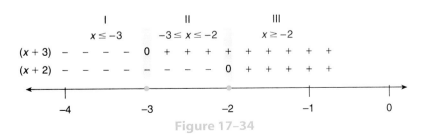

Figure 17–34

The product $(x + 3)(x + 2)$ is positive where the signs of the factors are alike, Regions I and III.

The product $(x + 3)(x + 2)$ is negative where the signs of the factors are unlike, Region II.

The factored form of the inequality is "less than or equal to" zero. Therefore, the **solution set includes the region or regions where the product is negative** since the inequality is less than zero (Fig. 17–35).

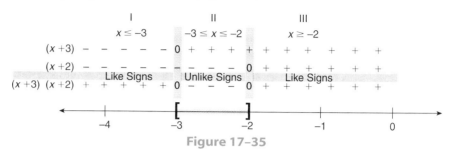

Figure 17–35

The solution set is $-3 \le x \le -2$. Interval notation: $[-3, -2]$.

(b) $2x^2 + x - 6 > 0$
$(2x - 3)(x + 2) > 0$

Critical values: $x = \dfrac{3}{2}$ and $x = -2$

The **solution set is in the region or regions where the product is positive**, or Regions I and III, since the inequality is greater than zero (Fig. 17–36).

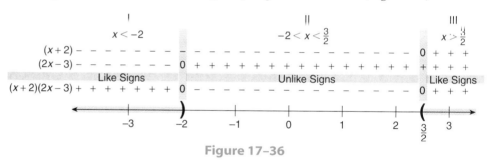

Figure 17–36

The solution set is $x < -2$ or $x > \frac{3}{2}$. Interval notation: $(-\infty, -2) \cup \left(\frac{3}{2}, \infty\right)$.

After examining several problems, you may notice a pattern for determining the solution set by inspection once the critical values are known. This pattern or generalization also can be applied to quadratic inequalities that do not factor.

TIP!

Finding Solution Sets by Inspection for Quadratic Inequalities

Once you understand how to identify the appropriate region or regions for the solution set of a quadratic inequality, you may observe a pattern that allows you to select appropriate regions for such inequalities by inspection. Patterns and generalizations are appropriate only under special conditions. Here, the inequality **must** be written in standard form ($ax^2 + bx + c < 0$ or $ax^2 + bx + c > 0$), where $a > 0$ (a is positive). The critical values are represented symbolically as s_1 and s_2, where $s_1 < s_2$. Figure 17–37 illustrates the critical values and signs of the two factors and the product.

Chapter 17 / Inequalities and Absolute Values

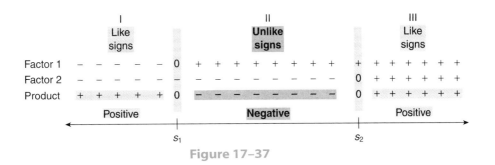

Figure 17–37

When the inequality $ax^2 + bx + c < 0$ is written in factored form, the two factors must have opposite signs to make a product that is less than zero (negative). From Figure 17–37, we see that the solution set is in Region II and is written symbolically as $s_1 < x < s_2$. The graphical representation of this solution appears in Fig. 17–38.

When the inequality $ax^2 + bx + c > 0$ is written in factored form, the two factors must have the same signs to make a product that is greater than zero (positive). From Figure 17–37, we see that the solution set is in Regions I and III and is written symbolically as $x < s_1$ or $x > s_2$. The graphical representation of this solution appears in Fig. 17–39.

Figure 17–38   Figure 17–39

Now, if you solve a quadratic equation by *any* method, including completing the square or using the quadratic formula, you can determine the solution set of the associated quadratic inequality by inspection after you have found the critical values. The solution can be checked by testing one point in the projected solution set. Similar generalizations can be made for $ax^2 + bx + c \leq 0$ and $ax^2 + bx + c \geq 0$.

2 Solve Rational Inequalities in One Variable.

When is a rational expression greater than zero? For solving a rational inequality, we use strategies similar to those used in solving quadratic inequalities. To solve a *rational inequality* like $\dfrac{x + 2}{x - 5} > 0$, we find the critical values by setting each factor equal to zero.

To solve a rational inequality like $\dfrac{x + a}{x + b} < 0$ or $\dfrac{x + a}{x + b} > 0$:

1. Find critical values by setting the numerator and denominator equal to zero.
2. Solve each equation from Step 1 and use the solutions (critical values) to divide the number line into three regions.
3. Regions that have like signs for both the numerator and denominator are in the solution set of $\dfrac{x + a}{x + b} > 0$ for $x \neq -b$ ($x = -b$ is an excluded value).
4. Regions that have unlike signs for the numerator and denominator are in the solution set of $\dfrac{x + a}{x + b} < 0$ for $x \neq -b$ ($x = -b$ is an excluded value).

EXAMPLE Solve the following inequalities.

(a) $\dfrac{x+2}{x-5} < 0$ (b) $\dfrac{x+2}{x-5} \geq 0$

$x + 2 = 0$ $x - 5 = 0$ Set numerator and denominator equal to zero.

$x = -2$ $x = 5$ Critical values for inequalities in both (a) and (b).

See Fig. 17–40. The quotient in Region I is positive (like signs). The quotient in Region II is negative (unlike signs). The quotient in Region III is positive (like signs).

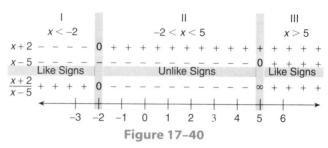

Figure 17–40

(a) Solution set (see Fig. 17–41):

For $\dfrac{x+2}{x-5} < 0$,

$-2 < x < 5$

or

$(-2, 5)$

Figure 17–41

(b) Solution set (see Fig. 17–42):

For $\dfrac{x+2}{x-5} \geq 0$,   The boundary point, 5, is an excluded value.

$x \leq -2$ or $x > 5$ Therefore, it is not

or   included in the

$(-\infty, -2] \cup (5, \infty)$ solution set.

Figure 17–42

SELF-STUDY EXERCISES 17–4

1 Graph the solution of the quadratic inequalities on a number line and write the solution in symbolic and interval notation.

1. $(x - 2)(x + 3) < 0$ 2. $(2x - 1)(x + 3) > 0$ 3. $(2x - 5)(3x - 2) < 0$

4. $(y + 6)(y - 2) \geq 0$ 5. $(a + 4)(a - 6) < 0$ 6. $x^2 - 7x + 10 \leq 0$

7. $x^2 - x - 6 \geq 0$ 8. $x^2 - 4x + 3 \leq 0$ 9. $4x^2 - 21x + 5 \leq 0$

10. $6x^2 + x > 1$ 11. $6x^2 + 2 \geq 7x$ 12. $2x^2 + 9x \leq 5$

2 Graph the solution of the rational inequalities on a number line and write the solution in symbolic and interval notation.

13. $\dfrac{x-3}{x+7} < 0$ 14. $\dfrac{x+8}{x-3} > 0$ 15. $\dfrac{2x-6}{3x-1} > 0$

16. $\dfrac{x-7}{x+7} \leq 0$ 17. $\dfrac{x}{x-1} < 0$ 18. $\dfrac{5x-10}{6x-18} \geq 0$

19. $\dfrac{3x-9}{6x-12} \leq 0$ 20. $\dfrac{3x}{2x-8} \geq 0$ 21. $\dfrac{5x-15}{4x+4} \geq 0$

Learning Outcome   **1**   Solve equations containing one absolute-value term.

1   **Solve Equations Containing One Absolute-Value Term.**

Earlier, absolute value was defined as the number of units from zero; that is, $|+7|$ or $|-7|$ is 7. If the *absolute value* of a term is indicated in an equation, then both possible values of the term must be considered in solving the equation.

EXAMPLE   Solve the equation $|x| = 5$.

By definition of absolute value, the solutions are $x = +5$ or $x = -5$ and are written symbolically as $x = \pm 5$.

The equation $|x| = 5$ has two roots, $+5$ and -5. To solve an equation that contains an absolute-value term, we must examine each of two cases. One case is when the expression within the absolute-value symbol is positive. The other case is when the expression within the absolute-value symbol is negative. Recall that the symbolic definition of the absolute value of a number is $|a| = a$ for $a \geq 0$, and $|a| = -a$ for $a < 0$, or $-|a| = a$ for $a < 0$. This gives us two ways to write the second case.

EXAMPLE   Find the roots for the equation $|x - 4| = 7$.

Two cases must be examined. We solve the equation given in each case to find the *two* roots.

Case 1: $x - 4 = 7$	*Case 2:* $(x - 4) = -7$	or	$-(x - 4) = 7$
$x - 4 = 7$	$(x - 4) = -7$	or	$-(x - 4) = 7$
$x = 7 + 4$	$x - 4 = -7$		$-x + 4 = 7$
$x = 11$	$x = -7 + 4$		$-x = 7 - 4$
	$x = -3$		$-x = 3$
			$x = -3$

The roots of the equation are 11 and −3. To check each root, substitute each root separately into the original equation.

If $x = 11,$ If $x = -3,$

$$|x - 4| = 7 \qquad |x - 4| = 7$$
$$|11 - 4| = 7 \qquad |-3 - 4| = 7$$
$$|7| = 7 \qquad |-7| = 7$$
$$7 = 7 \qquad 7 = 7 \quad \text{Each root checks.}$$

To solve an equation containing one absolute-value term:

1. Isolate the absolute-value term on one side of the equation.
2. Separate the equation into two cases. One case considers the expression within the absolute-value symbol to be positive. The other case considers it to be negative.

3. Solve each case to obtain the *two* roots of the equation.
4. $|x| = b$ has no solution if b is negative ($b < 0$).
5. $|x| = b$ has one root when $b = 0$.

EXAMPLE   Find the roots of the equation $|y + 3| - 5 = 6$.

$$|y + 3| - 5 = 6 \qquad \text{Isolate the absolute-value term.}$$
$$|y + 3| = 6 + 5 \qquad \text{Combine like terms.}$$
$$|y + 3| = 11 \qquad \text{Separate into two cases.}$$

Case 1: $y + 3 = 11$ \qquad *Case 2:* $(y + 3) = -11$ or $-(y + 3) = 11$

$$y = 11 - 3 \qquad\qquad y + 3 = -11 \qquad\qquad -y - 3 = 11$$
$$y = 8 \qquad\qquad\qquad y = -11 - 3 \qquad\qquad -y = 11 + 3$$
$$y = -14 \qquad\qquad\qquad -y = 14$$
$$y = -14$$

The roots of the equation are 8 and −14. Each root checks to be a true root.

TIP!

The Importance of Isolating the Absolute-Value Term

In finding the root for the case when the absolute-value term is negative (case 2), we are very careful to isolate the absolute-value term first. Let's examine what happens if we don't isolate the absolute-value term first. Let's look at both correct and incorrect procedures.

Find the root of $|y + 3| - 5 = 6$ when $(y + 3)$ is negative (case 2).

Correct Procedure With Isolating	**Correct Procedure Without Isolating**	**Incorrect Procedure Without Isolating**
$|y + 3| - 5 = 6$	$|y + 3| - 5 = 6$	$|y + 3| - 5 = 6$
$|y + 3| = 6 + 5$	Case 2: $-(y + 3) - 5 = 6$	Case 2: $y + 3 - 5 = -6$
$|y + 3| = 11$	$-y - 3 - 5 = 6$	$y - 2 = -6$
Case 2: $y + 3 = -11$	$-y - 8 = 6$	$y = -6 + 2$
$y = -11 - 3$	$-y = 6 + 8$	$y = -4$
$y = -14$	$-y = 14$	
	$y = -14$	

Check: $y = -14$ \qquad\qquad\qquad\qquad Check: $y = -4$

$$|y + 3| - 5 = 6 \qquad\qquad\qquad\qquad |y + 3| - 5 = 6$$
$$|-14 + 3| - 5 = 6 \qquad\qquad\qquad\qquad |-4 + 3| - 5 = 6$$
$$|-11| - 5 = 6 \qquad\qquad\qquad\qquad\qquad |-1| - 5 = 6$$
$$11 - 5 = 6 \qquad\qquad\qquad\qquad\qquad\qquad 1 - 5 = 6$$
$$6 = 6 \quad \text{Correct.} \qquad\qquad\qquad\qquad -4 = 6 \quad \text{Incorrect.}$$

Therefore, we must be very careful when applying the absolute-value properties for case 2 (when $a < 0$). If we choose *not* to isolate the absolute-value term first, then we must take the opposite of the absolute-value grouping rather than the value on the other side of the equation.

Solve the equation $|2x - 5| + 14 = 7$.

$|2x - 5| + 14 = 7$ Isolate the absolute-value term.

$|2x - 5| = 7 - 14$ Combine like terms.

$|2x - 5| = -7$ Cannot continue because the absolute value must be positive.

$|2x - 5| + 14 = 7$ **has no solution.**

SELF-STUDY EXERCISES 17–5

1 Solve the equations containing absolute values.

1. $|x| = 8$
2. $|x| = 15$
3. $|x - 3| = 5$
4. $|x - 7| = 2$
5. $|2x - 3| = 9$
6. $|3x - 7| = 2$
7. $|x| - 4 = 7$
8. $|x - 2| + 5 = 3$
9. $-5 + |x - 4| = -2$
10. $|2x - 1| - 3 = 4$
11. $|3x - 2| + 1 = -6$
12. $|3x - 5| + 12 = 18$
13. $|6 - 3x| + 2 = 8$
14. $|5 - 2x| = 15$
15. $|3x - 8| + 2 = 12$
16. $|7x - 9| - 1 = 8$
17. $|4x - 3| + 11 = 6$
18. $|5x - 3| - 7 = 8$
19. $|5x - 1| - 8 = 7$
20. $|2x - 3| - 8 = -5$
21. $|3x - 2| + 2 = 2$

17–6 | *Absolute-Value Inequalities*

Learning Outcomes

1 Solve absolute-value inequalities using the "less than" relationship.
2 Solve absolute-value inequalities using the "greater than" relationship.

1 **Solve Absolute-Value Inequalities Using the "Less Than" Relationship.**

An inequality that contains an absolute-value term is interpreted by the sense of the inequality. If the inequality is a "less than" or "less than or equal to" relationship, the following property is used.

"Less than" relationships for absolute-value inequalities:

If $|x| < b$ and $b > 0$, then $-b < x < b$.

or

If $|x| \leq b$ and $b > 0$, then $-b \leq x \leq b$.

Absolute-value inequalities with the "less than" or "less than or equal to" relationship have a solution set that is a *conjunction* (*and*).

On a number line (Fig. 17–43), the solution set is a continuous set of values. The values in the solution set of $|x| < b$ are between b and $-b$ when b is positive. Why are zero and negative values of b excluded in the property for "less than" relationships of absolute-value inequalities? Try some test points, $b = 0$ and $b = -1$. If $b = 0$, $|x| = 0$. Since zero is unsigned, we do not create an interval from -0 to 0. The solution is a single value, 0. If $b = -1$, $|x| < -1$. The absolute value of a number cannot be negative.

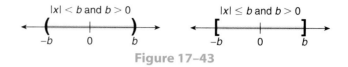

Figure 17–43

EXAMPLE   Find the solution set for each inequality.

(a) $|x| < 3$ (b) $|x + 5| < 8$ (c) $|x - 1| \leq -5$ (d) $|3x| + 2 \leq 14$

(a) $|x| < 3$ Apply the property of absolute-value inequalities having $<$ relationship.

Figure 17–44

$-3 < x < 3$ or $(-3, 3)$. See Fig. 17–44.

(b) $|x + 5| < 8$

		Apply the property of absolute-value inequalities having $<$ relationship.
$-8 < x + 5 < 8$		Separate into two simple inequalities.
$-8 < x + 5$ and	$x + 5 < 8$	Solve each inequality.
$-8 - 5 < x$	$x < 8 - 5$	
$-13 < x$ or $x > -13$	$x < 3$	Shade the overlap.

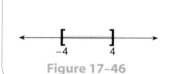

Figure 17–45

$-13 < x < 3$ or $(-13, 3)$. See Fig. 17–45.

(c) $|x - 1| \leq -5$ Apply the property of absolute-value inequalities having $\leq$ relationship.

An absolute-value term cannot be negative. **So $|x - 1| \leq -5$ has no solution.**

(d) $|3x| + 2 \leq 14$

$|3x| \leq 14 - 2$	Isolate the absolute-value term.
	Combine like terms.
$|3x| \leq 12$	Apply the property of absolute-value inequalities having $\leq$ relationship.
$-12 \leq 3x \leq 12$	Separate into two simple inequalities.
$-12 \leq 3x$ and $3x \leq 12$	Divide.
$-4 \leq x$ $x \leq 4$	Find the overlap.

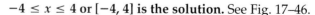

Figure 17–46

$-4 \leq x \leq 4$ or $[-4, 4]$ is the solution. See Fig. 17–46.

2 Solve Absolute-Value Inequalities Using the "Greater Than" Relationship.

The solution set for an absolute-value inequality with a "greater than" or "greater than or equal to" relationship is represented by the extreme values on the number line.

If $|x| > b$ and $b > 0$, then $x < -b$ or $x > b$.

or

If $|x| \geq b$ and $b > 0$, then $x \leq -b$ or $x \geq b$.

Absolute-value inequalities with the "greater than" or "greater than or equal to" relationship have a solution set that is a *disjunction (or)*.

The solution set for an absolute-value inequality with a $>$ or $\geq$ relationship is represented on the number lines in Figs. 17–47 and 17–48.

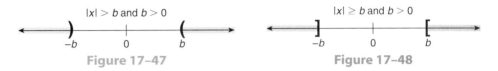

|x| > b and b > 0 |x| ≥ b and b > 0

Figure 17–47   Figure 17–48

EXAMPLE   Find the solution set for each inequality.

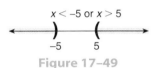

$x < -5$ or $x > 5$

Figure 17–49

(a) $|x| > 5$ (b) $|x| + 1 \geq 4$ (c) $|2x - 2| \geq 7$

(a) $|x| > 5$ Apply the property of absolute-value inequalities having $>$ relationship.

    $x < -5$ or $x > 5$; $(-\infty, -5) \cup (5, \infty)$. See Fig. 17–49.

(b) $|x| + 1 \geq 4$ Isolate the absolute-value term.

    $|x| \geq 4 - 1$ Combine like terms.

    $|x| \geq 3$ Apply the property of inequalities having $\geq$ relationship.

    $x \leq -3$ or $x \geq 3$

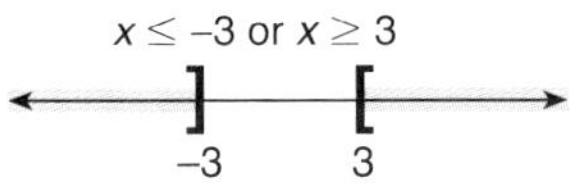

$x \leq -3$ or $x \geq 3$; $(-\infty, -3] \cup [3, \infty)$. See Fig. 17–50.

$x \leq -3$ or $x \geq 3$

Figure 17–50

(c) $|2x - 2| \geq 7$ Apply the property of absolute-value inequalities having $\geq$ relationship.

    $2x - 2 \leq -7$ or $2x - 2 \geq 7$ Solve each inequality.

    $2x \leq -7 + 2$ $2x \geq 7 + 2$

    $2x \leq -5$ $2x \geq 9$

    $x \leq -\dfrac{5}{2}$ $x \geq \dfrac{9}{2}$ Graph results.

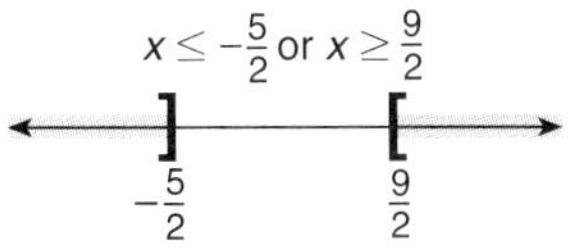

$x \leq -\frac{5}{2}$ or $x \geq \frac{9}{2}$

Figure 17–51

$x \leq -\frac{5}{2}$ or $x \geq \frac{9}{2}$, $\left(-\infty, -\frac{5}{2}\right] \cup \left[\frac{9}{2}, \infty\right)$. See Fig. 17–51.

Learning Strategy   *Develop Your Own Memory Flags.*

This chapter introduces many new concepts. Yet most of these concepts build on previously learned similar concepts.

    Develop your own list of "flags" or memory joggers to get you started when trying to recall a process. Often that is all you need to get yourself on the right track during an exam or other situation that requires recalling the process from memory.

Some examples of memory "flags" are

Concepts	Memory Flags
union	join all
intersection	overlap
inequalities	solution set
conjunction	"and," overlap
disjunction	"or," separate pieces
quadratic inequalities	critical values or boundaries, test points or signs test
rational inequalities	critical values or boundaries, test points or signs test
absolute-value equation	two cases, isolate absolute-value term
absolute-value "less than" relationship	overlap
absolute-value "greater than" relationship	separate pieces

SELF-STUDY EXERCISES 17–6

1 Find the solution set for each inequality. Graph the solution set on a number line and write it in symbolic and interval notation.

1. $|x| < 5$
2. $|x| < 1$
3. $|x| < 2$
4. $|x| \leq 7$
5. $|x + 3| < 2$
6. $|x + 4| < 3$
7. $|x - 3| < 4$
8. $|x - 2| < 3$
9. $|x - 5| \leq 6$
10. $|x - 1| \leq 7$
11. $|3x - 5| < 7$
12. $|2x - 4| < 3$
13. $|5x + 2| \leq 3$
14. $|4x + 7| \leq 5$
15. $|3x - 1| + 2 \leq 5$

2 Find the solution set for each inequality. Graph the solution set on a number line and write it in symbolic and interval notation.

16. $|x| > 2$
17. $|x| > 3$
18. $|x| \geq 5$
19. $|x| \geq 6$
20. $|x - 5| > 2$
21. $|x - 4| > 6$
22. $|x + 3| \geq 5$
23. $|x + 4| \geq 6$
24. $|2x - 5| \geq 0$
25. $|3x - 2| \geq 4$
26. $|5x + 8| \geq 2$
27. $|4x - 3| \geq 9$
28. $|3x - 1| + 3 \geq 5$
29. $|2x - 3| + 1 \geq 5$
30. $|4x - 2| - 3 \geq 7$
31. $|3x - 2| + 4 \leq 5$
32. $|2x - 5| - 3 \leq 2$
33. $|5x - 3| - 6 \geq 1$
34. $|4x - 6| + 4 \geq 5$
35. $|3x - 8| - 2 \leq -6$
36. $|5x + 7| + 3 \leq -6$

Find the solution set for each inequality and write in symbolic notation.

37. $|x - 3| \geq 3$
38. $|4x + 1| > 1$
39. $|x - 4| \leq 3$
40. $|2x - 1| < 0$
41. $|x| < 7$
42. $|x + 8| < 3$
43. $|x| > 12$
44. $|x + 8| > 2$
45. $|2x - 7| < -5$

ASSIGNMENT EXERCISES

Section 17–1

1. Describe the empty set.
2. Write the symbols used to indicate the empty set.
3. Use symbols to write the statement: 5 is an element of the set of whole numbers.

Represent each set of numbers on the number line and with interval notation.

4. $x \leq 3$
5. $x > -7$
6. $-3 < x < 5$
7. $-4 \leq x < 2$
8. All real numbers
9. $-2 < x$
10. $-5 > x$

Solve. Show the solution set on a number line and in symbolic and interval notation.

11. $42 > 8m - 2m$
12. $3m - 2m < 3$
13. $0 < 2x - x$
14. $1 \le x - 7$
15. $10 - 2x \ge 4$
16. $3x + 4 > x$
17. $10x + 18 > 8x$
18. $4x \le 5x + 8$
19. $12 + 5x > 6 - x$
20. $8 - 7y < y + 24$
21. $15 \ge 5(2 - y)$
22. $4t > 2(7 + 3t)$
23. $6x - 2(x - 3) \le 30$
24. $8x - (3x - 2) > 12$

25. A box of diskettes costs $5.60 more than a drafting pencil. The total cost of two boxes of diskettes and a dozen pencils must cost no more than $59.80. Write inequalities to express the cost range for a box of diskettes and a single pencil.

26. A shirt costs $3 less than a certain tie. If the total cost for six ties and two shirts is less than $130, what is the most each shirt and tie can cost?

Use sets U, A, B, C, and D to give the elements in the sets in Exercises 27–31.

$$U = \{-2, -1, 0, 1, 2, 3, 4, 5\}, \quad A = \{-2, -1\}, \quad B = \{0\}, \quad C = \{0, 1, 2, 3, 4\}, \quad D = \{-1, 0, 1\}$$

27. $A \cap B$
28. $B \cup D$
29. $A \cap D$
30. A'
31. $(A \cup C)'$

Solve. Show the solution on the number line and with interval notation.

32. $x - 7 < 2$
33. $x + 4 > 2$
34. $3x - 1 \le 8$
35. $3x - 2 \le 4x + 1$
36. $5 < x - 3 < 8$
37. $-3 < 2x - 4 < 5$
38. $-7 < x - 5 < 7$

Solve each compound inequality. Show the solution set on a number line and in symbolic and interval notation.

39. $x + 1 < 5 < 2x + 1$
40. $3x - 8 < x < 5x + 24$
41. $x + 2 < 7 < 2x - 15$
42. $x + 3 \le 7 \le 2x - 1$
43. $2x + 3 < 15 < 3x + 9$
44. $2 < 3x - 4 < 8$
45. $-5 \le -3x + 1 < 10$
46. $2x + 3 \le 5x + 6 < -3x - 7$
47. $-3 \le 4x + 5 \le 2$
48. $4x - 2 < x + 8 < 9x + 1$
49. $x + 3 < 5$ or $x > 8$
50. $x - 7 < 4$ or $x + 3 > 8$
51. $x - 3 < -12$ or $x + 1 > 9$
52. $3x < -4$ or $2x - 1 > 9$

53. The price of a particular brand of PC whiteboard is $1,365, and the stand costs $199. A college will spend between $15,000 and $30,000 for the equipment. Assuming an equal number of boards and stands are purchased, what is the range for the number of boards and stands to be purchased?

54. The blueprint for a modular desk shows the front edge to be 48 in. with a tolerance of 0.25 in. If the width of the desk is 8 in. less than the length and has the same tolerance, what is the allowable range for the width of the desk?

Solve each inequality and graph the solution on a number line and in symbolic and interval notation.

55. $(x - 5)(x - 2) > 0$
56. $(3x + 2)(x - 2) < 0$
57. $(3x + 1)(2x - 3) < 0$
58. $(5x - 6)(x + 1) \ge 0$
59. $(x + 1)(x - 2) \le 0$
60. $x^2 + x - 12 < 0$
61. $2x^2 \le 5x + 3$
62. $2x^2 - 3x \ge -1$
63. $2x^2 + 7x - 15 < 0$
64. $x^2 - 2x - 8 \ge 0$
65. $\dfrac{x - 7}{x + 1} < 0$
66. $\dfrac{x + 1}{x - 3} > 0$
67. $\dfrac{x}{x + 8} > 0$
68. $\dfrac{x - 8}{x + 1} < 0$

Solve each equation containing an absolute value.

69. $|x| = 12$
70. $|x - 9| = 2$
71. $|x + 3| = 7$
72. $|x + 4| = 11$
73. $|x - 8| = 12$
74. $|x + 7| = 3$
75. $|4x - 7| = 17$
76. $|2x + 3| = 5$
77. $|7x + 8| = 15$
78. $|4x + 1| = 9$
79. $|7x - 4| = 17$
80. $|6x - 2| = 3$
81. $|3x - 9| = 2$
82. $|x| + 8 = 10$
83. $|x| + 12 = 19$
84. $|2x| - 1 = 9$
85. $|x| - 9 = 7$
86. $|x - 4| - 10 = 6$

87. $-5 + |x - 3| = 2$ **88.** $|4x - 2| + 1 = 5$ **89.** $|4x - 3| - 12 = -7$
90. $|3x + 4| - 5 = 17$

Section 17-6

Graph the solution set for each inequality and write in symbolic and interval notation.
91. $|x - 3| < 4$ **92.** $|x + 7| > 3$ **93.** $|x - 4| - 3 < 5$
94. $|x - 1| < 9$ **95.** $|x - 3| < -4$

Write inequality statements.
96. You earn more than \$35,000 annually.

97. The income for Riddle's market exceeds that of Smith's market and falls short of the income of Duke's market. Smith's income is \$108,000, and Duke's income is \$250,000. Write an inequality expressing Riddle's income compared to Smith's and Duke's.

CHALLENGE PROBLEMS

You are a travel agent and have been asked to plan a trip for the high school math club. The math club has raised \$8,700 for the trip. Your agency charges a one-time setup fee of \$300. The cost per person is \$620.

98. Write an inequality that shows the relationship of the maximum number of club members who can go on the trip.

99. Solve the inequality in Exercise 98 to determine how many club members can go on the trip.

CONCEPTS ANALYSIS

1. The symmetric property of equations states that if $a = b$ then $b = a$. Is an inequality symmetric? Illustrate your answer with an example.

2. In equations, if $a = b$, then $ac = bc$, if c equals any real number. Is the statement, if $a < b$ then $ac < bc$, true if c is any real number? For what values of c is the statement false?

3. Write in your own words two differences in the procedures for solving equations and for solving simple inequalities.

4. Explain the difference between the statements $x < 2$ and $x \leq 2$.

5. If $x < 2$ or $x > 10$, is it correct to write the statement as $2 > x > 10$? Why or why not? Explain your answer.

6. Explain the difference between an *and* relationship and an *or* relationship in inequalities.

7. In your own words, write a procedure for solving a compound inequality such as $5 < x + 2 < 9$.

8. Find the mistake in the inequality. Correct and briefly explain the mistake.

$$3x + 5 < 5x - 7$$
$$3x - 5x < -7 - 5$$
$$-2x < -12$$
$$x < 6$$

9. In the inequality $x^2 + 6x + 5 < 0$, what roles do the numbers -5 and -1 play? Explain the solution for the inequality in words.

10. If p_1 and p_2 are real numbers that solve the equation $ax^2 + bx + c = 0$ and $p_1 < p_2$, when will the solution of $ax^2 + bx + c < 0$ be $p_1 < x < p_2$? When will the solution be $x < p_1$ or $x > p_2$? Give an example to illustrate each answer.

CHAPTER SUMMARY

Learning Outcomes	What to Remember with Examples

Section 17–1

1 Use set terminology (pp. 652–653).

Set-builder notation is used to show sets.

Use set-builder notation to show the following set: the set of integers between -2 and 3, including 3.

$$\{x \mid x \in Z \text{ and } -2 < x \leq 3\}$$

2 Show inequalities on a number line and write inequalities in interval notation (pp. 653–654).	Solutions to inequalities can be shown on the number line or by using interval notation. Boundaries that are included are indicated with brackets. Boundaries that are not included are indicated with parentheses.

The solution of $-2 < x \leq 3$ is shown on the number line in Fig. 17–52 and by using interval notation. $(-2, 3]$

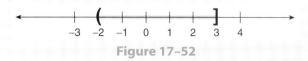

Figure 17–52

Section 17–2

1 Solve a simple linear inequality (pp. 655–658).	To solve a simple linear inequality, isolate the variable as in solving equations with two exceptions. When interchanging the sides of an inequality, reverse the sense of the inequality, and when multiplying or dividing by a negative number, reverse the sense of the inequality. Show the solution set symbolically, graphically, and using interval notation.

Solve.

$$4x - 1 < 7$$
$$4x < 8$$
$$x < 2 \quad \text{or} \quad (-\infty, 2)$$

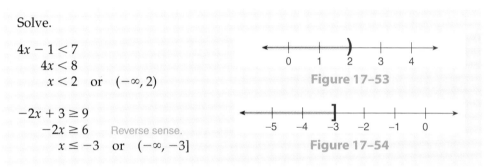

Figure 17–53

$$-2x + 3 \geq 9$$
$$-2x \geq 6 \qquad \text{Reverse sense.}$$
$$x \leq -3 \quad \text{or} \quad (-\infty, -3]$$

Figure 17–54

Section 17–3

1 Identify subsets of sets and perform set operations (pp. 659–661).	Every set has the set itself and the empty set (ϕ) as subsets. Set operations include union ($\cup$), intersection ($\cap$), and complements (A').

List the subsets of the set $\{3, 5, 8\}$:

$$\phi, \{3, 5, 8\} \text{ (the set itself)}, \{3\}, \{5\}, \{8\}, \{3, 5\}, \{3, 8\}, \{5, 8\}$$

If $U = \{1, 2, 3, 4, 5\}$, $A = \{1, 3\}$, and $B = \{2, 4\}$, find $A \cup B$, $A \cap B$, and A'.

$$A \cup B = \{1, 2, 3, 4\} \qquad A \cap B = \phi \qquad A' = \{2, 4, 5\}$$

2 Solve compound inequalities with the conjunction condition (pp. 661–663).	To solve a compound inequality that is a conjunction, separate the compound inequality into two simple inequalities using the conditions of the conjunction. Solve each simple inequality. Determine the solution set that includes the *intersection* of the solution sets of the two simple inequalities. *Note:* If there is no overlap in the sets, the solution is the empty set.

Indicate the solution using a number line and using interval notation.

$$-3 < x + 2 \leq 2$$

$$
\begin{array}{lcl}
-3 < x + 2 & \text{and} & x + 2 \leq 2 \\
-3 - 2 < x & & x \leq 2 - 2 \\
-5 < x \text{ or } x > -5 & & x \leq 0
\end{array}
$$

Solution: $-5 < x \leq 0$ **Interval notation:** $(-5, 0]$

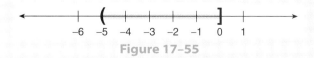

Figure 17–55

Chapter Summary   679

3 Solve compound inequalities with the disjunction condition (pp. 663–664).

To solve a compound inequality that is a disjunction: Solve each simple inequality. Determine the solution set that includes the union of the solution sets of the two simple inequalities.

Solve and indicate the solution set on the number line and by using interval notation.

$$\begin{aligned} x - 1 &< 3 \quad &\text{or} \quad x + 2 &> 8 \\ x &< 3 + 1 & x &> 8 - 2 \\ x &< 4 \quad &\text{or} \quad x &> 6 \end{aligned}$$

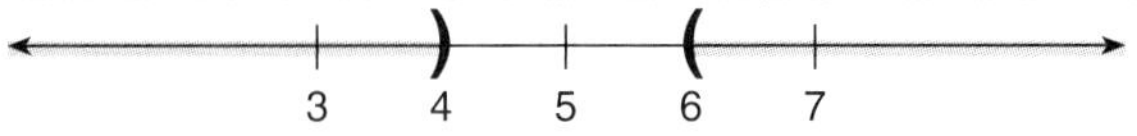

Figure 17–56

Interval notation: $(-\infty, 4) \cup (6, \infty)$

Section 17-4

1 Solve quadratic inequalities in one variable (pp. 665–669).

To solve a quadratic inequality by factoring, rearrange the inequality so that the right side of the inequality is zero and the leading coefficient is positive. Write the left side of the inequality in factored form. Determine the critical values that make each factor equal to zero. Test each region of values. The solution set for the quadratic inequality is the region or regions that produce a true statement.

Solve $x^2 + 4x - 21 < 0$.

$(x + 7)(x - 3) = 0$ Set each factor equal to zero.

$x + 7 = 0 \qquad x - 3 = 0$ Solve each equation.

$x = -7 \qquad x = 3$ Critical values.

Test Region I. Let $x = -8$.
$(-8)^2 + 4(-8) - 21 < 0?$
$64 - 32 - 21 < 0?$
$11 < 0$. *False.*

Test Region II. Let $x = 0$.
$0^2 + 4(0) - 21 < 0?$
$0 + 0 - 21 < 0?$
$-21 < 0$. *True.*

Test Region III. Let $x = 4$.
$4^2 + 4(4) - 21 < 0?$
$16 + 16 - 21 < 0?$
$11 < 0$. *False.*

Figure 17–57

The solution set is Region II.
$-7 < x < 3$ or $(-7, 3)$.

2 Solve rational inequalities in one variable (pp. 669–670).

To solve a rational inequality, set both the numerator and denominator equal to zero and solve for the variable. These solutions are the boundaries for the solution regions of the inequality solution. The region or regions that have like signs for the numerator and denominator are solutions for the $>$ and $\geq$ inequalities. The region or regions that have unlike signs for the numerator and denominator are solutions for the $<$ and $\leq$ inequalities.

Solve the rational inequality $\dfrac{x + 3}{x - 2} \leq 0$.

$x = 2$ is an excluded value. Set numerator and denominator equal to zero.

$x + 3 = 0 \qquad x - 2 = 0$ Solve each equation.

$x = -3 \qquad x = 2$ Critical values.

Figure 17–58

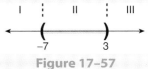

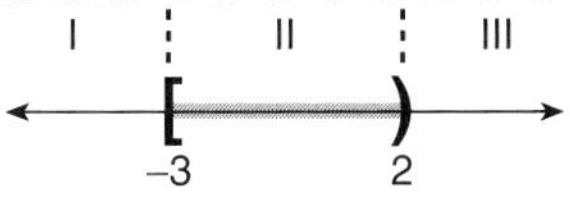

The critical value 2 is not included in the solution set because it is an excluded value.

Region I has like signs ($\frac{-}{-}$).

Region II has unlike signs ($\frac{+}{-}$).   Unlike signs indicate < 0.

Region III has like signs ($\frac{+}{+}$).

The solution set is Region II.
$-3 \le x < 2; [-3, 2)$

Section 17-5

1 Solve equations containing one absolute-value term (pp. 671–672).

To solve an absolute-value equation, isolate the absolute-value expression, then form two equations. One equation considers the expression within the absolute-value symbol to be positive; the other equation considers the expression to be negative. Solve each case to obtain the two roots of the equation.

Solve $|x - 3| + 1 = 5$.

$|x - 3| + 1 = 5$ Isolate the absolute-value term.
$|x - 3| = 4$

$Case\ 1: x - 3 = 4$ $Case\ 2: x - 3 = -4$ or $-(x - 3) = 4$
$x = 7$ $x = -4 + 3$ $-x + 3 = 4$
$x = -1$ $-x = 1$
$x = -1$

Section 17-6

1 Solve absolute-value inequalities using the "less than" relationship (pp. 673–674).

To solve an absolute-value inequality using the "less than" relationship, use the property, if $|x| < b$ and $b > 0$, then $-b < x < b$.

Solve $|x - 3| < 4$.   Show the solution set graphically, in symbolic and in interval notation.

$-4 < x - 3 < 4$ Separate into two simple inequalities connected by *and*.
$-4 < x - 3$ and $x - 3 < 4$ Solve each inequality.
$-1 < x$ or $x > -1$ $x < 7$
$-1 < x < 7; (-1, 7)$

2 Solve absolute-value inequalities using the "greater than" relationship (pp. 674–676).

To solve an absolute-value inequality using the "greater than" relationship, use the property, if $|x| > b$ and $b > 0$, then $x < -b$ or $x > b$.

Solve $|x + 7| > 1$.   Show the solution set graphically, in symbolic and in interval notation.

$x + 7 < -1$ or $x + 7 > 1$ Separate into two simple inequalities connected by *or*.
$x < -8$ $x > -6; (-\infty, -8) \cup (-6, \infty)$

CHAPTER TRIAL TEST

Represent each set of numbers on the number line and by using interval notation.
1. $x \ge -12$ 2. $-3 < x \le -2$

Graph the solution on a number line and write the solution in symbolic and interval notation.
3. $3x - 1 > 8$ 4. $2 - 3x \ge 14$ 5. $10 < 2 + 4x$
6. $-5b > -30$ 7. $\frac{1}{3}x + 5 \le 3$ 8. $2(1 - y) + 3(2y - 2) \ge 12$
9. $5 - 3x < 3 - (2x - 4)$ 10. $7 + 4x \le 2x - 1$ 11. $-5 < x + 3 < 7$
12. $\frac{1}{4}x + 2 < 3 < \frac{1}{3}x + 9$ 13. $3x - 1 \le 5 \le x - 5$ 14. $(x + 4)(x - 2) < 0$
15. $(2x + 3)(x - 1) > 0$ 16. $2y^2 + y < 15$ 17. $2x - 3 < 1$ or $x + 1 > 7$
18. $2(x - 1) < 3$ or $x + 7 > 15$

Use the following sets to give the elements in the indicated sets in Exercises 19–22.

$U = \{1, 2, 3, 4, 5, 6, 7, 8, 9\}$ $A = \{5, 8, 9\}$, $B = \{1, 2, 3, 4, 5, 6, 7, 8\}$

19. $A \cup B$ **20.** $A \cap B$ **21.** $A \cap B'$ **22.** List all the subsets of set A.

Solve. Graph the solution set on a number line and write the solution set in interval notation.

23. $\dfrac{x - 2}{x + 5} < 0$ **24.** $\dfrac{x - 3}{x} > 0$

Solve.

25. $|x| = 15$ **26.** $|x + 8| = 7$ **27.** $|x| + 8 = 10$

Solve and graph the solution set on a number line and write the solution set in interval notation.

28. $|4x - 7| < 17$ **29.** $|x + 8| > 10$ **30.** $|x + 1| - 3 < 2$

31. An airplane manufacturer orders a part from a supplier that measures 96.8 cm and has a tolerance of ±0.1 cm according to the specifications. Write an inequality that shows the range of acceptable measures for the part.

32. An electronically controlled thermostat is set to cause the cooling unit to operate when the temperature is above 68°F. Write an inequality that expresses the temperature range in Celsius degrees. The formula $°F = \frac{9}{5}°C + 32$ may be used.

Graphing Functions

18

18–1 | Graphical Representation of Linear Equations and Functions

Learning Outcomes
1 Represent the solutions of a linear function in a table of solutions.
2 Represent the solutions of a linear function on a graph.
3 Check the solution of an equation (or function) with two variables.
4 Find a specific solution of an equation (or function) with two variables when given the value of one variable.

Linear equations in one variable have at most one solution. The solution can be represented graphically as a point on a number line. A linear equation in two variables has an unlimited number of solutions. The solutions of a linear equation in two variables can be represented graphically as a line on a rectangular coordinate system. To graph a linear equation, we will review function notation and plotting points on a rectangular coordinate system.

1 Represent the Solutions of a Linear Function in a Table of Solutions.

The fixed monthly cost of a cellular phone is $15 plus $0.25 for each minute of use above the plan limits. An equation to represent this situation must have two variables. One variable represents the number of minutes of phone use (m) that are above the plan limits, and the other variable represents the total monthly cost (c). One variable is dependent on the other. For instance, the total monthly cost is dependent on the excess number of minutes of use: $c = \$15 + \$0.25m$. We can also write equations with two variables in function notation: $C(m) = \$15 + \$0.25m$ or $f(x) = 15 + 0.25x$. This type of function is often referred to as a *cost function*.

For every value of the independent variable x, there is a corresponding value of the dependent variable $f(x)$. These two values form a pair of values called an *ordered pair*. A representative selection of these solutions can be written in a *table of solutions*. The *domain* of the function is a set of suitable values determined by the context of the problem that can be selected for the independent variable of the function. The set of corresponding values of the dependent variable that is formed by considering each value in the domain is called the *range*.

To prepare a table of solutions for an equation or function with two variables:

1. Select any appropriate value from the domain of the independent variable.
2. Substitute the selected value for the independent variable and solve the resulting equation for the dependent variable.
3. Repeat Steps 1 and 2 until the desired number of solutions are obtained.
4. Write the solution as an ordered pair.

To make a table of selected solutions for the monthly cost of a cellular phone, consider the domain to be values that are zero or greater than zero.

EXAMPLE Make a table of solutions and graph the function $f(x) = 15 + 0.25x$ for five values of x in 50-min intervals. Discuss the characteristics of the domain and range.

x	$f(x)$
0	$15.00
50	$27.50
100	$40.00
150	$52.50
200	$65.00

$f(0) = 15 + 0.25(0)$
$f(0) = 15 + 0$
$f(0) = \textbf{15, or \$15.00}$

$f(50) = 15 + 0.25(50)$
$f(50) = 15 + 12.5$
$f(50) = \textbf{27.5, or \$27.50}$

$f(100) = 15 + 0.25(100)$
$f(100) = 15 + 25$
$f(100) = \textbf{40, or \$40.00}$

$f(150) = 15 + 0.25(150)$
$f(150) = 15 + 37.5$
$f(150) = \textbf{52.5, or \$52.50}$

$f(200) = 15 + 0.25(200)$
$f(200) = 15 + 50$
$f(200) = \textbf{65, or \$65.00}$

To represent this model graphically, use the x- or horizontal axis to represent the values of the independent variable. Use the y- or vertical axis to represent the values of the dependent variable (Fig. 18–1).

Each pair of values in the table of solutions is a point on the graph, and the ordered pair is written in point notation as (independent variable, dependent variable) or (x, $f(x)$). Each point is plotted on the graph as we did in Chapter 2, Section 1 Outcome 5.

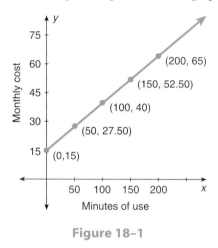

Figure 18–1

The domain is all whole numbers or possibly 0 and all positive rational numbers if fractional minutes are allowed. The range is numbers greater than or equal to 15.

Let's examine the graphical solution. What does the point (0, $15) represent? This is a *fixed cost*. Even if no minutes are used, the monthly cost will be $15. What if the monthly phone use is more than 200 min? When the table of values was represented on the graph, a pattern was established. We can use this pattern or model to project other values of x.

EXAMPLE Find the monthly cost from the graph of $f(x) = 15 + 0.25x$ (Fig. 18–2) for 250 min and 300 min.

From the graph find the corresponding values for $f(x)$ when $x = 250$ and when $x = 300$.

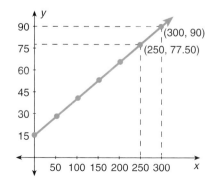

Figure 18–2

For $x = 250$, $f(x) = $77.50 (slightly more than $75). For $x = 300$, $f(x) = $90.00.

18–1 Graphical Representation of Linear Equations and Functions

685

Exact values are difficult and sometimes impossible to determine from a graph. To find the exact cost for 250 min of phone use we evaluate the function $f(x) = 15 + 0.25x$ for $x = 250$. $f(250) = 15 + 0.25(250); f(250) = 15 + 62.50; f(250) = \$77.50$.

2 Represent the Solutions of a Linear Function on a Graph.

In the preceding two examples, we were looking at a solution that restricted the values of x and $f(x)$ to zero or positive values. In many cases, we want to view the pattern established by considering all types of numbers. After viewing all values, some of these values may still be disregarded because they are impractical or impossible within the context of the situation.

To graph a linear function using a table of solutions:

1. Prepare a table of values by evaluating the function at different values (at least three) in the domain of the independent variable. Each pair of values is represented by a point on the graph.
2. Plot the points in the table of solutions on the rectangular coordinate system, with the independent variable on the x-axis and the dependent variable on the y-axis.
3. Connect the points with a straight line. Extend the graph beyond the three points and place an arrow on each end as appropriate to indicate that the line extends indefinitely.

The *table-of-solutions* or *table-of-values* procedure for graphing an equation works for *all* types of equations. Other methods for graphing equations focus on specific properties of each type of equation; these methods are generally less time-consuming and more efficient than the table-of-solutions procedure. However, the table-of-solutions method can be used as a backup or check if uncertainty arises with the other methods.

Linear functions can also be written as equations with two variables. In general, the independent variable is represented by the letter x and the dependent variable is represented by the letter y. Then, solutions are represented by the ordered pair (x, y) and are plotted on the rectangular coordinate system.

EXAMPLE Make a table of solutions and graph the equation $y = -2x + 5$. Discuss the characteristics of the domain and range.

Select at least three values for the independent variable: $-1, 0, 1$.

When $x = -1$, $y = -2(-1) + 5$	Substitute -1 for x.
$= 2 + 5$	Evaluate.
$= 7$	Solution: $x = -1$ and $y = 7$ or $(-1, 7)$.
When $x = 0$, $y = -2(0) + 5$	Substitute.
$= 0 + 5$	Evaluate.
$= 5$	Solution: $(0, 5)$.
When $x = 1$, $y = -2(1) + 5$	Substitute.
$= -2 + 5$	Evaluate.
$= 3$	Solution: $(1, 3)$.

Make a table of the three solutions.

x	y	Points on Graph
-1	7	$(-1, 7)$
0	5	$(0, 5)$
1	3	$(1, 3)$

Plot the solutions in the table of solutions as points on a rectangular coordinate system (Fig. 18–3).

Chapter 18 / Graphing Functions

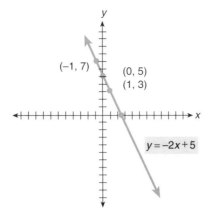

The domain is all real numbers and the range is all real numbers.

Figure 18–3

3 **Check the Solution of an Equation (or Function) with Two Variables.**

Many of the equations we have studied in this text have had only one variable, such as x or y, and took the form $3x - 2 = 7$ or $5y + 3 = 9$. However, we have seen that equations may have two variables, both x *and* y, such as $2x + 3y = 12$. Both types of equations are *linear equations*. A *linear equation* in two variables is an equation for which the graph is a straight line. It can be written in the standard form

$$ax + by = c$$

where a, b, and c are real numbers and a and b are not both zero. The degree of any term in the equation is 1 or zero.

The solutions of a linear equation in two variables are *ordered pairs* of numbers. There is one number for each variable. The numbers are generally ordered in the alphabetical order of the two variables rather than in the order the letters appear in the equation. Thus, in the equation $2y + x = 8$, a solution of $(4, 2)$ indicates that $x = 4$ and $y = 2$.

Solutions of equations in two variables are ordered pairs of numbers that make a true statement when used to evaluate the original equation. An equation in two variables has many solutions. We can check each solution of an equation in two variables by substituting the values for the respective variables and performing the operations indicated.

To check the solution of an equation with two variables:

1. Substitute the values for the respective variables into the equation (or function).
2. Evaluate each side of the equation.
3. If the ordered pair is a solution of the equation, the two sides of the equation will be equal.

EXAMPLE   Check the solution $(-2, 16)$ for the equation $2x + y = 12$.

$2x + y = 12$ Substitute −2 for *x* and 16 for *y*.
$2(-2) + 16 = 12$ Evaluate.
$-4 + 16 = 12$
$12 = 12$ True.

The ordered pair makes the equation true. So $(-2, 16)$ is a solution.

EXAMPLE   Check the ordered pair $(3, -2)$ for the equation $y = 2x - 5$.

$$y = 2x - 5 \qquad \text{Substitute 3 for } x \text{ and } -2 \text{ for } y.$$
$$-2 = 2(3) - 5 \qquad \text{Evaluate.}$$
$$-2 = 6 - 5$$
$$-2 = 1 \qquad \text{False.}$$

The ordered pair does not make the equation true. So $(3, -2)$ is not a solution.

4 **Find a Specific Solution of an Equation (or Function) with Two Variables When Given the Value of One Variable.**

If we are given the value of one variable for an equation with two variables, we can substitute the given value in the equation and solve the equation for the other variable.

To find a specific solution of an equation (or function) with two variables when given the value of one variable:

1. Substitute the given value of the variable in each place that variable occurs in the equation.
2. Solve the equation for the other variable.

EXAMPLE   Solve the equation $2x - 4y = 14$, if $x = 3$.

$$2x - 4y = 14 \qquad \text{Substitute 3 for } x.$$
$$2(3) - 4y = 14$$
$$6 - 4y = 14 \qquad \text{Sort.}$$
$$-4y = 14 - 6$$
$$-4y = 8 \qquad \text{Divide both sides by } -4.$$
$$\frac{-4y}{-4} = \frac{8}{-4}$$
$$y = -2$$

The solution is $(3, -2)$; $x = 3$ and $y = -2$.

The standard form of *a linear equation in two variables*, $ax + by = c$, is such that both a and b are not zero. However, either a or b may be zero.

When a or b is zero, the variable for which zero is the coefficient is eliminated from the equation.

$$ax + by = 15 \qquad \text{Let } a = 5 \text{ and } b = 0.$$
$$5x + 0(y) = 15 \qquad 0(y) = 0.$$
$$5x + 0 = 15 \qquad \text{Linear equation in one variable. } n + 0 = n.$$
$$5x = 15 \qquad \text{Divide both sides by 5.}$$
$$\frac{5x}{5} = \frac{15}{5}$$
$$x = 3$$

An equation like $x = 3$ is a linear equation and we think of the coefficient of the missing variable term as zero:

$$x = 3 \qquad \text{is considered as} \qquad x + 0y = 3$$

This means that, for any value of y, x is 3.

EXAMPLE For the equation $x = 6$, complete the given ordered pairs: (, 2); (, −3); (, 1).

Equation	Ordered Pairs			
$x = 6$	(, 2)	(, −3)	(, 1)	*x* is 6 for *any* value of *y*.
	(6, 2)	**(6, −3)**	**(6, 1)**	

Numerous application problems can be solved with equations in two variables.

EXAMPLE A small business photocopied a report that included 12 black-and-white pages and 25 color pages. The cost was $23.70. Letting b = the cost for a black-and-white page and c = the cost for a color page, set up an equation with two variables.

Cost of black-and-white copies: $12b$ (12 times cost per page)

Cost of color copies: $25c$ (25 times cost per page)

Total cost: black-and-white plus color = $23.70

Equation: $12b + 25c = 23.70$

An equation containing two variables shows the relationship between the two unknown amounts. If a value is known for either amount, the other amount can be found by solving the equation for the missing amount.

EXAMPLE Solve the equation from the preceding example to find the cost for each black-and-white copy if color copies cost $0.90 each.

$$12b + 25c = 23.70 \qquad \text{Substitute 0.90 for } c.$$
$$12b + 25(0.90) = 23.70$$
$$12b + 22.50 = 23.70 \qquad \text{Transpose.}$$
$$12b = 23.70 - 22.50$$
$$12b = 1.20 \qquad \text{Divide both sides by 12.}$$
$$\frac{12b}{12} = \frac{1.20}{12}$$
$$b = 0.10$$

Each black-and-white copy costs $0.10.

SELF-STUDY EXERCISES 18–1

1

1. Make a table of at least three solutions for the function $f(x) = 3x - 7$.
2. Make a table of three solutions for the function $f(x) = -2x + 3$.
3. Rental cars cost $39 for a day plus $0.10 per mile. Write a function that represents the total cost of the car for a single day if x represents the number of miles and $f(x)$ represents the total car rental cost.
4. Make a table of solutions for the function in Exercise 3.

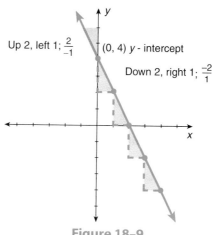

Up 2, left 1; $\dfrac{2}{-1}$ (0, 4) y - intercept

Down 2, right 1; $\dfrac{-2}{1}$

Figure 18–9

(b) $5x - y = -2$ Solve for y.

$\qquad\quad y = 5x + 2$

$\qquad$ y-intercept = (0, 2)   Locate this point on the y-axis.

$\qquad$ slope $= \dfrac{5}{1}$ or $\dfrac{-5}{-1}$

$\dfrac{5}{1}$ indicates vertical movement of $+5$ and horizontal movement of $+1$ from the y-intercept. $\dfrac{-5}{-1}$ indicates vertical movement of -5 and horizontal movement of -1. See Fig. 18–10.

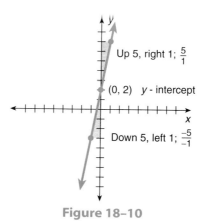

Up 5, right 1; $\dfrac{5}{1}$

(0, 2)   y - intercept

Down 5, left 1; $\dfrac{-5}{-1}$

Figure 18–10

(c) $3x + 4y = -12$ Solve for y.

$\qquad\quad y = -\dfrac{3}{4}x - \dfrac{12}{4}$

$\qquad\quad y = -\dfrac{3}{4}x - 3$

$\qquad$ y-intercept = (0, −3)   Locate this point on the y-axis.

$\qquad$ slope $= \dfrac{-3}{4}$ or $\dfrac{3}{-4}$

$\dfrac{-3}{4}$ indicates vertical movement of -3 and horizontal movement of $+4$ from the y-intercept. $\dfrac{3}{-4}$ indicates vertical movement of $+3$ and horizontal movement of -4. See Fig. 18–11.

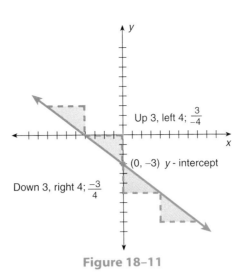

Up 3, left 4; $\frac{3}{-4}$

(0, –3) y - intercept

Down 3, right 4; $\frac{-3}{4}$

Figure 18–11

3 **Graph Linear Equations Using a Graphing Calculator.**

A graphing calculator can be used to graph linear equations written in the form $y = mx + b$. Also, computer software can be used to graph linear equations. These tools free us from the tedious task of graphing equations and allow us to focus on the patterns and properties of graphs.

Every model of graphing calculator and every brand of computer software may have a different set of keystrokes for graphing equations. You will need to refer to the owner's manual to adapt to a particular model or brand. However, we will illustrate the usefulness of these tools by showing a few features common to most calculators.

Feature	Purpose
$y=$	Screen for entering equation or function that is solved for y
Window	Setting the range (high and low values of each axis) for the viewing window
Zoom	Zooming to get a closer or wider view or a view to "fit" the full range of y-values for the selected x-values
Trace	Tracing to locate specific points and determine their coordinates
Graph	Displays the graph of all equations that have been entered

EXAMPLE Graph $y = 5x + 2$ using a graphing calculator.

Perform the appropriate function on the graphing calculator of your choice.

Clear graphing screen.	Erase previous equations from the $y=$ screen.
Set or initialize range.	Define the viewing range of the graph window. One option is to select the standard option from the Zoom screen. That means the range will be reset at a predetermined, factory-set domain and range.
Enter equation.	Equation must be solved for y. On the $y=$ screen enter the equation using the appropriate key for the variable. This key is often labeled $\boxed{X, T, \theta, n}$.
View graph.	Use the Graph key. Determine a different viewing window by setting the values on the Window screen or using the Zoom feature.
Determine key points on the graph.	Using the trace function, move the cursor using the left or right arrow keys. The x- and y-coordinates of the point at the cursor will appear at the bottom of the screen.

Find the corresponding *y*-value for a given *x*-value by using the *value* function from the CALC screen.

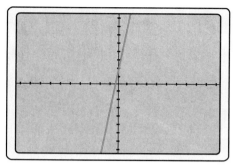

Figure 18–12

EXAMPLE   Use the function $y = 5x + 2$ and the graphing calculator to find the value for *y* when *x* is (a) 3 and (b) 2.7.

(a) $\boxed{Y=}$ $\boxed{\text{CLEAR}}$ 5 $\boxed{\text{X, T, }\theta, n}$ + 2   Enter the function.

           CALC

     $\boxed{\text{2nd}}$ $\boxed{\text{TRACE}}$ 1   Choose the value option from the CALC screen, option 1.

     3 $\boxed{\text{ENTER}}$ $\Rightarrow$ ***y = 17***   Let *x* = 3. The value entered must be on view screen.

           CALC

(b) $\boxed{\text{2nd}}$ $\boxed{\text{TRACE}}$ 1   Choose the value option from the CALC screen.

     2.7 $\boxed{\text{ENTER}}$ $\Rightarrow$ ***y = 15.5***   Let *x* = 2.7.

4   **Solve a Linear Equation Using a Graph.**

An equation in one variable, such as $2x + 5 = 3$, can be written in the form of a function by first rewriting all terms on either side of the equation.

$$2x + 5 = 3 \quad\quad \text{or} \quad\quad 2x + 5 = 3$$
$$2x + 5 - 3 = 0 \quad\quad\quad\quad\quad 0 = 3 - 2x - 5$$
$$2x + 2 = 0 \quad\quad\quad\quad\quad\quad 0 = -2x - 2$$

Then, write the nonzero side of each equation as a function of *x*.

$$f(x) = 2x + 2 \quad\quad \text{or} \quad\quad g(x) = -2x - 2$$

To write an equation in one variable as a function:

1. Rewrite the equation so that all terms are on one side of the equation.
2. Write the equation in function notation using $f(x) =$ the nonzero side of the equation from Step 1.

The solution of the equation is the value of x that makes the value of either function zero.

Solution	Functions Evaluated Using Solution		
$2x + 5 = 3$	$f(x) = 2x + 2$	or	$g(x) = -2x - 2$
$2x = 3 - 5$	$f(-1) = 2(-1) + 2$		$g(-1) = -2(-1) - 2$
$2x = -2$	$f(-1) = -2 + 2$		$g(-1) = 2 - 2$
$x = -1$	$f(-1) = 0$		$g(-1) = 0$

Now, suppose we want to solve an equation graphically *instead of* by traditional methods of isolating the variable. That is, suppose we want to use a graphing calculator or computer graphing software to solve an equation. We can graph the equation and find the solution by examining the point where the graph crosses the x-axis. Let's solve the equation $2x + 5 = 3$ graphically by using either function $f(x) = 2x + 2$ or $f(x) = -2x - 2$.

Use the intercepts method.

x	f(x)
-1	0
0	2

$f(x) = 2x + 2$

Let $f(x) = 0$.
$$0 = 2x + 2$$
$$-2 = 2x$$
$$-1 = x$$

Let $x = 0$.
$$f(0) = 2(0) + 2$$
$$= 2$$

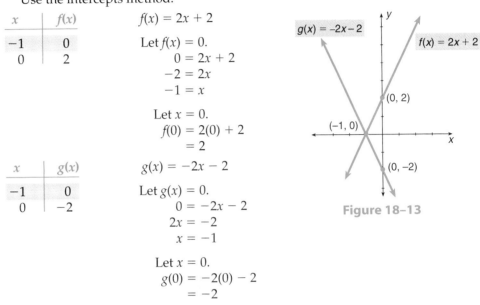

x	g(x)
-1	0
0	-2

$g(x) = -2x - 2$

Let $g(x) = 0$.
$$0 = -2x - 2$$
$$2x = -2$$
$$x = -1$$

Let $x = 0$.
$$g(0) = -2(0) - 2$$
$$= -2$$

Figure 18–13

If an equation in one variable is written in function notation, the solution of the equation is shown on the graph at the point where the graph crosses the x-axis. In Fig. 18–13, both functions cross the x-axis at $(-1, 0)$ and the solution of the equation $2x + 5 = 3$ is $x = -1$. Thus, we can solve equations in one variable by using the graphing function on a calculator or computer.

To solve an equation in one variable using a graph:

1. Write the equation as a function.
2. Graph the function using a calculator or computer.
3. The solution is the x-intercept. This point is also referred to as the *zero point* $(y = 0)$ of the function.

TIP!

Find the Zero of a Function Using a Calculator

On most graphing calculators there is a feature to find the zero point or points of a function. On one popular model, the TI-83 Plus, the feature is a selection on the CALC screen.

The option for finding the zero point requires several steps that are also used for finding other properties such as the maximum or minimum of a function.

EXAMPLE Solve the equation $2x + 5 = 3$ using the TI-83 Plus.

$$2x + 5 = 3$$
$$f(x) = 2x + 2 \qquad \text{Function form of the equation.}$$

$\boxed{Y=}$ Enter function: 2 $\boxed{\text{X, T, θ, } n}$ + 2

CALC

CALC $\boxed{\text{2nd}}\boxed{\text{TRACE}}$

Zero Option 2.

Left bound? Move cursor to any point to the left of where the graph crosses the *x*-axis.
Press $\boxed{\text{ENTER}}$.

Right bound? Move cursor to any point to the right of where the graph crosses the *x*-axis.
Press $\boxed{\text{ENTER}}$.

Guess? Move the cursor close to the *x*-intercept. Press $\boxed{\text{ENTER}}$.

$x = -1$ Read solution from *x* value at the bottom of the screen.

EXAMPLE Write $3x + 7 = x - 3$ as a function, graph the function, and find the solution from the graph.

$$3x + 7 = x - 3 \qquad \text{Write all terms on the left side of the equation.}$$
$$3x + 7 - x + 3 = 0 \qquad \text{Combine like terms.}$$
$$2x + 10 = 0 \qquad \text{Write as a function.}$$
$$f(x) = 2x + 10 \qquad \text{Graph the function.}$$

Find the *x*-coordinate of the *x*-intercept of the graph

$\boxed{Y=}$ 2 $\boxed{\text{X, T, θ, } n}$ + 10 Enter function.

CALC

$\boxed{\text{2nd}}\boxed{\text{TRACE}}$ 2 Find the *x*-intercept. Choose ZERO which is option 2.

Mark left bound, Read *x*-coordinate from bottom of screen.
right bound,
and guess.

$x = -5$

SELF-STUDY EXERCISES 18–2

1 Graph the equations using the intercepts procedure.

1. $x + y = 5$
2. $x + 3y = 5$
3. $\dfrac{1}{2}y = 4 + x$
4. $y = 3x - 1$
5. $5x = y + 2$
6. $x = -2y + 3$

2 Graph the equations using the slope-intercept procedure.

7. $y = 2x - 3$
8. $y = -\dfrac{1}{2}x - 2$
9. $y = -\dfrac{3}{5}x$
10. $x - 2y = 3$
11. $2x + y = 1$
12. $y = \dfrac{3}{4}x + 2$

Graph the equations using a graphing calculator.

13. $y = 0.3x + 0.5$

15. $y = -2.6x - 1.7$

17. $y = 212x + 757$

19. $y = 48x - 200$

14. Find y when $x = -0.4$ in $y = 0.3x + 0.5$.

16. Find y when $x = 1.3$ in $y = -2.6x - 1.7$.

18. Find y when $x = 5$ in $y = 212x + 757$.

20. Find y when $x = 7$ in $y = 48x - 200$.

21. The cost of printing a magazine is $5,000 to typeset and prepare copy for printing and $3 per copy to print. This is expressed as an equation $f(x) = 3x + 5,000$, where x is the number of copies printed and $f(x)$ is the total cost of printing x copies. Find the cost of printing 10,000 copies. Find the cost of printing 30,000 copies.

22. Income is expressed by the function $f(x) = \$8.30x$, where x is the number of hours per week an employee works. Find the income for an employee who works 40 hours in a week. Find the annual income if a person works for an entire year (52 weeks).

23. Some researchers say a person's maximum heart rate in beats per minute can be found by subtracting age from 220. This can be expressed as the function $f(x) = 220 - x$. What is the maximum heart rate for a person 25 years old? What is the maximum heart rate for a person who is 65 years old?

24. An inventory shows 196 jigsaw puzzles in stock. These puzzles can be produced at a rate of 15 per hour. The number of puzzles on hand can be expressed as a function $f(x) = 15x + 196$, where x represents the number of hours of production and $f(x)$ represents the total inventory. How many puzzles are in inventory if they have been produced for 60 hr?

25. The function $f(x) = 24x + 800$ is used to represent the cost of a party if x people attend. Find the cost for 100 people to attend the party.

Solve the following equations by writing them as functions and using a graphing calculator.

26. $3x + 2 = 7$

27. $5x - 4 = 2$

28. $4 = 6x - 1$

29. $3x + 7 = 4x + 6$

30. $3x - 1 = 4(x + 2)$

18-3 | Graphing Linear Inequalities with Two Variables

Learning Outcome   **1** Graph linear inequalities with two variables using test points.

When we solved inequalities in one variable the solution was a set of numbers that we can represent graphically. The *solution of linear inequalities with two variables* is a set of coordinate points.

1   **Graph Linear Inequalities with Two Variables Using Test Points.**

The graphical solution for a linear inequality ($<$ or $>$) includes all points on one side of the line representing the graph of the similar equation. If either the "less than or equal" ($\leq$) or "greater than or equal" ($\geq$) symbol is used, the points on the boundary line are also included.

To graph a linear inequality with two variables:

1. Find the boundary by graphing an equation that substitutes an equal sign for the inequality symbol. Make a solid line if the boundary is included ($\leq$ or $\geq$) or a dashed line if the boundary is not included ($<$ or $>$) in the solution set.
2. Test any point that is *not* on the boundary line.
3. If the test point makes a true statement with the original inequality, shade the side containing the test point.
4. If the test point makes a false statement with the original inequality, shade the side opposite the side containing the test point.

EXAMPLE Graph the inequality $2x + y < 3$.

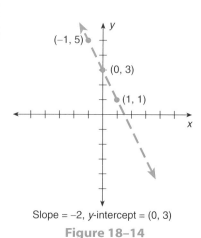

(−1, 5)

(0, 3)

(1, 1)

Slope = −2, y-intercept = (0, 3)

Figure 18–14

Graph the equation $2x + y = 3$ to establish the boundary. The boundary will not be included in the solution set and is represented with a dashed line.

$$2x + y = 3 \qquad \text{Solve the equation for } y.$$
$$y = -2x + 3$$

$$y\text{-intercept} = 3 \text{ or } (0, 3) \qquad \text{Constant term.}$$

$$\text{slope} = -2 \text{ or } \frac{-2}{1} \text{ or } \frac{2}{-1} \qquad \text{Coefficient of } x.$$

Select one point, on either side of the boundary, to use as a test point. Suppose that we choose (2, 2) as the point above the boundary.

$$\text{Is } 2x + y < 3 \quad \text{when} \quad x = 2 \text{ and } y = 2? \qquad \text{Substitute values.}$$
$$2(2) + 2 < 3 \qquad \text{Simplify.}$$
$$4 + 2 < 3$$
$$6 < 3 \qquad \text{False statement.}$$

Thus, the side of the line including the point (2, 2) is *not* in the solution set. Shade the opposite side. See Fig. 18–15.

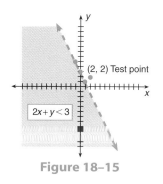

(2, 2) Test point

$2x + y < 3$

Figure 18–15

TIP!

Solid or Dashed Boundary Lines

In inequalities with two variables, we represent boundaries that are included in the solution set with solid lines and boundaries that are excluded from the solution set with dashed lines. Figure 18–16 shows the graph of $2x + y \leq 3$. Figure 18–17 shows the graph of $2x + y > 3$, and Figure 18–18 shows $2x + y \geq 3$.

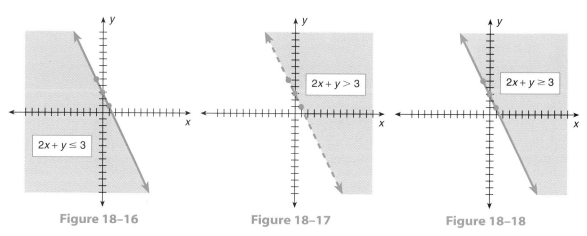

$2x + y \leq 3$

$2x + y > 3$

$2x + y \geq 3$

Figure 18–16   **Figure 18–17**   **Figure 18–18**

Chapter 18 / Graphing Functions

TIP!

Selecting Test Points

Choose numbers that are easy to work with when selecting a test point. If the boundary line does not pass through the origin, a good point to use is $(0, 0)$.

EXAMPLE Graph the inequality $y \geq 3x + 1$.

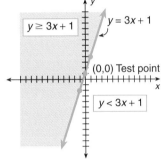

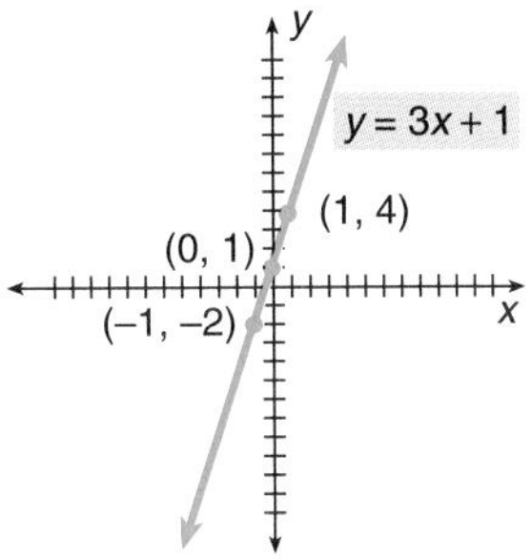

Figure 18–19

Graph the line $y = 3x + 1$.

y-intercept $= (0, 1)$ Constant term.

$\text{slope} = 3 \text{ or } \dfrac{3}{1} \text{ or } \dfrac{-3}{-1}$ Coefficient of x.

The boundary line is shown in Fig. 18–19 as a solid line because the inequality $y \geq 3x + 1$ *does* include the boundary.

The point $(0, 0)$ is not on the boundary line and it is on the right side of the boundary. We use $(0, 0)$ as our test point.

Is $y \geq 3x + 1$ when $x = 0$ and $y = 0$?   Substitute values.

$0 \geq 3(0) + 1$ Simplify.

$0 \geq 0 + 1$

$0 \geq 1$ False statement.

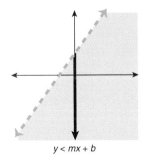

Thus, as shown in Fig. 18–20, the boundary and all points on the side of the boundary opposite the test point $(0, 0)$ are in the solution set.

Figure 18–20

TIP!

General Observation About Inequalities

When the inequality is in the form $y < mx + b$ or $y \leq mx + b$, the solution set includes the portion of the y-axis and all points *below* the boundary line. When an inequality is in the form $y > mx + b$ or $y \geq mx + b$, the solution set includes the portion of the y-axis and all points *above* the boundary line.

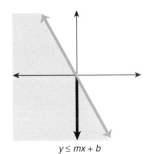 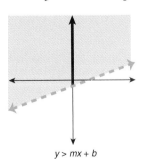 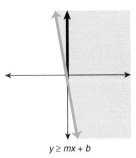

Figure 18–21

18-3 Graphing Linear Inequalities with Two Variables

SELF-STUDY EXERCISES 18–3

1 Graph the following linear inequalities using test points.

1. $x - y < 8$
2. $x + y > 4$
3. $3x + y < 2$
4. $2x + y \leq 1$
5. $x + 2y < 3$
6. $2x + 2y \geq 3$
7. $y \geq 2x - 3$
8. $y \geq -3x + 1$
9. $y > \frac{1}{2}x - 3$
10. $y > -\frac{3}{2}x + \frac{1}{2}$

18–4 | *Graphing Quadratic Equations and Inequalities*

Learning Outcomes

1 Graph quadratic equations using the table-of-solutions method.

2 Graph quadratic equations by examining properties.

3 Solve a quadratic equation from a graph.

4 Graph quadratic equations and inequalities using a graphing calculator.

1 Graph Quadratic Equations Using the Table-of-Solutions Method.

The graphs of linear equations are *straight* lines. We will examine the graphs of some equations that have a degree higher than 1, which are not linear equations.

The graph of an equation of a degree higher than 1 is a *curved* line. The curved line can be a parabola, hyperbola, circle, ellipse, or an irregular curved line. We do not define these terms at this time, but Fig. 18–22 illustrates them. The equation for a *parabola* that opens up or down is distinguished from other quadratic equations. The y variable has degree 1. The x variable must have one term with degree 2 and no term with a degree higher than 2. The quadratic equation for a parabola can be written in function notation: $f(x) = ax^2 + bx + c$.

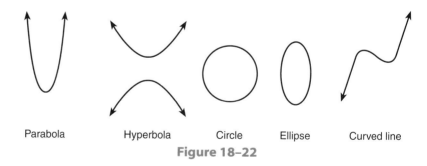

Parabola   Hyperbola   Circle   Ellipse   Curved line

Figure 18–22

One method of graphing quadratic equations is to form a table of solutions and plot the points like we did in Section 18–1. This method requires *more* points than we customarily use in graphing linear equations. Other methods of graphing quadratic equations involve examining characteristics or properties of the equation.

EXAMPLE Prepare a table of solutions and graph the function $f(x) = x^2 + x - 6$ using integral values of x between -3 and 3, inclusive. Discuss the characteristics of the domain and range.

$f(x) = x^2 + x - 6$

x	$f(x)$
-3	0
-2	-4
-1	-6
0	-6
1	-4
2	0
3	6

$f(-3) = (-3)^2 + (-3) - 6$
$f(-3) = 9 - 3 - 6$
$f(-3) = 0$

$f(0) = 0^2 + 0 - 6$
$f(0) = 0 + 0 - 6$
$f(0) = -6$

$f(3) = 3^2 + 3 - 6$
$f(3) = 9 + 3 - 6$
$f(3) = 6$

$f(-2) = (-2)^2 + (-2) - 6$
$f(-2) = 4 - 2 - 6$
$f(-2) = -4$

$f(1) = 1^2 + 1 - 6$
$f(1) = 1 + 1 - 6$
$f(1) = -4$

$f(-1) = (-1)^2 + (-1) - 6$
$f(-1) = 1 - 1 - 6$
$f(-1) = -6$

$f(2) = 2^2 + 2 - 6$
$f(2) = 4 + 2 - 6$
$f(2) = 0$

The domain is all real numbers. The range is real numbers greater than or equal to the lowest point of the graph (approximately -6).

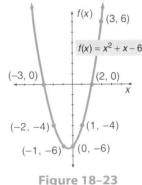

Plot the points indicated in the table of solutions and connect the points with a smooth, continuous curve. The curve is symmetrical.

Figure 18–23

2 Graph Quadratic Equations by Examining Properties.

As with linear equations, graphing quadratic equations that form a parabola by the table-of-solutions method can be time-consuming. The graph can be drawn and other important characteristics determined by examining some key properties of the parabola.

A parabola is symmetrical; that is, it can be folded in half and the two halves match. The fold line is called the *axis of symmetry*. For a parabola in the form $y = ax^2 + bx + c$, the equation of the axis of symmetry is $x = -\dfrac{b}{2a}$.

The point of the graph that crosses the axis of symmetry is the *vertex* of the parabola. Thus, the x coordinate of the vertex of the parabola is $-\dfrac{b}{2a}$.

To graph quadratic equations in the form $y = ax^2 + bx + c$:

1. Find the axis of symmetry: $x = -\dfrac{b}{2a}$.
2. Find the vertex: x-coordinate of vertex $= -\dfrac{b}{2a}$. To find the y-coordinate of the vertex, substitute the x-coordinate into the original equation and solve for y.
3. Find one or two additional points that are to the right of the axis of symmetry.
4. Apply the property of symmetry to find additional points to the left of the axis of symmetry.
5. Connect the plotted points with a smooth, continuous curved line.

EXAMPLE Graph the equation $y = x^2$ by finding the vertex, the axis of symmetry, and some additional points. Discuss the characteristics of the domain and range.

$y = x^2$

$y = x^2 + 0x + 0$ Standard form: $a = 1$, $b = 0$, $c = 0$.

$x = -\dfrac{b}{2a}$ Substitute values.

$x = -\dfrac{0}{2(1)}$ Simplify.

18-4 Graphing Quadratic Equations and Inequalities

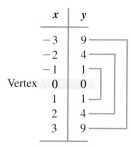

x	y
-3	9
-2	4
-1	1
Vertex 0	0
1	1
2	4
3	9

Figure 18–24

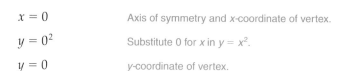

$x = 0$ Axis of symmetry and x-coordinate of vertex.

$y = 0^2$ Substitute 0 for x in $y = x^2$.

$y = 0$ y-coordinate of vertex.

Vertex: (0, 0)

Axis of symmetry: $x = 0$

Find y for $x = 1$: $y = 1^2$ Find y for $x = 2$: $y = 2^2$ Find y for $x = 3$: $y = 3^2$

$\qquad\qquad\qquad\quad y = 1$ $y = 4$ $y = 9$

Apply the principle of symmetry to complete the table. Plot the points and connect them with a smooth, continuous curve.

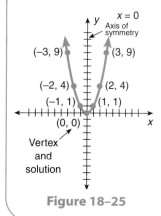

Figure 18–25

The domain is the set of all real numbers. The range is the set of real numbers greater than or equal to 0.

3 Solve a Quadratic Equation from a Graph.

As with linear equations, the real-number solutions of a quadratic equation are at the x-intercepts of the graph of the equation written as a function.

To find the solution of a quadratic function from the graph:

1. Graph the function.
2. Determine the x-coordinate of all x-intercepts.
3. If there are no x-intercepts, there are no real solutions.

If the equation $y = x^2$ is written in function notation as $f(x) = x^2$, *the solutions of the equation are the values or value of x that make the function zero. If $0 = x^2$ or $x^2 = 0$,* then $x = 0$ is the only solution of the equation. The solutions of the equation are found where the graph crosses or meets the x-axis. When we graph a quadratic equation, we can visually determine the nature of the solutions of the equation. *If the graph does not cross or meet the x-axis, the equation does not have any real-number solutions.*

The solutions can be found algebraically and used to graph the equation, or the equation can be graphed to find the solutions.

EXAMPLE Find the real-number solutions of the equations by examining the graphs.

(a) $y = x^2 - 2x - 3$ (b) $y = -3x^2$ (c) $y = x^2 + 5$

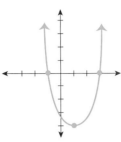

Figure 18–26

Solutions: $x = -1$
$x = 3$

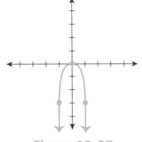

Figure 18–27

Solution: $x = 0$

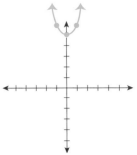

Figure 18–28

No real solutions

EXAMPLE Graph the equation $y = x^2 + x - 6$ from the example on p. 704 by using the axis of symmetry, the vertex, and the solutions. Discuss the characteristics of the domain and range.

Axis of symmetry: $x = -\dfrac{b}{2a}$ Substitute values $a = 1$ and $b = 1$.

$x = -\dfrac{1}{2(1)}$ Simplify.

$x = -\dfrac{1}{2}$ Also, x-coordinate of vertex.

Vertex: $\left(-\dfrac{1}{2}, y\right)$ Use the x-coordinate, $-\dfrac{1}{2}$, to find the y-coordinate of the vertex.

$y = x^2 + x - 6$ for $x = -\dfrac{1}{2}$ Substitute.

$y = \left(-\dfrac{1}{2}\right)^2 + \left(-\dfrac{1}{2}\right) - 6$ Raise to power.

$y = \dfrac{1}{4} - \dfrac{1}{2} - 6$ Write as equivalent fractions with a common denominator.

$y = \dfrac{1}{4} - \dfrac{2}{4} - \dfrac{24}{4}$ Combine fractions.

$\left(-\dfrac{1}{2}, -6\dfrac{1}{4}\right)$ $y = -\dfrac{25}{4}$ or $-6\dfrac{1}{4}$ y-coordinate of vertex.

Solutions: $0 = x^2 + x - 6$ Substitute $y = 0$ and solve for x.

$0 = (x + 3)(x - 2)$ Set factors equal to 0.

$x + 3 = 0$ $x - 2 = 0$ Solve each equation.

$x = -3$ $x = 2$ x-coordinates of x-intercepts.

$(-3, 0); (2, 0)$ Point notation of solutions.

Using the axis of symmetry, vertex, and two solutions, we can get a general idea of the shape of the graph.

The domain is all real numbers. The range is real numbers greater than or equal to $-6\frac{1}{4}$.

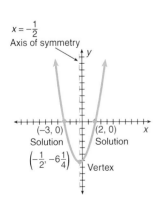

$x = -\dfrac{1}{2}$
Axis of symmetry

$(-3, 0)$ $(2, 0)$
Solution   Solution

$\left(-\dfrac{1}{2}, -6\dfrac{1}{4}\right)$ Vertex

Figure 18–29

18–4 Graphing Quadratic Equations and Inequalities

707

EXAMPLE Graph the equation $y = -x^2 + 3$ by using the axis of symmetry, vertex, and the solutions. Discuss the characteristics of the domain and range.

Axis of symmetry: $x = -\dfrac{b}{2a}$ $y = -x^2 + 0x + 3$. Substitute values. $a = -1$, $b = 0$.

$x = -\dfrac{0}{2(-1)}$ Simplify.

$x = 0$ Also, x-coordinate of vertex.

Vertex: $(0, y)$ Find y-coordinate of vertex.

$y = -0^2 + 0 + 3$ Substitute $x = 0$ and solve for y.

$y = 3$ y-coordinate of vertex.

$(0, 3)$

Solutions: $0 = -x^2 + 3$ Substitute $y = 0$ and solve for x.

$x^2 = 3$ Use square-root principle.

$x = \pm\sqrt{3}$ Exact solutions.

$x \approx \pm 1.7$ Approximate solutions.

$(1.7, 0); (-1.7, 0)$ Point notation of solutions.

Plot the points and connect them with a smooth, continuous curve. Two additional points are plotted to give a more complete view of the parabola.

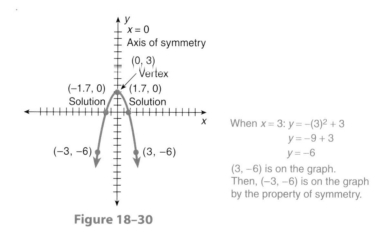

When $x = 3$: $y = -(3)^2 + 3$
$y = -9 + 3$
$y = -6$

$(3, -6)$ is on the graph.
Then, $(-3, -6)$ is on the graph by the property of symmetry.

Figure 18–30

The domain is all real numbers. The range is real numbers less than or equal to 3.

TIP!

Tips for Sketching Curves

When the coefficient of the squared letter term is negative as in the preceding example, the graph of the parabola opens downward. When the coefficient of the squared letter term is positive, the graph of the parabola opens upward. Think of the parabola as a cup or glass: positive holds water; negative spills water (Fig. 18–31).

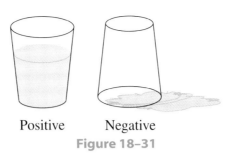

Positive   Negative

Figure 18–31

Graph Quadratic Equations and Inequalities Using a Graphing Calculator.

Quadratic equations are graphed on graphing calculators using the same procedure as for linear equations. The equation must first be solved for y. Critical values on the graph such as the vertex, x-intercepts or solutions, and the y-intercepts can be determined using calculator functions.

To find the vertex of a quadratic equation from the calculator display of the graph:

1. Graph the equation on the calculator.
2. Adjust the view of the graph by using the Window or Zoom features so that the vertex is visible. If the graph turns upward, the vertex is a low point or **minimum**. If the graph turns downward, the vertex is a high point or **maximum**.

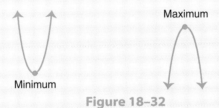

Maximum

Minimum

Figure 18–32

3. Select the appropriate minimum or maximum option from the CALC menu.
4. Move the cursor and press ENTER to mark points to the left of the vertex (left bound?), to the right of the vertex (right bound?), and near the vertex (guess?).
5. Read the coordinates of the vertex at the bottom of the display screen.

To find the real solutions of a quadratic equation from the calculator display of the graph:

1. Graph the equation on the calculator.
2. Adjust the view of the graph by using the Window or Zoom features so that one or two x-intercepts are visible or so that it can be determined that there are no x-intercepts.
3. If there are no x-intercepts, there are no real solutions.
4. If there are one or two intercepts, select the zero option from the CALC menu.
5. Move the cursor and press ENTER to mark points to the left (left bound?), right (right bound?), and near each x-intercept (guess?). Each x-intercept is one real solution.
6. Read the coordinates of each x-intercept at the bottom of the display screen.

EXAMPLE Graph the equation $y = 2x^2 - 5x - 3$ using a graphing calculator. Find the vertex and the solutions of the equation from the graph.

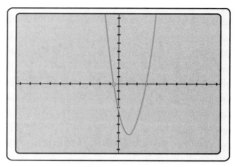

Figure 18–33

Find the vertex: The vertex is a minimum.

Vertex = (1.25, −6.125)

CALC
[2nd] [TRACE] 3

Mark left, right, and guess. $x = 1.25$ (rounded), $y = -6.125$.

Find the solutions: There are two real solutions.

CALC
[2nd] [TRACE] 2

Mark left, right, and guess for leftmost solution.
$x = -0.5, y = 0$.

Solution 1: $x = -0.5$

CALC
[2nd] [TRACE] 2

Mark left, right, and guess for rightmost solution.
$x = 3, y = 0$

Solution 2: $x = 3$

To graph a quadratic inequality:

1. Graph the boundary by substituting an equal sign for the inequality sign.
2. Select a test point that is *not* on the boundary.
3. Substitute the values from the test point into the original inequality and evaluate. The test point will be either *inside* or *outside* the boundary.
4. If the evaluated inequality makes a *true* statement, the solution set is the portion of the graph (inside or outside) that includes the test point.
5. If the evaluated inequality makes a *false* statement, the solution set is the portion of the graph (inside or outside) that does *not* include the test point.
6. The boundary *is not* included if the inequality is < or >. The boundary *is* included if the inequality is ≤ or ≥.

EXAMPLE Graph the inequality $y \geq 2x^2 - 5x - 3$.

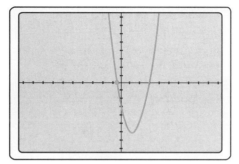

Figure 18–34

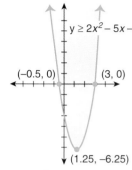

Figure 18–35

Test point: $(0, 0)$ Select a test point that is *not* on the boundary.

$y \geq 2x^2 - 5x - 3$ Substitute $x = 0$ and $y = 0$.

$0 \geq 2(0)^2 - 5(0) - 3$ Simplify.

$0 \geq 0 - 0 - 3$

$0 \geq -3$ True. Test point is included in solution set.

Sketch the graph and shade inside the graph. The boundary is included.

SELF-STUDY EXERCISES 18–4

1 Use a table of solutions to graph the quadratic equations. Discuss the characteristics of the domain and range.

1. $y = x^2$

2. $y = 3x^2$

3. $y = \dfrac{1}{3}x^2$

4. $y = -4x^2$

5. $y = -\dfrac{1}{4}x^2$

6. $y = x^2 - 4$

7. $y = x^2 + 4$

8. $y = x^2 - 6x + 9$

9. $y = -x^2 + 6x - 9$

2 Graph the equations using the vertex, solutions, and one other point on the graph. Discuss the characteristics of the domain and range.

10. $y = x^2 - 4x + 4$

11. $y = x^2 + 4x - 2$

12. $y = 2x^2 + 10x + 8$

13. $y = -3x^2 - 6x + 9$

14. $y = -3x^2 - 6x - 6$

15. $y = \dfrac{1}{2}x^2 - 6x + 3$

3 Find the real-number solutions of the equations by writing them as functions and graphing the equations.

16. $x^2 - 5x + 6 = 0$

17. $4x^2 = 0$

18. $3x^2 = -5$

19. The revenue R for a leather wallet is based on the price P of the item and is given by the formula $R = -4p^2 + 36p$. What prices generate \$0 revenue?

20. The distance y that an object falls in a vacuum because of gravity in t seconds after it is released is given by the formula $d = 4.9t^2$. How much time is elapsed when $d = 0$?

18–4 Graphing Quadratic Equations and Inequalities

Graph using a graphing calculator. Adjust the window if necessary to show the vertex of the parabola.

21. $y = 3x^2 + 5x - 2$ **22.** $y = (2x - 3)(x - 1)$ **23.** $y = 2x^2 - 9x - 5$ **24.** $y = -2x^2 + 9x + 5$
25. $y < 3x^2 + 5x - 2$ **26.** $y > (2x - 3)(x - 1)$ **27.** $y \leq 2x^2 - 9x - 5$ **28.** $y \geq -2x^2 + 9x + 5$

29. The revenue range for a leather wallet is based on the price of the item given by the formula $R \leq -4x^2 + 52x$, where R is the revenue and x is the price. Graph the revenue function and determine the greatest possible revenue.

30. The formula $d \geq 4.9t^2$ gives the distance d an object falls in t seconds. Graph the solution set to show the possible range of solutions.

18–5 | Graphing Other Nonlinear Equations

Learning Outcomes

1 Graph nonlinear equations using the table-of-solutions method.
2 Distinguish between a function and a relation using the vertical line test.
3 Find the domain and range of a relation from a graph.

1 Graph Nonlinear Equations Using the Table-of-Solutions Method.

Any nonlinear equation can be graphed using the table-of-solutions method. It is important to include enough points to get a complete view of the graph.

EXAMPLE Prepare a table of solutions and graph the function $f(x) = x^3$ using integral values of x between -2 and 2, inclusive. Discuss the characteristics of the domain and range.

x	$f(x)$
-2	-8
-1	-1
0	0
1	1
2	8

$f(-2) = (-2)^3$ $f(-1) = (-1)^3$ $f(0) = 0^3$
$\mathbf{f(-2) = -8}$ $\mathbf{f(-1) = -1}$ $\mathbf{f(0) = 0}$

$f(1) = 1^3$ $f(2) = 2^3$
$f(1) = 1$ $f(2) = 8$

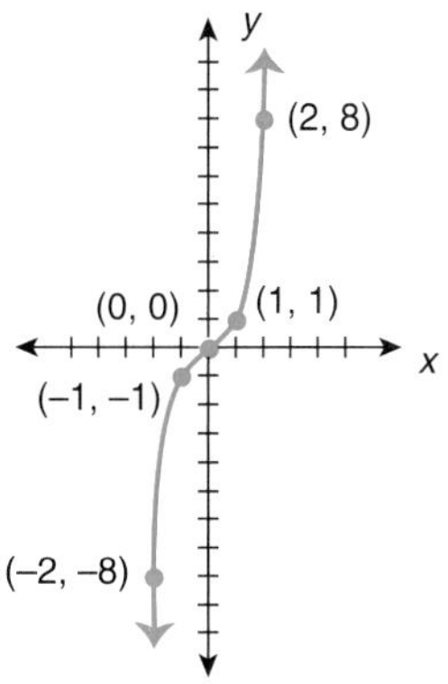

Plot the points indicated in the table of solutions and connect the points with a smooth, continuous curve.

The domain and range are all real numbers.

Figure 18–36

EXAMPLE Prepare a table of solutions and graph the function $f(x) = 2^x$ using integral values of x between -3 and 3, inclusive. Discuss the characteristics of the domain and range.

     Chapter 18 / Graphing Functions

$f(x) = 2^x$

x	$f(x)$
−3	0.125
−2	0.25
−1	0.5
0	1
1	2
2	4
3	8

$f(-3) = 2^{-3}$
$f(-3) = \dfrac{1}{2^3}$
$f(-3) = \dfrac{1}{8}$, or 0.125

$f(-2) = 2^{-2}$
$f(-2) = \dfrac{1}{2^2}$
$f(-2) = \dfrac{1}{4}$, or 0.25

$f(-1) = 2^{-1}$
$f(-1) = \dfrac{1}{2^1}$
$f(-1) = \dfrac{1}{2}$, or 0.5

$f(0) = 2^0$
$f(0) = 1$

$f(1) = 2^1$
$f(1) = 2$

$f(2) = 2^2$
$f(2) = 4$

$f(3) = 2^3$
$f(3) = 8$

Plot the points indicated in the table of solutions and connect the points with a smooth, continuous curve.

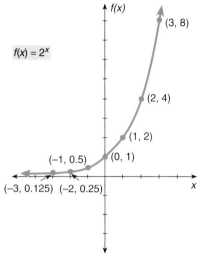

The domain is all real numbers. The range is real numbers greater than 0.

Figure 18–37

EXAMPLE Prepare a table of solutions and graph the function $f(x) = \ln 4x$ using the values of x: 0.125, 0.25, 0.5, 1, 2, 3, 4. Discuss the characteristics of the domain and range.

$f(x) = \ln 4x$

x	$f(x)$
0.125	−0.7
0.25	0
0.5	0.7
1	1.4
2	2.1
3	2.5
4	2.8

Using a calculator:

$f(0.125) = \ln 4(0.125)$
$f(0.125) = \ln 0.5$
$f(0.125) = -0.6931471806$

$f(0.25) = \ln 4(0.25)$
$f(0.25) = \ln 1$
$f(0.25) = 0$

$f(0.5) = \ln 4(0.5)$
$f(0.5) = \ln 2$
$f(0.5) = 0.6931471806$

$f(1) = \ln 4(1)$
$f(1) = \ln 4$
$f(1) = 1.386294361$

$f(2) = \ln 4(2)$
$f(2) = \ln 8$
$f(2) = 2.079441542$

$f(3) = \ln 4(3)$
$f(3) = \ln 12$
$f(3) = 2.48490665$

$f(4) = \ln 4(4)$
$f(4) = \ln 16$
$f(4) = 2.772588722$

Plot the points indicated in the table of solutions and connect the points with a smooth, continuous curve.

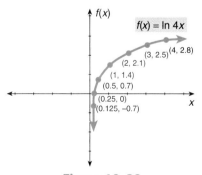

The domain is real numbers greater than 0. The range is all real numbers.

Figure 18–38

As you can see from these examples, all functions can be expressed graphically by using a table of values.

2 Distinguish Between a Function and a Relation Using the Vertical Line Test.

Equations in two variables describe a *relation* between the two variables. This connection between the two variables defines a set of ordered pairs. The equations in two variables that we have examined have been functions. A *function* is a relation that has exactly one value of the dependent variable (y-value) for every value of the independent variable (x-value). In other words, all functions are also relations, but not all relations are functions. A relation that is not a function will have at least one case where a value of the independent variable (x-value) corresponds to *more than one* value of the dependent variable (y-value).

To distinguish between a function and a relation using the vertical line test:

1. Graph the relation on a coordinate system.
2. Apply the vertical line test.
   (a) If a vertical line can be drawn so that it intersects a graph at more than one point, then the graph is the graph of a relation that is not a function.
   (b) If no vertical line can be drawn so that it intersects a graph at more than one point, then the graph is the graph of a function.

EXAMPLE Determine which graphs are graphs of functions.

(a) 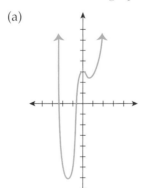 (b) 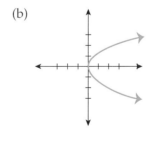 (c)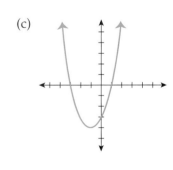

Figure 18–39

(a) **The graph is a graph of a function.** No vertical line can be drawn so that it intersects the graph at more than one point.
(b) **The graph is not a graph of a function.** For all values of x greater than zero, a vertical line will intersect the graph at two points.

(c) **The graph is a graph of a function.** No vertical line can be drawn so that it intersects the graph at more than one point.

3 Find the Domain and Range of a Relation from a Graph.

The *domain* of a relation is the set of all values that are appropriate for the independent variable and is the first component of the ordered pairs of the relation. The *range* of a relation is the set of all resulting values of the dependent variable that correspond to the independent variable and is the second component of the ordered pairs of the relation.

TIP!

Relation Versus Function

- For every value of the domain of a relation there is at least one corresponding value of the range.
- For every value of the domain of a function there is exactly one corresponding value of the range.

$x = y^2$

x	y
4	-2
1	-1
0	0
1	1
4	2

Domain   Range

0
*1
*4

-2
-1
0
1
2

*x-values with two different y-values
$x = y^2$ **is a relation, but not a function.**

$y = x^2$

x	y
-2	4
-1	1
0	0
1	1
2	4

Domain   Range

-2
-1
0
1
2

0
1

4

Each x-value only has one y-value.
Two different x-values can have the same y-value.
$y = x^2$ **is a function.**

To find the domain and range of a relation from a graph:

1. Graph the relation on a coordinate system.
2. The domain is all the values on the graph for the independent variable (x-values).
3. The range is all the values on the graph for the dependent variable (y-values).

EXAMPLE Determine the domain and range of each relation.

(a)

Range

Domain

(b)

(0, 2)

(−3, 0)   (3, 0)

Range

(0, −2)

Domain

(c)

(5, 6)

Range

(−3, −2)

(2, −4)

Domain

Figure 18–40

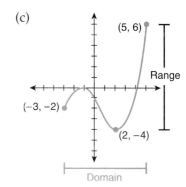

(a) Domain: $(-\infty, \infty)$
Range: $[0, \infty)$

(b) Domain: $[-3, 3]$
Range: $[-2, 2]$

(c) Domain: $[-3, 5]$
Range: $[-4, 6]$

SELF-STUDY EXERCISES 18–5

1 Graph the equations using the table-of-solutions method. Discuss the characteristics of the domain and range.

1. $y = 2x^3$

2. $y = \frac{1}{2}x^3$

3. $y = x^3 - x^2 - 4x + 4$

4. $y = x^3 + 2$

5. $y = 3^x$

6. $y = -3^x$

7. $y = 2^x - 3$

8. $y = -2^x - 3$

9. $y = \ln 2x$

10. $y = -\ln 2x$

2 Use the vertical line test to determine which are graphs of functions.

11. $y = x^3 - x^2 - 8x + 8$

12. $y = 3x^2 - 2x - 5$

13. $x = y^3 - y^2 - 3y - 3$

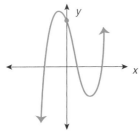

Figure 18–41

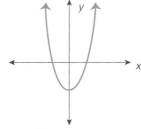

Figure 18–42

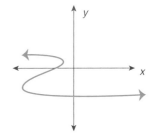

Figure 18–43

14. $x = y^2 - y - 5$

15. $y = x^4 - 3x^3 + x^2$

16. $x = 2y + 3$

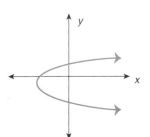

Figure 18–44

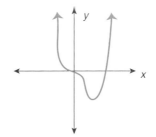

Figure 18–45

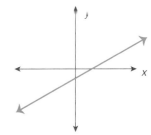

Figure 18–46

3 Find the domain and range of the relations from the graph.

17.

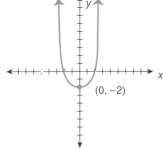

Figure 18–47

18.

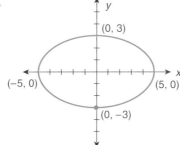

Figure 18–48

19.

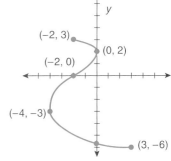

Figure 18–49

20.

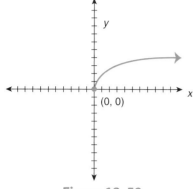

Figure 18–50

Section 18–1

Represent the solutions of the functions in a table of solutions and on a graph.

1. $f(x) = 2x - 3$
2. $f(x) = -4x + 1$
3. $f(x) = 3x$
4. $f(x) = 4x$
5. $f(x) = -3x$
6. $f(x) = -4x$
7. $f(x) = 2x + 1$
8. $f(x) = 2x + 5$
9. $f(x) = 4x - 2$
10. $f(x) = -5x + 1$
11. $f(x) = \frac{1}{2}x - 2$
12. $f(x) = -\frac{1}{4}x + 1$

Which of the coordinate pairs are solutions for the equation $2x - 3y = 12$?

13. $(-2, -3)$
14. $(1, -3)$
15. $(3, -2)$
16. $(0, 4)$
17. $(6, 0)$
18. In the equation $2x + y = 8$, find x when $y = 10$.
19. Find y in the equation $x - 3y = 5$ when $x = 8$.
20. Find y if $x = 7$ in the equation $3x - 4 = -2$.

Section 18–2

Graph the equations using the intercepts procedure.

21. $x = -4y - 1$
22. $x + y = -4$
23. $3x - y = 1$
24. $x = -4y$
25. $\frac{1}{2}x + \frac{1}{3}y = 1$
26. $2x + 3y = 6$

Graph using the slope and y-intercept procedure.

27. $y = 5x - 2$
28. $y = -x$
29. $y = -3x - 1$
30. $y = \frac{1}{2}x + 3$
31. $x - y = 4$
32. $2y + 4 = -3$
33. $x - 2y = -1$
34. $x + y = 5$
35. $y - 2x = -2$
36. $y + 3x = 4$

Graph the equations using a graphing calculator.

37. $y = 0.5x - 3$
38. $y = 2.1x + 0.5$
39. $y = 20x - 15$
40. $y = -30x + 10$

41. The function $f(x) = 5x + 8{,}000$ is used to express the cost of manufacturing a widget, where $8{,}000$ is the fixed cost of constructing molds and $5 is the cost of materials to make each widget. What is the cost of manufacturing 10,000 widgets?

42. A salesperson earns a salary of $200 weekly plus $12.50 commission for each photo sitting. Weekly income is expressed by the function $f(x) = 12.50x + 200$, where x represents the number of sittings in a week. Find the weekly income if 28 sittings are sold.

Write each equation as a function and find the solution graphically.

43. $x + 2 = 8$
44. $3x - 7 = 1$
45. $2x + 1 = 5x + 7$
46. $-14 = 2(x - 7)$
47. $5(x + 2) = 3(x + 4)$
48. $2x + 1 = 7x + 2 - 8$

Section 18–3

Graph the linear inequalities using test points and verify with a graphing calculator.

49. $4x + y < 2$
50. $x + y > 6$
51. $3x + y \le 2$
52. $x + 3y > 4$
53. $x - 2y < 8$
54. $3x + 2y \ge 4$
55. $y \ge 3x - 2$
56. $y \le -2x + 1$
57. $y > \frac{2}{3}x - 2$

Graph the quadratic equations. Discuss the characteristics of the domain and range.

58. $y = x^2 - 1$ **59.** $y = -x^2 - 1$ **60.** $y = x^2 + 3x - 10$

61. $y = x^2 - 6x + 8$ **62.** $y = x^2 - 2x + 1$ **63.** $y = -x^2 + 2x - 1$

64. $y = x^2 - 4x + 4$ **65.** $y = -x^2 + 4x - 4$

Find the real-number solutions of the equations by writing them as functions and graphing.

66. $x^2 + x - 12 = 0$ **67.** $x^2 - 4x - 12 = 0$

68. $x^2 - 5x = 14$ **69.** $x^2 + 8x = -16$

Use a calculator to graph the following. Adjust the window to display the vertex.

70. $y < x^2 - 2x + 1$ **71.** $y \geq -2x^2$ **72.** $y \leq \frac{1}{2}x^2 - 3$

Graph the equations using the table-of-solutions method. Discuss the characteristics of the domain and range.

73. $y = 5x^3$ **74.** $y = x^3 - x^2 - 8x + 1$ **75.** $y = 5^x$

76. $y = 5^x + 2$ **77.** $y = \ln 4x$ **78.** $y = -\ln 4x$

Use the vertical line test to determine which are graphs of functions.

79. $x = -y^3 + 2y^2 + 2y + 3$ **80.** $x = 6 - 3y$

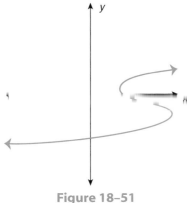

Figure 18–51

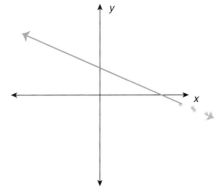

Figure 18–52

81. $\dfrac{x^2}{25} + \dfrac{y^2}{4} = 1$ **82.** $y = x^4 - 3x^3 + x$

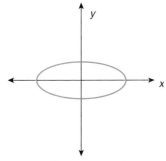

.**Figure 18–53**

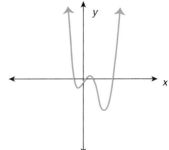

Figure 18–54

Find the domain and range of the relations from the graph.

83.

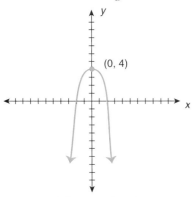

Figure 18–55

84.

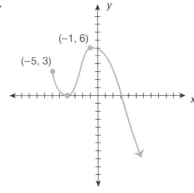

Figure 18–56

85.

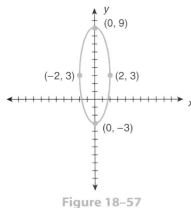

Figure 18–57

86.

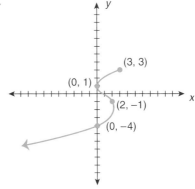

Figure 18–58

CHALLENGE PROBLEMS

87. A 1-gal can of indoor house paint is advertised to cover 400 ft^2 of wall surface.
   (a) Make a table to show the amount of wall surface area that can be covered by 1, 2, 3, ... , or 10 gallons of paint.
   (c) Use the graph to decide how many 1-gal cans of paint it would take to cover 5,500 ft^2 of wall surface.
   (e) The sales tax rate is 8.25%. Calculate the total cost of the paint.

   (b) Graph these data.

   (d) The paint being used for this job can be purchased for $19.95 a gallon. Find the cost of the paint for the 5,500 ft^2 of wall surface.

88. Using column 1 of the following table for the horizontal scale and one of the other columns for the vertical scale, which graph in Fig. 18–59 illustrates the function pattern for the column you selected?

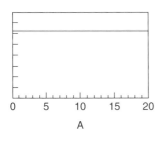
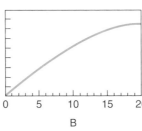
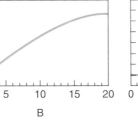
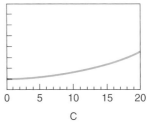

Figure 18–59

Challenge Problems

Payment Number	Payment Amount	Applied to Interest	Applied to Principal	Balance Owed	Total Interest
1	611.09	580.00	31.09	69,568.91	580.00
2	611.09	579.74	31.35	69,537.56	1,159.74
3	611.09	579.48	31.61	69,505.95	1,739.22
4	611.09	579.22	31.87	69,474.08	2,318.44
5	611.09	578.95	32.14	69,441.94	2,897.39
6	611.09	578.68	32.41	69,409.53	3,476.07
7	611.09	578.41	32.68	69,376.85	4,054.48
8	611.09	578.14	32.95	69,343.90	4,632.62
9	611.09	577.87	33.22	69,310.68	5,210.49
10	611.09	577.59	33.50	69,277.18	5,788.08

89. Sketch graphs to describe the pattern shown by the numbers in the two columns not illustrated by graphs A, B, and C in Fig. 18–59.

CONCEPTS ANALYSIS

1. What do we mean when we say a point is represented by an ordered pair of numbers?

2. What does the graph of an equation represent?

3. Explain the procedure for plotting the point $(5, -2)$ on a grid representing the rectangular coordinate system.

4. How can you tell from an equation if the graph will be a straight line?

5. Discuss the similarities and differences in the table-of-solutions method, the intercepts method, and the slope and y-intercept method for graphing a linear equation.

6. Why is it generally helpful to solve an equation for y before using the table-of-solutions method for graphing?

7. How is the graph of a linear inequality different from the graph of a linear equation?

8. When graphing linear inequalities, how do you determine which portion of the graph to shade?

9. How is the graph of a quadratic equation different from the graph of a linear equation?

10. What do *axis of symmetry* and *vertex* refer to on the graph of a quadratic equation that represents a parabola?

CHAPTER SUMMARY

Learning Outcomes

What to Remember with Examples

Section 18–1

1 Represent the solutions of a linear function in a table of solutions (pp. 684–686).

To prepare a table of solutions: (1) select any appropriate value from the domain of the independent variable, (2) substitute the selected value for the independent variable and solve the resulting equation, (3) repeat Step 1 and Step 2 until the desired number of solutions are obtained.

Make a table of solutions for the function $f(x) = -x + 3$. The domain is all real numbers.

$f(-2) = -(-2) + 3$ $f(0) = -0 + 3$ $f(2) = -2 + 3$
$f(-2) = 2 + 3$ $f(0) = 3$ $f(2) = 1$
$f(-2) = 5$

x	y
-2	5
0	3
2	1

Each solution is an ordered pair that can be written in point notation: $(-2, 5)$, $(0, 3)$, and $(2, 1)$. The range is all real numbers.

	2	Represent the solutions of a linear function on a graph (pp. 686–687).

2 Represent the solutions of a linear function on a graph (pp. 686–687).

To graph a function using a table of solutions: (1) Prepare a table of solutions. (2) Plot the points on a rectangular coordinate system. (3) Connect the points with a straight line.

Prepare a table of values and graph $f(x) = -x + 3$.

x	$f(x)$
-2	5
0	3
2	1

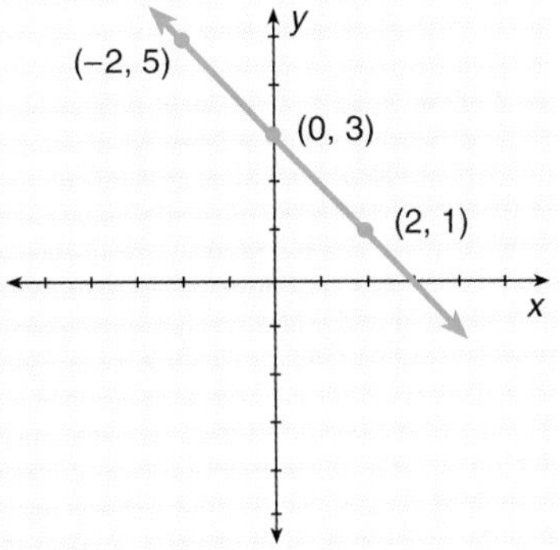

(−2, 5)

(0, 3)

(2, 1)

Figure 18–60

3 Check the solution of an equation (or function) with two variables (pp. 687–688).

To check the solution of an equation in two variables: (1) Substitute the value for x in each place it occurs and substitute the value for y in each place it occurs. (2) Simplify each side of the equation using the order of operations. (3) The solution makes a true statement; that is, both sides of the equation will be equal to the same number.

A solution for the equation $3x - y = 1$ is (1, 2). Check to verify the solution.

$3x - y = 1$ Substitute values.
$3(1) - 2 = 1$ Evaluate.
$3 - 2 = 1$
$1 = 1$ True. Solution checks.

4 Find a specific solution of an equation (or function) with two variables when given the value of one variable (pp. 688–689).

To evaluate an equation when the value of one variable is given: (1) Substitute the given value of the variable in each place that variable occurs in the equation. (2) Solve the equation for the other variable.

Evaluate the equation $4x - y = 6$ for x, if $y = 14$.

$4x - y = 6$ Substitute 14 for y.
$4x - 14 = 6$ Solve for x.
$4x = 20$
$x = 5$

Solution: (5, 14)

Section 18–2

1 Graph linear equations using intercepts (pp. 691–692).

1. Find the x-intercept by letting $y = 0$ and solving for x.
2. Find the y-intercept by letting $x = 0$ and solving for y.
3. Plot the two points and graph the equation.
4. Find a third point on the graph as a check point.

Graph the equation $y = 2x - 1$ by using the intercepts method.

x-intercept: $0 = 2x - 1$ Substitute $y = 0$. y-intercept: $y = 2(0) - 1$ Substitute $x = 0$.
$1 = 2x$ Solve for x. $y = 0 - 1$ Solve for y.
$\dfrac{1}{2} = x$ $y = -1$

$\left(\dfrac{1}{2}, 0 \right)$ $(0, -1)$

Chapter Summary

Plot the two points: then draw the graph.

Check point: For $x = 3$

$y = 2x - 1$
$y = 2(3) - 1$
$y = 6 - 1$
$y = 5$
$(3, 5)$

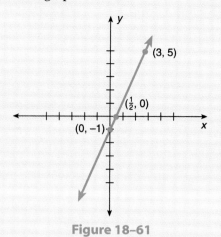

Figure 18–61

2 Graph linear equations using the slope and y-intercept (pp. 693–697).

(1) Write the equation in the form $y = mx + b$. (2) The slope is m. In fraction form the numerator of m is the vertical movement and the denominator of m is the horizontal movement. (3) The y-coordinate of the y-intercept is b.

To graph using the slope-intercept method: (1) Locate the y-intercept or b by counting vertically along the y-axis. (2) From this point, count the slope (vertical, then horizontal) and locate the second point. (3) Draw the line through the two points.

Use the slope-intercept method to graph $y = 2x - 1$. The y-intercept is -1, so count down 1 from the origin. The coordinates of this point are $(0, -1)$. From this point, move $+2$ vertically and $+1$ horizontally. The coordinates of the second point are $(1, 1)$. Connect the two points.

y-intercept $= (0, -1)$

$m = 2 = \dfrac{2}{1}$ or $\dfrac{-2}{-1}$

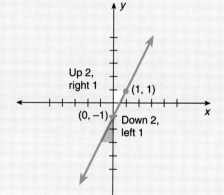

Figure 18–62

3 Graph linear equations using a graphing calculator (pp. 697–698).

To graph a linear equation using a graphing caculator: (1) Solve the equation for y. (2) Press the function key $\boxed{Y=}$; then enter the right side of the equation. (3) Press the graph key to show the graph on the screen. Adjust the view with the Window or Zoom keys if appropriate.

Graph the equation $y = 2x - 1$.

Using a TI-83 Plus:

$\boxed{Y=}$ $\boxed{\text{CLEAR}}$ 2 $\boxed{X, T, \theta, n}$ $\boxed{-}$ 1

$\boxed{\text{GRAPH}}$

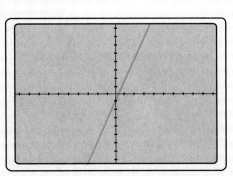

Figure 18–63

4 Solve a linear equation using a graph (pp. 698–700).

To write an equation in one variable as a function: (1) Rewrite the equation so that all terms are on one side of the equation. (2) Write the equation in function notation using $f(x) =$ the nonzero side of the equation. (3) Evaluate the function for at least two values and graph. The solution is the x-coordinate of the point where the line crosses the x-axis.

Write $2x + 3 = 6$ as a function, graph the function, and find the solution from the graph.

$$2x + 3 = 6$$
$$2x + 3 - 6 = 0$$
$$2x - 3 = 0$$

$$f(x) = 2x - 3$$
$$f(0) = 2(0) - 3$$
$$f(0) = -3$$

$$f(1) = 2(1) - 3$$
$$f(1) = -1$$

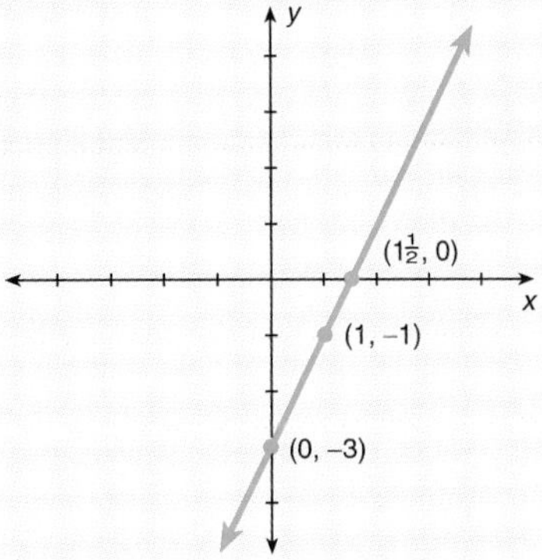

(graph with points labeled $(1\frac{1}{2}, 0)$, $(1, -1)$, $(0, -3)$)

Figure 18–64

Solution: $x = \frac{3}{2}$ or $1\frac{1}{2}$, the x-coordinate of the point where the line crosses the x-axis.

Section 18–3

1 Graph linear inequalities with two variables using test points (pp. 701–703).

To graph a linear inequality: (1) Graph the corresponding equation. (2) Select a point on one side of the boundary line and substitute each coordinate in place of the appropriate letter in the inequality. (3) If the statement is true, shade the side of the boundary that contains the point. (4) If false, shade the other side of the boundary. For $>$ or $<$ inequalities, make the boundary line dashed. For $\geq$ or $\leq$ inequalities, make the boundary line solid.

Graph $y \geq 2x - 1$.
First graph $y = 2x - 1$. Make the boundary line solid.

Chapter Summary

Test (1, 3):

$3 \geq 2(1) - 1$ Substitute $x = 1$ and $y = 3$.
$3 \geq 2 - 1$ Simplify.
$3 \geq 1$ True.

Shade the side of the boundary line that contains the point since the coordinates of the test point made a true statement in the inequality.

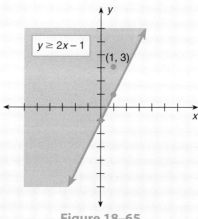

$y \geq 2x - 1$
(1, 3)

Figure 18–65

Section 18–4

1 Graph quadratic equations using the table-of-solutions method (pp. 704–705).

(1) Prepare a table of solutions of approximately five ordered pairs. (2) Plot the points on a rectangular coordinate system. (3) Connect the points with a smooth, continuous curve.

Graph $y = x^2 - 6x + 8$ using a table of solutions. Start with three values of x, -2, 0, and 2.

at $x = -2$:
$y = (-2)^2 - 6(-2) + 8$
$y = 4 + 12 + 8$
$y = 24$

at $x = 0$:
$y = 0^2 - 6(0) + 8$
$y = 0 - 0 + 8$
$y = 8$

at $x = 2$:
$y = 2^2 - 6(2) + 8$
$y = 4 - 12 + 8$
$y = 0$

Try two more, $x = 4$ and $x = 6$.

at $x = 4$
$y = 4^2 - 6(4) + 8$
$y = 16 - 24 + 8$
$y = 0$

at $x = 6$
$y = 6^2 - 6(6) + 8$
$y = 36 - 36 + 8$
$y = 8$

Finally, see what happens between $x = 2$ and $x = 4$.

at $x = 3$
$y = 3^2 - 6(3) + 8$
$y = 9 - 18 + 8$
$y = -1$

x	y
-2	24
0	8
2	0
3	-1
4	0

(0, 8)   (6, 8)
(2, 0) (4, 0)
(3, −1)

Figure 18–66

2 Graph quadratic equations by examining properties (pp. 705–706).	(1) Determine the x-coordinate of the vertex of the parabola by finding the value of $-\frac{b}{2a}$. Use the x-coordinate of the vertex to write an equation $x = -\frac{b}{2a}$, which is the axis of symmetry. (2) Substitute the value as the x-coordinate into the equation to find the corresponding y-value. (3) Replace y with zero and solve the resulting equation for x. Plot these points if any, which are the solutions for the equation. (4) Use symmetry to find additional points. (5) Connect the plotted points and the vertex with a smooth, continuous curved line.

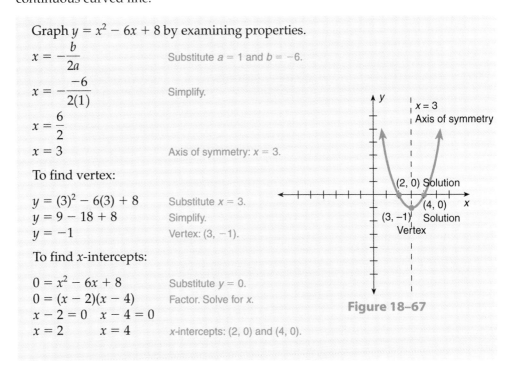

Graph $y = x^2 - 6x + 8$ by examining properties.

$x = -\dfrac{b}{2a}$ Substitute $a = 1$ and $b = -6$.

$x = -\dfrac{-6}{2(1)}$ Simplify.

$x = \dfrac{6}{2}$

$x = 3$ Axis of symmetry: $x = 3$.

To find vertex:

$y = (3)^2 - 6(3) + 8$ Substitute $x = 3$.
$y = 9 - 18 + 8$ Simplify.
$y = -1$ Vertex: $(3, -1)$.

To find x-intercepts:

$0 = x^2 - 6x + 8$ Substitute $y = 0$.
$0 = (x - 2)(x - 4)$ Factor. Solve for x.
$x - 2 = 0 \quad x - 4 = 0$
$x = 2 \qquad x = 4$ x-intercepts: $(2, 0)$ and $(4, 0)$.

Figure 18–67

3 Solve a quadratic equation from a graph (pp. 706–709).	(1) Find the x-intercepts on the graph. (2) The solutions of the equation are the x-coordinates of the x-intercepts.

Find the solutions for the equation $y = x^2 - 6x + 8$ from the graph (Fig. 18–68).

The x-intercepts are $(2, 0)$ and $(4, 0)$.
The solutions are $x = 2$ and $x = 4$.

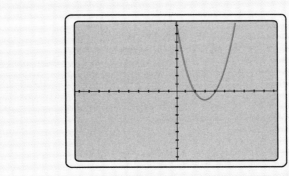

Figure 18–68

4 Graph quadratic equations and inequalities using a graphing calculator (pp. 709–711).	To graph quadratic equations using a graphing calculator: (1) Clear the graphing screen. (2) Enter the equation and press the graph function. To graph quadratic inequalities. (1) Graph the boundary by substituting an equal sign for the inequality sign. (2) Use a test point to determine if the solution set is inside or outside the boundary.

Graph $y = x^2 - 6x + 8$.

Clear the graphing screen.
Enter equation.
Graph.
Change viewing window if appropriate.

Graph $y \le x^2 - 6x + 8$.

Graph the boundary $y = x^2 - 6x + 8$.
Select a test point such as $(0, 0)$.
$y \le x^2 - 6x + 8$
$0 \le 0^2 - 6(0) + 8$
$0 \le 8$ True.

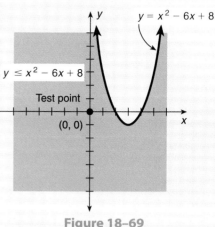

$y \le x^2 - 6x + 8$

Test point

$(0, 0)$

$y = x^2 - 6x + 8$

Figure 18–69

Shade the portion of the graph that includes the test point. The boundary is included in the solution set.

Section 18–5

1 Graph nonlinear equations using the table-of-solutions method (pp. 712–714).

Prepare a table of solutions with enough points to get a complete view of the graph.

Prepare a table of solutions and graph the function $f(x) = 3^x$.

x	y	
-2	$\frac{1}{9}$	$f(-2) = 3^{-2}$ or $\frac{1}{9}$
-1	$\frac{1}{3}$	$f(-1) = 3^{-1}$ or $\frac{1}{3}$
0	1	$f(0) = 3^0$ or 1
1	3	$f(1) = 3^1$ or 3
2	9	$f(2) = 3^2$ or 9

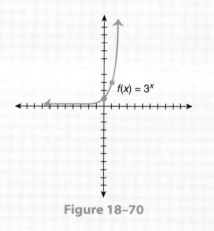

$f(x) = 3^x$

Figure 18–70

2 Distinguish between a function and a relation using the vertical line test (pp. 714–715).

1. Graph the relation on a coordinate system.
2. Apply the vertical line test.
   (a) If a vertical line can be drawn so that it intersects a graph at more than one point, then the graph is the graph of a relation that is not a function.
   (b) If no vertical line can be drawn so that it intersects a graph in more than one point, then the graph is the graph of a function.

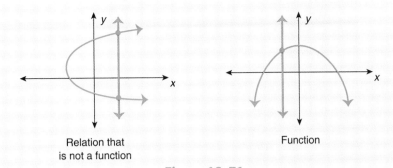

| Relation that is not a function | Function |

Figure 18–71

3 Find the domain and range of a relation from a graph (p. 715).

1. Graph the relation on a coordinate system.
2. The domain is all the values on the graph for the independent variable (x-values).
3. The range is all the values on the graph for the dependent variable (y-values).

Determine the domain and range.

(a)

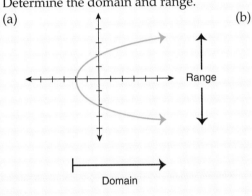

Domain: $[-2, \infty)$
Range: $(-\infty, \infty)$

(b)

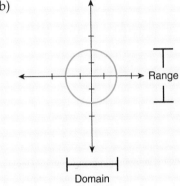

Domain: $[-2, 2]$
Range: $[-2, 2]$

Figure 18–72

CHAPTER TRIAL TEST

Make a table of values to represent the solutions to the following equations and show the solutions on a graph.

1. $f(x) = \dfrac{1}{2}x$

2. $f(x) = \dfrac{1}{2}x + 1$

3. $f(x) = 2x - 4$

4. $f(x) = 5 - x$

Write as functions and find the solution graphically.

5. $3x + 2 = 5$

6. $2(x + 1) = 3x$

Write the specific solution for each equation in ordered-pair form.

7. $x + y = 7$, if $x = 2$

8. $2x - y = 1$, if $y = -3$

9. Plot the graph for the relationship between horse-power (hp) and revolutions per minute (rpm) in a test engine. Use the same intervals as those in the data table.

10. A cost function is known to be $f(x) = 10x + 250$, where x is the number of units produced and \$250 is the fixed cost. Find the total cost of producing 5,000 units.

x-axis (hp)	y-axis (rpm)
30	500
45	1,000
60	1,500
75	2,000

Chapter Trial Test

121

11. A cake bakery estimates the profit function to be $f(x) = 3x - 800$, where x is the number of cakes produced in a month and \$800 is cost of overhead such as utilities. Find the profit if 8,000 cakes are produced.

Graph using the intercepts method.

12. $2x - 3y = 6$ **13.** $x + 2y = 8$

Graph using the slope-intercept method.

14. $y = -3x + 1$ **15.** $2x + y = -3$ **16.** $x + 2y = 1$ **17.** $y = x - 5$

Graph the linear inequalities.

18. $x + 2y < 4$ **19.** $2x - y \leq 2$ **20.** $y \leq 2x + 2$ **21.** $x + y < 1$

Graph the quadratic equations by examining properties. Show the axis of symmetry, vertex, and solutions.

22. $y = x^2 + 2$ **23.** $y = x^2 + 2x + 1$ **24.** $y = x^2 + 4x + 4$

Graph the quadratic inequalities.

25. $y \leq x^2 - 6x + 8$ **26.** $y > 2x^2$ **27.** $y < -\dfrac{1}{2}x^2$

Use the vertical line test to identify the graphs as functions or relations. State the domain and range.

28. $y = x^5 + 2x^4 - x^3$ **29.** $x = y^3 - 2y^2 - y$

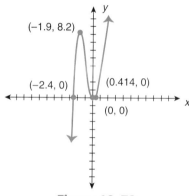

Figure 18–73

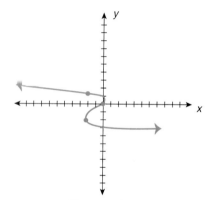

Figure 18–74

30. $\dfrac{x^2}{36} + \dfrac{x^2}{25} = 1$

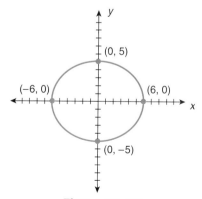

Figure 18–75

Slope and Distance

19

The concept of the rate of change is important in real-world applications. This concept is often referred to as *slope* and is used when we consider the slant of a roof or the grade of a roadway. We use the rate of change in applications, like changes in temperature, in the quality of a product, or in sales.

19-1 | Slope

Learning Outcomes

1 Calculate the slope of a line, given two points on the line.

2 Determine the slope of a horizontal or vertical line.

1 **Calculate the Slope of a Line, Given Two Points on the Line.**

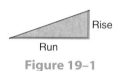

Figure 19-1

The *slope* of a straight line is the rate of change of the line or the ratio of the vertical rise of a line to the horizontal run of the line (see Fig. 19–1).

$$\text{Slope} = \frac{\text{Rise}}{\text{Run}} \quad \text{or} \quad \frac{\text{Vertical change}}{\text{Horizontal change}}$$

To find the slope of a line from two given points on the line:

1. Designate one point as point 1 with coordinates (x_1, y_1). Designate the other point as point 2 with coordinates (x_2, y_2).
2. Calculate the change (*difference*) in the y-coordinates to find the vertical *rise* $(y_2 - y_1)$ and in the x-coordinates to find the horizontal *run* $(x_2 - x_1)$.
3. Write a ratio of the rise to the run and reduce the ratio to lowest terms.

Symbolically, if P_1 and P_2 are any two points on the line,

$$\text{Slope} = \frac{\Delta y}{\Delta x} = \frac{y_2 - y_1}{x_2 - x_1}$$

where $P_1 = (x_1, y_1)$ and $P_2 = (x_2, y_2)$.

The Greek capital letter delta (Δ) is used to indicate a change and to write the slope definition symbolically.

EXAMPLE Find the slope of a line if the points $(2, -1)$ and $(5, 3)$ are on the line (see Fig. 19–2).

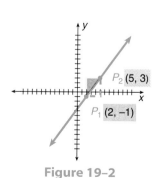

Figure 19-2

Chapter 19 / Slope and Distance

Identify (2, −1) as point 1 (P_1) and (5, 3) as point 2 (P_2). The coordinates of P_1 are (x_1, y_1) and the coordinates of P_2 are (x_2, y_2). Write the slope definition symbolically and substitute known values.

$$\text{Change in } y = \Delta y = \text{rise} = y_2 - y_1 = 3 - (-1) = 4$$
$$\text{Change in } x = \Delta x = \text{run} = x_2 - x_1 = 5 - 2 = 3$$
$$\text{Slope} = \frac{\Delta y}{\Delta x} = \frac{\text{rise}}{\text{run}} = \frac{y_2 - y_1}{x_2 - x_1} = \frac{3 - (-1)}{5 - 2} = \frac{4}{3}$$

The slope of the line through points (2, −1) and (5, 3) is $\frac{4}{3}$.

TIP!

What Points on a Line Are Used? In What Order Are They Used?

Any line has an infinite number of points. *Any* two points on the line can be used to find the slope. Also, the designation of P_1 and P_2 is not critical. Let's find the slope of the line in the preceding example by designating P_1 as (5, 3) and P_2 as (2, −1).

$$\Delta y = \text{Rise} = y_2 - y_1 = -1 - 3 = -4$$
$$\Delta y = \text{Run} = x_2 - x_1 = 2 - 5 = -3$$
$$\frac{\Delta y}{\Delta x} = \frac{\text{Rise}}{\text{Run}} = \frac{-4}{-3} = \frac{4}{3}$$

Let's compare the slope of a line with a positive slope and the slope of a line with a negative slope (Fig. 19–3). The slope of a line passing through the points (1, 6) and (4, 2) is $-\frac{4}{3}$.

$$\text{Slope} = \frac{\Delta y}{\Delta x} = \frac{2 - 6}{4 - 1} = \frac{-4}{3} = -\frac{4}{3}$$

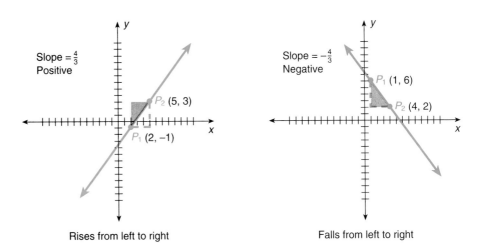

Figure 19–3

19–1 Slope

EXAMPLE

EXAMPLE Find the slope of the line passing through the pairs of points.

(a) $(-5, -1)$ and $(3, -2)$ (b) $(0, 0)$ and $(-5, 3)$

(a) Visualize the line on a coordinate system (Fig. 19–4).

Let $P_1 = (-5, -1)$ and $P_2 = (3, -2)$.   Line falls. Expect negative slope.

$$\text{Slope} = \frac{\Delta y}{\Delta x} = \frac{-2 - (-1)}{3 - (-5)} = \frac{-2 + 1}{3 + 5} = \frac{-1}{8} = -\frac{1}{8}$$

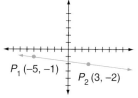

Figure 19–4

The slope is $-\dfrac{1}{8}$.

(b) Visualize the line on a coordinate system (Fig. 19–5).

Let $P_1 = (0, 0)$ and $P_2 = (-5, 3)$.   Line falls. Expect negative slope.

$$\text{Slope} = \frac{\Delta y}{\Delta x} = \frac{3 - 0}{-5 - 0} = \frac{3}{-5} = -\frac{3}{5}$$

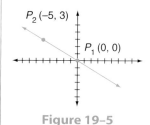

Figure 19–5

The slope is $-\dfrac{3}{5}$.

EXAMPLE Table 19-1 lists the annual cost in dollars of tuition and fees at public two-year colleges for selected years. Find the rate of change in tuition and fees from 1998 to 1999 and from 1997 to 2002.

Find the rate of change of tuition and fees from 1998 to 1999.

(1998, 1,554)   and   (1999, 1,649)   Write the data as ordered pairs.

$$m = \frac{y_2 - y_1}{x_2 - x_1}$$ Use the slope formula to find the rate of change.

$$\text{Rate of change} = \frac{1,649 - 1,554}{1999 - 1998}$$

$$\text{Rate of change} = \frac{95}{1} = \textbf{\$95 per year}$$

Find the rate of change from 1997 to 2002.
(1997, 1,567)   and   (2002, 1,735)

$$m = \frac{y_2 - y_1}{x_2 - x_1}$$

$$\text{Rate of change} = \frac{1,735 - 1,567}{2002 - 1997}$$

$$\text{Rate of change} = \frac{168}{5} = \frac{33.6}{1} = \textbf{\$33.60 per year}$$

The rate of change in tuition for one year from 1998 to 1999 was \$95. But the rate of change from 1997 to 2002 was \$33.60 per year.

2 **Determine the Slope of a Horizontal or Vertical Line.**

Let's examine the slopes of a horizontal line and a vertical line (Figs. 19–6 and 19–7). First look at the horizontal line that passes through the points $(3, 2)$ and $(-3, 2)$.

Chapter 19 / Slope and Distance

EXAMPLE   Find the slope of the horizontal line that passes through (3, 2) and (−3, 2).

Let $P_1 = (3, 2)$ and $P_2 = (-3, 2)$.

$$\text{Slope} = \frac{\Delta y}{\Delta x} = \frac{y_2 - y_1}{x_2 - x_1} = \frac{2 - 2}{-3 - 3} = \frac{0}{-6} = 0$$

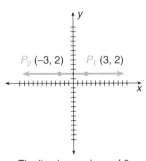

The line has a slope of 0.

Figure 19–6

Slope of horizontal lines:

The slope of a horizontal line is zero.

- Two points are on the same horizontal line if their y-coordinates are equal.
- A horizontal line has no rise or no change in y-values.

Next, let's look at the vertical line that passes through the points (3, 2) and (3, −2).

EXAMPLE   Find the slope of the vertical line that passes through the points (3, 2) and (3, −2).

Let $P_1 = (3, 2)$ and $P_2 = (3, -2)$.

$$\text{Slope} = \frac{\Delta y}{\Delta x} = \frac{y_2 - y_1}{x_2 - x_1} = \frac{-2 - 2}{3 - 3} = \frac{-4}{0}$$

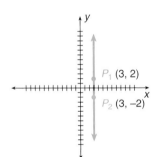

Division by zero is undefined; **therefore, the slope of the vertical line passing through (3, 2) and (3, −2) is undefined.**

The line has no slope.

Figure 19–7

Slope of vertical lines:

A vertical line has no slope; that is, the slope of a vertical line is undefined.

- Two points are on the same vertical line if their x-coordinates are equal.
- A vertical line has no change in x-values.

The necessary facts for writing the equation of a line are often obtained from the graph of the equation.

EXAMPLE Figure 19–8 shows the cost of producing picture frames.

(a) Write an equation that represents the graph.
(b) Using the equation, find the cost of producing 20 picture frames.

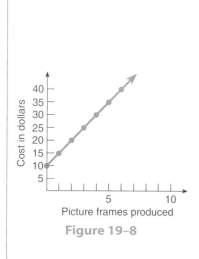

Figure 19–8

(a) The y-intercept represents the fixed cost of producing picture frames. $b = \$10$. The slope m is 5 units vertical change for every 1 unit horizontal change, which is $\frac{5}{1} = 5$.

$$y = 5x + 10 \qquad \text{Equation of the line.}$$

This type of equation is a *cost function*. The function notation used for the relationship described is $C(x) = 5x + 10$.

(b) Use the equation of the line found in part (a) to find y or $C(x)$ when $x = 20$.

$y = 5x + 10$	or	$C(x) = 5x + 10$	Substitute $x = 20$.
$y = 5(20) + 10$		$C(20) = 5(20) + 10$	Evaluate.
$y = 100 + 10$		$C(20) = 100 + 10$	
$y = \$110$		$C(20) = \$110$	Interpret.

The cost of producing 20 frames is \$110.

SELF-STUDY EXERCISES 19–3

1 Determine the slope and y-intercept by inspection.

1. $y = 4x + 3$
2. $y = -5x + 6$
3. $y = -\frac{7}{8}x - 3$
4. $y = 3$

Rewrite the following equations in slope-intercept form and determine the slope and y-intercept.

5. $4x - 2y = 10$
6. $2y = 5$
7. $\frac{1}{2}y + x = 3$

8. $2.1y - 4.2x = 10.5$
9. $2x = 8$
10. $x - 5 = 4$

2 Write the equation of the line with the given slope and y-intercept.

11. $m = \frac{1}{4}, b = 7$
12. $m = -8, b = -4$

13. A local business rents computer time for a \$3 setup charge and \$0.20 for every minute the computer is used. Write an equation to represent the cost y of using the computer for x minutes. How much does 45 min of computer time cost?

14. The cost of producing fine china plates is \$8 per plate plus a one-time equipment charge of \$12,000. Write an equation that represents the total cost y of producing x plates. What is the total cost of producing 8,000 plates?

15. A snowplow has a maximum speed of 40 miles per hour on a dry, flat road surface. Its speed decreases 0.9 miles per hour for every inch of snow on the highway. Write an equation that represents the speed y of the snowplow when moving x in. of snow. What is the maximum speed of the plow in moving 10 in. of snow?

Write the equations of the lines that are graphed in Figures 19–9 through 19–11.

16.   **17.**   **18.**

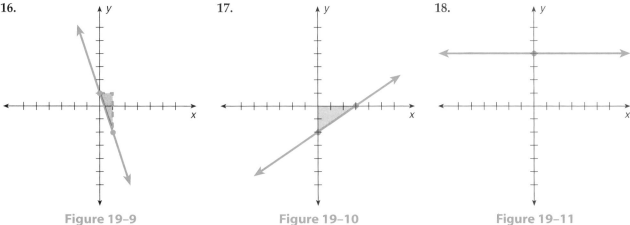

Figure 19–9   Figure 19–10   Figure 19–11

Solve Exercises 19 and 20 using the information given in Fig. 19–12, which shows the cost of producing widgets in dollars.

19. Write an equation that represents the cost of producing widgets shown in the graph.
20. Use the equation to find the cost of producing 10 widgets.

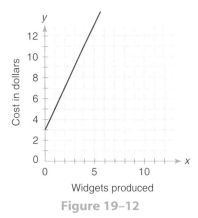

Figure 19–12

19–4 | *Parallel and Perpendicular Lines*

Learning Outcomes   **1** Find the equation of a line parallel to a given line when at least one point on the parallel line is known.

2 Find the equation of a line perpendicular to a given line when at least one point on the perpendicular line is known.

1 **Find the Equation of a Line Parallel to a Given Line When at Least One Point on the Parallel Line Is Known.**

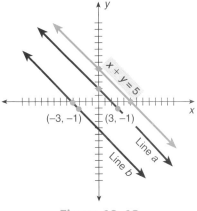

Figure 19–13

Parallel lines are two or more lines that are the same distance apart everywhere. They have no points in common.

Slope of parallel lines:

The slopes of parallel lines are equal.

If we write $x + y = 5$ in slope-intercept form, we see that $y = -x + 5$ and the slope is -1. *Any* equation that has a slope of -1 is either parallel to or coincides with the line formed by the equation $x + y = 5$. *Coincides* means one line lies on top of the other; that is, they are the same line.

We examined the similarities of parallel lines, but what are the differences? Parallel lines have different x- and y-intercepts.

EXAMPLE Write the equations for lines a and b in Figure 19–13.

$x + y = 5$ Solve for y.
$y = -x + 5$ Slope $= -1$.

The y-intercepts can be determined from the graph.

Line a: $y = -x + 2$ **Substitute $m = -1$ and $b = 2$**
Line b: $y = -x - 4$ **Substitute $m = -1$ and $b = -4$**

To find the equation of a line that is parallel to a given line when at least one point on the parallel line is known:

1. Solve the equation of the given line for y.
2. Determine the slope m from $y = mx + b$.
3. The slope of the parallel line is the same as the slope of the given line.
4. Use the point-slope form of a straight line $y - y_1 = m(x - x_1)$ and substitute values for m, x_1, and y_1.
5. Rearrange the equation to be in standard form ($ax + by = c$) or solved for y ($y = mx + b$).

EXAMPLE Find the equation of a line that is parallel to $2y = 3x + 8$ and passes through the point $(2, 1)$.

$2y = 3x + 8$ Solve for y.

$$\frac{2y}{2} = \frac{3x + 8}{2}$$

$$y = \frac{3}{2}x + 4$$ Identify the slope.

The slope of $2y = 3x + 8$ is $\dfrac{3}{2}$.

$$y - y_1 = m(x - x_1)$$

Substitute. The slope of a parallel line is equal to the slope of the given line. $m = \dfrac{3}{2}$; $x_1 = 2$; $y_1 = 1$.

$$y - 1 = \frac{3}{2}(x - 2)$$

Distribute.

$$y - 1 = \frac{3}{2}x - 3$$

Solve for y.

$$y = \frac{3}{2}x - 3 + 1$$

Combine like terms.

$$y = \frac{3}{2}x - 2$$

Slope-intercept form.

Write the equation in standard form by clearing fractions.

$$2y = 2\left(\frac{3}{2}x - 2\right)$$

Clear fractions. Multiply by 2.

$$2y = 3x - 4$$

Transpose variable term to the left.

$$-3x + 2y = -4$$

Multiply each term by -1 so that the coefficient of x is positive.

$$3x - 2y = 4$$

Standard form.

2 **Find the Equation of a Line Perpendicular to a Given Line When at Least One Point on the Perpendicular Line Is Known.**

Perpendicular lines are lines that intersect and make a square corner. Any line can have several lines perpendicular to it, but only one line perpendicular to a *given* line passes through a *given* point. Figure 19–14 shows a line that is perpendicular to the line $y = -2x + 7$ and passes through the point $(6, 8)$. In some cases, the given point may lie on the given line.

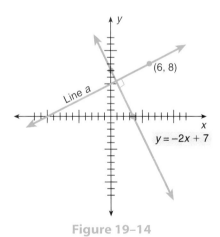

Figure 19–14

Perpendicular lines are two lines that intersect to form right angles ($90°$ angles). Another term for *perpendicular* is *normal*. In Fig. 19–14, the perpendicular (or normal) line that passes through the given point is indicated by the symbol ⌐, which means that a right or $90°$ angle is formed by the lines.

The slope of the equation $y = -2x + 7$ is -2. Two points on line a are the x-intercept $(-10, 0)$ and the y-intercept $(0, 5)$. The slope of line a is

$$m = \frac{5 - 0}{0 - (-10)} = \frac{5}{10} = \frac{1}{2}$$

$P_1 = (-10, 0)$ and $P_2 = (0, 5)$

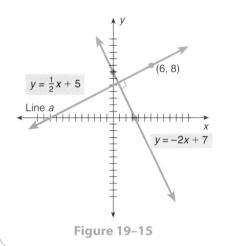

EXAMPLE   Find the equation of line *a* in Fig. 19–15.

Line *a* passes through point (6, 8).

$$y - y_1 = m(x - x_1)$$ Substitute. $m = \dfrac{1}{2}$, $P_1 = (6, 8)$.

$$y - 8 = \frac{1}{2}(x - 6)$$ Distribute.

$$y - 8 = \frac{1}{2}x - 3$$ Solve for *y*.

$$y = \frac{1}{2}x - 3 + 8$$

$$y = \frac{1}{2}x + 5$$ Slope-intercept form.

Figure 19–15

To further examine the relationship of the slopes of perpendicular lines, examine the graphs in Fig. 19–16.

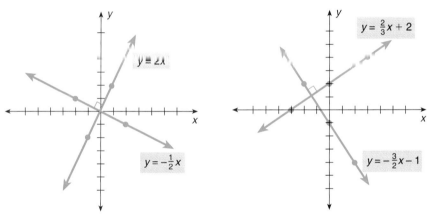

Figure 19–16

The slope of *any* line that is perpendicular to a given line is the negative reciprocal of the slope of the given line.

TIP!

Negative Reciprocals

The negative reciprocal of a number is not necessarily a negative value. It is the opposite of the given number.

To find the negative reciprocal:

1. Interchange the numerator and denominator.
2. Give the reciprocal the opposite sign.

The negative reciprocal of -5 is $+\dfrac{1}{5}$. The negative reciprocal of $-\dfrac{3}{4}$ is $+\dfrac{4}{3}$.

The negative reciprocal of $\dfrac{4}{5}$ is $-\dfrac{5}{4}$. The negative reciprocal of 3 is $-\dfrac{1}{3}$.

Chapter 19 / Slope and Distance

To find the equation of a line that is perpendicular (normal) to a given line and passes through a given point:

1. Determine the slope of the *given* line.
2. Find the *negative reciprocal* of this slope. The negative reciprocal is the slope of the perpendicular line.
3. Write the equation for the perpendicular line by substituting the coordinates of the *given* point for x_1 and y_1 and the slope of the *perpendicular* line for m into the point-slope form of the equation $y - y_1 = m(x - x_1)$.
4. Write the equation in slope-intercept or standard form.

EXAMPLE   Find the equation of a line that is perpendicular to the line $y = \frac{2}{3}x + 8$ and passes through the point (3, 10).

The slope of the given line $y = \frac{2}{3}x + 8$ is $\frac{2}{3}$. The negative reciprocal of $\frac{2}{3}$ is $-\frac{3}{2}$.

$$y - y_1 = m\,(x - x_1)$$ Substitute values: $m = -\dfrac{3}{2}$; $x_1 = 3$; $y_1 = 10$.

$$y - 10 = -\frac{3}{2}(x - 3)$$ Distribute.

$$y - 10 = -\frac{3}{2}x + \frac{9}{2}$$ Solve for y.

$$y = -\frac{3}{2}x + \frac{9}{2} + 10$$ Combine like terms.

$$y = -\frac{3}{2}x + \frac{9}{2} + \frac{20}{2}$$

$$y = -\frac{3}{2}x + \frac{29}{2}$$ Slope-intercept form. For standard form clear the fraction.

or

$$2y = -3x + 29$$ Put all variable terms on the left.

$$3x + 2y = 29$$ Standard form.

TIP!

Words as Subscripts

In the next example, we clarify the notation with subscripts. The phrase *slope of a given line* is notated as "slope$_{given}$." The slope of the perpendicular line is notated as "slope$_{perpendicular}$."

EXAMPLE   Find the equation of the line perpendicular to $4x + y = -3$ and passing through $(0, -3)$.

$$4x + y = -3$$ Given equation. Solve for y.

$$y = -4x - 3$$ Identify the slope.

$$\text{Slope}_{given} = -4$$

$$\text{Slope}_{perpendicular} = +\frac{1}{4}$$ Negative reciprocal of -4.

$$y - y_1 = m(x - x_1)$$ Point-slope form. Substitute $m = \dfrac{1}{4}$, $P_1 = (0, -3)$.

$$y - (-3) = \frac{1}{4}(x - 0)$$ Simplify.

19–4 Parallel and Perpendicular Lines

$$y + 3 = \frac{1}{4}x$$

Solve for y.

$$y = \frac{1}{4}x - 3$$

Slope-intercept form. For standard form clear the fraction and rearrange.

or

$$4y = x - 12$$

Move the x-variable term to the left and make the leading coefficient positive.

$$x - 4y = 12$$

Standard form.

EXAMPLE Which of the equations represents a line that is perpendicular to $2x - 3y = 1$ passing through $(2, -1)$?

(a) $y = \frac{2}{3}x - \frac{1}{3}$ (b) $y = \frac{2}{3}x - \frac{7}{3}$

(c) $y = -\frac{3}{2}x - \frac{1}{3}$ (d) $y = -\frac{3}{2}x + 2$

$$2x - 3y = 1$$

Rewrite the given equation in slope-intercept form. Solve for y.

$$-3y = -2x + 1$$

$$\frac{-3y}{-3} = \frac{-2x}{-3} + \frac{1}{-3}$$

$$y = \frac{2}{3}x - \frac{1}{3}$$

Slope-intercept form.

$$\text{Slope}_{\text{given}} = \frac{2}{3}$$

From $y = \frac{2}{3}x - \frac{1}{3}$.

$$\text{Slope}_{\text{perpendicular}} = -\frac{3}{2}$$

Negative reciprocal

Choices are now limited to (c) or (d) since (a) and (b) have slopes of $\frac{2}{3}$.

$$y - y_1 = m(x - x_1)$$

Substitute $m = -\frac{3}{2}$, $x_1 = 2$, $y_1 = -1$.

$$y - (-1) = -\frac{3}{2}(x - 2)$$

Distribute and simplify.

$$y + 1 = -\frac{3}{2}x + 3$$

Solve for y.

$$y = -\frac{3}{2}x + 3 - 1$$

$$y = -\frac{3}{2}x + 2$$

The correct equation is $y = -\frac{3}{2}x + 2$, or choice (d).

SELF-STUDY EXERCISES 19–4

1 Write the equations in standard form. Verify your answers using a graphing calculator.

1. Find the equation of the line that is parallel to the line $x + y = 6$ and passes through the point $(2, -3)$.

2. Find the equation of the line that is parallel to the line $2x + y = 5$ and passes through the point $(1, 7)$.

3. Find the equation of the line that is parallel to the line $3y = x - 2$ and passes through the point $(4, 0)$.

4. Find the equation of the line that is parallel to the line $3x - y = -2$ and passes through the point $(-3, -2)$.

5. Find the equation of the line that is parallel to the line $x + 3y = 7$ and passes through the point $(4, 1)$.
7. Find the equation of the line that is parallel to the line $2x + 3y = 5$ and passes through the point $(1, 1)$.
9. Find the equation of the line that is parallel to the line $2x - 5y = 0$ and passes through the point $(3, -1)$.

6. Find the equation of the line that is parallel to the line $3x - y = 4$ and passes through the point $(0, 3)$.
8. Find the equation of the line that is parallel to the line $3x + 2y = 1$ and passes through the point $(2, 0)$.
10. Find the equation of the line that is parallel to the line $-3x + 4y = -1$ and passes through the point $(\frac{1}{2}, 0)$.

11. The cost of producing tires on an assembly line includes a fixed cost of $45,000 and a variable cost of $12 per tire. This can be written as a cost function $C(x) = 12x + 45{,}000$, where x is the number of tires produced and $C(x)$ is the total cost of x tires. Research shows that 10,000 tires can be produced on a newly installed assembly line at a total cost of $140,000 with the same variable cost per tire as the original assembly line. Write a function that represents the total cost of producing tires on the new assembly line.

12. A car rental agency uses the function $C(x) = 0.5x + 25$, where x is the number of miles driven and $C(x)$ is the total cost of car rental. Another car rental agency has the same variable cost of $0.50 per mile and quotes the total cost of driving 500 mi as $295. Write a cost function to represent the second agency's rental charges.

2 Write the equation for each exercise in standard form. Verify your results using a graphing calculator. Be sure the range is set so the vertical and horizontal increments are equal.

13. Find the equation of the line that is perpendicular to the line $x + y = 6$ and passes through the point $(2, 3)$.
15. Find the equation of the line that is perpendicular to the line $3y = x - 2$ and passes through the point $(4, 0)$.
17. Find the equation of the line that is perpendicular to the line $x + 3y = 7$ and passes through the point $(4, 1)$.
19. Find the equation of the line that is perpendicular to the line $2x + 3y = 4$ and passes through the point $(3, -1)$.
21. Find the equation of the line that is perpendicular to the line $2x + 2y = 3$ and passes through the point $(\frac{1}{2}, 2)$.

14. Find the equation of the line that is perpendicular to the line $2x + y = 5$ and passes through the point $(1, 7)$.
16. Find the equation of the line that is perpendicular to the line $3x - y = 2$ and passes through the point $(-3, -2)$.
18. Find the equation of the line that is perpendicular to the line $x + 2y = 7$ and passes through the point $(-2, 3)$.
20. Find the equation of the line that is perpendicular to the line $4x + y = 1$ and passes through the point $(0, 0)$.
22. Find the equation of the line that is perpendicular to the line $5x - y = 6$ and passes through the point $(5, -\frac{1}{5})$.

19–5 | Distance and Midpoints

Learning Outcomes

1 Find the distance between two points on the rectangular coordinate system.

2 Find the coordinates of the midpoint of a line segment if given the coordinates of the end points.

1 **Find the Distance Between Two Points on the Rectangular Coordinate System.**

In Chapter 5 we found the distance between two points on a line. To expand this concept to the distance between two points on the rectangular coordinate system, we will apply the Pythagorean theorem.

$$c^2 = a^2 + b^2 \qquad \text{or} \qquad c = \sqrt{a^2 + b^2}$$

The distance between two points on the rectangular coordinate system is the shortest distance between the points.

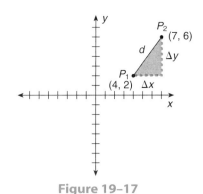

Figure 19–17

To find the distance between two points on the rectangular coordinate system:

1. Use the formula

$$d = \sqrt{(\Delta x)^2 + (\Delta y)^2} \quad \text{or} \quad \sqrt{(x_2 - x_1)^2 + (y_2 - y_1)^2}$$

   where P_1 and P_2 are two points with coordinates (x_1, y_1) and (x_2, y_2), respectively.
2. Substitute values for x_1, x_2, y_1, and y_2.
3. Evaluate to find d.

EXAMPLE   Find the distance from $(4, 2)$ to $(7, 6)$.

These points are shown in Fig. 19–17. Point $(4, 2)$ is labeled P_1, and point $(7, 6)$ is labeled P_2.

$d = \sqrt{(x_2 - x_1)^2 + (y_2 - y_1)^2}$ Substitute: $x_1 = 4$, $x_2 = 7$, $y_1 = 2$, and $y_2 = 6$.

$d = \sqrt{(7 - 4)^2 + (6 - 2)^2}$ Combine in each grouping.

$d = \sqrt{3^2 + 4^2}$ Square each term in the radicand.

$d = \sqrt{9 + 16}$ Add terms in the radicand.

$d = \sqrt{25}$ Find the principal square root.

$d = 5$

The distance from $(4, 2)$ to $(7, 6)$ is 5 units.

TIP!

Point Selection Does Not Matter.

The distance formula can be used to calculate the distance between two points, no matter which points we designate as P_1 and P_2. Let's rework the preceding example, letting $P_1 = (7, 6)$ and $P_2 = (4, 2)$. Thus, $x_1 = 7$, $x_2 = 4$, $y_1 = 6$, and $y_2 = 2$.

$d = \sqrt{(4 - 7)^2 + (2 - 6)^2}$ Substitute.

$d = \sqrt{(-3)^2 + (-4)^2}$ $(-3)^2 = +9; (-4)^2 = +16$

$d = \sqrt{9 + 16}$

$d = \sqrt{25}$

$d = 5$

The change in vertical movement has the same *absolute value*, regardless of which point is used first. The result after squaring is positive whether the difference is positive or negative. For similar reasons, the change in horizontal movement squared is the same, regardless of which point is used first.

Chapter 19 / Slope and Distance

2 **Find the Coordinates of the Midpoint of a Line Segment if Given the Coordinates of the End Points.**

In Chapter 5 we found the midpoint between two points on a line. We will expand that concept to find the midpoint between two points on the rectangular coordinate system.

To find the coordinates of the midpoint of a line segment if given the coordinates of the end points:

1. Average the respective coordinates of the end points of the segment using the formula

$$\text{Midpoint} = \left(\frac{x_1 + x_2}{2}, \frac{y_1 + y_2}{2} \right)$$

where P_1 and P_2 are end points of the segment, $P_1 = (x_1, y_1)$ and $P_2 = (x_2, y_2)$.
2. Substitute values for x_1, x_2, y_1, and y_2.
3. Evaluate to find the coordinates of the midpoint.

EXAMPLE Find the midpoint of each segment with the given end points.

(a) $(2, 4)$ and $(6, 10)$
(b) $(3, -5)$ and origin

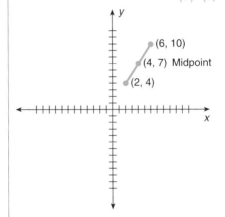

Figure 19–18

(a) $(2, 4)$ and $(6, 10)$. See Fig. 19–18.

$$\text{Midpoint} = \left(\frac{x_1 + x_2}{2}, \frac{y_1 + y_2}{2} \right)$$

Substitute: $x_1 = 2$, $x_2 = 6$, $y_1 = 4$, $y_2 = 10$.

$$\text{Midpoint} = \left(\frac{2 + 6}{2}, \frac{4 + 10}{2} \right)$$

Simplify.

$$\text{Midpoint} = \left(\frac{8}{2}, \frac{14}{2} \right)$$

Reduce.

$$\textbf{Midpoint} = \textbf{(4, 7)}$$

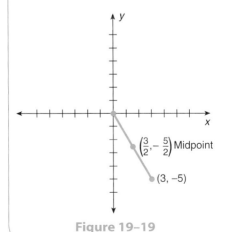

Figure 19–19

(b) $(3, -5)$ and origin. See Fig. 19–19.

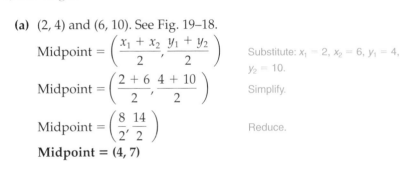

$$\text{Midpoint} = \left(\frac{x_1 + x_2}{2}, \frac{y_1 + y_2}{2} \right)$$

Substitute: $x_1 = 3$, $x_2 = 0$, $y_1 = -5$, $y_2 = 0$.

$$\text{Midpoint} = \left(\frac{3 + 0}{2}, \frac{-5 + 0}{2} \right)$$

$$\textbf{Midpoint} = \left(\frac{3}{2}, -\frac{5}{2} \right) \ \textbf{or} \ \left(1\frac{1}{2}, -2\frac{1}{2} \right)$$

TIP!

Midpoint Between the Origin and Another Point

To find the coordinates of the midpoint of a segment from the origin to any point, take one-half of each coordinate of the point that is not at the origin. Apply this tip to part (b) of the preceding example.

Learning Strategy   *Visualize and Estimate Whenever Practical.*

When formulas are developed to find desired information, we often rely totally on our calculation skills and ability to manipulate formulas. We can strengthen our understanding of the concepts and avoid mistakes by visualizing the problem and by estimating.

In some of the examples in this section, we provided a visualization of the problem. Now, let's examine some estimates. To estimate the distance between points $(4, 2)$ and $(7, 6)$, we find the vertical and horizontal changes. The vertical change is $4 (6 - 2)$ and the horizontal change is $3 (7 - 4)$. Applying properties of the Pythagorean theorem, the distance between the points (hypotenuse) is more than the longest leg (4) and less than the sum of the legs, $3 + 4 = 7$. So, we estimate the distance to be between 4 and 7 units.

To estimate the coordinates of the midpoint between these points, find the range of values. The x-coordinate of the midpoint is halfway between 4 and 7. The y-coordinate of the midpoint is halfway between 2 and 6.

SELF-STUDY EXERCISES 19–5

1   Graph the line segment determined by the given end points and find the distance between each of the pairs of points. Express the answer to the nearest thousandth when necessary.

1. $(7, 10)$ and $(1, 2)$
2. $(7, -7)$ and $(2, 5)$
3. $(5, 0)$ and $(0, 5)$
4. $(-2, -2)$ and $(3, -4)$
5. $(5, 7)$ and $(0, -3)$
6. $(8, -2)$ and $(-4, 3)$
7. $(-4, 6)$ and $(2, -2)$
8. $(3, 5)$ and $(0, 1)$
9. $(5, 4)$ and $(-7, -5)$
10. $(7, 2)$ and $(-2, -3)$

2   Find the coordinates of the midpoints of the segments determined by the given end points.

11. $(7, 10)$ and $(1, 2)$
12. $(7, -7)$ and $(2, 5)$
13. $(5, 0)$ and $(0, 5)$
14. $(-2, -2)$ and $(3, -4)$
15. $(5, 7)$ and $(0, -3)$
16. $(8, -2)$ and $(-4, 3)$
17. $(-4, 6)$ and $(2, -2)$
18. $(3, 5)$ and $(0, 1)$
19. $(5, 4)$ and $(-7, -5)$
20. $(7, 2)$ and $(-2, -3)$

ASSIGNMENT EXERCISES

Section 19–1

Find the slope of the line passing through the given pairs of points.

1. $(-2, 2)$ and $(1, 3)$
2. $(3, -1)$ and $(1, 3)$
3. $(3, 2)$ and $(5, 6)$
4. $(-1, -1)$ and $(2, 2)$
5. $(4, 3)$ and $(-4, -2)$
6. $(6, 2)$ and $(-3, 2)$
7. $(3, -4)$ and $(0, 0)$
8. $(1, -1)$ and $(5, -5)$
9. $(-4, 1)$ and $(-4, 3)$

10. $(4, -4)$ and $(1, 3)$

11. $(5, 0)$ and $(-2, 4)$

12. $(-2, 1)$ and $(0, 3)$

13. $(-4, -8)$ and $(-2, -1)$

14. $(3, 3)$ and $(3, 0)$

15. $(5, -3)$ and $(-1, -3)$

16. $(-5, -1)$ and $(-7, -3)$

17. $(-7, 0)$ and $(-7, 5)$

18. $(3, 5)$ and $(2, 5)$

19. $(5, 9)$ and $(7, 11)$

20. $(3, 5)$ and $(5, 3)$

21. Write the coordinates of two points that lie on the same horizontal line.

22. Write the coordinates of two points that lie on the same vertical line.

Use Table 19–2 for Exercises 23–28.

23. What is the rate of change in tuition and fees at public two-year colleges from 1998 to 1999?

24. What is the rate of change in tuition and fees at public two-year colleges from 2002 to 2003?

25. What is the rate of change in tuition and fees at public two-year colleges from 1993 to 1994?

26. What is the rate of change in tuition and fees at public two-year colleges from 1992 to 2003?

27. Explain why the rate of change found in Exercises 24, 25, and 26 are different.

28. If the data in Table 19–2 formed a straight line when graphed, what would you expect to find for the rate of change for the data in Exercises 24, 25, and 26?

Section 19–2

Find the equation of a line passing through the given point with the given slope. Solve the equation for y if necessary.

29. $(-6, 2)$, $m = \dfrac{1}{3}$

30. $(3, 2)$, $m = -\dfrac{2}{5}$

31. $(4, 0)$, $m = \dfrac{3}{4}$

32. $(0, -2)$, $m = 2$

33. $(2, 3)$, $m = 4$

34. $(6, 0)$, $m = -1$

35. $(5, -4)$, $m = -\dfrac{2}{3}$

36. $(-1, -5)$, $m = -3$

Find the equation of a line passing through the given pairs of points. Solve the equation for y if necessary.

37. $(-5, 2)$ and $(6, 1)$

38. $(1, 4)$ and $(-1, 3)$

39. $(-1, -3)$ and $(3, 4)$

40. $(-3, 0)$ and $(4, 0)$

41. $(-2, -3)$ and $(3, 6)$

42. $(2, -4)$ and $(3, -4)$

43. $(5, 2)$ and $(6, 3)$

44. $(4, 6)$ and $(1, -1)$

45. $(-1, -2)$ and $(-3, -4)$

46. $(4, 0)$ and $(4, -3)$

47. $(5, -2)$ and $(3, -2)$

48. $(5, 4)$ and $(0, 4)$

49. A salesperson sells 80 items and earns $3,800 in one month, and in another month 120 items are sold resulting in $4,200 earnings. If we assume this is a linear function, write a function that expresses the salesperson's monthly salary plus commission, S, as a function of the number of items sold, x. Salary is y-intercept.

50. Simple interest is a linear function. Write an equation to express interest earned if $500 is earned in 2 years and $2,000 is earned in 8 years.

Section 19–3

Determine the slope and y-intercept of the given equations by inspection.

51. $y = 3x + \dfrac{1}{4}$

52. $y = \dfrac{2}{3}x - \dfrac{3}{5}$

53. $y = -5x + 4$

54. $y = 7$

55. $x = 8$

56. $y = \dfrac{1}{3}x - \dfrac{5}{8}$

57. $y = \dfrac{x}{8} - 5$

58. $y = -\dfrac{x}{5} + 2$

59. The cost function for producing a particular calculator is represented by $C(x) = 18x + 5,050$. What is the fixed cost (y-intercept)?

60. In Exercise 59, what is the variable cost (cost per calculator or slope)?

Write the given equations in slope-intercept form and determine the slope and y-intercept.

61. $2x + y = 8$

62. $4x + y = 5$

63. $3x - 2y = 6$

64. $5x - 3y = 15$

65. $\dfrac{3}{5}x - y = 4$

66. $2.2y - 6.6x = 4.4$

67. $3y = 5$

68. $3x - 6y = 12$

Write the equations using the given slope and y-intercept.

69. $m = 3$, $b = -2$

70. Slope $= \dfrac{3}{5}$; y-intercept $= -7$

Find the distance between each pair of points and find the coordinates of the midpoint of the line segment made by the pairs of points.

30. $(5, 2)$ and $(-7, -3)$

31. $(-2, 1)$ and $(3, 3)$

32. A cellular phone provider uses the function $C(x) = 0.08x + 30$, where x is the number of minutes per month and $C(x)$ is the total monthly cost. A new system is being installed that requires the company to restructure its monthly fees. The variable cost will remain the same and the company has provided examples of the new pricing that show that 500 min would have a total cost of $140 and 2,000 min would have a total cost of $260. Write a new cost function that represents the new pricing.

Systems of Linear Equations and Inequalities

20

20-1 Solving Systems of Linear Equations and Inequalities Graphically

- **1** Solve a system of linear equations by graphing.
- **2** Solve a system of linear inequalities by graphing.

20-2 Solving Systems of Linear Equations Using the Addition Method

- **1** Use the addition method to solve a system of linear equations that contains opposite variable terms.
- **2** Use the addition method to solve a system of linear equations that does not contain opposite variable terms.
- **3** Apply the addition method to a system of linear equations with no solution or with many solutions.

20-3 Solving Systems of Linear Equations Using the Substitution Method

- **1** Use the substitution method to solve a system of linear equations.

20-4 Problem Solving Using Systems of Linear Equations

- **1** Use a system of linear equations to solve application problems.

Many real-world situations involve problems in which several conditions or constraints have to be considered. These conditions can be written in separate equations or inequalities that form a system of equations or inequalities. The solution of the system will be the value or values that satisfy all conditions.

20–1 | Solving Systems of Linear Equations and Inequalities Graphically

Learning Outcomes **1** Solve a system of linear equations by graphing.
2 Solve a system of linear inequalities by graphing.

1 Solve a System of Linear Equations by Graphing.

A *system of two linear equations*, each having two variables, is solved when we find the one ordered pair of solutions that satisfies *both* equations. One method of solving systems of two equations is to graph each equation and find the intersection of these graphs. The point where the two graphs intersect represents the ordered pair of solutions that the two graphs have in common.

To solve a system of two linear equations with two variables by graphing:

1. Graph each equation on the same pair of axes.
2. The solution will be the common point or points.

EXAMPLE A board is 20 ft long. It needs to be cut so that one piece is 2 ft longer than the other. What should be the length of each piece?

Write two equations to describe all the conditions of the problem.
  Since the board is not cut into equal pieces, we let the letter *l* represent the *longer* piece and the letter *s* represent the *shorter* piece.

Condition 1: The total length of the board is 20 ft. Thus, the two pieces (*l* and *s*) total 20 ft: $l + s = 20$.
Condition 2: One piece is 2 ft longer than the other. Thus, the shorter piece plus 2 ft equals the longer piece: $s + 2 = l$.

The two equations become a *system of equations*.

$l + s = 20$ Condition 1.
$s + 2 = l$ Condition 2.

  Graph each equation on the same set of axes and examine the intersection of the graphs.

Condition 1		**Condition 2**	Make a table of solutions for each equation.

$l + s = 20$ Intercepts: (0,20) and (20,0). $s + 2 = l$ Intercepts: (0, 2) and (-2, 0).
   or   or
$l = 20 - s$ Domain: [0, 20]. $l = s + 2$ Domain: (0, ∞).

s	l		s	l	
8	12	Choose values that are near the middle of the domain.	0	2	Choose only zero and positive values.
10	10		5	7	
12	8		10	12	

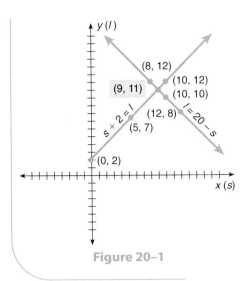

Figure 20–1

The point of intersection (Fig. 20–1) is (9, 11), which means that $s = 9$, and $l = 11$. **The short length is 9 ft and the longer length is 11 ft.** Check to see if the solution ($s = 9, l = 11$) satisfies both equations.

$l + s = 20$	$s + 2 = l$	
$11 + 9 = 20$	$9 + 2 = 11$	The ordered pair checks
$20 = 20$	$11 = 11$	in both equations.

Line relationships:

When graphing two straight lines, three possibilities can occur.

1. The two lines can intersect in **just one point.** This means the system is *independent* and one pair of values satisfy both equations.
2. The two lines **do not intersect** at all. This means that the system is *inconsistent* and no pair of values satisfies both equations.
3. The two lines **coincide** or fall exactly in the same place. This means that the system is *dependent* and the equations are identical or are multiples, and any pair of values that satisfies one equation satisfies both equations.

TIP!

Solving a System of Equations Using a Graphing Calculator

Systems of linear equations can be graphed on a graphing calculator or computer.

1. Graph the first equation.
2. Graph the second equation without clearing the graph screen.
3. Use the Trace feature or the CALC/Intersect feature to determine the approximate coordinates of the intersection of the graphs.
4. Use the Zoom or Box (Window) feature to get a closer view and thus a more accurate approximation of the intersection of the graphs.

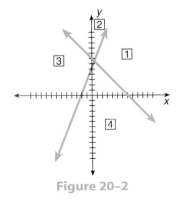

Figure 20–2

2 **Solve a System of Linear Inequalities by Graphing.**

When two intersecting lines are plotted on the same set of axes, the rectangular coordinate system is separated into four regions. The solution set of a system of linear inequalities will include all the points in one of the regions.

1. Graph each inequality on the same pair of axes.
2. The solution set of the system will be the overlapping portion of the solution sets of the two inequalities.

EXAMPLE Shade the portion on the graph that is represented by the following conditions: $y \leq 3x + 5$ and $x + y > 7$.

Graph the inequality $y \leq 3x + 5$:

The boundary $y = 3x + 5$ has a slope of 3 and a y-intercept of 5. The boundary will be included in the solution set.

Because $(0, 0)$ does not fall on the graph of $y = 3x + 5$, we use it as our test point in the original inequality, $y \leq 3x + 5$.

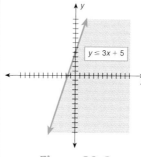

Figure 20–3

Test point: $(0, 0)$

$y \leq 3x + 5$	Substitute values.
$0 \leq 3(0) + 5$	Multiply.
$0 \leq 0 + 5$	Add.
$0 \leq 5$	True.

Thus, the point $(0, 0)$ is included in the solution set (see Fig. 20–3).

Graph the inequality, $x + y > 7$:

The boundary $x + y = 7$ in slope-intercept form is $y = -x + 7$. The slope is -1 and the y-intercept is 7. The boundary will not be included.

The point $(0, 0)$ does not fall on the graph of $x + y = 7$. Let's use it as our test point in the original inequality $x + y > 7$.

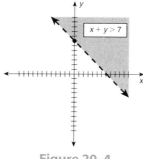

Figure 20–4

Test point: $(0, 0)$

$x + y > 7$	Substitute values.
$0 + 0 > 7$	Add.
$0 > 7$	False.

Thus, the point $(0, 0)$ is not included in the solution set. The solution set will be the opposite side of the boundary (see Fig. 20–4).

Find the solution set common to both inequalities. Visualize both graphs on the same axes. The portion that has overlapping shading represents the points that satisfy *both* conditions and forms the solution set for the system of inequalities (see Fig. 20–5).

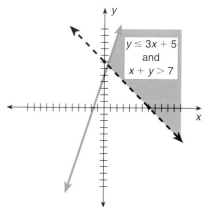

Figure 20–5

TIP!

Color-Code Your Graph

Not all graphing calculators or graphing software shade the intersecting region for the solution of a system of inequalities. However, the calculator does graph the boundaries. Through practice and watching for patterns, you will be able to shade the appropriate region from a mental analysis of the conditions of the problem.

Highlighters and your knowledge of primary colors and color combinations can be useful in finding overlapping regions of a graph.

- Choose two colors of highlighters such as yellow and blue.
- Shade the region represented by one inequality with one color.
- Shade the region represented by the other inequality with the second color.
- The overlap will be the combined color. Yellow and blue combine to be green. Yellow and pink combine to be orange.

SELF-STUDY EXERCISES 20–1

1 Solve the systems of equations by graphing.

1. $x + y = 12$
$x - y = \ 2$

2. $2x + y = 9$
$3x - y = 6$

3. $2x - y = 5$
$4x - 2y = 8$

4. $x - y = \ 9$
$3x - 3y = 27$

5. $3x + 2y = 8$
$x + y = 2$

2 Graph each system inequalities and shade the portion on the graph represented by the solution set.

6. $y < 2x + 1$ and $x + y > 5$

7. $x + y \geq 3$ and $x - y \geq 2$

8. $x - 2y \leq -1$ and $x + 2y \geq 3$

9. $x + y > 4$ and $x - y > -3$

10. $3x - 2y < 8$ and $2x + y \leq -4$

20–2 | Solving Systems of Linear Equations Using the Addition Method

Learning Outcomes

1 Use the addition method to solve a system of linear equations that contains opposite variable terms.

2 Use the addition method to solve a system of linear equations that does not contain opposite variable terms.

3 Apply the addition method to a system of linear equations with no solution or with many solutions.

We can see that solving systems of equations graphically is tedious and time-consuming. Also, many solutions to systems of equations are not whole numbers. Graphically, it is difficult to plot fractions or to read fractional intersection points. Therefore, we need a convenient algebraic procedure for solving systems of equations. We examine the two methods most commonly used to solve a system of equations with two variables, but there are other methods. The first method we introduce is the *addition method*.

20–2 Solving Systems of Linear Equations Using the Addition Method   765

The addition method incorporates the concept that *equals added to equals give equals.* Thus, when we add two equations, the result is still an equation. However, the addition method is effective *only* if one variable is *eliminated* in the addition process. This method is also called the *elimination method.*

To solve a system of linear equations that contains opposite variable terms by the addition (elimination) method:

1. Write each equation in standard form ($ax + by = c$).
2. Add the two equations.
3. Solve the equation from Step 2. This is one part of the solution ordered pair.
4. Substitute the solution from Step 3 in either equation and solve for the remaining variable. This completes the solution ordered pair.
5. Check the solution in each original equation.

EXAMPLE Solve the systems of equations by the addition method:

(a) $x + y = 7$
$\quad x - y = 5$

(b) $3x + 8y = 7$
$\quad -3x - 3y = 8$

(a) $x + y = 7$ Equation 1. Add the equations to eliminate one variable.

$\underline{x - y = 5}$ Equation 2.

$2x = 12$ Solve for x.

$x = 6$ Substitute 6 in place of x in *either* of the given equations.

$x + y = 7$ Equation 1. Substitute $x = 6$.

$6 + y = 7$ Solve for y.

$y = 7 - 6$ Simplify.

$y = 1$

The solution is $x = 6$ and $y = 1$, or (6, 1).

Check in the *other* equation:

$x - y = 5$ Equation 2. Substitute values from the ordered-pair solution.

$6 - 1 = 5$ Simplify.

$5 = 5$ The solution checks.

Checking in both equations helps ensure that no errors have been made in the addition process or in the substitution process.

(b) $3x + 8y = 7$ Equation 1. Add the equations to eliminate one variable.

$\underline{-3x - 3y = 8}$ Equation 2.

$5y = 15$ Solve for y.

$y = 3$ Substitute 3 for y in Equation 1.

$3x + 8y = 7$ Equation 1. Substitute $y = 3$.

$3x + 8(3) = 7$ Solve for x.

$3x + 24 = 7$

$3x = 7 - 24$

$3x = -17$

$x = \dfrac{-17}{3}$

The solution is $\left(-\dfrac{17}{3}, 3\right)$.

Check in equation 2:

$$-3x - 3y = 8 \qquad \text{Equation 2. Substitute } x = -\frac{17}{3} \text{ and } y = 3.$$

$$-3\left(\frac{-17}{3}\right) - 3(3) = 8 \qquad \text{Multiply.}$$

$$17 - 9 = 8 \qquad \text{Subtract.}$$

$$8 = 8 \qquad \text{The solution checks.}$$

2 **Use the Addition Method to Solve a System of Linear Equations That Does Not Contain Opposite Variable Terms.**

In the addition method one unknown must be eliminated; that is, the terms must add to 0. Thus, if neither pair of variable terms in a system of equations is opposite terms that will add to 0, we multiply one or both of the equations by numbers that *cause* the terms of one variable to add to 0. This is an application of the multiplication axiom (see p. 364).

To solve a system of linear equations using the addition method (elimination):

1. Write each equation in standard form ($ax + by = c$).
2. If necessary, multiply one or both equations by numbers that cause the terms of one variable to add to zero.
3. Add the two equations to eliminate a variable.
4. Solve the equation from Step 3 for the remaining variable.
5. Substitute the solution from Step 4 in either equation and solve for the remaining variable.
6. Check the solution in one or both original equations.

EXAMPLE Solve the system of equations: $2x + y = 7$ and $x + y = 3$.

Choice 1.
$$\begin{aligned} -2x - y &= -7 \qquad \text{Multiply the first equation by } -1 \text{ to eliminate the } y \text{ term.}\\ x + y &= 3 \qquad \text{Add the equations.}\\ \hline -x &= -4 \qquad \text{Solve for } x.\\ x &= 4 \end{aligned}$$

Choice 2.
$$\begin{aligned} 2x + y &= 7\\ -x - y &= -3\\ \hline x &= 4 \end{aligned}$$
Multiply the second equation by -1 and add the equations to eliminate the y term.

Choice 3.
$$\begin{aligned} 2x + y &= 7\\ -2x - 2y &= -6\\ \hline -y &= 1\\ y &= -1 \end{aligned}$$
Multiply the second equation by -2 and add the equations to eliminate the x term.

In choices 1 and 2, substitute $x = 4$:

$$-2x - y = -7$$
$$-2(4) - y = -7$$
$$-8 - y = -7$$
$$-y = -7 + 8$$
$$-y = 1$$
$$y = -1$$

In choice 3, substitute $y = -1$:

$$2x + y = 7$$
$$2x + (-1) = 7$$
$$2x - 1 = 7$$
$$2x = 7 + 1$$
$$2x = 8$$
$$x = 4$$

The solution is (4, −1).

Check the solution in the first original equation:

$$2x + y = 7$$
$$2(4) + (-1) = 7$$
$$8 + (-1) = 7$$
$$7 = 7$$

Check the solution in the second original equation:

$$x + y = 3$$
$$4 + (-1) = 3$$
$$3 = 3$$

TIP!

Importance of Checking Solutions

When an original equation is altered, it is very important to check your solution in *both* original equations. This enables you to identify mistakes such as forgetting to multiply *each* term in the equation by a number.

EXAMPLE Solve the system of equations: $x = 3y + 7$ and $2x + 3y = 2$.

$$x = 3y + 7$$ Write Equation 1 in standard form.
$$x - 3y = 7$$

$$x - 3y = 7$$ Equation 1. Add to eliminate the y terms.
$$2x + 3y = 2$$ Equation 2.
$$3x \quad\quad = 9$$ Solve for x.

$$x = 3$$ x-value of solution.

$$x - 3y = 7$$ Substitute $x = 3$.
$$3 - 3y = 7$$ Solve for y.
$$-3y = 7 - 3$$
$$-3y = 4$$

$$y = -\frac{4}{3}$$ y-value of solution.

The solution is $(3, -\frac{4}{3})$.

Check the solution in the original equations:

Equation 1:

$$x = 3y + 7$$

$$3 = 3\left(-\frac{4}{3}\right) + 7$$

$$3 = -4 + 7$$

$$3 = 3$$

Equation 2:

$$2x + 3y = 2$$ Substitute $x = 3$ and $y = -\frac{4}{3}$.

$$2(3) + 3\left(-\frac{4}{3}\right) = 2$$

$$6 + (-4) = 2$$

$$2 = 2$$ Solution checks.

EXAMPLE Solve the system of equations: $2x + 3y = 1$ and $3x + 4y = 2$.

In this system, no number can be multiplied by just one equation to eliminate a letter. Therefore, we need to multiply each equation by some number. There are several possibilities; we examine just two.

Choice 1.

$$
\begin{aligned}
-3(2x + 3y) &= -3(1) &&\text{Multiply Equation 1 by } -3.\\
2(3x + 4y) &= 2(2) &&\text{Multiply Equation 2 by } +2.\\
-6x - 9y &= -3 &&\text{Add equations to eliminate } x.\\
\underline{6x + 8y} &= 4\\
-y &= 1 &&\text{Solve for } y.\\
y &= -1 &&\text{y-value of solution.}\\
2x + 3y &= 1 &&\text{Substitute } y = -1.\\
2x + 3(-1) &= 1 &&\text{Solve for } x.\\
2x - 3 &= 1\\
2x &= 1 + 3\\
2x &= 4\\
x &= 2 &&\text{x-value of solution.}
\end{aligned}
$$

Choice 2.

$$
\begin{aligned}
4(2x + 3y) &= 4(1) &&\text{Multiply Equation 1 by 4.}\\
-3(3x + 4y) &= -3(2) &&\text{Multiply Equation 2 by } -3.\\
8x + 12y &= 4 &&\text{Add equations to eliminate } y.\\
\underline{-9x - 12y} &= -6\\
-x &= -2 &&\text{Solve for } x.\\
x &= 2 &&\text{x-value of solution.}\\
2x + 3y &= 1 &&\text{Substitute } x = 2.\\
2(2) + 3y &= 1 &&\text{Solve for } y.\\
3y &= 1 - 4\\
3y &= -3\\
y &= -1
\end{aligned}
$$

The solution is (2, −1).

Check the solution in the original equations:

$$
\begin{array}{lll}
2x + 3y = 1 & 3x + 4y = 2 & \text{Substitute } x = 2 \text{ and } y = -1.\\
2(2) + 3(-1) = 1 & 3(2) + 4(-1) = 2\\
4 + (-3) = 1 & 6 + (-4) = 2\\
1 = 1 & 2 = 2 & \text{Solution checks.}
\end{array}
$$

3 Apply the Addition Method to a System of Linear Equations with No Solution or with Many Solutions.

There are instances when a system of equations has no solution; the graphs of the two equations do not intersect. There are also instances when a system of equations has all solutions in common; the graphs of the two equations coincide. Let's look at examples of these situations and the addition method.

EXAMPLE Solve the system $x + y = 7$ and $x + y = 5$.

$$
\begin{aligned}
x + y &= 7 &&\text{Multiply Equation 2 by } -1.\\
-1(x + y) &= -1(5)\\
x + y &= 7 &&\text{Add the equations to eliminate a variable.}\\
\underline{-x - y} &= -5\\
0 &= 2 &&\text{Both variables are eliminated.}
\end{aligned}
$$

Notice, both variables are eliminated and the resulting equation, $0 = 2$, is *false*.
There are no solutions to this system.

Inconsistent Equations—No Solution

When solving a system of equations, if both variables are eliminated and the resulting statement is false, the equations are *inconsistent* and have no solution. The graphs of the equations are parallel lines.

EXAMPLE   Solve the system $2x - y = 7$ and $4x - 2y = 14$.

$$-2(2x - y) = -2(7)$$ Multiply Equation 1 by -2.

$$-4x + 2y = -14$$ Add the equations to eliminate a variable.

$$\underline{4x - 2y = 14}$$

$$0 = 0$$ Both variables are eliminated.

Again, both variables are eliminated; however, this time the result is a *true* statement ($0 = 0$). In this situation, **all solutions of one equation are also solutions of the other equation.**

Dependent Equations—All Solutions in Common

When solving a system of equations, if both variables are eliminated and the resulting statement is true, then the equations are *dependent* and have many solutions. The graphs of the equations coincide.

SELF-STUDY EXERCISES 20–2

1 Solve the systems of equations using the addition method.

1. $a - 2b = 7$
$3a + 2b = 13$

2. $3m + 4n = 8$
$2m - 4n = 12$

3. $x - 4y = 5$
$-x - 3y = 2$

4. $a - b = 6$
$2a + b = 3$

5. $x + 2y = 5$
$3x - 2y = 3$

2 Solve the systems of equations using the addition method.

6. $3x + y = 9$
$x + y = 3$

7. $7x + 2y = 17$
$y = 3x + 2$

8. $a + 6b = 18$
$4a - 3b = 0$

9. $3x + y = -1$
$4x - 2y = -8$

10. $a = 6y$
$2a - y = 11$

3 Solve the systems of equations using the addition method.

11. $x + y = 8$
$x + y = 3$

12. $x + y = 9$
$x + y = 3$

13. $3x - 2y = 6$
$9x - 6y = 18$

14. $2a + 4b = 10$
$a + 2b = 5$

15. $3a - 2b = 14$
$3a = 2b + 2$

Learning Outcome **1** Use the substitution method to solve a system of linear equations.

1 Use the Substitution Method to Solve a System of Linear Equations.

Another method for solving systems of equations is by substitution. Recall that in formula rearrangement (Chapter 12, Section 2), whenever more than one variable is used in an equation or formula, we can rearrange the equation or solve for a particular variable. In the *substitution method* for solving systems of equations, we solve one equation for one variable and then substitute the equivalent expression in place of the variable in the other equation.

To solve a system of equations by substitution:

1. Rearrange either equation to isolate one variable.
2. Substitute the equivalent expression from Step 1 into the *other* equation and solve for the remaining variable.
3. Substitute the portion of the solution from Step 2 into the equation from Step 1 to find the remaining value of the solution.
4. Check the solution in both original equations.

EXAMPLE Solve the system of equations $x + y = 15$ and $y = 2x$ using the substitution method.

$$x + y = 15$$
$$y = 2x \qquad \text{Equation 2 is already solved for } y.$$
$$x + y = 15 \qquad \text{Substitute } 2x \text{ for } y \text{ in Equation 1 and solve.}$$
$$x + 2x = 15 \qquad \text{Solve for } x.$$
$$3x = 15$$
$$x = 5 \qquad x\text{-value of solution.}$$
$$y = 2x \qquad \text{Substitute the solution for } x \text{ in Equation 2 to find } y.$$
$$y = 2(5) \qquad \text{Simplify.}$$
$$y = 10 \qquad y\text{-value of the solution.}$$

The solution is (5, 10).

Check:

$$x + y = 15 \qquad y = 2x \qquad \text{Substitute } x = 5 \text{ and } y = 10 \text{ in both equations.}$$
$$5 + 10 = 15 \qquad 10 = 2(5) \qquad \text{Simplify.}$$
$$15 = 15 \qquad 10 = 10 \qquad \text{The solution checks in both equations.}$$

EXAMPLE Solve the following system of equations using the substitution method.

$$2x - 3y = -14$$
$$x + 5y = 19$$

When using the substitution method, either equation can be solved for either unknown. In this example, the x-term in the second equation has a coefficient of 1, so the simplest choice would be to solve the second equation for x.

	Step 1	Step 2	Step 3

Step 1

$$x + 5y = 19$$
$$x = 19 - 5y$$

Step 2

$$2x - 3y = -14$$
$$2(19 - 5y) - 3y = -14$$
$$38 - 10y - 3y = -14$$
$$38 - 13y = -14$$
$$-13y = -14 - 38$$
$$-13y = -52$$
$$\frac{-13y}{-13} = \frac{-52}{-13}$$
$$y = 4$$

Step 3

$$x = 19 - 5y$$
$$x = 19 - 5(4)$$
$$x = 19 - 20$$
$$x = -1$$

The solution is $(-1, 4)$.

Check the roots $x = -1$, $y = 4$ in both original equations:

Step 4

$$2x - 3y = -14 \qquad x + 5y = 19$$
$$2(-1) - 3(4) = -14 \qquad -1 + 5(4) = 19$$
$$-2 - 12 = -14 \qquad -1 + 20 = 19$$
$$-14 = -14 \qquad 19 = 19$$

Learning Strategy   *Long Problems Don't Have to Be Difficult.*

Sometimes we let ourselves become overwhelmed by the mere length of a problem. Look at the previous example. Each step of the solution involves skills that we have previously used many times. Here are some tips to help you manage longer problems.

- Get a global or overall understanding of the problem you are solving.
- Make a prediction or estimate of the solution.
- Get a global or overall understanding of the process you are using to solve the problem.
- List in your own words (as briefly as possible) the steps of the process.
- Focus on one step at a time.
- Examine the solution to see if it matches your prediction or estimate.

SELF-STUDY EXERCISES 20–3

1   Solve the systems of equations using the substitution method.

1. $2a + 2b = 60$
    $a = 10 + b$

2. $7r + c = 42$
    $3r - 8 = c$

3. $x - 35 = -2y$
    $3x - 2y = 17$

4. $x + y = 12$
    $x = 2 + y$

5. $2p + 3k = 2$
    $2p - 3k = 0$

6. $x + 2y = 5$
    $x = 3y$

7. $x - 3 = 2y$
    $x = 3y - 2$

8. $a = 3x - 1$
    $x = a + 5$

Learning Outcome   **1** Use a system of linear equations to solve application problems.

1 **Use a System of Linear Equations to Solve Application Problems.**

Many job-related problems can be solved by setting up and solving systems of equations.

EXAMPLE   A restaurant ordered three hampers of blue crabs and one hamper of shrimp that together weighed 89 lb. In another order, two hampers of shrimp and five hampers of blue crabs weighed 160 lb. How much did one hamper of crabs weigh and how much did one hamper of shrimp weigh?

Known facts   3 hampers of blue crabs and 1 hamper of shrimp weigh 89 lb.
5 hampers of blue crabs and 2 hampers of shrimp weigh 160 lb.

Unknown facts   What is the weight of one hamper of blue crabs (c)?
What is the weight of one hamper of shrimp (s)?

Relationships   Total weight = Number of containers × Weight of each container

$3c + 1s = 89$ Equation 1.
$5c + 2s = 160$ Equation 2.

Estimation   In Equation 1, 4 hampers weigh 89 lb, which averages to more than 20 lb per hamper. In Equation 2, 7 hampers weigh 160 lb, which also averages to more than 20 lb per hamper. One type of seafood is expected to weigh more than 20 lb and the other type is expected to weigh 20 lb or less.

Calculations
$\begin{aligned} 3c + 1s &= 89 \end{aligned}$ Eliminate s by multiplying Equation 1 by -2.
$5c + 2s = 160$

$\begin{aligned} -6c - 2s &= -178 \\ 5c + 2s &= 160 \end{aligned}$ Add equations to eliminate s.

$-c = -18$ Solve for c.
$c = 18$ c-value of solution.

$5c + 2s = 160$ Substitute for c in Equation 2.
$5(18) + 2s = 160$ Solve for s.
$90 + 2s = 160$
$2s = 160 - 90$
$2s = 70$
$s = \dfrac{70}{2}$
$s = 35$ s-value of solution.

Interpretation   **One hamper of blue crabs weighed 18 lb; one hamper of shrimp weighed 35 lb.**

EXAMPLE   Two dry cells connected in series have a total internal resistance of 0.09 ohm. The difference between the internal resistance of each dry cell is 0.03 ohm. How much is each internal resistance?

Known facts   The total internal resistance of the two dry cells is 0.09 ohm.
The difference in internal resistance of the two dry cells is 0.03 ohm.

Unknown facts	What is the resistance of dry cell 1 (r_1)? What is the resistance of dry cell 2 (r_2)?
Relationships	$r_1 + r_2 = 0.09$ Equation 1. $r_1 - r_2 = 0.03$ Equation 2.
Estimation	If resistances were the same, they would each be 0.045 ohm. Because they are not the same, one will be more than 0.045 and one will be less than 0.045.

Calculations

$$r_1 + r_2 = 0.09 \qquad \text{Add the equations to eliminate } r_2.$$
$$\underline{r_1 - r_2 = 0.03}$$
$$2r_1 \quad\;\; = 0.12 \qquad \text{Solve for } r_1.$$
$$r_1 = \frac{0.12}{2}$$
$$r_1 = 0.06 \qquad r_1\text{-value of solution.}$$
$$r_1 + r_2 = 0.09 \qquad \text{Substitute for } r_1 \text{ in Equation 1.}$$
$$0.06 + r_2 = 0.09 \qquad \text{Solve for } r_2.$$
$$r_2 = 0.09 - 0.06$$
$$r_2 = 0.03 \qquad r_2\text{-value of solution.}$$

Interpretation	**Thus, the larger internal resistance is 0.06 ohm and the smaller internal resistance is 0.03 ohm.**

EXAMPLE	A tank holds a solution that is 10% herbicide. Another tank holds a solution that is 50% herbicide. If a farmer wants to mix the two solutions to get 200 gal of a solution that is 25% herbicide, how many gallons of each solution should be mixed?
Known facts	There are two strengths of herbicide, 10% and 50%. 200 gal of 25% herbicide are needed.
Unknown facts	How many gallons of 10% herbicide (h) are needed? How many gallons of 50% herbicide (H) are needed?
Relationships	200 gal of the new herbicide are needed: $h + H = 200$ Amount of pure herbicide in h gal of 10% herbicide: $0.1h$ Amount of pure herbicide in H gal of 50% herbicide: $0.5H$ Amount of pure herbicide in 200 gal of 25% herbicide: $0.25(200)$

$$\begin{aligned} h + \quad H &= 200 & \text{Equation 1 (total gallons).} \\ 0.1h + 0.5H &= 0.25(200) & \text{Equation 2 (gallons of pure herbicide).} \end{aligned}$$

Estimation	If equal amounts of herbicide were needed, we would need 100 gal of each solution. However, since the desired solution strength is not exactly halfway between the two original herbicide strengths, we will need unequal amounts of herbicide. One amount will be less than 100 gal and the other will be more than 100 gal.
Calculations	Solve by the substitution method.

$$h + H = 200 \qquad \text{Solve Equation 1 for } h.$$
$$h = 200 - H \qquad \text{Equivalent expression for } h.$$
$$0.1h + 0.5H = 0.25(200) \qquad \text{Substitute } (200 - H) \text{ for } h \text{ in Equation 2.}$$
$$0.1(200 - H) + 0.5H = 0.25(200) \qquad \text{Solve for } H.$$
$$20 - 0.1H + 0.5H = 50$$
$$20 + 0.4H = 50$$
$$0.4H = 50 - 20$$

Chapter 20 / Systems of Linear Equations and Inequalities

$$0.40H = 30$$

$$H = \frac{30}{0.40}$$

$$H = \boxed{75} \text{ gal} \qquad \text{\textit{H}-value of solution.}$$

$$h + \boxed{H} = 200 \qquad \text{Substitute 75 for \textit{H} in Equation 1.}$$

$$h + \boxed{75} = 200 \qquad \text{Solve for \textit{h}.}$$

$$h = 200 - 75$$

$$h = 125 \text{ gal} \qquad \text{\textit{h}-value of solution.}$$

Interpretation   **The farmer must mix 75 gal of the 50% solution and 125 gal of the 10% solution to make 200 gal of a 25% solution.**

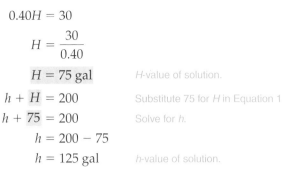

EXAMPLE   Rosita has $5,500 to invest and for tax purposes wants to earn exactly $500 interest for 1 year. She wants to invest part at 10% and the remainder at 5%. How much must she invest at each interest rate to earn exactly $500 interest in 1 year?

Let x = the amount invested at 10%. Let y = the amount invested at 5%. Interest for 1 year = rate × amount invested. Remember to convert percents to decimals. Using these relationships, we derive a system of equations.

Known facts   Total of $5,500 to be invested
$500 interest to be earned in one year

Unknown facts   How much should be invested at 10%?
How much should be invested at 5%?

Relationships   Amount invested at 10%: x
Interest earned at 10%: $0.1x$
Amount invested at 5%: y
Interest earned at 5%: $0.05y$

$$x + \quad y = 5{,}500 \qquad \text{Equation 1 (total investment).}$$
$$0.1x + 0.05y = 500 \qquad \text{Equation 2 (total interest in 1 year).}$$

Estimation   If the total amount were invested at 10%, the interest (in 1 year) would be $550 ($0.1 \times \$5{,}500$). Since we want $500 in interest, most of the money will need to be invested at 10%.

Calculations   Solve by the substitution method.

$$x + y = 5{,}500 \qquad \text{Solve Equation 1 for \textit{x}.}$$

$$x \quad = 5{,}500 - y \qquad \text{Equivalent expression for \textit{x}.}$$

$$0.1\,x + 0.05y = 500 \qquad \text{Substitute } (5{,}500 - y) \text{ for \textit{x} into Equation 2.}$$

$$0.1(5{,}500 - y) + 0.05y = 500 \qquad \text{Solve for \textit{x}.}$$

$$550 - 0.1y + 0.05y = 500 \qquad -0.10y + 0.05y = 0.05y.$$

$$550 - 0.05y = 500$$

$$-0.05y = 500 - 550$$

$$-0.05y = -50$$

$$y = \frac{-50}{-0.05}$$

$$y = \boxed{\$1{,}000} \qquad \text{Amount invested at 5\%.}$$

$$x + \boxed{y} = \$5{,}500 \qquad \text{Substitute \$1,000 for \textit{y} in Equation 1.}$$

$$x + \boxed{1{,}000} = 5{,}500$$

$$x = 5{,}500 - 1{,}000$$

$$x = \$4{,}500 \qquad \text{Amount invested at 10\%.}$$

Interpretation   **Rosita must invest $4,500 at 10% and $1,000 at 5% for 1 year to earn $500 interest.**

SELF-STUDY EXERCISES 20–4

1 Solve the problems using systems of equations with two unknowns.

1. Two boards together are 48 in. If one board is 17 in. shorter than the other, find the length of each board.

2. A broker invested $35,000 in two different stocks. One earned dividends at 4% and the other at 5%. If a $1,570 dividend was earned on both stocks together, how much was invested in each? (*Reminder:* Change 4% to 0.04 and 5% to 0.05.)

3. A department store buyer ordered 12 shirts and 8 hats for $380 one month and 24 shirts and 10 hats for $664 the following month. What was the cost of each shirt and each hat?

4. A mechanic makes $105 on each 8-cylinder engine tune-up and $85 on each 4-cylinder engine tune-up. If the mechanic did 10 tune-ups and made a total of $990, how many 8-cylinder jobs and how many 4-cylinder jobs were completed?

5. 30 resistors and 15 capacitors cost $12. And 10 resistors and 20 capacitors cost $8.50. How much does each capacitor and resistor cost?

6. A private airplane flew 420 mi in 3 hr with the wind. The return trip against the wind took 3.5 hr. Find the rate of the plane in calm air and the rate of the wind.

7. In 1 year, Melissa Deskin earned $660 in interest on two investments totaling $8,000. If she received 7% and 9% rates of return, how much did she invest at each rate?

8. A motorboat went 40 mi with the current in 3 hr. The return trip against the current took 4 hr. How fast was the current? What would have been the speed of the boat in calm water?

9. A visitor to south Louisiana purchased 3 lb of dark-roast pure coffee and 4 lb of coffee with chicory for $27.30 in a local supermarket. Another visitor at the same store purchased 2 lb of coffee with chicory and 5 lb of dark-roast pure coffee for $28. How much did each coffee cost per pound?

10. A lawn-care technician wants to spread a 200-lb seed mixture that is 50% bluegrass. If the technician has on hand a mixture that is 75% bluegrass and a mixture that is 10% bluegrass, how many pounds of each mixture are needed to make 200 lb of the 50% mixture? Round to the nearest whole lb.

11. A college bookstore received a partial shipment of 50 scientific calculators and 25 graphing calculators at a total cost of $2,200. Later the bookstore received the balance of the calculators: 25 scientific and 50 graphing at a cost of $3,800. Find the cost of each calculator.

12. A consumer received two 1-yr loans totaling $10,000 at interest rates of 10% and 15%. If the consumer paid $1,300 interest, how much money was borrowed at each rate?

13. For the first performance at the Overton Park Shell, 40 reserved seats and 80 general admission seats were sold for $2,000. For the second performance, 50 reserved seats and 90 general admission seats were sold for $2,350. What was the cost for a reserved seat and for a general admission seat?

14. A photographer has a container with a solution of 75% developer and a container with a solution of 25% developer. If she wants to mix the solutions to get 8 pt of solution with 50% developer, how many pints of each solution does she need to mix?

ASSIGNMENT EXERCISES

Section 20–1

Solve the systems of equations by graphing.

1. $x + y = 8$
   $x - y = 2$

2. $3x + 2y = 13$
   $x - 2y = 7$

3. $2x + 2y = 10$
   $3x + 3y = 15$

4. $x + y = 1$
   $3x - 4y = 10$

5. $2x - y = 5$
   $4x - 2y = 2$

Shade the portion of the graph represented by the sets of conditions.

6. $y \geq x + 3$ and $x + y < 4$ **7.** $2x + y < 6$ and $x - y < 1$

8. $x + 2y < -1$ and $x + 2y > 3$ **9.** $2x + y > 3$ and $x - y \leq 1$

10. $x + 2y > -4$ and $x - 2y > -1$

Section 20–2

Solve the systems of equations using the addition method.

11. $3x + y = 9$
$2x - y = 6$

12. $2a + 3b = 8$
$a - b = 4$

13. $\quad Q = 2P + 8$
$2Q + 3P = 2$

14. $4j + k = 3$
$8j + 2k = 6$

15. $\quad r = 2y + 6$
$2r + y = 2$

16. $3a + 3b = 3$
$2a - 2b = 6$

17. $\quad c = 2y$
$2c + 3y = 21$

18. $2x + 4y = 9$
$x + 2y = 3$

19. $3R - 2S = 7$
$-14 = -6R + 4S$

20. $x - 3 = -y$
$2y = 9 - x$

21. $\quad c = 2 + 3d$
$3c - 14 = d$

22. $Q - 10 = T$
$T = 2 - 2Q$

23. $x - 18 = -6y$
$4x - 0 = 3y$

24. $R + S = 3$
$S - 9 = -3R$

25. $3a - 2b = 6$
$6a - 12 = b$

26. $a - b = 2$
$a + b = 12$

27. $x + 2y = 7$
$x - y = 1$

28. $2c + 3b = 2$
$2c - 3b = 0$

29. $x + 2r = 5.5$
$2x = 1.5r$

30. $7x + 2y = 6$
$4y = 12 - x$

Section 20–3

Solve the systems of equations using either the addition or the substitution method.

31. $a + 7b = 32$
$3a - b = 8$

32. $x + y = 1$
$4x + 3y = 0$

33. $c - d = 2$
$c = 12 - d$

34. $3a + 4b = 0$
$a + 3b = 5$

35. $7x - 4 = -4y$
$3x + y = 6$

36. $5Q - 4R = -1$
$R + 3Q = -38$

37. $\quad a = 2b + 11$
$3a + 11 = -5b$

38. $\quad y = 5 - 2x$
$3x - 2y = 4$

39. $\quad c = 2q$
$2c + q = 2$

40. $3x + 2y = 10$
$\quad y = 6 - x$

41. $4x - 2.5y = 2$
$2x \quad 1.5y - -10$

42. $2a - c = 4$
$\quad a = 2 + c$

43. $4d - 7 = -c$
$3c - 6 = -6d$

44. $\quad x = 10 - y$
$5x + 2y = 11$

45. $3.5a + 2b = 2$
$\quad 0.5b = 3 - 1.5a$

46. $c + d = 12$
$c - d = 2$

47. $x + 4y = 20$
$4x + 5y = 58$

48. $a + 5y = 7$
$a + 4y = 8$

49. $3a + 1 = -2b$
$4b + 23 = 15a$

50. $6y + 0 = -5p$
$4y - 3p = 38$

Section 20–4

Solve the problems using systems of equations with two unknowns.

51. Three electricians and four apprentices earned a total of $365 on one job. At the same rate of pay, one electrician and two apprentices earned a total of $145. How much pay did each apprentice and electrician receive?

52. 6 bushels of bran and 2 bushels of corn weigh 182 lb. If 2 bushels of bran and 4 bushels of corn weigh 154 lb, how much does 1 bushel of bran and 1 bushel of corn each weigh?

53. A painter paid $22.50 for 2 qt of white shellac and 5 qt of thinner. If 3 qt of shellac and 2 qt of thinner cost the painter's helper $14.50, what is the cost of each quart of shellac and thinner?

54. A main current of electricity is the sum of two smaller currents whose difference is 0.8 A. What are the two smaller currents if the main current is 10 A?

55. The sum of two angles is 175°. Their difference is 63°. What is the measure of each angle?

56. A johnboat traveled 20 mi in 2 hr with the current. The return trip took 3 hr. Find the rate of the boat in calm water and the rate of the current.

| 2 Solve a system of linear inequalities by graphing (pp. 763–765). | Rewrite each inequality as an equation. Graph each equation and shade the area that satisfies each inequality. Where shaded areas overlap is the set of solutions to the system of inequalities. |

Graph $x + y \le -2$ and $x - y > 3$.
For $x + y = -2$,

when $x = 0, y = -2$
when $y = 0, x = -2$
when $x = 1, y = -3$
shade *below* the line.

For $x - y = 3$,

when $x = 0, y = -3$
when $y = 0, x = 3$
when $x = 1, y = -2$
shade *below* the line.

The overlapping shaded area meets both conditions of the system of inequalities.

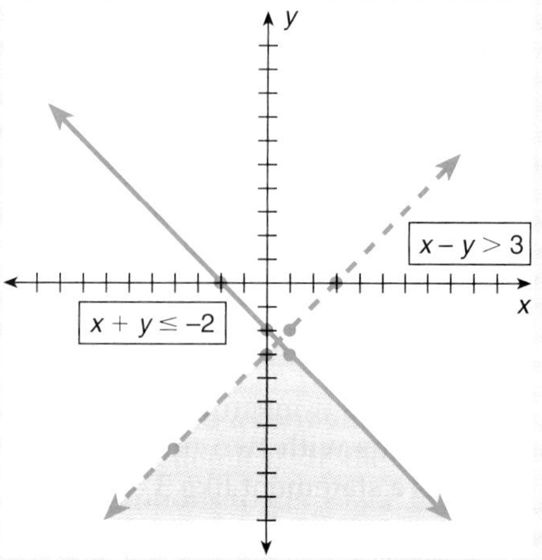

Figure 20–7

Section 20–2

| 1 Use the addition method to solve a system of linear equations that contains opposite variable terms (pp. 766–767). | To solve a system of equations with two opposite variable terms, add the equations so that the opposite terms add to zero. Solve the new equation and substitute the root in an original equation to find the value of the variable. |

Solve $x - y = 6$ and $x + y = 4$ ($-y$ and $+y$ are opposites).

$x - y = 6$ Equation 1. Add the equations to eliminate y.
$x + y = 4$ Equation 2.
$2x \quad = 10$ Solve for x.   Check: $x - y = 6$

$\dfrac{2x}{2} = \dfrac{10}{2}$ $5 - (-1) = 6$

$x = 5$ x-value of solution. $5 + 1 = 6$

                                         $6 = 6$

$x - y = 6$ Equation 1. Substitute $x = 5$. $x + y = 4$

                                   $5 + (-1) = 4$

$5 - y = 6$ Solve for y. $4 = 4$

$-y = 6 - 5$
$-y = 1$ Solution checks.
$y = -1$ y-value of solution.

The solution is (5, −1).

| 2 Use the addition method to solve a system of linear equations that does not contain opposite variable terms (pp. 767–769). | To solve a system of equations that does not contain opposite variables:

1. Multiply one or both equations by signed numbers that cause the terms of one variable to add to zero.
2. Add the equations.
3. Solve the equation from Step 2.
4. Substitute the root in an original equation and solve for the remaining variable.
5. Check solutions. |

Chapter 20 / Systems of Linear Equations and Inequalities

Solve $2x - y = 6$ and $4x + 2y = 4$.

$$2(2x - y) = 2(6)$$ Multiply Equation 1 by 2. Add equations to eliminate y.

$$4x - 2y = 12$$
$$\underline{4x + 2y = 4}$$
$$8x \qquad = 16$$ Solve for x.

$$\frac{8x}{8} = \frac{16}{8}$$

$$x = 2$$ x-value of solution.

$$4x + 2y = 4$$ Equation 2. Substitute $x = 2$.

$$4(2) + 2y = 4$$

$$8 + 2y = 4$$ Solve for y.

$$2y = 4 - 8$$

$$2y = -4$$

$$\frac{2y}{2} = \frac{-4}{2}$$

$$y = -2$$ y-value of solution.

Check:
$$2x - y = 6$$
$$2(2) - (-2) = 6$$
$$4 + 2 = 6$$
$$6 = 6$$
$$4(x) + 2(-2) = 4$$
$$4(2) + 2(-2) = 4$$
$$8 - 4 = 4$$
$$4 = 4$$
Solution checks.

Solution: $(2, -2)$

---

3 Apply the addition method to a system of linear equations with no solution or with many solutions (pp. 769–770).

Apply the addition rule. But note that *both* variables are eliminated. If the resulting statement is false, there is no solution. If the resulting statement is true, there are many solutions.

Solve $a + b = 2$ and $a + b = 4$.

$$-1 \, (a + b) = -1(2)$$ Multiply the first equation by -1. Add to eliminate variables.

$$-a - b = -2$$
$$\underline{a + b = 4}$$
$$0 = 2$$ False. No solution.

Section 20–3

1 Use the substitution method to solve a system of linear equations (pp. 771–772).

To solve a system of equations by substitution:

1. Rearrange either equation to isolate one of the variables.
2. Substitute the equivalent expression from Step 1 in the *other* equation and solve.
3. Substitute the root from Step 2 in one of the equations and solve for the other variable.
4. Check.

Solve $2x + y = 6$ and $2x - 2y = 8$.

$$y = 6 - 2x$$ Isolate y in the first equation.

$$2x - 2y = 8$$ Substitute $6 - 2x$ for y in second equation.

$$2x - 2(6 - 2x) = 8$$

$$2x - 12 + 4x = 8$$ Solve for x.

$$6x - 12 = 8$$

$$6x = 8 + 12$$

$$6x = 20$$

$$\frac{6x}{6} = \frac{20}{6}$$

$$x = \frac{10}{3} \qquad \text{\textit{x}-value of solution.}$$

Check: $2x + y = 6$ $\qquad\qquad$ $2x - 2y = 8$

$$y = 6 - 2\left(\frac{10}{3}\right) \qquad 2\left(\frac{10}{3}\right) + \left(\frac{-2}{3}\right) = 6 \qquad 2\left(\frac{10}{3}\right) - 2\left(-\frac{2}{3}\right) = 8$$

$$y = \frac{18}{3} - \frac{20}{3} \qquad\qquad \frac{20}{3} + \frac{-2}{3} = 6 \qquad\qquad \frac{20}{3} + \frac{4}{3} = 8$$

$$y = -\frac{2}{3} \quad \text{\textit{y}-value of solution.} \qquad \frac{18}{3} = 6 \qquad\qquad \frac{24}{3} = 8$$

$$6 = 6 \qquad\qquad 8 = 8$$

$$\textbf{Solution: } \left(\frac{10}{3}, -\frac{2}{3}\right) \qquad \text{Solution checks.}$$

Section 20–4

1 Use a system of linear equations to solve application problems (pp. 773–775).

Let two variables represent the unknown values. Use numbers and the variables to represent the conditions of the problem. Set up a system of equations and solve.

A taxidermist bought two pairs of glass deer eyes and five pairs of glass duck eyes for $15. She later purchased three pairs of deer eyes and five pairs of duck eyes for $20. Find the price per pair of each type of glass eye.

Let x = the price of a pair of deer eyes. Let y = the price of a pair of duck eyes. Price = the number of pairs of eyes times price of eye type:

$$2x + 5y = 15 \qquad \text{2 pair of deer eyes plus 5 pair of duck eyes cost \$15.}$$

$$3x + 5y = 20 \qquad \text{3 pair of deer eyes plus 5 pair of duck eyes cost \$20.}$$

$$-1(2x + 5y) = -1(15) \qquad \text{Multiply first equation by } -1. \text{ Add equations to eliminate } y.$$

$$\underline{\begin{aligned} -2x - 5y &= -15 \\ 3x + 5y &= 20 \end{aligned}}$$

$$x \quad\;\; = 5 \qquad \text{\textit{x}-value of solution.}$$

$$2x + 5y = 15 \qquad \text{Use first equation to find } y. \text{ Substitute } x = 5.$$

$$2(5) + 5y = 15$$

$$10 + 5y = 15 \qquad \text{Solve for } y.$$

$$5y = 15 - 10$$

$$5y = 5$$

$$\frac{5y}{5} = \frac{5}{5}$$

$$y = 1 \qquad \text{\textit{y}-value of solution.}$$

The deer eyes cost $5 a pair and the duck eyes cost $1 a pair.

CHAPTER TRIAL TEST

Solve the systems of equations graphically.

1. $2a + b = 10$
$\qquad a - b = 5$

2. $x + y = 8$
$\qquad 3x + 2y = 12$

3. $3x + 4y = 6$
$\qquad x + y = 5$

4. $a - 3b = 7$
$\qquad a - 5 = b$

5. $2c - 3d = 6$
$\qquad c - 12 = 3d$

Shade the portion on the graph represented by the conditions.

6. $y \leq 2x + 1$ and $x + y \geq 5$ **7.** $x + y < 4$ and $y > 3x + 2$

Solve the systems of equations using the addition method.

8. $x + y = 6$
$x - y = 2$

9. $p + 2m = 0$
$2p = -m$

10. $6p + 5t = -16$
$3p - 3 = 3t$

11. $3x + y = 5$
$2x - y = 0$

12. $7c - 2b = -2$
$c - 4b = -4$

Solve the systems of equations with two unknowns using the substitution method.

13. $4x + 3y = 14$
$x - y = 0$

14. $a + 2y = 6$
$a + 3y = 3$

15. $7p + r = -6$
$3p + r = 6$

Solve the systems of equations with two unknowns using either the addition or the substitution method.

16. $4x + 4 = -4y$
$6 + y = -6x$

17. $38 + d = -3a$
$5a + 1 = 4d$

Solve the problems using systems of equations with two unknowns.

18. Two lengths of stereo speaker wire total 32.5 ft. One length is 2.9 ft longer than the other. How long is each length of speaker wire?

19. Two currents add to 35 A and their difference is 5 A. How many amperes are in each current?

20. Six packages of common nails and four packages of finishing nails weigh 6.5 lb. If two packages of common nails and three packages of finishing nails weigh 3.0 lb, how much does one package of each kind of nail weigh?

21. The length of a piece of sheet metal is $1\frac{1}{2}$ times the width. The difference between the length and the width is 17 in. Find the length and the width.

22. A mixture of fieldstone is needed for a construction job and will cost $954 for the 27 tons of stone. The stone is of two types, one costing $38 per ton and one costing $32 per ton. How many tons of each are required?

23. A broker invested $25,000 in two different stocks. One stock earned dividends at 3.5% and the other at 4%. If a dividend of $900 was earned on both stocks together, how much was invested in each stock?

24. A mason purchased 2-in. cold-rolled channels and $\frac{3}{4}$-in. cold-rolled channels whose total weight was 820 lb. The difference in weight between the heavier 2-in. and lighter $\frac{3}{4}$-in. channels was 280 lb. How many pounds of each type of channel did the mason purchase?

25. A total capacitance in parallel is the sum of two capacitances. If the capacitance totals 0.00027 farad (F) and the difference between the two capacitances is 0.00016 F, what is the value of each capacitance in the system?

Selected Concepts of Geometry

21

784

21–6 *Inscribed and Circumscribed Regular Polygons and Circles*

1. Use the properties of inscribed and circumscribed equilateral triangles to find missing amounts.

2. Use the properties of inscribed and circumscribed squares to find missing amounts.

3. Use the properties of inscribed and circumscribed regular hexagons to find missing amounts.

Geometry is one of the oldest and most useful of the mathematical sciences. *Geometry* involves the study and measurement of shapes according to their sizes, volumes, and positions. A knowledge of geometry is necessary in many careers. Notice how the meanings of common words you are already familiar with change when they are precisely defined for geometry.

21–1 | *Basic Terminology and Notation*

Learning Outcomes

1. Use various notations to represent points, lines, line segments, rays, and planes.
2. Distinguish among lines that intersect, that coincide, and that are parallel.
3. Use various notations to represent angles.
4. Classify angles according to size.

1 Use Various Notations to Represent Points, Lines, Line Segments, Rays, and Planes.

Geometry is the study of size, shape, position, and other properties of the objects around us. The basic terms used in geometry are *point*, *line*, and *plane*. Generally, these terms are not defined. Instead, they are only described. Once described, they are used in definitions of other terms and concepts.

A *point* is a location or position that has no size or dimension. A dot is used to represent a point, and a capital letter may be used to label the point (Fig. 21–1).

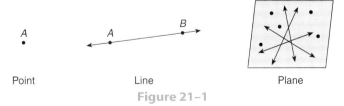

Point   Line   Plane

Figure 21–1

A *line* extends indefinitely in both directions and contains an infinite number of points. It has length but no width. In our discussions, the word *line* always refers to a straight line otherwise specified. In Fig. 21–1, the line can be identified by naming any two points on the line (such as *A* and *B*).

21–1 Basic Terminology and Notation   785

Figure 21–2

Figure 21–3

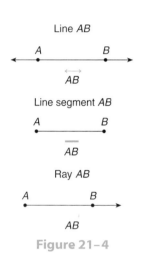

Line *AB*

$\overleftrightarrow{AB}$

Line segment *AB*

$\overline{AB}$

Ray *AB*

$\overrightarrow{AB}$

Figure 21–4

A *plane* is a flat, smooth surface that extends indefinitely in all directions. A plane contains an infinite number of points and lines (Fig. 21–1).

Since a line extends indefinitely in both directions, most geometric applications deal with parts of lines. A part of a line is called a *line segment* or *segment*. A line segment starts and stops at distinct points that we call *end points*. A *line segment*, or *segment*, consists of all points on the line between and including two points that are called *end points* (Fig. 21–2).

The notation for a line that extends through points *A* and *B* is $\overleftrightarrow{AB}$ (read "line *AB*"). The notation for the line segment including points *A* and *B* and all the points between is $\overline{AB}$ (read "line segment *AB*").

Another term used in connection with parts of a line is *ray*. Before we give the definition of a ray, consider the beam of light from a flashlight. The beam is like a ray. It seems to continue indefinitely in only one direction. A *ray* consists of a point on a line and all points of the line on one side of the point (Fig. 21–3).

The point from which the ray originates is called the *end point*, and all other points on the ray are called *interior points* of the ray. A ray is named by its end point and any interior point on the ray. In Fig. 21–3, we use the notation $\overrightarrow{RS}$ to denote the ray whose end point is *R* and that passes through *S*.

To see the contrast in the notation used for a line, line segment, and ray, look at Fig. 21–4.

To illustrate appropriate notations for lines, segments, and rays, consider a line with several points designated on the line (Fig. 21–5).

Figure 21–5

Any two points can be used to name the line in Fig. 21–5. For example $\overleftrightarrow{AB}$, $\overleftrightarrow{AC}$, $\overleftrightarrow{CE}$, $\overleftrightarrow{BD}$, and $\overleftrightarrow{BC}$ are some of the possible ways to name the line. However, a segment is named *only* by its end points. Thus, in Fig. 21–5, $\overline{AB}$ is not the same segment as $\overline{AC}$, but $\overleftrightarrow{AB}$ and $\overleftrightarrow{AC}$ represent the same line.

In Fig. 21–5, $\overrightarrow{BC}$ and $\overrightarrow{BD}$ represent the same ray, but $\overline{BC}$ and $\overline{BD}$ do not represent the same segment.

Learning Strategy   *Using Notation as Clues*

Notation is important in geometry because it allows us to shorten the amount of writing needed to describe a situation, and it allows us to recognize or recall certain properties at a glance.

Use geometric notations as visual clues for understanding the notation.

$\overleftrightarrow{AB}$ line continues indefinitely in both directions.

$\overrightarrow{AB}$ ray continues indefinitely in one direction.

$\overline{AB}$ line segment starts and stops at specific points.

2 Distinguish Among Lines That Intersect, That Coincide, and That Are Parallel.

A line can be extended indefinitely in either direction. If two lines are drawn in the same plane, three things can happen:

1. The two lines *intersect* in *one and only one point*. In Fig. 21–6, $\overleftrightarrow{AB}$ and $\overleftrightarrow{CD}$ intersect at point *E*.

2. The two lines *coincide*; that is, one line fits exactly on the other. In Fig. 21–6, $\overleftrightarrow{EF}$ and $\overleftrightarrow{GH}$ coincide.

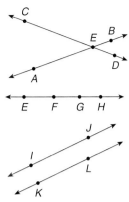

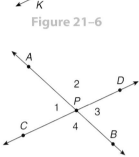

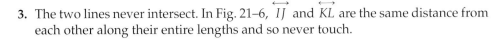

3. The two lines never intersect. In Fig. 21–6, $\overleftrightarrow{IJ}$ and $\overleftrightarrow{KL}$ are the same distance from each other along their entire lengths and so never touch.

The relationship described in the third situation has a special name, *parallel lines.* The symbol ∥ is used for parallel lines. (Chapter 19, Section 4, Outcome 1.)

Figure 21–6

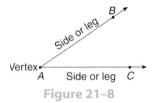

Figure 21–7

3   Use Various Notations to Represent Angles.

When two lines intersect in a point, four *angles* are formed, as shown in Fig. 21–7. An *angle* is a geometric figure formed by two rays that intersect in a point, and the point of intersection is the end point of each ray.

In Fig. 21–8, rays $\overrightarrow{AB}$ and $\overrightarrow{AC}$ intersect at point A. Point A is the end point of $\overrightarrow{AB}$ and $\overrightarrow{AC}$. $\overrightarrow{AB}$ and $\overrightarrow{AC}$ are called the *sides* or *legs* of the angle. Point A is called the *vertex* of the angle.

Angles can be named in several ways. An angle can be named using a number or lowercase letter, by the capital letter that names the vertex point, or by using three capital letters. If three capital letters are used, two of the letters name interior points of each of the two rays, and the middle letter names the vertex point of the angle. Using the symbol $\angle$ for angle, the angle in Fig. 21–9 can be named $\angle 1$, $\angle KLM$, $\angle MLK$, or $\angle L$. One capital letter is used only when it is perfectly clear which angle is designated by the letter. To name the angle in Fig. 21–10 with three letters, we write $\angle XZY$ or $\angle YZX$. This angle can also be named $\angle a$ or $\angle Z$.

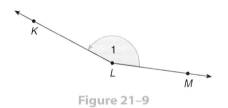

Figure 21–8

Figure 21–9

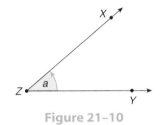

Figure 21–10

Figure 21–11 illustrates how the intersection of two rays actually forms two angles. In this text, we refer to the smaller of the two angles formed by two rays unless the other angle is specifically indicated. Arcs (curved lines) and arrows are often used to clarify which angle we are considering.

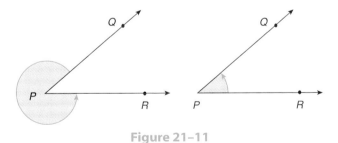

Figure 21–11

4   Classify Angles According to Size.

The measure of an angle is determined by the amount of opening between the two sides of the angle. The length of the sides does not affect the angle measure. Two units are commonly used to measure angles, *degrees* and *radians*. In this chapter, only degrees are used to measure angles. Radians are discussed in Chapter 22.

Consider the hands of a clock as the sides of an angle. When the two hands both point to the same number, the measure of the angle formed is 0 degrees (0°). An angle of 0 degrees is used in trigonometry but is seldom used in geometric applications. During 1 hour, the minute hand makes one complete revolution. Ignoring the movement of the hour hand, this revolution of the minute hand contains 360 degrees (360°). Figure 21–12

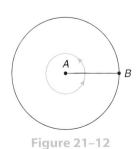

Figure 21–12

21–1 Basic Terminology and Notation

shows a revolution or rotation of 360°. Note that *A* is kept as a fixed point and *B* rotates around point *A*. This rotation can be either clockwise or counterclockwise. A complete rotation counterclockwise is +360°. A complete rotation clockwise is −360°.

Clockwise and Counterclockwise

The customary rotation around a point progresses *opposite* to the normal rotation of the hands of a clock. The direction is called *counterclockwise* (Fig. 21–13).

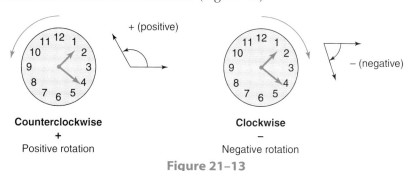

Counterclockwise
+
Positive rotation

Clockwise
−
Negative rotation

Figure 21–13

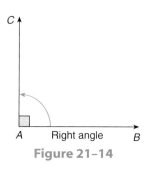

Figure 21–14

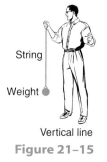

String

Weight

Vertical line

Figure 21–15

A *degree* is a unit for measuring angles. It represents $\frac{1}{360}$ of a complete rotation about the vertex. Suppose, in Fig. 21–14, that $\overrightarrow{AC}$ rotates from $\overrightarrow{AB}$ through one-fourth of a circle. Then $\overrightarrow{AC}$ and $\overrightarrow{AB}$ form a 90° angle ($\frac{1}{4}$ of 360 = 90). This angle is a *right angle*. The symbol for a right angle is ⌞.

If two lines intersect so that right angles are formed (90° angles), the lines are perpendicular to each other. (See Section 19–5.) The symbol for "perpendicular" is ⊥. However, the right-angle symbol also implies the lines forming the angle are perpendicular.

If a string is suspended at one end and weighted at the other (Fig. 21–15), the line it forms is a *vertical line*. A line that is perpendicular to the vertical line is a *horizontal line*. In Fig. 21–16, $\overleftrightarrow{AB}$ is a vertical line, and $\overleftrightarrow{AB}$ and $\overleftrightarrow{CD}$ form right angles. Thus, $\overleftrightarrow{AB} \perp \overleftrightarrow{CD}$, and $\overleftrightarrow{CD}$ is a horizontal line.

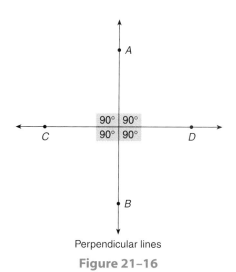

Perpendicular lines

Figure 21–16

Now look at Fig. 21–17. When $\overrightarrow{ML}$ rotates one-half a circle from $\overrightarrow{MN}$, an angle of 180° is formed ($\frac{1}{2}$ of 360 = 180). This angle is a *straight angle*.

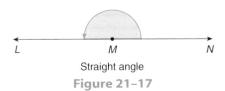

L   M   N

Straight angle

Figure 21–17

An angle that is less than 90° but more than 0° is an *acute angle*. An angle that is more than 90° but less than 180° is an *obtuse angle* (see Fig. 21–18).

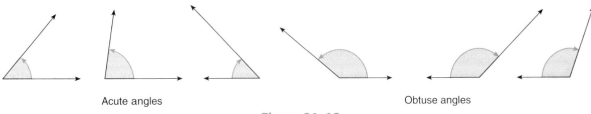

Acute angles   Obtuse angles

Figure 21–18

To classify angles according to size:

1. Examine the angle to determine the number of degrees it measures.
2. Classify the angle as a straight, right, obtuse, or acute angle based on its degree measure.

EXAMPLE   Classify the angles according to size (Fig. 21–19).

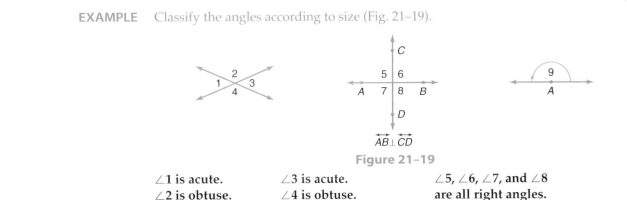

$\overleftrightarrow{AB} \perp \overleftrightarrow{CD}$

Figure 21–19

∠1 is acute.   ∠3 is acute.   ∠5, ∠6, ∠7, and ∠8
∠2 is obtuse.   ∠4 is obtuse.   are all right angles.
                                            ∠9 is a straight angle.

The angle classifications used so far—right, straight, acute, and obtuse—deal with one angle at a time. If two angles together form a right angle or if their measures total 90°, they are *complementary angles*. If two angles together form a straight angle or if their measures total 180°, they are *supplementary angles*.

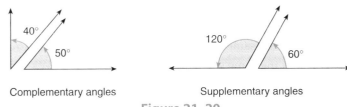

Complementary angles   Supplementary angles

Figure 21–20

21–1 Basic Terminology and Notation

6. A swimming pool has an octagon shape 15 ft on each side. If the distance from the center of the pool to the midpoint of a side is 18.1 ft, what is the area of the pool? The pool is protected by an octagonal cover. What is the measure of each angle of the cover?

7. A no. 5 soccer ball covering is made of 12 colored pentagons 1.75 in. on a side and 20 white hexagons 3.19 in. on a side. If the distance from the midpoint of a side of a colored pentagon to the center of the pentagon is 1.20 in., how many square inches of the covering are colored?

8. Find the area of the layout shown in Fig. 21–74.

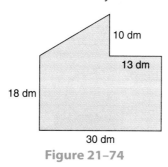

Figure 21–74

9. Find the area of the layout shown in Fig. 21–75.

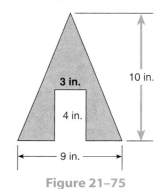

Figure 21–75

2 Give the specific name of each shape in Figs. 21–76 through 21–79 and the number of degrees in each angle.

10.

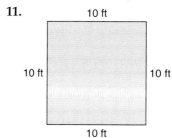

Figure 21–76

11.

10 ft

10 ft   10 ft

10 ft

Figure 21–77

12.

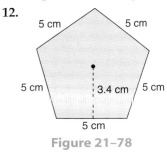

Figure 21–78

13.

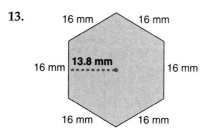

Figure 21–79

3

14. Find the area of Fig. 21–78.

15. Find the area of Fig. 21–79.

21–5 | *Sectors and Segments of a Circle*

Learning Outcomes

1 Find the area of a sector.

2 Find the arc length of a sector.

3 Find the area of a segment.

We often work with figures that are less than a whole circle. For example, earlier we worked with the semicircle in composite figures. *Sectors* and *segments* are both parts of a circle.

To

1. Arrar
2. Add
3. In sut
   on the
4. Write

792

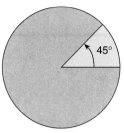

Figure 21–80

1 Find the Area of a Sector.

A *sector* of a circle is the portion of the area of a circle cut off by two radii. The sector is formed by a central angle of the circle and the arc connecting the sides of the angle (radii). See Fig. 21–80.

We use the Greek letter *theta* (θ) to represent the number of degrees in the angle of the sector.

To find the area of a sector:

1. Calculate the portion of the circle included in the sector. θ is a central angle measured in degrees.

$$\frac{\theta}{360} = \text{fractional part of circle}$$

2. Find the fractional part of the area of the circle.

$$A = \frac{\theta}{360} \pi r^2 \qquad \text{where } \pi r^2 = \text{Area of circle}$$

EXAMPLE Find the area of a sector with a central angle of 45° in a circle with a radius of 10 in. Round to hundredths.

$$A = \frac{\theta}{360} \pi r^2 \qquad\qquad \text{Substitute for } \theta \text{ and } r.$$

$$A = \frac{45}{360} (\pi)(10)^2 \qquad \text{Perform the indicated operations.}$$

$$A = 0.125 \, (\pi)(100)$$

$$A = 39.26990817 \text{ in}^2$$

The area of the sector is 39.27 in².

EXAMPLE A cone is made from sheet metal. To form a cone, a sector with a central angle of 40°20′ is cut from a metal circle whose diameter is 20 in. Find the area of the stretchout (portion of the circle) that is used to form the cone to the nearest hundredth (Fig. 21–81).

$$\text{Area used for cone} = \text{Area of circle} - \text{Area of sector}$$

Circle	**Sector**	
$A_1 = \pi r^2$	$A_2 = \dfrac{\theta}{360} \pi r^2$	Substitute known values. Convert 40°20′ to 40.333333°.
$A_1 = \pi(10)^2$	$A_2 = \dfrac{40.3\overline{3}}{360}(314.1592654)$	Substitute A_1 for πr^2.

21–5 Sectors and Segments of a Circle   811

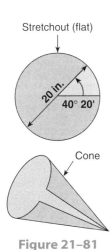

Stretchout (flat)

20 in.

40° 20'

Cone

Figure 21–81

$A_1 = \pi(100)$ $A_2 = 0.112037037(314.1592654)$

$A_1 = 314.1592654 \text{ in}^2$ $A_2 = 35.19747324 \text{ in}^2$

Area used for cone $= A_1 - A_2$

$A_3 = 314.1592654 - 35.19747324$

$A_3 = 278.96 \text{ in}^2$ Rounded.

The area of the metal sector used to form the cone is 278.96 in².

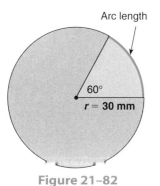

Arc length

60°

$r = 30 \text{ mm}$

Figure 21–82

2 **Find the Arc Length of a Sector.**

The *arc length* of a sector is the portion of the circumference intercepted by the sides of the sector (see Fig. 21–82).

To find the arc length of a sector:

1. Substitute known values into the formula.

$$s = \frac{\theta}{360}(2\pi r) \quad \text{or} \quad s = \frac{\theta}{360}(\pi d)$$

where θ is a central angle measured in degrees.

2. Evaluate.

EXAMPLE Find the arc length of the sector formed by a 60° central angle if the radius of the circle is 30 mm (Fig. 21–82).

$$s = \frac{\theta}{360}(2\pi r) \qquad \text{Substitute values.}$$

$$s = \frac{60}{360}(2)(\pi)(30) \qquad \text{Evaluate.}$$

$$s = 31.41592654$$

$$s = 31.42 \text{ mm.} \qquad \text{Rounded.}$$

Chapter 21 / Selected Concepts of Geometry

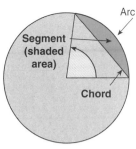

Figure 21-83

3 Find the Area of a Segment.

If a line segment (called a *chord*) joins the end points of the radii that form a sector, the sector is divided into two figures, a triangle and a *segment* (Fig. 21–83). A *chord* is a line segment joining two points on the circumference of a circle. The portion of the circumference cut off by a chord is an *arc*. A *segment* is the portion of the area of a circle bounded by a chord and an arc.

Because the chord divides the sector into a triangle and a segment, we can calculate the area of the segment by subtracting the area of the triangle from the area of the sector. Expressed symbolically, we have the formula for the area of a segment.

To find the area of a segment:

1. Substitute known values into the formula.

$$A = \frac{\theta}{360}\pi r^2 - \frac{1}{2}bh$$

where $\frac{\theta}{360}\pi r^2$ = Area of sector and $\frac{1}{2}bh$ = Area of triangle

2. Evaluate.

EXAMPLE Find the area to the nearest hundredth of the segment in Fig. 21–84.

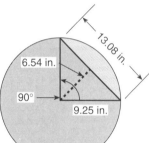

$$A = \frac{\theta}{360}\pi r^2 - \frac{1}{2}bh \qquad \text{Substitute in formula.}$$

$$A = \frac{90}{360}(\pi)(9.25)^2 - \frac{1}{2}(13.08)(6.54) \qquad \text{Evaluate.}$$

$$A = 67.20063036 - 42.7716$$

$$A = 24.43 \text{ in}^2 \qquad \text{Rounded.}$$

Figure 21-84

The area of the segment is 24.43 in².

EXAMPLE A segment in Fig. 21–85 is removed so that a template for a cam is made from the rest of the circle. What is the area of the template? Give the answer to the nearest hundredth.

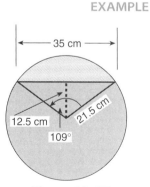

Area of template = Area of circle − Area of segment

Circle

$$A_1 = \pi r^2$$

$$A_1 = \pi(21.5)^2$$

$$A_1 = 1{,}452.201204 \text{ cm}^2$$

Segment

$$A_2 = \frac{\theta}{360}\pi r^2 - \frac{1}{2}bh$$

$$A_2 = \frac{109}{360}(1{,}452.201204) - \frac{1}{2}(35)(12.5)$$

$$A_2 = 220.9442535 \text{ cm}^2$$

Figure 21-85

Area₃ (template) = $A_1 - A_2$

$$A_3 = 1{,}452.201204 - 220.9442535$$

$$A_3 = 1{,}231.26 \text{ cm}^2 \qquad \text{Rounded.}$$

The area of the template is 1,231.26 cm².

SELF-STUDY EXERCISES 21–5

1 Find the area of the sectors of a circle using Fig. 21–86. Round to hundredths.

1. ∠ = 54°
   r = 16 cm

2. ∠ = 25°16′
   r = 30 mm

3. ∠ = 120°30′
   r = 1.52 ft

4. ∠ = 65°
   r = 5 in.

5. ∠ = 150°
   r = 1.45 m

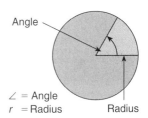

∠ = Angle
r = Radius

Figure 21–86

6. The library of a contemporary elementary school is circular (Fig. 21–87). The floor plan includes sectors reserved for science materials, literary materials, reference materials, and so on. Find the area of the reference section excluding its storage area.

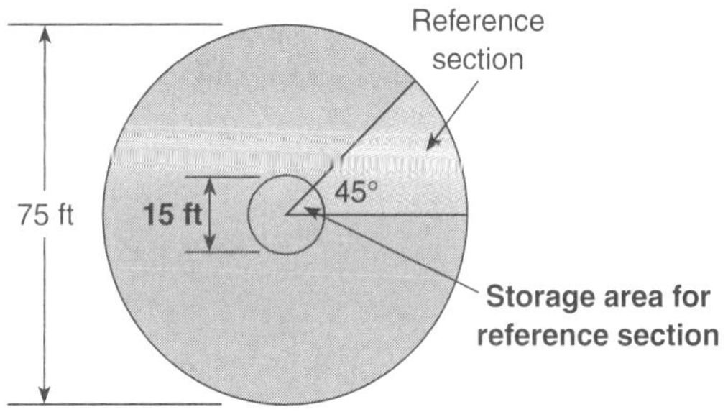

7. A mason lays a tile mosaic featuring a four-sector design (color portion of Fig. 21–88). What is the area of the design to the nearest hundredth?

Figure 21–87

Figure 21–88

8. If the angle formed by a pendulum swing is 19°30′ and the pendulum is 10 in. long (Fig. 21–89), what area does the pendulum move across as it swings? Express your answer to the nearest tenth. (60′ = 1°.)

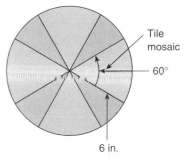

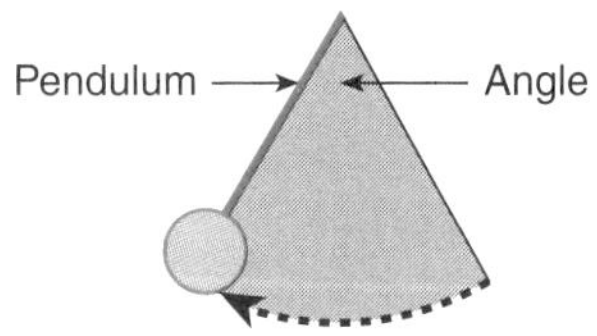

Pendulum → ← Angle

Figure 21–89

2 Find the arc length of the sectors of a circle. Round to hundredths.

9. ∠ = 54°
   r = 30 mm

10. ∠ = 150°
   r = 5 in.

11. ∠ = 120°30′
   r = 1.52 ft

12. ∠ = 25°16′
   r = 16 cm

13. ∠ = 65°
   r = 1.45 m

14. ∠ = 90°
   r = 10 in.

3 Find the area of the segments of a circle using Fig. 21–90. Round to hundredths.

15. ∠ = 60°
   r = 13.3 cm
   h = 11.52 cm
   b = 13.3 cm

16. ∠ = 110°
   r = 10 in.
   h = 5.74 in.
   b = 16.38 in.

17. ∠ = 105°
   r = 11.25 in.
   h = 6.85 in.
   b = 17.85 in.

18. ∠ = 60°
   r = 24 cm
   h = 20.8 cm
   b = 24 cm

19. $\angle = 108°$
   $r = 14$ in.
   $h = 8.25$ in.
   $b = 22.65$ in.

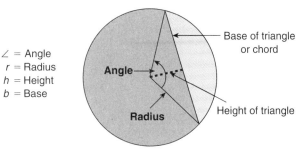

$\angle$ = Angle
r = Radius
h = Height
b = Base

Base of triangle
or chord

Angle

Radius

Height of triangle

Figure 21–90

20. A motor shaft has milled on it a flat for a setscrew to rest so that it can hold a pulley on the shaft (Fig. 21–91). What is the cross-sectional area of the shaft after being milled? Round to hundredths.

20 mm

Flat

14 mm

10 mm

Shaft

90°

Figure 21–91

21. A contractor pours a concrete patio in the shape of a circle except where the patio touches the exterior wall of the house (Fig. 21–92). What is the area of the patio in square feet? Round to hundredths.

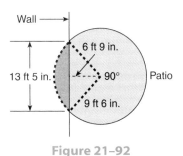

Wall

6 ft 9 in.

13 ft 5 in.

90°

Patio

9 ft 6 in.

Figure 21–92

21–6 | Inscribed and Circumscribed Regular Polygons and Circles

Learning Outcomes

1 Use the properties of inscribed and circumscribed equilateral triangles to find missing amounts.

2 Use the properties of inscribed and circumscribed squares to find missing amounts.

3 Use the properties of inscribed and circumscribed regular hexagons to find missing amounts.

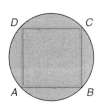

Inscribed square:
Vertices lie on circle
at points *A*, *B*, *C*, and *D*.

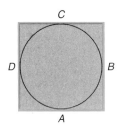

Circumscribed square:
Sides of square are tangent
at points *A*, *B*, *C*, and *D*.

Figure 21–93

Previously, we studied polygons and their areas and perimeters and the area and circumference of a circle. In this section, we examine regular polygons *inscribed* in a circle and *circumscribed* about a circle.

A polygon is *inscribed in a circle* when it is inside the circle and all its *vertices* (points where sides of each angle meet) are on the circle. The circle is said to be *circumscribed about the polygon* (Fig. 21–93).

A polygon is *circumscribed about a circle* when it is outside the circle and all its sides are *tangent* to (intersecting in exactly one point) the circumference. The circle is said to be *inscribed in the polygon* (Fig. 21–93).

In this section, we examine the three most common polygons inscribed in or circumscribed about a circle: the triangle, the square, and the hexagon.

1 Use the Properties of Inscribed and Circumscribed Equilateral Triangles to Find Missing Amounts.

An equilateral triangle can be inscribed in a circle or circumscribed about a circle.

The height (or altitude) of the equilateral triangle has a special relationship with the radius of the circle. The height of an equilateral triangle divides the triangle into two

21–6 Inscribed and Circumscribed Regular Polygons and Circles   **815**

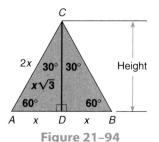

Figure 21–94

30°, 60°, 90° triangles (Fig. 21–94). By relating the radius of a circle to the height, we can then determine the length of the side of the equilateral triangle. Examine the relationships in Figure 21–95.

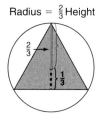

Radius = $\frac{2}{3}$ Height   Radius = $\frac{1}{3}$ Height

Inscribed triangle   Circumscribed triangle

Figure 21–95

Triangle–circle relationships

Equilateral Triangle Inscribed in a Circle	**Equilateral Triangle Circumscribed about a Circle**
Radius of circle = $\frac{2}{3}$(Height of triangle)	Radius of circle = $\frac{1}{3}$(Height of triangle)
Height of triangle = $\frac{3}{2}$(Radius of circle)	Height of triangle = 3(Radius of circle)
Side of triangle = $\dfrac{2(\text{Height of triangle})}{\sqrt{3}}$	Side of triangle = $\dfrac{2(\text{Height of triangle})}{\sqrt{3}}$
Side of triangle = $\dfrac{3(\text{Radius of circle})}{\sqrt{3}}$	Side of triangle = $\dfrac{6(\text{Radius of circle})}{\sqrt{3}}$
Radius of circle = $\dfrac{\sqrt{3}\,(\text{Side of triangle})}{3}$	Radius of circle = $\dfrac{\sqrt{3}(\text{Side of triangle})}{6}$

EXAMPLE Find the dimensions for the inscribed equilateral triangle in Fig. 21–96, where $CD = 13$ mm and $AO = 15$ mm.

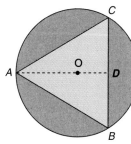

Figure 21–96

(a) Height of $\triangle ABC$ (b) BD (c) BC (d) $\angle CAD$
(e) $\angle ACD$ (f) Area of $\triangle ABC$

(a) Height of triangle = $\frac{3}{2}$ (radius of circle)

$AD = \frac{3}{2}(AO)$ Substitute $AO = 15$ mm.

$AD = \frac{3}{2}(15)$ Simplify.

$AD = \frac{45}{2}$

$AD = 22.5$ mm

(b) $BD = CD$ Corresponding sides of congruent triangles.
  $BD = 13$ mm   Substitute $CD = 13$ mm.

(c) $BC = BD + CD$ Substitute 13 mm for BD and CD.
  $BC = 13$ mm $+ 13$ mm   Simplify.
  $BC = 26$ mm

(d) $\angle CAD = 30°$ $\triangle CAD$ is a 30°, 60°, 90° $\triangle$, and $\angle CAD$ is opposite shortest side.

Chapter 21 / Selected Concepts of Geometry

(e) $\angle ACD = 60°$ $\angle ACD$ is opposite height or other leg of $\triangle$.

(f) $A = \dfrac{1}{2}bh$ Formula for area of triangle. Substitute for base and height.

$A = \dfrac{1}{2}(BC)(AD)$ $BC = 26$, $AD = 22.5$

$A = \dfrac{1}{2}(26)(22.5)$ Simplify.

$A = \textbf{292.5 mm}^2$ Area of $\triangle ABC$.

EXAMPLE   One end of a shaft 35 mm in diameter is milled as shown in Fig. 21–97. Find the radius of the smaller circle.

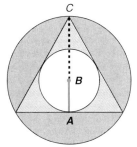

Figure 21–97

$BC = \text{Radius} = \dfrac{1}{2}\,\text{Diameter of large circle}$

$BC = \dfrac{1}{2}(35)$ Substitute 35 for diameter.

$BC = 17.5$ mm   Radius of large circle.

$AB = \dfrac{1}{2}BC$ From $AB = \dfrac{1}{3}AC$ and $BC = \dfrac{2}{3}AC$.

$AB = \dfrac{1}{2}(17.5)$

$AB = 8.75$ mm   Radius of small circle.

The radius of the smaller circle is 8.75 mm.

EXAMPLE   Find the area of the larger circle in Fig. 21–97. Round to hundredths.

$A = \pi r^2 = \pi(17.5)^2 = 962.11 \text{ mm}^2$ Substitute 17.5 for radius.

The area of the larger circle is 962.11 mm^2.

---

2 Use the Properties of Inscribed and Circumscribed Squares to Find Missing Amounts.

A square has four equal sides and four equal angles, each 90°. If a diagonal is drawn, the result is two congruent right triangles that have angles of 45°, 45°, 90°. A second diagonal divides the square into four 45°, 45°, 90° congruent right triangles, as shown in Fig. 21–98. The special 45°, 45°, 90° triangle and its properties may be reviewed in Section 3, Outcome 4. Diagonals AC and BD are equal in length and bisect each other. Point O, where the diagonals intersect, is the center of the square and the center of any inscribed or circumscribed circle (Fig. 21–99).

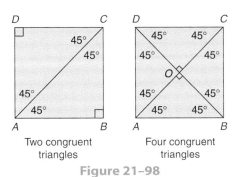

Two congruent triangles   Four congruent triangles

Figure 21–98

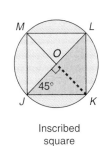

Inscribed square

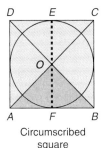

Circumscribed square

Figure 21–99

21–6 Inscribed and Circumscribed Regular Polygons and Circles   817

Square–circle relationships

Square Inscribed in Circle	Square Circumscribed About a Circle
Radius of circle = $\frac{1}{2}$ Diagonal of square	Side of square (across flats) = Diameter of circle
Diameter of circle = Diagonal of square (across corners)	Radius of circle = $\frac{1}{2}$ Side of square
$\dfrac{\text{Side of square}}{\text{(across flats)}} = \dfrac{\text{Diameter of circle}}{\sqrt{2}}$	Diagonal of square (across corners) = $\sqrt{2}$ (Diameter of circle)
Diameter of circle = $\sqrt{2}$(Side of square)	Diagonal of square (across corners) = $2\sqrt{2}$ (Radius of circle)

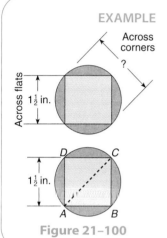

Figure 21–100

EXAMPLE To mill a square bolt head $1\frac{1}{2}$ in. across flats, round stock of what diameter is needed? Answer to the nearest hundredth (see Fig. 21–100).

The distance across corners is the diameter of the circle and the diagonal of the square. The distance across flats is the side of a square inscribed in a circle.

Diameter = $\sqrt{2}$ (Side of square)   Substitute 1.5 for the side of the square. $1\frac{1}{2}$ in. = 1.5 in.

Diameter = $\sqrt{2}$ (1.5)

Diameter = 2.12 in.   Rounded.

The diameter of the round stock is 2.12 in.

Figure 21–101

EXAMPLE Find the diagonal of an inscribed square if the area of the circle is 22 in². Round to hundredths (Fig. 21–101).

$$A = \pi r^2$$ Formula for the area of circle. Substitute $A = 22$.

$$22 = \pi r^2$$ Solve for r.

$$\frac{22}{\pi} = r^2$$

$$\sqrt{\frac{22}{\pi}} = r$$

$$2.646283714 = r$$ Radius of the circle.

Diagonal = Diameter = $2 \times$ Radius

$$= 2(2.646283714) = 5.29 \text{ in.}$$ Rounded.

The diagonal of the inscribed square is 5.29 in.

EXAMPLE What is the minimum depth of cut required to mill a circle on the end of a square piece of stock with a side measuring 4 cm? Answer to hundredths (Fig. 21–102).

The side of the square is 4 cm, so the diameter of the circle is 4 cm and the radius is 2 cm. If the radius is drawn perpendicular to the top side, a right triangle is formed as indicated (shading).

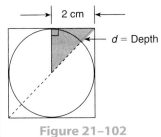

Figure 21–102

Let d = depth of milling, which equals the hypotenuse of the shaded triangle minus the radius of the circle.

$$c^2 = a^2 + b^2$$ — Find the hypotenuse using the Pythagorean theorem. Substitute $a = 2$ and $b = 2$..

$$c^2 = 2^2 + 2^2$$ — Solve for c.
$$c^2 = 4 + 4$$
$$c^2 = 8$$
$$c = \sqrt{8}$$
$$c = 2.828427125 \text{ cm}$$ — Hypotenuse.

Depth of milling = Hypotenuse − Radius
$$d = c - r$$ — Substitute $c = 2.828427125$ and $r = 2$.
$$d = 2.828427125 - 2$$
$$d = 0.83 \text{ cm}$$ — Rounded.

The minimum depth of milling is 0.83 cm.

3 **Use the Properties of Inscribed and Circumscribed Regular Hexagons to Find Missing Amounts.**

A regular hexagon is a figure with six equal sides and six equal angles of 120° each. Three diagonals joining pairs of opposite vertices divide the hexagon into six congruent equilateral triangles (all angles 60°) (see Fig. 21–103).

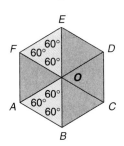

Figure 21–103

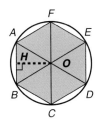
Inscribed hexagon

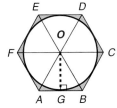
Circumscribed hexagon

Across flats

Across corners

Figure 21–104

Hexagon-circle relationships (Figure 21–104)

Hexagon Inscribed in Circle

Diameter of circle = Diagonal of hexagon (across corners)

Radius of circle = Side of hexagon

Across flats = (side of hexagon) $\left(\sqrt{3}\right)$

Hexagon Circumscribed About a Circle

Across flats = Diameter of circle

Across corners = 2 (side of hexagon)

Radius of circle = Height of one triangle formed by the three diagonals

$$= \frac{\text{side of hexagon}\left(\sqrt{3}\right)}{2}$$

EXAMPLE Find the indicated dimensions. When appropriate, round to hundredths. Use Fig. 21–105.

(a) $\angle EOD$ (b) $\angle COH$ (c) $\angle BHO$ (d) Radius
(e) Distance across corners   (f) FO

21-6 Inscribed and Circumscribed Regular Polygons and Circles

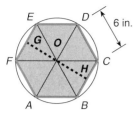

Figure 21–105

(a) $\angle EOD = 60°$ This is an angle of an equilateral triangle.

(b) $\angle COH = 30°$ Height *OH* bisects the 60° angle to form a 30°, 60°, 90° triangle.

(c) $\angle BHO = 90°$ The height *OH* forms a right angle with the base of $\triangle BOC$.

(d) Radius = 6 in.   The radius is a side of an equilateral triangle or a side of the hexagon.

(e) Distance across corners = 12 in.   The distance across corners is the diameter.

(f) $FO = 6$ in.   *FO* is a radius or a side of an equilateral triangle.

EXAMPLE   Find the indicated dimensions. When appropriate, round to hundredths. Use Fig. 21–106.

Figure 21–106

(a) Distance across flats   (b) *EO*   (c) Diagonal   (d) $\angle FED$

(a) Distance across flats = 8 cm   This is the diameter of the circle.

(b) $EO = 4.62$ cm   *EO* is a side of an equilateral triangle that equals the side of the hexagon.

(c) Diagonal = 9.24 cm   The diagonal is twice the side of an equilateral triangle.

(d) $\angle FED = 120°$ This is the angle at the vertex of a hexagon.

EXAMPLE   If a regular hexagon is cut from a circle with a radius of 6.45 in., how much of the circle is not used? Round to hundredths (Fig. 21–107)

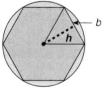

Figure 21–107

To solve, we find the area of the circle and the area of the hexagon and subtract to find the difference (portion not used).

$A_{circle} = \pi r^2$ Substitute $r = 6.45$ in.

$A_{circle} = \pi(6.45)^2$ Simplify.

$A_{circle} = \pi(41.6025)$

$A_{circle} = 130.6981084$ in^2 Area of circle.

$A_{hexagon} = 6\left(\dfrac{1}{2}bh\right)$ $A_{hexagon}$ = area of six triangles. The height of an equilateral triangle forms a 30°, 60°, 90° triangle, $h = \dfrac{b}{2}\sqrt{3} = 3.225 \sqrt{3} = 5.585863854$ in.

$A_{hexagon} = 6\left[\dfrac{1}{2}(6.45)(5.585863854)\right]$ Substitute $b = 6.45$ and $h = 5.585863854$.

$A_{hexagon} = 108.0864656$ in^2

Waste = $A_{circle} - A_{hexagon}$ Difference in areas.

Waste = $130.6981084 - 108.0864656$

Waste = 22.61 in^2 Rounded.

There are 22.61 in^2 of waste from the circle.

EXAMPLE   Use Fig. 21–108 to answer the questions. Round answers to hundredths.

(a) What is the smallest-diameter round stock from which a hex-bolt head $\frac{1}{4}$ in. on a side can be milled?

(b) What is the distance across the corners of the hex-bolt head?

(c) What is the distance across the flats?

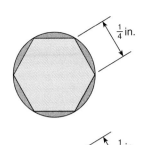

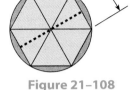

Figure 21–108

(a) Diameter $= 2 \times$ Side $=$ Diagonal

Smallest-diameter round stock $= \dfrac{1}{2}$ in.

Side $=$ radius. $2 \times \dfrac{1}{4} = \dfrac{1}{2}$

(b) Distance across corners $=$ Diagonal

Distance across corners $= \dfrac{1}{2}$ in.

(c) Distance across flats $=$ (side of hexagon)$(\sqrt{3})$

$\qquad = 0.25\left(\sqrt{3}\right)$

$\qquad = 0.43$

The distance across the flats is 0.43 in.

Side of hexagon $= \dfrac{1}{4}$ in. $= 0.25$ in.

To nearest hundredth.

Learning Strategy *Dealing with Relationship Overload*

Wow! How will I ever remember all these new words and relationships? It is important to remember the reason we examine relationships is to broaden our experience to have more potential solutions to pursue when finding missing information. In the workplace and in everyday life, you can look up and review the specifics of relationships. Remembering a few key words such as inscribed or circumscribed enables you to look up the words to locate examples, explanations, and exercises that can refresh your skills.

SELF-STUDY EXERCISES 21–6

1 Use Fig. 21–109 to find the indicated dimensions for an equilateral triangle inscribed in a circle. Round to hundredths.

1. $\angle CAB$ **2.** $\angle BCD$ **3.** $\angle ADC$
4. CB **5.** AC **6.** Diameter
7. Area of $\triangle ABC$ **8.** AD **9.** Radius
10. Circumference

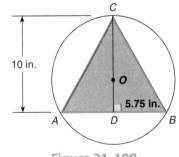

Figure 21–109

Solve the problems involving inscribed or circumscribed equilateral triangles. Round to hundredths.

11. The area of an equilateral triangle with a base of 10 cm is 43.5 cm². The triangle is circumscribed about a circle. What is the radius of the inscribed circle?

12. One end of a circular steel rod is milled into an equilateral triangle whose height is $\frac{3}{4}$ in. What is the diameter of the rod if the smallest diameter possible was used for the job?

2 Find the measures of the inscribed square in Fig. 21–110.

13. ∠DOC
14. ∠BCO
15. BO
16. ∠BOE
17. CE

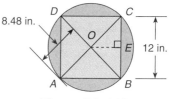

Figure 21–110

Find the measures of the circumscribed square in Fig. 21–111.

18. DO
19. Radius of circle
20. ∠AOB
21. BE

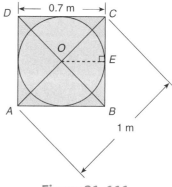

Figure 21–111

Solve the problems involving squares and circles. Round the answers to hundredths.

22. To what depth must a 2-in. shaft (Fig. 21–112) be milled on an end to form a square 1.414 in. on a side?
23. What is the smallest-diameter round stock (Fig. 21–113) needed to mill a square $\frac{3}{4}$ in. on a side on the end of the stock?
24. What is the area of the largest circle inscribed in a square with a perimeter of 36 dkm?

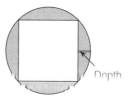

Figure 21–112

Figure 21–113

3 Find the dimensions for Fig. 21–114 if GO = 17.32 mm and FO = 20 mm.

25. Distance across flats
26. ∠CDE
27. ∠EFO
28. Distance across corners
29. ∠FGO
30. GA
31. AB

Figure 21–114

Find the dimension for Fig. 21–115 if FO = 3 in.

32. ∠ODC
33. ∠EOG
34. OC

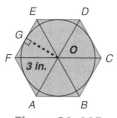

Figure 21–115

35. What is the distance across the flats of a hexagon cut from a circular blank with a circumference of 145 mm (Fig. 21–116)? Round to hundredths.
36. The end of a hexagonal rod is milled into the largest circle possible (Fig. 21–117). What is the distance across the corners of the hexagon if the area of the circle is 4.75 in²? Round to hundredths.

Figure 21–116

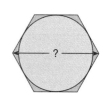

Figure 21–117

Section 21–1

Give the proper notation.
1. Line *AB*
2. Segment *AB*
3. Ray *AB*

Use Fig. 21–118 for Exercises 4–10.
4. Name the line in two different ways.
5. Name the ray with end point *N* and with an interior point *O*.
6. Name the ray with end point *M* and with an interior point *L*.
7. Is $\overline{MN}$ the same as $\overline{MO}$?
8. Is $\overleftrightarrow{NO}$ the same as $\overleftrightarrow{NM}$?
9. Is $\overrightarrow{NO}$ the same as $\overrightarrow{NM}$?
10. Is $\overrightarrow{MN}$ the same as $\overrightarrow{MO}$?

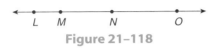

Figure 21–118

Use Fig. 21–119 for Exercises 11–15.
11. Which lines are parallel?
12. Which lines coincide?
13. Which lines intersect?
14. When will $\overleftrightarrow{AB}$ and $\overleftrightarrow{CD}$ meet?
15. Is $\overleftrightarrow{IJ}$ the same as $\overleftrightarrow{KJ}$?

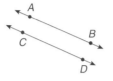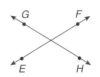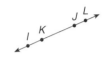

Figure 21–119

Use Fig. 21–120 for Exercises 16–21.
16. Name ∠*a* using three capital letters.
17. Name ∠*b* using three capital letters.
18. Name ∠*c* using three capital letters.
19. Name ∠*a* using one capital letter.
20. Name ∠*b* using one capital letter.
21. Name ∠*c* using one capital letter.

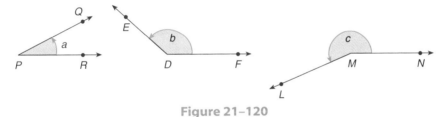

Figure 21–120

Classify the angle measures using the terms *right, straight, acute,* or *obtuse.*
22. 50°
23. 90°
24. 120°
25. 18°
26. 75°
27. 180°
28. 89°
29. 179°
30. 30°

State whether the angle pairs are complementary, supplementary, or neither.
31. 63°, 37°
32. 98°, 62°
33. 135°, 45°
34. 45°, 35°
35. 21°, 79°
36. 90°, 90°
37. Congruent angles have _____ measures.
38.–39. Perpendicular lines are formed when a _____ line and a _____ line intersect.

Section 21–2

Add or subtract as indicated. Write in standard notation.

40. 34° 23′ 41″
   + 18° 37′ 50″

41. 115° 34′ 29″
   − 84° 26′ 18″

42. 34° 29′ 35″
   + 19° 30′ 25″

43. 64° 15′ 37″
   − 29° 37′ 41″

44. 80°
   − 28° 14′ 28″

45. 74°
   − 13° 19′ 42″

46. Find the complement of an angle whose measure is 35° 29′ 14″.

Change to decimal degree equivalents. Express the decimals to the nearest ten-thousandth.
47. 29′
48. 47″
49. 7′34″

50. An angle of 34° 36′ 48″ is divided into two equal angles. Express the measure of each angle in decimal degrees. Round to the nearest ten-thousandth.

Change to equivalent minutes and seconds. Round to the nearest second when necessary.

51. 0.75°

52. 0.46°

53. 0.2176°

54. A right angle is divided into eight equal parts. Find the measure of each angle in degrees and minutes.

55. An angle of 140° is divided into six equal parts. Find the measure of each angle in degrees and minutes.

56. An angle of 18° 52′ 48″ needs to be 3 times as large. Find the measure of the new angle in decimal degrees.

57. An angle of 37° 19′ 41″ is divided into two equal angles. Find the measure of each equal angle in degrees, minutes, and seconds.

Section 21–3

Classify the triangles in Figs. 21–121 through 21–123 according to their sides.

58.

Figure 21–121

59.

Figure 21–122

60.

Figure 21–123

Identify the longest and shortest sides in Figs. 21–124 and 21–125.

61.

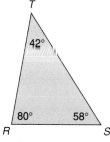

Figure 21–124

62.

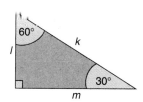

Figure 21–125

List the angles in order of size from largest to smallest in Figs. 21–126 and 21–127.

63.

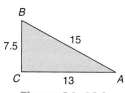

Figure 21–126

64.

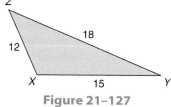

Figure 21–127

Write the corresponding parts not given for the congruent triangles in Figs. 21–128 through 21–130.

65.

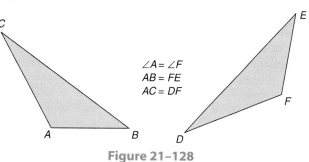

∠A = ∠F
AB = FE
AC = DF

Figure 21–128

66.

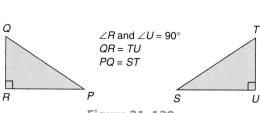

∠R and ∠U = 90°
QR = TU
PQ = ST

Figure 21–129

67.

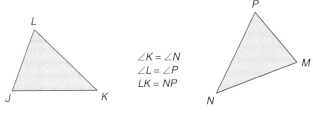

$\angle K = \angle N$
$\angle L = \angle P$
$LK = NP$

Figure 21–130

Use Fig. 21–131 to solve the following exercises. Round your final answers to the nearest thousandth if necessary.

68. $RS = 8$ mm; find ST and RT.
69. $RT = 15$ cm; find RS and ST.
70. $ST = 7.2$ ft; find RS and RT.
71. $RT = 9\sqrt{2}$ hm; find RS and ST.
72. $RT = 8\sqrt{5}$ dkm; find RS and ST.

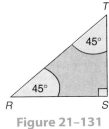

Figure 21–131

Use Fig. 21–132 to solve the following exercises. Round your final answers to the nearest thousandth if necessary.

73. $AC = 12$ dm; find AB and BC.
74. $AB = 15$ km; find AC and BC.
75. $BC = 10$ in.; find AC and AB.
76. $BC = 8\sqrt{2}$ hm; find AC and AB.
77. $AC = 40$ ft 7 in.; find AB and BC to the nearest inch.

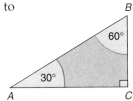

Figure 21–132

Solve. Round your final answers to the nearest thousandth if necessary.

78. The sides of an equilateral triangle are 4 dm in length. Find the altitude of the triangle. Round the answer to the nearest thousandth of a decimeter.

79. Find the total length to the nearest inch of the conduit $ABCD$ if $XY = 8$ ft, $CE = 14$ in., and angle $CBE = 30°$ (Fig. 21–133).

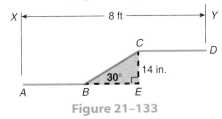

Figure 21–133

80. A piece of round steel is milled to a point at one end. The angle formed by the point is the angle of taper. If the steel has a 16-mm diameter and a 60° angle of taper, find the length c of the taper (Fig. 21–134).

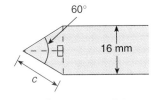

Figure 21–134

81. A V-slot forms a triangle whose angle at the vertex is 60°. The depth of the slot is 17 mm (see Fig. 21–135). Find the width of the V-slot.

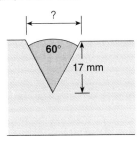

Figure 21–135

82. A manufacturer recommends attaching a guy wire to its 30-ft antenna at a 45° angle. If the antenna is installed on a flat surface, how long must the guy wire be (to the nearest foot) if it is attached to the antenna 4 ft from the top?

Assignment Exercises

Find the area of Figs. 21–136 and 21–137.

83.

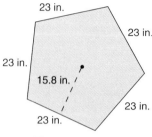

23 in.

23 in.

23 in.

15.8 in.

23 in.

23 in.

Figure 21–136

84.

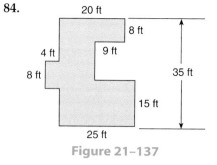

20 ft

8 ft

4 ft

9 ft

8 ft

35 ft

15 ft

25 ft

Figure 21–137

85. Identify the figure in Exercise 83 by giving its specific name and the number of degrees in each angle.

Solve the problems involving the perimeter and area of composite figures.

86. How many sections of 2 ft × 4 ft ceiling tiles will be required for the layout in Fig. 21–138?

87. How many feet of baseboard are needed to install around the room in Fig. 21–138 if we make no allowance for openings or doorways?

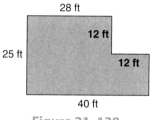

28 ft

12 ft

25 ft

12 ft

40 ft

Figure 21–138

88. A gable end of a gambrel roof (Fig. 21–139) is to be covered with 12-in. bevel siding at an $8\frac{1}{2}$-in. exposure to the weather. Estimate the square footage of siding needed if 18% is allotted for lap and waste. Answer to the nearest foot.

89. Find the square feet of floor space inside the walls of the plan of Fig. 21–140. Answer to the nearest square foot.

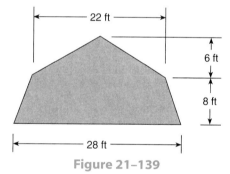

22 ft

6 ft

8 ft

28 ft

Figure 21–139

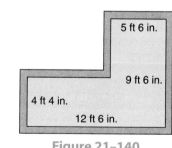

5 ft 6 in.

9 ft 6 in.

4 ft 4 in.

12 ft 6 in.

Figure 21–140

90. If one 2 in. × 4 in. stud is estimated for every linear foot of a wall when studs are 16 in. on center, how many studs are needed for the outside walls of the floor plan in Fig. 21–141?

91. A home has concrete steps in the rear (Fig. 21–142). The two sides are covered with a brick facing. How many bricks are needed if $\frac{1}{4}$-in. mortar joints are used? (Six bricks cover 1 ft².) Round any part of a square foot to the next-highest square foot. (144 in² = 1 ft².)

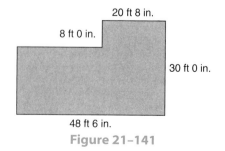

20 ft 8 in.

8 ft 0 in.

30 ft 0 in.

48 ft 6 in.

Figure 21–141

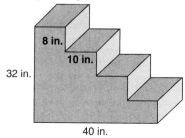

8 in.

10 in.

32 in.

40 in.

Figure 21–142

92. A mason lays tile to cover an area in the form of a regular octagon with eight sides each 18 ft. If the distance from the midpoint of a side to the center of the octagon is 21.7 ft, how many tiles are required if $7\frac{1}{2}$ tiles will cover 1 ft^2 and 36 tiles are added for waste?

Section 21–5

Find the area of the sectors of a circle using Fig. 21–143. Round to hundredths.

93. $\angle = 45° 9'$
$r = 2.58$ cm

94. $\angle = 165°$
$r = 15$ in.

95. $\angle = 15° 15'$
$r = 110$ mm

96. $\angle = 75°$
$r = 1$ m

97. $\angle = 40°$
$r = 2\frac{1}{2}$ ft

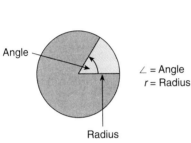

$\angle$ = Angle
r = Radius

Figure 21–143

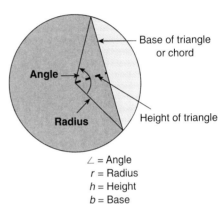

Base of triangle or chord

Angle

Radius

Height of triangle

$\angle$ = Angle
r = Radius
h = Height
b = Base

Figure 21–144

Find the area of the segments of a circle using Fig. 21–144. Round to hundredths.

98. $\angle = 52° 30'$
$r = 11.25$ in.
$h = 10.09$ in.
$b = 10$ in.

99. $\angle = 45°$
$r = 14$ mm
$h = 12.9$ mm
$b = 10.9$ mm

100. $\angle = 55°$
$r = 10''$
$h = 8.9''$
$b = 9.2''$

101. $\angle = 30°$
$r = 24$ cm
$h = 23.2$ cm
$b = 12.4$ cm

102. $\angle = 60°$
$r = 13.3$ cm
$h = 11.5$ cm
$b = 13.3$ cm

Solve.

103. What is the area (to the nearest tenth) of the flat-top bifocal lens shown in gray shading (Fig. 21–145)?

104. The minute hand on a grandfather clock makes an angle of 30° as it moves from 12 to 1 on the clock face. If the minute hand is 27.5 cm long, what is the area of the clock face over which the minute hand moves from 12 to 1? Round to tenths.

105. What is the area of a patio (Fig. 21–146) that has one rounded corner? Round to hundredths.

106. A machine cuts a 12-in.-diameter frozen pizza into slices with sides that form 72° angles at the center of the pizza. What is the surface area of each slice to the nearest square inch?

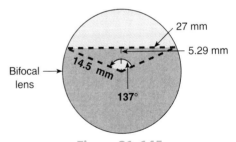

27 mm

5.29 mm

Bifocal lens

14.5 mm

137°

Figure 21–145

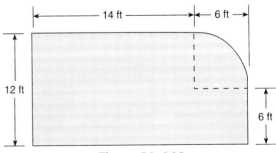

14 ft

6 ft

12 ft

6 ft

Figure 21–146

107. A drain pipe with a 20-in diameter has 4 in. of water in it (Fig. 21–147). What is the cross-sectional area of the water in the pipe to the nearest hundredth?

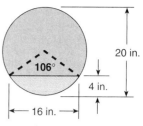

Figure 21–147

Find the arc length of the sectors of a circle using Fig. 21–143. Round to hundredths.

108. ∠ = 40°
   r = 1.45 ft

109. ∠ = 180°
   r = 10 in.

110. ∠ = 30°15′
   r = 20 cm

111. ∠ = 70°10′
   r = 30 mm

Section 21–6

For each figure, supply the missing dimensions.
For an inscribed triangle, see Fig. 21–148.
112. If FB = 9, FO = ————
113. If AB = 10, AE = ————
114. ∠ABF = ————

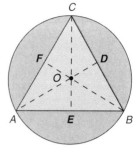

Figure 21–148

For a circumscribed square, see Fig. 21–149.
115. ∠GJO = ————
116. If GO = 7, HO = ————
117. If KO = 10, IJ = ————

For an inscribed hexagon, see Fig. 21–150.
118. If QO = 15, MN = ————
119. ∠MOP = ————
120. ∠MNO = ————

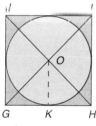

Figure 21–149

Figure 21–150

Solve the problems involving polygons and circles. Round the answers to hundredths.

121. A shaft with a 20-mm diameter is milled at one end, as illustrated in Fig. 21–151. What is the radius of the inscribed circle?

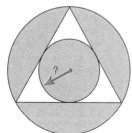

Figure 21–151

122. A hex nut 0.87 in. on a side is milled from the smallest-diameter round stock possible (Fig. 21–152). What is the diameter of the stock?

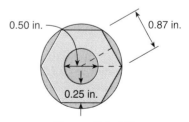

Figure 21–152

123. A square is milled on the end of a 5-cm shaft (Fig. 21–153). If the square is the largest that can be milled on the 5-cm shaft, what is the length of a side to hundredths?

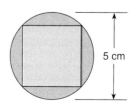

Figure 21–153

124. Compare the area of a circle with a radius of 5 cm to the area of an inscribed equilateral triangle. Compare the area of a circle with a radius of 5 cm to the area of an inscribed square. Compare the area of a circle with a radius of 5 cm to the area of the inscribed polygon in Fig. 21–154.

125. Estimate the area of the inscribed regular pentagon in Fig. 21–155 by giving two values the area is between. Present a convincing argument for your estimate. Similarly, estimate the area of an inscribed regular octagon and give an argument for your estimate.

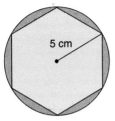

Figure 21–154

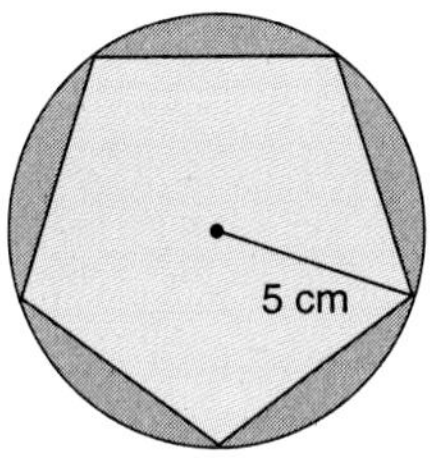

Figure 21–155

1. Can the measures 8 cm, 12 cm, and 25 cm represent the sides of a triangle? Illustrate and explain your answer.

2. The Pythagorean theorem, $a^2 + b^2 = c^2$, applies to right triangles where a and b are legs and c is the hypotenuse. For what type of triangle is the statement $a^2 + b^2 > c^2$ true? Illustrate your answer.

3. For what type of triangle is the statement $a^2 + b^2 < c^2$ true? Illustrate your answer.

4. A triangle has 6, 3, and $3\sqrt{3}$ as the measures of its sides. Sketch the triangle, place the measures on the appropriate sides, and show the measures of each of the three angles of the triangle.

5. Explain the process for finding the perimeter of a composite figure.

6. Explain the process for finding the area of a composite figure.

7. Draw a composite figure for which the area can be found by using either addition or subtraction and explain each approach.

8. Write in words the formula for finding the degrees in each angle of a regular polygon.

9. Describe a shortcut for finding the perimeter of an L-shaped figure. Illustrate your procedure with a problem.

10. Two pieces of property have the same area (acreage) and equal desirability for development. Both require expensive fencing. One piece is a square 600 ft on each side. The other piece is a rectangle 400 ft by 900 ft. Which piece of property requires the least amount of fencing and is thus more desirable?

Learning Outcomes

What to Remember with Examples

Section 21–1

1 Use various notations to represent points, lines, line segments, rays, and planes (pp. 785–786).

A dot represents a point and a capital letter names the point. A line extends in both directions and is named by any two points on the line (with double arrow above the letters, such as $\overleftrightarrow{AB}$). A line segment has a beginning point and an ending point, which are named by letters (with a bar above the letters, such as $\overline{CD}$). Line segments are sometimes named by letters only, such as CD. A ray extends from a point on a line and includes all points on one side of the point and is named by the end point and any other point on the line forming the ray (with an arrow above the letters, such as $\overrightarrow{AC}$). A plane contains an infinite number of points and lines on a flat surface.

Use proper notation for the following:

(a) Line *GH*   (b) Line segment *OP*   (c) Ray *ST*

(a) $\overleftrightarrow{GH}$ (b) $\overline{OP}$ (c) $\overrightarrow{ST}$

2 Distinguish among lines that intersect, that coincide, and that are parallel (pp. 786–787).

Lines that intersect meet at only one point. Lines that coincide fit exactly on top of one another. Lines that are parallel are the same distance from one another and never intersect.

Classify the pairs of lines in Fig. 21–156.

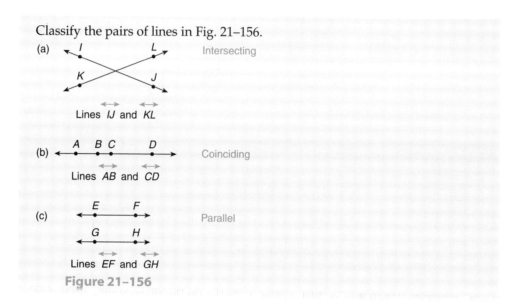

Figure 21–156

3 Use various notations to represent angles (p. 787).

When two rays intersect in a point (end point of rays), an angle is formed. Angles may be named with three capital letters (end point in middle and one point on each ray), such as ∠*ABC*. They may be named by only the middle letter (vertex of the angle), such as ∠*B*. They may be assigned a number or lowercase letter placed within the vertex, such as ∠2 or ∠*d*.

Name the angle in Fig. 21–157 three ways.

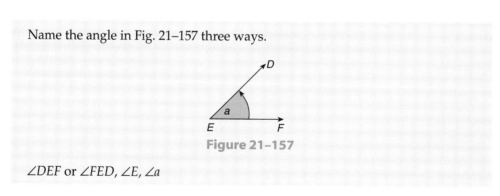

Figure 21–157

∠*DEF* or ∠*FED*, ∠*E*, ∠*a*

4 Classify angles according to size (pp. 787–790).

A right angle contains 90° or one-fourth a rotation. A straight angle contains 180° or one-half a rotation. An acute angle contains less than 90° but more than 0°. An obtuse angle contains more than 90° but less than 180°.

Identify the following angles:

(a) 30°   (b) 100°   (c) 90°   (d) 180°

(a) acute   (b) obtuse   (c) right   (d) straight

Section 21-2

1 Add and subtract angle measures (pp. 792–793).

Only like measures can be added or subtracted. Arrange measures in columns of like measures. Add or subtract each column separately. Write in standard notation.

Perform the following:

(a)
$$\begin{array}{r} 35° \quad 15' \ 10'' \\ + \ 10° \quad 55' \ \ 5'' \\ \hline 45° \quad 70' \ 15'' \\ + \ \ 1° - 60' \\ \hline 46° \quad 10' \ 15'' \end{array}$$

$60' = 1°$. Standard notation.

Borrowing and writing answers in standard form are also similar to operations with U.S. customary measures.

(b) Subtract $35°15'10'' - 10°55'5''$.

$$\begin{array}{r} \overset{34}{\cancel{35°}} \ \overset{75}{\cancel{15'}} \ 10'' \\ - \ 10° \ 55' \ \ 5'' \\ \hline 24° \ 20' \ \ 5'' \end{array}$$

Borrow. $1° = 60'$, $60' + 15' = 75'$.

2 Change minutes and seconds to a decimal part of a degree (pp. 793–795).

To change minutes to a decimal degree, multiply by $\dfrac{1°}{60 \text{ min}}$ or divide minutes by 60. To change seconds to a decimal degree, multiply by $\dfrac{1°}{3,600 \text{ sec}}$ or divide seconds by 3,600.

Change $15'36''$ to a decimal degree.

$15' \div 60 = 0.25$ Minutes to degrees.
$36'' \div 3,600 = 0.01$ Seconds to degrees.
$0.25 + 0.01 = 0.26°$ Combine.

3 Change the decimal part of a degree to minutes and seconds (p. 795).

To change a decimal degree to minutes, multiply the decimal degree by 60. To change a decimal part of a minute to seconds, multiply the decimal part of a minute by 60. To change a decimal part of a degree to seconds, multiply by 3,600.

Change $0.48°$ to minutes and seconds.

$0.48° \times 60 = 28.8'$ Degrees to minutes.
$0.8' \times 60 = 48''$ Minutes to seconds.

Thus, $0.48° = 28'48''$.

4 Multiply and divide angle measures (pp. 795–796).

To multiply angle measures, multiply each angle unit by the number and write in standard notation. To divide angle measures by a number, start with the largest angle unit. Convert remainders to the next smaller unit and continue.

Perform the following operations:

(a)
$$\begin{array}{r} 25° \quad 40' \ 10'' \\ \times \qquad\qquad 3 \\ \hline 75° \quad 120' \ 30'' \\ + \ \ 2° - 120' \\ \hline 77° \qquad\qquad 30'' \end{array}$$

Standard notation.

(b)
$$8° \quad 33' \quad 23\tfrac{1}{3}''$$
$$3\overline{)25° \quad 40' \quad 10''}$$
$$\underline{24°}$$
$$1° = \underline{60'} \quad \text{Convert to minutes.}$$
$$100'$$
$$\underline{99'}$$
$$1' = \underline{60''} \quad \text{Convert to seconds.}$$
$$70''$$
$$\underline{69''}$$
$$1''$$

Section 21–3

1 Classify triangles by sides (pp. 797–798).

An equilateral triangle has three equal sides. An isosceles triangle has exactly two equal sides. A scalene triangle has three unequal sides.

Identify the following triangles by the measures of their sides:

(a) 6 cm, 6 cm, 6 cm	Equilateral
(b) 12 in., 10 in., 12 in.	Isosceles
(c) 10 cm, 12 cm, 15 cm	Scalene

2 Relate the sides and angles of a triangle (p. 798).

An equilateral triangle has three equal angles. The equal sides of an isosceles triangle have opposite angles that are equal. If the three sides of a triangle are unequal, the largest angle is opposite the longest side, and the smallest angle is opposite the shortest side. In a right triangle, the hypotenuse is always the longest side.

Identify the largest and the smallest angles in triangles with the following sides: 12 in., 10 in., 9 in.
The largest angle is opposite the 12-in. side; the smallest angle is opposite the 9-in. side.

3 Determine if two triangles are congruent using inductive and deductive reasoning (pp. 798–800).

Triangles are congruent (one can fit exactly over the other) (a) if the three sides of one triangle equal the corresponding sides of the other triangle (SSS), (b) if two sides and the included angle of one triangle are equal to two sides and the included angle of the other triangle (SAS), or (c) if two angles and the common side of one triangle are equal to two angles and the common side of the other triangle (ASA).

Determine whether the pairs of triangles in Fig. 21–158 are congruent.

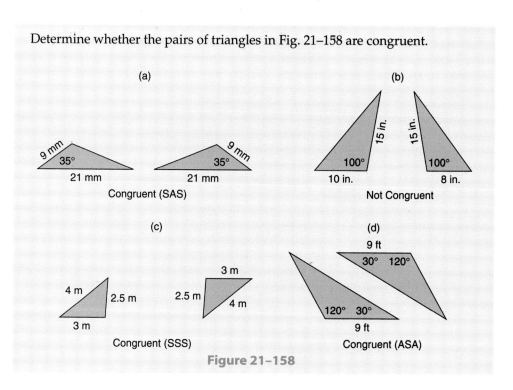

Figure 21–158

4 Use the properties of a 45°, 45°, 90° triangle to find missing parts (pp. 800–801).

To find the hypotenuse of a 45°, 45°, 90° triangle, multiply a leg by $\sqrt{2}$. To find a leg of a 45°, 45°, 90° triangle, divide the product of the hypotenuse and $\sqrt{2}$ by 2.

Given a 45°, 45°, 90° triangle, find

(a) the hypotenuse if a leg is 10 cm.
(b) a leg if the hypotenuse is 15 in.

(a) Hypotenuse = Leg ($\sqrt{2}$)

$\qquad$ = $10\sqrt{2}$

$\qquad$ = 14.14213562 or 14.1 cm $\qquad$ Rounded.

(b) Leg = $\dfrac{\text{Hypotenuse }(\sqrt{2})}{2}$

$\qquad$ = $\dfrac{15\sqrt{2}}{2}$

$\qquad$ = 10.60660172 or 10.6 in. $\qquad$ Rounded.

5 Use the properties of a 30°, 60°, 90° triangle to find missing parts (pp. 802–803).

To find the hypotenuse of a 30°, 60°, 90° triangle, multiply the side opposite the 30° angle by 2. To find the side opposite the 30° angle in a 30°, 60°, 90° triangle, divide the hypotenuse by 2. To find the side opposite the 60° angle of a 30°, 60°, 90° triangle, multiply the side opposite the 30° angle by $\sqrt{3}$.

Given a 30°, 60°, 90° triangle, find

(a) the hypotenuse if the side opposite the 30° angle = 4.5 cm.
(b) the side opposite the 30° angle if the hypotenuse is 20 in.
(c) the side opposite the 60° angle if the side opposite the 30° angle is 35 mm.

(a) Hypotenuse $\quad$ = 2 (Side opposite 30°)

$\qquad$ = 2(4.5)

$\qquad$ = 9 cm

(b) Side opposite 30° = $\dfrac{\text{Hypotenuse}}{2}$

$\qquad$ = $\dfrac{20}{2}$

$\qquad$ = 10 in.

(c) Side opposite 60° = Side opposite 30° ($\sqrt{3}$)

$\qquad$ = $35(\sqrt{3})$

$\qquad$ = 60.62177826 or 60.6 mm. $\qquad$ Rounded.

Section 21–4

1 Find the area of a composite figure (pp. 806–807).

Use the area formulas for each polygon. Add the areas.

Find the area of the layout in Fig. 21–159.

$A_{\text{square}} = s^2$ $\qquad$ Area of square. Substitute $s = 6$.

$A_{\text{square}} = (6 \text{ in.})^2$

$\qquad$ = 36 in^2

Before finding the area of the triangle, find the height of the triangle. The height divides the triangle into two congruent right triangles. The base of each right triangle is $\frac{1}{2}$ the base of the original triangle. $b = \frac{1}{2}$ (6 in.) = 3 in.

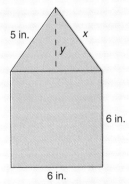

5 in. x

y

6 in.

6 in.

Figure 21–159

Use the Pythagorean theorem to find the height of the triangle.

$$c^2 = a^2 + b^2 \qquad \text{Substitute } c = 5, b = 3, \text{ and } y = a.$$
$$5^2 = y^2 + 3^2 \qquad \text{Solve for } y.$$
$$25 = y^2 + 9$$
$$25 - 9 = y^2$$
$$16 = y^2$$
$$4 = y \qquad \text{Height of triangle.}$$
$$A_{\text{triangle}} = \frac{1}{2}bh \qquad \text{Area of triangle. Substitute } b = 6 \text{ and } h = 4.$$
$$A_{\text{triangle}} = \frac{1}{2}(6)(4) \qquad \text{Evaluate.}$$
$$A_{\text{triangle}} = 12 \text{ in}^2 \qquad \text{Total area of triangle.}$$
$$A_{\text{layout}} = A_{\text{square}} + A_{\text{triangle}} \qquad \text{Substitute.}$$
$$A_{\text{layout}} = 36 + 12 \qquad \text{Simplify.}$$
$$A_{\text{layout}} = 48 \text{ in}^2 \qquad \text{Area of layout.}$$

2 Find the number of degrees in each angle of a regular polygon (pp. 807–808).

To find the number of degrees in each angle of a regular polygon: Multiply the number of sides less 2 by 180° and divide the product by the number of sides.

Find the number of degrees in each angle of a pentagon.
A pentagon has five sides.

$$\text{Degrees per angle} = \frac{180(5 - 2)}{5} = \frac{180(3)}{5} = \frac{540}{5} = 108°$$

3 Find the area of a regular polygon (pp. 808–809).

A regular polygon has all sides equal. Regular polygons with 3, 4, 5, 6, and 8 sides are equilateral triangles, squares, pentagons, hexagons, and octagons, respectively. Lines drawn from the center of a regular polygon to each vertex will form as many congruent triangles as the polygon has sides.

To find the area of a regular polygon, find the area of one of the congruent triangles and multiply by the number of triangles.

Find the area of the regular hexagon in Fig. 21–160.
Each congruent triangle formed is an equilateral triangle (Fig. 21–161). Six congruent central angles make each central angle 60°: $\frac{360°}{6} = 60°$. Each base angle of the triangles equals one-half the angle at each vertex of the hexagon: $\frac{1}{2}(120°) = 60°$. The congruent triangles are 60°, 60°, 60° triangles, or equilateral.
Find the area of one congruent triangle of the hexagon.

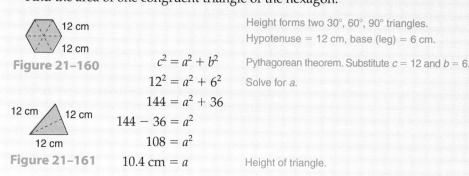

Figure 21–160

Figure 21–161

Height forms two 30°, 60°, 90° triangles.
Hypotenuse = 12 cm, base (leg) = 6 cm.

$$c^2 = a^2 + b^2 \qquad \text{Pythagorean theorem. Substitute } c = 12 \text{ and } b = 6.$$
$$12^2 = a^2 + 6^2 \qquad \text{Solve for } a.$$
$$144 = a^2 + 36$$
$$144 - 36 = a^2$$
$$108 = a^2$$
$$10.4 \text{ cm} = a \qquad \text{Height of triangle.}$$

$$A_{\text{triangle}} = \frac{1}{2}bh$$ Substitute $b = 12$ and $h = 10.4$.

$$A_{\text{triangle}} = \frac{1}{2}(12)(10.4)$$ Evaluate.

$$A_{\text{triangle}} = 62.4 \text{ cm}^2$$ Area of one congruent triangle.

$$A_{\text{hexagon}} = 6(A_{\text{triangle}})$$ Substitute.

$$A_{\text{hexagon}} = 6(62.4)$$

$$A_{\text{hexagon}} = 374.4 \text{ cm}^2$$

Section 21–5

1 Find the area of a sector (pp. 811–812).

A sector is a portion of a circle cut off by two radii.

Area of a sector $= \frac{\theta}{360}\pi r^2$, where θ is the degrees of the central angle formed by the radii.

Find the area of a sector that has a central angle of 50° and a radius of 20 cm (Fig. 21–162).

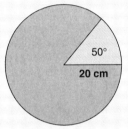

Figure 21–162

$$\text{Area} = \frac{\theta}{360}\pi r^2$$ Formula for area of a sector of a circle. Substitute $\theta = 50°$ and $r = 20$ cm.

$$\text{Area} = \frac{50}{360}(\pi)(20)^2$$ Evaluate.

$$\text{Area} = 174.5329252 \quad \text{or} \quad 174.53 \text{ cm}^2 \quad \text{Rounded.}$$

2 Find the arc length of a sector (p. 812).

Arc length is the portion of the circumference formed by the sides of a sector.

$$\text{Arc length } (s) = \frac{\theta}{360}(2\pi r) \quad \text{or} \quad s = \frac{\theta}{360}(\pi d)$$

Find the arc length of a sector formed by a 50° central angle if the radius is 20 cm (Fig. 21–163).

$$s = \frac{\theta}{360}(2\pi r)$$ Substitute $\theta = 50°$ and $r = 20$ cm.

$$s = \frac{50}{360}(2)(\pi)(20)$$ Evaluate.

$$s = 17.45329252 \quad \text{or} \quad 17.45 \text{ cm}$$

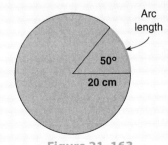

Figure 21–163

Chapter Summary

A chord is a line segment joining two points on a circle. An arc is the portion of the circumference cut off by a chord. A segment is the portion of a circle bounded by a chord and an arc.

The area of a segment is the area of the sector minus the area of the triangle formed by the chord and the central angle of the sector:

$$A = \frac{\theta}{360}\pi r^2 - \frac{1}{2}bh$$

Find the area of the segment formed by a 12-cm chord if the height of the triangle part of the sector is 6 cm. The central angle is 90° (Fig. 21–164).

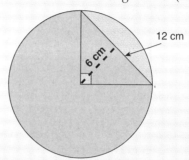

12 cm

6 cm

Figure 21–164

Find r using the Pythagorean theorem.

$$c^2 = a^2 + b^2 \qquad \text{Substitute } c = 12, a = r, \text{ and } b = r.$$
$$12^2 = r^2 + r^2$$
$$144 = 2r^2$$
$$\frac{144}{2} = r^2$$
$$72 = r^2$$
$$8.485281374 = r$$

$$A_{\text{segment}} = \frac{\theta}{360}\pi r^2 - \frac{1}{2}bh \qquad \text{Substitute } \theta = 90°, r = 8.485281374,$$
$$b = 12, h = 6.$$

$$A_{\text{segment}} = \frac{90}{360}(\pi)(8.485281374)^2 - \frac{1}{2}(12)(6) \quad \text{Evaluate.}$$

$$A_{\text{segment}} = 0.25(\pi)(72) - 0.5(72)$$

$$A_{\text{segment}} = 20.54866776 \qquad \text{or} \qquad 20.55 \text{ cm}^2 \quad \text{Rounded.}$$

Section 21–6

The height of an equilateral triangle forms two congruent 30°, 60°, 90° triangles.

The radius of a circle circumscribed about an equilateral triangle is two-thirds the height of the triangle or twice the distance from the center of the circle to the base of the triangle.

The radius of a circle inscribed in an equilateral triangle is one-third the height of the triangle or one-half the distance from the vertex of the triangle to the center of the circle.

One end of a shaft 30 mm in diameter is milled as shown in Fig. 21–165. Find the diameter of the smaller circle.

BC is the radius and half the diameter of the large circle.

$$BC = \frac{1}{2}(30)$$

Chapter 21 / Selected Concepts of Geometry

$$BC = 15 \text{ mm} \qquad \text{Radius of large circle.}$$

$$AB = \frac{1}{2}BC \qquad \text{Radius of small circle.}$$

$$AB = \frac{1}{2}(15)$$

$$AB = 7.5 \text{ mm}$$

The diameter of the small circle = 2(7.5) = 15 mm.

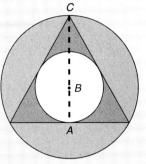

Figure 21–165

2 Use the properties of inscribed and circumscribed squares to find missing amounts (pp. 817–819).

The diagonal of a square inscribed in or circumscribed about a circle forms congruent 45°, 45°, 90° triangles.

The diagonal of a square inscribed in a circle equals the diameter of the circle.

The side of a square circumscribed about a circle equals the radius of the circle.

The radius of a circle when a square is inscribed in the circle equals the height of a 45°, 45°, 90° triangle formed by one diagonal.

The radius of a circle when a square is circumscribed about the circle equals the height of a 45°, 45°, 90° triangle formed by two diagonals or it equals one-half the length of a side of the square.

The end of a 20-mm diameter round rod is milled as shown in Fig. 21–166. Find the area of the square.

The radius of the circle equals the height of a right triangle formed by one diagonal. The diameter is the base. Find the area of the triangle and double it.

$$A_{\text{triangle}} = \frac{1}{2}bh$$

$$A_{\text{triangle}} = \frac{1}{2}(20)(10)$$

$$A_{\text{triangle}} = 100 \text{ mm}^2$$

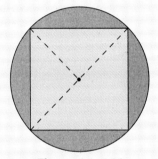

Figure 21–166

Area of square end: $100 \times 2 = 200 \text{ mm}^2$

An alternative approach would be to find the side of the square using the proportions of a 45°, 45°, 90° triangle and then to find the area of the square.

3 Use the properties of inscribed and circumscribed regular hexagons to find missing amounts (pp. 819–821).

The three diagonals that join opposite vertices of a regular hexagon form six congruent equilateral triangles.

The height of each of the equilateral triangles forms two congruent 30°, 60°, 90° triangles. The diagonal of a regular hexagon inscribed in a circle is the diameter of the circle (distance across corners).

The radius of the circle when the regular hexagon is inscribed in the circle is a side of an equilateral triangle formed by the three diagonals and is one-half the distance across corners.

The radius of the circle when a regular hexagon is circumscribed about the circle is the height of an equilateral triangle formed by the three diagonals.

The diameter of the circle when a regular hexagon is circumscribed about the circle is the distance across flats or twice the height of an equilateral triangle formed by the three diagonals.

Find the following dimensions for Fig. 21–167 if the diameter is 10 cm.

(a) ∠AOF
(b) Radius
(c) Distance across corners

(a) ∠AOF = 60° (angle of equilateral triangle)
(b) 5 cm (radius of a circle is one-half the diameter)
(c) 10 cm (diameter of a circle is the
    distance across corners of the hexagon)

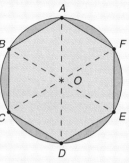

Figure 21–167

CHAPTER TRIAL TEST

1. Show the proper notation to name the line segment from B to C in Fig. 21–168.

Figure 21–168

2. Name the angle in Fig. 21–169 in four different ways.

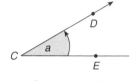

Figure 21–169

Classify the following angles as right, straight, acute, or obtuse.

3. 42°   **4.** 158°

Identify the lines in Figs. 21–170 through 21–172 as intersecting, parallel, or perpendicular.

5.

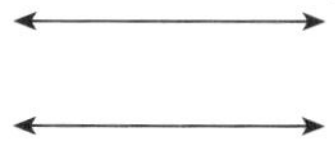

Figure 21–170

6.

Figure 21–171

7.

Figure 21–172

8. Subtract 40° 37′ 26″ from 75°.

9. An angle of 47° 16′ 28″ is divided into three equal angles. Find the measure of each angle in degrees, minutes, and seconds (to the nearest second).

10. Change 15′ 32″ to a decimal degree to the nearest ten-thousandth.

11. Change 0.3125° to minutes and seconds.

12. Name the following triangles: (a) all sides unequal, (b) exactly two sides equal, and (c) all sides equal.

13. Identify the largest and the smallest angles in Fig. 21–173.

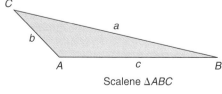

Scalene △ABC

Figure 21–173

14. In Fig. 21–174, identify (a) the angle corresponding to ∠DEF and (b) the side corresponding to HI.

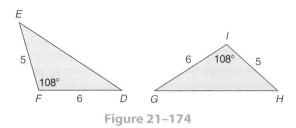

Figure 21–174

15. A conduit *ABCD* is made so that ∠*CBE* is 45° (Fig. 21–175). If *BE* = 4 cm and *AK* = 12 cm, find the length of the conduit to the nearest thousandth centimeter.

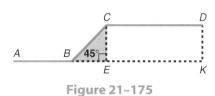

Figure 21–175

16. The rafters of a house make a 30° angle with the joists (Fig. 21–176). If the rafters have an 18-in. overhang and the center of the roof is 10 ft above the joists, how long must the rafters be cut?

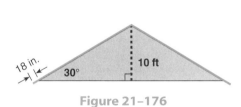

Figure 21–176

17. Find the area of the composite figure in Fig. 21–177.

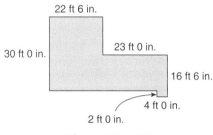

Figure 21–177

18. A section of a hip roof is a trapezoid measuring 35 ft at the bottom, 15 ft at the top, and 10 ft high. Find the area of this section of the roof in square feet.

20. Find the area of the sector in Fig. 21–178 if the central angle is 55° and the radius is 45 mm. Round to hundredths.

21. Find the arc length cut off by the sector in Fig. 21–178. Round to hundredths.

19. Find the number of degrees in each angle of a regular octagon (eight sides).

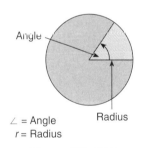

∠ = Angle
r = Radius

Figure 21–178

22. A segment is removed from a flat metal circle so that the piece rests on a horizontal base (Fig. 21–179). Find the area of the segment that is removed.

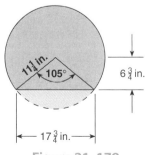

Figure 21–179

23. Find the circumference of a circle inscribed in an equilateral triangle if the height of the triangle is 12 in. (Fig. 21–180).

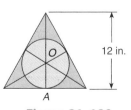

Figure 21–180

24. Ken Bennett milled a square metal rod so that a circle is formed at the end (see Fig. 21–181). If the square cross-section is 1.8 in. across the corners, what is the diameter of the circle?

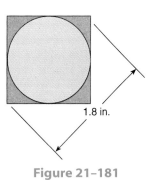

Figure 21–181

25. Laura Deskin milled a circle on the cross-sectional end of a hexagonal rod (Fig. 21–182). If the distance across the flats of the cross-section is $\frac{1}{2}$ in., what is the circumference of the circle?

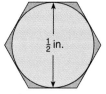

Figure 21–182

Introduction to Trigonometry

22

One of the most important uses of *trigonometry* is to find by indirect measurement lengths or distances that are difficult or impossible to measure directly. Although trigonometry, which means "triangle measurement," involves more than a study of just the triangle, we limit our study to trigonometry of the triangle.

22-1 | *Radians and Degrees*

Learning Outcomes

1 Find the arc length, given a central angle in radians.
2 Find the area of a sector, given a central angle in radians.
3 Convert angle measures from degrees to radians.
4 Convert angle measures from radians to degrees.

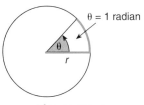

$\theta = 1$ radian

Figure 22–1

In geometry, we learned that angles can be measured in units called *degrees*. Angles can also be measured in *radians*. A *radian* is the measure of a central angle of a circle whose intercepted arc is equal in length to the radius of the circle (Fig. 22–1). The abbreviation for radian is *rad*.

The circumference of a circle is related to the radius by the formula $C = 2\pi r$. Thus, the ratio of the circumference to the radius of any circle is $\frac{C}{r} = 2\pi$; that is, the radius could be measured off 2π times (about 6.28 times) along the circumference. A complete rotation is 2π radians (see Fig. 22–2).

1 radian

3 radians

3.14 or π radians

5 radians

6.28 or 2π radians

Figure 22–2

How are radians and degrees related? A central angle measuring 1 radian makes an arc length equal to the radius, so we use the arc-length formula from Chapter 21 to find the equivalent degree measure for 1 radian.

$$\text{Arc length} = \frac{\theta}{360} 2\pi r; \theta \text{ in degrees}$$

If the radius and arc length are equal to 1,

$$1 = \frac{\theta}{360} 2\pi(1) \qquad \text{Substitute arc length} = 1 \text{ and } r = 1. \text{ Multiply both sides by 360.}$$

$$360 = \theta(2\pi) \qquad \text{Solve for } \theta. \text{ Divide both sides by } 2\pi.$$

$$\frac{360}{2\pi} = \frac{\theta(2\pi)}{2\pi}$$

$$57.29577951 = \theta$$

One radian is approximately $57.3°$.

TIP!

Degree and Radian Notation

Angles measured in degrees always require the word *degree* or the degree symbol to be written. No comparable symbol exists for the radian. The abbreviation *rad* or no unit at all indicates that radian is the unit.

1 Find the Arc Length, Given a Central Angle in Radians.

Radians are used frequently in physics and engineering mechanics. One application that uses radians is in determining *arc length*, the length of the arc intercepted by a central angle.

The circumference of a circle is $2\pi r$, where r is the radius of the circle. 2π is the radian measure of a complete rotation.

An arc length is a portion of the circumference. An angle measured in radians identifies the portion of the circumference being considered. We use the Greek lowercase letter theta (θ) to represent the central angle measured in radians in this formula for arc length.

To find the arc length, given a central angle in radians:

1. Substitute the given radian measure of the central angle and the radius in the formula

$$s = \theta r$$

   where θ is the central angle measured in radians and r is the radius of the circle.
2. Simplify the expression.

EXAMPLE Find the arc length intercepted on the circumference of the circle in Fig. 22–3 by a central angle of $\dfrac{\pi}{3}$ radians (rad) if the radius of the circle is 10 cm.

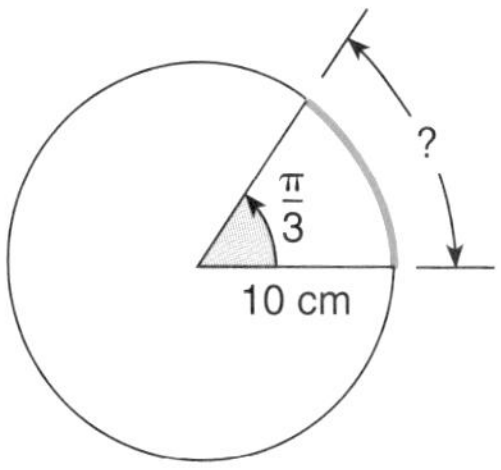

Figure 22–3

$s = \theta r$ Substitute $\frac{\pi}{3}$ for θ and 10 cm for r.

$s = \dfrac{\pi}{3}(10 \text{ cm})$ Evaluate.

$s = \dfrac{\pi(10 \text{ cm})}{3}$ Use calculator value for π.

$s = 10.47 \text{ cm}$ To the nearest hundredth.

Thus, the arc length of the intercepted arc is 10.47 cm.

TIP!

Analyzing Arc Length Dimensions

In the preceding example, $\left(\dfrac{\pi}{3} \text{ rad}\right)(10 \text{ cm}) = 10.47 \text{ cm}$, what happened to the radians? In the definition of a radian, we relate the measure of an arc connecting the end points of a central angle to the measure of the radius of the circle. Therefore, the arc length will have the same measuring unit as the radius.

To find the central angle or radius of a sector given the arc length:

1. Substitute given values in the formula $s = \theta r$, where s is the arc length, θ is the central angle measured in radians, and r is the radius of the circle.
2. Solve for the missing value.

EXAMPLE Find the radian measure of an angle at the center of a circle of radius 5 m. The angle intercepts an arc length of 12.5 m (see Fig. 22–4).

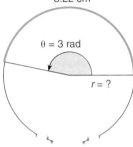

$s = \theta r$ — Formula for arc length. Solve for θ.

$\theta = \dfrac{s}{r}$ — Substitute 12.5 m for s and 5 m for r. Arc length and radius measuring units are compatible.

$\theta = \dfrac{12.5 \text{ m}}{5 \text{ m}}$

$\theta = 2.5 \text{ rad}$

The angle is 2.5 rad.

Figure 22–4

EXAMPLE Find the radius of an arc if the length of the arc is 8.22 cm and the intercepted central angle is 3 rad (see Fig. 22–5).

$s = \theta r$ — Formula for arc length. Solve for r.

$r = \dfrac{s}{\theta}$ — Substitute 8.22 cm for s and 3 rad for θ.

$r = \dfrac{8.22 \text{ cm}}{3}$ — Simplify.

$r = 2.74 \text{ cm}$ — Same measuring unit as arc length.

The radius of the arc is 2.74 cm.

Figure 22–5

2 Find the Area of a Sector, Given a Central Angle in Radians.

In Chapter 21, we examined the formula for finding the area of a sector when the angle measure is expressed in degrees. Since the radian measure for angles is based on a complete rotation of 2π radians, the formula for the area of a sector formed by a central angle θ is $A = \frac{1}{2}\theta r^2$. The relationship between the formula for the area of a sector given the degree measure of the central angle and the formula given the radian measure is presented in the Tip on p. 845.

To find the area of a sector, given a central angle in radians:

1. Substitute known values into the formula $A = \dfrac{1}{2}\theta r^2$, where θ is the central angle measured in radians and r is the radius of the circle.
2. Solve for the missing value.

EXAMPLE Find the area of a sector that has a central angle of 5 rad and has a radius of 7.2 in.

$A = \dfrac{1}{2}\theta r^2$ — Substitute $\theta = 5$ and $r = 7.2$.

$A = \dfrac{1}{2}(5)(7.2 \text{ in.})^2$ — Simplify.

$A = 129.6 \text{ in}^2$ — Area of sector.

The area of the sector is 129.6 in^2.

TIP!

Two Formulas for the Area of a Sector

The formula for the area of a circle is $A_{circle} = \pi r^2$. To find the area of a sector, the formula is multiplied by a fractional portion of the complete rotation of a circle, $A_{sector} = \dfrac{\theta}{360°}\,\pi r^2$. In radians, a complete rotation is 2π radians, so a sector with a central angle expressed in radians can be expressed as $A_{sector} = \dfrac{\theta}{2\pi}\,\pi r^2$. Since the π factors will reduce, the resulting formula is $A_{sector} = \dfrac{\theta}{2}r^2$, or $A_{sector} = \dfrac{1}{2}\theta r^2$.

3 **Convert Angle Measures from Degrees to Radians.**

A complete rotation is $360°$, or 2π rad. To convert from one unit of angle measure to another, we multiply by a *unity ratio* that relates degrees and radians. Because $360° = 2\pi$ rad, we can simplify the relationship to $180° = \pi$ rad $\left(\dfrac{360°}{2} = \dfrac{2\pi}{2}\text{ rad}\right)$.

To convert degrees to radians:

1. Multiply degrees by the unity ratio $\dfrac{\pi \text{ rad}}{180°}$.
2. Write the product in simplest form.

EXAMPLE Convert the angle measures to radians. Use the calculator value for π to change to a decimal equivalent rounded to the nearest hundredth.

(a) $20°$ (b) $175°$ (c) $270°$

(a) $20° \times \dfrac{\pi \text{ rad}}{180°} = \dfrac{20(\pi)}{180} = \textbf{0.35 rad}$ Multiply by $\dfrac{\pi \text{ rad}}{180°}$.

(b) $175° \times \dfrac{\pi \text{ rad}}{180°} = \dfrac{175(\pi)}{180} = \textbf{3.05 rad}$

(c) $270° \times \dfrac{\pi \text{ rad}}{180°} = \dfrac{270(\pi)}{180} = \textbf{4.71 rad}$

When angle measures are expressed in decimal degrees, we use the same procedure as before to convert to radians. When angle measures are expressed in degrees, minutes, and seconds, we first convert the minutes and seconds to a decimal degree before converting to radians.

EXAMPLE Convert the following measures to radians. Round the decimal equivalent of degrees, minutes, and seconds to the nearest ten-thousandth; then round the radians to the nearest hundredth.

(a) $15.25°$ (b) $25°20'$ (c) $110°25'30''$

(a) $15.25° \times \dfrac{\pi \text{ rad}}{180°} = \dfrac{15.25(\pi)}{180} = \textbf{0.27 rad}$ Multiply by $\dfrac{\pi \text{ rad}}{180°}$. Rounded to hundredths.

(b) $25°20' = 25.3333°$

    Convert minutes to a decimal degree:

    $20' \times \dfrac{1°}{60'} = \dfrac{20°}{60} = 0.3333°$.

$25.3333° \times \dfrac{\pi \text{ rad}}{180°} = \dfrac{25.3333(\pi)}{180} = \textbf{0.44 rad}$ Multiply by $\dfrac{\pi \text{ rad}}{180°}$. Round.

(c) $110°25'30'' = 110.425°$

Convert minutes and seconds to a decimal degree: $110 + \dfrac{25}{60} + \dfrac{30}{3,600} = 110.425$.

$$110.425° \times \frac{\pi \text{ rad}}{180°} = \frac{110.425(\pi)}{180} = \textbf{1.93 rad}$$

Multiply by $\dfrac{\pi \text{ rad}}{180°}$. Round.

4 Convert Angle Measures from Radians to Degrees.

When converting from radians to degrees, we multiply by a unity ratio so that the radians cancel and are replaced by degrees.

To convert radians to degrees:

1. Multiply radians by the unity ratio $\dfrac{180°}{\pi \text{ rad}}$.
2. Write the product in simplest form.

EXAMPLE Convert to degrees. Round to the nearest ten-thousandth of a degree.

(a) 2 rad   (b) $\dfrac{\pi}{2}$ rad   (c) 0.25 rad

(a) $2 \text{ rad} \times \dfrac{180°}{\pi \text{ rad}} = \dfrac{360°}{\pi} = \textbf{114.5916°}$

Multiply by $\dfrac{180°}{\pi \text{ rad}}$. Use calculator value of π and round.

(b) $\dfrac{\pi}{2} \text{ rad} \times \dfrac{180°}{\pi \text{ rad}} = \dfrac{180°}{2} = \textbf{90°}$

Multiply by $\dfrac{180°}{\pi \text{ rad}}$. Because π canceled, we do not substitute for π.

(c) $0.25 \text{ rad} \times \dfrac{180°}{\pi \text{ rad}} = \dfrac{(0.25)(180°)}{\pi} = \textbf{14.3239°}$

Multiply by $\dfrac{180°}{\pi \text{ rad}}$. Use calculator value for π and round.

EXAMPLE Convert to degrees, minutes, and seconds. Round to the nearest second.

(a) 1 rad   (b) 3.2 rad

(a) $1 \text{ rad} \times \dfrac{180°}{\pi \text{ rad}} = \dfrac{180°}{\pi} = 57.29577951°$

Multiply by $\dfrac{180°}{\pi \text{ rad}}$. Continue with the decimal part of the degree measure.

$0.29577951° \times \dfrac{60'}{1°} = 17.74677077'$

Multiply by $\dfrac{60'}{1°}$. Continue with the decimal part of the minute measure.

$0.74677077' \times \dfrac{60''}{1'} = 45''$

Multiply by $\dfrac{60''}{1'}$. Convert the decimal part of minute to seconds, to the nearest second.

Thus, **1 rad = 57°17'45''.**

(b) $3.2 \text{ rad} \times \dfrac{180°}{\pi \text{ rad}} = \dfrac{3.2(180°)}{\pi} = 183.3464944°$

Multiply by $\dfrac{180°}{\pi \text{ rad}}$. Continue with the decimal part of the degree measure.

$0.3464944° \times \dfrac{60'}{1°} = 20.7896664'$

Multiply by $\dfrac{60'}{1°}$. Convert the decimal part of degree to minutes.

$0.7896664' \times \dfrac{60''}{1'} = 47''$

Multiply by $\dfrac{60''}{1'}$. Convert the decimal part of minute to seconds, to the nearest second.

Thus, **3.2 rad = 183°20'47''.**

Chapter 22 / Introduction to Trigonometry

Using Only the Decimal Part of a Calculator Result

When a procedure requires us to continue a calculation with only the decimal part of a result, *first subtract the whole-number part*. Look at the calculator sequence for the preceding example, part (a):

$\boxed{180}$ $\boxed{\div}$ $\boxed{\pi}$ $\boxed{=}$ $\boxed{-}$ 57 $\boxed{=}$ $\boxed{\times}$ 60 $\boxed{=}$ [17.74677078 in display] $\boxed{-}$ 17 $\boxed{=}$ $\boxed{\times}$ 60 $\boxed{=}$ $\Rightarrow$ 44.8062471

There is a slight rounding descrepancy since more decimal places appear in 57.29577951 when 57 is subtracted.

SELF-STUDY EXERCISES 22–1

1 Solve the following problems. Round answers to hundredths if necessary.

1. Find the arc length intercepted on the circumference of a circle by a central angle of 2.15 rad if the radius of the circle is 3 in.

2. Find the arc length intercepted on the circumference of a circle by a central angle of 4 rad if the radius of the circle is 3.5 cm.

3. Use radians to find an angle at the center of a circle of radius 2 in. if the angle intercepts an arc length of 8.5 in.

4. Use radians to find an angle at the center of a circle of radius 4.3 cm if the angle intercepts an arc length of 15 cm.

5. Find the radius of an arc if the length of the arc is 14.7 cm and the intercepted central angle is 2.1 rad.

6. Find the radius of an arc if the length of the arc is 12.375 in. and the intercepting central angle is 2.75 rad.

7. Find the arc length of an arc whose intercepting angle is .66 rad and whose radius is 2.3 cm. Round to hundredths.

8. Find the number of radians in a central angle whose arc length is 3.2 in. and whose radius is 3 in. Round to the nearest tenth radian.

2

9. Find the area of a sector whose central angle is 2.14 rad and whose radius is 4 in.

10. Find the area of a sector that has a central angle of 6 rad and a radius of 1.2 cm.

11. Find the radius of a sector if the area of the sector is 7.5 cm^2 and the central angle is 3 rad.

12. How many radians does the central angle of a sector measure if its area is 1.7 in^2 and its radius is 2 in.?

13. Find the area of a sector whose central angle is 1.83 rad and whose radius is 7.2 cm. Round to hundredths.

14. Find the number of radians of a central angle of a sector whose area is 5.6 cm^2 and whose radius is 4 cm.

3 Convert the measures to radians rounded to the nearest hundredth.

15. 45°   16. 56°   17. 78°   18. 140°

Convert the measures to degrees rounded to the nearest ten-thousandth. Then convert to radians to the nearest hundredth.

19. 21°45'   20. 177°33'   21. 44°54'12"   22. 10°31'15"

4 Convert the measures to degrees. Round to the nearest ten-thousandth of a degree.

23. $\dfrac{\pi}{4}$ rad   24. $\dfrac{\pi}{6}$ rad   25. 2.5 rad   26. 1.4 rad

Convert the measures to degrees, minutes, and seconds. Round to the nearest second.

27. 0.5 rad   28. $\dfrac{\pi}{8}$ rad   29. 0.75 rad   30. 1.1 rad

1 Find the sine, cosine, and tangent of angles of right triangles, given the measures of at least two sides.

2 Find the cosecant, secant, and cotangent of angles of right triangles, given the measures of at least two sides.

1 **Find the Sine, Cosine, and Tangent of Angles of Right Triangles, Given the Measures of at Least Two Sides.**

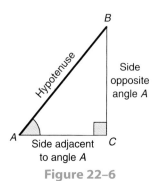

Figure 22–6

In geometry, we studied the basic properties of similar triangles and right triangles that allowed us to find missing measures of the sides of the triangle. Using trigonometry, we can determine the measure of either acute angle of a right triangle if we know the measure of at least two sides of the right triangle. We will define several functions that are ratios of various sides of a triangle, and these ratios will be used later to determine the angles of a triangle.

Fig. 22–6 shows a right triangle, *ABC*, with the sides of the triangle labeled according to their relationship to angle *A*. The *hypotenuse* is the side opposite the right angle of the triangle and the hypotenuse forms one side of angle *A*. The other side that forms angle *A* is the adjacent side of angle *A*. The third side of the triangle is the opposite side of angle *A*.

In Fig. 22–7, the sides of the right triangle *ABC* are labeled according to their relationship to angle *B*. The hypotenuse forms one side of angle *B*, the other side that forms angle *B* is the adjacent side of angle *B*, and the third side is the opposite side of angle *B*. In Chapters 11 and 21, we used the term hypotenuse, but the terms adjacent side and opposite side are also very important in understanding trigonometric functions. The *adjacent side* of an acute angle of a right triangle is the side that forms the angle with the hypotenuse. The *opposite side* of an acute angle of a right triangle is the side that does *not* form the given angle.

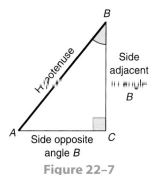

Figure 22–7

The three most commonly used trigonometric functions are the *sine, cosine,* and *tangent.* The sine, cosine, and tangent of angle *A* in Fig. 22–8 are defined as ratios of the sides of the right triangle.

$$\text{sine of angle } A = \frac{\text{side opposite angle } A}{\text{hypotenuse}}$$

$$\text{cosine of angle } A = \frac{\text{side adjacent to angle } A}{\text{hypotenuse}}$$

$$\text{tangent of angle } A = \frac{\text{side opposite angle } A}{\text{side adjacent to angle } A}$$

For convenience, the sine, cosine, and tangent functions are abbreviated as *sin, cos,* and *tan,* respectively. Furthermore, we will use letters to designate the sides of the *standard right triangle* in Fig. 22–8. In the standard right triangle, side *a* is opposite angle *A*, side *b* is opposite angle *B*, and side *c* (hypotenuse) is opposite angle *C* (right angle). We can identify the functions of angle *A* as follows.

Trigonometric functions of angle *A* in a standard right triangle:

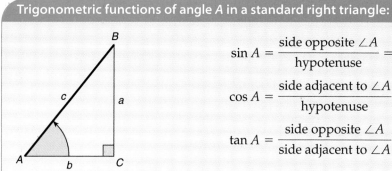

$$\sin A = \frac{\text{side opposite } \angle A}{\text{hypotenuse}} = \frac{a}{c}$$

$$\cos A = \frac{\text{side adjacent to } \angle A}{\text{hypotenuse}} = \frac{b}{c}$$

$$\tan A = \frac{\text{side opposite } \angle A}{\text{side adjacent to } \angle A} = \frac{a}{b}$$

Figure 22–8

Similarly, we can identify the sine, cosine, and tangent relationship for the other acute angle, angle B (Fig. 22–9).

Trigonometric functions of angle B in a standard right triangle:

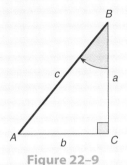

$$\sin B = \frac{\text{side opposite } \angle B}{\text{hypotenuse}} = \frac{b}{c}$$

$$\cos B = \frac{\text{side adjacent to } \angle B}{\text{hypotenuse}} = \frac{a}{c}$$

$$\tan B = \frac{\text{side opposite } \angle B}{\text{side adjacent to } \angle B} = \frac{b}{a}$$

Figure 22–9

Learning Strategy   *Words Are Easier to Remember Than Letters.*

We often substitute letters for words in relationships because it shortens the amount of information we need to write. However, letters sometimes are arbitrary and meaningless and words are easier to understand and remember. Here are two common tips for remembering the trigonometric relationships of sin, cos, and tan.

• "Oscar had a heap of apples."

$$\sin = \frac{\text{opposite}}{\text{hypotenuse}} = \frac{\text{Oscar}}{\text{had}}$$

$$\cos = \frac{\text{adjacent}}{\text{hypotenuse}} = \frac{\text{a}}{\text{heap}}$$

$$\tan = \frac{\text{opposite}}{\text{adjacent}} = \frac{\text{of}}{\text{apples}}$$

• "Chief Soh-Cah-Toa" or "Some old hogs come around here tasting our apples."

$$\sin = \frac{\text{opposite}}{\text{hypotenuse}} \qquad \cos = \frac{\text{adjacent}}{\text{hypotenuse}} \qquad \tan = \frac{\text{opposite}}{\text{adjacent}}$$

To write the trigonometric ratios for a right triangle given the measures of two sides:

1. Identify the acute angle that is being used.
2. Identify the hypotenuse, opposite side, and adjacent side in relation to the acute angle selected in Step 1.
3. Write the appropriate trigonometric ratio.
4. Simplify the ratio or convert to decimal equivalent.

22-2 Trigonometric Functions

Find the sine, cosine, and tangent of angles A and B in Fig. 22–10. Express answers as fractions in lowest terms.

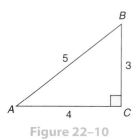

Figure 22–10

$$\sin A = \frac{\text{opposite}}{\text{hypotenuse}} = \frac{3}{5}$$

$$\cos A = \frac{\text{adjacent}}{\text{hypotenuse}} = \frac{4}{5}$$

$$\tan A = \frac{\text{opposite}}{\text{adjacent}} = \frac{3}{4}$$

$$\sin B = \frac{\text{opposite}}{\text{hypotenuse}} = \frac{4}{5}$$

$$\cos B = \frac{\text{adjacent}}{\text{hypotenuse}} = \frac{3}{5}$$

$$\tan B = \frac{\text{opposite}}{\text{adjacent}} = \frac{4}{3}$$

EXAMPLE Find the sine, cosine, and tangent of angles A and B in Fig. 22–11. Express answers as decimals. Round to the nearest ten-thousandth.

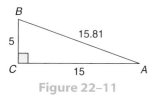

Figure 22–11

$$\sin A = \frac{\text{opposite}}{\text{hypotenuse}} = \frac{5}{15.81} = 0.3163$$

$$\cos A = \frac{\text{adjacent}}{\text{hypotenuse}} = \frac{15}{15.81} = 0.9488$$

$$\tan A = \frac{\text{opposite}}{\text{adjacent}} = \frac{5}{15} = 0.3333$$

$$\sin B = \frac{\text{opposite}}{\text{hypotenuse}} = \frac{15}{15.81} = 0.9488$$

$$\cos B = \frac{\text{adjacent}}{\text{hypotenuse}} = \frac{5}{15.81} = 0.3163$$

$$\tan B = \frac{\text{opposite}}{\text{adjacent}} = \frac{15}{5} = 3$$

Is a Trigonometric Ratio Expressed as a Unit of Measure?

When both terms of a ratio are expressed in the same unit of measure, that is, *like* units, then the ratio itself is unitless because the common units cancel, for example, $\frac{3 \text{ in.}}{5 \text{ in.}} = \frac{3}{5}$. Thus, trigonometric ratios are numerical values with no unit of measure.

EXAMPLE Find the sine, cosine, and tangent of angles A and B in Fig. 22–12. Leave the answers as fractions in lowest terms.

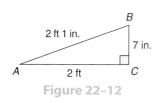

Figure 22–12

$$\sin A = \frac{\text{opposite}}{\text{hypotenuse}} = \frac{7 \text{ in.}}{2 \text{ ft 1 in.}} = \frac{7 \text{ in.}}{25 \text{ in.}} = \frac{7}{25}$$

$$\cos A = \frac{\text{adjacent}}{\text{hypotenuse}} = \frac{2 \text{ ft}}{2 \text{ ft 1 in.}} = \frac{24 \text{ in.}}{25 \text{ in.}} = \frac{24}{25}$$

$$\tan A = \frac{\text{opposite}}{\text{adjacent}} = \frac{7 \text{ in.}}{2 \text{ ft}} = \frac{7 \text{ in.}}{24 \text{ in.}} = \frac{7}{24}$$

$$\sin B = \frac{\text{opposite}}{\text{hypotenuse}} = \frac{2 \text{ ft}}{2 \text{ ft 1 in.}} = \frac{24 \text{ in.}}{25 \text{ in.}} = \frac{24}{25}$$

$$\cos B = \frac{\text{adjacent}}{\text{hypotenuse}} = \frac{7 \text{ in.}}{2 \text{ ft 1 in.}} = \frac{7 \text{ in.}}{25 \text{ in.}} = \frac{7}{25}$$

$$\tan B = \frac{\text{opposite}}{\text{adjacent}} = \frac{2 \text{ ft}}{7 \text{ in.}} = \frac{24 \text{ in.}}{7 \text{ in.}} = \frac{24}{7}$$

2 **Find the Cosecant, Secant, and Cotangent of Angles of Right Triangles, Given the Measures of at Least Two Sides.**

Each basic trigonometric function (sine, cosine, and tangent) has a *reciprocal function.* The *cosecant* (*csc*) function is the reciprocal of the sine function, the *secant* (*sec*) function is the reciprocal of the cosine function, and the *cotangent* (*cot*) function is the reciprocal of the tangent function.

Using Fig. 22–13, we can write the reciprocal trigonometric functions of angle *A*.

Reciprocal trigonometric functions of angle A in a standard right triangle:

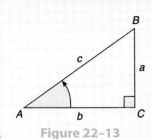

Figure 22–13

$$\csc A = \frac{\text{hypotenuse}}{\text{opposite}} = \frac{c}{a}$$

$$\sec A = \frac{\text{hypotenuse}}{\text{adjacent}} = \frac{c}{b}$$

$$\cot A = \frac{\text{adjacent}}{\text{opposite}} = \frac{b}{a}$$

TIP!

Reciprocal Trigonometric Functions

$$\sin A = \frac{1}{\csc A} \qquad \cos A = \frac{1}{\sec A} \qquad \tan A = \frac{1}{\cot A}$$

$$\csc A = \frac{1}{\sin A} \qquad \sec A = \frac{1}{\cos A} \qquad \cot A = \frac{1}{\tan A}$$

For example, if $\sin A = \dfrac{a}{c}$, then $\csc A = \dfrac{c}{a}$ $\qquad \dfrac{1}{\frac{a}{c}} = 1 \div \dfrac{a}{c} = \dfrac{c}{a}$

Similarly, we can write the reciprocal trigonometric functions of angle *B* in Fig. 22–14.

Reciprocal trigonometric functions of angle B in a standard right triangle:

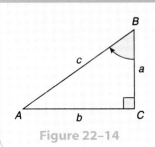

Figure 22–14

$$\csc B = \frac{\text{hypotenuse}}{\text{opposite}} = \frac{c}{b}$$

$$\sec B = \frac{\text{hypotenuse}}{\text{adjacent}} = \frac{c}{a}$$

$$\cot B = \frac{\text{adjacent}}{\text{opposite}} = \frac{a}{b}$$

EXAMPLE Write the six trigonometric ratios for angles A and B in Fig. 22–15. For convenience, *opposite*, *hypotenuse*, and *adjacent* are now abbreviated.

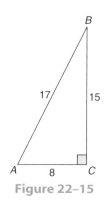

Figure 22–15

Angle A

$$\sin A = \frac{\text{opp}}{\text{hyp}} = \frac{15}{17} \qquad \csc A = \frac{\text{hyp}}{\text{opp}} = \frac{17}{15}$$

$$\cos A = \frac{\text{adj}}{\text{hyp}} = \frac{8}{17} \qquad \sec A = \frac{\text{hyp}}{\text{adj}} = \frac{17}{8}$$

$$\tan A = \frac{\text{opp}}{\text{adj}} = \frac{15}{8} \qquad \cot A = \frac{\text{adj}}{\text{opp}} = \frac{8}{15}$$

Angle B

$$\cos B = \frac{\text{adj}}{\text{hyp}} = \frac{15}{17} \qquad \sec B = \frac{\text{hyp}}{\text{adj}} = \frac{17}{15}$$

$$\sin B = \frac{\text{opp}}{\text{hyp}} = \frac{8}{17} \qquad \csc B = \frac{\text{hyp}}{\text{opp}} = \frac{17}{8}$$

$$\tan B = \frac{\text{opp}}{\text{adj}} = \frac{8}{15} \qquad \cot B = \frac{\text{adj}}{\text{opp}} = \frac{15}{8}$$

EXAMPLE Find the six trigonometric functions for angles A and B in Fig. 22–16. Express ratios to the nearest ten-thousandth.

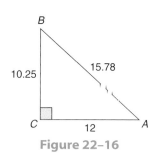

Figure 22–16

$$\sin A = \frac{10.25}{15.78} = \mathbf{0.6496} \qquad \sin B = \frac{12}{15.78} = \mathbf{0.7605}$$

$$\cos A = \frac{12}{15.78} = \mathbf{0.7605} \qquad \cos B = \frac{10.25}{15.78} = \mathbf{0.6496}$$

$$\tan A = \frac{10.25}{12} = \mathbf{0.8542} \qquad \tan B = \frac{12}{10.25} = \mathbf{1.1707}$$

$$\cot A = \frac{12}{10.25} = \mathbf{1.1707} \qquad \cot B = \frac{10.25}{12} = \mathbf{0.8542}$$

$$\csc A = \frac{15.78}{10.25} = \mathbf{1.5395} \qquad \csc B = \frac{15.78}{12} = \mathbf{1.3150}$$

$$\sec A = \frac{15.78}{12} = \mathbf{1.3150} \qquad \sec B = \frac{15.78}{10.25} = \mathbf{1.5395}$$

TIP!

Patterns and Relationships

Examine the boxes in the two previous examples. The ratios are arranged and highlighted in the first example to illustrate *reciprocal relationships*. The arrangement and highlighting in the second example illustrate the relationships of ratios for angle A and angle B.

$$\sin A = \cos B \qquad \csc A = \sec B \qquad \cos A = \sin B \qquad \sec A = \csc B \qquad \tan A = \cot B \qquad \cot A = \tan B$$

The relationships are referred to as *cofunctions*. Cofunctions are discussed in Section 3, Outcome 4 (pp. 856–858).

1 Use Fig. 22–17 to find the indicated trigonometric ratios.

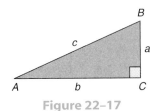

Figure 22–17

Use $a = 5$, $b = 12$, and $c = 13$ in Exercises 1–6 and express the ratios as fractions in lowest terms.

1. $\sin A$ **2.** $\cos A$ **3.** $\tan A$ **4.** $\sin B$ **5.** $\cos B$ **6.** $\tan B$

Use $a = 9$, $b = 12$, and $c = 15$ in Exercises 7–12 and express the ratios as fractions in lowest terms.

7. $\sin A$ **8.** $\cos A$ **9.** $\tan A$ **10.** $\sin B$ **11.** $\cos B$ **12.** $\tan B$

Use $a = 16$, $b = 30$, and $c = 34$ in Exercises 13–18 and express the ratios as fractions in lowest terms.

13. $\sin A$ **14.** $\cos A$ **15.** $\tan A$ **16.** $\sin B$ **17.** $\cos B$ **18.** $\tan B$

Use $a = 9$, $b = 14$, and $c = 16.64$ in Exercises 19–24 and express the ratios as decimals to the nearest ten-thousandth.

19. $\sin A$ **20.** $\cos A$ **21.** $\tan A$ **22.** $\sin B$ **23.** $\cos B$ **24.** $\tan B$

2 Find the six trigonometric ratios for angles A and B in Fig. 22–17 using the values given in Exercises 25 and 26.

25. $a = 12$, $b = 16$, $c = 20$

26. $a = 8.15$, $b = 5.32$, $c = 9.73$. Express the ratios as decimals to the nearest ten-thousandth.

22–3 | Using a Calculator to Find Trigonometric Values

Learning Outcomes

1 Find trigonometric values for the sine, cosine, and tangent using a calculator.

2 Find the angle measure, given a trigonometric value.

3 Find the trigonometric values for the cosecant, secant, and cotangent using the reciprocal relationship.

4 Find the trigonometric values for the cosecant, secant, and cotangent using the cofunction relationship.

In studying similar triangles, we found that corresponding angles of similar triangles are equal and that corresponding sides are proportional. These properties of similar triangles lead us to a very important property of trigonometric functions.

Values of trigonometric functions:

Every angle of a specified measure has a specific set of values for its trigonometric functions.

For centuries, trigonometric values were recorded in tables and were readily available in mathematics textbooks to facilitate calculations. Today, scientific and graphing calculators and software packages have the values of the trigonometric functions already programmed. These calculator values are expressed to more decimal places than even the most accurate of written tables, thus producing more accurate results.

1 Find Trigonometric Values for the Sine, Cosine, and Tangent Using a Calculator.

We recommend that you use your calculator to find trigonometric values. To find trigonometric values on your calculator, the keys SIN, COS, and TAN are used most often. The angle measure can be entered in degrees or radians, depending on the selected mode.

To find trigonometric values using the calculator:

1. Set your calculator to the desired mode of angle measure. Calculators generally have at least two angle modes: degrees and radians.
2. Press the appropriate function key (sin, cos, tan) and then enter the angle measure.
3. Display the result by pressing the = or ENTER key.

EXAMPLE Using a calculator, find the trigonometric values.

(a) sin 27°   (b) sin 20°30′

(c) cos 1.34   (d) $\tan \dfrac{\pi}{4}$

(a) Be sure your calculator is in degree mode.

SIN 27 = ⇒ **0.4539904997**   Keystrokes may vary.

(b) Be sure your calculator is in degree mode.

SIN 20.5 = ⇒ **0.3502073813**   20°30′ = 20.5°.

With some calculators, angle measures can be entered using degrees, minutes, and seconds *or* using the decimal equivalent.

(c) Reset your calculator to radian mode.

COS 1.34 = ⇒ **0.2287528078**

(d) Be sure your calculator is in radian mode.

TAN (π ÷ 4) = ⇒ **1**

TIP!

Automatic Parentheses in Calculator Functions

Some calculators automatically insert an open parenthesis after certain functions. If a close parenthesis is not entered before the calculation is executed, the close parenthesis is automatically interpreted at the end of the entry. Test your calculator with and without the close parenthesis to verify the interpretation made by your calculator.

Entry: SIN 27 ENTER
Display: SIN (27
                    .4539904997
Entry: SIN 27) ENTER
Display: SIN (27)
                    .4539904997

2 **Find the Angle Measure, Given a Trigonometric Value.**

When the measures of two sides of a right triangle are known, we can find the angle measure from a trigonometric value.

Finding one of the acute angle measures of a right triangle when given the trigonometric value is an *inverse operation* to finding the trigonometric value when given the angle measure. The notation most commonly used is $\sin^{-1}$, $\cos^{-1}$, or $\tan^{-1}$. Another notation is arcsin, arccos, or arctan.

To find angle measures using the calculator:

1. Set your calculator to the desired mode of angle measure (degrees or radians).
2. Select the appropriate inverse trigonometric function key or menu option ($\boxed{\text{SIN}^{-1}}$, $\boxed{\text{COS}^{-1}}$, or $\boxed{\text{TAN}^{-1}}$). Enter the trigonometric value and $\boxed{=}$ or $\boxed{\text{ENTER}}$.

Some calculators use an inverse key $\boxed{\text{INV}}$, shift key $\boxed{\text{SHIFT}}$ or second $\boxed{\text{2nd}}$ key to access the inverse keys. Test your calculator with a known value. For example, in the previous example we found sin 27° = 0.4539904997. $\sin^{-1} 0.4539904997$ should be 27°.

EXAMPLE Find the angle in degrees given the trigonometric values in parts (a) and (b). θ represents the unknown angle measure. Round to the nearest tenth of a degree. For part (c), find the radians to the nearest thousandth.

(a) sin θ = 0.6561   (b) cos θ = 0.4226   (c) tan θ = 2.825

(a) Be sure your calculator is in degree mode.

$\boxed{\text{SIN}^{-1}}.6561\boxed{=}$ ⟹ 41.0031105 ≈ **41.0°**   Rounded.

(b) Be sure your calculator is in degree mode.

$\boxed{\text{COS}^{-1}}.4226\boxed{=}$ ⟹ 65.00115448 ≈ **65.0°**   Rounded.

(c) Be sure your calculator is in radian mode.

$\boxed{\text{TAN}^{-1}}2.825\boxed{=}$ ⟹ 1.230578215 ≈ **1.231 rad**   Rounded.

---

3 **Find the Trigonometric Values for the Cosecant, Secant, and Cotangent Using the Reciprocal Relationship.**

The reciprocal key ($\boxed{x^{-1}}$ or $1/x$) on calculators can be used to find the value of *reciprocal trigonometric functions*. For instance, after the sine of an angle is found, the cosecant can be determined by pressing the reciprocal key.

The following statements indicate the reciprocal relationships of the tangent and cotangent, the sine and cosecant, and the cosine and secant functions.

$$\cot \theta = \frac{1}{\tan \theta} \qquad \csc \theta = \frac{1}{\sin \theta} \qquad \sec \theta = \frac{1}{\cos \theta}$$

A relationship statement such as $\cot \theta = \dfrac{1}{\tan \theta}$ is a true statement or fact. A relationship statement such as this is a *trigonometric identity*.

To find the value of the cosecant, secant, or cotangent using the reciprocal relationship:

1. Set your calculator to the desired mode of angle measure (degrees or radians).
2. Select the appropriate reciprocal relationship (sin, cos, or tan). Enter the angle measure and $\boxed{=}$ or $\boxed{\text{ENTER}}$.
3. Find the reciprocal of the answer. Select the reciprocal key $\boxed{x^{-1}}$ or $\boxed{1/x}$ and press $\boxed{=}$ or $\boxed{\text{ENTER}}$.

EXAMPLE Determine the value of the trigonometric functions.

(a) csc 18.5°   (b) sec 1.0821

(a) $\csc 18.5° = \dfrac{1}{\sin 18.5°} = \dfrac{1}{0.3173046564} = \textbf{3.1515}$ To nearest ten-thousandth.

In degree mode:

$\boxed{\text{SIN}}\, 18.5 \boxed{=} \boxed{x^{-1}} \boxed{=} \Rightarrow 3.151545305$

(b) $\sec 1.0821 = \dfrac{1}{\cos 1.0821} = \dfrac{1}{0.4694752149} = \textbf{2.1300}$ To nearest ten-thousandth.

In radian mode:

$\boxed{\text{COS}}\, 1.0821 \boxed{=} \boxed{x^{-1}} \boxed{=} \Rightarrow 2.130037898$

To find the angle measure for a given value of the cosecant, secant, or cotangent using reciprocal functions:

1. Find the reciprocal of the trigonometric value using the $\boxed{x^{-1}}$ or $\boxed{1/x}$ key.
2. Select the appropriate inverse function key ($\boxed{\text{SIN}^{-1}}$, $\boxed{\text{COS}^{-1}}$, or $\boxed{\text{TAN}^{-1}}$) and enter the reciprocal found in Step 1 ($\boxed{\text{ANS}}$).
3. Display the result of the calculation by entering $\boxed{=}$ or $\boxed{\text{ENTER}}$.

EXAMPLE Find an angle in degrees whose secant is 2.7320. Express the measure to the nearest tenth of a degree. Find the radians of an angle to the nearest hundredth radian whose cotangent is 0.2167.

In degree mode:

$2.732\, \boxed{x^{-1}} \boxed{=} \boxed{\text{COS}^{-1}} \boxed{\text{ANS}} \boxed{=}$ $\boxed{\text{ANS}}$ enters the previous answer.

$\theta = \textbf{68.5°}$

In radian mode:

$.2167\, \boxed{x^{-1}} \boxed{=} \boxed{\text{TAN}^{-1}} \boxed{\text{ANS}} \boxed{=}$

$\theta = \textbf{1.36 rad}$

---

4 **Find the Trigonometric Values for Cosecant, Secant, and Cotangent Using the Cofunction Relationship.**

We can use calculators to investigate relationships among trigonometric functions. Because the sum of all three angles of a triangle is 180° and the right angle in a right triangle is 90°, the other two angles in a right triangle are both acute angles (less than 90°) and their sum is 90°. That makes the two acute angles of any right triangle *complementary angles*. Let's examine the relationship between the sine of an angle and the cosine of its complement.

$$\sin 40° = 0.6428 \qquad \cos 50° = 0.6428$$
$$\sin 30° = 0.5 \qquad \cos 60° = 0.5$$
$$\sin 80° = 0.9848 \qquad \cos 10° = 0.9848$$
$$\sin 90° = 1 \qquad \cos 0° = 1$$

We can generalize our observations.

The relationship of the sine of an angle and the cosine of its complement:

The sine of an angle equals the cosine of its complement.

Now, let's look at the relationship between the tangent of an angle and the cotangent of its complement.

$\tan 0° = 0$	$\cot 90° = 0$	
$\tan 30° = 0.5774$	$\cot 60° = 0.5774$	
$\tan 10° = 0.1763$	$\cot 80° = 0.1763$	
$\tan 45° = 1$	$\cot 45° = 1$	
$\tan 90° = \text{error}$ Undefined.	$\cot 0° = \text{error}$ Undefined.	

The relationship of the tangent of an angle and the cotangent of its complement:

The tangent of an angle equals the cotangent of its complement.

Now let's compare the secant of an angle with the cosecant of its complement.

$\sec 0° = 1$	$\csc 90° = 1$
$\sec 30° = 1.1547$	$\csc 60° = 1.1547$
$\sec 10° = 1.0154$	$\csc 80° = 1.0154$
$\sec 45° = 1.4142$	$\csc 45° = 1.4142$
$\sec 90° = \text{error}$ Undefined.	$\csc 0° = \text{error}$ Undefined.

The relationship of the secant of an angle and the cosecant of its complement:

The secant of an angle equals the cosecant of its complement.

TIP!

The pairs of functions given in these three properties are referred to as trigonometric *cofunctions*.

Cofunction Versus Reciprocal Function

Trigonometric functions can be paired as cofunctions or as reciprocal functions.

Cofunctions

sine θ and cosine θ   secant θ and cosecant θ   tangent θ and cotangent θ

The values of a function of an angle and of the cofunction of the angle's complement are equal.

We can state these relationships as **trigonometric identities**.

For Degrees:

$$\sin \theta = \cos (90° - \theta)$$

$$\sec \theta = \csc (90° - \theta)$$

$$\tan \theta = \cot (90° - \theta)$$

For Radians:

$$\sin \theta = \cos \left(\frac{\pi}{2} - \theta \right)$$

$$\sec \theta = \csc \left(\frac{\pi}{2} - \theta \right)$$

$$\tan \theta = \cot \left(\frac{\pi}{2} - \theta \right)$$

Reciprocal Functions

$$\text{sine } \theta = \frac{1}{\text{cosecant } \theta} \qquad \text{cosine } \theta = \frac{1}{\text{secant } \theta} \qquad \text{tangent } \theta = \frac{1}{\text{cotangent } \theta}$$

1. Find the complement of the given angle measure.
2. Select the appropriate cofunction.
3. Evaluate the cofunction for the complement of the given angle.

EXAMPLE Find the values to the nearest ten-thousandth using the cofunction relationship.

(a) cot 25°   (b) cot 1.414

(a) $\cot 25° = \tan (90° - 25°) = \tan 65° = 2.144506921 = \mathbf{2.1445}$

(b) $\cot 1.414 = \tan \left(\dfrac{\pi}{2} - 1.414 \right) = \tan 0.1567963268 = 0.1580940406 = \mathbf{0.1581}$

$\boxed{\text{TAN}}\ (\ (\ \boxed{\pi}\ \boxed{\div}\ 2\ \boxed{-}\ 1.414\ )\ )\ \boxed{=}\ \Rightarrow\ 0.1580940406$

---

1. Use the appropriate inverse cofunction ($\boxed{\text{SIN}^{-1}}, \boxed{\text{COS}^{-1}}$, or $\boxed{\text{TAN}^{-1}}$) and the given value to find the angle measure of the cofunction.
2. Find the complement of the angle found in Step 1.

EXAMPLE Find the angle measure of (a) cot θ = 2.6, in degrees, (b) cot θ = 0.24, in radians.

(a) $\cot \theta = \tan (90° - \theta) = 2.6$

       $\boxed{\text{TAN}^{-1}}\ 2.6\ \boxed{=}\ \Rightarrow\ 68.96248897°$

       $90° - \theta = 68.96248897°$

             $\theta = 90° - 68.96248897°$

             $\boldsymbol{\theta = 21.03751103°}$ **or**   **21.0°**

Find $\boxed{\text{TAN}^{-1}}$ of 2.6.

90° − θ in degrees = 68.96248897°.
Solve for θ to find the complement of 68.96248897°.

Use $\boxed{\text{ANS}}$ to enter 68.96248897.

(b) $\cot \theta = \tan \left(\dfrac{\pi}{2} - \theta \right) = 0.24$

       $\boxed{\text{TAN}^{-1}}\ 0.24\ \boxed{=}\ \Rightarrow\ 0.2355449807$

       $\dfrac{\pi}{2} - \theta = 0.2355449807$

           $\theta = \dfrac{\pi}{2} - 0.2355449807$

           $\boldsymbol{\theta = 1.335251346}$ **or**   **1.3353 rad**

Find $\boxed{\text{TAN}^{-1}}$ of 0.24.

$\dfrac{\pi}{2} - \theta$ in radians = 0.2355449807.

Solve for θ.

        Chapter 22 / Introduction to Trigonometry

1 Use your calculator to find the trigonometric values. Express your answers in ten-thousandths.

1. sin 21°	**2.** cos 3.5°	**3.** tan 47°	**4.** cos 52.5°
5. sin 0.5498	**6.** cos 21°30′	**7.** cos 1.1519	**8.** cos 0.3665
9. sin 53°30′	**10.** tan 42.5°	**11.** tan 47.7°	**12.** sin 62°10′
13. cos 12°40′	**14.** cos 1.0530	**15.** tan 73°14′	**16.** sin 1.2363
17. cos 46.8°	**18.** cos 0.3549	**19.** cos 1.1636	**20.** tan 12.4°

2 Find the angles of the trigonometric values in degrees. θ represents the unknown angle measure. Express each answer to the nearest tenth of a degree.

21. sin θ = 0.3420	**22.** cos θ = 0.9239	**23.** tan θ = 2.356
24. cos θ = 0.4617	**25.** cos θ = 0.540	**26.** sin θ = 0.5712
27. tan θ = 1.265	**28.** cos θ = 0.137	**29.** sin θ = 0.6298
30. cos θ = 0.9325		

Find the angles of the trigonometric values in radians. Express each answer to the nearest ten-thousandth.

31. sin θ = 0.5299	**32.** tan θ = 0.8098	**33.** cos θ = 0.6947
34. cos θ = 0.3907	**35.** cos θ = 0.968	**36.** sin θ = 0.9959
37. tan θ = 0.3160	**38.** tan θ = 2.430	**39.** cos θ = 0.9610
40. cos θ = 0.4210		

3 Find the indicated trigonometric values using the reciprocal relationship. Round to the nearest ten-thousandth.

41. cot 24.5°	**42.** csc 42°	**43.** sec 0.2443	**44.** csc 1.3788
45. sec 1.2165	**46.** cot 28.6°	**47.** cot 87°	**48.** sec 42°30′
49. csc 0.2136	**50.** sec 1.0372		

Find θ to the nearest tenth of a degree.

51. cot θ = 0.4238	**52.** sec θ = 1.8291	**53.** csc θ = 3.7129
54. sec θ = 8.2156	**55.** cot θ = 1.7318	

4 Find the indicated trigonometric value using the cofunction relationship. Round to the nearest ten-thousandth.

56. sin 82°	**57.** cos 15°	**58.** tan 1.023	**59.** cot 0.896

Find θ to the nearest tenth of a degree.

60. sin θ = 0.8724	**61.** cos θ = 0.4687	**62.** csc θ = 3.1067	**63.** cot θ = 5.213

ASSIGNMENT EXERCISES

Section 22–1

Convert the degree measures to radians rounded to the nearest hundredth.

1. 60°	**2.** 212°	**3.** 300°

Convert the degree, minute, and second measures to degrees rounded to the nearest ten-thousandth. Then convert to radians to the nearest hundredth.

4. 25°30′	**5.** 99°45′	**6.** 120°20′40″

Convert the radian measures to degrees. Round to the nearest ten-thousandth of a degree.

7. $\dfrac{5\pi}{6}$ rad	**8.** 2.4 rad	**9.** 1.7 rad

Convert the radian measures to degrees, minutes, and seconds. Round to the nearest second.

10. 0.9 rad

11. $\dfrac{3\pi}{8}$ rad

12. 1.2 rad

Find the arc length, radius, or central angle (in radians) for each of the following. Round to hundredths when necessary.

13. Find s if $\theta = 0.7$ rad and $r = 2.3$ cm.

14. Find θ if $s = 6.2$ cm and $r = 5$ cm.

15. Find r if $\theta = 2.1$ rad and $s = 3.6$ ft.

Find the area of the sector, the radius, or the central angle (in radians) of the sector for each of the following. Round to hundredths when necessary.

16. Find A if $\theta = 0.88$ rad and $r = 1.5$ m.

17. Find r if $\theta = 4.2$ rad and $A = 24$ in^2.

18. Find θ if $r = 4$ cm and $A = 12$ cm^2.

19. A pendulum 6 in. long swings through an angle of 0.35 rad. Find the arc length the pendulum swings over from one extreme position to the other. Round to hundredths.

20. A movable part on a machine swings through an angle of 0.7 rad with an arc length of 8 in. What is the length of the part? Round to hundredths.

21. Find the area of a sector with a central angle of 1.48 rad and whose radius is 4.6 cm. ($A = \frac{1}{2}\theta r^2$, where θ is in radians.) Round to hundredths.

22. Find the number of radians to the nearest ten-thousandth of a central angle of a sector that has an area of 8.4 cm^2 and a radius of 6 cm.

Section 22–2

Find the trigonometric ratios using Fig. 22–18.

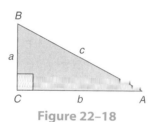

Figure 22–18

Use $a = 15$, $b = 20$, $c = 25$ in Exercises 23–34 and express the ratios as fractions in lowest terms.

23. sin A **24.** cos A **25.** tan A **26.** csc A **27.** sec A **28.** cot A

29. sin B **30.** cos B **31.** tan B **32.** csc B **33.** sec B **34.** cot B

Use $a = 2$ ft, $b = 10$ in., $c = 2$ ft 2 in. in Exercises 35–40 and express the ratios as fractions in lowest terms.

35. sin A **36.** sec B **37.** cot B **38.** csc A **39.** tan B **40.** cos A

Use $a = 7$, $b = 10.5$, $c = 12.62$ in Exercises 41–47 and express the ratios in decimals to the nearest ten-thousandth.

41. cos B **42.** csc A **43.** tan A **44.** sin B **45.** sec A **46.** sin A

47. cot A

Section 22–3

Use your calculator to find the trigonometric values. Round to the nearest ten-thousandth.

48. cos 42.5°   **49.** sin 0.4712   **50.** sin 65.5°   **51.** cot 73°

52. tan 1.0210   **53.** tan 47°   **54.** tan 15.6°   **55.** sin 0.8610

56. tan 25°40′   **57.** cos 32°50′   **58.** cot 0.7510   **59.** cos 80°10′

Find an angle (in degrees) having the trigonometric values. θ represents the unknown angle measure. Express answers to the nearest tenth of a degree.

60. sin $\theta = 0.5446$ **61.** cos $\theta = 0.6088$ **62.** tan $\theta = 0.8720$

63. cot $\theta = 0.9884$ **64.** cos $\theta = 0.8897$ **65.** cot $\theta = 3.340$

Find an angle (in radians) having the trigonometric values. Round to the nearest ten-thousandth.

66. sin $\theta = 0.9205$ **67.** tan $\theta = 2.723$ **68.** cos $\theta = 0.9450$

69. cot $\theta = 0.3772$ **70.** sin $\theta = 0.2896$ **71.** tan $\theta = 0.3440$

Use reciprocal functions to find the trigonometric values. Round to the nearest ten-thousandth.

72. sec 15.5°   **73.** csc 71°   **74.** sec 0.4363

75. csc 1.082   **76.** sec 1.2886   **77.** csc 0.4829

Use cofunctions to evaluate the following.

78. cot 45°

79. csc 82°20′

80. sec 38°

81. csc 1.235

82. cot 0.1823

83. sec 4.23

CHALLENGE PROBLEMS

84. Is there an angle between 0° and 360° for which the sine and cosine of the angle are equal? Justify your answer by making a table of values for the sine and cosine of angles between 0° and 360° and by graphing sin x and cos x for values between 0° and 360°.

85. Investigate the relationship described in Problem 84 for the sine and tangent functions and justify your answer.

CONCEPTS ANALYSIS

1. Explain how the sine function is a ratio.

2. Use your calculator to find the value of the sine of several angles in each of the four quadrants and make a general statement about the greatest and least values the sine function can have.

3. Draw three right triangles. The acute angles are 30° and 60° for triangle 1; 25° and 65° for triangle 2; and 80° and 10° for triangle 3. For each triangle, find the sine of the first angle and the cosine of the second angle. Compare the sine and cosine values for each triangle. What generalization can you make from these three triangles?

4. Draw three right triangles of your choice and verify the generalization you made in Exercise 3.

5. Use your calculator to find the sine of the angles given below and make a table. List the given angle in column 1 and the sine of the angle in column 2. What is the angle measure at which the sine of the angle begins to repeat?
30°, 60°, 90°, 120°, 150°, 180°, 210°, 240°, 270°, 300°, 330°, 360°, 390°, 420°

6. Find the sine of several angles larger than those given in Exercise 5 to determine where, if anywhere, the sine begins to repeat again.

7. Make a general statement about the sine function as it repeats. Compare your general statement with the graph of the sine function.

8. Use a graphing calculator to graph the following functions on the same grid: sin x; 2 sin x; 3 sin x; $\frac{1}{2}$ sin x; $\frac{1}{3}$ sin x. Describe how the coefficient of the sine function impacts the graph of the function. This coefficient is called the amplitude. Look up the definition of the word *amplitude*. Does the dictionary definition make sense in view of your description? Set your graphing window for x as 0° to 360° and for y as −3 to +3.

9. Graph sin x and sin 2x on the same graph. Compare these graphs. How do they differ and how are they alike?

10. Graph sin x and sin($\frac{1}{2}$)x on the same graph. Compare these two graphs to the graphs for Exercise 9. What effect does the coefficient of x have on the graph of the sine function? Graph other sine functions in which the coefficient of x is different to verify your description.

CHAPTER SUMMARY

Learning Outcomes

What to Remember with Examples

Section 22–1

1 Find the arc length, given a central angle in radians (pp. 843–844).

To find the length of an arc of a circle, multiply the radians of the central angle by the radius of the circle.

Use $s = \theta r$, where θ is given in radians.

Find the arc length intercepted on the circumference of a circle by a central angle of $\frac{\pi}{4}$ if the radius of the circle is 24 m.

$$s = \theta r$$
$$s = \frac{\pi}{4}(24)$$
$$s = \pi(6)$$
$$s = 6\pi$$
$$s = 18.850 \text{ m}$$

2 Find the area of a sector, given a central angle in radians (pp. 844–845).

The area of a sector is $\frac{1}{2}\theta r^2$, where θ is given in radians.

Find the area of a sector that has a central angle of 1.4 rad and a radius of 5.2 ft.

$$A = \frac{1}{2}\theta r^2$$
$$A = 0.5(1.4)(5.2)^2$$
$$A = 18.928 \text{ ft}^2$$

3 Convert angle measures from degrees to radians (pp. 845–846).

Angle measures in degrees are converted to radians by multiplying by the unity ratio $\frac{\pi \text{ rad}}{180°}$.

Write 82° in radians and round to the nearest thousandth.

$$82° \times \frac{\pi \text{ rad}}{180°} = 1.431 \text{ rad}$$

4 Convert angle measures from radians to degrees (pp. 846–847).

Angle measures in radians are converted to degrees by multiplying by the unity ratio $\frac{180°}{\pi \text{ rad}}$.

Write 0.45 radians in degrees and round to the nearest tenth of a degree.

$$0.45 \text{ rad} \times \frac{180°}{\pi \text{ rad}} = 25.8°$$

Section 22–2

1 Find the sine, cosine, and tangent of angles of right triangles, given the measures of at least two sides (pp. 848–850).

Use the three trigonometric functions to calculate the value of the function when the appropriate sides are given. The Pythagorean theorem may be needed to find the length of a third side in some instances.

$$\text{sine } A = \frac{\text{opposite}}{\text{hypotenuse}} \qquad \text{cosine } A = \frac{\text{adjacent}}{\text{hypotenuse}} \qquad \text{tangent } A = \frac{\text{opposite}}{\text{adjacent}}$$

Find the sine, cosine, and tangent of angle A in Fig. 22–19 and round to the nearest ten-thousandth.

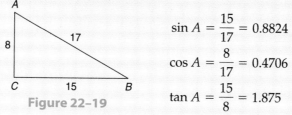

Figure 22–19

$$\sin A = \frac{15}{17} = 0.8824$$
$$\cos A = \frac{8}{17} = 0.4706$$
$$\tan A = \frac{15}{8} = 1.875$$

Chapter 22 / Introduction to Trigonometry

2 Find the cosecant, secant, and cotangent of angles of right triangles, given the measures of at least two sides (pp. 851–852).

The cosecant, secant, and cotangent are reciprocal functions of the sine, cosine, and tangent, respectively.

$$\text{cosecant } A = \frac{\text{hypotenuse}}{\text{opposite}} \qquad \csc = \frac{1}{\sin}$$

$$\text{secant } A = \frac{\text{hypotenuse}}{\text{adjacent}} \qquad \sec = \frac{1}{\cos}$$

$$\text{cotangent } A = \frac{\text{adjacent}}{\text{opposite}} \qquad \cot = \frac{1}{\tan}$$

Find the cosecant, secant, and cotangent of angle A in Fig. 22–19 and round to the nearest ten-thousandth.

$$\csc A = \frac{17}{15} = 1.1333$$

$$\sec A = \frac{17}{8} = 2.125$$

$$\cot A = \frac{8}{15} = 0.5333$$

Section 22–3
1 Find trigonometric values for the sine, cosine, and tangent using a calculator (p. 854).

Use the $\boxed{\text{SIN}}$, $\boxed{\text{COS}}$, and $\boxed{\text{TAN}}$ keys to find the trigonometric value of a specified angle. Be sure your calculator is set to the appropriate mode: degree or radian.

Use your calculator to find the following: sin 35°, cos 198°, tan 125°, sin 2.8, cos 2.98, tan 4.2.

Most calculators:
Be sure your calculator is in degree mode:

$$\boxed{\text{SIN}} \quad 35 \boxed{=} \Rightarrow 0.5735764364$$
$$\boxed{\text{COS}} \quad 198 \boxed{=} \Rightarrow -0.9510565163$$
$$\boxed{\text{TAN}} \quad 125 \boxed{=} \Rightarrow -1.428148007$$

Change to radian mode:

$$\boxed{\text{SIN}} \quad 2.8 \boxed{=} \Rightarrow 0.3349881502$$
$$\boxed{\text{COS}} \quad 2.98 \boxed{=} \Rightarrow -0.9869722927$$
$$\boxed{\text{TAN}} \quad 4.2 \boxed{=} \Rightarrow 1.777779775$$

2 Find the angle measure, given a trigonometric value (p. 855).

To find the angle measure when the trigonometric value is given, use the inverse trigonometric function key, which may be accessed by pressing the $\boxed{\text{SHIFT}}$, $\boxed{\text{INV}}$ or $\boxed{\text{2nd}}$ key.

Find the value of x in degrees: $\cos x = 0.906307787$ or $x = \cos^{-1}(0.906307787)$.

$$\boxed{\text{COS}^{-1}} 0.906307787 \boxed{=}$$
$$x = 25°$$

3 Find the trigonometric values for the cosecant, secant, and cotangent using the reciprocal relationship (pp. 855–856).

Use the $\boxed{\text{SIN}}$, $\boxed{\text{COS}}$, and $\boxed{\text{TAN}}$ keys and then the reciprocal key $\boxed{x^{-1}}$ or $\boxed{1/x}$ to find the values of the inverse functions.

Use your calculator to find the following: csc 25°, sec 1.

$$\csc 25°: \boxed{\text{SIN}} \ 25 \boxed{=} \boxed{x^{-1}} \boxed{=} \Rightarrow 2.366201583 \quad \text{Use degree mode.}$$

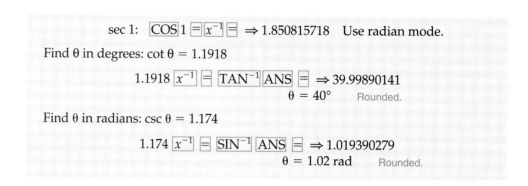

sec 1: COS 1 $=$ x^{-1} $=$ $\Rightarrow$ 1.850815718 Use radian mode.

Find θ in degrees: cot θ = 1.1918

1.1918 x^{-1} $=$ TAN^{-1} ANS $=$ $\Rightarrow$ 39.99890141
θ = 40°   Rounded.

Find θ in radians: csc θ = 1.174

1.174 x^{-1} $=$ SIN^{-1} ANS $=$ $\Rightarrow$ 1.019390279
θ = 1.02 rad   Rounded.

4 Find the trigonometric values for the cosecant, secant, and cotangent using the cofunction relationship (pp. 856–858).

To find the cosecant, secant, or cotangent of an angle, find its complement by subtracting the angle from 90°. Then find the sine, cosine, or tangent of the complement.

Find the cotangent of 35°.

In degree mode:

90° − 35° = 55°   Find the complement of 35°.
tan 55° = 1.428148007   Find the value of the cofunction of the complement.

Thus, cot 35° = 1.428148007.

Find θ for sec θ = 0.75.

In radian mode:

$\sin^{-1}$ 0.75 = 0.848062079 rad

$\dfrac{\pi}{2}$ − 0.848062079 = 0.7227342478

θ = 0.72 rad   Rounded.

CHAPTER TRIAL TEST

Convert the degree measures to radians rounded to the nearest hundredth.
1. 35°   **2.** 122°   **3.** 315°   **4.** 240°

Convert the degree, minute, and second measures to degrees rounded to the nearest ten-thousandth.
Then convert to radians to the nearest hundredth.
5. 15°25′   **6.** 142°32′15″   **7.** 16°12′   **8.** 32°18′37″

Convert the radian measures to degrees. Round to the nearest ten-thousandth of a degree.
9. $\dfrac{5\pi}{8}$ rad   **10.** 3.1 rad

Convert the radian measures to degrees, minutes, and seconds. Round to the nearest second.
11. 1.2 rad   **12.** $\dfrac{\pi}{6}$ rad

Use the relationships of arc length, area, central angle (in radians), and radius to solve the problems relating to sectors. Round to hundredths if necessary.
13. Find s if θ = 0.5 and r = 2 in.   **14.** Find θ if s = 5.3 m and r = 7 m.
15. Find r if θ = 1.7 and s = 2.9 m.   **16.** Find A if θ = 0.6 and r = 7.3 cm.

Write the trigonometric ratios as fractions in lowest terms for the following trigonometric functions using triangle ABC in Fig. 22–20.
$a = 10, b = 24, c = 26$.

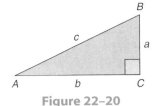

Figure 22–20

17. $\sin A$
18. $\tan B$
19. $\csc A$

Express the trigonometric values in decimals to the nearest ten-thousandth. Refer to Fig. 22–20 and use $a = 5, b = 11.5, c = 12.54$.

20. $\cot A$ 　　　　　　21. $\cos A$ 　　　　　　22. $\sin B$

Use your calculator to find the trigonometric values.

23. $\sin 53°$ 　　　　24. $\sin 61°10'$ 　　　　25. $\sin 1.1519$ 　　　　26. $\cos 1.0297$

Find an angle (in degrees) having the indicated trigonometric values; θ represents the unknown angle measure. Express your answers to the nearest tenth of a degree.

27. $\sin \theta = 0.2756$ 　　28. $\tan \theta = 1.280$ 　　29. $\cos \theta = 0.9426$ 　　30. $\cot \theta = 1.540$

Find an angle (in radians) having the indicated trigonometric values. Round to the nearest ten-thousandth.

31. $\sin \theta = 0.7660$ 　　32. $\cos \theta = 0.8387$ 　　33. $\tan \theta = 0.3259$

Find the indicated trigonometric values. Round to the nearest ten-thousandth.

34. $\sec 25.5°$ 　　　　35. $\csc 47°$ 　　　　36. $\cot 0.3316$

Right-Triangle Trigonometry

23

23–1 Sine, Cosine, and Tangent Functions

1 Find the missing parts of a right triangle using the sine function.

2 Find the missing parts of a right triangle using the cosine function.

3 Find the missing parts of a right triangle using the tangent function.

23–2 Applied Problems Using Right-Triangle Trigonometry

1 Select the most direct method for solving right triangles.

2 Solve applied problems using right-triangle trigonometry.

We use right triangles extensively in real-world applications. By using the trigonometric relationships we learned in Chapter 22, we can find all the angles and sides of a right triangle if we know the measure of one side and any other part.

23–1 | *Sine, Cosine, and Tangent Functions*

Learning Outcomes

1 Find the missing parts of a right triangle using the sine function.
2 Find the missing parts of a right triangle using the cosine function.
3 Find the missing parts of a right triangle using the tangent function.

We often use the sine, cosine, and tangent functions to find unknown parts of a right triangle. To do this, we manipulate formulas and use other algebraic principles depending on what information is given and what information needs to be found.

1 **Find the Missing Parts of a Right Triangle Using the Sine Function.**

We abbreviate the sine function to read $\sin \theta = \frac{\text{opp}}{\text{hyp}}$. Using this relationship, we can find parts of right triangles when we know any two parts that involve the sine function: one acute angle, the side opposite the known acute angle, and the hypotenuse.

Use the sine function to find unknown parts of a right triangle:

1. Two of these three parts must be known:
   a. One acute angle
   b. The side opposite the known acute angle
   c. Hypotenuse
2. Substitute the two known values in $\sin \theta = \frac{\text{opp}}{\text{hyp}}$.
3. Solve for the missing part.

TIP!

Does It Matter Which Acute Angle Is Known?

Not really. The acute angles of a right triangle are complementary. Therefore, if we know either acute angle (a), we can find the other one ($90° - a$) or $\left(\frac{\pi}{2} - a\right)$. Thus, we can use the sine function if we know *either* acute angle and any side.

EXAMPLE   Find angle A to the nearest tenth of a degree if $a = 7$ and $c = 21$ (see Fig. 23–1).

The two known values are the side opposite angle A and the hypotenuse, so we use the sine function for the acute angle A.

$$\sin A = \frac{\text{opp}}{\text{hyp}}$$

Substitute the known values: opp = 7, hyp = 21.

$$\sin A = \frac{7}{21}$$

Convert ratio to a decimal equivalent.

$$\sin A = 0.3333333333$$

Find $\sin^{-1}$ of 0.3333333333.

$$A = 19.47122063° \quad \text{or} \quad 19.5°$$

Round to the nearest tenth of a degree.

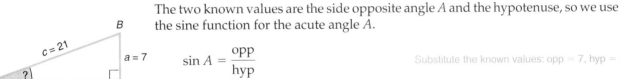

Figure 23–1

TIP!

How Do I Round My Answers?

Rounding practices are generally dictated by the context of the problem or industry standards; however, **for consistency, we round all trigonometric ratios to four significant digits and all angle values to the nearest 0.1° throughout Chapters 23 and 24 unless otherwise indicated.**

EXAMPLE   Find side a in triangle ABC (see Fig. 23–2).

We are given $\angle A$ and the hypotenuse and are asked to find the measure of the side opposite $\angle A$, so we use the sine function.

$$\sin A = \frac{\text{opp}}{\text{hyp}}$$

Substitute known values: $\angle A = 53.5°$, hyp = 15 m.

$$\sin 53.5° = \frac{a}{15}$$

$\sin 53.5° = 0.8038568606$.

$$0.8038568606 = \frac{a}{15}$$

Solve for a.

$$15(0.8038568606) = a$$

$$a = 12.05785291 \qquad \text{or} \qquad 12.06 \text{ m} \qquad \text{Rounded to four significant digits.}$$

(Figure 23–2: right triangle ABC with B at top, right angle at C. Side $AB = 15$ m, angle at $A = 53.5°$, $a = ?$ is side BC.)

Figure 23–2

TIP!

Rearrange the Formula Before You Calculate.

In the preceding example and those to follow, we can visualize a continuous sequence of calculator steps if we rearrange the formula for the missing part *before* we make any calculations.

$$\sin 53.5° = \frac{a}{15} \qquad \text{Rearrange for } a.$$

$$15(\sin 53.5°) = a \qquad \text{Evaluate.}$$

Then a continuous series of calculations is made. In degree mode,

$$15 \times \boxed{\text{SIN}} \, 53.5 \, \boxed{=} \Rightarrow 12.05785291$$

EXAMPLE   Find the hypotenuse in triangle RST (see Fig. 23–3).

We are given an acute angle and the side opposite the angle and are asked to find the hypotenuse, so we use the sine function.

$$\sin R = \frac{\text{opp}}{\text{hyp}}$$

Substitute known values: $\angle R = 32°$, opp = 25 ft.

$$\sin 32° = \frac{25}{t}$$

Rearrange for t.

$$t(\sin 32°) = 25$$

$$t = \frac{25}{\sin 32°} \qquad \text{Evaluate.}$$

$$t = 47.17699787$$

$$t = 47.18 \text{ ft} \qquad \text{Rounded to four significant digits.}$$

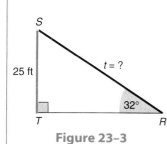

Figure 23–3

Choose Given Values over Calculated Values Whenever Possible.

It is best to use given values rather than calculated values when finding missing parts of a triangle. Because the rounded value for one missing part is sometimes used to find other missing parts, final answers may vary slightly due to rounding discrepancies. For instance, the angles of a triangle may add to as little as 179° or as much as 181°. Or the length of a side may be slightly different in the last significant digit. If the full calculator value of a side or angle is used to find other missing parts, the rounding discrepancy is reduced.

To *solve* a triangle means to find the measures of all sides and all angles. A right triangle can be solved if we know one side and any other part besides the right angle. In the next example, we find values of all sides and angles of the triangle using the sine function.

EXAMPLE Solve triangle *DEF* (see Fig. 23–4). One side and one other part besides the right angle are known.

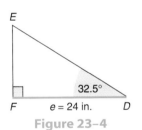

Figure 23–4

Because we have one acute angle and a side that is not opposite the known angle, we must find the other angle of the triangle.

$$\angle E = 90° - 32.5° = \mathbf{57.5°} \quad \text{Find the complement of } \angle D.$$

To find the hypotenuse f:

$$\sin E = \frac{\text{opp}}{\text{hyp}}$$

Use the sine function and substitute $E = 57.5°$ and opp = 24.

$$\sin 57.5° = \frac{24}{f} \qquad \text{Solve for } f.$$

$$f(\sin 57.5°) = 24$$

$$f = \frac{24}{\sin 57.5°} \qquad \text{Evaluate.}$$

$$f = 28.45653714$$

$$\mathbf{f = 28.46 \text{ in.}} \qquad \text{Rounded to four significant digits.}$$

To find side d:

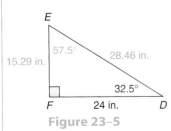

Figure 23–5

$$\sin D = \frac{\text{opp}}{\text{hyp}}$$

Substitute $D = 32.5°$ and hyp = 28.45653714. Use full calculator value for f to get the most accurate result.

$$\sin 32.5° = \frac{d}{28.45653714} \qquad \text{Solve for } d.$$

$$28.45653714(\sin 32.5°) = d \qquad \text{Evaluate.}$$

$$15.28968626 = d$$

$$\mathbf{15.29 \text{ in.} = d} \qquad \text{Rounded to four significant digits.}$$

The solved triangle is shown in Fig. 23–5.

Check Computations Using the Pythagorean Theorem.

In this and other problems involving right triangles, you can check your computations using the Pythagorean theorem, $(\text{hyp})^2 = (\text{leg})^2 + (\text{leg})^2$. Let's check the solution to the preceding example.

With Values to the Nearest Hundredth

$$(28.46)^2 = (15.29)^2 + (24)^2$$
$$809.9716 = 233.7841 + 576$$
$$809.9716 = 809.7841$$

With Full Calculator Values

$$(28.45653714)^2 = (15.28968626)^2 + (24)^2$$
$$809.774506 = 233.7745059 + 576$$
$$809.774506 = 809.7745059$$

Rounding discrepancies are minimized when more significant digits are used.

Learning Strategy   *See the BIG Picture, Then Focus on the Little Parts.*

In solving a right triangle, we are generally given the values of three parts and are asked to find the values of the three missing parts. Look at the big picture first.

• Identify the three given parts, one of which is the right angle.
• Identify the missing parts.
• Plan a strategy to find each missing part.
• Focus on one part at a time.
• Check by using the Pythagorean theorem and the property that the three angles of a triangle add to 180°.

2 **Find the Missing Parts of a Right Triangle Using the Cosine Function.**

In some of the previous problems, when we found a side not opposite the given angle, we had to find the other angle first by subtracting the given acute angle from 90°. If we use the cosine function, however, we can find the same side by using the given angle, rather than its complement. The abbreviated cosine ratio is $\cos \theta = \dfrac{\text{adj}}{\text{hyp}}$.

Use the cosine function to find unknown parts of a right triangle:

1. Two of these three parts of a right triangle must be known:
   a. One acute angle
   b. The side adjacent to the known acute angle
   c. Hypotenuse
2. Substitute two known values in the ratio $\cos \theta = \dfrac{\text{adj}}{\text{hyp}}$.
3. Solve for the missing part.

EXAMPLE   Find angle *A* of Fig. 23–6.

We are asked to find an angle and we are given the hypotenuse and the side adjacent to the angle, so we use the cosine function.

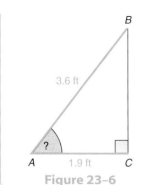

Figure 23–6

$$\cos A = \frac{\text{adj}}{\text{hyp}} \qquad \text{Substitute known values.}$$

$$\cos A = \frac{1.9}{3.6} \qquad \text{Use the inverse cosine function. Evaluate.}$$

$$\cos^{-1}\left(\frac{1.9}{3.6}\right) = A$$

$$A = 58.14456918$$

$$A = \mathbf{58.1°} \qquad \text{Rounded to nearest } 0.1°.$$

EXAMPLE Find side b of Fig. 23–7.

We can use either the sine or the cosine function because we are given the hypotenuse and an angle. However, we do not have to find the complement of the given angle if we use the cosine function.

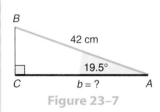

Figure 23–7

$$\cos A = \frac{\text{adj}}{\text{hyp}} \qquad \text{Substitute known values.}$$

$$\cos 19.5° = \frac{b}{42} \qquad \text{Solve for } b.$$

$$42(\cos 19.5°) = b \qquad \text{Evaluate.}$$

$$39.59094263 = b$$

$$\mathbf{39.59 \text{ cm}} = b \qquad \text{Four significant digits.}$$

EXAMPLE Find side t of Fig. 23–8.

The cosine function is the most efficient function to use because we are finding the hypotenuse and given an acute angle and its adjacent side.

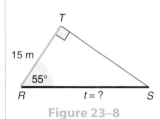

Figure 23–8

$$\cos R = \frac{\text{adj}}{\text{hyp}} \qquad \text{Substitute known values.}$$

$$\cos 55° = \frac{15}{t} \qquad \text{Solve for } t.$$

$$t(\cos 55°) = 15$$

$$t = \frac{15}{\cos 55°} \qquad \text{Evaluate.}$$

$$t = 26.15170193$$

$$t = \mathbf{26.15 \text{ m}} \qquad \text{Four significant digits.}$$

3 Find the Missing Parts of a Right Triangle Using the Tangent Function.

If we have a right triangle in which we know only the length of the two legs, we cannot use the sine or cosine function to solve the triangle unless we use the Pythagorean theorem to find the length of the hypotenuse. However, we can use the tangent function directly: $\tan \theta = \frac{\text{opp}}{\text{adj}}$.

1. Two of the following three parts of a right triangle must be known:
   a. An acute angle
   b. The side opposite the acute angle
   c. The side adjacent to the acute angle
2. Substitute two known values in the formula $\tan \theta = \dfrac{\text{opp}}{\text{adj}}$.
3. Solve for the missing part.

Whenever we wish to check our calculations of the sides, we can use the Pythagorean theorem as we did earlier in the section.

EXAMPLE Find angle A of Fig. 23–9.

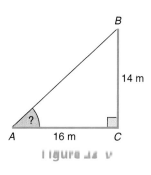

We are looking for an angle and given the side opposite and the side adjacent to the angle, so we use the tangent function.

$$\tan A = \frac{\text{opp}}{\text{adj}} \qquad \text{Substitute known values.}$$

$$\tan A = \frac{14}{16} \qquad \text{Use the inverse tangent function. Evaluate.}$$

$$\tan^{-1}\left(\frac{14}{16}\right) = A$$

$$A = 41.18592516^{\circ}$$

$$A = \mathbf{41.2^{\circ}} \qquad \text{Rounded to nearest } 0.1^{\circ}.$$

EXAMPLE Use the tangent function to find a of Fig. 23–10.

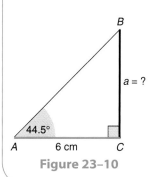

Figure 23–10

We are given an acute angle and the side adjacent to the acute angle. We are looking for the opposite side so use the tangent function.

$$\tan A = \frac{\text{opp}}{\text{adj}} \qquad \text{Substitute known values.}$$

$$\tan 44.5^{\circ} = \frac{a}{6} \qquad \text{Solve for } a.$$

$$6(\tan 44.5^{\circ}) = a \qquad \text{Evaluate.}$$

$$5.896183579 = a$$

$$\mathbf{5.896\ cm} = \boldsymbol{a} \qquad \text{Four significant digits.}$$

EXAMPLE Find side b of Fig 23–11.

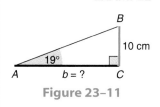

Figure 23–11

We know an acute angle and its opposite side. We are looking for the side adjacent to the acute angle.

$$\tan A = \frac{\text{opp}}{\text{adj}} \qquad \text{Substitute known values.}$$

$$\tan 19^{\circ} = \frac{10}{b} \qquad \text{Solve for } b.$$

$$b(\tan 19°) = 10$$

$$b = \frac{10}{\tan 19°} \qquad \text{Evaluate.}$$

$$b = 29.04210878$$

$$\mathbf{b = 29.04 \text{ cm}} \qquad \text{Four significant digits.}$$

SELF-STUDY EXERCISES 23–1

1 Use the sine function to find the indicated parts of the triangle *LMN* in Fig. 23–12. Round lengths of sides to four significant digits and angles to the nearest 0.1°.

1. Find *M* if $n = 15$ m and $m = 7$ m.
2. Find *l* if $n = 13$ in. and $L = 32°$.
3. Find *m* if $l = 15$ m and $L = 28°$.
4. Find *n* if $l = 12$ ft and $M = 42°$.
5. Find *M* if $m = 13$ cm and $n = 19$ cm.
6. Find *M* if $n = 3.7$ in. and $l = 2.4$ in.

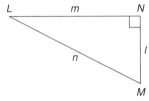

Figure 23–12

Solve triangle *STU* in Fig. 23–13 for the given values using the sine function. Check the measures of the sides by using the Pythagorean theorem. Round as above.

7. Solve if $t = 18$ yd and $s = 14$ yd.
8. Solve if $U = 45°$ and $u = 4.7$ m.
9. Solve if $S = 34.5°$ and $t = 8.5$ mm.
10. Solve if $S = 16°$ and $s = 14$ m.

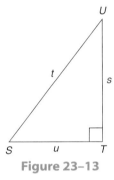

Figure 23–13

2 Use the cosine function to find the indicated parts of triangle *KLM* in Fig. 23–14. Round sides to four significant digits and angles to the nearest 0.1°.

11. Find *M* if $k = 13$ m and $l = 16$ m.
12. Find *k* if $l = 11$ cm and $M = 24°$.
13. Find *l* if $M = 31°$ and $k = 27$ ft.
14. Find *l* if $m = 15$ dm and $M = 25°$.
15. Find *k* if $K = 72°$ and $l = 16.7$ mm.
16. Find *l* if $K = 67°$ and $k = 13$ yd.

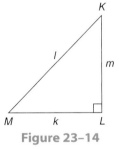

Figure 23–14

Use the sine *or* cosine function to solve triangle *QRS* in Fig. 23–15. Round as above.

17. Solve if $s = 23$ ft and $q = 16$ ft.
18. Solve if $s = 17$ cm and $R = 46°$.
19. Solve if $q - 14$ dkm and $Q = 73.5°$.
20. Solve if $R = 59.5°$ and $q = 8$ m.

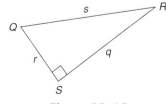

Figure 23–15

3 Use the tangent function to find the indicated parts of triangle *ABC* in Fig. 23–16. Round sides to four significant digits and angles to the nearest 0.1°.

21. Find *A* if $b = 11$ cm and $a = 6$ cm.
22. Find *b* if $a = 1.9$ m and $A = 25°$.
23. Find *a* if $A = 40.5°$ and $b = 7$ ft.
24. Find *A* if $b = 10.8$ m and $a = 4.7$ m.
25. Find *a* if $A = 43°$ and $b = 0.05$ cm.
26. Find *a* if $B = 68°$ and $b = 0.03$ m.

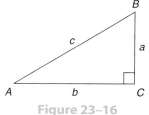

Figure 23–16

Solve triangle *DEF* in Fig. 23–17. Round sides to four significant digits and angles to the nearest 0.1°.

27. Solve if $e = 4.6$ m and $d = 3.2$ m.
28. Solve if $D = 42°$ and $e = 7$ ft.
29. Solve if $E = 73.5°$ and $e = 20.13$ in.
30. Solve if $d = 11$ ft and $e = 8$ ft.

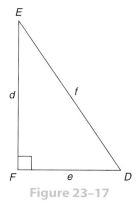

Figure 23–17

23–2 | Applied Problems Using Right-Triangle Trigonometry

Learning Outcomes

1 Select the most direct method for solving right triangles.
2 Solve applied problems using right-triangle trigonometry.

1 Select the Most Direct Method for Solving Right Triangles.

Our first task in solving any problem involving right triangles is to *select the most convenient and efficient function.*

To select the most direct method for solving a right triangle:

1. Where possible, choose the function that uses given parts rather than unknown parts that must be calculated.
2. Where possible, choose the function that gives the desired part directly, that is, without having to find other parts first.

EXAMPLE   In △*ABC* of Fig. 23–18, find *c*.

The most direct way of finding side *c* is to use the *Pythagorean theorem*. The two legs of a right triangle are given.

$c^2 = a^2 + b^2$ Substitute known values.

$c^2 = 6^2 + 9^2$ Solve for *c*.

$c^2 = 36 + 81$

$c^2 = 117$ Take square root of both sides.

Figure 23–18

$$c = 10.81665383$$
$$c = \mathbf{10.82}$$ Four significant digits.

Alternative method: Trigonometric functions can be used to find c.

First find one of the acute angles.

$$\tan A = \frac{\text{opp}}{\text{adj}}$$ Opposite and adjacent sides are given for this function.

$$\tan A = \frac{6}{9}$$ Use inverse tangent function.

$$A = \tan^{-1}\frac{6}{9}$$ Evaluate.

$$A = 33.69006753$$

$$A = 33.7°$$ Rounded to nearest 0.1°

Next find the hypotenuse using the sine function.

$$\sin A = \frac{\text{opp}}{\text{hyp}}$$ Substitute 33.69006753 for A and 6 for the opposite side.

$$\sin 33.69006753 = \frac{6}{c}$$ Solve for the hypotenuse.

$$c = \frac{6}{\sin 33.69006753}$$ Evaluate.

$$c = 10.81665383$$

$$c = \mathbf{10.82}$$ Four significant digits.

As you see in the alternative method, more steps are required.

EXAMPLE In $\triangle ABC$ of Fig. 23–19, find angle B.

We are given the side adjacent to and the side opposite angle B, so we use the tangent function for a direct solution.

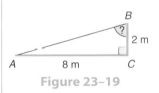

Figure 23–19

$$\tan \theta = \frac{\text{opp}}{\text{adj}}$$ Substitute known values.

$$\tan B = \frac{8}{2}$$ Reduce.

$$\tan B = 4$$ Use the inverse tangent function.

$$\tan^{-1}4 = B$$ Evaluate.

$$B = 75.96375653$$

$$B = \mathbf{76.0°}$$ Round to nearest 0.1°.

2 Solve Applied Problems Using Right-Triangle Trigonometry.

In solving technical problems, it's a good idea to draw diagrams or pictures to visualize the various relationships.

EXAMPLE A jet takes off at a 30° angle (see Fig. 23–20). If the runway (from takeoff) is 875 ft long, find the altitude of the airplane as it flies over the end of the runway.

Known facts One acute angle = 30°; adjacent side = 875 ft

Unknown fact Plane's altitude at end of runway or opposite side

Relationship $\tan \theta = \dfrac{\text{opp}}{\text{adj}}$

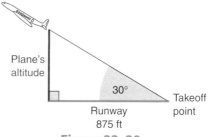

Figure 23–20

Estimation   In a 45° 45° 90° right triangle, the legs are equal. Because 30° is less than 45°, the side opposite the 30° angle should be less than 875 ft.

Calculations

$$\tan \theta = \frac{\text{opp}}{\text{adj}}$$ Substitute known values.

$$\tan 30° = \frac{a}{875}$$ Solve for *a*.

$$875(\tan 30°) = a$$ Evaluate.

$$505.1814855 = a$$

Interpretation   **505.2 ft = plane's altitude**   Four significant digits.

Many right-triangle applications use the terminology *angle of elevation* and *angle of depression*. See Fig. 23–21. The angle of elevation is generally used when we are looking *up* at an object. We use the angle of depression to describe the location of an objective below our eye level. Both angles are formed by a line of sight and a horizontal line from the point of sight.

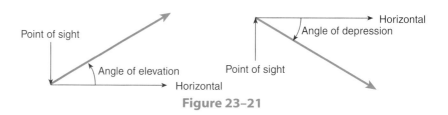

Figure 23–21

EXAMPLE   A stretch of roadway drops 30 ft for every 300 ft of road (see Fig. 23–22). Find the *angle of declination* of the road.

The *angle of declination* is the angle of depression. The opposite side and the hypotenuse are given.

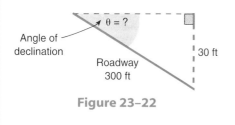

Figure 23–22

$$\sin \theta = \frac{\text{opp}}{\text{hyp}}$$ Substitute known values.

$$\sin \theta = \frac{30}{300}$$ Reduce.

$$\sin \theta = 0.1$$ Use the inverse sine function.

$$\sin^{-1} 0.1 = \theta$$

$$\theta = 5.739170477°$$

The angle of declination of the road is 5.7° rounded to the nearest 0.1°.

EXAMPLE A surveyor locates two points on a steel column so that it can be set plumb (perpendicular to the horizon). If the angle of elevation is 15° and the surveyor's transit is 175 ft from the column (see Fig. 23–23), find the distance from the transit to the upper point (point B) on the column. (A transit is a surveying instrument used for measuring angles.)

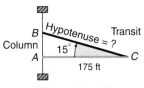

Figure 23–23

An acute angle and the adjacent side are given. To find the distance from the transit to point *B* on the column (hypotenuse), we use the *cosine function.*

$$\cos \theta = \frac{adj}{hyp}$$ — Substitute known values.

$$\cos 15° = \frac{175}{hyp}$$ — Solve for hypotenuse.

$$hyp(\cos 15°) = 175$$

$$hyp = \frac{175}{\cos 15°}$$ — Evaluate.

$$hyp = 181.1733316$$

$$hyp = 181.2 \text{ ft}$$ — Four significant digits.

Point *B* is 181.2 ft from the transit.

EXAMPLE Find the angle a rafter makes with a joist of a house if the rise is 12 ft and the span is 30 ft (see Fig. 23–24). Also, find the length of the rafter.

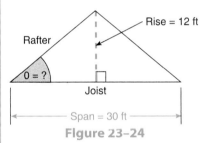

Figure 23–24

The span is twice the distance from the outside end to the center point of the joist. Therefore, to solve the right triangle for the desired angle, we draw the triangle shown in Fig. 23–25. Because the legs of a right triangle are given, the tangent function is used.

Find θ.

$$\tan \theta - \frac{opp}{adj}$$ — Substitute known values.

$$\tan \theta = \frac{12}{15}$$ — Use the inverse tangent function.

$$\tan^{-1} \frac{12}{15} = \theta$$

$$\theta = 38.65980825$$ — Evaluate.

$$\theta = 38.7°$$ — Round to nearest 0.1°

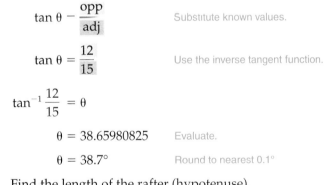

Figure 23–25

Find the length of the rafter (hypotenuse).

We use the *Pythagorean theorem* to find the length of the rafter directly.

$$(hyp)^2 = (leg)^2 + (leg)^2$$ — Substitute known values.

$$(hyp)^2 = 12^2 + 15^2$$ — Evaluate.

$$(hyp)^2 = 144 + 225$$

$$(hyp)^2 = 369$$ — Take square root of both sides.

$$hyp = 19.20937271$$

$$hyp = 19.21 \text{ ft}$$ — Four significant digits.

The angle the rafter makes with the joist is 38.7° and the rafter is 19.21 ft.

EXAMPLE   Find the angle formed by the rod in the mechanical assembly shown in Fig. 23–26.

Given the hypotenuse and the side opposite the desired angle, we use the sine function.

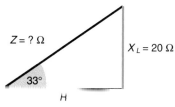

$\theta = ?$

59 cm

10.6 cm

Figure 23–26

$$\sin \theta = \frac{\text{opp}}{\text{hyp}}$$ Substitute known values.

$$\sin \theta = \frac{10.6}{59}$$ Use the inverse sine function. Evaluate.

$$\sin^{-1} \frac{10.6}{59} = \theta$$

$$\theta = 10.35001563$$

$$\theta = 10.4°$$ Rounded.

The angle formed by the rod is 10.4°.

EXAMPLE   Find the impedance Z of a circuit with 20 Ω of reactance X_L represented by the vector diagram in Fig. 23–27.

Because we are looking for the hypotenuse Z and are given an acute angle and the opposite side, we use the sine function.

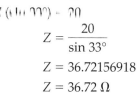

$Z = ?\ \Omega$

$X_L = 20\ \Omega$

$33°$

H

Figure 23–27

$$\sin \theta = \frac{\text{opp}}{\text{hyp}}$$ Substitute known values.

$$\sin 33° = \frac{20}{Z}$$ Solve for Z.

$$Z(\sin 33°) = 20$$

$$Z = \frac{20}{\sin 33°}$$ Evaluate.

$$Z = 36.72156918$$

$$Z = 36.72\ \Omega$$

The impedance, Z, is 36.72 Ω.

SELF-STUDY EXERCISES 23–2

1 Find the indicated part of the right triangles in Figs. 23–28 through 23–32 by the most direct method.

1.
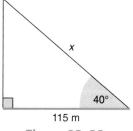

x

$40°$

115 m

Figure 23–28

2.
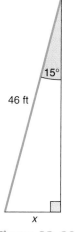

$15°$

46 ft

x

Figure 23–29

3.
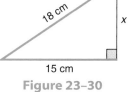

18 cm

x

15 cm

Figure 23–30

4.

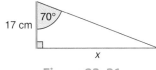

17 cm 70°

x

Figure 23–31

5.

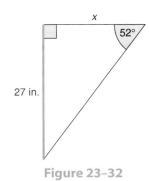

x

52°

27 in.

Figure 23–32

2 Use the trigonometric functions to solve the problems. Round side lengths to four significant digits and angles to 0.1°.

6. A sign is attached to a building by a triangular brace. If the horizontal length of the brace is 48 in. and the angle at the sign is 25° (see Fig. 23–33), what is the length of the wall support piece?

7. A surveyor uses right triangles to measure inaccessible property lines. To measure a property line that crosses a pond, a surveyor sights to a point across the pond, then makes a right angle, measures 50 ft, and sights the point across the pond with a 47° angle (see Fig. 23–34). Find the distance across the pond from the initial point.

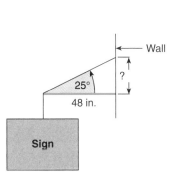

Wall

?

25°

48 in.

Sign

Figure 23–33

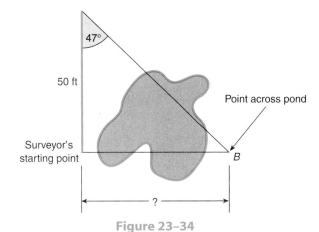

47°

50 ft

Point across pond

Surveyor's
starting point

B

?

Figure 23–34

8. At what angle must a jet descend if it is 900 ft above the end of the runway and must touch down 1,500 ft from the runway's end?

9. A 50-ft wire is used to brace a utility pole. If the wire is attached 4 ft from the top of the 35-ft pole, how far from the base of the pole will the wire be attached to the ground?

10. A roadway rises 4 ft for every 15 ft along the road. What is the angle of inclination of the roadway?

11. A shadow cast by a tree is 32 ft long when the angle of inclination of the sun is 36°. How tall is the tree?

12. The vector diagram of the circuit in Fig. 23–35 has a known impedance Z. Find the reactance X_L. All units are in ohms.

13. Using Fig. 23–35, find resistance R.

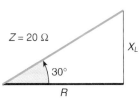

$Z = 20\ \Omega$

X_L

30°

R

Figure 23–35

14. A piston assembly at the midpoint of its stroke forms a right triangle (see Fig. 23–36). Find the length of R.

15. Find the angle a rafter makes with a joist of a house if the rise is 18 ft and the span is 50 ft. Refer to Fig. 23–24 on page 877.

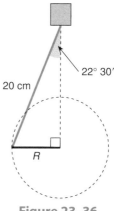

Figure 23–36

Section 23–1

Find the indicated parts of the triangles in Figs. 23–37 through 23–46. Round sides to four significant digits and round angles to the nearest 0.1°.

1. Find A.

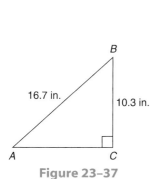

Figure 23–37

2. Find r.

Figure 23–38

3. Find h.

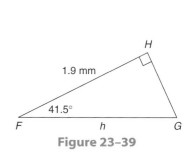

Figure 23–39

4. Find c.

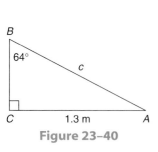

Figure 23–40

5. Find y.

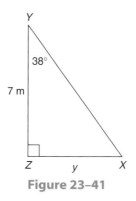

Figure 23–41

6. Find B.

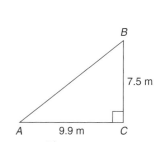

Figure 23–42

Solve the triangles in Figs. 23–43 through 23–46. Round sides to four significant digits and round angles to the nearest 0.1°.

7.

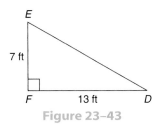

Figure 23–43

8.

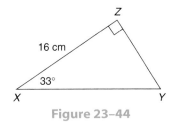

Figure 23–44

9.

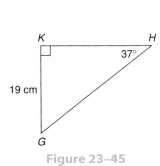

Figure 23–45

10.

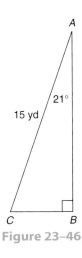

Figure 23–46

Section 23–2

Find the indicated parts of the triangles in Figs. 23–47 through 23–49. Round sides to four significant digits and angle measures to the nearest 0.1°.

11. Find A.

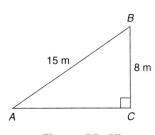

Figure 23–47

12. Find A.

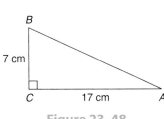

Figure 23–48

13. Find a.

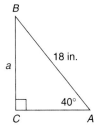

Figure 23–49

Solve the triangles in Figs. 23–50 and 23–51.

14.

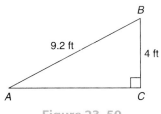

Figure 23–50

15.

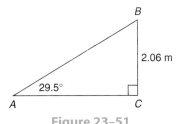

Figure 23–51

Assignment Exercises

16. A railway inclines 14°. How many feet of track must be laid if the hill is 15 ft high?

17. A corner shelf is cut so that the sides placed on the wall are 37 in. and 42 in. What are the measures of the acute angles?

18. From a point 5 ft above the ground and 20 ft from the base of a building, a surveyor uses a transit to sight an angle of 38° to the top of the building. Find the height of the building.

19. A surveyor makes the measures indicated in Fig. 23–52. Solve the triangle. All parts of the triangle should be included in a report.

20. What length of rafter is needed for a roof if the rafters form an angle of 35.5° with a joist and the rise is 8 ft?

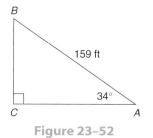

Figure 23–52

21. Find the angle formed by the rod and the horizontal in the mechanical assembly shown in Fig. 23–53.

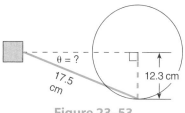

Figure 23–53

22. A utility pole is 40 ft above ground level. A wire must be attached to the pole 3 ft from the top to give it support. If the wire forms a 20° angle with the ground, how long is the wire? Disregard the length needed for attaching the wire to the pole or ground. Round to hundredths.

23. Solve for reactance X_L and resistance R in Fig. 23–54. All units are in ohms. Round to hundredths.

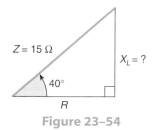

Figure 23–54

CHALLENGE PROBLEMS

24. You are designing a stairway for a home and need to determine how many steps the stairway will have. If the stairway is 8 ft high and the floor space allotted is 14 ft, determine how many steps are needed and determine the optimum measure (rise and run) for each step.

25. Draw triangle ABC so that angle C is 90° and angle A is 70°. Point E is on side $\overline{AB}$ and is 10 cm from A. Point D is on $\overline{BC}$, and $\overline{DE}$ is parallel to $\overline{CA}$. If $\overline{CA}$ is 24 cm, find the length of $\overline{BD}$.

CONCEPTS ANALYSIS

1. Use a calculator to find sin 45° and cos 45°. Using your knowledge of geometry, explain the relationship between sin 45° and cos 45°. Does it follow that the sine and cosine of the same angle are always equal? Are there other angles for which the two functions are equal? If so, discuss them.

2. If you are given an angle and an adjacent side of a right triangle, explain how you would find the hypotenuse of the triangle.

3. You know the measure of both legs of a right triangle and the measure of one of the acute angles. Describe two different methods you could use to find the length of the hypotenuse.
5. Explain the differences and similarities between angle of inclination and angle of declination.

7. Devise a right triangle that has an angle (other than the right angle) and a side given. Then use trigonometric functions or other methods to find the measures of all the other sides and angles.
9. Use your calculator to discover which of the three trigonometric functions—sine, cosine, tangent—can have values greater than 1. For what angle measures is the value of this function greater than 1?

4. What is the least number of parts of a triangle that must be known to find the measures of all the other parts? What are the parts required?

6. Using a calculator, compare the sine of an acute angle with the sine of its complement for several angles. Compare the sine of an acute angle with the cosine of its complement for several angles. Generalize your findings.
8. Describe the calculator steps required to find the measure of an angle whose sine function is given.

10. Use your graphing calculator to graph the sine and cosine trigonometric functions. Describe how the graphs are similar and how they are different.

CHAPTER SUMMARY

Learning Outcomes   **What to Remember with Examples**

Section 23–1

1 Find the missing parts of a right triangle using the sine function (pp. 867–870).

$$\sin \theta = \frac{\text{opposite side}}{\text{hypotenuse}}$$

Use the sine function to find side x (Fig. 23–55).

$$\sin \theta = \frac{\text{opp}}{\text{hyp}}$$ Substitute known values.

$$\sin 25° = \frac{x}{12}$$ Solve for x.

$$x = 12 \sin 25°$$ Evaluate.

$$x = 5.071419141$$

$$x = 5.071 \text{ in.}$$ Four significant digits.

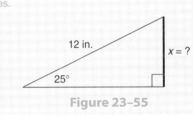

Figure 23–55

12 in. $x = ?$ 25°

2 Find the missing parts of a right triangle using the cosine function (pp. 870–871).

$$\cos \theta = \frac{\text{adjacent side}}{\text{hypotenuse}}$$

Use the cosine function to find side x (Fig. 23–56).

$$\cos \theta = \frac{\text{adj}}{\text{hyp}}$$ Substitute known values.

$$\cos 37° = \frac{26}{x}$$ Solve for x.

$$x = \frac{26}{\cos 37°}$$ Evaluate.

$$x = 32.55552711$$

$$x = 32.56 \text{ m}$$

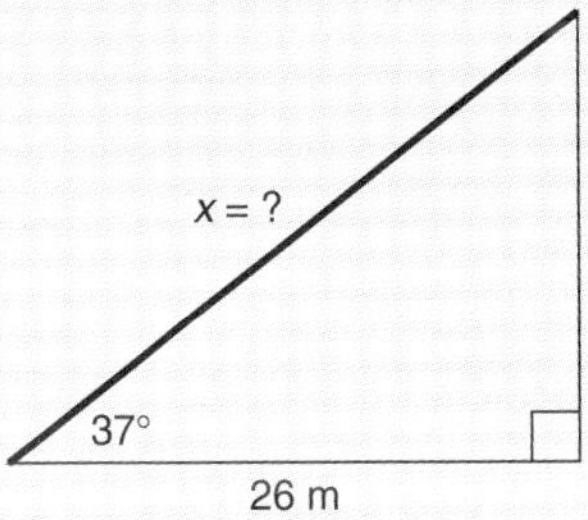

$x = ?$ 37°   26 m

Figure 23–56

3 Find the missing parts of a right triangle using the tangent function (pp. 871–873).

$$\tan \theta = \frac{\text{opposite side}}{\text{adjacent side}}$$

Use the tangent function to find θ in degrees (Fig. 23–57).

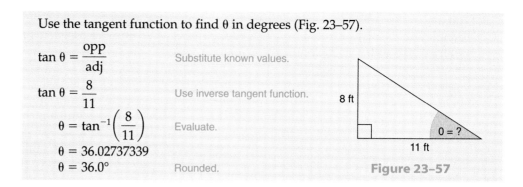

$$\tan \theta = \frac{\text{opp}}{\text{adj}}$$ Substitute known values.

$$\tan \theta = \frac{8}{11}$$ Use inverse tangent function.

$$\theta = \tan^{-1}\left(\frac{8}{11}\right)$$ Evaluate.

$$\theta = 36.02737339$$

$$\theta = 36.0°$$ Rounded.

Figure 23–57

Section 23–2

1 Select the most direct method for solving right triangles (pp. 874–875).

Whenever possible, choose the function that uses parts given in the problem rather than parts that are not given and must be calculated.

Whenever possible, choose the function that gives the desired part directly; that is, find the desired part without having to find other parts first.

Find the parts indicated in Fig. 23–58.

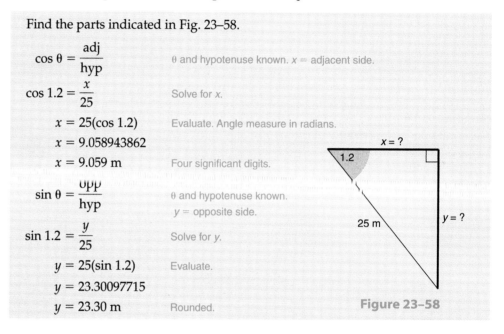

$$\cos \theta = \frac{\text{adj}}{\text{hyp}}$$ θ and hypotenuse known. x = adjacent side.

$$\cos 1.2 = \frac{x}{25}$$ Solve for x.

$$x = 25(\cos 1.2)$$ Evaluate. Angle measure in radians.

$$x = 9.058943862$$

$$x = 9.059 \text{ m}$$ Four significant digits.

$$\sin \theta = \frac{\text{opp}}{\text{hyp}}$$ θ and hypotenuse known. y = opposite side.

$$\sin 1.2 = \frac{y}{25}$$ Solve for y.

$$y = 25(\sin 1.2)$$ Evaluate.

$$y = 23.30097715$$

$$y = 23.30 \text{ m}$$ Rounded.

Figure 23–58

2 Solve applied problems using right-triangle trigonometry (pp. 875–878).

To solve a problem using the trigonometric functions, first determine the given parts and the missing part or parts; then identify the trigonometric function that relates the given and missing parts. Use this function and the given information to find the missing information.

A jet takes off at a 25° angle. Find the distance traveled by the plane from the takeoff point to the end of the runway if the runway is 950 ft long (Fig 23–59).

$$\cos \theta = \frac{\text{adj}}{\text{hyp}}$$ θ and adjacent side known. Find the hypotenuse.

$$\cos 25° = \frac{950}{x}$$ Solve for x.

$$x = \frac{950}{\cos 25°}$$ Evaluate.

$$x = 1,048.209023$$

$$x = 1,048 \text{ ft}$$ Four significant digits.

Figure 23–59

The measures are indicated for a standard right triangle, *ABC*, where *C* is the right angle. Draw the figures and use the sine, cosine, or tangent function to find the indicated parts of the triangle. Round side lengths to four significant digits and angles to the nearest 0.1°.

1. $a = 16$ m, $b = 14$ m, find A.
2. $a = 7$ in., $A = 33°$, find c.
3. $c = 17$ ft, $B = 25°$, find a.
4. $a = 21$ m, $A = 48.5°$, find b.
5. $a = 32$ cm, $c = 47$ cm, find A.
6. $c = 12$ m, $A = 35°$, find a.
7. $b = 21$ cm, $A = 17°$, find a.
8. $b = 1$ cm, $A = 87°$, find c.
9. $b = 3.1$ m, $c = 6.8$ m, find A.
10. $a = 0.15$ m, $c = 0.46$ m, find A.
11. Solve triangle *ABC* in Fig. 23–60.
12. Solve triangle *DEF* in Fig. 23–61. Round as above.

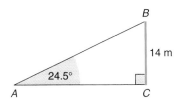

Figure 23–60

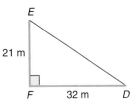

Figure 23–61

13. A stair has a rise of 4 in. for every 5 in. of run. What is the angle of inclination of the stair?
14. A surveyor uses indirect measurement to find the width of a river at a certain point. The surveyor marks off 50 ft along the river bank at right angles with the river and then sights an angle of 43° to point *A* across the river (Fig. 23–62). Find the width of the river.

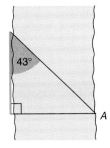

Figure 23–62

15. Steel girders are reinforced by placing steel supports between two runners so that right triangles are formed (see Fig. 23–63). If the runners are 24 in. apart and a 30° angle is desired between the support and a runner, find the length of the support to be placed at a 30° angle.

16. In the circuit represented by the diagram of Fig. 23–64, find the total current I_t. All units are in amps.

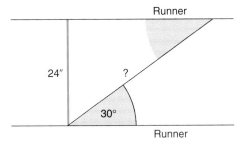

Figure 23–63

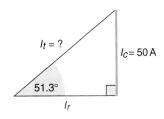

Figure 23–64

17. Find the current in the resistance branch I_r of the circuit in Exercise 16.
18. The minimum clearances for the installation of a metal chimney pipe are shown in Fig. 23–65. How far from the ridge should the hole be cut for the pipe to pass through the roof?

19. Refer to Fig. 23–65. What angle is formed by the chimney pipe and the roof where the pipe passes through the roof?

20. Elbows are used to form bends in rigid pipe and are measured in degrees. What is the angle of bend of the elbow in the installation shown in Fig. 23–66 to the nearest whole degree?

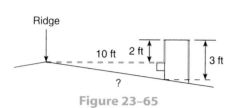

Figure 23–65

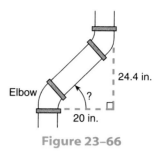

Figure 23–66

Oblique Triangles

24

Although right triangles are perhaps the most frequently used triangles, we often work with other kinds of triangles. In this chapter, we apply the laws of sines and cosines to oblique triangles, find the area of oblique triangles when only selected parts are given, and use *vectors* for solving triangles that have an angle greater than 90°.

24–1 | Vectors

Learning Outcomes

1 Find the magnitude of a vector in standard position, given the coordinates of the end point.

2 Find the direction of a vector in standard position, given the coordinates of the end point.

3 Find the sum of vectors.

1 Find the Magnitude of a Vector in Standard Position, Given the Coordinates of the End Point.

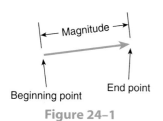

Figure 24–1

Quantities that we have discussed so far in this text have been described by specifying their size or magnitude. Quantities such as area, volume, length, and temperature, which are characterized by magnitude only, are called *scalars*. There are many other quantities such as electrical current, force, velocity, and acceleration that are called *vectors*. A quantity described by magnitude (or length) and direction is a *vector*. A vector represents a shift from one point to another.

When we consider the speed of a plane to be 500 mph, we are considering a scalar quantity. However, when we consider the speed of a plane *traveling northeast from a given location*, we are concerned with both the distance traveled and the direction. This is a vector quantity. We have seen such quantities in the vector diagrams used to solve electronic problems by means of right triangles in Chapter 23 and elsewhere.

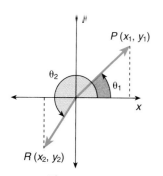

Figure 24–2

Vectors are represented by straight arrows. The length of the arrow represents the *magnitude* of the vector (see Fig. 24–1). The curved arrow shows the counterclockwise *direction* of the vector (see Fig. 24–2), with the end of the arrow being the beginning point and the arrowhead being the end point.

In relating trigonometric functions to vector quantities, we will place the vectors in *standard position*. Standard position places the beginning point of the vector at the origin of a rectangular coordinate system. The direction of the vector is the counterclockwise angle measured from the positive x-axis (horizontal axis).

In Fig. 24–2, vector P is a first-quadrant vector that has a direction of θ_1 and an end point at the point P. The magnitude of a vector can be determined if the vector is in standard position and the x- and y-coordinates of the end point are known. Vector R is a third-quadrant vector that has a direction of θ_2 and an end point at point R. Its magnitude may be determined similarly.

Magnitude of a vector in standard position:

The *magnitude* of a vector in standard position is the length of the hypotenuse of the right triangle formed by the vector, the x-axis, and the vertical line from the end point of the vector to the x-axis.

To find the magnitude of a vector in standard position:

1. Substitute into the Pythagorean theorem the coordinates of the end points of the vector (x_1, y_1) as the legs of a right triangle.
2. Solve for the hypotenuse.

EXAMPLE Find the magnitude of vector p in Fig. 24–3 if the coordinates of P are $(4, 3)$.

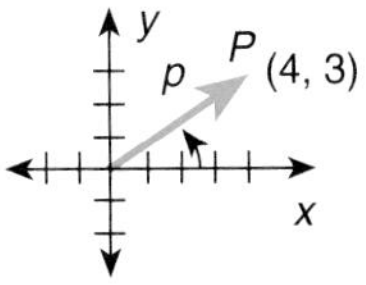

$p^2 = x^2 + y^2$ Substitute $x = 4$ and $y = 3$.

$p^2 = 4^2 + 3^2$ Solve for p.

$p^2 = 16 + 9$

$p^2 = 25$

$p = \pm\sqrt{25}$

$p = \pm 5$ We use only $+5$ because length or magnitude is positive.

The magnitude of vector p is 5.

Figure 24–3

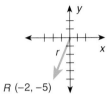

EXAMPLE Find the magnitude of vector r in Fig. 24–4 if the coordinates of point R are $(-2, -5)$.

The magnitude of a vector is always positive; however, the x- and y-coordinates can be negative.

$r^2 = x^2 + y^2$ Substitute $x = -2$ and $y = -5$.

$r^2 = (-2)^2 + (-5)^2$ Solve for r.

$r^2 = 4 + 25$

$r^2 = 29$

$r = \sqrt{29}$

$r = 5.385$ Rounded.

The magnitude of vector r is 5.385.

Figure 24–4

2 **Find the Direction of a Vector in Standard Position, Given the Coordinates of the End Point.**

If the vector in standard position falls in quadrant I, we can use right-triangle trigonometry to find the direction of the vector. For a vector in standard position with an end point in quadrant I, the tangent function can be used to find the angle or direction of the vector.

To find the direction of a quadrant I vector in standard position:

1. Identify the coordinates of the end point of the vector.
2. Substitute the coordinates of the end point into the tangent function: $\tan\theta = \dfrac{\text{opp}}{\text{adj}}$ or $\tan\theta = \dfrac{y}{x}$.
3. Solve for θ.

EXAMPLE Find the direction of vector p in Fig. 24–5 if the coordinates of P are $(4, 3)$.

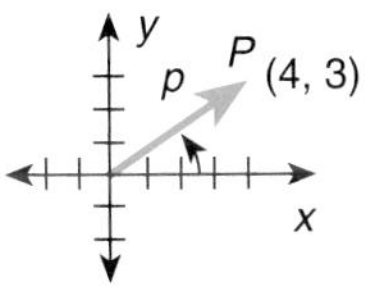

$\tan\theta = \dfrac{y}{x}$ Substitute values for end point of vector P. $x = 4$ and $y = 3$.

$\tan\theta = \dfrac{3}{4}$ Use the inverse tangent function and the decimal equivalent of $\dfrac{3}{4}$.

$\tan^{-1} 0.75 = \theta$ Evaluate.

$\theta = 36.9°$ Nearest 0.1°.

The direction of vector p is 36.9°.

Figure 24–5

3 Find the Sum of Vectors.

Two vectors with the same direction can be added by aligning the beginning point of one vector with the end point of the other. The vector represented by the sum is called the *resultant vector,* or *resultant*. The resultant has the same direction as the vectors being added and a magnitude that is the sum of the two magnitudes.

To add vectors of the same direction:

1. Add the magnitudes of each vector.
2. The resultant vector will have the same direction as the original vectors.

EXAMPLE Add two vectors with a direction of 35° if the magnitudes of the vectors are 4 and 5, respectively (see Fig. 24–6).

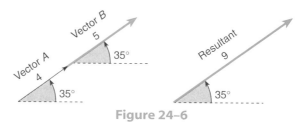

Figure 24–6

The resultant vector has a magnitude of 9 and a direction of 35°.

Two vectors have opposite directions if the directions of the vectors differ by 180°. Two vectors with opposite directions can be added by aligning the beginning point of the vector with the smaller magnitude, with the end point of the vector with the larger magnitude. The resultant has the same direction as the vector with the larger magnitude and a magnitude that is the difference of the two magnitudes.

To add vectors of opposite direction:

1. Subtract the magnitude of the smaller vector from the magnitude of the larger vector.
2. The resultant vector has the direction of the vector with the larger magnitude.

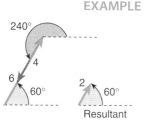

Figure 24–7

EXAMPLE Add a vector with a direction of 60° and a magnitude of 6 to a vector with a direction of 240° and a magnitude of 4 (see Fig. 24–7).

Subtract the magnitudes: $6 - 4 = 2$.

The resultant has a direction of 60° and a magnitude of 2.

Chapter 24 / Oblique Triangles

Adding any two vectors that have different directions is accomplished by performing the shifts of each vector in succession.

EXAMPLE Show a graphical representation of the sum of a 45° vector with a magnitude of 5 (vector A) and a 60° vector with a magnitude of 6 (vector B) (see Fig. 24–8).

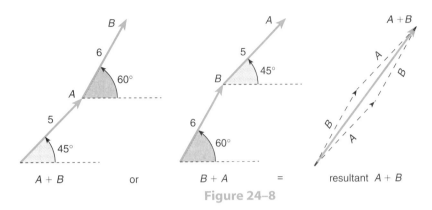

Figure 24–8

To add vectors with different directions:

1. Determine the x- and y-coordinates of the end points of each vector in standard position.
2. Add the x-coordinates of the end points for the x-coordinate of the end point of the resultant vector.
3. Add the y-coordinates of the end points for the y-coordinate of the end point of the resultant vector.
4. Find the magnitude using the Pythagorean theorem.
5. Find the direction using the tangent function.

Complex numbers in the form $a + bi$ are sometimes used to represent vectors. The x-coordinate of a vector in standard position is represented as the real component a. The y-coordinate of a vector in standard position is represented as the imaginary component bi.

To write a vector in complex notation:

1. Find the x-coordinate of the end point of the vector (Figure 24–9). Use the relationship

$$\cos \theta = \frac{\text{adj}}{\text{hyp}} = \frac{x}{r} \text{ and solve for } x. \quad \text{\small{r = magnitude}}$$

2. Find the y-coordinate of the end point of the vector. Use the relationship

$$\sin \theta = \frac{\text{opp}}{\text{hyp}} = \frac{y}{r} \text{ and solve for } y. \quad \text{\small{r = magnitude}}$$

3. Write the vector in complex notation as $x + yi$.

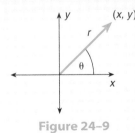

Figure 24–9

EXAMPLE Write the vectors A, B, and $A + B$ from Fig. 24–8 in complex notation.

Vector A:	**x-coordinate**	**y-coordinate**	
Magnitude = 5	$\cos 45° = \dfrac{x}{5}$	$\sin 45° = \dfrac{y}{5}$	Solve for x or y.
Direction = 45°	$5(\cos 45°) = x$	$5(\sin 45°) = y$	Evaluate.
	$3.535533906 = x$	$3.535533906 = y$	

Vector A in complex form: $3.535533906 + 3.535533906i$

Vector B:	**x-coordinate**	**y-coordinate**	
Magnitude = 6	$\cos 60° = \dfrac{x}{6}$	$\sin 60° = \dfrac{y}{6}$	Solve for x or y.
Direction = 60°	$6(\cos 60°) = x$	$6(\sin 60°) = y$	Evaluate.
	$3 = x$	$5.196152423 = y$	

Vector B in complex form: $3 + 5.196152423i$

Resultant $A + B = (3.535533906 + 3.535533906i)$ Add x values.
$+ (3 + 5.196152423i)$ Add y values.

Resultant $A + B = 6.535533906 + 8.731686329i$ or   **$6.536 + 8.732i$**

EXAMPLE Find the magnitude and direction of the resultant $A + B$ in the preceding example.

The resultant in complex form is $6.535533906 + 8.731686329i$; thus, the x-coordinate of the end point is 6.535533906 and the y-coordinate is 8.731686329.

Magnitude $= \sqrt{x^2 + y^2}$ Substitute for x and y.

Magnitude $= \sqrt{(6.535533906)^2 + (8.731686329)^2}$ Evaluate.

Magnitude $= \sqrt{118.9555496}$

Magnitude = 10.91   Round to 4 significant digits.

$\tan \theta = \dfrac{y}{x}$ Substitute for x and y. Evaluate.

$\tan \theta = \dfrac{8.731686329}{6.535533906}$

$\tan \theta = 1.336032596$ Use inverse tangent function.

$\tan^{-1} 1.336032596 = \theta$ Evaluate.

$\theta = 53.18570658$

Direction: $\theta = 53.2°$ Nearest 0.1°.

TIP!

Vector Notation in Electronics

Vectors written in complex form are often used in electronics. The imaginary part is referred to as the *j*-factor.

Resultant vector $A + B$ from the first example on the preceding page would be written as $6.536 + j8.732$. The customary notation is for *j* to be followed by the coefficient.

SELF-STUDY EXERCISES 24-1

1 Find the magnitude of the vectors in standard position with end points at the indicated points. Round to the nearest thousandth if necessary.

1. $(5, 12)$ **2.** $(-12, 9)$ **3.** $(2, -7)$ **4.** $(-8, -3)$ **5.** $(1.5, 2.3)$

2 Find the direction of the vectors in standard position with end points at the indicated points. Round to the nearest hundredth.

6. $(5, 12)$ **7.** $(6, 8)$ **8.** $(8, 3)$ **9.** $(2, 5)$ **10.** $(1, 4)$

3

11. Find the resultant vector of two vectors that have a direction of 42° and magnitudes of 7 and 12, respectively.

12. Two vectors have a direction of 72°. Find the sum of the vectors if their magnitudes are 1 and 7, respectively.

13. Find the sum of two vectors if one has a direction of 45° and a magnitude of 7 and the other has a direction of 225° and a magnitude of 8.

14. Find the sum of two vectors if one has a direction of 75° and a magnitude of 15 and the other has a direction of 255° and a magnitude of 9.

Find the magnitude and direction of the resultant of the sum of the two given vectors in complex notation.

15. $5 + 3i$ and $7 + 2i$

16. $1 + 2i$ and $5 + 2i$

24-2 | Trigonometric Functions for Any Angle

Learning Outcomes

1 Find related acute angles for angles or vectors in quadrants II, III, and IV.

2 Determine the signs of trigonometric values of angles of more than 90°.

3 Find the trigonometric values of angles of more than 90° using a calculator.

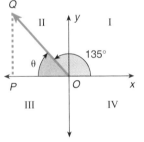

Figure 24-10

1 **Find Related Acute Angles for Angles or Vectors in Quadrants II, III, and IV.**

The direction of any vector in standard position can be determined, if the coordinates of the end point are known, by applying our knowledge of trigonometric functions. A right triangle that we will refer to as our *reference triangle* can be formed by drawing a vertical line from the end point of the vector to the *x*-axis. See the example of a reference triangle in Fig. 24–10. The angle θ is called the *related angle*. The *related angle* is the acute angle formed by the *x*-axis and the vector.

In quadrant I, the related angle is the same as the direction of the vector. Therefore, the direction of the vector in quadrant I is always less than 90°. For vectors in quadrants II, III, and IV, see Figs. 24–10, 24–11, and 24–12.

In Fig. 24–10, $\overline{PQ}$, the *x*-axis, and vector Q form a right triangle. The direction of the vector is 135°; therefore, the related angle is 45° (180° − 135°). 135° is a second-quadrant angle. Second-quadrant angles are more than 90° and less than 180°, or more than

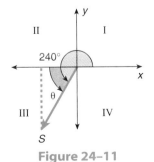

Figure 24-11

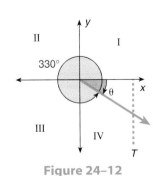

Figure 24-12

$\frac{\pi}{2}$ radians (1.57) and less than π radians (3.14). The related angle for any second-quadrant vector can be found by subtracting the direction of the vector from 180° (or π radians).

Third-quadrant angles are more than 180° and less than 270°, or more than π radians (3.14) and less than $\frac{3\pi}{2}$ radians (4.71). In Fig. 24–11, vector S is a third-quadrant vector. The related angle θ is 60° (240° − 180°). The related angle for any third-quadrant vector is found by subtracting 180° (or π radians) from the direction of the vector.

Vector T in Fig. 24–12 is a fourth-quadrant vector. Fourth-quadrant angles are more than 270° and less than 360°, or more than $\frac{3\pi}{2}$ radians (4.71) and less than 2π radians (6.28). The related angle θ is 30° (360° − 330°). The related angle for any fourth-quadrant vector is found by subtracting the direction of the vector from 360° (or 2π radians).

To find related angles for vectors that are more than 90°:

The related angle for quadrant I angles and vectors is equal to the direction (angle) of the vector. The related angle for angles and vectors more than 90° can be found as follows (Fig. 24–13):

Quadrant II angle (90° < θ < 180°   or $\frac{\pi}{2}$ rad < θ < π rad):   180° − θ_2 or π − θ_2

Quadrant III angle (180° < θ < 270°   or π rad < θ < $\frac{3\pi}{2}$ rad): θ_3 − 180°   or θ_3 − π

Quadrant IV angle (270° < θ < 360°   or $\frac{3\pi}{2}$ rad < θ < 2π rad):   360° − θ_4 or 2π − θ_4

Related angles are always angles less than 90° or $\frac{\pi}{2}$ radians (1.57 rad).

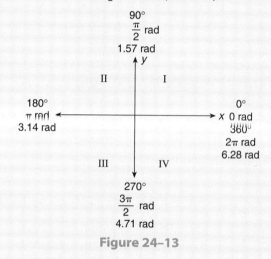

Figure 24–13

EXAMPLE   Find the related angle for the following angles.
(a) 210°   (b) 1.93 rad

(a) 210° is between 180° and 270° and is a quadrant III angle.

$$\theta_3 - 180° =$$
$$210° - 180° = 30°$$

The related angle is 30°.

(b) 1.93 rad is between $\frac{\pi}{2}$ and π rad and is a quadrant II angle.

$$\pi - 1.93 = 1.211592654 \text{ rad.}$$

The related angle is 1.21 rad to the nearest hundredth.

2   **Determine the Signs of Trigonometric Values of Angles of More Than 90°.**

To determine the appropriate sign of trigonometric functions for angles more than 90°, we examine the trigonometric functions in each quadrant. The sign of the function indicates the direction or quadrant of the vector.

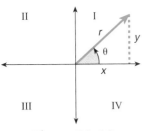

Figure 24–14

The vector in Fig. 24–14 has a magnitude of r and direction of θ. To relate our trigonometric functions to vectors, the magnitude of the vector is the hypotenuse, the side opposite θ is y, and the side adjacent to θ is x. The magnitude of a vector (r) is always positive. In quadrant I, the x value and the y value are both positive.

Trigonometric Functions in Quadrant I

$$\sin \theta_1 = \frac{y}{r} \qquad \cos \theta_1 = \frac{x}{r} \qquad \tan \theta_1 = \frac{y}{x}$$

$$\csc \theta_1 = \frac{r}{y} \qquad \sec \theta_1 = \frac{r}{x} \qquad \cot \theta_1 = \frac{x}{y}$$

In quadrant I, the sign of all six trigonometric functions is positive.

For quadrant II vectors (Fig. 24–15), again the magnitude (r) is positive and the y value is positive, but the x value is negative.

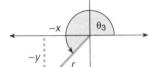

Figure 24–15

Trigonometric Functions in Quadrant II

$$\sin \theta_2 = \frac{y}{r} \qquad \cos \theta_2 = \frac{-x}{r} = -\frac{x}{r} \qquad \tan \theta_2 = \frac{y}{-x} = -\frac{y}{x}$$

$$\csc \theta_2 = \frac{r}{y} \qquad \sec \theta_2 = -\frac{r}{x} \qquad \cot \theta_2 = -\frac{x}{y}$$

In quadrant II the sine and cosecant functions are positive, and the remaining functions are negative.

Quadrant III vectors (Fig. 24–16) have a positive magnitude (r), negative x value, and negative y value.

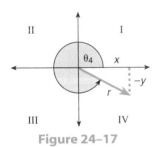

Figure 24–16

Trigonometric Functions in Quadrant III

$$\sin \theta_3 = \frac{-y}{r} = -\frac{y}{r} \qquad \cos \theta_3 = \frac{-x}{r} = -\frac{x}{r} \qquad \tan \theta_3 = \frac{-y}{-x} = \frac{y}{x}$$

$$\csc \theta_3 = -\frac{r}{y} \qquad \sec \theta_3 = -\frac{r}{x} \qquad \cot \theta_3 = \frac{x}{y}$$

In quadrant III the tangent and cotangent functions are positive and the remaining functions are negative.

Quadrant IV vectors (Fig. 24–17) have a positive magnitude (r), positive x value, and negative y value.

Figure 24–17

Trigonometric Functions in Quadrant IV

$$\sin \theta_4 = \frac{-y}{r} = -\frac{y}{r} \qquad \cos \theta_4 = \frac{x}{r} \qquad \tan \theta_4 = \frac{-y}{x} = -\frac{y}{x}$$

$$\csc \theta_4 = -\frac{r}{y} \qquad \sec \theta_4 = \frac{r}{x} \qquad \cot \theta_4 = -\frac{x}{y}$$

In quadrant IV the cosine and secant functions are positive and the remaining functions are negative.

Learning Strategy   *Understanding Beats Memorizing.*

We have said it before but it is worth saying again, *Understand rather than memorize.*

To understand the signs of the trigonometric functions in each quadrant, we fit together information we already know:

r or the hypotenuse is always positive.
Two negatives in division make a positive quotient.
One negative and one positive make a negative quotient.

Use these three tools along with the fractional relationships for sine, cosine, and tangent to determine the sign of the sine, cosine, and tangent functions for each of the four quadrants (Fig. 24–18).

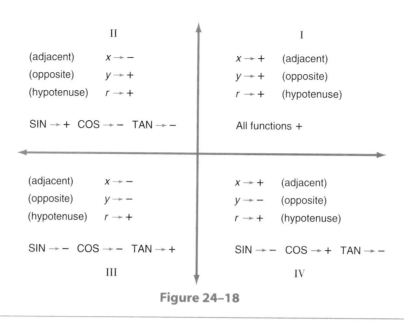

Figure 24–18

Trigonometric functions, just like algebraic functions, can be represented graphically. This graphical representation can help you to visualize the sign patterns of the various functions and draws attention to other properties of trigonometric functions. To graph these functions, we make a table of values from 0° to 360°, or from 0 rad to 2π rad. See Figs. 24–19 through 24–21.

		Quadrant I				Quadrant II				Quadrant III				Quadrant IV			
sin θ	0	0.5	0.71	0.87	1	0.87	0.71	0.5	0	−0.5	−0.71	−0.87	−1	−0.87	−0.71	−0.5	0
θ	0	$\frac{\pi}{6}$	$\frac{\pi}{4}$	$\frac{\pi}{3}$	$\frac{\pi}{2}$	$\frac{2\pi}{3}$	$\frac{3\pi}{4}$	$\frac{5\pi}{6}$	π	$\frac{7\pi}{6}$	$\frac{5\pi}{4}$	$\frac{4\pi}{3}$	$\frac{3\pi}{2}$	$\frac{5\pi}{3}$	$\frac{7\pi}{4}$	$\frac{11\pi}{6}$	2π
θ°	0	30	45	60	90	120	135	150	180	210	225	240	270	300	315	330	360

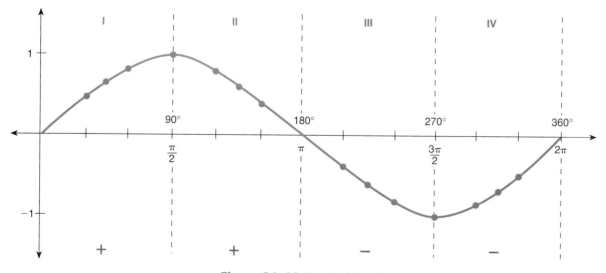

Figure 24–19 Graph of sine θ.

Chapter 24 / Oblique Triangles

		Quadrant I				Quadrant II				Quadrant III				Quadrant IV			
cos θ	1	0.87	0.71	0.5	0	−0.5	−0.71	−0.87	−1	−0.87	−0.71	−0.5	0	0.5	0.71	0.87	1
θ	0	$\frac{\pi}{6}$	$\frac{\pi}{4}$	$\frac{\pi}{3}$	$\frac{\pi}{2}$	$\frac{2\pi}{3}$	$\frac{3\pi}{4}$	$\frac{5\pi}{6}$	π	$\frac{7\pi}{6}$	$\frac{5\pi}{4}$	$\frac{4\pi}{3}$	$\frac{3\pi}{2}$	$\frac{5\pi}{3}$	$\frac{7\pi}{4}$	$\frac{11\pi}{6}$	2π
θ°	0	30	45	60	90	120	135	150	180	210	225	240	270	300	315	330	360

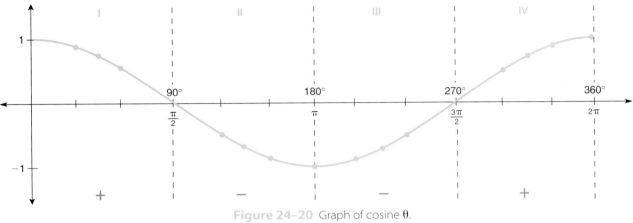

Figure 24–20 Graph of cosine θ.

		Quadrant I				Quadrant II				Quadrant III				Quadrant IV			
tan θ	0	0.58	1	1.7	∞	−1.7	−1	−0.58	0	0.58	1	1.7	∞	−1.7	−1	−0.58	0
θ	0	$\frac{\pi}{6}$	$\frac{\pi}{4}$	$\frac{\pi}{3}$	$\frac{\pi}{2}$	$\frac{2\pi}{3}$	$\frac{3\pi}{4}$	$\frac{5\pi}{6}$	π	$\frac{7\pi}{6}$	$\frac{5\pi}{4}$	$\frac{4\pi}{3}$	$\frac{3\pi}{2}$	$\frac{5\pi}{3}$	$\frac{7\pi}{4}$	$\frac{11\pi}{6}$	2π
θ°	0	30	45	60	90	120	135	150	180	210	225	240	270	300	315	330	360

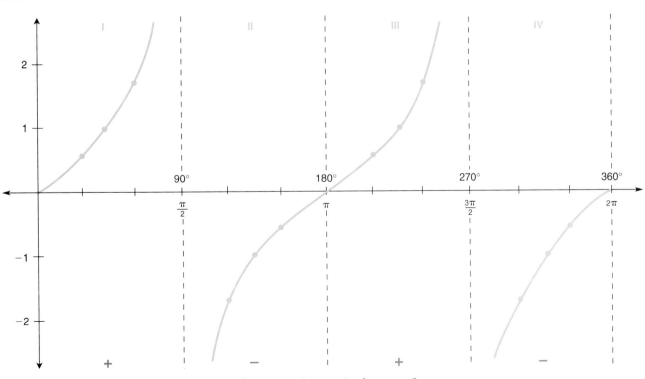

Figure 24–21 Graph of tangent θ.

Signs of Trigonometric Functions

To help remember the signs of the trigonometric functions in the various quadrants, we may find the following reminder helpful.

ALL-SIN-TAN-COS

This reminder gives the positive functions in the quadrants I to IV, respectively. Also, the functions cosecant, cotangent, and secant have the same signs as their respective reciprocal functions.

3 **Find the Trigonometric Values of Angles of More Than 90° Using a Calculator.**

General Tips for Using the Calculator

To find the value of the trigonometric function of an angle that is more than 90° using a calculator:

1. Set the calculator to degree or radian mode as desired.
2. Enter the trigonometric function.
3. Enter the angle measure in degrees or radians.
4. Press $=$ or ENTER .

To find the related angle, given the end point of a vector:

1. Set the calculator to degree or radian mode as desired.
2. Press the inverse tangent TAN⁻¹ or ARCTAN function.
3. Enter the coordinates of the end point as appropriate in the tangent ratio $\left(\dfrac{y}{x}\right)$.
4. Press $=$ or ENTER .

EXAMPLE   Using a calculator and the π key, find the values of the trigonometric functions. Round the final answer to four significant digits.

(a) sin 155°   (b) cos 3   (c) tan 208°   (d) sin 4.2

(e) cos 304.5°   (f) tan $\dfrac{5\pi}{3}$

(a) sin 155° = **0.4226**   degree mode   **(b)** cos 3 = **−0.9900**   radian mode
(c) tan 208° = **0.5317**   degree mode   **(d)** sin 4.2 = **−0.8716**   radian mode

(e) cos 304.5° = **0.5664**   degree mode   **(f)** tan $\dfrac{5\pi}{3}$ = **−1.732**   radian mode

---

EXAMPLE   Find the direction in degrees of a vector in standard position if the coordinates of its end point are (6, −8) (Fig. 24–22).

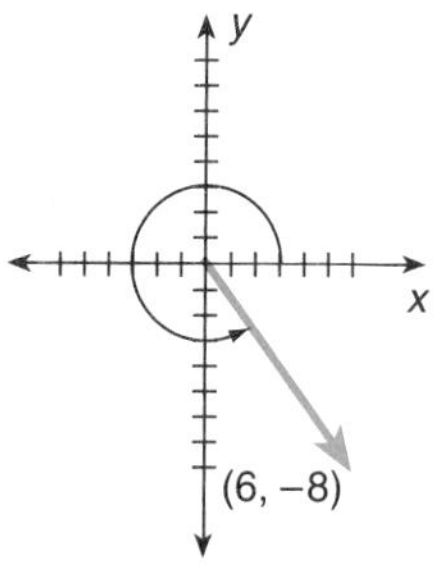

$$\tan \theta = \frac{y}{x} \qquad \text{Substitute values of the end points.}$$

$$\tan \theta = \frac{-8}{6}$$

$$\tan \theta = -1.333333333 \qquad \text{Use inverse tangent function.}$$

$$\tan^{-1}(-1.333333333) = \theta \qquad \text{Evaluate.}$$

$$\theta = -53.1° \qquad \text{To the nearest tenth degree.}$$

Figure 24–22

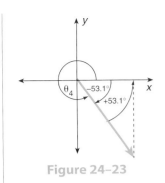

Figure 24–23

Related angle = 53.1°

$360° - \theta_4 = 53.1°$

$360° - 53.1° = \theta_4$

$306.9° = \theta_4$

Fig. 24–23. Positive direction from vector to *x* axis.

Solve for θ_4, direction of vector.

Thus, the direction of the vector is 306.9°.

TIP!

When Do We Use the Cosecant, Secant, and Cotangent Functions?

As you have probably noticed, we don't often use the cosecant, secant, and cotangent functions. What causes a fraction or ratio to be undefined? A denominator of zero. One use of reciprocal functions would be to avoid undefined terms. For example, if $x = 0$ and $y = 3$, the tangent function is undefined: $\tan \theta = \dfrac{y}{x} = \dfrac{3}{0} = \infty$. However, the cotangent function is defined: $\cot \theta = \dfrac{x}{y} = \dfrac{0}{3} = 0$.

SELF-STUDY EXERCISES 24–2

1 Find the related angle for the angles. (Use the calculator value for π and round to hundredths.)

1. 120°   **2.** 195°   **3.** 290°   **4.** 345°   **5.** 148°

6. 250°   **7.** 212°   **8.** 118°   **9.** 2.18 rad   **10.** 5.84 rad

2 Give the signs of all six trigonometric functions of vectors with the indicated end points.

11. $(3, 5)$ **12.** $(-2, 6)$ **13.** $(-4, -2)$ **14.** $(5, -3)$ **15.** $(5, 0)$

3 Using a calculator and the π key, evaluate the trigonometric functions. Round to four significant digits.

16. $\sin 210°$ **17.** $\tan 140°$ **18.** $\cos 2.5$ **19.** $\cos 4$ **20.** $\sin 300°$

21. $\tan 6$ **22.** $\cos 100°$ **23.** $\sin \dfrac{5\pi}{6}$

24. Find the direction in degrees of a vector in standard position if the coordinates of its end point are $(-3, 2)$. Round to the nearest 0.1°.

25. Find the direction in radians of a vector in standard position if the coordinates of its end point are $(-2, -1)$. Round to the nearest hundredth.

24–3   *Law of Sines*

Learning Outcomes

1 Find the missing parts of an oblique triangle, given two angles and a side.

2 Find the missing parts of an oblique triangle, given two sides and an angle opposite one of them.

24-3 Law of Sines

An *oblique triangle* is a triangle that does not contain a right angle. Because these triangles do not have right angles, we cannot use the Pythagorean theorem or trigonometric functions directly as we did previously to find sides and angles of right triangles. However, two formulas based on the trigonometric functions of right triangles can be used to solve oblique triangles. In this section, we study one of these formulas, the *law of sines*.

Law of sines:

The **law of sines** states that the ratios of the sides of a triangle to the sines of the angles opposite these respective sides are equal (see Fig. 24–24).

$$\frac{a}{\sin A} = \frac{b}{\sin B} = \frac{c}{\sin C}$$

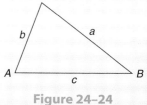

Figure 24–24

TIP!

Conditions for Using the Law of Sines

The law of sines is used to solve triangles when either of the conditions exists:

1. Two angles and a side are known.
2. Two sides and the angle opposite one of them are known.

When using the law of sines you must have at least one complete ratio—an angle and side opposite. However, if we know two angles of a triangle, we can always find the third angle. Then, we can form a complete ratio.

1 **Find the Missing Parts of an Oblique Triangle, Given Two Angles and a Side.**

To find a missing part of an oblique triangle given two angles and a side:

1. Find the third angle of the triangle by adding the two known angles and subtracting the sum from 180° or 2π radians.
2. Set up a ratio with the given side and sine of the opposite angle.
3. Set up a second ratio with a missing side and the sine of the opposite angle.
4. Form a proportion using the ratios in Steps 1 and 2 and solve the proportion for the missing side.
5. Repeat Steps 2–4 if the value of the third side is needed.

In the examples that follow, all digits of the calculated value that show in the calculator display will be given. Rounding should be done after the last calculation has been made.

EXAMPLE Solve triangle ABC if $A = 50°$, $B = 75°$, and $b = 12$ ft.

Sketch the triangle and label the parts, as shown in Fig. 24–25.

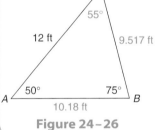

Figure 24–25

Find $\angle C$.

$$A + B + C = 180°$$ Substitute $A = 50°$ and $B = 75°$.
$$50° + 75° + C = 180°$$ Solve for C.
$$C = 180° - 50° - 75°$$
$$\mathbf{C = 55°}$$

Find a.

$$\frac{a}{\sin A} = \frac{b}{\sin B}$$ Side b and $\angle B$ form the ratio with both values known.
Substitute $b = 12$, $A = 50°$, $B = 75°$.

$$\frac{a}{\sin 50°} = \frac{12}{\sin 75°}$$ Solve for a.
$$a \sin 75° = 12 \sin 50°$$
$$a = \frac{12 \sin 50°}{\sin 75°}$$ Perform calculations.
$$a = 9.516810781$$
$$\mathbf{a = 9.517 \ ft}$$ Four significant digits.

Find c.

$$\frac{b}{\sin B} = \frac{c}{\sin C}$$ Substitute $b = 12$, $B = 75°$, and $C = 55°$.

$$\frac{12}{\sin 75°} = \frac{c}{\sin 55°}$$ Solve for c.
$$12 \sin 55° = c \sin 75°$$
$$\frac{12 \sin 55°}{\sin 75°} = c$$ Perform calculations.
$$c = 10.1765832$$
$$\mathbf{c = 10.18 \ ft}$$ Four significant digits.

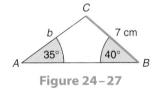

Figure 24–26

As a check for our work, the longest side should be opposite the largest angle, and the shortest side should be opposite the smallest angle (see Fig. 24–26).

EXAMPLE Solve triangle ABC if $A = 35°$, $a = 7$ cm, and $B = 40°$.

Sketch the triangle and label its parts (see Fig. 24–27).

Figure 24–27

Find $\angle C$.

$$A + B + C = 180°$$ Substitute $A = 35°$ and $B = 40°$.
$$35° + 40° + C = 180°$$ Solve for C.
$$C = 180° - 35° - 40°$$
$$\mathbf{C = 105°}$$

Find b.

$$\frac{a}{\sin A} = \frac{b}{\sin B}$$ Side a and $\angle A$ form the ratio with both values known.
Substitute $a = 7$, $A = 35°$, and $B = 40°$.

$$\frac{7}{\sin 35°} = \frac{b}{\sin 40°}$$ Solve for b.
$$7 \sin 40° = b \sin 35°$$

24–3 Law of Sines

$$\frac{7 \sin 40°}{\sin 35°} = b$$
Evaluate.

$$b = 7.844661989$$

$$b = \textbf{7.845 cm}$$
Four significant digits.

Find c.

$$\frac{a}{\sin A} = \frac{c}{\sin C}$$
Substitute $a = 7$, $A = 35°$, and $C = 105°$.

$$\frac{7}{\sin 35°} = \frac{c}{\sin 105°}$$
Solve for c.

$$7 \sin 105° = c \sin 35°$$

$$\frac{7 \sin 105°}{\sin 35°} = c$$
Evaluate.

$$c = 11.78828201$$

$$c = \textbf{11.79 cm}$$
Four significant digits.

As a check for our work, the longest side should be opposite the largest angle, and the shortest side should be opposite the smallest angle (see Fig. 24–28).

C
7.845 cm   105°   7 cm
35°   40°
A   11.79 cm   B

Figure 24–28

2 Find the Missing Parts of an Oblique Triangle, Given Two Sides and an Angle Opposite One of Them.

When two sides of a triangle and an angle opposite one of the sides are known, we do not always have a single triangle. If the given sides are a and b and the given angle is B, three possibilities may exist (see Fig. 24–29).

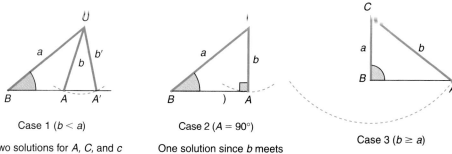

Case 1 ($b < a$)

Two solutions for A, C, and c since b can meet side c in either of two points, A and A'.

Case 2 ($A = 90°$)

One solution since b meets side c in exactly one point.

Case 3 ($b \geq a$)

One solution since b meets side c in only one point.

Figure 24–29

- If $b < a$ and $\angle A \neq 90°$, we have two possible solutions (case 1).
- If $b < a$ and $\angle A = 90°$, we have one solution (case 2).
- If $b \geq a$ and $\angle A \neq 90°$, we have one solution (case 3).

Note that side b above could be any side of the triangle that is opposite the given angle. Side a is then the other given side. Because of the lack of clarity when two sides and an angle opposite one of them are given, the situation is called the *ambiguous case*. (*Ambiguous* means that the given information can be interpreted in more than one way.)

To find the missing parts of an oblique triangle, given two sides and an angle opposite one of them:

1. Use the law of sines to find the measure of the angle opposite the other given side.
2. Determine if there are two angle values that are possible for the solution.
3. Find the third angle by subtracting from 180° the sum of the two known angles.
4. Find the third side using the law of sines.
5. If two angles were found in Step 2, find the second possible solution by repeating Steps 3 and 4.

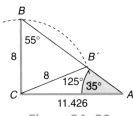

Figure 24–30

EXAMPLE Solve triangle ABC if $A = 35°$, $a = 8$, and $b = 11.426$ (see Fig. 24–30).

Find $\angle B$.

$$\frac{a}{\sin A} = \frac{b}{\sin B}$$

Substitute $A = 35°$, $a = 8$, and $b = 11.426$.

$$\frac{8}{\sin 35°} = \frac{11.426}{\sin B}$$

Solve for B.

$$8 \sin B = 11.426 \sin 35°$$

$$\sin B = \frac{11.426 \sin 35°}{8}$$

Evaluate.

$$\sin B = 0.8192105452$$

$$\sin^{-1} 0.8192105452 = B$$

Use inverse sine function.

$$B = 55.00584421°$$ or $$B' = 180° - 55.00584421°$$
$$B' = 124.99415579°$$
$$B' = 125.0°$$

$$\boldsymbol{B = 55.0°}$$

Nearest 0.1°.

Possible solution 1:

Find $\angle C$.

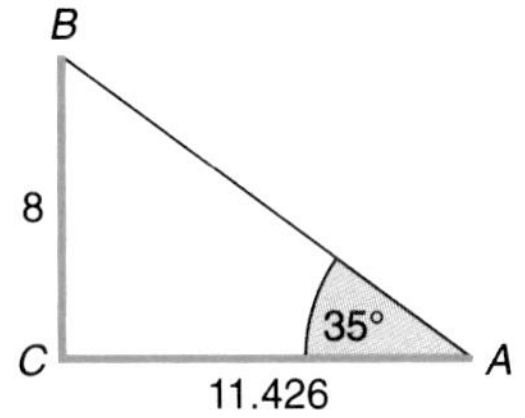

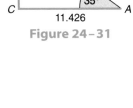

Figure 24–31

$$A + B + C = 180°$$

Substitute $A = 35°$ and $B = 55°$.

$$35° + 55° + C = 180°$$

Solve for C.

$$C = 180° - (55° + 35°)$$

$$\boldsymbol{C = 90°}$$

ABC is a right triangle (Fig. 24–31).

This is a right triangle, but we still have two possible solutions. The angle opposite the second given side is not the 90° angle. To find the third side of this right triangle, we can use the Pythagorean theorem or the law of sines.

Find c.

$$\frac{a}{\sin A} = \frac{c}{\sin C}$$

Substitute $a = 8$, $A = 35°$, and $C = 90°$.

$$\frac{8}{\sin 35°} = \frac{c}{\sin 90°}$$

Solve for c.

$$8 \sin 90° = c \sin 35°$$

$$\frac{8 \sin 90°}{\sin 35°} = c$$

Evaluate.

$$c = 13.94757436$$

$$\boldsymbol{c = 13.95}$$

Four significant digits.

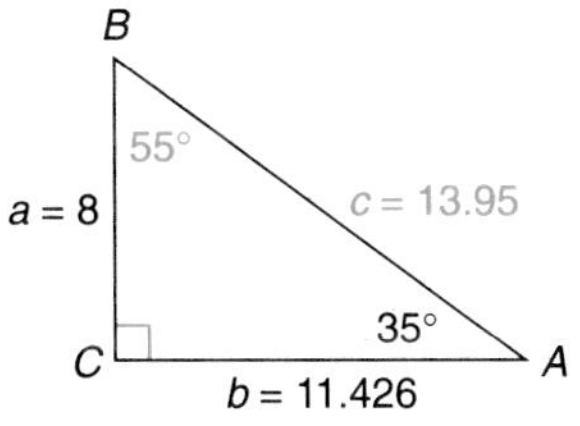

The solved triangle for this possibility is shown in Fig. 24–32.

Figure 24–32

Possible solution 2:

Find $\angle C'$.

$$C' = 180° - (35° + 124.99415579°) = 20.00584421°$$

$$\boldsymbol{C' = 20.0°}$$

See Fig. 24–33.

Find c'.

$$\frac{a}{\sin A} = \frac{c'}{\sin C'}$$

Substitute $a = 8$, $A = 35°$, and $C' = 20.00584421°$.

24–3 Law of Sines

$$\frac{8}{\sin 35°} = \frac{c'}{\sin 20.00584421°}$$ Solve for c'.

$$c' \sin 35° = 8 \sin 20.00584421°$$

$$c' = \frac{8 \sin 20.00584421°}{\sin 35°}$$

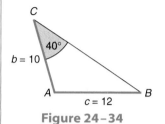

Figure 24–33

$$c' = 4.771688224 \qquad \text{or} \qquad \mathbf{c' = 4.772}$$

The solved triangle for this possibility is shown in Fig. 24–33.

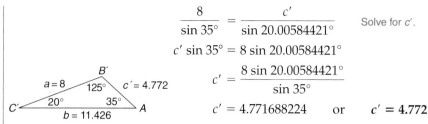

EXAMPLE   Solve triangle ABC if $b = 10$, $c = 12$, and $C = 40°$ (Fig. 24–34).

Find B.

Because c, the side opposite the given angle C, is *longer* than the other given side, b, we expect *one* solution.

$$\frac{c}{\sin C} = \frac{b}{\sin B}$$ Substitute $c = 12$, $C = 40°$, and $b = 10$.

$$\frac{12}{\sin 40°} = \frac{10}{\sin B}$$ Solve for B.

$$12 \sin B = 10 \sin 40°$$

$$\sin B = \frac{10 \sin 40°}{12}$$ Evaluate.

$$\sin B = 0.5356563414$$ Use inverse sine function.

$$\sin^{-1} 0.5356563414 = B$$

$$B = 32.38843382°$$

$$\mathbf{B = 32.4°}$$ Nearest 0.1.

Both $32.38843382°$ and $147.6115662°$ have a sine of approximately 0.5356563414; however, we cannot have an angle of $147.6115662°$ in this triangle because the triangle already has a $40°$ angle. $147.6115662° + 40° = 187.6115662°$. The *three* angles of a triangle total only $180°$. Therefore, we have only *one* solution.

Find A.

$$A = 180° - 40° - 32.38843382°$$ Use full calculator value for B.

$$A = 107.6115662°$$

$$\mathbf{A = 107.6°}$$

Find a.

$$\frac{a}{\sin A} = \frac{c}{\sin C}$$ Substitute full calculator value for A.

$$\frac{a}{\sin 107.6115662°} = \frac{12}{\sin 40°}$$ Solve for a.

$$a \sin 40° = 12 \sin 107.6115662°$$

$$a = \frac{12 \sin 107.6115662°}{\sin 40°}$$ Evaluate.

$$a = 17.79367733$$

$$\mathbf{a = 17.79}$$ Four significant digits.

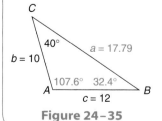

Figure 24–35

The solved triangle is shown in Fig. 24–35.

Chapter 24 / Oblique Triangles

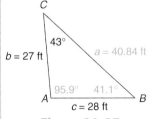

EXAMPLE A technician checking a surveyor's report is given the information shown in Fig. 24–36. Calculate the missing information.

Find B.

$$\frac{c}{\sin C} = \frac{b}{\sin B}$$ Substitute given values.

$$\frac{28}{\sin 43°} = \frac{27}{\sin B}$$ Solve for B.

$$28 \sin B = 27 \sin 43°$$

$$\sin B = \frac{27 \sin 43°}{28}$$

$$\sin B = 0.6576412758$$ Use the inverse sine function.

$$\sin^{-1} 0.6576412758 = B$$

$$B = 41.12023021°$$

$$\boldsymbol{B = 41.1°}$$ Nearest 0.1°.

There is only one case here because we are given the triangle. Also, the other angle whose sine is 0.6576412758 is 138.8797698°, which we exclude because 138.8797698° + 43° = 181.8797698°.

Find angle A.

$$A = 180° - (43° + 41.12023021°)$$

$$A = 95.87976979°$$

$$\boldsymbol{A = 95.9°}$$ Nearest 0.1°

Find a.

$$\frac{a}{\sin A} = \frac{c}{\sin C}$$ Substitute.

$$\frac{a}{\sin 95.87976979°} = \frac{28}{\sin 43°}$$

$$a \sin 43° = 28 \sin 95.87976979°$$

$$a = \frac{28 \sin 95.87976979°}{\sin 43°}$$ Evaluate.

$$a = 40.83982458$$

$$\boldsymbol{a = 40.84 \text{ ft}}$$ Four significant digits.

The completed survey should have the measures shown in Fig. 24–37.

Figure 24–37

1 Solve each of the oblique triangles using the law of sines. Round the final answer for sides to four significant digits and angles to the nearest 0.1°.

1. $a = 46, A = 65°, B = 52°$
2. $b = 7.2, B = 58°, C = 72°$
3. $a = 65, B = 60°, A = 87°$
4. $c = 3.2, A = 120°, C = 30°$
5. $b = 12, A = 95°, B = 35°$

2 Use the law of sines to solve the triangles. If a triangle has two possibilities, find both solutions. Round sides to four significant digits and angles to the nearest 0.1°.

6. $a = 42, b = 24, A = 40°$
7. $b = 15, c = 3, B = 70°$
8. $a = 18, c = 9, C = 20°$
9. $a = 8, b = 4, A = 30°$
10. Find the missing angle and sides of the plot of land described by Fig. 24–38.
11. Find the distance from A to B on the surveyed plot shown in Fig. 24–39.

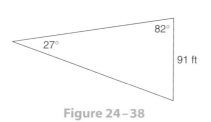

Figure 24–38

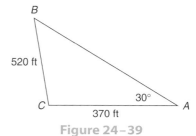

Figure 24–39

24–4 | Law of Cosines

Learning Outcomes

1 Find the missing parts of an oblique triangle, given three sides of the triangle.

2 Find the missing parts of an oblique triangle, given two sides and the included angle of the triangle.

In some cases, our given information does not allow us to use the law of sines. For example, if we know all three sides of a triangle, we cannot use the law of sines to find the angles. In a case such as this, however, we can use the *law of cosines*. This law is based on the trigonometric functions just as the law of sines is.

Law of cosines:

The square of any side of a triangle equals the sum of the squares of the other sides minus twice the product of the other two sides and the cosine of the angle opposite the first side. For triangle *ABC*:

$$a^2 = b^2 + c^2 - 2bc \cos A$$
$$b^2 = a^2 + c^2 - 2ac \cos B$$
$$c^2 = a^2 + b^2 - 2ab \cos C$$

TIP!

When Do You Use the Law of Cosines?

To use the law of cosines efficiently, we are given

1. Three sides of a triangle

   or

2. Two sides and the included angle of a triangle.

We can sometimes use *either* the law of sines or the law of cosines to solve certain triangles. However, whenever possible, the law of sines is generally preferred because it involves fewer calculations.

To find the missing parts of an oblique triangle, given three sides of the triangle:

1. Substitute the given values of the three sides into any version of the law of cosines.
2. Solve for the unknown angle.
3. Use the angle measure found in Step 2 and find a second angle using a different version of the law of cosines.
4. Find the third angle of the triangle by subtracting from 180° the sum of the two angles found in Steps 2 and 3.

EXAMPLE   Find the angles in triangle ABC (see Fig. 24–40).

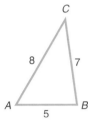

Figure 24–40

We may find any one of the angles first. If we choose to find angle A first, we must use the formula that contains $\cos A$: $a^2 = b^2 + c^2 - 2bc \cos A$. Rearrange the formula to solve for $\cos A$.

Find A.

$$a^2 - b^2 - c^2 = -2bc \cos A$$

Multiply each term on both sides by -1 to reduce the number of negative signs.

$$-a^2 + b^2 + c^2 = 2bc \cos A$$

Solve for $\cos A$.

$$\frac{-a^2 + b^2 + c^2}{2bc} = \cos A$$

Substitute $a = 7$, $b = 8$, and $c = 5$.

$$\frac{-(7)^2 + 8^2 + 5^2}{2(8)(5)} = \cos A$$

Solve for A.

$$\frac{-49 + 64 + 25}{80} = \cos A$$

$$\frac{40}{80} = \cos A$$

$$0.5 = \cos A$$

$$\cos^{-1} 0.5 = A$$

Use inverse cosine function.

$$\mathbf{60° = A}$$

Find B.

To find angle B, we use the law of cosines and only given values. The law of sines could be used but would involve using a calculated value.

$$b^2 = a^2 + c^2 - 2ac \cos B$$

Solve for $\cos B$.

$$2ac \cos B = a^2 + c^2 - b^2$$

$$\cos B = \frac{a^2 + c^2 - b^2}{2ac}$$

Substitute $a = 7$, $b = 8$, and $c = 5$.

$$\cos B = \frac{7^2 + 5^2 - (8)^2}{2(7)(5)}$$

Solve for B.

$$\cos B = \frac{49 + 25 - 64}{70}$$

$$\cos B = \frac{10}{70}$$

$$\cos B = 0.1428571429 \qquad \text{\small Use inverse cosine function.}$$
$$\cos^{-1} 0.1428571429 = B$$
$$B = 81.7867893°$$
$$\mathbf{B = 81.8°} \qquad \text{\small Nearest 0.1°.}$$

Find C.

The law of cosines or the law of sines can be used to find the third angle. However, the quickest way to find this angle is to subtract the sum of A and B from 180°.

$$C = 180° - (A + B)$$
$$C = 180° - 60° - 81.7867893°$$
$$C = 180° - 141.7867893°$$
$$C = 38.2132107°$$
$$\mathbf{C = 38.2°} \qquad \text{\small Nearest 0.1°.}$$

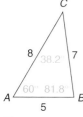

Figure 24–41

The solved triangle is shown in Fig. 24–41.

EXAMPLE Find the angles in triangle ABC (see Fig. 24–42).

Find A.

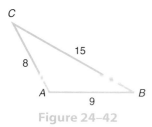

Figure 24–42

$$a^2 = b^2 + c^2 - 2bc \cos A \qquad \text{\small Solve for cos A.}$$
$$\cos A = \frac{b^2 + c^2 - a^2}{2bc} \qquad \text{\small Substitute.}$$
$$\cos A = \frac{8^2 + 9^2 - 15^2}{2(8)(9)} \qquad \text{\small Evaluate.}$$
$$\cos A = \frac{64 + 81 - 225}{144}$$
$$\cos A = \frac{-80}{144}$$
$$\cos A = -0.5555555556 \qquad \text{\small Use inverse cosine function.}$$
$$\cos^{-1} -0.5555555556 = A$$
$$A = 123.7489886°$$
$$\mathbf{A = 123.7°} \qquad \text{\small Nearest 0.1°.}$$

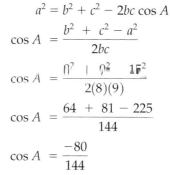

Figure 24–43

Recall from Section 24–2 that the cosine is *negative* in the second and third quadrants. Because A is either acute ($<90°$) or obtuse ($>90°$ but $<180°$) and because its cosine is negative (see Fig. 24–43), it must be in the second quadrant. If a calculator is used and -0.5555555556 is entered, 123.7489886° will appear in the display. We can use the law of sines to find the second angle.

Find B.

$$\frac{a}{\sin A} = \frac{b}{\sin B} \qquad \text{\small Substitute.}$$
$$\frac{15}{\sin 123.7489886°} = \frac{8}{\sin B} \qquad \text{\small Solve for B.}$$
$$15 \sin B = 8 \sin 123.7489886°$$
$$\sin B = \frac{8 \sin 123.7489886°}{15}$$
$$\sin B = 0.4434556903 \qquad \text{\small Use the inverse sine function.}$$
$$\sin^{-1} 0.4434556903 = B$$

Chapter 24 / Oblique Triangles

$$B = 26.32457654°$$
$$\mathbf{B = 26.3°}$$

C

29.9°   15

8

123.7° 26.3°
A   9   B

Figure 24–44

Find C.

$$C = 180° - A - B$$
Substitute.

$$C = 180° - 123.7489886° - 26.32457654°$$
Evaluate.

$$\mathbf{C = 29.9°}$$

The solved triangle is given in Fig. 24–44.

2 **Find the Missing Parts of an Oblique Triangle, Given Two Sides and the Included Angle of the Triangle.**

The law of cosines is needed to solve oblique triangles if two sides and the included angle are given.

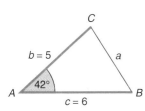

To find the missing parts of an oblique triangle, given two sides and the included angle of the triangle:

1. Find the side opposite the given angle using the law of cosines.
2. Use the law of sines to find one of the two unknown angles.
3. Find the third angle by subtracting from 180° the sum of the two known angles.

EXAMPLE   Solve triangle ABC in Fig. 24–45.

Find a.

C

b = 5   a

42°
A   c = 6   B

Figure 24–45

$$a^2 = b^2 + c^2 - 2bc \cos A$$
Substitute into the appropriate version of the law of cosines.

$$a^2 = 5^2 + 6^2 - 2(5)(6) \cos 42°$$
Evaluate.

$$a^2 = 25 + 36 - 60 \cos 42°$$
$\cos 42° = 0.7431448255$

$$a^2 = 25 + 36 - 44.58868953$$
Simplify.

$$a^2 = 16.41131047$$
Solve for a.

$$a = 4.051087566$$

$$\mathbf{a = 4.051}$$
Four significant digits.

Find B.

$$\frac{a}{\sin A} = \frac{b}{\sin B}$$
Substitute $a = 4.051087566$, $b = 5$, and $A = 42°$.

$$\frac{4.051087566}{\sin 42°} = \frac{5}{\sin B}$$
Solve for B.

$$4.051087566 \sin B = 5 \sin 42°$$

$$\sin B = \frac{5 \sin 42°}{4.051087566}$$
Evaluate.

$$\sin B = 0.8258653946$$
Use the inverse sine function.

$$B = 55.67632738°$$

$$\mathbf{B = 55.7°}$$
Nearest 0.1°.

Find C.

$$\frac{a}{\sin A} = \frac{c}{\sin C}$$
Substitute $a = 4.051087566$, $c = 6$, and $A = 42°$.

24–4 Law of Cosines

$$\frac{4.051087566}{\sin 42°} = \frac{6}{\sin C}$$ Solve for C.

$$4.051087566 \sin C = 6 \sin 42°$$

$$\sin C = \frac{6 \sin 42°}{4.051087566}$$ Evaluate.

$$\sin C = 0.9910384737$$ Use inverse sine function.

$$\sin^{-1} 0.9910384737 = C$$

$$C = 82.32367267°$$

$$\mathbf{C = 82.3°}$$ Nearest 0.1°.

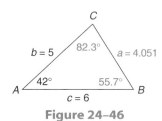

Figure 24–46

The solved triangle is given in Fig. 24–46.

To check, verify that the sum of the angles adds to 180°, the longest side is opposite the largest angle, and the shortest side is opposite the smallest angle.

EXAMPLE A vertical 45-ft pole is placed on a hill that is inclined 17° to the horizontal (see Fig. 24–47). How long a wire is needed if the wire is placed 5 ft from the top of the pole and attached to the ground at a point 32 ft uphill from the base of the pole?

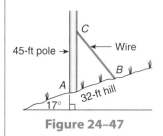

Figure 24–47

Point A is where the pole enters the ground. Point B is where the wire is attached to the ground. Point C is where the wire is attached to the pole (5 ft from the top of the pole). Side b is the length of the pole from the wire to the ground ($45 - 5 = 40$ ft). The length of the wire is side a in the triangle. Because AC makes a 90° angle with the horizontal, we know that angle A is $90° - 17°$, or $73°$. Using the law of cosines, we have

$$a^2 = b^2 + c^2 - 2bc \cos A$$ Substitute $b = 40$, $c = 32$, and $A = 73°$.

$$a^2 = 40^2 + 32^2 - 2(40)(32) \cos 73°$$ Evaluate.

$$a^2 = 1{,}600 + 1{,}024 - 2{,}560(0.2923717047)$$

$$a^2 = 1{,}600 + 1{,}024 - 748.4715641$$

$$a^2 = 1{,}875.528436$$

$$a = 43.30737161$$

$$a = 43.31 \text{ ft}$$ Four significant digits.

The length of the wire is 43.31 ft.

SELF-STUDY EXERCISES 24–4

 Solve the triangles in Figs. 24–48 and 24–49. Round side lengths to four significant digits and angles to the nearest 0.1°.

1.

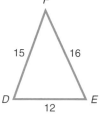

Figure 24–48

2.

Figure 24–49

Chapter 24 / Oblique Triangles

3.

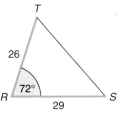

Figure 24–50

4.

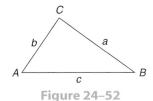

L

12

108.5°

J 11 K

Figure 24–51

Solve the problems. Round as above.

5. A hill is inclined 20° to the horizontal. A pole stands vertically on the side of the hill with 35 ft above the ground. How much wire will it take to reach from a point 2 ft from the top of the

pole to a point on the ground 27 ft downhill from the base of the pole?

6. A triangular tabletop is to be 8.4 ft by 6.7 ft by 9.3 ft. What angles must be cut?

24–5 | Area of Triangles

Learning Outcomes

1 Find the area of a triangle when the height is unknown and at least three parts of the triangle are known.

C

b   a

A   c   B

Figure 24–52

1 **Find the Area of a Triangle When the Height Is Unknown and at Least Three Parts of the Triangle Are Known.**

We can use trigonometric relationships to find the area of *any* triangle if we know the measure of any two sides and the included angle. In Fig. 24–52, the area of triangle *ABC* can be found by using one of three formulas.

Formula for the area of a triangle using trigonometric functions:

$$\text{Area} = \frac{1}{2} ab \sin C$$

$$\text{Area} = \frac{1}{2} ac \sin B$$

$$\text{Area} = \frac{1}{2} bc \sin A$$

These three formulas may be stated as a rule.

To find the area of a triangle when the height is unknown:

1. Use the law of sines or law of cosines to find two sides and the included angle of the triangle.
2. Select the appropriate version of the formula for the area of a triangle to match the known two sides and included angle.
3. Evaluate the formula.

EXAMPLE Find the area of triangle ABC in Fig. 24–53.

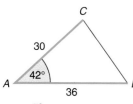

Figure 24–53

$$\text{Area} = \frac{1}{2}bc \sin A$$

Two sides and included angle are given. Substitute $b = 30$, $c = 36$, and $A = 42°$.

$$\text{Area} = \frac{1}{2}(30)(36)(\sin 42°)$$

Evaluate.

$$\text{Area} = 361.3305274$$

Area = 361.3 square units

Four significant digits.

EXAMPLE Find the area of triangle DEF in Fig. 24–54.

We are not given two sides and the included angle. Find side d or side e to have two sides and an included angle.

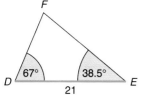

Figure 24–54

Find $\angle F$.

$$67° + 38.5° + F = 180°$$

Solve for F.

$$F = 180° - 67° - 38.5°$$

Evaluate.

$$F = 74.5°$$

Find side d.

$$\frac{f}{\sin F} = \frac{d}{\sin D}$$

Substitute $f = 21$, $F = 74.5°$, and $D = 67°$.

$$\frac{21}{\sin 74.5°} = \frac{d}{\sin 67°}$$

Solve for d.

$$21 \sin 67° = d \sin 74.5°$$

$$\frac{21 \sin 67°}{\sin 74.5°} = d$$

Evaluate.

$$d = 20.06018164$$

$$d = 20.06$$

Four significant digits.

Using the formula for area, we have

$$\text{Area} = \frac{1}{2}df \sin E$$

Angle E is included between d and f. Substitute $d = 20.06018164$, $f = 21$, and $E = 38.5°$.

$$\text{Area} = \frac{1}{2}(20.06018164)(21) \sin 38.5°$$

Evaluate.

$$\text{Area} = 131.1214452$$

Area = 131.1 square units

Four significant digits.

EXAMPLE Find the area of triangle ABC (Fig. 24–55).

To find the area of $\triangle ABC$, we must find the measure of any angle.

Find A.

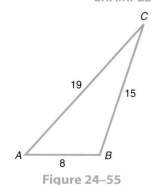

Figure 24–55

$$a^2 = b^2 + c^2 - 2bc \cos A$$

Use the law of cosines. Solve for $\cos A$.

$$\cos A = \frac{b^2 + c^2 - a^2}{2bc}$$

Substitute.

$$\cos A = \frac{19^2 + 8^2 - 15^2}{2(19)(8)}$$

$$\cos A = 0.6578947368$$

Use inverse cosine function.

$$\cos^{-1} 0.6578947368 = A$$

Chapter 24 / Oblique Triangles

$$A = 48.86048959° \text{ or } \mathbf{48.9°} \quad \text{Nearest } 0.1°.$$

Using the area formula, we have

Area $= \frac{1}{2}bc \sin A$ Substitute $b = 19$, $c = 8$, and $A = 48.86048959°$.

Area $= \frac{1}{2}(19)(8) \sin 48.86048959°$ Evaluate.

Area $= 57.23635209$ or $\mathbf{57.24}$ **square units**   Four significant digits.

When the measures of three sides of a triangle are known, Heron's formula can also be used (Chapter 6, Section 1, Outcome 2).

SELF-STUDY EXERCISES 24–5

1   Find the area of the triangles shown in Figs. 24–56 though 24–67. Use the law of sines or the law of cosines when necessary. Round answers to four significant digits.

1.

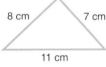

Figure 24–56

2.

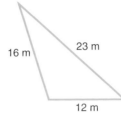

Figure 24–57

3.

Figure 24–58

4.

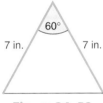

Figure 24–59

5.

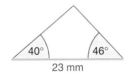

Figure 24–60

6.

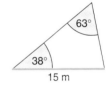

Figure 24–61

7.

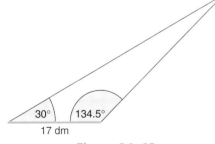

Figure 24–62

8.

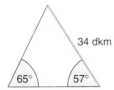

Figure 24–63

9.

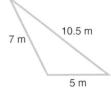

Figure 24–64

10.

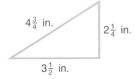

Figure 24–65

11.

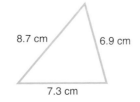

Figure 24–66

12.

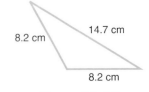

Figure 24–67

24–5   Area of Triangles

913

13. A tool and die maker for a medical implant manufacturer must calculate the area of a triangle whose sides are 2.06 cm, 3.27 cm, and 4.16 cm to make a die. What is the area of the triangle to three significant digits?

14. What is the area of a piece of carpeting that has sides measuring 14.3 ft, 12.8 ft, and 11.6 ft? Round to the nearest tenth of a square foot.

15. What is the area of the irregularly shaped property in Fig. 24–68? Round to the nearest square foot.

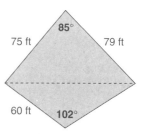

Figure 24–68

ASSIGNMENT EXERCISES

Section 24–1

Find the direction and magnitude of vectors in standard position with the indicated end points. Round lengths to four significant digits and angles to the nearest tenth of a degree.

1. $(4, 6)$ 2. $(2, 8)$ 3. $(6, 1)$

Find the magnitude and direction of the resultant of the sum of the two given vectors in complex notation. Round as above.

4. $3 + 5i$ and $2 + 3i$ 5. $1 + 4i$ and $6 + 8i$

Find the related angle for the angles.

6. $115°$ 7. 3.04 rad   8. 4.75 rad   9. $221°$

10. $305°$ 11. 5.4 rad   12. $138.5°$ 13. $212°15'10''$

Using the calculator, evaluate the trigonometric functions. Round to four significant digits.

14. $\cos 250°$ 15. $\sin 2.1$ 16. $\tan 175°$ 17. $\sin 340°$

18. $\tan 4.5$ 19. $\cos 290°$ 20. $\sin \dfrac{3\pi}{4}$ 21. $\tan \dfrac{5\pi}{4}$

22. Find the magnitude and direction of a vector of an electrical current in standard position if the coordinates of the end point of the vector are $(2, -3)$. Round the magnitude (in amps) to the nearest hundredth. Express the direction in degrees to the nearest $0.1°$.

23. Find the direction and magnitude of a vector in standard position if the coordinates of the end point of the vector are $(-2, 2)$. Express its direction in radians and round its direction and its magnitude to the nearest hundredth.

Section 24–3

Solve the triangles. If a triangle has two possible solutions, find both solutions. Round sides to tenths and angles to the nearest $0.1°$.

24. $A = 60°, B = 40°, b = 20$ 25. $B = 120°, C = 20°, a = 8$

26. $A = 60°, B = 60°, a = 10$ 27. $a = 5, c = 7, C = 45°$

28. $b = 10, c = 8, B = 52°$ 29. $a = 9.2, b = 6.8, B = 28°$

30. A surveyor needs the measure of JK in Fig. 24–69. Find JK to the nearest foot.

31. In Fig. 24–70, find RS.

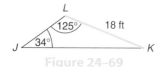

Figure 24–69

Figure 24–70

Solve the triangles in Figs. 24–71 to 24–76. Round sides to four significant digits and angles to nearest 0.1°.

32.

7   8.5

9

Figure 24–71

33.

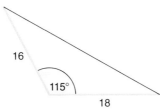

16

115°

18

Figure 24–72

34.

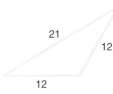

21

12

12

Figure 24–73

35.

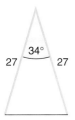

34°

27   27

Figure 24–74

36.

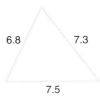

6.8   7.3

7.5

Figure 24–75

37.

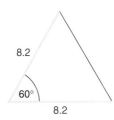

8.2

60°

8.2

Figure 24–76

Solve the problems. Round sides to four significant digits and angles to nearest 0.1°.

38. A triangular lot has sides 180 ft long, 160 ft long, and 123.5 ft long. Find the angles of the lot.

39. A vertical 50-ft pole stands on top of a hill inclined 18° to the horizontal. What length of wire is needed to reach from a point 6 ft from the pole's top to a point 75 ft downhill from the base of the pole?

40. A ship sails from a harbor 35 nautical miles east, then 42 nautical miles in a direction 32° south of east (Fig. 24–77). How far is the ship from the harbor?

41. A hill with a 35° grade (inclined to the horizontal) is cut down for a roadbed to a 10° grade. If the distance from the base to the top of the original hill is 800 ft (Fig. 24–78), how many vertical feet will be removed from the top of the hill, and what is the distance from the bottom to the top of the hill for the roadbed?

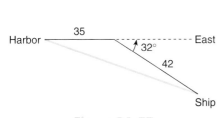

Harbor   35   East

32°

42

Ship

Figure 24–77

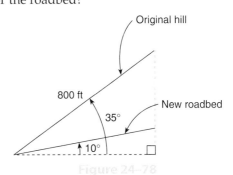

Original hill

800 ft

New roadbed

35°

10°

Figure 24–78

Find the area of the triangles shown in Figs. 24–79 to 24–87. Round the answers to four significant digits.

42.

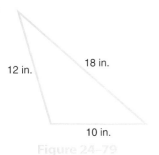

12 in.   18 in.

10 in.

Figure 24–79

43.

14 m   17 m

15 m

Figure 24–80

44.

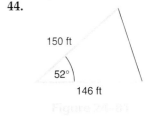

150 ft

52°

146 ft

Figure 24–81

Assignment Exercises

915

45.

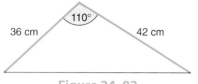

Figure 24–82

46.

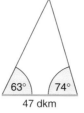

Figure 24–83

47.

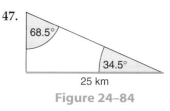

Figure 24–84

48.

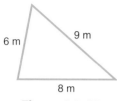

Figure 24–85

49.

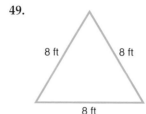

Figure 24–86

50.

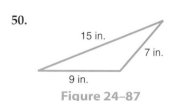

Figure 24–87

51. Find the area of a triangular tabletop that has sides measuring 46 in., 37 in., and 40 in.

53. Find the area of the irregularly shaped property in Fig. 24–88.

52. What is the area of a triangular plot of ground with sides that measure 146 ft, 85 ft, and 195 ft?

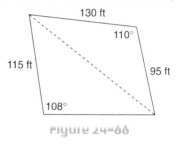

Figure 24–88

54. Cy Pipkin needs to finish his surveying report by finding the area of the triangular plot of land he has surveyed. Read his survey report and finish it. Be sure to show all calculations so he can check them.

Survey report

From a corner of the property, one property line measures 196 ft and the other property line measures 216 ft. The angle formed by these two property lines measures 42°.

1. Speed is a scalar measure, whereas velocity is a vector measure. What are the differences and similarities in these two measures?

3. Locate a point in the second quadrant by a set of coordinates. Find the magnitude and direction of the vector determined by the point, and find the related angle for the vector.

5. Give the sign of the tangent function in each of the four quadrants. Use your calculator to find the value of an angle in each of the four quadrants to validate your answer.

7. Draw a triangle and assign values to three parts of the triangle, then use the law of sines and other methods to find the remaining angles and sides.

9. Write the law of sines in words.

2. Explain in your own words what a related angle is, and explain how to find the related angle for an angle located in the third quadrant.

4. Give the sign of the sine function in each of the four quadrants. Use your calculator to find the value of an angle in each of the four quadrants to validate your answer.

6. What parts of a triangle must be given to use the law of sines?

8. Write the law of cosines in words.

10. How would you know whether to select the law of sines or the law of cosines to find missing angles or sides of a triangle?

Learning Outcomes

What to Remember with Examples

Section 24–1

1 Find the magnitude of a vector in standard position, given the coordinates of the end point (pp. 888–889).

Use the Pythagorean theorem to find the magnitude of a vector in standard position: $r = \sqrt{x^2 + y^2}$, where r = magnitude, x = x-coordinate of the end point, and y = y-coordinate of the end point.

Find the magnitude of a vector in standard position with the end point at $(4, 2)$.

$$r = \sqrt{x^2 + y^2}$$ Substitute values of x and y.
$$r = \sqrt{4^2 + 2^2}$$ Evaluate.
$$r = \sqrt{16 + 4}$$
$$r = \sqrt{20}$$
Magnitude $= 4.472$ Rounded.

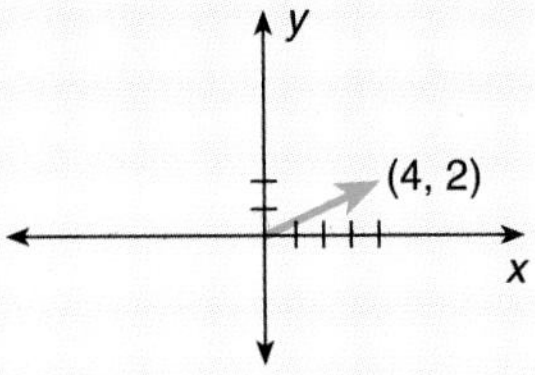

Figure 24–89

2 Find the direction of a vector in standard position, given the coordinates of the end point (p. 889).

Use the tangent function to find the direction of a vector in standard position: $\tan \theta = \frac{y}{x}$, where θ is the direction of the vector, x is the x-coordinate of the end point and y is the y-coordinate of the end point.

Find the direction of a vector in standard position with the end point at $(2, 4)$.

$$\tan \theta = \frac{y}{x}$$ Substitute values of x and y.

$$\tan \theta = \frac{4}{2}$$ Simplify.

$$\tan \theta = 2$$ Use inverse tangent function.
$$\tan^{-1} 2 = 63.43494882°$$
$$\theta = 63.4°$$ Rounded.
Direction $= 63.4°$

3 Find the sum of vectors (pp. 890–892).

1. Represent the vectors in the form of complex numbers, where the x-coordinate of the end point of a vector in standard position is the real component and the y-coordinate of the end point of a vector is the imaginary component.
2. Add the like components.

Add vectors A and B. See Fig. 24–90.

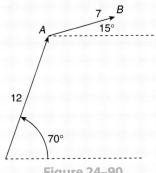

Figure 24–90

Vector A:

x-coordinate: $\cos \theta = \dfrac{x}{r}$ | r = magnitude. Substitute θ = 70° and r = 12.

$$\cos 70° = \dfrac{x}{12}$$ Solve for x.

$$12(\cos 70°) = x$$

$$4.10424172 = x$$

$$x = 4.104$$ Rounded.

y-coordinate: $\sin \theta = \dfrac{y}{r}$ | r = magnitude.

$$\sin 70° = \dfrac{y}{12}$$ Solve for y.

$$12(\sin 70°) = y$$

$$11.27631145 = y$$

$$y = 11.28$$ Rounded.

Vector $A = 4.104 + 11.28i$

Vector B:

x-coordinate: $\cos \theta = \dfrac{x}{r}$ | Substitute θ = 15° and r = 7.

$$\cos 15° = \dfrac{x}{7}$$ Solve for x.

$$7(\cos 15°) = x$$

$$6.761480784 = x$$

$$x = 6.761$$ Rounded.

y-coordinate: $\sin \theta = \dfrac{y}{r}$ | Substitute θ = 15° and r = 7.

$$\sin 15° = \dfrac{y}{7}$$ Solve for y.

$$7(\sin 15°) = y$$

$$1.811733316 = y$$

$$y = 1.812$$ Rounded.

Vector $B = 6.761 + 1.812i$

Vector $A + B = (4.104 + 11.28i) + (6.761 + 1.812i)$ Add like vector components.

$$= 10.87 + 13.09i$$

Section 24–2

1 Find related acute angles for angles or vectors in quadrants II, III, and IV (pp. 893–894).

To find a related acute angle for an angle (θ) in any quadrant, draw the angle and form the third side of a triangle by drawing a line perpendicular to the x-axis from the line forming the angle. The related angle is the angle formed by the ray forming the angle and the x-axis portion of the triangle.

Quadrant II angle (Fig. 24–91):

$$\theta = 180° - \theta_2 \quad \text{or} \quad \theta = \pi - \theta_2$$

Quadrant III angle (Fig. 24–92):

$$\theta = \theta_3 - 180° \quad \text{or} \quad \theta = \theta_3 - \pi$$

Figure 24–91

Figure 24–92

Chapter 24 / Oblique Triangles

Quadrant IV angle (Fig. 24–93):

$$\theta = 360° - \theta_4 \quad \text{or} \quad \theta = 2\pi - \theta_4$$

Figure 24–93

Find the related angle for an angle of 156°.

The angle is in quadrant II, thus, 180° − 156° = 24°.

2 Determine the signs of trigonometric values of angles of more than 90° (pp. 894–898).

The memory device ALL-SIN-TAN-COS can be used to remember the signs of the trigonometric functions in the various quadrants. The signs of *all* trigonometric functions are positive in the first quadrant. The sign of the SIN function is positive in the second quadrant (COS and TAN are negative). The sign of the TAN function is positive in the third quadrant. The sign of the COS function is positive in the fourth quadrant.

What is the sign of cos 145°?

145° is a quadrant II angle and the cosine function is negative in quadrant II. Thus, the sign of cos 145° is negative.

3 Find the trigonometric values of angles of more than 90° using a calculator (pp. 898–899).

Most calculators give the trigonometric value of an angle regardless of the quadrant in which it is found.

Use a calculator to find the value of the following trigonometric functions. Round to four significant digits.

$$\sin 125° = 0.8192$$
$$\cos 215° = -0.8192$$
$$\tan 335° = -0.4663$$

Section 24–3

1 Find the missing parts of an oblique triangle, given two angles and a side (pp. 900–902).

Use the law of sines to find the missing part of an oblique triangle when two angles and a side opposite one of them are given.

$$\frac{a}{\sin A} = \frac{b}{\sin B} = \frac{c}{\sin C}$$

Find side AB in the triangle of Fig. 24–94.

$$\frac{a}{\sin A} = \frac{c}{\sin C}$$ Substitute $a = 13$, $A = 22°$, and $C = 97°$.

$$\frac{13}{\sin 22°} = \frac{c}{\sin 97°}$$ Solve for c.

$$c \sin 22° = 13 \sin 97°$$

$$c = \frac{13 \sin 97°}{\sin 22°}$$ Evaluate.

$$c = 34.44440167 \text{ cm}$$

$$AB = c = 34.44 \text{ cm}$$ Rounded.

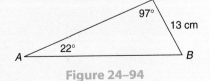

Figure 24–94

2 Find the missing parts of an oblique triangle, given two sides and an angle opposite one of them (pp. 902–905).

Use the law of sines to find the missing part of an oblique triangle when two sides and an angle opposite one of them are given.

Find the measure of angle A in the triangle of Fig. 24–95.

$$\frac{a}{\sin A} = \frac{b}{\sin B}$$ Substitute $a = 18$, $b = 16$, and $B = 32°$.

$$\frac{18}{\sin A} = \frac{16}{\sin 32°}$$ Solve for A.

$$16 \sin A = 18 \sin 32°$$

$$\sin A = \frac{18 \sin 32°}{16}$$ Evaluate.

$$\sin A = 0.5961591723$$ Use the inverse sine function.

$$\sin^{-1} 0.5961591723 = 36.59531105°$$

$$A = 36.6°$$ Rounded.

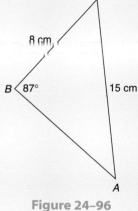

Figure 24–95

Given $\triangle ABC$ and b is a known side and a is a known side opposite a known angle, if $b < a$ and $\angle A \neq 90°$, we have two possible solutions. If $b < a$ and $\angle A = 90°$, we have one solution. If $b \geq a$, we have one solution.

Determine how many solutions exist for the triangle in Fig. 24–96 based on the given facts. Explain your response.

$B = 87°$

$b = 15$ cm (known side opposite known angle)

$a = 8$ cm (known side opposite unknown angle)

Because $b > a$, there is one possible solution.

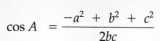

Figure 24–96

Section 24–4

1 Find the missing parts of an oblique triangle, given three sides of the triangle (pp. 907–909).

To find an angle when three sides are given, we need to use the law of cosines. One variation of the law of cosines is $a^2 = b^2 + c^2 - 2bc \cos A$. Rearrange the formula to solve for $\cos A$. The rearranged formula is

$$\cos A = \frac{-a^2 + b^2 + c^2}{2bc}$$

Find the measure of angle A in the triangle of Fig. 24–97.

$a = 20$ cm
$b = 9$ cm
$c = 15$ cm

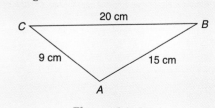

Figure 24–97

$$\cos A = \frac{-a^2 + b^2 + c^2}{2bc}$$ Substitute known values.

$$\cos A = \frac{-20^2 + 9^2 + 15^2}{2(9)(15)}$$ Evaluate.

$$\cos A = \frac{-400 + 81 + 225}{270}$$

$$\cos A = -0.3481481481$$ Use inverse cosine function.

$$\cos^{-1}(-0.3481481481) = 110.3740893°$$

$$A = 110.4°$$ Rounded.

2 Find the missing parts of an oblique triangle, given two sides and the included angle of the triangle (pp. 909–910).

Use the law of cosines to find the missing parts of an oblique triangle when given two sides of the triangle and the included angle. Solve a variation of the law of cosines for the side opposite the given angle.

$$a = \sqrt{b^2 + c^2 - 2bc \cos A}$$

Find the side opposite the given angle of the triangle in Fig. 24–98.

$$a = \sqrt{b^2 + c^2 - 2bc \cos A}$$ Substitute $A = 70°$, $b = 17$, and $c = 23$.

$$a = \sqrt{17^2 + 23^2 - 2(17)(23) \cos 70°}$$

$$a = \sqrt{289 + 529 - 267.4597521}$$

$$a = \sqrt{550.5402479}$$

$$a = 23.4635941$$

$$a = 23.46 \text{ cm}$$ Rounded.

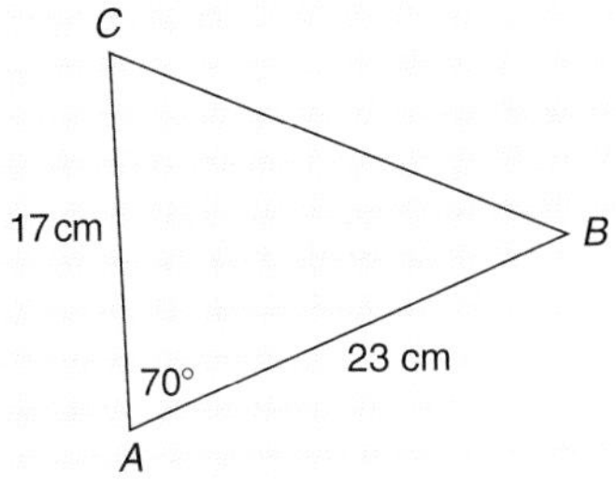

Figure 24–98

Find as many parts as possible without using information that you have calculated.

Find the measures of all the angles and sides of the triangle in Fig. 24–99.

$$\frac{a}{\sin A} = \frac{c}{\sin C}$$ Substitute $a = 6.8$, $A = 48°$, and $c = 7.5$.

$$\frac{6.8}{\sin 48°} = \frac{7.5}{\sin C}$$ Solve for C.

$$\sin C = \frac{7.5 \sin 48°}{6.8}$$

$$\sin C = 0.8196450281$$ Use inverse sine function.

$$\sin^{-1} 0.8196450281 = 55.04927548°$$

$$C = 55.0°$$ Rounded.

$$\angle B = 180° - \angle A - \angle C$$

$$\angle B = 180° - 48° - 55.04927548°$$

$$\angle B = 76.95072452°$$

$$\angle B = 77.0°$$ Rounded.

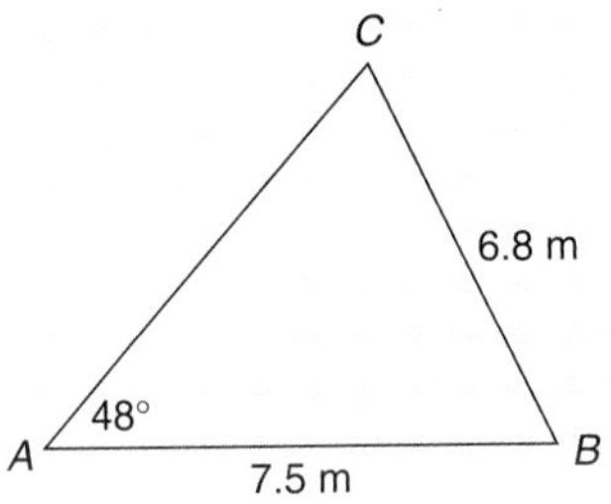

Figure 24–99

Use the law of sines or law of cosines to find b or AC.

$$\frac{a}{\sin A} = \frac{b}{\sin B}$$ Using the law of sines, solve for b.

$$b = \frac{a \sin B}{\sin A}$$ Substitute $a = 6.8$, $B = 76.95072452°$, and $A = 48°$.

$$b = \frac{6.8 \sin 76.95072452°}{\sin 48°}$$ Evaluate.

$$b = 8.914007364$$

$$b = 8.914 \text{ m}$$

Section 24-5

[1] Find the area of a triangle when the height is unknown and at least three parts of the triangle are known (pp. 911–913).

Area $= \frac{1}{2}ab \sin C$, where a and b are two adjacent sides of a triangle and angle C is the angle between sides a and b.

Find the area of triangle ABC in Fig. 24–100.

Area $= \dfrac{1}{2}ab \sin C$ Substitute $a = 4$, $b = 3$, and $C = 37°$.

Area $= 0.5(4)(3)(\sin 37°)$ Evaluate.

Area $= 3.610890139$

Area $= 3.611 \text{ in}^2$

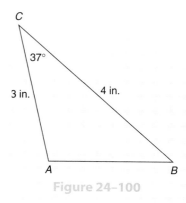

Figure 24-100

CHAPTER TRIAL TEST

Find the values. Round to four significant digits.

1. $\sin 125°$
2. $\tan 140°$
3. $\cos 160°$
4. Find the length of the vector that has end point coordinates of $(8, 15)$.
5. Find the angle of the vector whose end point coordinates are $(-8, 8)$.

Use the law of sines or the law of cosines to find the side or angle indicated in Figs. 24–101 to 24–108. Round sides to four significant digits and angles to the nearest $0.1°$.

6.

Figure 24-101

7.

Figure 24-102

8.
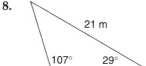
Figure 24-103

9.
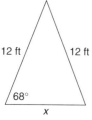
Figure 24-104

10.
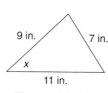
Figure 24-105

11.
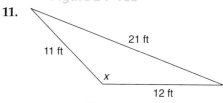
Figure 24-106

12.
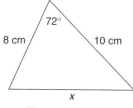
Figure 24-107

13. 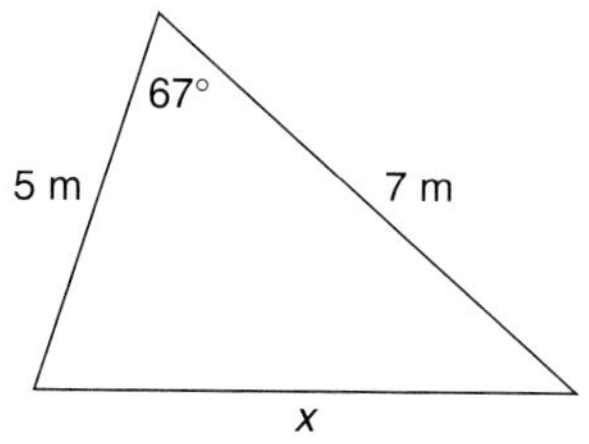
Figure 24-108

Chapter 24 / Oblique Triangles

14. Find the area of triangle *ABC* in Fig. 24–109. Round to four significant digits.

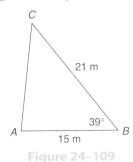

Figure 24–109

16. Find the direction in degrees (to the nearest 0.1°) of the vector I_t (total current) in Fig. 24–111.

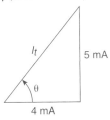

Figure 24–111

18. What is the area of a triangular cast if the sides measure 31 mm, 42 mm, and 27 mm?

20. A plane is flown from an airport due east for 32 mi, and then turns 15° north of east and travels 72 mi. How far is the plane from the airport?

22. A rod 11 cm long joins a crank 20 cm long to form a triangle with a third, imaginary line (Fig. 24–112). If the crank and the rod form an angle of 150°, find the angle formed by the rod and the imaginary line.

23. Find the magnitude of vector *Z* in Fig. 24–113 if vectors *R* and X_L form a right angle. All units are in ohms.

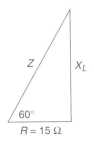

Figure 24–113

15. Find the area of triangle *RST* in Fig. 24–110. Round to four significant digits.

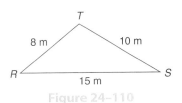

Figure 24–110

17. A surveyor measures two sides of a triangular lot and the angle formed by these two sides. What is the length of the third side if the two measured sides are 76 ft and 110 ft and the angle formed by these two sides is 107°?

19. Find the area of a triangular flower bed if two sides measure 12 ft and 15 ft and the included angle measures 48°.

21. Find the magnitude to the nearest tenth of the vector I_t in Exercise 16.

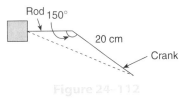

Figure 24–112

24. Find the area of the lot described in Problem 17. (Round to four significant digits.)

Selected Answers to Student Exercise Material

Answers to all Self-Study Exercises and answers to odd-numbered exercises in the Assignment Exercises and Chapter Trial Tests are included here. Solutions for the assignment exercises and Chapter Trial Tests are available separately in the Student Solutions Manual (odd-numbered) or the Instructor's Resource Manual (even-numbered).

Self-Study Exercises 1–1

1   **1.** hundred millions   **2.** ten thousands   **3.** hundreds   **4.** ten millions   **5.** billions
6. hundred thousands   **7.** tens   **8.** ten millions   **9.** billions   **10.** ten thousands

2   **11.** six thousand, seven hundred, four   **12.** eighty-nine thousand, twenty-one   **13.** six hundred
sixty-two million, nine hundred thousand, seven hundred fourteen   **14.** three million, one hundred one
15. fifteen billion, four hundred seven million, two hundred ninety-four thousand, three hundred seventy-six.
16. one hundred fifty   **17.** 7,000,000,400   **18.** 1,627,106   **19.** 58,201   **20.** $1,006   **21.** $(7 \times 100) + (1 \times 10) + (8 \times 1)$
22. $(4 \times 10) + (2 \times 1)$ **23.** $(1 \times 1,000) + (9 \times 100) + (8 \times 10) + (3 \times 1)$ **24.** $(8 \times 1,000) + (0 \times 100) + (2 \times 10) +$
(1×1) **25.** $(5 \times 10,000) + (2 \times 1,000) + (0 \times 100) + (1 \times 10) + (0 \times 1)$ **26.** $(7 \times 100) + (0 \times 10) + (0 \times 1)$
27. 537   **28.** 9,008   **29.** 65   **30.** 40,217

3   **31.** $6 < 8; 8 > 6$ **32.** $32 < 42; 42 > 32$ **33.** $148 < 196; 196 > 148$ **34.** $2,517 < 2,802; 2,802 > 2,517$
35. $7,809 < 8,902; 8,902 > 7,809$ **36.** $42,999 < 44,000; 44,000 > 42,999$

4   **37.** thousandths   **38.** ones   **39.** hundred-thousandths   **40.** millionths   **41.** ten-thousandths
42. ten-thousandths   **43.** 4   **44.** 0   **45.** 6   **46.** 3   **47.** 7   **48.** 3

5   **49.** twenty-one and three hundred eighty-seven thousandths   **50.** four hundred twenty and fifty-nine
thousandths   **51.** eighty-nine hundredths   **52.** five hundred sixty-eight ten-thousandths   **53.** thirty and
two thousand three hundred seventy-nine hundred-thousandths   **54.** twenty-one and two hundred five
thousand eighty-five millionths   **55.** 3.42   **56.** 78.195   **57.** 500.0005   **58.** 0.75034

6   **59.** 0.5   **60.** 0.23   **61.** 0.07   **62.** 6.83   **63.** 0.079   **64.** 0.468   **65.** 5.87   **66.** 0.108
67. 6.03   **68.** 4.00 or 4

7   **69.** 3.72   **70.** 7.08   **71.** 0.3   **72.** 0.56   **73.** 2.75   **74.** 0.2   **75.** 8.88   **76.** 0.25   **77.** 0.913
78. 0.76   **79.** 5.983   **80.** 1.972   **81.** 0.179, 0.23, 0.314   **82.** 1.87, 1.9, 1.92   **83.** 72.07, 72.1, 73   **84.** 0.837 in.
85. the micrometer reading   **86.** yes   **87.** 0.04 in.   **88.** No. 10 wire   **89.** 16.2 kilograms   **90.** $4.2 > 3.8; 3.8 < 4.2$
91. $1.68 > 1.6; 1.6 < 1.68$ **92.** $7.09 < 7.21; 7.21 > 7.09$ **93.** $52.97 < 98.4; 98.4 > 52.97$
94. $2.742 < 27.420; 27.420 > 2.742$ **95.** $15,000 > 13,481.2; 13,481.2 < 15,000$ **96.** $119 > 62; 62 < 119$; The distance
from Cleveland to Toledo is greater than the distance from Detroit to Toledo.   **97.** $96°F > 82°F; 82°F < 96°F$; The

average temperature in July at Laredo is higher than the average temperature at El Paso.   **98.** 26°F < 70°F; 70°F > 26°F   **99.** 57°F < 71°F; 71°F > 57°F   **100.** San Francisco 1906, 8.3   **101.** 1.14 in 1980 > 1.02 in 1990; falling growth rate   **102.** 1.7 in Asia < 2.0 in dev. world; Asia is growing slower than dev. world   **103.** 0.139 > 0.111; Rolling a sum of 6 has the greater probability.   **104.** 0.026 kW > 0.003 kW; 800 watt toaster   **105.** 0.394 < 0.621; 0.621

8   **106.** 500   **107.** 6,200   **108.** 8,000   **109.** 430,000   **110.** 40,000   **111.** 40,000   **112.** 285,000,000   **113.** 470   **114.** 83,000,000,000   **115.** 300,000,000   **116.** Corpus Christi 40 ft, Jacksonville 20 ft; Corpus Christi is higher; tens   **117.** Denver 5,000, Salt Lake City 4,000; thousands   **118.** Indianapolis 700, Salt Lake City 4,200; hundreds   **119.** Answers may vary. Compare the number of digits for the altitude in the pair of cities. Round to the same place when comparing each pair.   **120.** 43   **121.** 367   **122.** 8   **123.** 103   **124.** 3   **125.** 8.1   **126.** 12.9   **127.** 42.6   **128.** 83.2   **129.** 6.0   **130.** 7.04   **131.** 42.07   **132.** 0.79   **133.** 3.20   **134.** 7.77   **135.** 0.217   **136.** 0.020   **137.** 1.509   **138.** 4.238   **139.** 7.004   **140.** $219   **141.** $83   **142.** $507   **143.** $3   **144.** $6   **145.** $8.24   **146.** $0.29   **147.** $0.53   **148.** $5.80   **149.** $238.92   **150.** 0.784 in.   **151.** 2.8 A   **152.** 3.82 in.   **153.** 87

9   **154.** 500   **155.** 8   **156.** 60   **157.** 0.5   **158.** 0.009   **159.** 0.1   **160.** 3   **161.** 2   **162.** 80   **163.** 50   **164.** $3   **165.** 20   **166.** March and December   **167.** Drivers A, B, and D

Self-Study Exercises 1–2

1   **1.** 15   **2.** 27   **3.** 22   **4.** 36   **5.** 37 bolts   **6.** 9,192   **7.** 16,956   **8.** 106,285   **9.** 616, 060   **10.** 2,191,062   **11.** 2,310   **12.** 97,614   **13.** 7,007   **14.** 30,133   **15.** $456   **16.** 24,349 lb   **17.** 298 cartridges   **18.** 4,553 bricks   **19.** 50 gal   **20.** 14,325 students

2   **21.** 15.7   **22.** 34.18   **23.** 8.13   **24.** 87.4   **25.** 129.97   **26.** 26.03   **27.** 78.2   **28.** 45.7   **29.** 261.335   **30.** 356.612   **31.** 6.984   **32.** 9.525   **33.** 1,126.6   **34.** 413.6   **35.** 18.1   **36.** 18.8   **37.** 0.81805   **38.** 1.17642   **39.** 55.512   **40.** 85.411   **41.** 93,012   **42.** 69.987   **43.** 55.3   **44.** 13.58   **45.** 55.17   **46.** 3.077 in   **47.** 15.503 A   **48.** $181.25   **49.** 391.4 ft   **50.** $19.93   **51.** $5.04   **52.** 100.9'   **53.** $28,334.01   **54.** $224.2800   **55.** 5.8 million   **56.** $2,189.45   **57.** 36,901.9   **58.** $16.26   **59.** $304.97   **60.** 31,000; 30,986.41   **61.** 4,700; 4,740.80   **62.** 13,100; 13,157.42   **63.** 70; 72.3   **64.** $112.17

3   **65.** 17,100; 17,018.21   **66.** 0; 56.365   **67.** 84,200; 84,213   **68.** 200; 183.405   **69.** 402,300; 402,199   **70.** 0; 81.401   **71.** $3,440; $3,443.60   **72.** $3,400; $3,365   **73.** 8,900; 9,239   **74.** 600; 624.18   **75.** 9,000; 8,449   **76.** 800; 801.39   **77.** 60,000; 52,801   **78.** 50,000; 49,241   **79.** 159 lb   **80.** Yes, total capacity available is 115 gal.   **81.** Yes, 479 pages are needed.   **82.** 350 ft   **83.** 190,000; 190,786   **84.** 1,900,000; 1,859,867   **85.** 50,000; 54,359.65   **86.** 51,000,000; 47,520,014   **87.** $17,039.04   **88.** $258,540.18

Self-Study Exercises 1–3

1   **1.** 5   **2.** 5   **3.** 1   **4.** 7   **5.** 2   **6.** 3   **7.** 7   **8.** 5   **9.** 2   **10.** 4   **11.** 1   **12.** 1   **13.** 24   **14.** 401   **15.** 1,020   **16.** 115   **17.** 53,036   **18.** 4,078   **19.** 21,879   **20.** 13,567   **21.** 1,278   **22.** 22 bags   **23.** 341 boxes   **24.** 321 ft   **25.** $237

2   **26.** 6.93   **27.** 15.834   **28.** 803.693   **29.** 56.14   **30.** 4.094   **31.** 3.9   **32.** 291.82   **33.** 7.5   **34.** 310.8   **35.** 4.4   **36.** 598.473   **37.** 12.7 claims per 1,000   **38.** 5°F   **39.** 15.1 lb   **40.** 12.08 in., 12.10 in.   **41.** 59.83 cm   **42.** 4.189 in., 4.201 in.   **43.** 0.22 dm   **44.** 11.55 A   **45.** $8.75   **46.** $2.25   **47.** $4.75   **48.** $4,647.75

3   **49.** 45   **50.** 608   **51.** 1,573   **52.** 22,205   **53.** 100,000; 91,034   **54.** 2,000; 2,088.884   **55.** 0; 174   **56.** $200; $599.15   **57.** 4,000; 4,397   **58.** 1,000; 1,187.251   **59.** 500,000; 508,275   **60.** 17 L   **61.** 186 bricks   **62.** 213.8 in.   **63.** 125.5 in.   **64.** 45 in.

Self-Study Exercises 1–4

1 **1.** 15 **2.** 0 **3.** 378 **4.** 84 **5.** 0 **6.** 224 **7.** 581 **8.** 630 **9.** 102,612 **10.** commutative property of multiplication **11.** $42 **12.** Any two numbers may be grouped for multiplication.

$$3[(5)(2)] = [3(5)]2$$
$$3(10) \;\;= (15)2$$
$$30 \;\;\;\;\;= \;\;30 \qquad \text{Answers may vary.}$$

13. 864 pieces **14.** 4,096 washers **15.** 84 books **16.** 700 tickets **17.** $288 **18.** 249 students **19.** $560 **20.** $23,780

2 **21.** 15.486 **22.** 56.55 **23.** 3.2445 **24.** 0.05805 **25.** 0.08672 **26.** 7.141 **27.** 0.0834 **28.** 170.12283 **29.** 0.38381871 **30.** 4.9386274 **31.** 30.66 **32.** 596.97 **33.** 50.7357 **34.** 38.6232 **35.** 339.04 **36.** 540.27 **37.** 254 **38.** 184.2 **39.** 0.9307 **40.** 0.7602 **41.** 0.58635 **42.** 0.73265 **43.** 9.21702 **44.** 11.42356 **45.** 0.0176 **46.** 0.027045 **47.** 0.915371 **48.** 0.390483 **49.** 0.00015 **50.** 0.00056 **51.** 3.957 in. **52.** 5.25 in. **53.** $27.48 **54.** 151.2 in. **55.** $64.20 **56.** $10,728 **57.** 0.375 in. **58.** $1,960.50 **59.** 15 A **60.** $3,276

3 **61.** 55 **62.** 60 **63.** 196 **64.** 12 **65.** 15.6 **66.** 5.6 **67.** $P = 40$ ft **68.** $P = 61$ in. **69.** 56 ft **70.** 170 ft

4 **71.** $600; $768 **72.** $3,000; $4,080 **73.** 123.54 ft^2 **74.** 1,164 cm^2 **75.** 1,200 ft^2; 1,374.75 ft^2 **76.** 96 ft^2 **77.** 3,578,040.664 **78.** 23,379,045 **79.** 561,500,160 **80.** 0 **81.** $36,180 **82.** $22,006.40 **83.** 147,000,000,000 **84.** 16,755,200,000 **85.** 420,000,000

Self-Study Exercises 1–5

1 **1.** $8 \div 4; 4\overline{|8}; \dfrac{8}{4}$ **2.** $9 \div 3; 3\overline{|9}; \dfrac{9}{3}$ **3.** $24 \div 6; 6\overline{|24}; \dfrac{24}{6}$ **4.** $30 \div 7; 7\overline{|30}; \dfrac{30}{7}$ **5.** $6 \div 2; 2\overline{|6}; \dfrac{6}{2}$

6. $8 \div 4; 4\overline{|8}; \dfrac{8}{4}$

2 **7.** 75 **8.** 23 **9.** 20 **10.** 3 **11.** 16 **12.** 12 **13.** 43 **14.** 47 **15.** 12 **16.** 23R1 **17.** 124R30 **18.** 56R80 **19.** 32 ft **20.** $1,245 **21.** 6 in.

3 **22.** 1.26 **23.** 3.09 **24.** 0.063 **25.** 285 **26.** 5.9 **27.** 45 **28.** 10.9 **29.** 0.19 **30.** 1.06 **31.** 0.33 **32.** 25 **33.** 20,700 **34.** 90,200 **35.** 10,700 **36.** 1.8 **37.** 6 lb **38.** 23 rolls **39.** 600 revolutions **40.** 0.7 ft **41.** 93.75 volts **42.** A measure divided by a number equals a measure. **43.** 19.85 **44.** 1.82 **45.** 2,601.17 **46.** 1.99 **47.** 0.08 **48.** 169.3 **49.** 0.16 **50.** 9 **51.** $0.96 **52.** $1 **53.** $575 **54.** $1.98 **55.** 0.0812 in. **56.** 0.2 in. **57.** 54 **58.** 301 **59.** 40 **60.** 89,500 **61.** 40,200 **62.** 6,201 **63.** 1,070 bricks **64.** $10,500 **65.** $104.75 **66.** $1,950 **67.** $2,465 **68.** $4.45 **69.** 24 parts **70.** 10 lengths **71.** $78,505 **72.** $59.71 **73.** 0.8 in.

4 **74.** 100; 125.6 **75.** 4,000; 3,151.74 **76.** 4,000; 3,731.24 **77.** 8; 8.43 **78.** 128 loads **79.** 127 cords

5 **80.** 80.4° **81.** 85 **82.** 454 lb **83.** $765.32 **84.** 1.69 in. **85.** 3.5 A

Self-Study Exercises 1–6

1 **1.** 4; 3 **2.** 9; 4 **3.** 2.7; 9 **4.** 15; 2 **5.** 3.375 **6.** 49 **7.** 1,000 **8.** 16 **9.** 11.56 **10.** 15 **11.** 8 **12.** 1 **13.** 8^1 **14.** 14.5^1 **15.** 12^1 **16.** 23^1 **17.** leaves base unchanged **18.** Use the base as a factor 3 times.

2 **19.** 64 **20.** 4 **21.** 12.25 **22.** 324 **23.** 1.96 **24.** 169 **25.** 1 **26.** 625 **27.** 10,000 **28.** 14,641 **29.** 12,996 **30.** 14.44 **31.** 64 **32.** 81 **33.** 289 **34.** 144 **35.** 324 **36.** 10,201 **37.** 484 **38.** 225 **39.** 5 **40.** 7 **41.** 9 **42.** 6 **43.** 14 **44.** 15 **45.** 11 **46.** 12 **47.** Use the number as a factor two times. **48.** Find a number that is used as a factor two times to give the desired number.

3   **49.** 1,000   **50.** 32,000   **51.** 30,000   **52.** 1,200,000   **53.** 20,000   **54.** 10,200   **55.** 22,000   **56.** 13,600   **57.** 25   **58.** 21   **59.** 3   **60.** 42   **61.** 9   **62.** 250   **63.** 1.2   **64.** 23   **65.** Shift the decimal to the right in the number being multiplied by the power of 10 as indicated by the exponent.   **66.** Shift the decimal to the left in the number being multiplied by the power of 10 as indicated by the exponent.   **67.** 225   **68.** 343   **69.** 78,125   **70.** 20,736   **71.** 18   **72.** 28   **73.** 33   **74.** 14

Self-Study Exercises 1–7

1   **1.** 26   **2.** 18   **3.** 9   **4.** 12.5   **5.** 24   **6.** 32   **7.** 18   **8.** 10   **9.** 24   **10.** 9   **11.** 10   **12.** 9   **13.** 30   **14.** 15   **15.** 156   **16.** 109   **17.** 23   **18.** 145   **19.** 15.6   **20.** 25.04   **21.** 9   **22.** 5   **23.** 7   **24.** 15   **25.** 160   **26.** 1,458   **27.** 80   **28.** 483   **29.** 443   **30.** 70   **31.** 13   **32.** 11   **33.** 29   **34.** 5   **35.** 39   **36.** 7   **37.** 14   **38.** 20   **39.** 30   **40.** 20.52   **41.** undefined   **42.** 191.75   **43.** parentheses   **44.** addition or subtraction   **45.** 67   **46.** 86

2   **47.** 29 boxes   **48.** $26,178.44   **49.** 5,660 people   **50.** 1,080 boxes of cards

Assignment Exercises, Chapter 1

1. (a) 0.3   (b) 0.15   (c) 0.04   **3.** thousandths   **5.** 7   **7.** (a) tens   (b) ten-thousands   (c) millions   **9.** fifty-six million, one hundred nine thousand, one hundred ten   **11.** 1,265,401   **13.** six and eight hundred three thousandths   **15.** 0.625   **17.** (a) 40   (b) 70   (c) 24   (d) $43   (e) 80   (f) $8.94   (g) $1.00   (h) 0.09703   **19.** (a) 500   (b) 50,000   (c) 41.4   (d) 6.90   (e) 23.4610   **21.** 4.79   **23.** 0.02; 0.021; 0.0216   **25.** $\frac{7}{8}$ **27.** (a) 23   (b) 18   (c) 28   (d) 25   **29.** 34.9 kW   **31.** (a) 29,000; 29,092.09   (b) 26,000; 26,018   **33.** (a) 1.61   (b) 3.127   (c) 2   (d) 3   (e) 0   (f) 204.899   (g) 144   (h) 12,140   **35.** 8.291 in.; 8.301 in.   **37.** 200 miles, 190 miles   **39.** 0.43 in.   **41.** 8.930 in.; 8.940 in.   **43.** 84   **45.** 1,143   **47.** 13,725   **49.** 3,349,890   **51.** 394,254,080   **53.** $1,407   **55.** $43,920   **57.** 20   **59.** 60   **61.** 80   **63.** $328   **65.** $14,800, $13,140   **67.** 1,140,000 ft^2, 1,204,010.28 ft^2 **69.** 0.12096 in.   **71.** $5 \div 3, 3\overline{)5}, \frac{5}{3}$ **73.** 1   **75.** undefined   **77.** 5.26875   **79.** 13   **81.** 2,008.4   **83.** 23 envelopes, 11 leftover   **85.** 58.375   **87.** 5; 5.52   **89.** $10; $10.63   **91.** 48.79 ft   **93.** (a) 7, 3, 343   (b) 2.3, 4, 27.9841   (c) 8, 4, 4,096   **95.** (a) 0.9   (b) 35   (c) 1   **97.** (a) 1   (b) 1   (c) 1   (d) 1   **99.** (a) 1   (b) 15,625   (c) 31.36   (d) 441   **101.** (a) 10^1 (b) 10^3 (c) 10^4 (d) 10^5 **103.** (a) 7   (b) 0.04056   (c) 0.605   (d) 2.3079   (e) 44.582   **105.** 75   **107.** 3   **109.** 18   **111.** 54   **113.** 16   **115.** 15   **117.** 15   **119.** 13   **121.** 13.6   **123.** 113.608   **125.** 49 boxes   **127.** 16,309 people   **129.** $12,880   **131.** $52.60   **133.** $10.74   **135.** 45 yd × 48 yd   **137.** (a) 0.0009   (b) 0.0049   (c) 0.000025   (d) 0.000081   (e) 0.000008   (f) 0.000000064

Trial Test, Chapter 1

1. five million, thirty thousand, one hundred two   **3.** 7.027   **5.** 2,700   **7.** 5.09   **9.** 48.3   **11.** 1,007   **13.** $9,271,314   **15.** 134   **17.** 106   **19.** 0.0086   **21.** $310, $310   **23.** $10, $14   **25.** Commutative means we can add numbers in any order. 2 + 4 = 6; 4 + 2 = 6. Associative means we can group numbers differently. (2 + 1) + 3 = 6; 2 + (1 + 3) = 6.   **27.** 42,730   **29.** 11.6   **31.** 83   **33.** 0.6   **35.** 70   **37.** $17,500

Self-Study Exercises 2–1

1   **1.** right   **2.** left   **3.** infinity   **4.** 1   **5.** neither   **6.** −3, −2, −1, 0, 1, 2, 3   **7.** to the right   **8.** to the left

9.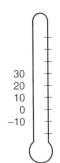

10.
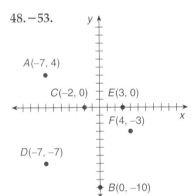

    a. New York City
    b. Greenwich
    c. Moscow

---

2   **11.** $<$ **12.** $>$ **13.** $<$ **14.** $<$ **15.** $>$ **16.** $<$ **17.** $>$ **18.** $<$ **19.** $x < y$ **20.** $a > b$ **21.** $3 + 5 > 6$
22. $9 < 18 - 6$ **23.** $k > t$ **24.** $r < s$

3   **25.** 23   **26.** 0   **27.** 10   **28.** -17 **29.** 13   **30.** 345   **31.** 67   **32.** 61

4   **33.** (number line: -7, 0, 7)   **34.** (number line: -8, 0, 8)   **35.** (number line: -4, 0, 4)

36. (number line: -12, 0, 12)   **37.** 0   **38.** a negative integer; a positive integer

5   **39.** R: horizontal, 4; vertical, 2   **40.** S: horizontal, -4; vertical, 3   **41.** T: horizontal, 5; vertical, -3
42. U: horizontal, -2; vertical, -5 **43.** V: horizontal, 7; vertical, 0   **44.** W: horizontal, 0; vertical, 5
45. X: horizontal, 0; vertical, 0   **46.** Y: horizontal, -3; vertical, 0
47. $A = (4, 2)$ $B = (-3, 2)$ $C = (-2, -1)$ $D = (3, -2)$ **48.–53.**
54. y-value $= 0$ **55.** x-value $= 0$ **56.** $(-x, +y)$ **57.** $(+x, -y)$

(graph plotting points $A(-7, 4)$, $C(-2, 0)$, $E(3, 0)$, $F(4, -3)$, $D(-7, -7)$, $B(0, -10)$)

Self-Study Exercises 2–2

1   **1.** 23   **2.** -16 **3.** -13 **4.** -29 **5.** -55 **6.** -124 **7.** 29   **8.** -20 **9.** 99   **10.** -59 **11.** -48
12. -161 **13.** -53 **14.** 232   **15.** $-1,005$

2   **16.** 2   **17.** 2   **18.** 4   **19.** -4 **20.** 6   **21.** -6 **22.** -2 **23.** 4   **24.** -2 **25.** 2   **26.** -10
27. -40 **28.** 7   **29.** 15   **30.** -8 **31.** -33 **32.** 9   **33.** -11 **34.** 13   **35.** 9   **36.** 3   **37.** 0   **38.** 0
39. 0   **40.** 0   **41.** 0   **42.** 9   **43.** -3 **44.** -7 **45.** 18   **46.** -8 **47.** -28 **48.** 0   **49.** $5 + (-5) = 0$
Answers may vary.   **50.** 0   **51.** Adding numbers of like signs increases the absolute value whether positive or
negative.   **52.** \$36   **53.** \$66   **54.** $+2:00$ **55.** $-8:00$ **56.** $-23°$

Self-Study Exercises 2–3

1   **1.** -12 **2.** 6   **3.** -6 **4.** 4   **5.** -25 **6.** -3 **7.** 8   **8.** -3 **9.** -9 **10.** 13
11. -1 **12.** -8

2   **13.** 15   **14.** -8 **15.** -12 **16.** 8   **17.** 7   **18.** 10   **19.** 56   **20.** -92 **21.** 14
22. -36 **23.** 0   **24.** 0

3 **25.** 4 **26.** 7 **27.** 6 **28.** −4 **29.** 1 **30.** 14 **31.** −3 **32.** −9 **33.** −5 **34.** 13 **35.** 2 **36.** −1 **37.** 107°F **38.** 107°F **39.** $115,054 **40.** Subtracting zero from a number results in the same number with the same sign. Subtracting a number from zero results in the opposite of the number. **41.** +3:00 **42.** −1:00

Self-Study Exercises 2–4

1 **1.** 40 **2.** 12 **3.** 35 **4.** 21 **5.** 24 **6.** 6 **7.** −15 **8.** −10 **9.** −32 **10.** −12 **11.** −56 **12.** −24 **13.** 4(−28) = −$112 **14.** (7)(−40) = −$280 **15.** When multiplying signed numbers, the absolute values are always multiplied. When numbers with like signs are multiplied, the result is positive. When numbers with unlike signs are multiplied, the result is negative.

2 **16.** −60 **17.** 36 **18.** 0 **19.** 90 **20.** 0 **21.** −42 **22.** 54 **23.** −210 **24.** 2,268 **25.** −20,160

3 **26.** 0 **27.** 0 **28.** 0 **29.** 0 **30.** 0 **31.** 0 **32.** 0 **33.** 0 **34.** 0 **35.** 0 **36.** −30 **37.** 168 **38.** 42 **39.** −8 **40.** 0 **41.** −16 **42.** The multiplicative inverse of a number is the number that, when multiplied by the original number, results in 1, the multiplicative identity. **43.** Answers will vary, 5(−3) = 3(−5) = −15.

4 **44.** 9 **45.** −8 **46.** 25 **47.** 0 **48.** −8 **49.** −512 **50.** 625 **51.** 81 **52.** 2,401 **53.** 1,764 **54.** 5 **55.** −28 **56.** 12 **57.** −8 **58.** −10 **59.** −8 **60.** −27 **61.** −121 **62.** 121 **63.** −125 **64.** 10^7 or 10,000,000 **65.** 20,000,000 **66.** No, we don't normally use 000 as the first 3 digits. **67.** 17,576,000 **68.** 10,000 **69.** 45°

Self-Study Exercises 2–5

1 **1.** 2 **2.** −3 **3.** −4 **4.** −4 **5.** −4 **6.** 8 **7.** 5 **8.** −6 **9.** 8 **10.** 5 **11.** −$1,800; $\dfrac{1,800}{6}$ = −$300 **12.** −6°C per hour

2 **13.** undefined **14.** undefined **15.** 0 **16.** 0 **17.** 0 **18.** undefined **19.** undefined **20.** undefined **21.** 0 **22.** undefined **23.** multiplication and division **24.** −10°F **25.** $3,472.50 **26.** The numerator must be zero and the denominator must be any number except zero.

Self-Study Exercises 2–6

1 **1.** −4 **2.** −30 **3.** −9 **4.** 17 **5.** 20 **6.** 24 **7.** −1 **8.** −13 **9.** 21 **10.** −13 **11.** 13 **12.** 15 **13.** 8 **14.** 24 **15.** −5 **16.** −7 **17.** −3 **18.** −28 **19.** −12 **20.** −1,260 **21.** 27 **22.** −14 **23.** 28 **24.** −35 **25.** −29 **26.** −132 **27.** −E−; indicating that division by zero is not possible. Calculator displays will vary. **28.** Numbers can be added in any order. **29.** When the order of subtraction is changed, the result is not the same. **30.** Numbers can be multiplied in any order. **31.** The order of division of two numbers cannot be changed. **32.** Zero divided byany nonzero number results in 0. Any number divided by 0 gives an error on the calculator because division by 0 is undefined.

Assignment Exercises, Chapter 2

1. G **3.** F **5.** E **7.** A **9.** L **11.** M **13.** 5, 21, 987 (Answers may vary.) **15.** 0 **17.** 1 **19.** > **21.** > **23.** < **25.** 5 **27.** 3 **29.** 7 **31.** 11 **33.** 12 **35.** −15 **37.** 2 **39.** 42 **41.** −87 **43.−48.** See figure to the right. **49.** (0, 0) **51.** A(3, 0); B(2, 2); C(2, −5); D(−4, −1); E(−3, 1) **53.** 17 **55.** −7 **57.** −13 **59.** 0 **61.** −8 **63.** −6 **65.** gain of 8 yd; +8 yd **67.** $569

43.−48.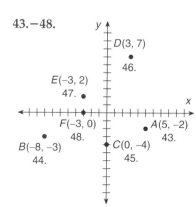

69. The additive identity, 0, if added to any number, will result in the same number. Numbers that are additive inverses, when added, result in 0.   **71.** -13 **73.** 14   **75.** -15 **77.** -5 **79.** 2   **81.** 36   **83.** 70°
85. -5.4°F   **87.** 36°C   **89.** -14 **91.** -12 **93.** 0   **95.** 0   **97.** -168 **99.** 343   **101.** -16 **103.** 25
105. -10°   **107.** higher, 20, $-4 \times 5 = -20$ points   **109.** 4   **111.** $+1$ **113.** 4   **115.** -4 **117.** 17
119. undefined   **121.** -3 **123.** -15°   **125.** 56   **127.** 1   **129.** 36   **131.** 73   **133.** -2 **135.** 391, 464
137. 9:00 A.M.: -1°C   10:00 A.M.:   0°C   11:00 A.M.:   0°C   12:00 P.M.:   1°C
1:00 P.M. :   1°C   2:00 P.M.:   4°C   3:00 P.M.:   0°C   4:00 P.M.: -7°C
5:00 P.M. : -15°C   6:00 P.M.: -27°C

Trial Test, Chapter 2

1. $<$ **3.** $>$ **5.** -8 **7.** 4   **9.** -48 **11.** 3   **13.** 8   **15.** 0   **17.** -7 **19.** undefined   **21.** -3 **23.** -2
25. 4   **27.** 33   **29.** 22   **31.** -35 **33.** If a number is multiplied by the multiplicative identity, or if a number is added to the additive identity, the result is the original number. $5 \times 1 = 5$ or $3 + 0 = 3$ Answers will vary.
35. 175°F

Self-Study Exercises 3–1

1   **1.** $\frac{5}{6}$ **2.** $\frac{3}{4}$ **3.** $\frac{3}{8}$ **4.** $\frac{7}{12}$ **5.** $\frac{1}{16}$ **6.** $\frac{3}{3}$ **7.** $\frac{4}{6}$ or $\frac{2}{3}$ **8.** (a) 6   (b) 5   (c) 5   (d) 6   (e) 5; 6
(f) 6   (g) 5   (h) proper   (i) 5; 6   (j) less than 1   **9.** (a) 8   (b) 8   (c) 8   (d) 8   (e) 8; 8   (f) 8
(g) 8   (h) improper   (i) 8; 8   (j) equal to 1   **10.** (a) 5   (b) 11   (c) 11   (d) 5   (e) 11; 5   (f) 5
(g) 11   (h) improper   (i) 11; 5   (j) greater than 1   **11.** (a) 3   (b) 12   (c) 12   (d) 3   (e) 12; 3   (f) 3
(g) 12   (h) improper   (i) 12; 3   (j) greater than 1   **12.** b   **13.** a   **14.** f   **15.** b   **16.** d   **17.** d
18. a   **19.** e   **20.** c   **21.** c   **22.** c

Self-Study Exercises 3–2

1   **1.** $5 = 5 \times 1, 10 = 5 \times 2, 15 = 5 \times 3, 20 = 5 \times 4, 25 = 5 \times 5, 30 = 5 \times 6$ **2.** $6 = 6 \times 1, 12 = 6 \times 2, 18 = 6 \times 3,$
$24 = 6 \times 4, 30 = 6 \times 5, 36 = 6 \times 6$ **3.** $8 = 8 \times 1, 16 = 8 \times 2, 24 = 8 \times 3, 32 = 8 \times 4, 40 = 8 \times 5, 48 = 8 \times 6$
4. $9 = 9 \times 1, 18 = 9 \times 2, 27 = 9 \times 3, 36 = 9 \times 4, 45 = 9 \times 5, 54 = 9 \times 6$ **5.** $10 = 10 \times 1, 20 = 10 \times 2, 30 = 10 \times 3,$
$40 = 10 \times 4, 50 = 10 \times 5, 60 = 10 \times 6$ **6.** $30 = 30 \times 1, 60 = 30 \times 2, 90 = 30 \times 3, 120 = 30 \times 4, 150 = 30 \times 5,$
$180 = 30 \times 6$ **7.** $6 \times 1 = 6, 6 \times 2 = 12, 6 \times 3 = 18, 6 \times 4 = 24, 6 \times 5 = 30$; Answers may vary.   **8.** $12 \times 1 = 12,$
$12 \times 2 = 24, 12 \times 3 = 36, 12 \times 4 = 48, 12 \times 5 = 60$; Answers may vary.   **9.** $13 \times 1 = 13, 13 \times 2 = 26, 13 \times 3 = 39,$
$13 \times 4 = 52, 13 \times 5 = 65, 13 \times 6 = 78$; Answers may vary.   **10.** $3 \times 1 = 3, 3 \times 2 = 6, 3 \times 3 = 9, 3 \times 4 = 12, 3 \times 5 = 15$;
Answers may vary.   **11.** $50 \times 1 = 50, 50 \times 2 = 100, 50 \times 3 = 150, 50 \times 4 = 200, 50 \times 5 = 250$; Answers may vary.
12. $4 \times 1 = 4, 4 \times 2 = 8, 4 \times 3 = 12, 4 \times 4 = 16, 4 \times 5 = 20$; Answers may vary.

2   **13.** no; not divisible by 3   **14.** yes; ends in zero   **15.** no, last two digits not divisible by 4   **16.** yes; sum of digits divisible by 3   **17.** yes; no remainder   **18.** yes; all three digits divisible by 8   **19.** yes; sum of digits divisible by 3   **20.** yes; divisible by both 2 and 3   **21.** yes; last digit even number

3   **22.** $1 \cdot 4, 2 \cdot 2$ **23.** $1 \cdot 8, 2 \cdot 4$ **24.** $1 \cdot 12, 2 \cdot 6, 3 \cdot 4$ **25.** $1 \cdot 15, 3 \cdot 5$ **26.** $1 \cdot 16, 2 \cdot 8, 4 \cdot 4$
27. $1 \cdot 20, 2 \cdot 10, 4 \cdot 5$ **28.** $1 \cdot 24, 2 \cdot 12, 3 \cdot 8, 4 \cdot 6$ **29.** $1 \cdot 30, 2 \cdot 15, 3 \cdot 10, 5 \cdot 6$ **30.** $1 \cdot 36, 2 \cdot 18, 3 \cdot 12, 4 \cdot 9, 6 \cdot 6$
31. $1 \cdot 38, 2 \cdot 19$ **32.** 1, 2, 4, 5, 8, 10, 20, 40   **33.** 1, 2, 23, 46   **34.** 1, 2, 4, 13, 26, 52   **35.** 1, 2, 4, 8, 16, 32, 64
36. 1, 2, 3, 4, 6, 8, 9, 12, 18, 24, 36, 72   **37.** 1, 3, 9, 27, 81   **38.** 1, 5, 17, 85   **39.** 1, 2, 4, 23, 46, 92   **40.** 1, 2, 3, 4, 6, 8, 12, 16, 24, 32, 48, 96   **41.** 1, 2, 7, 14, 49, 98

Self-Study Exercises 3–3

1   **1.** prime   **2.** composite   **3.** composite   **4.** prime   **5.** composite   **6.** composite   **7.** composite
8. composite   **9.** composite   **10.** prime   **11.** composite   **12.** prime

13. $2 \cdot 2 \cdot 3$ **14.** $2 \cdot 3 \cdot 3$ **15.** $2 \cdot 2 \cdot 5$ **16.** $2 \cdot 2 \cdot 2 \cdot 3$ **17.** $5 \cdot 5$ **18.** $3 \cdot 3 \cdot 3$ **19.** 29   **20.** $2 \cdot 3 \cdot 5$
21. $5 \cdot 7$ **22.** $2 \cdot 2 \cdot 2 \cdot 5$ **23.** 47   **24.** $7 \cdot 7$ **25.** $2 \cdot 5 \cdot 5$ **26.** $2 \cdot 2 \cdot 13$ **27.** $5 \cdot 13$ **28.** $3 \cdot 5 \cdot 5$
29. $2 \cdot 2 \cdot 5 \cdot 5$ **30.** $3 \cdot 5 \cdot 7$ **31.** $2 \cdot 2 \cdot 3 \cdot 3 \cdot 3$ **32.** $5 \cdot 23$ **33.** $11 \cdot 11$ **34.** $2 \cdot 2 \cdot 2 \cdot 2 \cdot 3 \cdot 3$ **35.** $2 \cdot 2 \cdot 3 \cdot 13$
36. 157   **37.** $2^3 \cdot 3^2$ **38.** $2^4 \cdot 7$ **39.** $2^2 \cdot 31$ **40.** $2^2 \cdot 41$ **41.** $2^3 \cdot 71$ **42.** $2^2 \cdot 3^2 \cdot 5^2$

Self-Study Exercises 3–4

1   **1.** 6   **2.** 30   **3.** 56   **4.** 12   **5.** 72   **6.** 60   **7.** 24   **8.** 18   **9.** 24   **10.** 700   **11.** 27   **12.** 16   **13.** 90
14. 120   **15.** 180   **16.** 30   **17.** 96   **18.** 72   **19.** 60   **20.** 300   **21.** 66   **22.** 312   **23.** 14   **24.** 144

2   **25.** 1   **26.** 1   **27.** 1   **28.** 6   **29.** 5   **30.** 12   **31.** 9   **32.** 5   **33.** 2   **34.** 6   **35.** 5   **36.** 3   **37.** 2
38. 2   **39.** 2

Self-Study Exercises 3–5

1   **1.** $\dfrac{4}{5}, \dfrac{8}{10}, \dfrac{12}{15}, \dfrac{16}{20}, \dfrac{20}{25}, \dfrac{24}{30}$ **2.** $\dfrac{7}{10}, \dfrac{14}{20}, \dfrac{21}{30}, \dfrac{28}{40}, \dfrac{35}{50}, \dfrac{42}{60}$ **3.** $\dfrac{3}{4} = \dfrac{18}{24}$ **4.** 6   **5.** 12   **6.** 36   **7.** 5   **8.** 20
9. 21   **10.** 12

2   **11.** $\dfrac{1}{2}$ **12.** $\dfrac{3}{5}$ **13.** $\dfrac{3}{4}$ **14.** $\dfrac{5}{16}$ **15.** $\dfrac{1}{2}$ **16.** $\dfrac{7}{8}$ **17.** $\dfrac{5}{16}$ **18.** $\dfrac{1}{5}$ **19.** $\dfrac{1}{5}$ **20.** $\dfrac{6}{25}$ **21.** $\dfrac{5}{8}$ **22.** $\dfrac{1}{4}$
23. $\dfrac{3}{4}$ **24.** $\dfrac{3}{16}$ **25.** $\dfrac{7}{32}$

3   **26.** $\dfrac{1}{2}$ **27.** $\dfrac{1}{10}$ **28.** $\dfrac{1}{5}$ **29.** $\dfrac{7}{10}$ **30.** $\dfrac{1}{4}$ **31.** $\dfrac{1}{40}$ **32.** $3\dfrac{9}{10}$ **33.** $4\dfrac{4}{5}$ **34.** $\dfrac{189}{500}$ **35.** $\dfrac{7}{8}$ **36.** $\dfrac{3}{8}$
37. $\dfrac{5}{8}$ **38.** $\dfrac{3}{4}$ in.   **39.** $\dfrac{3}{16}$ ft   **40.** $2\dfrac{3}{8}$ in.   **41.** $\dfrac{83}{100}$ lb   **42.** $\dfrac{5}{16}$ in.   **43.** $3\dfrac{1}{8}$ in.

4   **44.** 0.6   **45.** 0.3   **46.** 0.875   **47.** 0.375   **48.** 0.45   **49.** 0.98   **50.** 0.21   **51.** 3.875   **52.** 1.4375
53. 4.5625   **54.** 0.17   **55.** 0.44   **56.** 1.43   **57.** 3.45   **58.** 0.67   **59.** 0.27   **60.** 0.78   **61.** 0.38   **62.** 0.83
63. 0.58   **64.** 2.38   **65.** 5.57   **66.** 2.046875 in.   **67.** 0.045   **68.** 0.125 in.

Self-Study Exercises 3–6

1   **1.** $2\dfrac{2}{5}$ **2.** $1\dfrac{3}{7}$ **3.** 1   **4.** $4\dfrac{4}{7}$ **5.** 4   **6.** $2\dfrac{1}{7}$ **7.** $2\dfrac{5}{9}$ **8.** $9\dfrac{2}{5}$ **9.** $9\dfrac{5}{9}$ **10.** $1\dfrac{17}{21}$ **11.** $3\dfrac{4}{5}$ **12.** 16
13. $7\dfrac{1}{5}$ **14.** $9\dfrac{1}{2}$ **15.** 9

2   **16.** $\dfrac{7}{3}$ **17.** $\dfrac{25}{8}$ **18.** $\dfrac{15}{8}$ **19.** $\dfrac{77}{12}$ **20.** $\dfrac{77}{8}$ **21.** $\dfrac{31}{8}$ **22.** $\dfrac{89}{12}$ **23.** $\dfrac{103}{16}$ **24.** $\dfrac{257}{32}$ **25.** $\dfrac{69}{64}$ **26.** $\dfrac{73}{10}$
27. $\dfrac{26}{3}$ **28.** $\dfrac{100}{3}$ **29.** $\dfrac{200}{3}$ **30.** $\dfrac{25}{2}$ **31.** 15   **32.** 18   **33.** 56   **34.** 32   **35.** 48

Self-Study Exercises 3–7

1   **1.** 72   **2.** 30   **3.** 50   **4.** 48   **5.** 24

6. $\frac{2}{3}$ **7.** $\frac{7}{16}$ **8.** $\frac{8}{9}$ **9.** $\frac{11}{16}$ **10.** $\frac{15}{32}$ **11.** $\frac{7}{12}$ **12.** $\frac{4}{5}$ **13.** $\frac{9}{10}$ **14.** $\frac{4}{15}$ **15.** $\frac{1}{2}$ **16.** 0.37 **17.** $\frac{4}{5}$

18. $\frac{5}{12}$ **19.** $\frac{1}{2}$ **20.** 0.34 **21.** No, $\frac{15}{64}$ is greater **22.** Yes, $\frac{3}{8}$ is greater **23.** Yes, $\frac{7}{16}$ is greater **24.** Yes

25. $\frac{5}{8}$ in. end **26.** No **27.** No **28.** Yes **29.** No **30.** too small

Self-Study Exercises 3–8

1 **1.** $\frac{3}{8}$ **2.** $1\frac{3}{8}$ **3.** $\frac{5}{8}$ **4.** $\frac{25}{32}$ **5.** $\frac{9}{16}$ **6.** $1\frac{7}{16}$ **7.** $\frac{11}{64}$ **8.** $1\frac{19}{40}$ **9.** $1\frac{23}{36}$ **10.** $1\frac{1}{12}$ **11.** $1\frac{2}{5}$

12. $\frac{19}{21}$ **13.** $\frac{15}{16}$ in. **14.** $1\frac{27}{32}$ in. **15.** $1\frac{15}{16}$ in. **16.** $1\frac{7}{8}$ in. **17.** $1\frac{1}{4}$ in.

2 **18.** $6\frac{4}{5}$ **19.** $4\frac{1}{8}$ **20.** $16\frac{13}{16}$ **21.** $1\frac{11}{18}$ **22.** $4\frac{13}{16}$ **23.** $5\frac{17}{32}$ **24.** $4\frac{11}{16}$ **25.** $13\frac{1}{2}$ **26.** $4\frac{7}{15}$

27. $12\frac{11}{16}$ **28.** $8\frac{13}{16}$ in. **29.** $3\frac{15}{16}$ in. **30.** $11\frac{5}{8}$ gal **31.** $24\frac{5}{32}$ in. **32.** $22\frac{7}{8}$ in. **33.** $5\frac{7}{8}$ cups

Self-Study Exercises 3–9

1 **1.** $\frac{1}{4}$ **2.** $\frac{3}{16}$ **3.** $\frac{1}{16}$ **4.** $\frac{1}{8}$ **5.** $\frac{9}{64}$ **6.** $\frac{1}{8}$

2 **7.** $4\frac{11}{16}$ **8.** $17\frac{3}{4}$ **9.** $4\frac{15}{16}$ **10.** $5\frac{21}{32}$ **11.** $1\frac{5}{7}$ **12.** $1\frac{3}{4}$ **13.** $8\frac{13}{16}$ in. **14.** $10\frac{1}{8}$ in. **15.** $3\frac{1}{10}$ lb

16. $\frac{3}{16}$ in. **17.** $\frac{29}{32}$

Self-Study Exercises 3–10

1 **1.** $\frac{3}{32}$ **2.** $\frac{7}{32}$ **3.** $\frac{7}{16}$ **4.** $\frac{7}{12}$ **5.** $\frac{1}{3}$ **6.** $\frac{5}{32}$ **7.** $\frac{1}{15}$ **8.** $\frac{11}{25}$ **9.** $\frac{2}{25}$ **10.** $\frac{21}{32}$ **11.** $\frac{7}{15}$ **12.** $\frac{5}{6}$

2 **13.** $21\frac{7}{8}$ **14.** 75 **15.** $4\frac{1}{8}$ **16.** $36\frac{1}{10}$ **17.** $1\frac{21}{40}$ **18.** $2\frac{1}{6}$ **19.** $18\frac{3}{4}$ L **20.** 90 in. **21.** $15\frac{5}{8}$ in.

22. 264 kg copper, 84 kg tin, 36 kg zinc

3 **23.** $\frac{16}{25}$ **24.** $\frac{1}{16}$ **25.** $\frac{27}{125}$ **26.** $\frac{1}{27}$ **27.** $\frac{49}{81}$ **28.** $\frac{25}{49}$ **29.** $\frac{64}{81}$ **30.** $\frac{9}{100}$

Self-Study Exercises 3–11

1 **1.** $\frac{7}{3}$ **2.** $\frac{8}{5}$ **3.** $\frac{5}{11}$ **4.** $\frac{3}{11}$ **5.** $\frac{1}{7}$ **6.** $\frac{1}{8}$ **7.** $\frac{10}{9}$ **8.** 4 **9.** $\frac{10}{13}$ **10.** $\frac{5}{9}$

2 **11.** $\frac{6}{7}$ **12.** $\frac{9}{10}$ **13.** $\frac{11}{12}$ **14.** 2 **15.** $1\frac{1}{20}$ **16.** $2\frac{1}{12}$ **17.** $4\frac{2}{3}$ **18.** $\frac{2}{5}$ **19.** 8 segments **20.** 28 segments

3 **21.** 32   **22.** $13\frac{1}{3}$ **23.** $12\frac{1}{2}$ **24.** $\frac{5}{8}$ **25.** $14\frac{3}{4}$ **26.** 7   **27.** 4   **28.** $2\frac{1}{2}$ **29.** $1\frac{1}{3}$ **30.** $\frac{3}{5}$

31. 18—2 × 4's   **32.** 10 lengths   **33.** 12 shovels   **34.** $23\frac{11}{12}$ in.   **35.** 20 ft × 15 ft   **36.** 4 whole pieces

37. 7 strips   **38.** 23 straws, $3\frac{3}{4}$ in. left   **39.** 8 pieces

4 **40.** $\frac{1}{10}$ **41.** 3   **42.** $5\frac{3}{5}$ **43.** $\frac{12}{25}$ **44.** $\frac{1}{3}$ **45.** 5   **46.** 28   **47.** $\frac{2}{9}$ **48.** 3   **49.** $\frac{1}{3}$ **50.** $\frac{1}{6}$ **51.** $\frac{5}{6}$

52. $\frac{2}{3}$ **53.** $\frac{3}{8}$ **54.** $\frac{1}{8}$

Self-Study Exercises 3–12

1 **1.** $-\frac{-5}{8}$, $-\frac{5}{-8}$, $\frac{-5}{-8}$ **2.** $-\frac{-3}{-4}$, $\frac{-3}{4}$, $\frac{3}{-4}$ **3.** $-\frac{2}{-5}$, $-\frac{-2}{5}$, $\frac{2}{5}$ **4.** $-\frac{7}{8}$, $\frac{-7}{8}$, $\frac{7}{-8}$ **5.** $-\frac{-7}{8}$, $-\frac{7}{-8}$, $\frac{-7}{-8}$

2 **6.** $-\frac{2}{8}$ or $-\frac{1}{4}$ **7.** $-1\frac{1}{10}$ **8.** $1\frac{1}{10}$ **9.** $\frac{6}{11}$ **10.** $-\frac{25}{32}$

3 **11.** -26.297 **12.** -1.11 **13.** -91.44 **14.** -110.72 **15.** -59.04 **16.** 340.71 **17.** -27.73
18. 0.41   **19.** 4.6   **20.** 0.03   **21.** -2.3 **22.** -5.5

4 **23.** $-\frac{1}{5}$ **24.** $-\frac{47}{8}$ **25.** -7 **26.** $-\frac{18}{5}$ **27.** $\frac{37}{144}$ **28.** $-\frac{127}{245}$ **29.** -10.38 **30.** -3.992 **31.** -11.88

Assignment Exercises, Chapter 3

1. $\frac{3}{8}$ **3.** $\frac{1}{3}$ **5.** $\frac{13}{24}$ **7.** (a) 9   (b) 4   (c) 4   (d) 9   (e) 9, 4   (f) 4   (g) 9   (h) improper   (i) >1
(j) 9, 4   **9.** f   **11.** d   **13.** e   **15.** d   **17.** 20, 30, 40, 50, 60 Answers may vary.   **19.** 42, 63, 84, 105, 126
Answers may vary.   **21.** 14, 21, 28, 35, 42 Answers may vary.   **23.** 16, 24, 32, 40, 48 Answers may vary.
25. Yes; sum of digits divisible by 3   **27.** No, divisible by 2 but not by 3   **29.** Yes; ends in zero   **31.** No;
sum of digits not divisible by 9   **33.** $1 \times 48, 2 \times 24, 3 \times 16, 4 \times 12, 6 \times 8$; 1, 2, 3, 4, 6, 8, 12, 16, 24, 48
35. $1 \times 51, 3 \times 17$; 1, 3, 17, 51   **37.** $1 \times 74, 2 \times 37$; 1, 2, 37, 74   **39.** composite   **41.** composite   **43.** $2 \cdot 3 \cdot 7$
45. $2 \cdot 7 \cdot 7$ or $2 \cdot 7^2$ **47.** 360   **49.** 180   **51.** 2   **53.** 6   **55.** $\frac{15}{24}$ **57.** $\frac{25}{60}$ **59.** $\frac{10}{15}$ **61.** $\frac{24}{32}$ **63.** $\frac{11}{55}$
65. $\frac{16}{20}$ **67.** $\frac{1}{2}$ **69.** $\frac{1}{8}$ **71.** $\frac{1}{4}$ **73.** $\frac{17}{32}$ **75.** $\frac{3}{8}$ **77.** $\frac{3}{4}$ **79.** $\frac{7}{10}$ **81.** $\frac{19}{20}$ **83.** $\frac{109}{125}$ **85.** $\frac{1}{50}$ **87.** 0.2
89. 0.625   **91.** 0.818   **93.** $3\frac{3}{5}$ **95.** $4\frac{7}{8}$ **97.** $5\frac{3}{8}$ **99.** $87\frac{1}{2}$ **101.** $1\frac{1}{2}$ **103.** $\frac{8}{1}$ **105.** $\frac{57}{8}$ **107.** $\frac{147}{16}$
109. $\frac{23}{5}$ **111.** $\frac{12}{1}$ **113.** $\frac{16}{3}$ **115.** 20   **117.** 33   **119.** 60   **121.** 16   **123.** 60   **125.** larger   **127.** no
129. smaller   **131.** $\frac{3}{8}$ **133.** $\frac{3}{8}$ **135.** $\frac{3}{16}$ **137.** $\frac{27}{32}$ **139.** $\frac{9}{19}$ **141.** $\frac{3}{4}$ **143.** $\frac{3}{11}$ **145.** $\frac{21}{64}$ **147.** $1\frac{13}{30}$
149. $8\frac{19}{32}$ **151.** $10\frac{9}{32}$ **153.** $18\frac{1}{16}$ in.   **155.** $12\frac{19}{32}$ in.   **157.** $15\frac{25}{32}$ in.   **159.** $4\frac{3}{4}$ in.   **161.** $\frac{7}{8}$ in.   **163.** $\frac{1}{3}$
165. $1\frac{5}{8}$ **167.** $5\frac{31}{32}$ **169.** $7\frac{11}{16}$ **171.** $34\frac{3}{4}$ **173.** $\frac{7}{16}$ in.   **175.** $1\frac{29}{64}$ in.   **177.** $\frac{7}{24}$ **179.** $\frac{7}{24}$ **181.** $\frac{1}{2}$
183. $7\frac{7}{8}$ **185.** $1\frac{1}{5}$ **187.** 2   **189.** $100\frac{1}{2}$ in.   **191.** $2\frac{3}{4}$ cups   **193.** 150 cm   **195.** $\frac{9}{16}$ **197.** $\frac{16}{81}$ **199.** $\frac{1}{8}$
201. $\frac{81}{100}$ **203.** $\frac{1}{4}$ **205.** $\frac{10}{7}$ or $1\frac{3}{7}$ **207.** $1\frac{1}{6}$ **209.** $9\frac{1}{3}$ **211.** 24   **213.** 2   **215.** $\frac{3}{5}$ **217.** $2\frac{55}{64}$ in.

219. $\frac{3}{16}$ yd   **221.** 36 lengths   **223.** $\frac{1}{18}$ **225.** $5\frac{1}{3}$ **227.** $\frac{1}{4}$ **229.** $\frac{1}{8}$ **231.** $-\frac{-3}{-8}, +\frac{-3}{8}, \frac{3}{-8}$

233. $\frac{7}{8}, \frac{-7}{8}, \frac{7}{-8}$ **235.** $\frac{-2}{5}, \frac{2}{5}, \frac{-2}{-5}$ **237.** $\frac{62}{63}$ **239.** $1\frac{5}{7}$ **241.** 0.0351   **243.** -62 **245.** $-1\frac{7}{24}$

247. $-11\frac{13}{16}$ **249.** $-1\frac{1}{6}$ **251.** 23.76   **253.** 5.74   **255.** (a) page 17 (b) 74 pages (c) 3 blank pages

Trial Test, Chapter 3

1. $\frac{3}{4}$ **3.** 3   **5.** $\frac{34}{7}$ **7.** $2^5 \cdot 3$ **9.** $\frac{1}{4}$ **11.** $3\frac{8}{9}$ **13.** $13\frac{1}{2}$ **15.** $1\frac{5}{12}$ **17.** $7\frac{13}{14}$ **19.** $\frac{1}{9}$ **21.** $5\frac{1}{10}$

23. 1   **25.** $\frac{5}{16}$ **27.** 0.6   **29.** $\frac{2}{7}$ **31.** $3\frac{5}{6}$ cups   **33.** 8 yd

Self-Study Exercises 4–1

1   **1.** 40%   **2.** 70%   **3.** 62.5%   **4.** 77.8%   **5.** 0.7%   **6.** 0.3%   **7.** 20%   **8.** 14%   **9.** 0.7%   **10.** 1.25%
11. 500%   **12.** 800%   **13.** 133.3%   **14.** 350%   **15.** 430%   **16.** 220%   **17.** 305%   **18.** 720%   **19.** 1,510%
20. 3,625%

2   **21.** $\frac{9}{25}, 0.36$ **22.** $\frac{9}{20}, 0.45$ **23.** $\frac{1}{5}, 0.2$ **24.** $\frac{3}{4}, 0.75$ **25.** $\frac{1}{16}, 0.0625$ **26.** $\frac{5}{8}, 0.625$ **27.** $\frac{2}{3}, 0.6667$

28. $\frac{3}{500}, 0.006$ **29.** $\frac{1}{500}, 0.002$ **30.** $\frac{1}{2,000}, 0.0005$ **31.** $\frac{1}{12}, 0.0833$ **32.** $\frac{3}{16}, 0.1875$ **33.** 8   **34.** 4

35. $2\frac{1}{2}, 2.5$ **36.** $4\frac{1}{4}, 4.25$ **37.** $1\frac{19}{25}, 1.76$ **38.** $3\frac{4}{5}, 3.8$ **39.** $1\frac{3}{8}, 1.375$ **40.** $3\frac{7}{8}, 3.875$ **41.** $1\frac{2}{3}$ **42.** $3\frac{1}{6}$

43. 1.153   **44.** 2.125   **45.** 1.0625   **46.** $\frac{1}{10}, 0.1$ **47.** 25%, 0.25   **48.** 20%, $\frac{1}{5}$ **49.** $33\frac{1}{3}\%, 0.33\frac{1}{3}$ **50.** $\frac{1}{2}, 0.5$

51. 80%, 0.8   **52.** 75%, $\frac{3}{4}$ **53.** $\frac{2}{3}, 0.66\frac{2}{3}$ **54.** 100%, $\frac{1}{1}$ **55.** 30%, 0.3   **56.** $\frac{2}{5}, 0.4$ **57.** 70%, $\frac{7}{10}$ **58.** 90%, 0.9

59. $\frac{3}{5}, 0.6$ **60.** 100%, 1   **61.** 25%, $\frac{1}{4}$ **62.** $66\frac{2}{3}\%, 0.66\frac{2}{3}$ **63.** $\frac{7}{10}, 0.7$ **64.** 50%, $\frac{1}{2}$ **65.** 60%, 0.6

66. 10%, $\frac{1}{10}$ **67.** $\frac{1}{5}, 0.2$ **68.** $33\frac{1}{3}\%, \frac{1}{3}$ **69.** $66\frac{2}{3}\%, \frac{2}{3}$ **70.** 75%, 0.75   **71.** $\frac{3}{10}, 0.3$ **72.** 90%, $\frac{9}{10}$

73. 20%, 0.2   **74.** 60%, $\frac{3}{5}$ **75.** $\frac{1}{1}, 1$ **76.** 60%   **77.** 40%   **78.** 70%   **79.** 62.5%   **80.** 90%

Self-Study Exercises 4–2

1   **1.** $x = 5$ **2.** $a = 9$ **3.** $a = 8$ **4.** $b = 18$ **5.** R missing; $B = 10$; $P = 2$ **6.** $P = 2$; $R = 20\%$; B missing
7. $P = 3$; R missing; $B = 4$ **8.** R missing; $B = 25$; $P = 5$ **9.** $P = 6$; $B = 15$; R missing   **10.** P missing;
$R = 20\%$; $B = 15$ **11.** $R = 35\%$; B missing; $P = 70$ **12.** R missing; $B = \$45$; $P = \$3.15$

2   **13.** 5.4   **14.** 252   **15.** 1.8   **16.** 26   **17.** 50%   **18.** $333\frac{1}{3}\%$ **19.** 87.5%   **20.** 150%   **21.** 64
22. 70,000

3   **23.** 75   **24.** 63   **25.** 206   **26.** 115.92   **27.** 0.675   **28.** 0.94   **29.** 154.10   **30.** 231   **31.** 924
32. 345   **33.** 25%   **34.** 30%   **35.** $33\frac{1}{3}\%$ **36.** 32.9%   **37.** 15.75%   **38.** 16%   **39.** 0.8%   **40.** $0.66\frac{2}{3}\%$
41. 500%   **42.** 111.25%   **43.** 72   **44.** 50   **45.** 344   **46.** 46   **47.** 360   **48.** 275   **49.** 75   **50.** 250
51. 18.4   **52.** 261   **53.** 3.75   **54.** 0.625   **55.** $66\frac{2}{3}\%$ **56.** $37\frac{1}{2}\%$ **57.** 14.25   **58.** 350   **59.** 9.375

60. 220   **61.** 20%   **62.** 200   **63.** 1.0625 lb   **64.** 3%   **65.** 200 hp   **66.** 0.021 lb   **67.** 5%   **68.** $575
69. 1,200 lb   **70.** 11%   **71.** 9,625 swabs   **72.** 100,880 welds   **73.** $4.65   **74.** $6.45   **75.** 0.4   **76.** 0.4
77. $4,285.71   **78.** $1,364.16

4   **79.** $4.55, $80.38   **80.** $829.06   **81.** $23.30   **82.** 25%   **83.** 23%   **84.** $41.93

5   **85.** $11,832   **86.** $17.52   **87.** $3,866.00   **88.** $944.50   **89.** $434.84   **90.** $3,205.75

6   **91.** $9.75   **92.** $150   **93.** 1.75%   **94.** $171.50   **95.** $1,460   **96.** 2.5%

Self-Study Exercises 4–3

1   **1.** 108   **2.** 109.2   **3.** 10.2   **4.** 33.75   **5.** 672,830; 2,878,830   **6.** 630,504; 2,576,504

2   **7.** 60%   **8.** 82%   **9.** 13.7%   **10.** $66\frac{2}{3}$%   **11.** 99.91%   **12.** 99%   **13.** 27 in.   **14.** 1,815 board feet
15. 2,393 board feet   **16.** 25,908 bricks   **17.** $16,802.50   **18.** 1,366.4 yd^3 **19.** $53,350.20   **20.** $2,041.46
21. $3,961.82   **22.** $58.80

3   **23.** 7%   **24.** 45%   **25.** 20%   **26.** 5%   **27.** 350 hp   **28.** 10.5 yd^3 **29.** 3,019 board feet   **30.** 28.5 lb
31. 40 in.   **32.** 20%   **33.** 150 hp

Assignment Exercises, Chapter 4

1. 72%   **3.** 23%   **5.** 70%   **7.** 83%   **9.** 320%   **11.** 12,500%   **13.** 1,730%   **15.** 0.72 or $\frac{10}{25}$ **17.** 0.125 or $\frac{1}{8}$
19. 0.0066$\frac{2}{3}$ or $\frac{1}{150}$ **21.** 2.75 or $2\frac{3}{4}$ **23.** 1.125 or $1\frac{1}{8}$ **25.** 2.272   **27.** 3.4   **29.** 0.83
31. 0.125   **33.** $a = 4.5$ **35.** $y = 48$ **37.** $R = 5\%$; $B = 180$; P missing   **39.** $R = 45\%$; B missing; $P = \$36$
41. $P = 6$; R missing; $B = 25$ **43.** $R = 18\%$; $B = 150$; P missing   **45.** 24   **47.** 0.4375   **49.** 60%   **51.** 250%
53. 83   **55.** 152   **57.** 15.3%   **59.** 500%   **61.** 84   **63.** 37.5%   **65.** 1.14%   **67.** $266   **69.** 4%
71. 200 students   **73.** $204.35   **75.** $439.84   **77.** $365.66   **79.** $1,707.60   **81.** 1.75%   **83.** $3,935.63

85. $0.89   **87.** 8.5%   **89.** $369.69   **91.** $355   **93.** 87.1%   **95.** 78.5%   **97.** $67\frac{3}{5}$%   **99.** 8%   **101.** 7%

103. 14%   **105.** 142.1 kg   **107.** 62 cm, 63 cm   **109.** $1.69   **111.** 12.5%   **113.** 57.5 lb   **115.** $33\frac{1}{3}$%
117. 4.8%   **119.** 289 hp   **121.** 86%   **123.** 26.6%   **125.** 28.6%

Trial Test, Chapter 4

1. 80%   **3.** 0.003   **5.** $R = 40\%$; $B = 10$; P missing   **7.** $P = 9$; R missing; $B = 27$ **9.** $R = 12\%$; $B = 50$; P missing
11. 9   **13.** 305   **15.** 115   **17.** 67.10   **19.** 12%   **21.** $525   **23.** $2,500   **25.** 21.14%   **27.** 13.17%   **29.** 5%
31. $104.87

Self-Study Exercises 5–1

1   **1.** lb   **2.** oz   **3.** lb or oz   **4.** gal   **5.** qt   **6.** T   **7.** lb   **8.** qt   **9.** in.   **10.** c   **11.** in.
12. yd   **13.** $\frac{mi}{hr}$ **14.** oz   **15.** oz   **16.** mi   **17.** lb or oz   **18.** lb   **19.** lb and oz   **20.** oz

2 **21.** $\dfrac{2\text{ pt}}{1\text{ qt}}, \dfrac{1\text{ qt}}{2\text{ pt}}$ **22.** $\dfrac{5{,}280\text{ ft}}{1\text{ mi}}, \dfrac{1\text{ mi}}{5{,}280\text{ ft}}$ **23.** $\dfrac{12\text{ in.}}{1\text{ ft}}, \dfrac{1\text{ ft}}{12\text{ in.}}$ **24.** $\dfrac{3\text{ ft}}{1\text{ yd}}, \dfrac{1\text{ yd}}{3\text{ ft}}$ **25.** $\dfrac{7\text{ da}}{1\text{ wk}}, \dfrac{1\text{ wk}}{7\text{ da}}$

26. $\dfrac{4\text{ qt}}{1\text{ gal}}, \dfrac{1\text{ gal}}{4\text{ qt}}$ **27.** 48 in. **28.** 21 ft **29.** 4,400 yd **30.** $9\frac{1}{3}$ yd **31.** 2 mi **32.** 48 oz **33.** $2\frac{3}{10}$ or 2.3 lb

34. 732.8 oz **35.** 19.2 oz **36.** 408 oz, 25.5 lb **37.** 20 qt **38.** 13 pt **39.** $1\frac{1}{2}$ gal **40.** 60 pt **41.** 9 gal

42. 256 oz **43.** $1\frac{1}{2}$ or 1.5 yd **44.** 108 in. **45.** 7,040 ft **46.** 24 c

3 **47.** feet to inches = 12; inches to feet = $\dfrac{1}{12}$ or 0.083 **48.** quarts to gallons = $\dfrac{1}{4}$ or 0.25; gallons to quarts = 4

49. quarts to pints = 2; pints to quarts = $\dfrac{1}{2}$ or 0.5 **50.** ounces to cups = $\dfrac{1}{8}$ or 0.125; cups to ounces = 8

51. 28.75 pt **52.** 92 pt **53.** 16.67 yd **54.** 25 gal **55.** 36.25 lb **56.** 153 in. **57.** 720 in. **58.** $4\frac{2}{3}$ ft or 4.67 ft

59. 7,920 ft **60.** 28 qt

4 **61.** 3 ft 8 in. **62.** 2 mi 1,095 ft **63.** 3 lb $3\frac{1}{2}$ oz **64.** 2 gal 1 qt **65.** 2 gal **66.** 2 T 500 lb

67. 2 yd 2 ft 11 in. **68.** 2 qt 2 c 2 oz **69.** 2 ft 10 in. **70.** 6 lb 9 oz **71.** 4 gal 2 qt 16 oz **72.** 6 qt 20 oz

73. 2 mi 5 yd 2 ft 1 in. **74.** 5 ft 4 in. **75.** 3 T 600 lb 15 oz **76.** 4 lb 5 oz **77.** 2 lb 5 oz **78.** 2 yd 2 ft 4 in.

79. 2 yd 2 ft 11 in. **80.** 1 mi 875 ft 6 in.

Self-Study Exercises 5–2

1 **1.** 2 lb 5 oz **2.** 4 ft 7 in. **3.** 15 lb 11 oz **4.** 18 ft 3 in. **5.** 8 qt $\frac{1}{2}$ pt **6.** 14 gal 1 qt **7.** 9 yd 1 ft

8. 5 c 1 oz or 2 pt 1 c 1 oz **9.** 5 ft 4 in. or 1 yd 2 ft 4 in. **10.** 7 yd 1 ft 5 in. **11.** 2 ft 7 in. **12.** 11 ft

13. 5 lb 15 oz **14.** 12 lb

2 **15.** 6 in. **16.** 1 pt **17.** 2 ft **18.** 4 lb 14 oz **19.** 1 lb 6 oz **20.** 6 lb 11 oz **21.** 11 in. **22.** 10 lb 13 oz

23. 2 in. **24.** 3 gal 3 qt $1\frac{1}{2}$ pt or 3 gal 3 qt 1 pt 1 c **25.** 4 ft 2 in. **26.** 2 ft 3 in. **27.** 45 sec **28.** 39 sec

29. 69 lb 7 oz **30.** 16 lb 14 oz

Self-Study Exercises 5–3

1 **1.** 60 mi **2.** 108 gal **3.** 252 lb **4.** 14 qt or 3 gal 2 qt **5.** 57 lb 8 oz **6.** 38 gal 2 qt **7.** 43 gal 3 qt

8. 58 ft **9.** 36 lb **10.** 7 qt 1 pt or 1 gal 3 qt 1 pt

2 **11.** 35 in^2 **12.** 108 ft^2 **13.** 180 yd^2 **14.** 108 mi^2 **15.** 378 tiles **16.** 46 gal 18 oz

17. 7 qt 1 pt or 1 gal 3 qt 1 pt **18.** 6 lb 4 oz **19.** 7 lb 8 oz

3 **20.** 6 gal **21.** 1 day 15 hr **22.** 10 yd 1 ft 3 in. **23.** 1 yd 1 ft 7 in. **24.** 3 qt $\frac{1}{2}$ pt or 3 qt 1 c **25.** $6\frac{2}{3}$ gal

26. 1 hr 40 min **27.** $5\frac{1}{4}$ qt or 5 qt 1 c **28.** 7 ft 6 in. **29.** 3 gal 2 qt **30.** 9 pieces **31.** 5 ft

32. 3 gal 1 qt 5 oz **33.** 24 lb 3 oz

4 **34.** 3 **35.** 17 **36.** 8 **37.** 18 **38.** 3 **39.** 8 pieces **40.** 9 boxes **41.** 24 cans **42.** 9 tickets

43. 9 pieces

44. $15\dfrac{\text{gal}}{\text{sec}}$ **45.** $\dfrac{3}{4}\dfrac{\text{lb}}{\text{min}}$ **46.** $15{,}840\dfrac{\text{ft}}{\text{hr}}$ **47.** $2{,}304\dfrac{\text{oz}}{\text{min}}$ **48.** $2\dfrac{\text{qt}}{\text{sec}}$ **49.** $20\dfrac{\text{oz}}{\text{hr}}$ **50.** $40\dfrac{\text{ft}}{\text{min}}$ **51.** $440\dfrac{\text{ft}}{\text{sec}}$

52. $44\dfrac{\text{ft}}{\text{sec}}$ **53.** $3\dfrac{\text{qt}}{\text{min}}$ **54.** $53\dfrac{1}{3}\dfrac{\text{lb}}{\text{min}}$ **55.** $\dfrac{5}{6}\dfrac{\text{gal}}{\text{sec}}$

Self-Study Exercises 5–4

1 **1.** (a) 1,000 m (b) 10 L (c) $\dfrac{1}{10}$ of a gram (d) $\dfrac{1}{1{,}000}$ of a meter (e) 100 g (f) $\dfrac{1}{100}$ of a liter **2.** b
3. b **4.** c **5.** b **6.** a **7.** c **8.** b **9.** c **10.** a **11.** b **12.** c **13.** b **14.** a **15.** b **16.** b

2 **17.** 40 **18.** 70 **19.** 580 **20.** 80 **21.** 2.5 **22.** 210 **23.** 85 **24.** 142 **25.** 153 mL **26.** 460 m
27. 75 dkg **28.** 160 mm **29.** 400 **30.** 8,000 **31.** 58,000 **32.** 800 **33.** 250 **34.** 2,100 **35.** 102,500
36. 8,330 **37.** 2,000,000 **38.** 70 **39.** 236 L **40.** 467 cm **41.** 38,000 dg **42.** 13,000 cm **43.** 2.8
44. 23.8 **45.** 10.1 **46.** 6 **47.** 2.9 **48.** 19.25 **49.** 1.7 **50.** 438.9 dm **51.** 4.7 g **52.** 0.225 dL
53. 2.743 **54.** 0.385 **55.** 0.15 **56.** 0.08 **57.** 2,964.84 **58.** 0.2983 **59.** 0.0003 **60.** 0.004
61. 0.002857 **62.** 15.285 **63.** 0.0297 hm **64.** 0.00003 L

3 **65.** 11 m **66.** 12 hL **67.** 6 cg **68.** 2.4 dm or 24 cm **69.** 5.9 cL or 59 mL **70.** cannot add
71. 10.1 kL or 101 hL **72.** 0.55 g or 55 cg **73.** cannot subtract **74.** 7.002 km or 7,002 m **75.** 1,000 mL
76. 1.47 kL or 147 dkL **77.** 516 m **78.** 40.8 m **79.** 150.96 dm **80.** 969.5 m **81.** 13 m **82.** 9 cL
83. 163 g **84.** 0.4 m or 4 dm **85.** 5 **86.** 30 **87.** 16 prescriptions **88.** 80 containers **89.** 5.74 dL or 57.4 cL
90. 2.3 dkm or 23 m **91.** 165.7 hm or 16.57 km **92.** 9.1 kL or 91 hL **93.** 1.25 cL **94.** 70 mm **95.** 7.5 dm
96. 50 mL **97.** 21,250 containers **98.** 19 vials

Self-Study Exercises 5–5

1 **1.** 354.33 in. **2.** 131.232 yd **3.** 26.0988 mi **4.** 6.3402 liq qt **5.** 9.463 L **6.** 59.5242 lb **7.** 22.68 kg
8. 17.78 cm **9.** 5.4864 m **10.** 1.3188 oz **11.** 21.7649 L **12.** 23.9498 ft **13.** 5.0000 in. **14.** 0.4724 in.
15. 146.029 mi **16.** 310.7 mi **17.** 711.2 mm **18.** 91.44 cm **19.** 1.2192 m **20.** 1,609.344 m **21.** 91.44 m
22. 321.86 km **23.** 52.835 qt **24.** 94.63 L **25.** 0.8472 oz **26.** 44.092 lb **27.** 68.04 kg **28.** 28.412 kg
29. 56.6572 g **30.** 14.1643 g **31.** 6.6138 lb **32.** 30.48 m **33.** 27.216 kg **34.** 241.395 km **35.** 32.808 yd
36. 236.132 mi **37.** 7.4568 mi **38.** 11.3556 L **39.** 328.08 ft **40.** 22 **41.** 18

Self-Study Exercises 5–6

1 **1.** 1.5 days **2.** 9.67 min **3.** 150 min **4.** 318 sec **5.** 210 min **6.** 3 min 2 sec **7.** 432 min
8. 1 min 28 sec, 1.467 min

2 **9.** 4,320 lb/hr **10.** 2,400 gal/hr **11.** 10,950 lb/yr **12.** 1,680 vehicles/hr **13.** 0.0167 mi/sec
14. 2.0833 gal/min **15.** 60 lb/min **16.** 1.6 gal/sec

3 **17.** 8:00 A.M. **18.** 3:00 A.M. **19.** 7:00 P.M. **20.** 11:00 A.M. **21.** 7:00 P.M. **22.** 9:00 P.M. **23.** 4:00 P.M.
24. 12:00 A.M. **25.** 1:00 P.M. **26.** 8:00 P.M. **27.** 4:30 A.M. **28.** Wellington, New Zealand or other locations
that have GMT +12:00. **29.** 8 hr **30.** 5 hr **31.** 6:00 P.M. **32.** New York 9:00 A.M.−11:00 A.M.;
St. Louis 8:00 A.M.−10.00 A.M.; London 2:00 P.M.−4:00 P.M. **33.** Paris 4:00 P.M.; Stanford 8:00 A.M.
34. 1 hr later in Paris **35.** 11:00 P.M. **36.** 10 hr

Self-Study Exercises 5–7

1 **1.** 3 **2.** 4 **3.** 2 **4.** 4 **5.** 4 **6.** 2 **7.** 3 **8.** 5 **9.** 5 **10.** 3

11. $\frac{1}{32}$ in.   **12.** 0.05 mm   **13.** $\frac{1}{8}$ in.   **14.** $\frac{1}{2}$ or 0.5 oz   **15.** 0.5 L   **16.** $\frac{1}{16}$ in.   **17.** $\frac{1}{64}$ in.   **18.** $\frac{1}{16}$ oz
19. 0.05 mi   **20.** 0.05 cm   **21.** 0.005 dg   **22.** 0.05 cg   **23.** 0.05 cL   **24.** 0.05 km   **25.** 0.05 km   **26.** 0.05 km

27. $4\frac{9}{16}$ in.   **28.** $4\frac{1}{16}$ in.   **29.** $3\frac{13}{16}$ in.   **30.** $3\frac{3}{8}$ in.   **31.** $2\frac{1}{4}$ in.   **32.** 2 in.   **33.** $1\frac{3}{4}$ in.   **34.** $1\frac{3}{16}$ in.
35. $\frac{3}{4}$ in.   **36.** $\frac{3}{8}$ in.

37. 115 mm or 11.5 cm   **38.** 102 mm or 10.2 cm   **39.** 96 mm or 9.6 cm   **40.** 85 mm or 8.5 cm
41. 57 mm or 5.7 cm   **42.** 50 mm or 5 cm   **43.** 44 mm or 4.4 cm   **44.** 30.5 mm or 3.05 cm
45. 19 mm or 1.9 cm   **46.** 10 mm or 1 cm

47. 3 in.   **48.** $2\frac{3}{4}$ in.   **49.** 7.5 cm   **50.** 13.6 cm   **51.** $11\frac{1}{4}$ in.   **52.** $10\frac{7}{8}$ in.   **53.** 4.4 cm   **54.** 4.9 cm
55. 7.9 cm   **56.** 4.1 cm   **57.** $11\frac{3}{8}$ in.   **58.** $5\frac{7}{8}$ in.

Self-Study Exercises 5–8

1. 355   **2.** 165   **3.** 290   **4.** 0   **5.** -175 **6.** 344   **7.** 333   **8.** -81

9. 920   **10.** 558   **11.** 250   **12.** 460   **13.** 640   **14.** 672   **15.** -110 **16.** 140   **17.** -460 **18.** 492

19. 35   **20.** 0   **21.** 45   **22.** 5   **23.** 15   **24.** 10   **25.** 65   **26.** 50   **27.** 80   **28.** 120

29. 158   **30.** 59   **31.** 113   **32.** 122   **33.** 68   **34.** 419   **35.** 590   **36.** 770   **37.** 365   **38.** 32

Assignment Exercises, Chapter 5

1. lb or oz   **3.** qt   **5.** oz   **7.** T   **9.** qt   **11.** c   **13.** yd   **15.** $\frac{4 \text{ c}}{1 \text{ qt}}, \frac{1 \text{ qt}}{4 \text{ c}}$ **17.** $\frac{2,000 \text{ lb}}{1 \text{ T}}, \frac{1 \text{ T}}{2,000 \text{ lb}}$

19. 4 yd   **21.** 6,336 ft   **23.** 80 oz   **25.** $42\frac{1}{2}$ lb   **27.** 304 oz or 19 lb   **29.** 15 pt   **31.** 24 pt   **33.** 6,600 ft

35. 7 ft 5 in.   **37.** 13 lb $1\frac{1}{2}$ oz   **39.** 2 gal 1 pt   **41.** 4 yd 4 in.   **43.** 4 ft 1 in. or 1 yd 1 ft 1 in.   **45.** 2 gal 16 oz
or 2 gal 1 pt   **47.** 3 ft 11 in.   **49.** 10 ft 11 in. or 3 yd 1 ft 11 in.   **51.** 8 gal 2 qt   **53.** 14 oz   **55.** 1 ft 9 in.
57. 17 in. or 1 ft 5 in.   **59.** 504 ft^2 **61.** 63 in^2 **63.** 10 yd 1 ft 3 in.   **65.** $5\frac{5}{12}$ ft or 5.42 ft   **67.** 4 lb 8 oz

69. 3.5   **71.** $4\frac{4}{9}$ **73.** 300 mi/hr   **75.** 60 mi/hr   **77.** 1.25 gal/min   **79.** $1\frac{1}{2}$ qt   **81.** kilo-

83. milli-   **85.** centi-   **87.** 10 times   **89.** $\frac{1}{1,000}$ of   **91.** 1,000 times   **93.** a   **95.** a   **97.** c   **99.** b
101. 6.71 dkm   **103.** 2,300 mm   **105.** 12,300 mm   **107.** 230,000 mm   **109.** 413.27 km   **111.** 3.945 hg
113. 30.00974 kg   **115.** cannot add unlike measures   **117.** 748 cg or 7.48 g   **119.** 61.47 cg   **121.** 15   **123.** 8.5 hL
125. 18.9 m   **127.** 245 mL   **129.** 6 m   **131.** 100 servings   **133.** 40 shirts   **135.** 235.124 yd   **137.** 15.8505 liq qt
139. 70.5472 lb   **141.** 22.86 cm   **143.** 156.3916 qt   **145.** 60.96 m   **147.** 281.6275 km   **149.** 21   **151.** 3 da

153. 2 hr, 38 min   **155.** 4 da   **157.** 3 yr, 3 mo   **159.** 6   **161.** 4   **163.** $\frac{1}{4}$ **165.** $\frac{1}{32}$ **167.** 0.05 cm   **169.** 0.05 cm

171. $5\frac{1}{4}$ in.   **173.** $4\frac{7}{16}$ in.   **175.** $3\frac{15}{16}$ in.   **177.** $3\frac{9}{16}$ in.   **179.** $2\frac{3}{4}$ in.   **181.** 117 mm or 118 mm   **183.** 99 mm

185. 60 mm   **187.** 45 mm   **189.** 20 mm   **191.** $5\frac{1}{4}$ in.   **193.** $4\frac{27}{32}$ in.   **195.** 2.7 cm   **197.** $20\frac{1}{4}$ in.

199. 5.6 cm   **201.** 7.4 cm   **203.** 351 kelvins   **205.** 472°R   **207.** 203°F   **209.** 104°F   **211.** 185°C
213. Answers will vary.

Trial Test, Chapter 5

1. 36 in.   **3.** 8 gal   **5.** 4 qt/sec   **7.** 165 yd   **9.** $80\frac{2}{3}$ ft/sec   **11.** $1\frac{1}{4}$ in.   **13.** 2 gal 1 qt 4 oz   **15.** 1 hr
17. deci-   **19.** 0.298 km   **21.** 9.48 L or 94.8 dL   **23.** 120.6975 km   **25.** 8.4536 pt   **27.** 746.76 m/sec
29. 8.89°C   **31.** 4.85 cm

Self-Study Exercises 6–1

1   **1.** 38 in.   **2.** 21.2 cm   **3.** 156 in.   **4.** 124 in.   **5.** 160 in.   **6.** 108 in.   **7.** 10 ft   **8.** 43.8 ft   **9.** 930 ft
10. 54 ft   **11.** 59 ft   **12.** 156 ft   **13.** 17 ft   **14.** 130 in.   **15.** 12 cm   **16.** 35.6 cm   **17.** 590 ft   **18.** 48 tiles
19. 80 in.   **20.** 309.3 mm   **21.** 52.9 ft   **22.** 207 in.   **23.** 75 ft   **24.** 402 ft   **25.** 98 ft   **26.** 28.6 in.   **27.** 48 cm
28. 28.3 cm   **29.** 72 ft   **30.** 3 yd 1 ft 8 in.

2   **31.** 72 in^2 **32.** 13.8 cm^2 **33.** 9,000 ft^2 **34.** 504 in^2 **35.** 6 ft^2 **36.** 116.84 ft^2 **37.** 42,500 ft^2 **38.** 180 ft^2
39. 241 ft^2 **40.** 142.5 board feet   **41.** 9 cm^2 **42.** 79.21 cm^2 **43.** 193 ft^2 **44.** 900 yd^2 **45.** $\frac{1}{16}$ mi^2
46. 81 tiles   **47.** 108 tiles   **48.** 400 in^2 **49.** 18 tiles   **50.** 1,000 rectangles   **51.** 8 signs   **52.** 5,053.05 mm^2
53. 162 ft^2 **54.** 348 in^2 **55.** 260 ft^2 **56.** 8,220 ft^2 **57.** 33 in^2 **58.** 96 cm^2 **59.** 18.4 cm^2 **60.** 216 ft^2
61. 9.75 ft^2 **62.** 30 ft^2 **63.** 24.75 in^2 **64.** $175.50

Self-Study Exercises 6–2

1   **1.** 25.1 cm   **2.** 47.1 m.   **3.** 18.8 in.   **4.** 100.5 yd   **5.** 9.4 ft   **6.** 53.4 ft   **7.** 17.3 m   **8.** 16.5 in.

2   **9.** 50.3 cm^2 **10.** 176.7 m^2 **11.** 28.3 in.^{2} **12.** 804.2 yd^2 **13.** 7.1 ft^2 **14.** 227.0 ft^2 **15.** 23.8 m^2
16. 21.6 in.

3   **17.** $y = 22$ ft   **18.** $y = 25$ ft 6 in.   **19.** 0.6 m^2 **20.** 2.1 ft^2 **21.** 14.6 cm^2 **22.** 1.0 in^2 **23.** 245.5 ft
            $x = 16$ ft $x = 6$ ft
24. 1,942.5 ft^2 **25.** 58 ft 8 in.   **26.** $330.00   **27.** 24 lb   **28.** $69.50   **29.** 592.7 ft^2 **30.** 268.27 in.
31. 1,178.1 mm^2 **32.** 168.5 mm^2 **33.** Yes, the cross-section area of the third pipe is larger than the combined area
of the other two pipes.   **34.** Yes, the combined cross-sectional area of the two pipes is 25.1 in^2 which is greater than
20 in^2—the area of the large pipe.   **35.** 2,827.4 ft/min   **36.** 1,570.8 ft/min   **37.** 78.5 ft^2 **38.** 0.45 in.   **39.** 4.4 m

Self-Study Exercises 6–3

1   **1.** 576 in^3 **2.** 103.74 m^3 **3.** 3,817.04 cm^3 **4.** 233.80 cm^3 **5.** 320 cm^3 **6.** 55.36 in^2 **7.** 23.3 in^3
8. 18,850 ft^3 **9.** 236 ft^3 **10.** 447.84 cm^3

Assignment Exercises, Chapter 6

1   **1.** 57 cm   **3.** 210 mm   **5.** 28.8 m   **7.** 35.8 in.   **9.** 84 ft   **11.** 66 ft   **13.** 288 in.   **15.** 60 in.
17. 162 cm^2 **19.** 2,450 mm^2 **21.** 51.84 m^2 **23.** 66.5 in^2 **25.** 306 ft^2 **27.** 33,000 ft^2 **29.** 33 yd^2 **31.** $14.25
33. 235 ft^2 **35.** 3 rolls   **37.** $A = 50.27$ m^2 **39.** $A = 126.68$ m^2 **41.** $A = 12$ cm^2 **43.** 5.8 m^2 **45.** 2.4 in.
47. 33 ft/min   **49.** 15.0 in.   **51.** 1.4 yd^2 **53.** 59.1 in.   **55.** 600 cm^3 **57.** 7,854 cm^3 **59.** 102 yd^3
61. 348,962 gal   **63.** circle; 24 ft^2

Trial Test, Chapter 6

1. $P = 91$ ft, $A = 514.5\ ft^2$ **3.** $P = 12.2$ in., $A = 6.08\ in^2$ **5.** $P = 67.3$ cm, $A = 213.85\ cm^2$ **7.** $P = 37$ cm, $A = 43.5\ cm^2$ **9.** $C = 144.5$ m, $A = 1{,}661.9\ m^2$
11. $571.8\ cm^3$ **13.** $4{,}398\ cm^3$ **15.** \$52 **17.** $282\ ft^2$ **19.** 37.70 in. **21.** 385.62 ft **23.** $0.29\ in^2$ **25.** $0.15\ in^2$

Self-Study Exercises 7–1

1 **1.** 10% **2.** 36.8% **3.** 54.3% **4.** 20.8% **5.** 33.3% **6.** 75%

2 **7.** debt retirement **8.** misc. expenses and general government **9.** social projects and education costs **10.** 1970 **11.** 7,355,000 barrels; 5,834,000 barrels **12.** 445.8% **13.** 2000

3 **14.** 5 Amps **15.** 50 volts **16.** 35 volts **17.** 25 ohms **18.** 1980 **19.** motor gasoline **20.** 1960–1970 **21.** 1960–1970 **22.** 20.3%

Self-Study Exercises 7–2

1 **1.** 10 **2.** 2 **3.** $\frac{1}{3}$ **4.** $\frac{1}{7}$ **5.** 12% **6.** 28% **7.** 35–37 and 38–40 **8.** 20–22 and 23–25 **9.** 7 **10.** 16

	Midpoint	Tally	Class Frequency
11.	15	||	2
12.	12	||||	4
13.	9	𝍸 ||	7
14.	6	𝍸 𝍸 𝍸 𝍸	20

	Midpoint	Tally	Class Frequency
15.	93	||	2
16.	88	𝍸	5
17.	83	𝍸 ||	7
18.	78	𝍸 𝍸	10
19.	73	|||	3
20.	68	𝍸 |	6
21.	63	||	2
22.	58	𝍸	5

23.

Class Interval	Midpoint	Tally	Class Frequency
15–19	17	𝍸 ||	7
10–14	12	𝍸 |||	8
5–9	7	𝍸 ||||	9
0–4	2	𝍸 |	6

24.

Class Interval	Midpoint	Tally	Class Frequency
51–60	55.5	||	2
41–50	45.5	||||	4
31–40	35.5	||	2
21–30	25.5	𝍸 ||	7
11–20	15.5	𝍸	5

2 **25.**

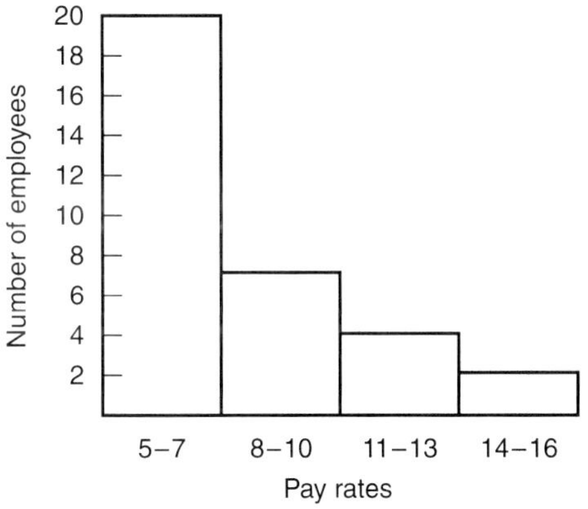

26.

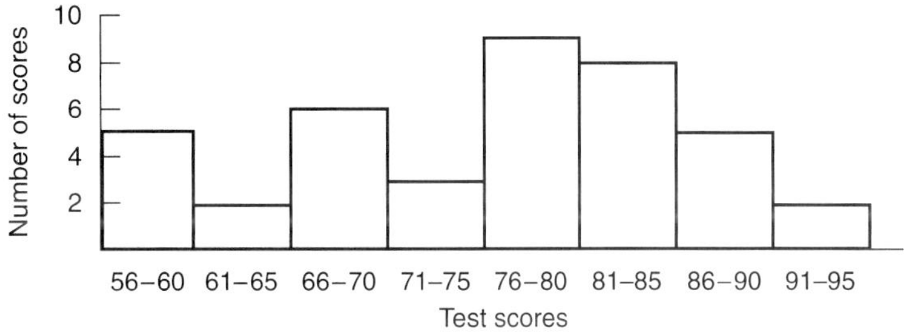

27.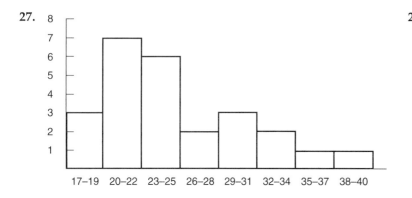

28.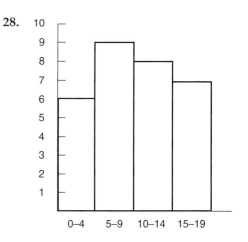

29.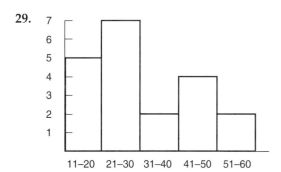

30.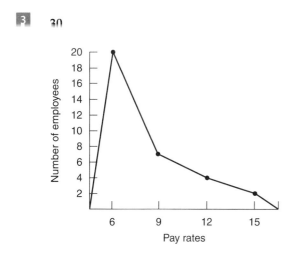

31.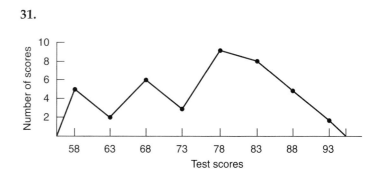

32.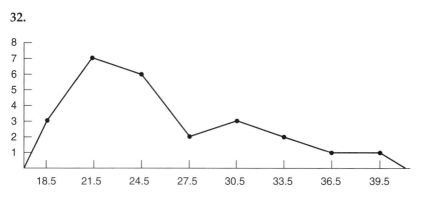

33.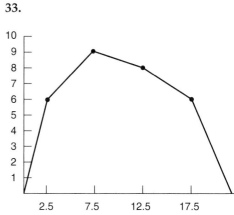

34.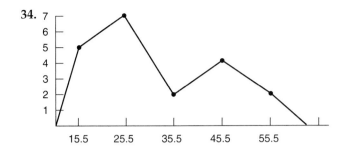

Self-Study Exercises 7–3

1 **1.** 16   **2.** 17   **3.** 66   **4.** 73.75   **5.** 33.7   **6.** 66.6   **7.** 42.33°F   **8.** 12.67°C   **9.** $34.80   **10.** $39.20
11. 14.67 in.   **12.** 8 in.   **13.** 20   **14.** 76   **15.** 17.3 runs   **16.** 102   **17.** 24.6 mpg   **18.** 10.2 mpg

2 **19.** 44   **20.** 43   **21.** 15.5   **22.** 26.5   **23.** $30   **24.** $66   **25.** $8.25   **26.** $8.85   **27.** 2   **28.** 5
29. no mode   **30.** no mode   **31.** $67   **32.** $32   **33.** 4 hours   **34.** $1.97

3 **35.** 24   **36.** 23   **37.** 13   **38.** 18   **39.** $25   **40.** $33   **41.** 48° F   **42.** 3.16   **43.** 8.29   **44.** 11.06
45. 17.36   **46.** 4   **47.** 8

Self-Study Exercises 7–4

1 **1.** Keaton Brienne Renee
      Keaton Renee Brienne
      Brienne Keaton Renee
      Brienne Renee Keaton
      Renee Keaton Brienne
      Renee Brienne Keaton
      $3 \cdot 2 \cdot 1 = 6$ ways

2. ABCD   BACD
   ABDC   BADC
   ACBD   BCAD
   ACDB   BCDA
   ADBC   BDAC
   ADCB   BDCA
   CABD   DABC
   CADB   DACB
   CBAD   DBAC
   CBDA   DBCA
   CDAB   DCAB
   CDBA   DCBA
   24 ways
   $4 \cdot 3 \cdot 2 \cdot 1 = 24$ ways

3. $4 \cdot 3 \cdot 2 \cdot 1 = 24$ ways   **4.** $5 \cdot 4 \cdot 3 \cdot 2 \cdot 1 = 120$ ways   **5.** $5 \cdot 4 \cdot 3 \cdot 2 \cdot 1 = 120$ ways

2 **6.** $\frac{1}{24}, \frac{1}{23}$ **7.** $\frac{1}{4}$ **8.** $\frac{11}{48}$ **9.** $\frac{2}{5}$ **10.** $\frac{1}{3}$ **11.** $\frac{1}{6}$

Assignment Exercises, Chapter 7

1. 1999, 2001, 2002   **3.** 2000, 2003, 2004   **5.** 12.9%   **7.** 17.4%   **9.** 7-10-04 @ 4:00 P.M.   **11.** 50 computers

13. 110 computers   **15.** $\frac{1}{2}$

	Midpoint	Tally	Class Frequency
17.	60.5	卌 卌	10
19.	40.5	卌 卌 ||	12
21.	20.5	卌 ||	7

23.

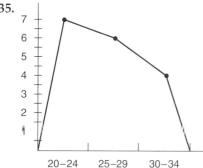

25. 36–45   **27.** 15   **29.** $\dfrac{1}{2}$ **31.** $\dfrac{10}{54} = 18.5\%$

33.

Miles per Gallon	Midpoint	Tally	MPG Frequency
20–24	22	卌 ||	7
25–29	27	卌		6
30–34	32	||||	4

(Answers may vary)

35.

37. 12.6 cars   **39.** 13.8 mpg   **41.** $6.03   **43.** $1.85   **45.** 3.67   **47.** 83

49. range: 59   **51.** 21.47   **53.** 24   **55.** $\dfrac{10}{13}, \dfrac{3}{4}$ **57.** 16   **59.** $\dfrac{1}{5}$ **61.** $\dfrac{1}{4}$
mean: 81.67
median: 89
mode: none

Trial Test, Chapter 7

1. bar   **3.** line   **5.** 2°C   **7.** $65   **9.** 25%   **11.** $\dfrac{15}{200} = \dfrac{3}{40}$ **13.** English dept., Electronics dept.   **15.** 104.2%

17. $\dfrac{42}{44} = \dfrac{21}{22}$ **19.** 6   **21.** $\dfrac{1}{2}$

	Midpoint	Tally	Class Frequency
23.	5	卌 ||||	9

25. 14   **27.** cannot be determined   **29.** range: 25   **31.** 8.49   **33.** 6   **35.** $\dfrac{3}{5}$
mean: 77.9
median: 78
mode: 81

Self-Study Exercises 8–1

1   **1.** 13   **2.** -12 **3.** -24 **4.** 22   **5.** -33 **6.** -42 **7.** 6   **8.** 4   **9.** -6 **10.** -35 **11.** -3
12. 5   **13.** -16 **14.** -27 **15.** -37 **16.** -60 **17.** -5 **18.** -11 **19.** -22 **20.** -9 **21.** -13
22. -22 **23.** 14   **24.** 25   **25.** $-21°$F   **26.** $-\$61$ **27.** -9 **28.** -10 **29.** 20   **30.** 2   **31.** -25 **32.** 2
33. 72   **34.** 72   **35.** 56   **36.** 135   **37.** -15 **38.** -60 **39.** -80 **40.** -210 **41.** -144 **42.** -15

43. 8   **44.** $\dfrac{5}{16}$ **45.** 31.92   **46.** -34.84 **47.** -7 **48.** -6 **49.** -9 **50.** -9 **51.** -8 **52.** 9   **53.** 7

54. 5   **55.** 8   **56.** $\dfrac{5}{6}$ **57.** 1.26   **58.** -11.67 **59.** -38 **60.** -7 **61.** -50 **62.** 22   **63.** -50

2   **64.** -23 **65.** 19   **66.** -6 **67.** 335   **68.** 869   **69.** -15 **70.** 25   **71.** 0

Self-Study Exercises 8–2

1   **1.** $-15 = -15$ **2.** $11 = 11$ **3.** $2 = 2$ **4.** $-7 = -7$ **5.** $n = 11$ **6.** $m = 6$ **7.** $y = -4$ **8.** $x = 6$

9. $p = 9$ **10.** $b = -9$ **11.** $\boxed{7} + \boxed{c}$ **12.** $\boxed{4a} - \boxed{7}$ **13.** $\boxed{3x} - \boxed{2(x+3)}$ **14.** $\boxed{\dfrac{a}{3}}$ **15.** $\boxed{7xy} + \boxed{3x} - \boxed{4} + \boxed{2(x+y)}$

16. $\boxed{14x} + \boxed{3}$ **17.** $\boxed{\dfrac{7}{(a+5)}}$ **18.** $\boxed{\dfrac{4x}{7}} + \boxed{5}$ **19.** $\boxed{11x} - \boxed{5y} + \boxed{15xy}$ (Answers may vary.)   **20.** 5   **21.** -4
22. $\dfrac{1}{5}$ **23.** $\dfrac{2}{7}$ **24.** 6   **25.** $-\dfrac{4}{5}$ **26.** $-15c$ (Answers may vary.)

2   **27.** Four more than a number is 7. (Answers may vary.)   **28.** Five less than a number is 2. (Answers may vary.)   **29.** Three times a number is 15. (Answers may vary.)   **30.** One more than 3 times a number is 7. (Answers may vary.)

3   **31.** $x + 5 = 12$ **32.** $\dfrac{x}{6} = 9$ **33.** $4(x - 3) = 12$ **34.** $4x - 3 = 12$ **35.** $12 + 7 + x = 17$ **36.** $2x + 7 = 21$

37. $x + 15° = 48°$ **38.** $x + 15 \text{ mL} = 45 \text{ mL}$ **39.** $x \cdot 5 = 45$ **40.** $\dfrac{18 - 3}{5} = x$

4   **41.** $-3a$ **42.** $-8x - 2y$ **43.** $7x - 2y$ **44.** $-3a + 6$ **45.** 10   **46.** $3a + 6b + 8c + 1$ **47.** $10x - 20$
48. $6x + 13$ **49.** $-4x + 5$ **50.** $-12a + 3$

Self-Study Exercises 8–3

1   **1.** $x = 8$ **2.** $x = 15$ **3.** $x = -3$ **4.** $x = -8$ **5.** $x = 4$ **6.** $x = -11$ **7.** $x = -26$ **8.** $x = 3$ **9.** $x = 5$
10. $x = -7$ **11.** $x = 16$ **12.** $x = 9$ **13.** $x = -2$ **14.** $x = 4$

2   **15.** $x = 8$ **16.** $x = 9$ **17.** $x = 5$ **18.** $n = 30$ **19.** $a = -9$ **20.** $b = -2$ **21.** $c = 7$ **22.** $x = -9$
23. $x = \dfrac{32}{5}$ **24.** $y = \dfrac{9}{2}$ **25.** $n = -\dfrac{5}{2}$ **26.** $b = \dfrac{28}{3}$ **27.** $x = -7$ **28.** $y = 2$ **29.** $x = 15$ **30.** $y = 8$
31. $x = \dfrac{10}{3}$ **32.** $y = -\dfrac{20}{3}$ **33.** $x = 12$ **34.** $n = 56$

3   **35.** $x = 6$ **36.** $m = 7$ **37.** $a = -3$ **38.** $m = -\dfrac{1}{2}$ **39.** $y = 8$ **40.** $x = 0$ **41.** $y = -7$ **42.** $y = -4$

43. $x = -6$ **44.** $b = -1$ **45.** $x = 8$ **46.** $t = 6$ **47.** $x = 3$ **48.** $y = 5$ **49.** $x = -\dfrac{7}{2}$ **50.** $a = -2$

51. all real numbers   **52.** no solution   **53.** $x - -8$ **54.** $t = 7$ **55.** $x = -1$ **56.** $y = 11$ **57.** $y = -2$

58. $x = 27$ **59.** $y = 13$ **60.** $R = -68$ **61.** $P = \dfrac{5}{6}$ **62.** $x = 14$ **63.** $x = -\dfrac{4}{15}$ **64.** $m = \dfrac{49}{18}$ **65.** $s = \dfrac{8}{5}$

66. $m = \dfrac{8}{3}$ **67.** $T = \dfrac{68}{9}$ **68.** $x = 2$ **69.** $R = 0.8$ **70.** $x = 6.03$ **71.** $x = 0.08$ **72.** $R = 0.34$

1   **1.** $x = 7$ **2.** $y = 2$ **3.** $x = -5$ **4.** $x = -2$ **5.** $x = 5$ **6.** $x = -6$ **7.** $x = 1$ **8.** $x = 8$ **9.** $x = 1$
10. $x = 0$ **11.** no solution   **12.** all real numbers   **13.** no solution   **14.** no solution   **15.** $x = 2$ **16.** $x = 5$
17. $x = \dfrac{1}{2}$ **18.** $x = -15$ **19.** $x = -9$ **20.** $x = 1$ **21.** $x = -2$ **22.** $x = -1$ **23.** $x = \dfrac{16}{5}$ **24.** $x = 5$
25. $x + 4 = 12; x = 8$ **26.** $2x - 4 = 6; 5$

27. $x + x + (x - 3) = 27$;
Two parts weigh 10 lb.
Third part weighs 7 lb.

28. $24 + x = 60$;
36 gal

29. $4.03 - 3.97 = x$;
0.06 kg

30. $x + 3x = 400$
100 ft of solid pipe
300 ft of perforated pipe

31. $x + 2x + \$70 = \$235$
Spanish text: \$55
Calculator: \$110

32. $C = \$14(4)(150) = \$8,400$

33. $x + 2x = 325$

tank 1: $216\dfrac{2}{3}$ gal

tank 2: $108\dfrac{1}{3}$ gal

34. $A = \dfrac{1}{2}bh$; 12 ft

$60 = \dfrac{1}{2}(b)(10)$

35. B14 = B5 + B6 + B7 + B8 + B9 + B10 + B11 + B12
D14 = D5 + D6 + D7 + D8 + D9 + D10 + D11 + D12
C5 = B5 ÷ B14 × 100
C6 = B6 ÷ B14 × 100
C7   D7 : B11)(100
⋮
C12 = B12 ÷ B14 × 100
E5 = D5 ÷ D14 × 100
E6 = D6 ÷ D14 × 100
E7 = D7 ÷ D14 × 100
⋮
E12 = D12 ÷ D14 × 100
F5 = (D5 − B5) ÷ B5 × 100
F6 = (D6 − B6) ÷ B6 × 100
F7 = (D7 − B7) ÷ B7 × 100
⋮
F12 = (D12 − B12) ÷ B12 × 100

36. See the figure.

	A	B	C	D	E	F
1		The 7th Inning: Budget Operating Expenses and Actual Expenses				
2						
3	Expense	Budget Amount	Percent of Total Budget	Actual Expenses	Percent of Actual Total Expense	Percent Difference from Budget
4						
5	Salaries	$45,000.00	17.93%	$42,000.00	16.57%	−6.7%
6	Rent	$37,000.00	14.74%	$36,000.00	14.20%	−2.7%
7	Depreciation	$12,000.00	4.78%	$14,000.00	5.52%	+16.7%
8	Utilities and phone	$13,000.00	5.18%	$10,862.56	4.28%	−16.4%
9	Taxes and Insurance	$15,000.00	5.98%	$13,583.29	5.36%	−9.4%
10	Advertising	$2,000.00	0.80%	$2,847.83	1.12%	+42.4%
11	Purchases	$125,000.00	49.80%	$132,894.64	52.41%	+6.3%
12	Other	$2,000.00	0.80%	$1,356.35	0.53%	−32.2%
13						
14	Total	$251,000.00	100.01%	$253,544.67	99.99%	

Self-Study Exercises 8–5

1. $4x - 3 = 9$

x	$4x - 3$	9	Observations
0	−3	9	$4(0) - 3 = -3$ Expression values differ by 12.
1	1	9	$4(1) - 3 = 1$ Expression values differ by 8.
2	5	9	$4(2) - 3 = 5$ Expression values differ by 4.
3	9	9	$4(3) - 3 = 9$ Expressions are equal.

The solution is 3.

2. $7x - 1 = 4x + 17$

x	$7x - 1$	$4x + 17$	Observations
0	−1	+17	Values differ by 18.
2	13	25	Values differ by 12.
4	27	33	Values differ by 6.
6	41	41	Values are equal.

The solution is 6.

3. $7x = 8x + 4$

x	$7x$	$8x + 4$	Observations
1	7	12	Values differ by 5.
2	14	20	Values differ by 6.
−1	−7	−4	Values differ by 3.
−3	−21	−20	Values differ by 1.
−4	−28	−28	Values are equal.

The solution is −4.

4. $4x - 13 = 2x + 15$

x	$4x - 13$	$2x + 15$	Observations
10	27	25	Values differ by 2.
11	31	37	Values differ by 6.
14	43	43	Values are equal.

The solution is 14.

5. $7(x + 2) = -6 + 2x$

x	$7(x + 2)$	$-6 + 2x$	Observations
10	84	14	Values differ by 70.
0	14	-6	Values differ by 20.
-4	-14	-14	Values are equal.

The solution is -4.

6. The shorter side of a carpenter's square in the shape of an L is 3 in. shorter than the longer side. If the total length of the L-shaped tool is 28 in., what is the measure of each side of the carpenter's square?

Longer Side	Shorter Side	Observations
t	$t - 3$	Total length = 28 in.
20	17	37
15	12	27
16	13	29
$15\frac{1}{2}$	$12\frac{1}{2}$	28

Length $= 15\frac{1}{2}$ in. Width $= 12\frac{1}{2}$ in.

7. Nurse Vance can prepare dosages for her patients in 30 min. If she gets help from assistants who work at the same rate and together they complete the preparation in 5 min, how many helpers did she get?

Helpers	Time Required to Complete Task	Observations
h	$\dfrac{30 \text{ min}}{h + 1}$	Nurse Vance + helpers complete the task in 5 min.
1	15	Difference is 10 min.
2	10	Difference is 5 min.
3	7.5	Difference is 2.5 min.
4	6	Difference is 1 min.
5	5	Five workers take 5 min.

She got 5 helpers.

8. A certain fabric sells at a rate of 3 yd for $8.00. How many yards can Kristin buy for $36?

Yards of Fabric	Total Cost	Observations
y	$\dfrac{\$8.00}{3} \times y$	Total cost should be $36.00
3	8	Difference is $28.
12	32	Difference is $4.
15	40	Difference is $4.
13.50	36	Two amounts are equal.

Kristin can buy 13.5 yds.

948

9. A group of investors purchase land in Rhode Island for $120,000. When four more investors joined the group, the cost dropped $8,000 per person for the original group. How many investors were in the original group?

Number of Original Investors	Cost per n Investors	Cost per $n + 4$ Investors	Observations
n	$\dfrac{\$120,000}{n}$	$\dfrac{\$120,000}{n+4}$	New cost is $8,000 less per investor than the original cost per investor.
8	15,000	10,000	Difference is $5,000.
4	30,000	15,000	Difference is $15,000.
6	20,000	12,000	Difference is $8,000.

There were 6 investors in the original group.

10. What are the dimensions of a rectangular room if the area is 84.5 m² and the room is twice as long as it is wide?

Width	Length	Observations
w	$2w$	Area is 84.5 m²
6 m	12 m	72 m²
7	14	98 m²
6.5	13	84.5 m²

The width is 6.5 meters and the length is 13 meters.

Self-Study Exercises 8–6

1 1. $f(2) = 7, f(-3) = 2, f(0) = 5$ 2. $f(-4) = -19, f(5) = 8, f\left(\dfrac{1}{6}\right) = -6\dfrac{1}{2}$ 3. $f(-0.2) = -1.4, f\left(\dfrac{3}{4}\right) = 2.4,$

$f\left(-\dfrac{3}{5}\right) = -3$ 4. $f(0) = 6, f(1) = 9, f(-5) = -9$

2 5. positive multiples of 5 6. odd integers 7. integers that are perfect squares 8. integers less than 8

3 9. $f(-2) = -2$
$f(-1) = -1$
$f(0) = 0$
$f(1) = 1$
$f(2) = 2$

$f(x)$

(2, 2)
(1, 1)
(-1, -1) (0, 0) x
(-2, -2)

10. $f(-2) = -4$
$f(-1) = -2$
$f(0) = 0$
$f(1) = 2$
$f(2) = 4$

$f(x)$

(2, 4)
(1, 2)
(-1, -2) (0, 0) x
(-2, -4)

11. x is the number of days and is the independent variable.

$f(x)$ is the cost of renting the wheelchair and is the dependent variable.

$f(x) = 12x + 20$

x	f(x)
0	20
1	32
2	44
3	56

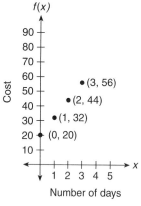

12. x is the number of minutes and the independent variable. $f(x)$ is the monthly cellphone cost and the dependent variable. $f(x) = 0.10x + 35$

x	f(x)
0	35
50	40
100	45
150	50
200	55

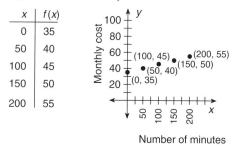

Assignment Exercises, Chapter 8

1. 12   **3.** -19 **5.** 107   **7.** -1 **9.** 6   **11.** -5 **13.** -3 **15.** -25 **17.** -2 **19.** 23   **21.** -1 **23.** 1   **25.** 15   **27.** 40   **29.** 33   **31.** 4   **33.** -12 **35.** -9 **37.** -37 **39.** 21   **41.** 2   **43.** -554 **45.** $12 = 12$

47. $8 = 8$ **49.** $x = 14$ **51.** $y = -8$ **53.** $\boxed{15x} - \boxed{\dfrac{3a}{7}} + \boxed{\dfrac{(x-7)}{5}}$ **55.** A number increased by 5 equals 2. Answers will vary.   **57.** The quotient of a number and 8 equals 7. Answers will vary.   **59.** Three times the sum of a number and 7 equals -3. Answers will vary.   **61.** $2x + 7 = 11$ **63.** $2(x + 8) = 40$ **65.** $5a$ **67.** $7y - 12$ **69.** $-3a - 11$ **71.** $-2x + 21$ **73.** $x = 13$ **75.** $x = 38$ **77.** $x = -2$ **79.** $x = 19$ **81.** $x = 7$ **83.** $y = -\dfrac{15}{2}$ **85.** $y = 7$ **87.** $x = -8$ **89.** $x = 15$ **91.** $x = 64$ **93.** $x = 10$ **95.** $x = 84$ **97.** $x = 5$ **99.** $b = -2$ **101.** $x = 7$ **103.** $x = -4$ **105.** $x = 3$ **107.** $x = -1$ **109.** $a = 5$ **111.** $x = 5$ **113.** $x = 7$ **115.** $x = 4$ **117.** $x = 2$ **119.** $x = -3$ **121.** no solution   **123.** $y = -12$ **125.** $y = -3$ **127.** $x = \dfrac{9}{2}$ **129.** $y = 7$ **131.** $x = 9$ **133.** $x = \dfrac{20}{13}$ **135.** $x = \dfrac{7}{20}$ **137.** $c = -\dfrac{9}{32}$ **139.** $y = 1.5$ **141.** $R = 0.61$ **143.** $y = -1$ **145.** $x = -9$ **147.** $x = 1$ **149.** $x = 3$ **151.** $x = 2$ **153.** $x = -4$ **155.** $x = 2$ **157.** $x = 0$ **159.** $x = 3$ **161.** $x = 3$ **163.** $x = -1$ **165.** $x = -\dfrac{1}{8}$ **167.** $x - 6 = 8; x = 14$

169. $5(x + 6) = 42 + x; x = 3$ **171.** $x + (x - 3) = 51; 27$ hr, 24 hr   **173.** $6 = \pi d; d = 1.9$ in.; no   **175.** $720 = 2(w + 2w); w = 120$ ft; $l = 2w = 240$ ft   **177.** $3x - 8 = 4$

x	3x − 8	4	Observations
0	−8	4	Difference of 12.
2	−2	4	Difference of 6.
4	4	4	4 = 4

The solution is 4.

179. $6x = 7x + 3$

x	$6x$	$7x + 3$	Observations
0	0	3	Difference of 3.
1	6	10	Difference of 4.
−1	−6	−4	Difference of 2.
−2	−12	−11	Difference of 1.
−3	−18	−18	−18 = −18

The solution is −3.

181. $5(x + 1) = 3x - 7$

x	$5(x + 1)$	$3x - 7$	Observations
2	15	−1	Difference of 16.
−2	−5	−13	Difference of 8.
−3	−10	−16	Difference of 6.
−5	−20	−22	Difference of 2.
−6	−25	−25	−25 = −25

The solution is −6.

183.

Additional/ Robots	Time Required to Complete Task		Observations
r	$\dfrac{12 \text{ hr}}{r + 1}$	4	First robot & additional robots complete the task in 4 hr.
0	12	4	Difference of 8 hr.
1	6	4	Difference of 2 hr.
2	4	4	4 = 4

The solution is 2 additional robots for a total of 3 robots.

185. $f(-2) = -5, f(-1) = -3, f(0) = -1, f(1) = 1, f(2) = 3$ **187.** $f(-3) = 4, f(0) = 1, f(3) = -2, f(5) = -4$
189. all integers   **191.** whole-number multiples of 3

193.

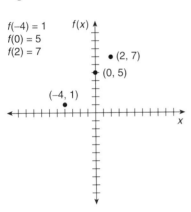

$f(-4) = 1$
$f(0) = 5$
$f(2) = 7$

195.

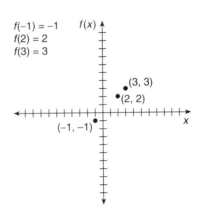

$f(-1) = -1$
$f(2) = 2$
$f(3) = 3$

197.

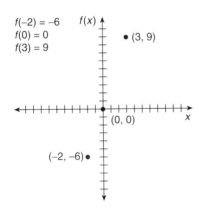

$f(-2) = -6$
$f(0) = 0$
$f(3) = 9$

199. $f(x) = 3x - 2; f(1) = 1; f(2) = 4$ Answers will vary.

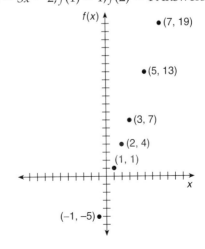

Trial Test, Chapter 8

1. 14   **3.** 6   **5.** 10   **7.** -3 **9.** -15 **11.** 3   **13.** 2   **15.** 2   **17.** 1.25   **19.** 2   **21.** $x + x + 4,980 = 11,034$; Ross WMA = 3,027 acres; Potts WMA = 8,007 acres   **23.** $f(-2) = 7; f(5) = -35$

Self-Study Exercises 9–1

1   **1.** 28   **2.** -11 **3.** $\dfrac{18}{7}$ **4.** 0   **5.** $-\dfrac{27}{5}$ **6.** $-\dfrac{5}{27}$ **7.** $\dfrac{8}{5}$ **8.** $\dfrac{11}{15}$ **9.** -350 **10.** 48   **11.** 27
12. $\dfrac{1}{3}$ **13.** 13   **14.** $\dfrac{25}{8}$ **15.** 0   **16.** $\dfrac{1}{9}$ **17.** 576   **18.** 4   **19.** 28   **20.** -68 **21.** $\dfrac{27}{35}$ **22.** $\dfrac{108}{5}$
23. $\dfrac{217}{24}$ **24.** 8   **25.** -5 **26.** $\dfrac{3}{10}$ **27.** 70   **28.** 0   **29.** $\dfrac{32}{63}$ **30.** $\dfrac{32}{5}$ **31.** $-\dfrac{36}{11}$ **32.** $-\dfrac{1}{108}$ **33.** 21
34. $\dfrac{4}{25}$ **35.** $\dfrac{2}{21}$ **36.** $\dfrac{9}{7}$

2   **37.** 80 vases   **38.** $1\dfrac{1}{3}$ hours   **39.** $3\dfrac{1}{13}$ hr   **40.** $3\dfrac{3}{7}$ days   **41.** $\dfrac{3}{5}$ hr or 36 min   **42.** 3 min   **43.** 15 min
44. $2\dfrac{1}{3}$ min   **45.** 8.57 ohms   **46.** 5 ohms   **47.** 7.38 ohms   **48.** 3.6 ohms

Self-Study Exercises 9–2

1   **1.** 2   **2.** 0.8   **3.** 6.03   **4.** 16   **5.** 4.7   **6.** 0.08   **7.** 0.035   **8.** 0.34   **9.** -3.3 **10.** 38.8   **11.** 3.3
12. -0.2 **13.** 0.6   **14.** 1.128   **15.** 16

2   **16.** 87.5 lb   **17.** 4.71 in.   **18.** 12 hr   **19.** 6.5 hr   **20.** 8.3 Ω **21.** $7.30   **22.** 156.25 V   **23.** $264
24. $136   **25.** $1,925   **26.** 12.25%   **27.** 1.2%   **28.** 18.5%   **29.** $2,777.78   **30.** $\dfrac{1}{2}$ yr or 6 mo

Self-Study Exercises 9–3

1   **1.** 3   **2.** 2   **3.** 29   **4.** $\dfrac{5}{8}$ **5.** $\dfrac{34}{5}$ **6.** $\dfrac{17}{14}$ **7.** 3   **8.** 21   **9.** 4   **10.** $-\dfrac{4}{5}$

2   **11.** $5.90   **12.** $74.13   **13.** 48 engines   **14.** 15 headpieces   **15.** 1,425 mi   **16.** 3,360 mi   **17.** 550 lb
18. 1.7 gal   **19.** $0.0019 per gram   **20.** $0.0022 per gram   **21.** larger can has lower cost per gram   **22.** 320 bottles per hour

3   **23.** 360   **24.** 960   **25.** 900 rpm   **26.** 200 in^3 **27.** 20 in.   **28.** 18 painters   **29.** 10 hr   **30.** 50 rpm
31. 5 in.   **32.** 4 hr   **33.** 5 machines   **34.** 120 rpm   **35.** 4 helpers (5 workers total)   **36.** 9 in.

4   **37.** $\angle A = \angle F, \angle B = \angle E, \angle C = \angle D$ **38.** $PQ = ST, \angle P = \angle S, \angle Q = \angle T$ **39.** $JL = MP, LK = NP, \angle L = \angle P$
40. $a = 20, d = 9$ **41.** $DC = 6, DE = 5\dfrac{1}{3}$ **42.** 65 ft

Assignment Exercises, Chapter 9

1. $\dfrac{1}{2}$ **3.** $\dfrac{5}{6}$ **5.** $-\dfrac{4}{15}$ **7.** $\dfrac{49}{18}$ **9.** $-\dfrac{60}{13}$ **11.** $\dfrac{8}{3}$ **13.** $\dfrac{9}{2}$ **15.** $-\dfrac{7}{4}$ **17.** $\dfrac{16}{7}$ **19.** $\dfrac{10}{7}$ **21.** 6   **23.** $\dfrac{31}{39}$

25. $2\frac{1}{10}$ hr **27.** 25 fixtures **29.** 1.33 ohms **31.** 1.9 **33.** 2 **35.** 77.2 **37.** -11.8 **39.** -0.8 **41.** 1.5

43. 0.44 **45.** 0.02 **47.** 3.4 **49.** 17 Ω **51.** 110 V **53.** 375 lb **55.** 2 **57.** $\frac{7}{6}$ **59.** $\frac{1}{2}$ **61.** $\frac{49}{12}$

63. $-\frac{21}{2}$ **65.** $\frac{3}{14}$ **67.** $-\frac{21}{23}$ **69.** 23 machines **71.** $2\frac{1}{2}$ hr **73.** 4,500 women **75.** 1,500 rpm **77.** 3 days

79. $4\frac{1}{5}$ ft **81.** 129.7 gal **83.** 5.0 hr **85.** 1,351 ft **87.** 60 rpm **89.** 3 machines **91.** $\frac{AB}{RT} = \frac{BC}{TS} = \frac{AC}{RS}$

93. 10 **95.** 100 teeth **97.** $2\frac{6}{7}$ gal water, $7\frac{1}{7}$ gal antifreeze

Trial Test, Chapter 9

1. 16 **3.** 21 **5.** $-\frac{11}{5}$ **7.** $-\frac{22}{7}$ **9.** $-\frac{56}{3}$ **11.** $\frac{7}{25}$ **13.** $\frac{10}{21}$ **15.** 6.17 **17.** 0.30 **19.** 254.24 **21.** 2.5
23. -0.15 **25.** 168.75 rpm **27.** 1.33 psi **29.** 54.7 L **31.** 11.429 Ω **33.** 107 lb **35.** 5

Self-Study Exercises 10–1

1. x^7 **2.** m^4 **3.** a^2 **4.** x^{10} **5.** y^6 **6.** a^2b **7.** $12a^5b^7$ **8.** $10x^{11}$

9. y^5 **10.** x^4 **11.** $\frac{1}{a}$ **12.** b **13.** 1 **14.** $\frac{1}{x^2}$ **15.** y^4 **16.** n^7 **17.** $\frac{1}{x^{10}}$ **18.** $\frac{2}{n}$ **19.** $\frac{3x^7}{2}$

20. $\frac{1}{x^5}$ **21.** $\frac{x}{y^6}$ **22.** $\frac{a^3}{b^2}$ **23.** xy^2 **24.** $\frac{a}{b}$

25. 4,096 **26.** x^8 **27.** 1 **28.** x^{40} **29.** a^{21} **30.** $-\frac{1}{8}$ **31.** $\frac{4}{49}$ **32.** $\frac{a^4}{b^4}$ **33.** $8m^6n^3$

34. $\frac{x^6}{y^3}$ **35.** x^6y^{12} **36.** $4a^2$ **37.** x^6y^3 **38.** $-27a^3b^6$ **39.** $-16,807x^{10}$ **40.** $16x^4y^{16}$

Self-Study Exercises 10–2

1. binomial **2.** binomial **3.** monomial **4.** monomial **5.** monomial **6.** monomial
7. trinomial **8.** trinomial **9.** monomial **10.** binomial **11.** binomial **12.** binomial

13. 1 **14.** 2 **15.** 2, 1, 0 **16.** 3, 1, 0 **17.** 1, 0 **18.** 2, 0 **19.** 0 **20.** 0 **21.** 1, 0 **22.** 2, 0
23. 5, 2 **24.** 5, 8 **25.** 2 **26.** 3 **27.** 6 **28.** 5 **29.** 2 **30.** 1 **31.** 6 **32.** 6

33. $-3x^2 + 5x$; 2; $-3x^2$; -3 **34.** $-x^3 + 7$; 3; $-x^3$; -1 **35.** $9x^2 + 4x - 8$; 2; $9x^2$; 9
36. $5x^2 - 3x + 8$; 2; $5x^2$; 5 **37.** $7x^3 + 8x^2 - x - 12$; 3; $7x^3$; 7 **38.** $-15x^4 + 12x + 7$; 4; $-15x^4$; -15
39. $8x^6 - 7x^3 - 7x$; 6; $8x^6$; 8 **40.** $-14x^8 + x + 15$; 8; $-14x^8$; -14 **41.** $8x^4y^5 + 15x^3y^7 + 12x$; 10; $8x^4y^5$; 8
42. $-7x^4y + 5x^3y + 7xy^3 - 8y^6$; 6; $-7x^4y$; -7

Self-Study Exercises 10–3

1. $7a^2$ **2.** $3x^3$ **3.** $-2a^2 + 3b^2$ **4.** $2a - 2b$ **5.** $-5x^2 - 2x$ **6.** $5a^2$ **7.** $-x^2 - 2y$ **8.** $2m^2 + n^2$
9. $9a + 2b + 10c$ **10.** $-2x + 3y + 2z$ **11.** -4 **12.** $3x + 1$

13. $14x^3$ **14.** $2m^3$ **15.** $-21m^2$ **16.** $-2y^6$ **17.** $8x^3 - 28x^2$ **18.** $-6a^3b^2 + 15a^2b^3$ **19.** $35x^4 - 21x^3 - 49x$
20. $6x^3y^2 - 10x^2y^3$ **21.** $3x^2 - 18x$ **22.** $12x^3 - 28x^2 + 32x$ **23.** $-8x^2 + 12x$ **24.** $10x^2 + 4x^3$

25. $2x^2$ **26.** $-\dfrac{a}{2}$ **27.** $\dfrac{1}{2x^2}$ **28.** $-\dfrac{3x^3}{4}$ **29.** $3x - 2$ **30.** $4x^3 - 2x - 1$ **31.** $7x - \dfrac{1}{x}$

32. $4x^2 + 3x$ **33.** $\dfrac{x^2}{3} + \dfrac{4}{9}$ **34.** $5ab^2 - 3b - \dfrac{7}{ab}$ **35.** $\dfrac{5ab^2}{2} - \dfrac{b}{2}$ **36.** $\dfrac{3x}{y} - \dfrac{2y}{x} - 1$ **37.** $5x - 14x^4$

38. $24x^9y^6 + 4$ **39.** $-4 - 4x^2$ **40.** $\dfrac{x}{5y} - \dfrac{y}{3x} - 2x^2$

Self-Study Exercises 10–4

1. 37 **2.** 1,820 **3.** 0.56 **4.** 1.42 **5.** 780,000 **6.** 62 **7.** 0.00046 **8.** 0.61 **9.** 0.72 **10.** 42
11. 10^{10} **12.** 10^{-7} or $\dfrac{1}{10^7}$ **13.** 10^{-3} or $\dfrac{1}{10^3}$ **14.** 10 **15.** 10^3 **16.** 10^2 **17.** $\dfrac{1}{10^3}$ **18.** 1 **19.** $\dfrac{1}{10^5}$ **20.** $\dfrac{1}{10}$

21. 430 **22.** 0.0065 **23.** 2.2 **24.** 73 **25.** 0.093 **26.** 83,000 **27.** 0.0058 **28.** 80,000 **29.** 6.732
30. 0.00589

31. 3.92×10^2 **32.** 2×10^{-2} **33.** 7.03×10^0 **34.** 4.2×10^4 **35.** 8.1×10^{-2} **36.** 2.1×10^{-3}
37. 2.392×10^1 **38.** 1.01×10^{-1} **39.** 1.002×10^0 **40.** 7.21×10^2 **41.** 4.2×10^5 **42.** 3.26×10^4
43. 2.13×10^1 **44.** 6.2×10^{-6} **45.** 5.6×10^1

46. 2.144×10^7 **47.** 5.6×10^3 **48.** 2.36×10^{-2} **49.** 4.73×10^{-3} **50.** 7×10^2 **51.** 6.5×10^{-5}
52. 7×10^2 **53.** 8×10^{-3} **54.** 3.2285×10^{13} miles **55.** 4.2×10^2 or 420 Å

Assignment Exercises, Chapter 10

1. x^{10} **3.** $21x^4$ **5.** x^5 **7.** $7x^5$ **9.** $\dfrac{y}{y^3}$ **11.** x^{11} **13.** x^{16} **15.** $-27x^6$ **17.** $\dfrac{1}{y^2}$ **19.** 5

21. $3x^3 + x^2 + 5x - 8; 3; 3x^3, 3$ **23.** $2x^4 + 5x^2 - 12x - 32; 4; 2x^4; 2$ **25.** $-2x^4 - 2x^3 + 3x$ **27.** $-3x^2 + 5y^2$

29. $21x^6$ **31.** $2x^3 + 6x^2 - 10x$ **33.** $-2x^4 + 14x^3 - 30x$ **35.** $-\dfrac{2x^3}{3}$ **37.** $-\dfrac{14}{5y^2}$ **39.** $2x^2 - 4x + 7$

41. $\dfrac{x^3}{2} - 6x^6$ **43.** 1,000,000,000,000 **45.** $\dfrac{1}{1,000}$ or 0.001 **47.** 8,730 **49.** 375,000 **51.** 0.0000387 **53.** 5.2×10^4
55. 6.7×10^2 **57.** 1.7×10^{-4} **59.** 5.6×10^3 **61.** 5.2×10^4 **63.** 5.83×10^6 **65.** 4.198×10^7 **67.** 3.3×10^{-2}
69. 8×10^{-9} **71.** 7.19×10^{-5} **73.** 6.75×10^{-4} **75.** 4×10^{-3} **77.** 8×10^{-7} **79.** 1.6×10^3 **81.** 9.7×10^7
83. 5.2×10^{-4} **85.** 1.053×10^1 **87.** 5.082×10^3 **89.** 7.8×10^3 **91.** 2.5×10^4 **93.** 4×10^3 **95.** 4.7×10^{-3}
97. 4.368×10^{126} **99.** 3.4×10^{10} (rounded to tenths) **101.** 2.5×10^8 **103.** 1.2×10^6 km or 1,200,000 km

Trial Test, Chapter 10

1. x^5 **3.** $\dfrac{16}{49}$ **5.** $\dfrac{x^4}{y^2}$ **7.** $12a^3 - 8a^2 + 20a$ **9.** 10^6 **11.** $x^2 - 5x$ **13.** 42,000 **15.** 0.059 **17.** 2.4×10^2

19. 2.1×10^{-2} **21.** 7.83×10^{-3} **23.** 3.5×10^2 **25.** 1.5×10^6 ohms

Self-Study Exercises 11–1

1. $\sqrt{36}; 36^{1/2}$ **2.** $\sqrt[3]{27}; 27^{1/3}$ **3.** $\sqrt[4]{81}; 81^{1/4}$ **4.** $\sqrt[5]{32}; 32^{1/5}$ **5.** $\sqrt{81}; 81^{1/2}$ **6.** $\sqrt[3]{1,331}; 1,331^{1/3}$

7. 6 and 7 **8.** 8 and 9 **9.** 3 and 4 **10.** 10 and 11 **11.** 2 and 3 **12.** 3 and 4 **13.** 3 and 4 **14.** 4 and 5

15.–22. See number line and answers that follow. **15.** 5 **16.** 4.1 **17.** 4.5 **18.** 3.2 **19.** 4.2
20. 5.7 **21.** 2.6 **22.** 6.3

23. $x^{3/2}; \sqrt{x^3}$ **24.** $x^{2/3}; \sqrt[3]{x^2}$ **25.** $x^{4/3}; \left(\sqrt[3]{x}\right)^4$ **26.** $x^{2/3}; \left(\sqrt[3]{x}\right)^2$ **27.** $x^{3/4}; \left(\sqrt[4]{x}\right)^3$ **28.** $x^{5/3}; \left(\sqrt[3]{x}\right)^5$
29. $\sqrt[5]{x^3}$ **30.** $\sqrt[5]{x^2}$ **31.** $\sqrt[4]{x^3}$ **32.** $\sqrt[8]{x^5}$ **33.** $\sqrt[5]{y}$ **34.** $\sqrt[8]{y^7}$ **35.** $\sqrt[8]{y^3}$ **36.** $\sqrt[5]{y^4}$ **37.** $x^{2/3}$
38. $x^{1/2}$ **39.** $x^{2/5}$ **40.** $x^{3/5}$ **41.** $x^{5/6}$ **42.** $x^{1/2}$ **43.** $x^{2/3}$ **44.** $x^{4/7}$

Self-Study Exercises 11–2

1 **1.** x^5 **2.** x^3 **3.** $\pm x^8$ **4.** $\dfrac{x^7}{y^{12}}$ **5.** $a^3 b^5$ **6.** $-ab^2 c^6$ **7.** $\dfrac{xy^2}{z^5}$ **8.** $\dfrac{a^2}{b^5 c^6}$

2 **9.** $y^{1/3}$ **10.** $(5x)^{1/3}$ or $5^{1/3} x^{1/3}$ **11.** $x^{5/2} y^{5/2}$ **12.** $r^{7/2}$ **13.** $3x^2$ **14.** $2r^3$ **15.** $16x^4 y^8$ **16.** $32x^{5/4} y^{25/4}$
17. $x^{1/4}$ **18.** $(7x)^{1/5}$ or $7^{1/5} x^{1/5}$ **19.** $(xy)^{4/3}$ or $x^{4/3} y^{4/3}$ **20.** $3b^5$ **21.** $4x^{2/3} y^{14/3}$ **22.** x^4 **23.** y^3 **24.** $64ab^6$
25. $8x^2 y^{5/2}$ **26.** $\dfrac{1}{x^{1/3}}$ **27.** $x^{3/5}$ **28.** $\dfrac{1}{y^{1/6}}$ **29.** $b^{1/6}$ **30.** $4x^{7/2}$ **31.** $5x^{7/3}$ **32.** 2 **33.** 192 **34.** $a^{5.2}$
35. $49a^8$ **36.** x^2 **37.** $3x^2$ **38.** $6x^6$

3 **39.** $2\sqrt{6}$ **40.** $7\sqrt{2}$ **41.** $4\sqrt{3}$ **42.** $x^4\sqrt{x}$ **43.** $y^7\sqrt{y}$ **44.** $2x\sqrt{3x}$ **45.** $2a^2\sqrt{14a}$
46. $6ax^2\sqrt{2a}$ **47.** $2x^2yz^3\sqrt{11xz}$ **48.** $\sqrt{8x^3} = 2x\sqrt{2x}$; Answers will vary.

Self-Study Exercises 11–3

1 **1.** $12\sqrt{3}$ **2.** $-4\sqrt{5}$ **3.** $2\sqrt{7}$ **4.** $9\sqrt{3} - 8\sqrt{5}$ **5.** $-3\sqrt{11} + \sqrt{6}$ **6.** $43\sqrt{2} + 6\sqrt{3}$
7. $12\sqrt{3}$ **8.** $13\sqrt{5}$ **9.** $11\sqrt{7}$ **10.** $-3\sqrt{6}$ **11.** $8\sqrt{2} - 6\sqrt{7}$ **12.** $6\sqrt{3}$ **13.** $27\sqrt{5}$ **14.** $47\sqrt{2}$
15. $6\sqrt{10}$ **16.** $-\sqrt{3}$ **17.** $\sqrt{42}$ **18.** 6

2 **19.** $15\sqrt{10}$ **20.** 240 **21.** $20x\sqrt{15x}$ **22.** $28x^5\sqrt{6x}$ **23.** $7\sqrt{2} + 35$ **24.** $48 - 4\sqrt{3}$
25. $\sqrt{10} - 3\sqrt{5}$ **26.** $4\sqrt{7} + \sqrt{21}$ **27.** $\sqrt{15} + \sqrt{6}$ **28.** $\sqrt{22} + \sqrt{33}$ **29.** $x\sqrt{3x}$
30. $\dfrac{2\sqrt{2}}{\sqrt{15y}}$ or $\dfrac{2\sqrt{30y}}{15y}$ **31.** $\dfrac{2\sqrt{2}}{x\sqrt{7}}$ or $\dfrac{2\sqrt{14}}{7x}$ **32.** $\dfrac{2}{3y\sqrt{2y}}$ or $\dfrac{\sqrt{2y}}{3y^2}$ **33.** $\dfrac{4x\sqrt{x}}{3}$ **34.** $x\sqrt{7}$ **35.** $4\sqrt{3}$
36. $\dfrac{x}{\sqrt{6}}$ or $\dfrac{x\sqrt{6}}{6}$ **37.** $\dfrac{y}{\sqrt{7}}$ or $\dfrac{y\sqrt{7}}{7}$ **38.** $2x\sqrt{2x}$

3 **39.** 3 **40.** $\dfrac{\sqrt{10}}{2}$ **41.** $2\sqrt{2}$ **42.** $\dfrac{10\sqrt{3}}{3}$ **43.** $\dfrac{8}{5\sqrt{3}}$ or $\dfrac{8\sqrt{3}}{15}$ **44.** $\dfrac{15x^3\sqrt{x}}{4}$

4 **45.** $\dfrac{5\sqrt{3}}{3}$ **46.** $\dfrac{6\sqrt{5}}{5}$ **47.** $\dfrac{\sqrt{2}}{4}$ **48.** $\dfrac{\sqrt{55}}{11}$ **49.** $\dfrac{\sqrt{6}}{3}$ **50.** $\dfrac{5\sqrt{3x}}{3}$ **51.** $\dfrac{2x\sqrt{7}}{7}$ **52.** $\dfrac{2x^3\sqrt{3x}}{3}$

Self-Study Exercises 11–4

1 **1.** $5i$ **2.** $6i$ **3.** $8xi$ **4.** $4y^2 i\sqrt{2y}$

2 **5.** i **6.** i **7.** 1 **8.** -1 **9.** 1 **10.** i **11.** 1 **12.** $-i$

3 **13.** $15 + 0i$ **14.** $17 + 0i$ **15.** $0 + 7i$ **16.** $0 + 9i$ **17.** $0 + 33i$ **18.** $0 + 7i$ **19.** $-4 + 0i$ **20.** $5 + 2i$
21. $8 + 4i\sqrt{2}$ **22.** $7 - i\sqrt{3}$

4 **23.** $11 + 5i$ **24.** $(2\sqrt{3} + 2\sqrt{2}) - 8i\sqrt{3}$ **25.** $4 + i$ **26.** $12 + 20i$ **27.** $1 + 9i$ **28.** $12 - 8i$

Self-Study Exercises 11–5

1 **1.** ± 5 **2.** ± 5 **3.** ± 3.464 **4.** ± 0.667 **5.** ± 3 **6.** ± 0.926 **7.** ± 1.155 **8.** ± 4 **9.** ± 2.449
10. $\pm 5i$ **11.** ± 3.207 **12.** ± 3

2 **13.** $x = 81$ **14.** $y = 192$ **15.** $y = 66$ **16.** $x = \pm 5.292$ **17.** $x = -1.979$ **18.** $x = \pm 2$ **19.** $x = 50$
20. $x = \pm 4.899$ **21.** $y = \pm 2.105$ **22.** $y = \pm 2.9$ **23.** $x = \pm 3.606$ **24.** $x = 9$

3 **25.** $AC = 24$ cm **26.** $BC = 10$ mm **27.** $AB = 17$ yd **28.** $c = 6.403$ cm **29.** $a = 6.325$ m **30.** $b = 12$ ft
31. 30 ft **32.** $AC = 39.783$ cm **33.** 26 ft 10 in. **34.** 12.806 ft **35.** 13 cm **36.** 45 dm **37.** 10.607 mm
38. 61.083 ft **39.** $E_a = 170.294$ V

Assignment Exercises, Chapter 11

1. $\sqrt{49}; 49^{1/2}$ **3.** $\sqrt[4]{16}; 16^{1/4}$ **5.** $\sqrt{121}; 121^{1/2}$ **7.** 6 and 7 **9.** 11 and 12 **11.** 3 and 4 **13.–15.**
See graph and answers that follow. **13.** 3.9 **15.** 2.2 **17.** $x^{7/2}; \sqrt{x^7}$ **19.** $x^{2/3}; (\sqrt[3]{x})^2$ **21.** $x^{1/2}$

23. $x^{4/5}$ **25.** $x^{4/3}y^{4/3}$ **27.** $7^{1/2}$ **29.** $\sqrt[5]{y^3}$ **31.** y^6 **33.** $-b^9$ **35.** $x^{1/3}$ **37.** $(4y)^{1/5}$ **39.** $2b^4$

41. a^2 **43.** y **45.** $27x^{3/4}y^6$ **47.** $64a^3x^{3/2}$ **49.** $x^{1/2}$ **51.** $a^{7/6}$ **53.** $\dfrac{1}{x^{1/8}}$ **55.** $a^{8/3}$ **57.** $2a^{7/2}; 22.627$

59. $\dfrac{3}{2a^{22/5}}; 0.071$ **61.** $a^{6.3}; 78.793$ **63.** 5 **65.** x **67.** $3p\sqrt{p}$ **69.** $3a\sqrt{2b}$ **71.** $4x^2y\sqrt{2x}$

73. $5x^5y^4\sqrt{3y}$ **75.** $-2\sqrt{3}$ **77.** $-\sqrt{7}$ **79.** $11\sqrt{6}$ **81.** $-28\sqrt{3}$ **83.** $-5\sqrt{2}$ **85.** $-17\sqrt{2}$

87. $24\sqrt{3}$ **89.** $40\sqrt{21}$ **91.** $-160\sqrt{6}$ **93.** $6 - 5\sqrt{3}$ **95.** $3\sqrt{2} - 3\sqrt{5}$ **97.** $\dfrac{3}{4}$ **99.** $\dfrac{3\sqrt{3}}{4\sqrt{2}}$ or $\dfrac{3\sqrt{6}}{8}$

101. $\dfrac{\sqrt{3}}{2\sqrt{5}}$ or $\dfrac{\sqrt{15}}{10}$ **103.** $\sqrt{3}$ **105.** $\dfrac{9}{16}$ **107.** $\dfrac{5\sqrt{17}}{17}$ **109.** $\dfrac{\sqrt{21}}{6}$ **111.** $\dfrac{\sqrt{6}}{4}$ **113.** $\dfrac{5\sqrt{2}}{4}$ **115.** $10i$

117. $\pm 2y^3 i\sqrt{6y}$ **119.** -1 **121.** i **123.** $0 + 15i$ **125.** $0 - 12i$ **127.** $7 - 4i$ **129.** $11 + i$ **131.** ± 6

133. ± 2 **135.** $\pm 2i$ **137.** ± 3 **139.** $\dfrac{3\sqrt{6}}{2}$ or 3.674 **141.** 46 **143.** ± 10.262 **145.** ± 8.888 **147.** ± 1

149. ± 7.874 **151.** 15 in. **153.** 7.141 ft **155.** 8 yd **157.** 20.248 mi **159.** 30 cm **161.** 21.633 in.
163. 4.243 in. **165.** $20 + 15i$

Trial Test, Chapter 11

1. $6\sqrt{14}$ **3.** $6\sqrt{3}$ **5.** $\dfrac{4\sqrt{6}}{3}$ **7.** $2\sqrt{6}$ **9.** $\dfrac{7\sqrt{15x}}{2x}$ **11.** ± 1 **13.** 49 **15.** 9 **17.** $\pm 7i$ or no real

solutions **19.** $3x^5$ **21.** $x^2y^3z^6$ **23.** $5x^{\frac{1}{6}}y^2$ **25.** $2x$ **27.** $-i$ **29.** $-3 + 5i$ **31.** $\sqrt{35} - 4\sqrt{7}$ **33.** 13

35. 6 **37.** 17 in.

Selected Answers to Student Exercise Material

Self-Study Exercises 12–1

1   **1.** 280 mi   **2.** $120   **3.** 2,826 in^2 **4.** 20.78 m   **5.** 904.47 cm^2 **6.** $310   **7.** 14.25%   **8.** 3 years
9. $2,600   **10.** 90   **11.** 8%   **12.** $3,378.38   **13.** 13.6   **14.** 17%   **15.** 66 in.   **16.** 9.0 cm   **17.** $10.50

18. $\frac{1}{4}$ mi   **19.** 153.9 in^2 **20.** 6.25 km^2 **21.** $48.75   **22.** 4.8 Ω   **23.** 61.6%   **24.** 35 miles per hour

25. 1.5 amperes   **26.** 6 ft^3 **27.** 3 amperes   **28.** 2 cylinders   **29.** 1,600 rpm   **30.** 71.71 in.   **31.** 4.4 Ω

2   **32.** $32,980   **33.** $149,800   **34.** $32.98   **35.** $14.98   **36.** $40,000; $400,000   **37.** (a) $7,020 (b) $250,200
38. (a) $7.02 (b) $25.02

Self-Study Exercises 12–2

1   **1.** $R = S_n - S$ **2.** $S = I - I_n$ **3.** $r = \dfrac{S}{2\pi h}$ **4.** $b = y - mx$ **5.** $m = \dfrac{y - b}{x}$ **6.** $T_2 = \dfrac{T_1 V_2}{V_1}$

7. $r = \sqrt{\dfrac{V}{\pi h}}$ **8.** $a = \sqrt{c^2 - b^2}$ **9.** $R = \dfrac{100P}{B}$ **10.** $b = \dfrac{P - 2s}{2}$ **11.** $r = \dfrac{C}{2\pi}$ **12.** $l = \dfrac{A}{w}$

13. $C = \dfrac{R}{A - B}$ **14.** $s = \sqrt{A}$ **15.** $R = \dfrac{D}{T}$ **16.** $D = P - S$ **17.** $r = \sqrt{\dfrac{A}{\pi}}$ **18.** $C = \sqrt{a^2 + b^2}$

19. $I = A - P$ **20.** $T = \dfrac{I}{PR}$

Self-Study Exercises 12–3

1   **1.** $LSA = 37.3$ in^2; $TSA = 47.1$ in^2 **2.** $LSA = 1,885$ ft^2; $TSA = 4,398$ ft^2 **3.** $LSA = 200$ cm^2;
$TSA = 264$ cm^2 **4.** $LSA = 96$ in^2; $TSA = 109.84$ in^2 **5.** 1,885 ft^2

2   **6.** 12,600 cm^3 **7.** 76,800 m^3 **8.** 415.6 m^3 **9.** 2,617.33 in^3 **10.** 2,428.20 cm^3 **11.** 3,660 in^3 **12.** 4,116 in^3

3   **13.** 314.2 cm^2 **14.** 904.8 in^3 **15.** 6,361.7 ft^2; 47,712.9 ft^3 **16.** 356,892.8 gal   **17.** 50.3 ft^2; 33.5 ft^3
18. 225.6 gal

4   **19.** $LSA = 188.5$ ft^2; $TSA = 301.6$ ft^2; $V = 301.6$ ft^3 **20.** 4,712.4 ft^3 **21.** 589.0 cm^2 **22.** 50.3 L
23. 1,649.3 ft^2 **24.** 10.0 ft   **25.** $347

Assignment Exercises, Chapter 12

1. $0.81   **3.** 9.6%   **5.** 10.8   **7.** $193.60   **9.** 2 years   **11.** 50 in.   **13.** $89.50   **15.** 95.0 in^2 **17.** 2.75 amperes
19. 10 ft^3 **21.** 3.2 amperes   **23.** 1,400 rpm   **25.** 2 cylinders   **27.** (a) $10,348 (b) $13,480 (c) $44,800

29. (a) $4,000 (b) $40,000 (c) $400,000   **31.** (a) −$63.48 (b) $26.52 (c) $35.52   **33.** $w = \dfrac{V}{lh}$ **35.** $r = \sqrt{\dfrac{B}{cx}}$

37. $B = \dfrac{A}{P}$ **39.** $r = \dfrac{I}{Pt}$ **41.** $r = s + d$ **43.** $t = \dfrac{v_0 - v}{32}$ **45.** $t = \dfrac{A - P}{Pr}$ **47.** $P = S + D$ **49.** 616 cm^2
51. 2,733.186 cm^2 **53.** 60 ft^2 **55.** 124,407 ft^2 **57.** 1,728 m^3 **59.** 123 ft^3 **61.** 1,017.9 m^2 **63.** 7,238.2 ft^3
65. 169.6 cm^2 **67.** 377.0 in^3 **69.** 2 gal   **71.** 116.6 ft^2 **73.** 243.3 in^2 **75.** 2,094 ft^3 **77.** Answers will vary.

Trial Test, Chapter 12

1. $L = \dfrac{RA}{P}$ **3.** $r = \sqrt{\dfrac{d}{\pi sn}}$ **5.** 50 in^2 **7.** 791.7 ft^2 **9.** 6,768 gal   **11.** 3,848.5 ft^3 **13.** 2,080.7 cm^2

15. 26.2 ft^2 **17.** 1,497.05 lb   **19.** 12,000 calories   **21.** 912 m^3

1 **1.** $7(a + b)$ **2.** $12(x + y)$ **3.** $m(m + 2)$ **4.** $y^2(5y + 8)$ **5.** $3x(2x + 1)$ **6.** $6y^3(2 + 3y)$ **7.** $6x^4(2x - 1)$ **8.** $5x(1 - 3y)$ **9.** prime **10.** prime **11.** $5(ab + 2a + 4b)$ **12.** $2ax(2x + 3a + 5ax)$ **13.** prime **14.** $3(4a^2 - 5a + 2)$ **15.** $3x(x^2 - 3x - 2)$ **16.** $2ab(4a + 7b^2 + 14a^2b^2)$ **17.** $3m^2(1 - 2m + 4m^2)$ **18.** $-1(x + 7)$ **19.** $-1(3x + 8)$ **20.** $-1(5x - 2)$ **21.** $-1(12x - 7)$ **22.** $-1(x^2 - 3x + 8)$ **23.** $-1(2x^2 + 7x + 11)$ **24.** $-2(x^2 - 3x + 4)$ **25.** $-3(x^2 + 3x - 5)$ **26.** $-7(x^2 + 3x - 2)$ **27.** $-6(2x^2 - 3x - 1)$ **28.** $(x + 3)(5x + 8y)$ **29.** $(2x - 1)(3x + 5)$ **30.** $(3y - 5)(4y + 7)$ **31.** $(7a + 2b)(a - b)$ **32.** $(2m - 3n)(5m - 7n)$ **33.** $(y - 2)(y - 3)$ **34.** $(2x - 7)(3x - 8)$ **35.** $(9y - 2)(7y - 5)$ **36.** $5(\sqrt{7} + 2)$ **37.** $4(2\sqrt{3} - 3)$ **38.** $3(\sqrt{2} - 3\sqrt{3})$

1 **1.** $x^2 + 10x + 21$ **2.** $x^2 + 13x + 40$ **3.** $2x^2 + 3x - 2$ **4.** $3x^2 - 8x - 35$ **5.** $x^3 + 7x^2 + 9x - 5$ **6.** $x^3 + 4x^2 - 19x + 14$ **7.** $x^3 - 12x^2 + 37x - 14$ **8.** $x^3 - 11x^2 + 25x - 3$ **9.** $2x^3 - 7x^2 + 7x - 5$ **10.** $3x^3 + x^2 - 11x + 6$ **11.** $6x^3 + 11x^2 - 31x + 14$ **12.** $20x^3 - 22x^2 - 9x + 9$ **13.** $x^3 - 8$ **14.** $8x^3 - 27$ **15.** $x^3 + 27$ **16.** $27x^3 + 8$ **17.** $x^3 + 125$ **18.** $1{,}331x^3 - 27$

2 **19.** $a^2 + 11a + 24$ **20.** $x^2 + x - 20$ **21.** $y^2 - 11y + 18$ **22.** $y^2 - 10y + 21$ **23.** $2a^2 + 11a + 12$ **24.** $3a^2 - 2a - 5$ **25.** $3a^2 - 8ab + 4b^2$ **26.** $5x^2 - 26xy + 5y^2$ **27.** $6x^2 - 17x + 12$ **28.** $2a^2 - 7ab + 5b^2$ **29.** $21 - 52m + 7m^2$ **30.** $40 - 21x + 2x^2$ **31.** $x^2 + 11x + 28$ **32.** $y^2 - 12y + 35$ **33.** $m^2 - 4m - 21$ **34.** $3bx + 18b - 2x - 12$ **35.** $12r^2 - 7r - 10$ **36.** $35 - 22x + 3x^2$ **37.** $4 - 14m + 6m^2$ **38.** $6 + 13x + 6x^2$ **39.** $2x^2 + x - 15$ **40.** $20x^2 - 13xy - 21y^2$ **41.** $14a^2 + 19ab - 3b^2$ **42.** $30a^2 - 13ab - 10b^2$ **43.** $27x^2 + 30xy - 8y^2$ **44.** $20x^2 - 47xy + 24y^2$ **45.** $21m^2 + 29mn - 10n^2$ **46.** $5 - 3\sqrt{3}$ **47.** $2\sqrt{15} + 5\sqrt{5} - 10\sqrt{3} - 25$

3 **48.** $a^2 - 9$ **49.** $4x^2 - 9$ **50.** $a^2 - y^2$ **51.** $16r^2 - 25$ **52.** $25x^2 - 4$ **53.** $49 - m^2$ **54.** $9y^2 - 25$ **55.** $64y^2 - 9$ **56.** $9a^2 - 121b^2$ **57.** $25y^2 - 9$ **58.** $x^2 - 49$ **59.** $r^2 - 121$ **60.** $4 - 12r + 9r^2$ **61.** $9x^2 + 24x + 16$ **62.** $Q^2 + 2QL + L^2$ **63.** $m^4 + 2m^2 + 1$ **64.** $4d^2 - 20d + 25$ **65.** $9a^2 + 12ax + 4x^2$ **66.** $9x^2 - 42x + 49$ **67.** $36 + 12Q + Q^2$ **68.** $y^2 + 10xy + 25x^2$ **69.** $16 - 24j + 9j^2$ **70.** $9m^2 - 12mp + 4p^2$ **71.** $m^4 + 2m^2p^2 + p^4$ **72.** $4a^2 - 28ac + 49c^2$ **73.** $81 - 234a + 169a^2$ **74.** -1 **75.** 5 **76.** 1 **77.** 65 **78.** 26 **79.** -2 **80.** 17 **81.** $x^3 + p^3$ **82.** $Q^3 + L^3$ **83.** $27 + a^3$ **84.** $8x^3 + 64p^3$ **85.** $27m^3 + 8$ **86.** $216 - p^3$ **87.** $125y^3 - p^3$ **88.** $x^3 - 8y^3$ **89.** $Q^3 - 216$ **90.** $27T^3 - 8$

4 **91.** $x + 4$, yes **92.** $x + 6$, yes **93.** $x^2 - 3x + 2$, yes **94.** $x^2 + x - 5$, yes **95.** $3x^2 - 2x + 3 - \dfrac{1}{x + 3}$, no **96.** $2x^2 + 7x + 2 + \dfrac{18}{x - 5}$, no **97.** $x^2 + x - 6$, yes **98.** $x^2 - x - 2$, yes **99.** $x^2 + 3x + 9$, yes **100.** $x^2 - 5x + 25$, yes

1 **1.** difference **2.** not difference **3.** difference **4.** difference **5.** not difference **6.** difference **7.** $(y + 7)(y - 7)$ **8.** $(4x + 1)(4x - 1)$ **9.** $(3a + 10)(3a - 10)$ **10.** $(2m + 9n)(2m - 9n)$ **11.** $(3x + 8y)(3x - 8y)$ **12.** $(5x + 8)(5x - 8)$ **13.** $(10 + 7x)(10 - 7x)$ **14.** $(2x + 7y)(2x - 7y)$ **15.** $(11m + 7n)(11m - 7n)$ **16.** $(9x + 13)(9x - 13)$ **17.** $(2a + 3)(2a - 3)$ **18.** $(5r + 4)(5r - 4)$ **19.** $(6x + 7y)(6x - 7y)$ **20.** $(7 - 12x)(7 + 12x)$ **21.** $(4x + 9y)(4x - 9y)$

2 **22.** not perfect square **23.** perfect square **24.** not perfect square **25.** not perfect square **26.** perfect square **27.** perfect square **28.** $(x + 3)^2$ **29.** $(x + 7)^2$ **30.** $(x - 6)^2$ **31.** $(x - 8)^2$ **32.** $(2a + 1)^2$ **33.** $(5x - 1)^2$ **34.** $(3m - 8)^2$ **35.** $(2x - 9)^2$ **36.** $(x - 6y)^2$ **37.** $(2a - 5b)^2$ **38.** $(y - 5)^2$ **39.** $(3x + 10y)^2$ **40.** $-1(x + 6)^2$ **41.** $-1(3x - 1)^2$ **42.** $-(x + 4)^2$

3 **43.** difference of two cubes **44.** not a special sum or difference of cubes **45.** sum of two cubes **46.** not a special sum or difference of cubes **47.** difference of two cubes **48.** sum of two cubes **49.** $(m - 2)(m^2 + 2m + 4)$ **50.** $(y - 5)(y^2 + 5y + 25)$ **51.** $(Q + 3)(Q^2 - 3Q + 9)$ **52.** $(c + 1)(c^2 - c + 1)$ **53.** $(5d - 2p)(25d^2 + 10dp + 4p^2)$ **54.** $(a + 4)(a^2 - 4a + 16)$ **55.** $(6a - b)(36a^2 + 6ab + b^2)$ **56.** $(x + Q)(x^2 - xQ + Q^2)$ **57.** $(2p - 5)(4p^2 + 10p + 25)$ **58.** $(3 - 2y)(9 + 6y + 4y^2)$ **59.** $-1(a + 2)(a^2 - 2a + 4)$ **60.** $-1(x + 3)(x^2 - 3x + 9)$

Self-Study Exercises 13–4

1 **1.** $(x+6)(x+1)$ **2.** $(x-6)(x-1)$ **3.** $(x-2)(x-3)$ **4.** $(x+3)(x+2)$ **5.** $(x-7)(x-4)$
6. $(x+6)(x+2)$ **7.** $(x-6)(x-2)$ **8.** $(x+12)(x+1)$ **9.** $(x-12)(x-1)$ **10.** $(x+4)(x+3)$
11. $(x-4)(x-3)$ **12.** $(x-3)(x-1)$ **13.** $(x+7)(x+1)$ **14.** $(x+5)(x+2)$ **15.** $(x-3)(x+2)$
16. $(x+3)(x-2)$ **17.** $(x-6)(x+1)$ **18.** $(x+6)(x-1)$ **19.** $(x-4)(x+3)$ **20.** $(x+4)(x-3)$
21. $(x+6)(x-2)$ **22.** $(x-6)(x+2)$ **23.** $(x-12)(x+1)$ **24.** $(x+12)(x-1)$ **25.** $(y-5)(y+2)$
26. $(y-3)(y+2)$ **27.** $(b+3)(b-1)$ **28.** $(b-7)(b+2)$ **29.** $(x-4)(x-3)$ **30.** $(x-6)(x+5)$
31. $(x+9)(x+2)$ **32.** $(x-6)(x-3)$ **33.** $(x-9)(x+2)$ **34.** $(x+18)(x-1)$ **35.** $(x+5)(x+4)$
36. $(x-10)(x-2)$ **37.** $(x-8)(x-2)$ **38.** $(x-16)(x-1)$ **39.** $(x-14)(x+1)$ **40.** $(x-7)(x+2)$

2 **41.** $(x+y)(x+4)$ **42.** $(3x+2)(2x-y)$ **43.** $(3x+5)(m-2n)$ **44.** $(6x-7)(5y-6)$ **45.** $(x-2)(x+8)$
46. $(3x-1)(2x-7)$ **47.** $(x-4)(x+1)$ **48.** $(2x-1)(4x+3)$ **49.** $(x-5)(x+4)$ **50.** $(x-2)(3x+5)$
51. $(x+2)(4x-3)$ **52.** $(x-2)(4x+3)$ **53.** $(x+2)(4x+3)$ **54.** $(x-2)(4x-3)$ **55.** $(2x+1)(4x-3)$

3 **56.** $(3x+1)(x+2)$ **57.** $(3x+2)(x+4)$ **58.** $(3x+2)(2x+3)$ **59.** $(4x+3)(2x-1)$ **60.** $(3x-4)(2x-3)$
61. $(2x-5)(x-2)$ **62.** $(3x-5)(2x-1)$ **63.** $(4x+3)(2x+1)$ **64.** $(6x-5)(x-1)$ **65.** $(4x+3)(2x+5)$
66. $(3x-5)(5x+1)$ **67.** $(2x-1)(4x-3)$ **68.** $(2x-7)(x+1)$ **69.** $(6x-5)(2x+3)$ **70.** $(5x+3)(2x-1)$
71. $(3x+2)(4x-1)$ **72.** $(3x-2)(4x+1)$ **73.** $(4x+1)(3x+2)$ **74.** $(4x-1)(3x-2)$ **75.** $(8x-1)(3x+1)$
76. $(3x-1)(8x-1)$ **77.** $(6x-5y)(x+2y)$ **78.** $(3a+2b)(2a-7b)$ **79.** $(6x-5)(3x+2)$ **80.** $(5x-4y)(4x+3y)$
81. $(7x+8)(x-1)$ **82.** $(2a+19)(a+1)$

4 **83.** $4(x-1)$ **84.** $(x+3)(x-2)$ **85.** $(2x+3)(x-1)$ **86.** $(x+3)(x-3)$ **87.** $4(x+2)(x-2)$
88. $(m+5)(m-3)$ **89.** $2(a+2)(a+1)$ **90.** $(b+3)^2$ **91.** $(4m-1)^2$ **92.** $(x+7)(x+1)$ **93.** $(2m+1)(m+2)$
94. $(2m+1)(m-3)$ **95.** $(2a-5)(a+1)$ **96.** $(3x-2)(x+4)$ **97.** $(3x+5)(2x-3)$ **98.** $(4x-1)(2x+3)$
99. $-2(x-2)(x-1)$ **100.** $(x^2+4)(x+2)(x-2)$ **101.** $(x^3+9)(x^2-9)$ **102.** $(x^2-3)(x^4+3x^2+9)$
103. $3(x^2+4)(x+2)(x-2)$ **104.** $6(x-2)(x^2+2x+4)$ **105.** $3(x-4)(x+2)$

Assignment Exercises, Chapter 13

1. $5(x+y)$ **3.** $4(3m^2-2n^2)$ **5.** $2a(a^2-7a-1)$ **7.** $5x(3x^2-x-4)$ **9.** $6a^2(3a+2)$ **11.** $-2x(3x+5)$
13. $x^2+11x+28$ **15.** $m^2-4m-21$ **17.** $12r^2-7r-10$ **19.** $4-14m+6m^2$ **21.** $2x^2+x-15$ **23.** $14a^2+19ab-3b^2$ **25.** $27x^2+30xy-8y^2$ **27.** $21m^2+29mn-10n^2$ **29.** $36x^2-25$ **31.** $49y^2-121$ **33.** $64a^2-25b^2$
35. 4 **37.** 10 **39.** $x^2+18x+81$ **41.** x^2-6x+9 **43.** $16x^2-120x+225$ **45.** $64+112m+49m^2$ **47.** $16x^2-88x+121$ **49.** g^3-h^3 **51.** $8H^3-27T^3$ **53.** $216+i^3$ **55.** z^3+8t^3 **57.** $343T^3+8$ **59.** $x-2$ **61.** $3x-2$
63. x^2+x-5 **65.** $3x+1$ **67.** not difference   **69.** difference   **71.** difference   **73.** not perfect-square trinomial
75. perfect-square trinomial   **77.** perfect-square trinomial   **79.** difference of two cubes   **81.** difference of two cubes   **83.** sum of two cubes   **85.** $(5y-2)(5y+2)$ **87.** *NSP*, this is a sum of two squares, not a difference.
89. $(a+1)^2$ **91.** $(4c-3b)^2$ **93.** $(n-13)^2$ **95.** $(6a+7b)^2$ **97.** $(7-x)^2$ **99.** *NSP*, this is a sum of two squares, not a difference.   **101.** *NSP*, the middle term needs a y factor.   **103.** $(7+9y)(7-9y)$ **105.** $(3x+10y)(3x-10y)$

107. $(3x-y)^2$ **109.** $(3xy+z)(3xy-z)$ **111.** $(x+2)^2$ **113.** $\left(\frac{2}{5}x+\frac{1}{4}y\right)\left(\frac{2}{5}x-\frac{1}{4}y\right)$ **115.** $(T-2)(T^2+2T+4)$

117. $(d+9)(d^2-9d+81)$ **119.** $(3k+4)(9k^2-12k+16)$ **121.** $(x+8)(x+3)$ **123.** $(x+10)(x+3)$
125. $(x-8)(x-1)$ **127.** $(x-13)(x+2)$ **129.** $(x+8)(x-3)$ **131.** $(6x+1)(x+4)$ **133.** $(5x-4)(x-6)$
135. $(3x+7)(2x-5)$ **137.** $(7x+8)(x-3)$ **139.** $(3a+10)(3a-10)$ **141.** $(2x+1)(x-2)$ **143.** $(a+9)(a-9)$
145. $(y-7)^2$ **147.** $(b+5)(b+3)$ **149.** $(13+m)(13-m)$ **151.** $(x-8)(x+4)$ **153.** $(x+20)(x-1)$
155. $2(x-4)(x+2)$ **157.** $2x(x-6)(x+1)$ **159.** This enables us to factor more rapidly and easily.
161. $(25+x)(15+x); 375+40x+x^2$ **163.** $(25+3)(15+7)=28\cdot22=616$ ft^2 Answers will vary.

Trial Test, Chapter 13

1. m^2-49 **3.** a^2+6a+9 **5.** $2x^2-11x+15$ **7.** x^3-8 **9.** $125a^3-27$ **11.** $2x+7$ **13.** $x(7x+8)$
15. $7ab(a-2)$ **17.** $(3x-5)(3x+5)$ **19.** $(2y+3x)(4y^2-6xy+9x^2)$ **21.** $(x-9)^2$ **23.** $(x-6)^2$
25. $(3x+2)(2x-3)$ **27.** $(a+8b)^2$ **29.** $(b-5)(b+2)$ **31.** $(3x+5)(x+6)$ **33.** $(3m-2)(m-1)$
35. $(3x-2)(x-3)$ **37.** $3(x-2)(x^2+2x+4)$ **39.** $5(x+2)(x-2)$ **41.** $3(x+2)^2$

Self-Study Exercises 14–1

1 1. $\dfrac{4}{9}$ 2. $\dfrac{3}{8}$ 3. $2ab^2$ 4. $\dfrac{3a}{2b^3}$ 5. $\dfrac{x}{x+2}$ 6. $\dfrac{3(3x+2)}{x(2x-1)}$ or $\dfrac{9x+6}{2x^2-x}$ 7. $\dfrac{x^2-3x-10}{x^2+3x-10}$ 8. $\dfrac{1}{2}$

9. $\dfrac{1}{a-b}$ 10. $\dfrac{x-2}{x+2}$ 11. $\dfrac{2x-4}{3}$ or $\dfrac{2(x-2)}{3}$ 12. -1 13. $-(x+2)$ 14. $-\dfrac{1}{2}$ 15. -3

16. $-\dfrac{1}{2}$ 17. -5 18. $-\dfrac{2}{(x-2)}$ 19. $-\dfrac{x-5}{x+1}$ 20. -1 21. $\dfrac{x+3}{x+2}$ 22. $\dfrac{x-3}{x+3}$ 23. $\dfrac{x+1}{x+4}$

24. $\dfrac{2x-1}{3x-2}$

Self-Study Exercises 14–2

1 1. $\dfrac{1}{2}$ 2. $\dfrac{1}{10}$ 3. $\dfrac{10x^3}{9y^2}$ 4. $\dfrac{7}{15}$ 5. $\dfrac{3}{x-2}$ 6. $\dfrac{4}{x-2}$ 7. $3(a-b)$ or $3a-3b$ 8. 4 9. $\dfrac{a-7}{b+5}$

10. $\dfrac{3(x+1)}{x-1}$ or $\dfrac{3x+3}{x-1}$ 11. $\dfrac{1}{12a}$ 12. $\dfrac{3}{2}$ 13. $\dfrac{2}{3}$ 14. $\dfrac{3y^3}{2x}$ 15. $\dfrac{3ab}{10}$ 16. $\dfrac{1}{2}$ 17. -2 18. 1

19. $\dfrac{x}{x-2}$ 20. $\dfrac{50(a-b)}{ab}$ or $\dfrac{50a-50b}{ab}$ 21. 1

2 22. $\dfrac{25}{27}$ 23. $1\dfrac{1}{20}$ 24. $\dfrac{x-5}{12x}$ 25. $\dfrac{2}{3x(x-2)}$ or $\dfrac{2}{3x^2-6x}$ 26. $\dfrac{x+3}{2}$ 27. $\dfrac{x-3}{3}$ 28. $x+y$

29. $-\dfrac{x-1}{x+1}$ or $\dfrac{-x+1}{x+1}$ 30. $3(x-6)$ 31. $\dfrac{3y^2}{4}$ 32. $-3(2+\sqrt{5})$ 33. $5+2\sqrt{3}$ 34. $\dfrac{10-9\sqrt{2}}{14}$

35. $\dfrac{14-\sqrt{7}}{9}$ 36. $\dfrac{14+4\sqrt{3}+21\sqrt{5}+6\sqrt{15}}{37}$ 37. $13+5\sqrt{6}$ 38. $-\dfrac{1}{14+7\sqrt{5}}$ 39. $\dfrac{14}{75+15\sqrt{11}}$

40. $\dfrac{1}{8-2\sqrt{6}}$ 41. $\dfrac{1}{3-\sqrt{2}}$ 42. $-\dfrac{2}{28-21\sqrt{2}}$ 43. $\dfrac{13}{40-16\sqrt{3}}$

Self-Study Exercises 14–3

1 1. $\dfrac{5}{7}$ 2. $\dfrac{13}{16}$ 3. $\dfrac{3x}{2}$ 4. $\dfrac{23x}{24}$ 5. $\dfrac{4+3x}{2x}$ 6. $\dfrac{55}{6x}$ 7. $\dfrac{9x-1}{(x+1)(x-1)}$ or $\dfrac{9x-1}{x^2-1}$

8. $\dfrac{10x+6}{(x+3)(x-1)}$ or $\dfrac{10x+6}{x^2+2x-3}$ 9. $\dfrac{x-7}{(2x+1)(x-1)}$ or $\dfrac{x-7}{2x^2-x-1}$ 10. $\dfrac{x+1}{(x-6)(x-5)}$

or $\dfrac{x+1}{x^2-11x+30}$

2 11. $\dfrac{5(2x+1)}{3x}$ or $\dfrac{10x+5}{3x}$ 12. $\dfrac{8(3x+2)}{5x}$ or $\dfrac{24x+16}{5x}$ 13. $-\dfrac{28}{9}$ or $-3\dfrac{1}{9}$ 14. $-\dfrac{1}{7}$ 15. $\dfrac{11}{3x^2}$

16. $\dfrac{10+9x}{3+10x}$ 17. $\dfrac{-2(x-2)}{x+2}$ or $\dfrac{4-2x}{x+2}$ 18. 9

Self-Study Exercises 14–4

1 1. 0 2. 0 3. $5, 0$ 4. $0, -9$ 5. $-8, 8$ 6. $2, -2$ 7. $\dfrac{3}{4}, 0$ 8. $0, -3$

9. 35   **10.** 5   **11.** 1   **12.** $\dfrac{1}{4}$ **13.** 4   **14.** $-\dfrac{4}{5}$ **15.** $1\dfrac{5}{7}$ days   **16.** 8 investors   **17.** 25 mph going

18. 3 min   **19.** 50 lb dark-roast, 75 lb medium-roast, \$2 per pound   **20.** 6 ohms

Assignment Exercises, Chapter 14

1. $\dfrac{3}{4}$ **3.** $\dfrac{b^2}{2ac}$ **5.** $\dfrac{x-3}{2(x+3)}$ **7.** -1 **9.** $\dfrac{m^2-n^2}{m^2+n^2}$ **11.** $\dfrac{1}{1+y}$ **13.** 5   **15.** $y+1$ **17.** $\dfrac{2}{x+6}$ **19.** 3

21. $\dfrac{5x^3}{4y^2}$ **23.** 15   **25.** $\dfrac{4(2y-1)}{y-1}$ **27.** 1   **29.** $\dfrac{x+3}{x+2}$ **31.** $\dfrac{1}{4}$ **33.** $\dfrac{(y-1)^3}{y}$ **35.** $\dfrac{y+3}{y+2}$ **37.** $\dfrac{3x^2}{2}$

39. $(y+4)(y-3)$ **41.** $\dfrac{5}{4x-12}$ **43.** $\dfrac{4x^2}{3}$ **45.** $\dfrac{72+12\sqrt{5}}{31}$ **47.** $\dfrac{26+7\sqrt{3}}{23}$

49. $\dfrac{35+15\sqrt{5}+7\sqrt{2}+3\sqrt{10}}{4}$ **51.** $\dfrac{26}{20-5\sqrt{3}}$ **53.** $\dfrac{1}{16+4\sqrt{13}}$ **55.** $\dfrac{9}{40+8\sqrt{7}}$ **57.** $\dfrac{2}{3}$

59. $\dfrac{4x}{7}$ **61.** $\dfrac{19x}{12}$ **63.** $\dfrac{15-7x}{3x}$ **65.** $\dfrac{13}{4x}$ **67.** $\dfrac{10x+5}{(x-3)(x+2)}$ **69.** $\dfrac{6x-38}{(x+3)(x-4)}$ **71.** $\dfrac{x+3}{x-5}$

73. $\dfrac{51}{28}$ **75.** $\dfrac{3x^2-30}{2x^2+12}$ **77.** $0,2$ **79.** $\dfrac{1}{2},0$ **81.** $\dfrac{-20}{3}$ **83.** $\dfrac{1}{7}$ **85.** 3 students   **87.** $7\dfrac{1}{2}$ hr

Trial Test, Chapter 14

1. $\dfrac{1}{2}$ **3.** $\dfrac{3x-4}{x+3}$ **5.** $-\dfrac{x-2}{x+2}$ or $\dfrac{2-x}{x+2}$ **7.** $\dfrac{3a}{y}$ **9.** $\dfrac{1}{x^2(x+2y)}$ **11.** $\dfrac{2x+1}{x}$ **13.** $-\dfrac{5}{(x+2)(x-3)}$

15. $\dfrac{12+x}{4x}$ **17.** $-\dfrac{2}{3x-2}$ or $\dfrac{2}{2-3x}$ **19.** $\dfrac{1}{x+2y}$ **21.** $\dfrac{2x^2}{x-3}$ **23.** $0,-3$ **25.** $\dfrac{11}{7}$

27. $\dfrac{1}{3}$ **29.** 5 persons

Self-Study Exercises 15–1

1. $7x^2-4x+5=0$ **2.** $8x^2-6x-3=0$ **3.** $7x^2-5=0$ **4.** $x^2-6x+8=0$ **5.** $x^2-9x+8=0$
6. $x^2-4x-8=0$ **7.** $3x^2-6x+5=0$ **8.** $x^2-6x-5=0$ **9.** $x^2-16=0$ **10.** $8x^2-7x-8=0$
11. $8x^2+8x-10=0$ **12.** $0.3x^2-0.4x-3=0$

13. $x^2-5x=0$
$a=1, b=-5, c=0$

14. $3x^2-7x+5=0$
$a=3, b=-7, c=5$

15. $7x^2-4x=0$
$a=7, b=-4, c=0$

16. $3x^2-5x+8=0$
$a=3, b=-5, c=8$

17. $x^2-5x+6=0$
$a=1, b=-5, c=6$

18. $11x^2-8x=0$
$a=11, b=-8, c=0$

19. $x^2-x=0$
$a=1, b=-1, c=0$

20. $9x^2-7x-12=0$
$a=9, b=-7, c=-12$

21. $x^2-5=0$
$a=1, b=0, c=-5$

22. $x^2+6x-3=0$
$a=1, b=6, c=-3$

23. $5x^2-0.2x+1.4=0$
$a=5, b=-0.2, c=1.4$

24. $\dfrac{2}{3}x^2-\dfrac{5}{6}x-\dfrac{1}{2}=0$

$a=\dfrac{2}{3}, b=-\dfrac{5}{6}, c=-\dfrac{1}{2}$

25. $1.3x^2-8=0$
$a=1.3, b=0, c=-8$

26. $\sqrt{3}x^2+\sqrt{5}x-2=0$
$a=\sqrt{3}, b=\sqrt{5}, c=-2$

27. $8x^2-2x-3=0$

28. $x^2+3x=0$

29. $5x^2+2x-7=0$

30. $2.5x^2-0.8=0$

Selected Answers to Student Exercise Material

31. $x = \pm 3$ **32.** $x = \pm 7$ **33.** $x = \pm \dfrac{8}{3}$ **34.** $x = \pm \dfrac{7}{4}$ **35.** $y = \pm 3$ **36.** $x = \pm \dfrac{9}{2}$

37. $x = \pm\sqrt{5}$ or ± 2.236 **38.** $x = \pm 1$ **39.** $x = \pm 2$ **40.** $x = \pm 1.732$ **41.** 23 ft   **42.** 81.759 yd
43. Isolate the squared letter, then take the square root of both sides.   **44.** opposites

Self-Study Exercises 15–2

1   **1.** $x = 3, 0$ **2.** $x = 0, 6$ **3.** $x = 0, 2$ **4.** $x = 0, -\dfrac{1}{2}$ **5.** $x = 0, \dfrac{1}{2}$ **6.** $x = 0, 3$ **7.** $x = 0, \dfrac{7}{3}$

8. $y = 0, -4$ **9.** $x = 0, -\dfrac{2}{3}$ **10.** $x = 0, \dfrac{4}{3}$ **11.** $x = 0, 3$ **12.** $x = 0, \dfrac{2}{3}$ **13.** 8 or 0   **14.** 15 units

15. An incomplete quadratic equation is missing the constant or number term while a pure quadratic equation is missing the linear term.   **16.** Yes, the common factor of x will be set equal to zero.

2   **17.** $-3, -2$ **18.** $3, 3$ (double root)   **19.** $7, -2$ **20.** $3, -6$ **21.** $-4, -3$ **22.** $5, 3$ **23.** $14, -1$

24. $3, 625.$ $\dfrac{1}{2}, 3$ **26.** $-\dfrac{1}{3}, -4$ **27.** $\dfrac{3}{5}, -\dfrac{1}{2}$ **28.** $-\dfrac{1}{3}, -\dfrac{3}{2}$ **29.** $-\dfrac{3}{2}, -5$ **30.** $\dfrac{4}{3}, 2$ **31.** $\dfrac{1}{6}, -3$ **32.** $-4, -\dfrac{2}{3}$

33. $5, \dfrac{3}{2}$ **34.** $\dfrac{1}{2}, \dfrac{2}{3}$ **35.** $\dfrac{3}{4}, \dfrac{1}{2}$ **36.** $\dfrac{2}{5}, -1$ **37.** $\dfrac{5}{3}, -\dfrac{3}{2}$ **38.** $-\dfrac{5}{3}, -\dfrac{1}{3}$ **39.** $\dfrac{1}{3}, \dfrac{3}{2}$ **40.** $-3, \dfrac{2}{5}$

41. $w = 5$ ft   **42.** $w = 18$ in.
$\quad\;\; x = 11$ ft $\qquad x = 21$ in.

Self-Study Exercises 15–3

1   **1.** $-8, 6$ **2.** $-1, 9$ **3.** $1, 9$ **4.** $4, 6$ **5.** $-6, 4$ **6.** $-14, -2$ **7.** $\dfrac{3 \pm \sqrt{29}}{2}$ **8.** $\dfrac{5 \pm \sqrt{33}}{2}$

9. $\dfrac{5 \pm \sqrt{37}}{2}$ **10.** $\dfrac{3 \pm \sqrt{13}}{2}$ **11.** $\dfrac{2 \pm \sqrt{10}}{2}$ **12.** $\dfrac{1 \pm \sqrt{3}}{2}$ **13.** $\dfrac{3 \pm \sqrt{3}}{3}$ **14.** $\dfrac{-6 \pm 2\sqrt{3}}{3}$ **15.** $-1, 3$

16. $\dfrac{3 \pm \sqrt{205}}{14}$

Self-Study Exercises 15–4

1   **1.** $3, -\dfrac{2}{3}$ **2.** $3, -4$ **3.** $\dfrac{11}{5}, -1$ **4.** $3, 3$ **5.** $3, -2$ **6.** $\dfrac{3}{4}, -\dfrac{1}{2}$ **7.** $1.84, -10.84$ **8.** $-0.18, -1.82$

9. $3.14, -0.64$ **10.** $2.39, 0.28$ **11.** $-0.75 \pm 0.97i$ **12.** $0.5 \pm 1.32i$ **13.** width $= 5.55$ cm, length $= 8.55$ cm
14. width $= 4$ in., length $= 10$ in.   **15.** 4 m   **16.** width $= 11$ ft, length $= 16$ ft   **17.** length $= 110$ ft,
width $= 70$ ft (nearest ft); See figure below.   **18.** 111.5 kg

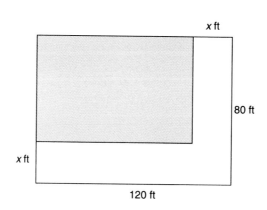

x ft

80 ft

x ft

120 ft

19. real, rational, $x = \dfrac{2}{3}, -1$ **20.** real, irrational, $x = \dfrac{3 \pm \sqrt{5}}{2}$ or 2.62, 0.38

21. real, irrational, $x = \dfrac{-1 \pm \sqrt{17}}{4}$ or 0.78, -1.28 **22.** no real solutions or $x = \dfrac{1 \pm i\sqrt{2}}{3}$ or $0.33 \pm 0.47i$

23. real, irrational, $x = \dfrac{3 \pm \sqrt{37}}{2}$ or 4.54, -1.54 **24.** real, irrational, $x = \dfrac{-5 \pm \sqrt{97}}{6}$ or 0.81 or -2.47

25. The discriminant must be greater than or equal to zero.   **26.** The discriminant must be greater than zero and a prefect square.

Self-Study Exercises 15–5

1. 2   **2.** 1   **3.** 4   **4.** 1   **5.** 4   **6.** 7

7. $x = 0, 2, -3$ **8.** $x = 0, \dfrac{1}{2}, -3$ **9.** $x = 0, \dfrac{5}{2}, \dfrac{2}{3}$ **10.** $x = 0, 2, 5$ **11.** $x = 0, 3, -2$ **12.** $x = 0, -\dfrac{1}{2}, -2$

13. $x = 0, 3, -3$ **14.** $x = 0, \dfrac{1}{2}, -\dfrac{1}{2}$ **15.** $x = 0, \dfrac{3}{4}, -\dfrac{3}{4}$ **16.** $3; x = 0, 4, -4$

17. real root: $x = 0$; imaginary roots: $x = \pm i\sqrt{5}$ **18.** $w(w + 3)(w + 7) = 421$; No; The factored form of the equation does not equal to zero. When put into factored form a cubic equation is formed that can not be factored using methods developed in this text.

Assignment Exercises, Chapter 15

1. pure   **3.** pure   **5.** incomplete   **7.** pure   **9.** complete   **11.** $2x^2 - 8x - 5 = 0$ **13.** $x^2 - 7x + 5 = 0$

15. $4x^2 + 3x - 1 = 0$ **17.** $x = \pm 10$ **19.** $x = \pm \dfrac{3}{2}$ **21.** $y = \pm 1.740$ **23.** $x = \pm 2.828$ **25.** $x = \pm 2.236$

27. $x = \pm 2$ **29.** $x = \pm 4.123$ **31.** $y = \pm 3.055$ **33.** $x = \pm 4$ **35.** $x = \pm 8$ **37.** 16.4 cm   **39.** 0, 5   **41.** 0, 2

43. $0, -\dfrac{1}{2}$ **45.** 0, 5   **47.** $0, -\dfrac{2}{3}$ **49.** $0, -3$ **51.** 0, 9   **53.** $0, -8$ **55.** $0, \dfrac{5}{3}$ **57.** $0, \dfrac{1}{2}$ **59.** $x = 0$ or 4

61. 3, 1   **63.** $-5, 2$ **65.** $-6, -1$ **67.** 2, 4   **69.** $-\dfrac{2}{3}, \dfrac{3}{2}$ **71.** $-\dfrac{2}{5}, \dfrac{5}{2}$ **73.** $-\dfrac{3}{4}, -1$ **75.** $\dfrac{3}{4}, -\dfrac{1}{3}$ **77.** $-21, 2$

79. $\dfrac{2}{3}, -1$ **81.** 3, 2   **83.** 6, -3 **85.** $-6, -5$ **87.** $-9, 2$ **89.** width $= 12$ ft   length $= 19$ ft

91. 2, 2 (double root)   **93.** 2, 6   **95.** $4 - \sqrt{2}, 4 + \sqrt{2}$ **97.** $3 - i\sqrt{3}, 3 + i\sqrt{3}$ **99.** 1, 4

101. $\dfrac{3 - \sqrt{37}}{2}, \dfrac{3 + \sqrt{37}}{2}$ **103.** $a = 1$, $b = -2$, $c = -8$ **105.** $a = 1$, $b = 3$, $c = -4$ **107.** $a = 1$, $b = -3$, $c = 2$ **109.** 9, -1 **111.** 2, -4 **113.** $2, -\dfrac{1}{2}$

115. 1.78, -0.28 **117.** $-0.23, -1.43$ **119.** $w = 11$ ft, $l = 22$ ft   **121.** $w = 14$ in., $l = 42$ in.   **123.** real, rational, equal

125. real, irrational, unequal   **127.** real, rational, unequal   **129.** 3.   **131.** 1   **133.** 8   **135.** $0, 2, \dfrac{2}{3}$

137. $0, -2, -3$ **139.** $0, -5, \dfrac{1}{2}$ **141.** 0, 2   **143.** $0, -2, -4$ **145.** $0, 5, -4$ **147.** $0, 3 \pm \sqrt{2}$ or 4.414, 1.586

149. $0, -1, 4$ **151.** $-2, -1, 0$ **153.** $0, \sqrt{17}, -\sqrt{17}$ or ± 4.123

Trial Test, Chapter 15

1. pure   **3.** incomplete   **5.** ± 9 **7.** $\pm \dfrac{4}{3}$ **9.** ± 2.33 **11.** 0, 2   **13.** 3, 2   **15.** $\dfrac{3}{2}, 4$ **17.** 169.79 mils

19. 1.98 amps   **21.** 6   **23.** $0, \dfrac{3}{2}, -5$ **25.** 0, 3

Self-Study Exercises 16–1

1   **1.** $2,415.77   **2.** $2,050.40   **3.** (a) 122.14 (b) 149.18 (c) 164.87 (d) 182.21   **4.** (a) $1,197.22 (b) $1,576.91
5. 64   **6.** 0.0041   **7.** 9,765,625   **8.** 0.002   **9.** 243   **10.** 0.0032

2   **11.** 2.08×10^{-87} **12.** 4.28×10^{-96} **13.** 5.58×10^{-44} **14.** 7.39   **15.** 0.05   **16.** 1.23   **17.** 0.03
18. $720.98   **19.** $14,448.88; $6,448.88   **20.** $15,373.05; $4,873.05   **21.** $14,414.30   **22.** $8\frac{1}{4}\%$ annually is
the better deal.   **23.** $90,305.50   **24.** $9.26   **25.** $7.72   **26.** $27.99   **27.** $321.89   **28.** 8.24%   **29.** 10.25%
30. 6.14%   **31.** 12.55%   **32.** $2,252.25   **33.** $7,906.98   **34.** $19,462.47   **35.** $6,462.60   **36.** $1,587.66
37. $1,254.59   **38.** $6,736.20   **39.** $13,611.60   **40.** $6,270; $270   **41.** $8,280; $280   **42.** $16,232
43. $9,786.68   **44.** $127,391.11; $91,000   **45.** $60,743.42   **46.** $25,129.02   **47.** $7,243.28   **48.** $5,866.60
49.   **50.** $908.92   **51.** $1,155.89

Years	Total Investment	Total Interest
Ten-year	$5,000	$2,243.28
Five-year	$5,000	$866.60

The 10-year investment earned more interest even
though half as much money was invested per year. At
the same period interest rate, investing for twice as
long gives a better yield on your investment than
investing the same amount for half as long. Thus, the
earlier you start saving, the better.

3   **52.** $x = 7$ **53.** $x = -3$ **54.** $x = 2$ **55.** $x = 10$ **56.** $x = 7$ **57.** $x = 3$ **58.** $x = 6$ **59.** $x = -4$
60. $x = -6$ **61.** $x = \frac{7}{6}$ **62.** $x = \frac{5}{2}$ **63.** $x = 5$ **64.** $x = 8$

Self-Study Exercises 16–2

1   **1.** $\log_3 9 = 2$ **2.** $\log_2 32 = 5$ **3.** $\log_9 3 = \frac{1}{2}$ **4.** $\log_{16} 2 = \frac{1}{4}$ **5.** $\log_4 \frac{1}{16} = -2$ **6.** $\log_3 \frac{1}{81} = -4$

2   **7.** $3^4 = 81$ **8.** $12^2 = 144$ **9.** $2^{-3} = \frac{1}{8}$ **10.** $5^{-2} = \frac{1}{25}$ **11.** $25^{-0.5} = \frac{1}{5}$ **12.** $4^{-0.5} = \frac{1}{2}$ **13.** $x = 3$
14. $x = \frac{1}{81}$ **15.** $x = 2$ **16.** $x = -2$ **17.** $x = 625$ **18.** $x = -4$

3   **19.** 0.4771   **20.** 0.7782   **21.** 0.3802   **22.** 0.6232   **23.** 2.1761   **24.** -2.9208 **25.** 1.3863   **26.** 0.9163
27. -1.8971 **28.** 5.6168   **29.** 4.6052

4   **30.** 3   **31.** 6   **32.** 5   **33.** 2   **34.** 1.9358

5   **35.** 3   **36.** 5   **37.** 8   **38.** 13.9 years   **39.** 2.0 years

6   **40.** 2.096   **41.** 1.893   **42.** 1.631

Assignment Exercises, Chapter 16

1. $x = 8$ **3.** $x = -2$ **5.** $x = 4$ **7.** $x = -\frac{5}{3}$ **9.** $x = 4$ **11.** $x = -5$ **13.** $x = 1$ **15.** $x = 4$

17. 0.0183   **19.** 0.0000454   **21.** $708.64   **23.** $524.95   **25.** $5,372.55   **27.** $2,189.90   **29.** $13,928.16
31. $1,356.25   **33.** 8.24%   **35.** $190.99   **37.** $2,913.80   **39.** $11,000 in 18 months is better.   **41.** $4,781.09
43. $26,361.59   **45.** $7,102.10   **47.** $2,294.19   **49.** $5,591.97   **51.** $3,407.89   **53.** $1,917.31

55. 1.25×10^{-6} **57.** 2.71×10^{-35} **59.** $\log_2 8 = 3$ **61.** $\log_3 81 = 4$ **63.** $\log_{27} 3 = \dfrac{1}{3}$ **65.** $\log_4 \dfrac{1}{64} = -3$

67. $\log_9 \dfrac{1}{3} = -\dfrac{1}{2}$ **69.** $\log_{12} \dfrac{1}{144} = -2$ **71.** $11^2 = 121$ **73.** $15^0 = 1$ **75.** $7^1 = 7$ **77.** $4^{-2} = \dfrac{1}{16}$ **79.** $9^{-0.5} = \dfrac{1}{3}$

81. $10^3 = 1,000$ **83.** 0.6990 **85.** 2.2553 **87.** -0.3979 **89.** 5.5984 **91.** -0.2231 **93.** 2.1133 **95.** 2
97. 343 **99.** $x = -2$ **101.** a. 2 b. 4 c. 8.1761 **103.** 3.17 **105.** 9.89 years **107.** (a) $C_t = 72,000,000(0.65)^x$
(b) 30,420,000 units per cubic meter after 60 days, 8,354,092.5 units per cubic meter after 150 days
(c) approximately 500 days or 16 to 17 months; Guess and check or logarithms are two possible methods.

Trial Test, Chapter 16

1. $3,657.26$ **3.** 58.09 **5.** $248,832$ **7.** $x = 9$ **9.** $x = 2$ **11.** $\log_4 \dfrac{1}{2} = -\dfrac{1}{2}$ **13.** $3^{-3} = \dfrac{1}{27}$ **15.** 3.4657

17. $x = 3$ **19.** 1.4641 **21.** $S = \$224.94$ thousands **23.** $\$5,596.82$ **25.** $\$19,350$ **27.** $\$5,727.50$ **29.** 12.55%
31. $\$680$ in 1 year is better **33.** Option 2 yields the greater return by $\$0.68$. **35.** $\$2,225.54$ **37.** $\$13,586.96$

Self-Study Exercises 17–1

[1] **1.** F **2.** F **3.** T **4.** T **5.** $\{6, 7, 8, 9, 10, 11\}$ **6.** $\{4, 6\}$ **7.** $\{15, 20, 25, 30, 35\}$ **8.** $\{-4, -3, -2, -1\}$
9. $\{x \mid x < 10 \text{ and } x \text{ is odd natural number}\}$ **10.** $\{x \mid -7 \le x \le -1 \text{ and } x \text{ is odd}\}$

11. $(5, 9)$ **12.** $(-7, -3)$ **13.** $[-5, -3]$

[2] **14.** $[8, 12]$ **15.** $(-\infty, 7)$ **16.** $(-\infty, -2)$

17. $[2, \infty)$ **18.** $(-\infty, 3]$ **19.** $(5, \infty)$

20. $[-3, \infty)$ **21.** $(-\infty, -2]$ **22.** $(-6, \infty)$

23. $(843,000, 1,000,000)$ **24.** $[4, 18]$ **25.** $n < 12$

Self-Study Exercises 17–2

[1] **1.** $y > 8$; $(8, \infty)$

2. $x < 1$; $(-\infty, 1)$

3. $x \le -3$; $(-\infty, 3]$

4. $x < 6$; $(-\infty, 6)$

5. $a \ge -3$; $[-3, \infty)$

6. $y \le 8$; $(-\infty, 8]$

7. $b > -1$; $(-1, \infty)$

$$b > -1$$

(number line with open paren at -1, shaded right; marks -3 -2 -1 0 1 2)

8. $t < 6$; $(-\infty, 6)$

$$t < 6$$

(number line with open paren at 6, shaded left; marks 2 3 4 5 6)

9. $y < 5$; $(-\infty, 5)$

$$y < 5$$

(number line with open paren at 5, shaded left; marks 3 4 5 6 7)

10. $a \le -2$; $(-\infty, -2]$

$$a \le -2$$

(number line with bracket at -2, shaded left; marks -4 -3 -2 -1 0 1)

11. $x \le \dfrac{1}{4}$; $\left(-\infty, \dfrac{1}{4}\right]$

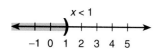

(number line with bracket, shaded left; marks -3 -2 -1 0$\frac{1}{4}$ 1 2)

12. $t \le 7$; $(-\infty, 7]$

$$t \le 7$$

(number line with bracket at 7, shaded left; marks 5 6 7 8 9 10)

13. $y > 11$; $(11, \infty)$

$$y > 11$$

(number line with open paren at 11, shaded right; marks 9 10 11 12 13 14)

14. $x \ge 3$; $[3, \infty)$

$$x \ge 3$$

(number line with bracket at 3, shaded right; marks 0 1 2 3 4 5)

15. $x < 1$; $(-\infty, 1)$

$$x < 1$$

(number line with open paren at 1, shaded left; marks -1 0 1 2 3 4 5)

16. $x > \dfrac{1}{3}$; $\left(\dfrac{1}{3}, \infty\right)$

$$x > \tfrac{1}{3}$$

(number line with open paren, shaded right; marks -1 0$\frac{1}{3}$ 1 2 3)

17. $a \le -3$, $(-\infty, -3]$

$$a \le -3$$

(number line with bracket at -3, shaded left; marks -4 -3 -2 -1 0 1)

18. $x < \dfrac{1}{2}$; $\left(-\infty, \dfrac{1}{2}\right)$

$$x < \tfrac{1}{2}$$

(number line with open paren, shaded left; marks -2 -1 0$\frac{1}{2}$ 1 2 3)

19. $2x + 196 \ge 52{,}800$
$\qquad x \ge 26{,}302$

20. $x > \$3.60 \cdot 2$
$\qquad x > \$7.20$

(number line with open paren, shaded right; marks 0 2 4 6 8 10)

Self-Study Exercises 17–3

[1] **1.** T; all elements in B also in U **2.** F; 2 is not in A **3.** T; ϕ is a subset of all sets.
4. F; No element of A is in B **5.** $\{\ \}$ or ϕ **6.** $\{2, 0, -2\}$ **7.** $\{-4, -2, 0, 1, 2, 3, 4, 5\}$
8. $\{-5, -3, -1, 1, 2, 3, 4, 5\}$ **9.** $\{-5, -4, -3, -2, -1, 1, 2, 3, 4, 5\}$ **10.** $\{-5, -4, -3, -1, 0\}$

[2] **11.** $3 \le x \le 4$; $[3, 4]$

(number line with brackets at 3 and 4, shaded between; marks 3 4)

12. $5 \le x \le 7$; $[5, 7]$

(number line with brackets at 5 and 7, shaded between; marks 1 2 3 4 5 6 7)

13. $-1 \le x \le 3$; $[-1, 3]$

(number line with brackets at -1 and 3, shaded between; marks -2 -1 0 1 2 3 4)

14. $-1 \le x \le 2$; $[-1, 2]$

(number line with brackets at -1 and 2, shaded between; marks -2 -1 0 1 2 3)

Selected Answers to Student Exercise Material

15. no solution; ϕ

16. $3 < x < 7$; $(3, 7)$

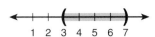

17. $0 < x < 3$; $(0, 3)$

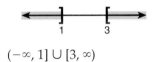

18. no solution; ϕ

19. $-6 < x < -3$; $(-6, -3)$

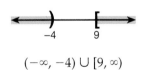

20. $3 < x < 5$; $(3, 5)$

21. $-2 \le x \le 2$; $[-2, 2]$

22. $0 \le x \le \dfrac{4}{7}$; $\left[0, \dfrac{4}{7}\right]$

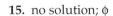 **23.** $x < 2$ or $x > 9$

$(-\infty, 2) \cup (9, \infty)$

24. $x \le 3$ or $x \ge 7$

$(-\infty, 3] \cup [7, \infty)$

25. $x \le 4$ or $x \ge 5$

$(-\infty, 4] \cup [5, \infty)$

26. $x \le 1$ or $x \ge 3$

$(-\infty, 1] \cup [3, \infty)$

27. $x \le 1\dfrac{1}{2}$ or $x \ge 3\dfrac{1}{2}$

$\left(-\infty, 1\dfrac{1}{2}\right] \cup \left[3\dfrac{1}{2}, \infty\right)$

28. $x < -4$ or $x \ge 9$

$(-\infty, -4) \cup [9, \infty)$

29. $5.22 \le 5.27 \le 5.32$;
$x - 0.05 \le x \le x + 0.05$

30. $14.7 \le 15.2 \le 15.7$

Self-Study Exercises 17–4

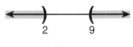 **1.** $-3 < x < 2$; $(-3, 2)$

2. $x < -3$ or $x > \dfrac{1}{2}$; $(-\infty, -3) \cup \left(\dfrac{1}{2}, \infty\right)$

3. $\dfrac{2}{3} < x < \dfrac{5}{2}; \left(\dfrac{2}{3}, \dfrac{5}{2}\right)$

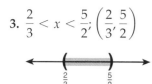

4. $y \le -6$ or $y \ge 2; (-\infty, -6] \cup [2, \infty)$

5. $-4 < a < 6; (-4, 6)$

6. $2 \le x \le 5; [2, 5]$

7. $x \le -2$ or $x \ge 3; (-\infty, -2] \cup [3, \infty)$

8. $1 \le x \le 3; [1, 3]$

9. $\dfrac{1}{4} \le x \le 5; \left[\dfrac{1}{4}, 5\right]$

10. $x < -\dfrac{1}{2}$ or $x > \dfrac{1}{3}; \left(-\infty, -\dfrac{1}{2}\right) \cup \left(\dfrac{1}{3}, \infty\right)$

11. $x \le \dfrac{1}{2}$ or $x \ge \dfrac{2}{3}; \left(-\infty, \dfrac{1}{2}\right] \cup \left[\dfrac{2}{3}, \infty\right)$

12. $-5 \le x \le \dfrac{1}{2}; \left[-5, \dfrac{1}{2}\right]$

[2]   **13.** $-7 < x < 3; (-7, 3)$

14. $x < -8$ or $x > 3; (-\infty, -8) \cup (3, \infty)$

15. $x < \dfrac{1}{3}$ or $x > 3; \left(-\infty, \dfrac{1}{3}\right) \cup (3, \infty)$

16. $-7 < x < 7; (-7, 7]$

17. $0 < x < 1; (0, 1)$

18. $x \le 2$ or $x > 3; (-\infty, 2] \cup (3, \infty)$

19. $2 < x \le 3; (2, 3]$

20. $x \le 0$ or $x > 4; (-\infty, 0] \cup (4, \infty)$

21. $x < -1$ or $x \ge 3; (-\infty, -1) \cup [3, \infty)$

Selected Answers to Student Exercise Material

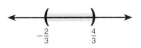

Self-Study Exercises 17–5

1. ± 8 **2.** ± 15 **3.** $8, -2$ **4.** $9, 5$ **5.** $6, -3$ **6.** $3, \dfrac{5}{3}$ **7.** ± 11 **8.** no solution   **9.** $7, 1$

10. $4, -3$ **11.** no solution   **12.** $-\dfrac{1}{3}, +3\dfrac{2}{3}$ **13.** $0, 4$ **14.** $-5, 10$ **15.** $-\dfrac{2}{3}, 6$ **16.** $0, \dfrac{18}{7}$ **17.** no solution

18. $-\dfrac{12}{5}, \dfrac{18}{5}$ **19.** $-\dfrac{14}{5}, \dfrac{16}{5}$ **20.** $0, 3$ **21.** $\dfrac{2}{3}$

Self-Study Exercises 17–6

1. $-5 < x < 5; (-5, 5)$

2. $-1 < x < 1; (-1, 1)$

3. $-2 < x < 2; (-2, 2)$

4. $-7 \le x \le 7; [-7, 7]$

5. $-5 < x < -1; (-5, -1)$

6. $-7 < x < -1; (-7, -1)$

7. $-1 < x < 7; (-1, 7)$

8. $-1 < x < 5; (-1, 5)$

9. $-1 \le x \le 11; [-1, 11]$

10. $-6 \le x \le 8; [-6, 8]$

11. $-\dfrac{2}{3} < x < 4; \left(-\dfrac{2}{3}, 4\right)$

12. $\dfrac{1}{2} < x < \dfrac{7}{2}; \left(\dfrac{1}{2}, \dfrac{7}{2}\right)$

13. $-1 \le x \le \dfrac{1}{5}; \left[-1, \dfrac{1}{5}\right]$

14. $-3 \le x \le -\dfrac{1}{2}; \left[-3, -\dfrac{1}{2}\right]$

15. $-\dfrac{2}{3} < x < \dfrac{4}{3}; \left(-\dfrac{2}{3}, \dfrac{4}{3}\right)$

16. $x < -2$ or $x > 2$; $(-\infty, -2) \cup (2, \infty)$

17. $x < -3$ or $x > 3$; $(-\infty, -3) \cup (3, \infty)$

18. $x \le -5$ or $x \ge 5$; $(-\infty, -5] \cup [5, \infty)$

19. $x \le -6$ or $x \ge 6$; $(-\infty, -6] \cup [6, \infty)$

20. $x < 3$ or $x > 7$; $(-\infty, 3) \cup (7, \infty)$

21. $x < -2$ or $x > 10$; $(-\infty, -2) \cup (10, \infty)$

22. $x \le -8$ or $x \ge 2$; $(-\infty, -8] \cup [2, \infty)$

23. $x \le -10$ or $x \ge 2$; $(-\infty, -10] \cup [2, \infty]$

24. $x \le \dfrac{5}{2}$ or $x \ge \dfrac{5}{2}$ or all real numbers; $(-\infty, \infty)$

25. $x \le -\dfrac{2}{3}$ or $x \ge 2$; $\left(-\infty, -\dfrac{2}{3}\right] \cup [2, \infty)$

26. $x \le -2$ or $x \ge -\dfrac{6}{5}$; $\left(-\infty, -2\right] \cup \left[-\dfrac{6}{5}, \infty\right)$

27. $x \le -\dfrac{3}{2}$ or $x \ge 3$; $\left(-\infty, -\dfrac{3}{2}\right] \cup [3, \infty)$

28. $x \le -\dfrac{1}{3}$ or $x \ge 1$; $\left(-\infty, -\dfrac{1}{3}\right] \cup [1, \infty)$

29. $x \le -\dfrac{1}{2}$ or $x \ge \dfrac{7}{2}$; $\left(-\infty, -\dfrac{1}{2}\right] \cup \left[\dfrac{7}{2}, \infty\right)$

30. $x \le -2$ or $x \ge 3$; $(-\infty, -2] \cup [3, \infty)$

31. $\dfrac{1}{3} \le x \le 1$; $\left[\dfrac{1}{3}, 1\right]$

32. $0 \le x \le 5$

33. $x \le -\dfrac{4}{5}$ or $x \ge 2$

34. $x \le 1\dfrac{1}{4}$ or $x \ge 1\dfrac{3}{4}$

35. no solution
$\{\,\}$ or ϕ

36. no solution; { } or φ   **37.** $x \le 0$ or $x \ge 6$ **38.** $x < -\dfrac{1}{2}$ or $x > 0$ **39.** $1 \le x \le 7$ **40.** no solution

41. $-7 < x < 7$ **42.** $-11 < x < -5$ **43.** $x < -12$ or $x > 12$ **44.** $x < -10$ or $x > -6$ **45.** no solution

Assignment Exercises, Chapter 17

1. a set with no elements { } or φ

3. $5 \in W$

5. $(-7, \infty)$

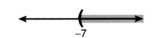

7. $[-4, 2)$

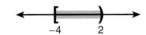

9. $(-2, \infty)$

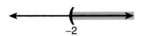

11. $m < 7; (-\infty, 7)$

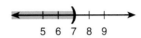

13. $x > 0; (0, \infty)$

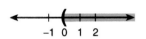

15. $x \le 3; (-\infty, 3]$

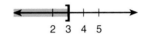

17. $x > -9; (-9, \infty)$

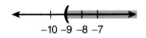

19. $x > -1; (-1, \infty)$

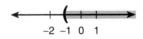

21. $y \ge -1; [-1, \infty)$

23. $x \le 6; (-\infty, 6]$

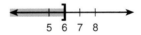

25. $D = P + \$5.60; 2D + 12P \le \$59.80$; therefore, $P \le \$3.47; D \le \$9.07$ **27.** { } or φ   **29.** $\{-1\}$ **31.** $\{5\}$

33. $x > -2; (-2, \infty)$

35. $x \ge -3; [-3, \infty)$

37. $\dfrac{1}{2} < x < 4\dfrac{1}{2}; \left(\dfrac{1}{2}, \dfrac{9}{2}\right)$

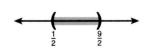

39. $2 < x < 4; (2, 4)$

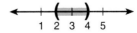

41. no solution; φ

43. $2 < x < 6; (2, 6)$

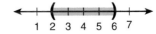

45. $-3 < x \le 2$; $(-3, 2]$

47. $-2 \le x \le -\frac{3}{4}$; $\left[-2, -\frac{3}{4}\right]$

49. $x < 2$ or $x > 8$; $(-\infty, 2) \cup (8, \infty)$

51. $x < -9$ or $x > 8$; $(-\infty, -9) \cup (8, \infty)$

53. $10 < x < 19$

55. $x < 2$ or $x > 5$; $(-\infty, 2) \cup (5, \infty)$

57. $-\frac{1}{3} < x < \frac{3}{2}$; $\left(\frac{1}{3}, \frac{3}{2}\right)$

59. $-1 \le x \le 2$; $[-1, 2]$

61. $-\frac{1}{2} \le x \le 3$; $\left[-\frac{1}{2}, 3\right]$

63. $-5 < x < \frac{3}{2}$; $\left(-5, \frac{3}{2}\right)$

65. $1 < x < 7$; $(1, 7)$

67. $x < -8 \cup x > 0$; $(-\infty, -8) \cup (0, \infty)$

69. ± 12 **71.** $4, -10$ **73.** $20, -4$ **75.** $6, -\dfrac{5}{2}$ **77.** $1, -\dfrac{23}{7}$ **79.** $3, -\dfrac{13}{7}$ **81.** $\dfrac{11}{3}, \dfrac{7}{3}$ **83.** ± 7 **85.** ± 16

87. $10, -4$ **89.** $2, -\dfrac{1}{2}$ **91.** $-1 < x < 7$; $(-1, 7)$ **93.** $-4 < x < 12$; $(-4, 12)$

95. no solution   **97.** $\$108,000 < I < \$250,000$ **99.** 13 club members with $\$340$ left

Trial Test, Chapter 17

1. $x \ge -12$; $[-12, \infty)$

3. $x > 3$; $(3, \infty)$

5. $x > 2$; $(2, \infty)$

7. $x \le -6$; $(-\infty, -6]$

9. $x > -2$; $(-2, \infty)$

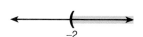

11. $-8 < x < 4$; $(-8, 4)$

Selected Answers to Student Exercise Material

13. no solution

15. $x < -\dfrac{3}{2} \cup x > 1; \left(-\infty, -\dfrac{3}{2}\right) \cup (1, \infty)$

17. $x < 2 \cup x > 6; (-\infty, 2) \cup (6, \infty)$

19. $A \cup B = \{1, 2, 3, 4, 5, 6, 7, 8, 9\}$

21. $A \cap B' = \{9\}$

23. $-5 < x < 2; (-5, 2)$

25. ± 15 **27.** ± 2 **29.** $x < -18 \cup x > 2; (-\infty, -18) \cup (2, \infty)$ **31.** $96.7 \le 96.8 \le 96.9$

Self-Study Exercises 18–1

1   Values chosen for table of values may vary.

1.

x	f(x)
−2	−13
−1	−10
0	−7
1	−4
2	−1

2.

x	f(x)
−1	5
0	3
1	1
2	−1

3. $f(x) = 39 + 0.1x$

4.

x	f(x)
0	39
50	44
100	49
150	54
200	59

5.

x	f(x)
−2	13
−1	9
0	5
1	1
2	−3
3	−7
4	−11

6.

x	f(x)
−6	0
−2	2
0	3
2	4
6	6

7.

x	f(x)
−6	3
−3	1
0	−1
3	−3
6	−5

8.

x	f(x)
−1	−3
0	2
1	7

9.

x	f(x)
−2	3
0	2
2	1
4	0

10.

x	f(x)
−6	0
−3	1
0	2
3	3
6	4

11.

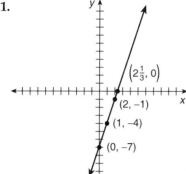

12.

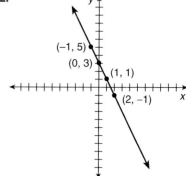

13.

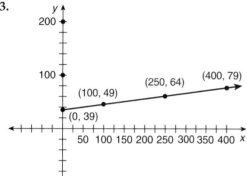

14. $79

15.

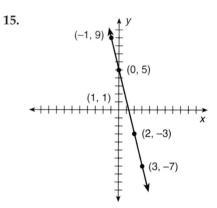

16.

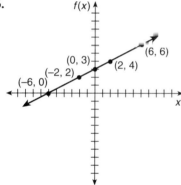

17.

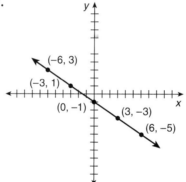

18.

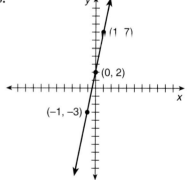

19.

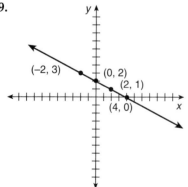

20.

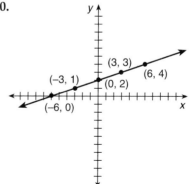

3   **21.** no   **22.** yes   **23.** yes   **24.** no   **25.** yes   **26.** yes   **27.** yes   **28.** no   **29.** yes   **30.** no
31. yes   **32.** yes   **33.** yes   **34.** no   **35.** no   **36.** no   **37.** yes   **38.** no

4   **39.** $(8, 5)$ **40.** $(2, 5)$ **41.** $(-6, 1)$ **42.** $(21, 3)$ **43.** $(5, 15)$ **44.** $(2, 0)$ **45.** $(8, -2)$
46. $(3, -3)$ **47.** $(6, 2)$ **48.** $(9, 1)$ **49.** $(12, 2)$ **50.** $(0, -12)$ **51.** $(18, -4)$ **52.** $(-2, 12)$
53. $(1, 3)$; $(1, -2)$; $(1, 0)$; $(1, 1)$ **54.** $(-1, 4)$; $(3, 4)$; $(0, 4)$; $(2, 4)$ **55.** $(2, -7)$; $(-3, -7)$; $(0, -7)$;
$(5, -7)$ **56.** $(9, -2)$; $(9, 3)$; $(9, 0)$; $(9, 2)$ **57.** $12   **58.** $2   **59.** $34   **60.** $13,800

1 1. x-intercept, (5, 0)
   y-intercept, (0, 5)

2. x-intercept, (5, 0)
   y-intercept, $\left(0, \dfrac{5}{3}\right)$

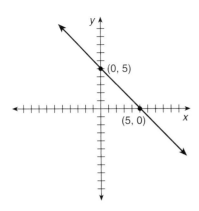

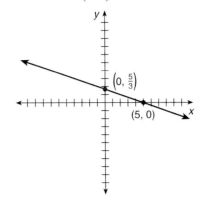

3. x-intercept, (−4, 0)
   y-intercept, (0, 8)

4. x-intercept, $\left(\dfrac{1}{3}, 0\right)$
   y-intercept, (0, −1)

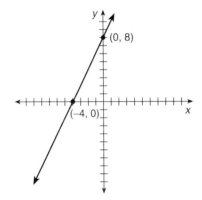

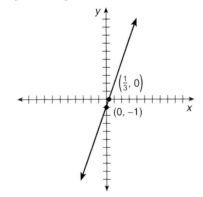

5. x-intercept, $\left(\dfrac{2}{5}, 0\right)$
   y-intercept, (0, −2)

6. x-intercept, (3, 0)
   y-intercept, $\left(0, \dfrac{3}{2}\right)$

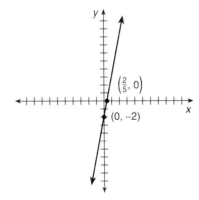

7. slope $= \dfrac{2}{1}$; y-intercept, $(0, -3)$

8. slope $= -\dfrac{1}{2}$; y-intercept, $(0, -2)$

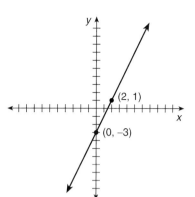

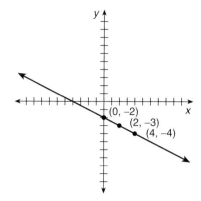

9. slope $= -\dfrac{3}{5}$; y-intercept, $(0, 0)$

10. slope $= \dfrac{1}{2}$; y-intercept, $\left(0, -1\dfrac{1}{2}\right)$

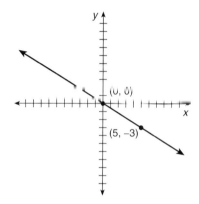

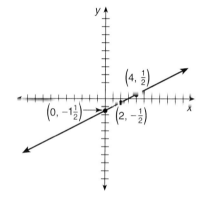

11. slope $= -\dfrac{2}{1}$; y-intercept, $(0, 1)$

12. slope $= \dfrac{3}{4}$; y-intercept, $(0, 2)$

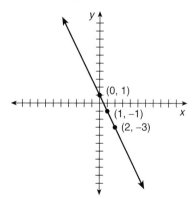

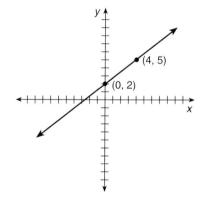

13.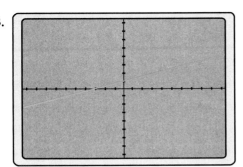

14. 0.38

15.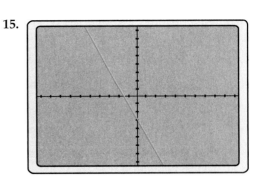

16. -5.08

17.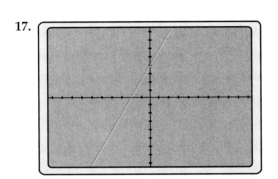

18. 1,817

19.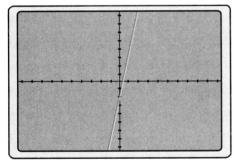

20. 136

21. \$35,000; \$95,000

22. \$332; \$17,264

23. 195; 155

24. 1,096 puzzles

25. \$3,200

26. $f(x) = 3x - 5$
$x = \dfrac{5}{3}$ or 1.66667

27. $f(x) = 5x - 6$
$x = 1.2$

28. $f(x) = 6x - 5$
$x = 0.83333$

29. $f(x) = x - 1$
$x = 1$

30. $f(x) = x + 9$
$x = -9$

Self-Study Exercises 18–3

1 1.

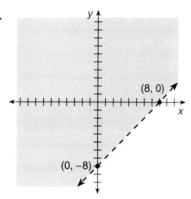

2.

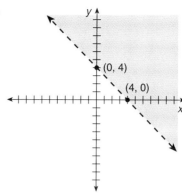

3.

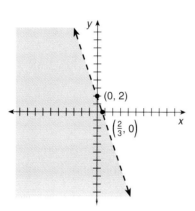

4.

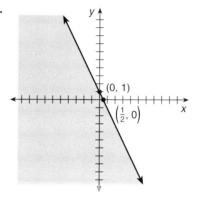

5.

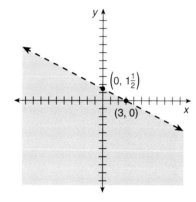

6.

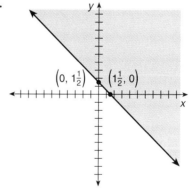

7.

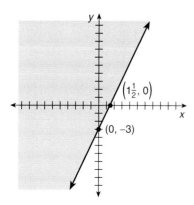

8.

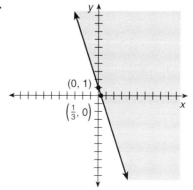

9.

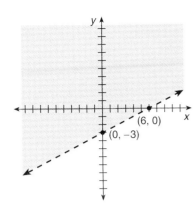

10.

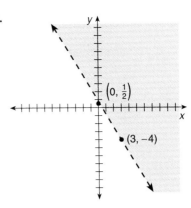

Self-Study Exercises 18–4

1. $y = x^2$. The domain is all real numbers. The range is all real numbers greater than or equal to zero.

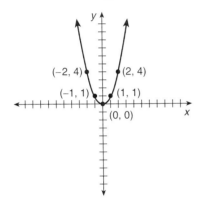

2. $y = 3x^2$. The domain is all real numbers. The range is all real numbers greater than or equal to zero.

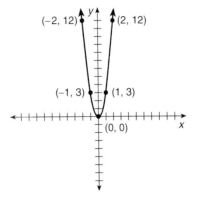

3. $y = \frac{1}{3}x^2$. The domain is all real numbers. The range is all real numbers greater than or equal to zero.

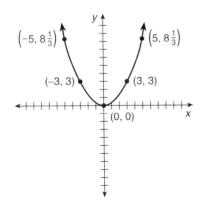

4. $y = -4x^2$. The domain is all real numbers. The range is all real numbers less than or equal to zero.

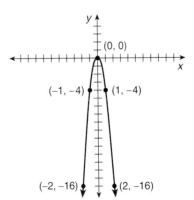

5. $y = -\dfrac{1}{4}x^2$. The domain is all real numbers. The range is all real numbers less than or equal to zero.

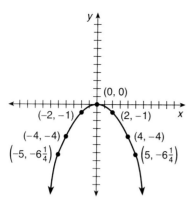

6. $y = x^2 - 4$. The domain is all real numbers. The range is all real numbers greater than or equal to -4.

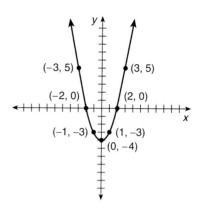

7. $y = x^2 + 4$. The domain is all real numbers. The range is all real numbers greater than or equal to 4.

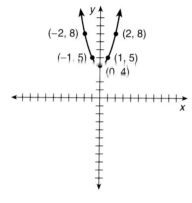

8. $y = x^2 - 6x + 9$. The domain is all real numbers. The range is all real numbers greater than or equal to zero.

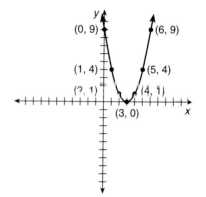

9. $y = -x^2 + 6x - 9$. The domain is all real numbers. The range is all real numbers less than or equal to 0.

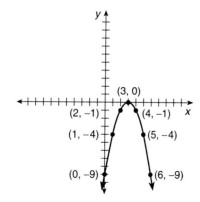

2

10. The domain is all real numbers. The range is all real numbers greater than or equal to 0.

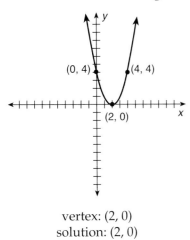

vertex: $(2, 0)$
solution: $(2, 0)$

11. The domain is all real numbers. The range is all real numbers greater than or equal to -6.

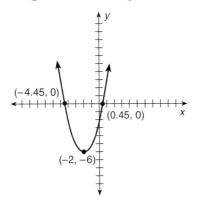

vertex: $(-2, -6)$
solutions: $(-4.45, 0)$; $(0.45, 0)$

12. The domain is all real numbers. The range is all real numbers greater than or equal to -4.5.

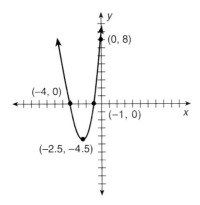

vertex: $(-2.5, -4.5)$
solutions: $(-4, 0)$; $(-1, 0)$

13. The domain is all real numbers. The range is all real numbers less than or equal to 12.

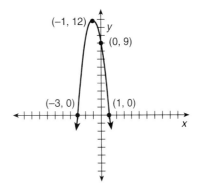

vertex: $(-1, 12)$
solutions: $(-3, 0)$; $(1, 0)$

14. The domain is all real numbers. The range is all real numbers less than or equal to -3.

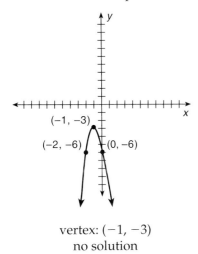

vertex: $(-1, -3)$
no solution

15. The domain is all real numbers. The range is all real numbers greater than or equal to -15.

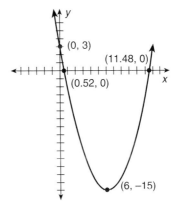

vertex: $(6, -15)$
solutions: $(0.52, 0)$; $(11.48, 0)$

16. $x = 2, x = 3$ **17.** $x = 0$ **18.** no real solution

19. $0, 9

20. $t = 0$ seconds

21. $y = 3x^2 + 5x - 2$

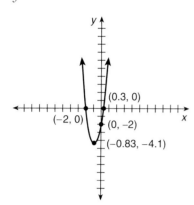

22. $y = (2x - 3)(x - 1)$

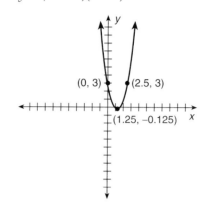

23. $y = 2x^2 - 9x - 5$

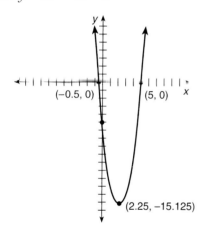

24. $y = -2x^2 + 9x + 5$

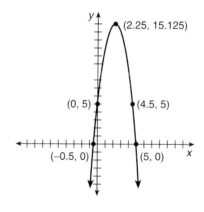

25.

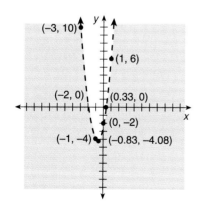

26.

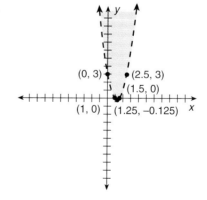

27.

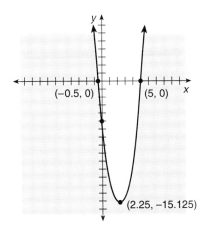

28.

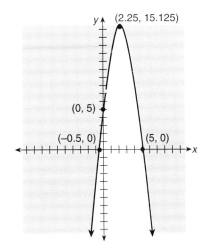

29.

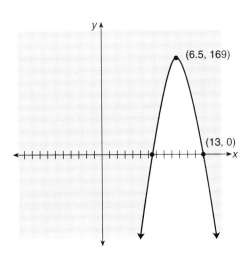

30.

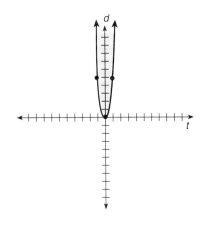

Self-Study Exercises 18–5

1️⃣   **1.** $y = 2x^3$

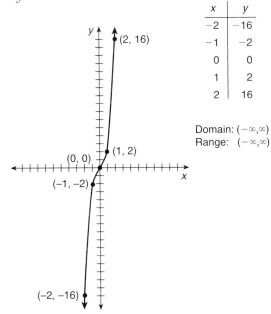

x	y
−2	−16
−1	−2
0	0
1	2
2	16

Domain: $(-\infty, \infty)$
Range: $(-\infty, \infty)$

2. $y = \dfrac{1}{2}x^3$

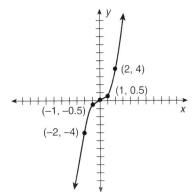

x	y
−2	−4
−1	−0.5
0	0
1	0.5
2	4

Domain: $(-\infty, \infty)$
Range: $(-\infty, \infty)$

3. $y = x^3 - x^2 - 4x + 4$

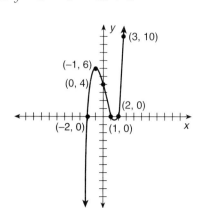

x	y
-3	-20
-2	0
-1	6
0	4
1	0
2	0
3	10

Domain: $(-\infty, \infty)$
Range: $(-\infty, \infty)$

4. $y = x^3 + 2$

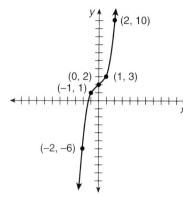

x	y
-2	-6
-1	1
0	2
1	3
2	10

Domain: $(-\infty, \infty)$
Range: $(-\infty, \infty)$

5. $y = 3^x$

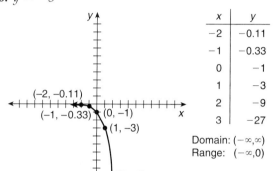

x	y
-2	0.11
-1	0.33
0	1
1	3
2	9
3	27

Domain: $(-\infty, \infty)$
Range: $(0, \infty)$

6. $y = -3^x$

x	y
-2	-0.11
-1	-0.33
0	-1
1	-3
2	-9
3	-27

Domain: $(-\infty, \infty)$
Range: $(-\infty, 0)$

7. $y = 2^x - 3$

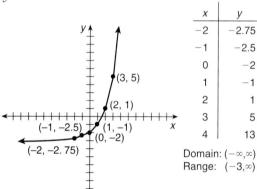

x	y
-2	-2.75
-1	-2.5
0	-2
1	-1
2	1
3	5
4	13

Domain: $(-\infty, \infty)$
Range: $(-3, \infty)$

8. $y = -2^x - 3$

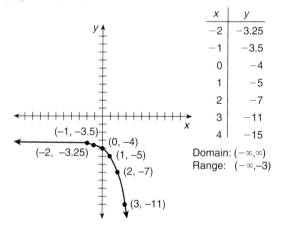

x	y
-2	-3.25
-1	-3.5
0	-4
1	-5
2	-7
3	-11
4	-15

Domain: $(-\infty, \infty)$
Range: $(-\infty, -3)$

9. $y = \ln 2x$

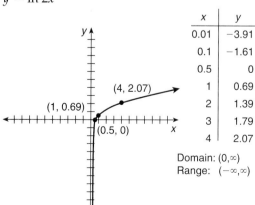

x	y
0.01	-3.91
0.1	-1.61
0.5	0
1	0.69
2	1.39
3	1.79
4	2.07

Domain: $(0, \infty)$
Range: $(-\infty, \infty)$

10. $y = -\ln 2x$

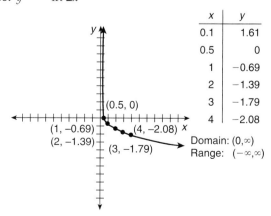

x	y
0.1	1.61
0.5	0
1	-0.69
2	-1.39
3	-1.79
4	-2.08

Domain: $(0, \infty)$
Range: $(-\infty, \infty)$

11. The graph is the graph of a function.

12. The graph is the graph of a function.

13. The graph is the graph of a relation but not a function.

14. The graph is the graph of a relation but not a function.

15. The graph is the graph of a function.

16. The graph is the graph of a function.

17. domain—all real numbers
range—all real numbers greater than or equal to -2

18. domain—the real numbers between and including -5 and 5
range—the real numbers between and including -3 and 3

19. domain—the real numbers between and including -4 and 3
range—the real numbers between and including -6 and 3

20. domain—the real numbers greater than or equal to 0
range—the real numbers greater than or equal to 0

Assignment Exercises, Chapter 18

Since plotted solutions will vary, check graphs by comparing x- and y-intercepts.

1.

x	$f(x)$
-1	-5
1	-1
3	3

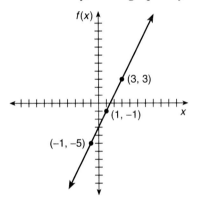

3.

x	$f(x)$
-1	-3
0	0
1	3

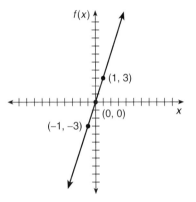

5.

x	$f(x)$
-1	3
0	0
1	-3

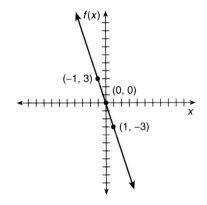

7.

x	$f(x)$
-1	-1
0	1
1	3

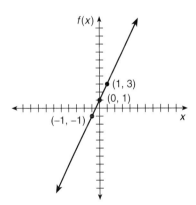

9.

x	f(x)
−2	−10
−1	−6
0	−2
1	2
3	10

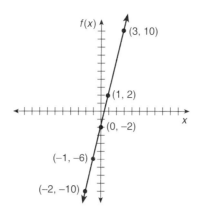

11.

x	f(x)
−2	−3
0	−2
2	−1
4	0

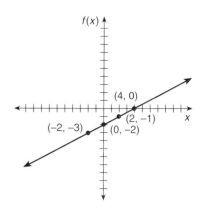

13. no   **15.** yes   **17.** yes   **19.** $y = 1$

21. x-intercept, $(−1, 0)$; y-intercept, $\left(0, -\dfrac{1}{4}\right)$

23. $3x − y = 1$; x-intercept, $\left(\dfrac{1}{3}, 0\right)$; y-intercept, $(0, −1)$

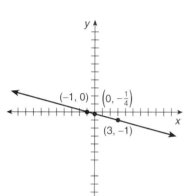

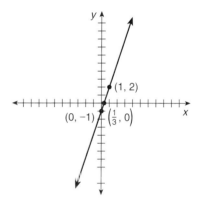

25. $\dfrac{1}{2}x + \dfrac{1}{3}y = 1$

x-intercept $(2, 0)$; y-intercept $(0, 3)$

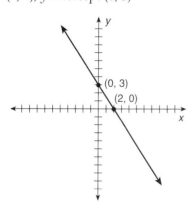

27. $y = 5x − 2$, $m = \dfrac{5}{1}$, $b = −2$

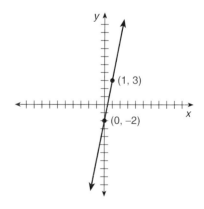

29. $y = -3x - 1$, $m = \dfrac{-3}{1}$, $b = -1$

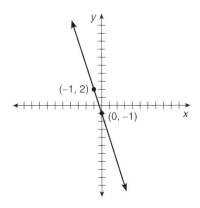

31. $y = x - 4$, $m = \dfrac{1}{1}$, $b = -4$

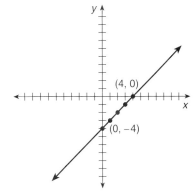

33. $y = \dfrac{1}{2}x + \dfrac{1}{2}$, $m = \dfrac{1}{2}$, $b = \dfrac{1}{2}$

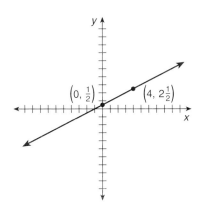

35. $y = 2x - 2$, $m = 2$, $b = -2$

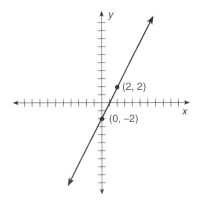

37.

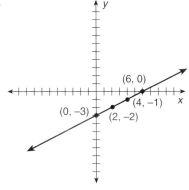

39.

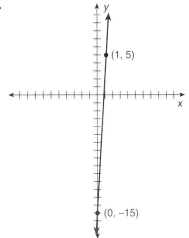

41. $58,000

43. $f(x) = x - 6$

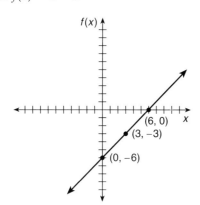

45. $f(x) = 3x + 6$

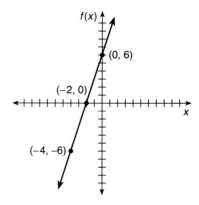

47. $f(x) = 2x - 2$

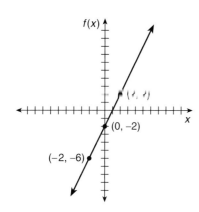

49. $4x + y < 2, \left(\frac{1}{2}, 0\right), (0, 2)$

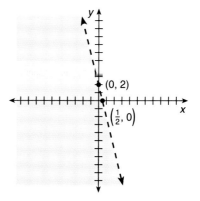

51. $3x + y \leq 2, \left(\frac{2}{3}, 0\right), (0, 2)$

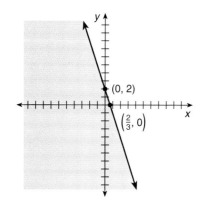

53. $x - 2y < 8, (8, 0), (0, -4)$

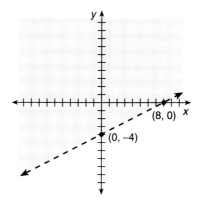

55. $y \geq 3x - 2$, $\left(\frac{2}{3}, 0\right)$, $(0, -2)$

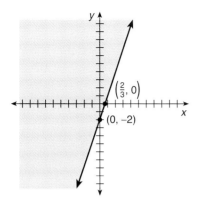

57. $y > \frac{2}{3}x - 2$, $(3, 0)$, $(0, -2)$

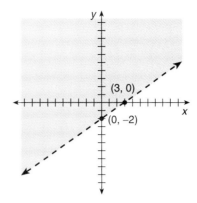

59. $y = -x^2 - 1$

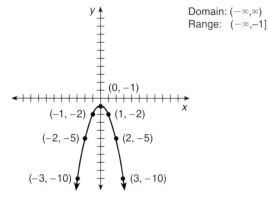

Domain: $(-\infty, \infty)$
Range: $(-\infty, -1]$

61. $y = x^2 - 6x + 8$

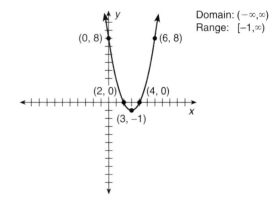

Domain: $(-\infty, \infty)$
Range: $[-1, \infty)$

63. $y = -x^2 + 2x - 1$

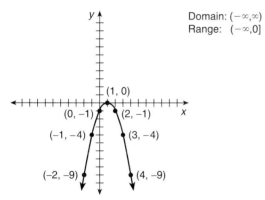

Domain: $(-\infty, \infty)$
Range: $(-\infty, 0]$

65. $y = -x^2 + 4x - 4$

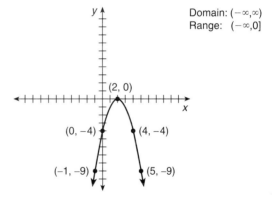

Domain: $(-\infty, \infty)$
Range: $(-\infty, 0]$

67. $x = -2, 6$

69. $x = -4$ (double root)

71. $y \geq -2x^2$

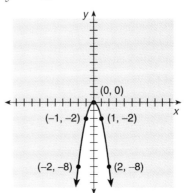

73. $y = 5x^3$

x	y
-2	-40
-1	-5
0	0
1	5
2	40
3	135

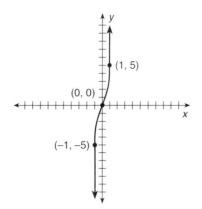

The domain is the set of all real numbers. The range is the set of all real numbers.

75. $y = 5^x$

x	y
-3	0.008
-2	0.04
-1	0.2
0	1
1	5
2	25
3	125

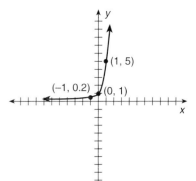

The domain is the set of all real numbers. The range is the set of all real numbers greater than zero.

77. $y = \ln 4x$

x	y
0.001	-5.52
0.1	-0.92
0.5	0.69
1	1.39
2	2.07
3	2.48
4	2.77

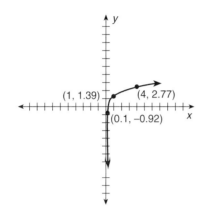

The domain is the set of all real numbers that are greater than zero. The range is the set of all real numbers.

79. The graph is the graph of a relation but not a function since at least one vertical line can be drawn that intersects the graph in more than one point.

81. The graph is the graph of a relation but not a function. A vertical line intersects the graph in two points everywhere except at $(-5, 0)$ and $(5, 0)$.

83. domain—all real numbers
range—all real numbers less than or equal to 4

85. domain—all real numbers between and including -2 and 2
range—all real numbers between and including -3 and 9

87. (a)

gal	ft²
1	400
2	800
3	1,200
4	1,600
5	2,000
6	2,400
7	2,800
8	3,200
9	3,600
10	4,000

(b)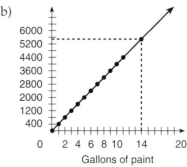

(c) 14 gal
(d) $279.30
(e) $302.34

89.

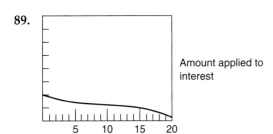

Amount applied to interest

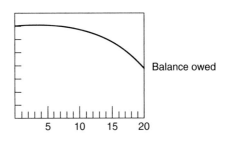

Balance owed

Trial Test, Chapter 18

Since plotted solutions will vary, check graphs by comparing x- and y-intercepts.

1.

x	f(x)
−2	−1
0	0
2	1

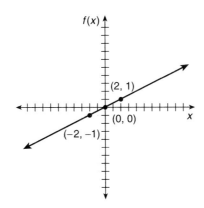

3.

x	f(x)
−1	−6
0	−4
1	−2
2	0

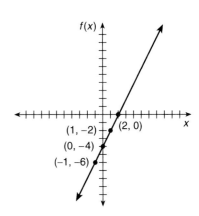

5. $f(x) = 3x - 3; x = 1$

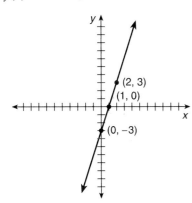

7. $(2, 5)$

9.

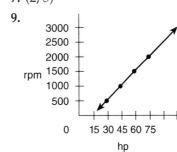

11. $23,200

13. x-intercept $(8, 0)$
y-intercept $(0, 4)$

15. y-intercept $= (0, -3)$

slope $= \dfrac{-2}{1}$

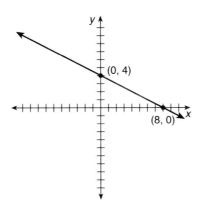

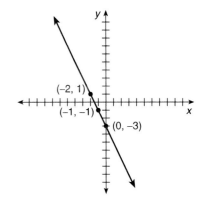

17. y-intercept $= (0, -5)$

slope $= \dfrac{1}{1}$

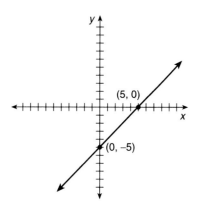

19. $2x - y \leq 2$

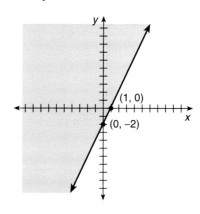

21. $x + y < 1$

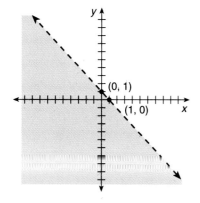

23. $y = x^2 + 2x + 1$

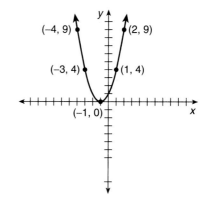

25. $y \leq x^2 - 6x + 8$

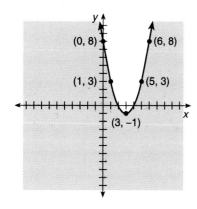

27. $y < -\dfrac{1}{2}x^2$

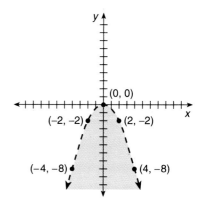

29. The graph is the graph of a relation but not a function. The domain is the set of all real numbers. The range is the set of all real numbers.

Self-Study Exercises 19–1

1   **1.** $\dfrac{1}{2}$ **2.** $-\dfrac{5}{3}$ **3.** 3   **4.** 1   **5.** $\dfrac{3}{4}$ **6.** 0   **7.** $-\dfrac{5}{4}$ **8.** -1 **9.** undefined   **10.** -3 **11.** 0

12. not defined   **13.** $1,400   **14.** $10   **15.** 2,000 ft/min   **16.** $113.20 per year   **17.** $123.40 per year   **18.** $316.00 per year

19. $32.50 per year

20. −$2.00 per year

21. The data in this table do not form a perfect straight line. If the slope (rate of change) remains the same for such data, the graph will be a straight line.

22. The rate of change is greater for public four-year colleges than it is for public two-year colleges.

2 **23.**

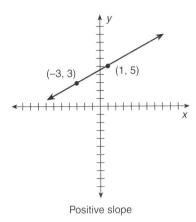

Positive slope

24.

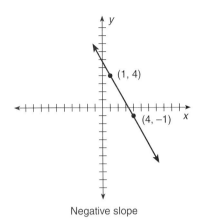

Negative slope

25.

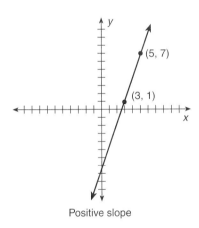

Positive slope

26.

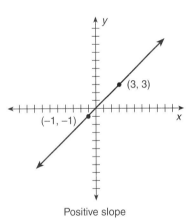

Positive slope

27.

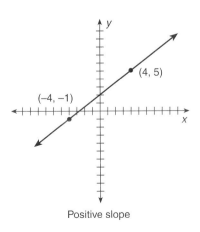

Positive slope

28.

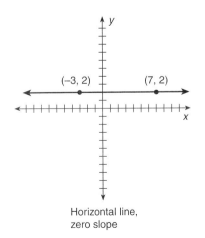

Horizontal line, zero slope

29.

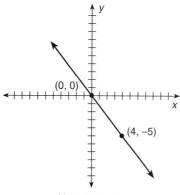

(0, 0)

(4, –5)

Negative slope

30.

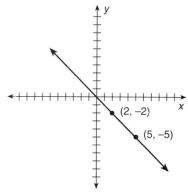

(2, –2)

(5, –5)

Negative slope

31.

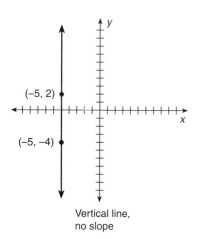

(–5, 2)

(–5, –4)

Vertical line,
no slope

32.

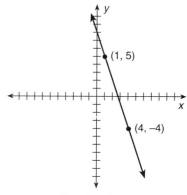

(1, 5)

(4, –4)

Negative slope

33.

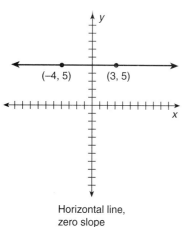

(–4, 5)   (3, 5)

Horizontal line,
zero slope

34.

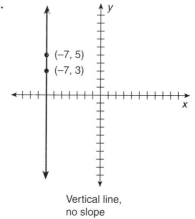

(–7, 5)

(–7, 3)

Vertical line,
no slope

Self-Study Exercises 19–2

1   **1.** $y = \dfrac{2}{3}x + \dfrac{25}{3}$ **2.** $y = -\dfrac{1}{2}x + 3$ **3.** $y = 2x + 1$ **4.** $y = x - 1$

2   **5.** $y = -\dfrac{5}{3}x + \dfrac{38}{3}$ **6.** $x = -1$ **7.** $y = -3$ **8.** $x = -4$ **9.** $y = 2x$ **10.** $y = 0$

11. $f(x) = 5x + 1{,}935$ **12.** $f(x) = 22x + 50{,}000$

Self-Study Exercises 19–3

1   **1.** slope $= 4$, y-intercept $= 3$ **2.** slope $= -5$, y-intercept $= 6$ **3.** slope $= -\dfrac{7}{8}$, y-intercept $= -3$

4. slope $= 0$, y-intercept $= 3$ **5.** $y = 2x - 5$, slope $= 2$, y-intercept $= -5$ **6.** $y = \dfrac{5}{2}$, slope $= 0$, y-intercept $= \dfrac{5}{2}$

7. $y = -2x + 6$, slope $= -2$, y-intercept $= 6$ **8.** $y = 2x + 5$, slope $= 2$, y-intercept $= 5$
9. $x = 4$, slope not defined, no y-intercept **10.** $x = 9$, slope not defined, no y-intercept

2 **11.** $y = \dfrac{1}{4}x + 7$ **12.** $y = -8x - 4$ **13.** $y = 0.2x + 3$; \$12 **14.** $y = 8x + 12{,}000$; \$76,000

15. $y = -0.9x + 40$; 31 mph **16.** $y = -3x + 1$ **17.** $y = \dfrac{2}{3}x - 2$ **18.** $y = 4$ **19.** $y = 2x + 3$

20. \$23

Self-Study Exercises 19–4

1 **1.** $x + y = -1$ **2.** $2x + y = 9$ **3.** $x - 3y = 4$ **4.** $3x - y = -7$ **5.** $x + 3y = 7$ **6.** $3x - y = -3$
7. $2x + 3y = 5$ **8.** $3x + 2y = 6$ **9.** $2x - 5y = 11$ **10.** $6x - 8y = 3$ **11.** $C(x) = 12x + 20{,}000$
12. $C(x) = 0.5x + 45$

2 **13.** $x - y = -1$ **14.** $x - 2y = -13$ **15.** $3x + y = 12$ **16.** $x + 3y = -9$ **17.** $3x - y = 11$
18. $2x - y = -7$ **19.** $3x - 2y = 11$ **20.** $x - 4y = 0$ **21.** $2x - 2y = -3$ **22.** $x + 5y = 4$

Self-Study Exercises 19–5

1 **1.** 10

2. 13

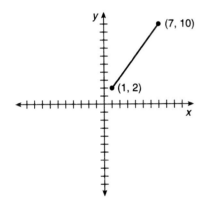

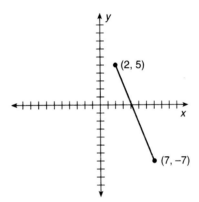

3. 7.071

4. 5.385

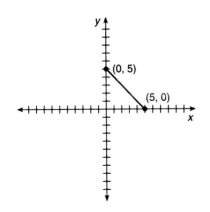

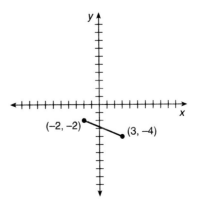

Selected Answers to Student Exercise Material

995

5. 11.180

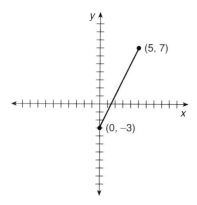

6. 13

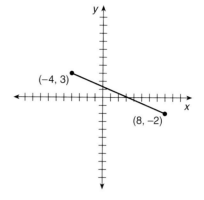

7. 10

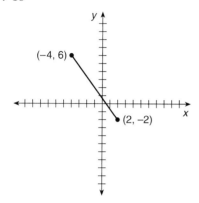

8. 5

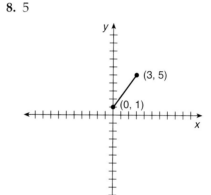

9. 15

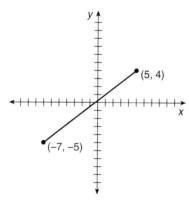

10. 10.290

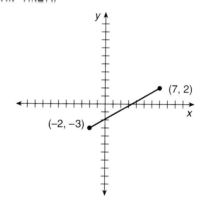

2   **11.** $(4, 6)$ **12.** $\left(\dfrac{9}{2}, -1\right)$ **13.** $\left(\dfrac{5}{2}, \dfrac{5}{2}\right)$ **14.** $\left(\dfrac{1}{2}, -3\right)$ **15.** $\left(\dfrac{5}{2}, 2\right)$ **16.** $\left(2, \dfrac{1}{2}\right)$ **17.** $(-1, 2)$

18. $\left(\dfrac{3}{2}, 3\right)$ **19.** $\left(-1, -\dfrac{1}{2}\right)$ **20.** $\left(\dfrac{5}{2}, -\dfrac{1}{2}\right)$

Assignment Exercises, Chapter 19

1. $\dfrac{1}{3}$ **3.** 2   **5.** $\dfrac{5}{8}$ **7.** $-\dfrac{4}{3}$ **9.** undefined   **11.** $-\dfrac{4}{7}$ **13.** $\dfrac{7}{2}$ **15.** 0   **17.** undefined   **19.** 1

21. $(-1, 2)$ $(3, 2)$; answers will vary.   **23.** $-\$43$ **25.** $\$124$ **27.** The table of values is not a perfect linear function.

29. $y = \frac{1}{3}x + 4$ **31.** $y = \frac{3}{4}x - 3$ **33.** $y = 4x - 5$ **35.** $y = -\frac{2}{3}x - \frac{2}{3}$ **37.** $y = -\frac{1}{11}x + \frac{17}{11}$

39. $y = \frac{7}{4}x - \frac{5}{4}$ **41.** $y = \frac{9}{5}x + \frac{3}{5}$ **43.** $y = x - 3$ **45.** $y = x - 1$ **47.** $y = -2$ **49.** $S = 10x + 3{,}000$

51. $m = 3, b = \frac{1}{4}$ **53.** $m = -5, b = 4$ **55.** no slope, no y-intercept   **57.** $m = \frac{1}{8}, b = -5$ **59.** \$5,050

61. $y = -2x + 8, m = -2, b = 8$ **63.** $y = \frac{3}{2}x - 3, m = \frac{3}{2}, b = -3$ **65.** $y = \frac{3}{5}x - 4, m = \frac{3}{5}, b = -4$

67. $y = \frac{5}{3}, m = 0, b = \frac{5}{3}$ **69.** $y = 3x - 2$ **71.** $y = 2x - 2$ **73.** $x + y = 7$ **75.** $x - 2y = 8$

77. $x - 3y = 20$ **79.** $x + 3y = -10$ **81.** $3x - 4y = -7$ **83.** $y = 0.03x + 35$ **85.** $x - y = -4$
87. $2x - y = -4$ **89.** $x - 5y = -11$ **91.** $2x + 10y = 31$ **93.** $x + 4y = 2$ **95.** 4.472   **97.** 10.440

99. 5.831   **101.** 9.434   **103.** 7.280   **105.** (1, 5)   **107.** $\left(1\frac{1}{2}, 2\right)$ **109.** $\left(-1\frac{1}{2}, 2\frac{1}{2}\right)$ **111.** $\left(1, -\frac{1}{2}\right)$

113. $\left(-1\frac{1}{2}, -3\right)$

115. (a)

Lipsticks	Cost
x	y
0	5,000
100	5,453
1,000	9,530
2,000	14,060

117.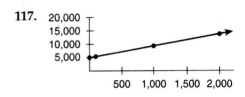

119. $4.53x + 5{,}000 = 8.99x$
$5{,}000 = 4.46x$
$x = 1{,}121.076$
or 1,122 lipsticks

Trial Test, Chapter 19

1. $-\frac{2}{3}$ **3.** $\frac{5}{2}$ **5.** 1   **7.** \$0.23   **9.** $y = \frac{2}{3}x - 7$ **11.** $y = \frac{2}{3}x + \frac{7}{3}$ **13.** $y = 2$ **15.** $m = 3, b = -22$

17. \$50; \$9.70   **19.** $y = 2x + 34, m = 2, b = 34$ **21.** $y = \frac{1}{4}x, m = \frac{1}{4}, b = 0$ **23.** $y = -6x + 3, m = -6, b = 3$

25. $y = \frac{3}{2}x + 3$ **27.** $2x + y = 5$ **29.** $x - 2y = 10$ **31.** $5.385, \left(\frac{1}{2}, 2\right)$

Self-Study Exercises 20–1

1   **1.** $x = 7, y = 5$

2. $x = 3, y = 3$

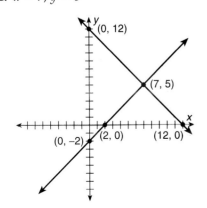

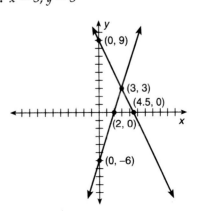

3. no solution; no intersection

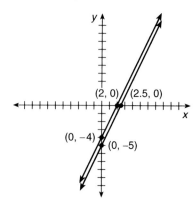

4. many solutions; lines coincide

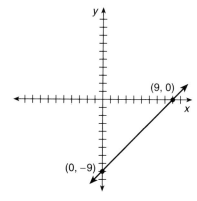

5. $x = 4, y = -2$

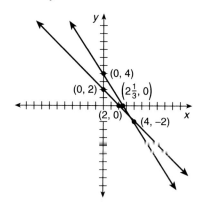

6.

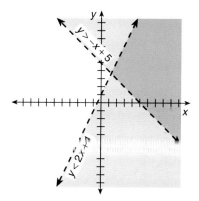

7.

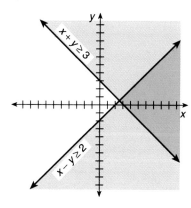

8. $x - 2y \leq -1$ $x + 2y \geq 3$

$$y \geq \frac{1}{2}x + \frac{1}{2} \qquad y \geq -\frac{1}{2}x + \frac{3}{2}$$

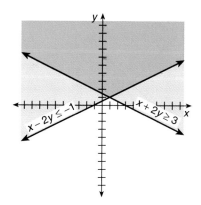

9. $x + y > 4$ $x - y > -3$

     $y > -x + 4$ $y > x + 3$

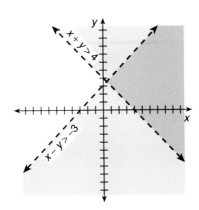

10. $3x - 2y < 8$ $2x + y \leq -4$

     $y > \dfrac{3}{2}x - 4$ $y \leq -2x - 4$

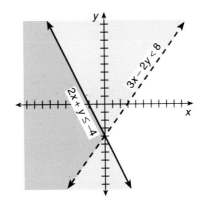

Self-Study Exercises 20–2

[1] **1.** $a = 5, b = -1$ **2.** $m = 4, n = -1$ **3.** $x = 1, y = -1$ **4.** $a = 3, b = -3$ **5.** $x = 2, y = \dfrac{3}{2}$

[2] **6.** $x = 3, y = 0$ **7.** $x = 1, y = 5$ **8.** $a = 2, b = \dfrac{8}{3}$ **9.** $x = -1, y = 2$ **10.** $a = 6, y = 1$

[3] **11.** inconsistent; no solution   **12.** inconsistent; no solution   **13.** dependent; many solutions
14. dependent; many solutions   **15.** inconsistent; no solution

Self-Study Exercises 20–3

[1] **1.** $a = 20, b = 10$ **2.** $r = 5, c = 7$ **3.** $x = 13, y = 11$ **4.** $x = 7, y = 5$ **5.** $p = \dfrac{1}{2}, k = \dfrac{1}{3}$
6. $x = 3, y = 1$ **7.** $x = 13, y = 5$ **8.** $x = -2, a = -7$

Self-Study Exercises 20–4

[1] **1.** short 15.5 in.   **2.** $18,000 at 4%   **3.** $21 per shirt   **4.** (7) 8-cylinder jobs
       long 32.5 in. $17,000 at 5% $16 per hat   (3) 4-cylinder jobs
5. resistor $0.25   **6.** rate of plane = 130 mph   **7.** $3,000 at 7%   **8.** rate of current = 1.67 mph
    capacitor $0.30   rate of wind = 10 mph $5,000 at 9%   rate of motorboat = 11.67 mph
9. dark roast, $4.10   **10.** 75% mixture, 123 lb   **11.** scientific $8.00   **12.** $4,000 at 10%
    with chicory, $3.75   10% mixture, 77 lb   graphing $72.00 $6,000 at 15%
13. reserved $20.00   **14.** 4 pints at 75%
    general $15.00   4 pints at 25%

1.

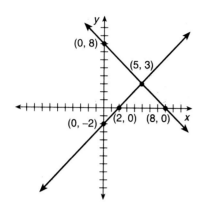

3.

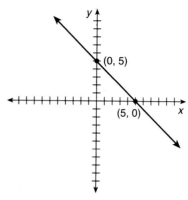

dependent; many solutions, lines coincide

5.

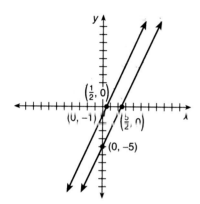

no solutions, lines parallel

7.

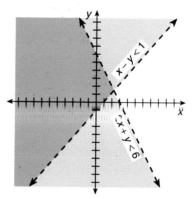

9.

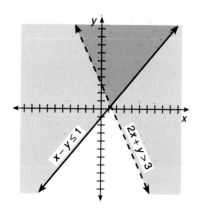

11. $(3, 0)$ **13.** $(-2, 4)$ **15.** $(2, -2)$ **17.** $(6, 3)$ **19.** dependent, many solutions   **21.** $(5, 1)$

23. $\left(2, \dfrac{8}{3}\right)$ **25.** $(2, 0)$ **27.** $(3, 2)$ **29.** $(2, 1.5)$ **31.** $(4, 4)$ **33.** $(7, 5)$

35. $(4, -6)$ **37.** $(3, -4)$ **39.** $\left(\dfrac{4}{5}, \dfrac{2}{5}\right)$ **41.** $(28, 44)$ **43.** $\left(-3, \dfrac{5}{2}\right)$ **45.** $(4, -6)$ **47.** $(12, 2)$ **49.** $(1, -2)$

51. electrician $75   **53.** shellac $2.50   **55.** 119°, 56°   **57.** $2,000 at 5%   **59.** Colombian = $5
    apprentice $35   thinner $3.50 $3,000 at 6%   blended = $4
61. Ohio = $1.95   **63.** name brand = $320   **65.** telephone = $15,000   **67.** 4 gal. 75%
    Alaska = $2.10   generic = $175   showroom = $25,000   4 gal. 25%
69. In line 4, $-6c$ should be $-9c$.   **71.** Answers will vary.
    $c = 3$
    $d = 1$

Trial Test, Chapter 20

1. $(5, 0)$ **3.** $(14, -9)$

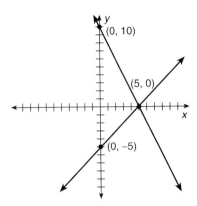

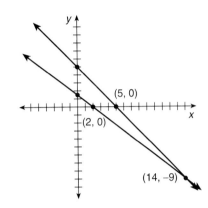

5. $(-6, -6)$

7. $\left(\dfrac{1}{2}, \dfrac{7}{2}\right)$

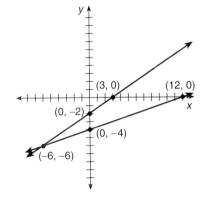

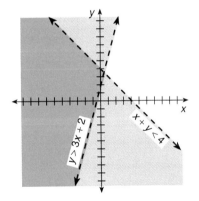

9. $(0, 0)$ **11.** $(1, 2)$ **13.** $(2, 2)$ **15.** $(-3, 15)$ **17.** $(-9, -11)$ **19.** $20A, 15A$
21. $L = 51$ in.   **23.** $20,000 at 3.5%   **25.** $0.000215\ F$
    $W = 34$ in. $\ 5,000 at 4% $0.000055\ F$

Self-Study Exercises 21–1

1   **1.** $\overleftrightarrow{PQ}, \overleftrightarrow{QR}, \&\ \overleftrightarrow{PR}$ **2.** $\overrightarrow{PQ}, \overrightarrow{PR}$ **3.** $\overline{QR}$ **4.** Yes   **5.** No, end points are different.

6. no   **7.** yes   **8.** yes   **9.** Yes. $\overline{XY}$ and $\overline{YX}$ contain same points.   **10.** no

2   **11.** parallel   **12.** intersect   **13.** coincide   **14.** intersect   **15.** $\overleftrightarrow{EF}$ and $\overleftrightarrow{GH}$

3   **16.** ∠DEF, ∠FED   **17.** ∠E   **18.** ∠1   **19.** ∠a   **20.** ∠ABC or ∠CBA   **21.** ∠B

4   **22.** 360°   **23.** 180°   **24.** 90°   **25.** acute   **26.** obtuse   **27.** right   **28.** obtuse   **29.** acute
30. straight   **31.** acute   **32.** obtuse   **33.** yes   **34.** no   **35.** no   **36.** no   **37.** yes   **38.** yes
39. no   **40.** 90°   **41.** congruent

Self-Study Exercises 21–2

1   **1.** 54°   **2.** 46° 8′ 18″   **3.** 37° 52′ 56″   **4.** 74° 3′ 12″   **5.** 16° 59′ 45″   **6.** 54° 44′ 12″
7. 64° 24′ 46″

2   **8.** 0.7833°   **9.** 0.01°   **10.** 0.0872°   **11.** 0.1708°   **12.** 29.7°

3   **13.** 21′   **14.** 12′   **15.** 7′ 12″   **16.** 12′ 47″   **17.** 18′ 54″

4   **18.** 7° 30′   **19.** 11° 25′ 43″   **20.** 38.1908°   **21.** 12° 38′ 50″

Self-Study Exercises 21–3

1   **1.** scalene   **2.** equilateral   **3.** isosceles

2   **4.** longest $\overline{TS}$, shortest $\overline{RS}$ **5.** longest k, shortest l **6.** ∠C, ∠B, ∠A   **7.** ∠X, ∠Z, ∠Y

?   **0.** ∠A, ∠F, ∠C   ∠D, ∠B   ∠E   **9.** ∠P = ∠C, ∠Q = ∠T, QP = TC   **10.** ∠L = ∠P, PM = LJ, LK = PN

4   **11.** $BC = 12$ cm, $AB = 16.971$ cm   **12.** $AC = 7.071$ m, $BC = 7.071$ m
13. $AC = 9.899$ m, $AB = 14.0$ m   **14.** $AC = 8$ m, $BC = 8$ m   **15.** $AC = 14.697$ mm, $BC = 14.697$ mm
16. 25′5″   **17.** 37.830 ft

5   **18.** $AB = 12$ cm, $BC = 10.392$ cm   **19.** $AC = 9.0$ mm, $BC = 15.588$ mm
20. $AC = 4.619$ in., $AB = 9.238$ in.   **21.** $AC = 5.715$ cm, $AB = 11.431$ cm   **22.** $AB = 5$ ft 6 in., $BC = 4$ ft 9 in.
23. 35′   **24.** rafter = 18 ft 0 in., run = 15 ft 7 in.   **25.** 4.5 cm

Self-Study Exercises 21–4

1   **1.** 912 ft^2 **2.** 304.5 ft^2 **3.** 1,942.5 ft^2 **4.** 20 yd^2 @ \$16.50 = \$330   **5.** 24 lb   **6.** 1,086 ft^2, ∠ = 135°
7. 63 in^2 **8.** 625 dm^2 **9.** 33 in^2

2   **10.** equilateral triangle, ∠60°   **11.** square, ∠90°   **12.** regular pentagon, ∠108°
13. regular hexagon, ∠120°   **14.** 42.5 cm^2 **15.** 662.4 mm^2

Self-Study Exercises 21–5

1   **1.** 120.64 cm^2 **2.** 198.44 mm^2 **3.** 2.43 ft^2 **4.** 14.18 in^2 **5.** 2.75 m^2 **6.** 530.14 ft^2
7. 75.40 in^2 **8.** 17.0 in^2

2   **9.** 28.27 mm   **10.** 13.09 in.   **11.** 3.20 ft   **12.** 7.06 cm   **13.** 1.64 m   **14.** 15.71 in.

3   **15.** 16.01 cm^2 **16.** 48.98 in^2 **17.** 54.83 in^2 **18.** 51.99 cm^2 **19.** 91.29 in^2
20. 561.81 mm^2 **21.** 257.93 ft^2

Self-Study Exercises 21–6

1   **1.** $60°$ **2.** $30°$ **3.** $90°$ **4.** 11.5 in.   **5.** 11.5 in.   **6.** 13.33 in.   **7.** 57.5 in^2 **8.** 5.75 in.
9. 6.67 in.   **10.** 41.89 in.   **11.** 2.9 cm   **12.** 1 in.

2   **13.** $90°$ **14.** $45°$ **15.** 8.49 in.   **16.** $45°$ **17.** 6 in.   **18.** 0.5 m   **19.** 0.35 m   **20.** $90°$
21. 0.35 m   **22.** 0.29 in.   **23.** 1.06 in.   **24.** 63.62 dkm^2

3   **25.** 34.64 mm   **26.** $120°$ **27.** $60°$ **28.** 40 mm   **29.** $90°$ **30.** 10 mm   **31.** 20 mm
32. $60°$ **33.** $30°$ **34.** 3 in.   **35.** 39.97 mm   **36.** 2.84 in.

Assignment Exercises, Chapter 21

1. $\overleftrightarrow{AB}$ **3.** $\overrightarrow{AB}$ **5.** $\overrightarrow{NO}$ **7.** no   **9.** no   **11.** $\overleftrightarrow{AB}$ & $\overleftrightarrow{CD}$ **13.** $\overleftrightarrow{EF}$ & $\overleftrightarrow{GH}$ **15.** yes
17. $\angle EDF$ or $\angle FDE$ **19.** $\angle P$ **21.** $\angle M$ **23.** right   **25.** acute   **27.** straight   **29.** obtuse
31. neither   **33.** supplementary   **35.** neither   **37.** equal   **39.** vertical   **41.** $31° 8' 11''$
43. $34° 37' 56''$ **45.** $60° 40' 18''$ **47.** $0.4833°$ **49.** $0.1261°$ **51.** $45'$ **53.** $13' 3''$ **55.** $23° 20'$
57. $18° 39' 50.5''$ **59.** isosceles   **61.** longest TS, shortest RS **63.** $\angle C, \angle B, \angle A$ **65.** $\angle C = \angle D$,
$\angle B = \angle E, CB = DE$ **67.** $\angle M = \angle J, JL = PM, JK = MN$ **69.** $RS = 10.607$ cm, $ST = 10.607$ cm
71. $RS = 9.0$ hm, $ST = 9.0$ hm   **73.** $AB = 13.856$ dm, $BC = 6.928$ dm   **75.** $AC = 17.321$ in., $AB = 20.0$ in.
77. $AB = 46$ ft 10 in., $BC = 23$ ft 5 in.   **79.** 8 ft 4 in.   **81.** 19.630 mm   **83.** 908.5 in^2
85. regular pentagon, $108°$ **87.** 130 ft   **89.** 83 ft^2 **91.** 72 bricks   **93.** 2.62 cm^2 **95.** $1{,}610.28 \text{ mm}^2$
97. 2.18 ft^2 **99.** 6.66 mm^2 **101.** 6.96 cm^2 **103.** 480.57 mm^2 **105.** 232.27 ft^2 **107.** 44.50 in^2
109. 31.42 in.   **111.** 36.74 mm   **113.** $AE = 5$ **115.** $\angle GJO = 45°$ **117.** $IJ = 20$ **119.** $\angle MOP = 30°$
121. 5 mm pen   **123.** 3.54 cm
125. area pentagon $<$ area circle with a radius of 5 cm; $\pi(5)^2 = 78.54 \text{ cm}^2$
    area pentagon $>$ area square with a side of 7.07 cm; $(7.07)^2 = 50 \text{ cm}^2$
    area octagon $<$ area circle with a radius of 5 cm; $\pi(5)^2 = 78.54 \text{ cm}^2$
    area octagon $>$ area hexagon with a side of 5 cm; $6(0.5)(5)(4.33) = 64.95 \text{ cm}^2$

Trial Test, Chapter 21

1. $\overline{BC}$ **3.** acute   **5.** parallel   **7.** perpendicular   **9.** $15°45'29''$ **11.** $18'45''$ **13.** largest $\angle A$,
smallest $\angle B$ **15.** 13.657 cm   **17.** 155 ft   **19.** $135°$ **21.** 43.20 mm   **23.** $25.13''$ **25.** 1.57 in.

Self-Study Exercises 22–1

1   **1.** 6.45 in.   **2.** 14 cm   **3.** 4.25 rad   **4.** 3.49 rad   **5.** 7 cm   **6.** 4.5 in.
7. 1.52 cm   **8.** 1.1 rad

2   **9.** 17.12 in^2 **10.** 4.32 cm^2 **11.** 2.24 cm   **12.** 0.85 rad   **13.** 47.43 cm^2 **14.** 0.7 rad

3   **15.** 0.79 rad   **16.** 0.98 rad   **17.** 1.36 rad   **18.** 2.44 rad   **19.** $21.75°$; 0.38 rad   **20.** $177.55°$; 3.10 rad
21. $44.9033°$; 0.78 rad   **22.** $10.5208°$; 0.18 rad

4   **23.** $45°$ **24.** $30°$ **25.** $143.2394°$ **26.** $80.2141°$ **27.** $28°38'52''$ **28.** $22°30'$
29. $42°58'19''$ **30.** $63°1'31''$

Self-Study Exercises 22–2

1   **1.** $\dfrac{5}{13}$ **2.** $\dfrac{12}{13}$ **3.** $\dfrac{5}{12}$ **4.** $\dfrac{12}{13}$ **5.** $\dfrac{5}{13}$ **6.** $\dfrac{12}{5}$ **7.** $\dfrac{9}{15} = \dfrac{3}{5}$ **8.** $\dfrac{12}{15} = \dfrac{4}{5}$ **9.** $\dfrac{9}{12} = \dfrac{3}{4}$
10. $\dfrac{12}{15} = \dfrac{4}{5}$ **11.** $\dfrac{9}{15} = \dfrac{3}{5}$ **12.** $\dfrac{12}{9} = \dfrac{4}{3}$ **13.** $\dfrac{16}{34} = \dfrac{8}{17}$ **14.** $\dfrac{30}{34} = \dfrac{15}{17}$ **15.** $\dfrac{16}{30} = \dfrac{8}{15}$

16. $\dfrac{30}{34} = \dfrac{15}{17}$ **17.** $\dfrac{16}{34} = \dfrac{8}{17}$ **18.** $\dfrac{30}{16} = \dfrac{15}{8}$ **19.** $\dfrac{9}{16.64} = 0.5409$ **20.** $\dfrac{14}{16.64} = 0.8413$

21. $\dfrac{9}{14} = 0.6429$ **22.** $\dfrac{14}{16.64} = 0.8413$ **23.** $\dfrac{9}{16.64} = 0.5409$ **24.** $\dfrac{14}{9} = 1.5556$

2 **25.** $\sin A = \dfrac{12}{20} = \dfrac{3}{5}$ $\sin B = \dfrac{16}{20} = \dfrac{4}{5}$ **26.** $\sin A = \dfrac{8.15}{9.73} = 0.8376$ $\sin B = \dfrac{5.32}{9.73} = 0.5468$

$\cos A = \dfrac{16}{20} = \dfrac{4}{5}$ $\cos B = \dfrac{12}{20} = \dfrac{3}{5}$ $\cos A = \dfrac{5.32}{9.73} = 0.5468$ $\cos B = \dfrac{8.15}{9.73} = 0.8376$

$\tan A = \dfrac{12}{16} = \dfrac{3}{4}$ $\tan B = \dfrac{16}{12} = \dfrac{4}{3}$ $\tan A = \dfrac{8.15}{5.32} = 1.5320$ $\tan B = \dfrac{5.32}{8.15} = 0.6528$

$\csc A = \dfrac{20}{12} = \dfrac{5}{3}$ $\csc B = \dfrac{20}{16} = \dfrac{5}{4}$ $\csc A = \dfrac{9.73}{8.15} = 1.1939$ $\csc B = \dfrac{9.73}{5.32} = 1.8289$

$\sec A = \dfrac{20}{16} = \dfrac{5}{4}$ $\sec B = \dfrac{20}{12} = \dfrac{5}{3}$ $\sec A = \dfrac{9.73}{5.32} = 1.8289$ $\sec B = \dfrac{9.73}{8.15} = 1.1939$

$\cot A = \dfrac{16}{12} = \dfrac{4}{3}$ $\cot B = \dfrac{12}{16} = \dfrac{3}{4}$ $\cot A = \dfrac{5.32}{8.15} = 0.6528$ $\cot B = \dfrac{8.15}{5.32} = 1.5320$

Self-Study Exercises 22–3

1 **1.** 0.3584 **2.** 0.9981 **3.** 1.0724 **4.** 0.6088 **5.** 0.5225 **6.** 0.9304 **7.** 0.4068
8. 0.9336 **9.** 0.8039 **10.** 0.9163 **11.** 1.0990 **12.** 0.8843 **13.** 0.9757 **14.** 0.4950
15. 3.3191 **16.** 0.9446 **17.** 0.6845 **18.** 0.9377 **19.** 0.3960 **20.** 0.2199

2 **21.** 20.0° **22.** 22.5° **23.** 67° **24.** 62.5° **25.** 57.3° **26.** 34.8° **27.** 51.7° **28.** 82.1°
29. 39.0° **30.** 21.2° **31.** 0.5585 rad **32.** 0.6807 rad **33.** 0.8028 rad **34.** 1.1694 rad
35. 0.2537 rad **36.** 1.4802 rad **37.** 0.3061 rad **38.** 1.1804 rad **39.** 0.2802 rad **40.** 1.1362 rad

3 **41.** 2.1943 **42.** 1.4945 **43.** 1.0306 **44.** 1.0187 **45.** 2.8824 **46.** 1.8341 **47.** 0.0524
48. 1.3563 **49.** 4.7174 **50.** 1.9661 **51.** 67.0° **52.** 56.9° **53.** 15.6° **54.** 83.0° **55.** 30.0° **56.** 0.9903
57. 0.9659 **58.** 1.6391 **59.** 0.8001 **60.** 60.7° **61.** 62.1° **62.** 18.8° **63.** 10.9°

Assignment Exercises, Chapter 22

1. 1.05 rad **3.** 5.24 rad **5.** 1.74 rad **7.** 150° **9.** 97.4028° **11.** 67°30′

13. 1.61 cm **15.** 1.71 ft **17.** 3.38 in. **19.** 2.1 in. **21.** 15.66 cm² **23.** $\dfrac{15}{25} = \dfrac{3}{5}$ **25.** $\dfrac{15}{20} = \dfrac{3}{4}$

27. $\dfrac{25}{20} = \dfrac{5}{4}$ **29.** $\dfrac{20}{25} = \dfrac{4}{5}$ **31.** $\dfrac{20}{15} = \dfrac{4}{3}$ **33.** $\dfrac{25}{15} = \dfrac{5}{3}$ **35.** $\dfrac{2 \text{ ft}}{2 \text{ ft 2 in.}} = \dfrac{24 \text{ in.}}{26 \text{ in.}} = \dfrac{12}{13}$

37. $\dfrac{2 \text{ ft}}{10 \text{ in.}} = \dfrac{24 \text{ in.}}{10 \text{ in.}} = \dfrac{12}{5}$ **39.** $\dfrac{10 \text{ in.}}{2 \text{ ft}} = \dfrac{10 \text{ in.}}{24 \text{ in.}} = \dfrac{5}{12}$ **41.** $\dfrac{7}{12.62} = 0.5547$ **43.** $\dfrac{7}{10.5} = 0.6667$

45. $\dfrac{12.62}{10.5} = 1.2019$ **47.** $\dfrac{10.5}{7} = 1.5000$ **49.** 0.4540 **51.** 0.3057 **53.** 1.0724 **55.** 0.7585

57. 0.8403 **59.** 0.1708 **61.** 52.5° **63.** 45.3° **65.** 16.7° **67.** 1.2188 rad **69.** 1.2101 rad
71. 0.3313 rad **73.** 1.0576 **75.** 1.1326 **77.** 2.1536 **79.** 1.0090 **81.** 1.0592 **83.** −2.1557
85. $\sin \theta = \tan \theta$ for $\theta = 0°, 180°,$ and $360°$.

1. 0.61 rad **3.** 5.50 rad **5.** 15.4167° = 0.27 rad **7.** 16.2° = 0.28 rad **9.** 112.5° **11.** 68°45′18″ **13.** 1 in.

15. 1.71 m **17.** $\dfrac{10}{26} = \dfrac{5}{13}$ **19.** $\dfrac{26}{10} = \dfrac{13}{5}$ **21.** $\dfrac{11.5}{12.54} = 0.9171$ **23.** 0.7986 **25.** 0.9135 **27.** 16.0°

29. 19.5° **31.** 0.8726 rad **33.** 0.3150 rad **35.** 1.3673

Self-Study Exercises 23–1

1 **1.** 27.8° **2.** 6.889 in. **3.** 28.21 m **4.** 16.15 ft **5.** 43.2° **6.** 49.6°

7. $S = 51.1°$

$U = 38.9°$

$u = 11.31$ yd

$18^2 = 14^2 + (11.31)^2$

$324 \approx 323.9161$

8. $S = 45°$

$t = 6.647$ m

$s = 4.700$ m

$(6.647)^2 = (4.7)^2 + (4.7)^2$

$44.18 = 44.18$

9. $U = 55.5°$

$s = 4.814$ mm

$u = 7.005$ mm

$(8.5)^2 \approx (4.814)^2 + (7.005)^2$

$72.25 \approx 23.17 + 49.07$

$72.25 \approx 72.24$

10. $U = 74°$

$t = 50.79$ m

$u = 48.82$ m

$(50.79)^2 \approx (48.82)^2 + (14)^2$

$2{,}579.62 \approx 2{,}383.39 + 196$

$2.579.62 \approx 2{,}579.39$

2 **11.** 35.7° **12.** 10.05 cm **13.** 31.50 ft **14.** 35.49 dm **15.** 15.88 mm

16. 14.12 yd **17.** $R = 45.9°$ **18.** $Q = 44°$ **19.** $R = 16.5°$ **20.** $Q = 30.5°$

$Q = 44.1°$ $r = 12.23$ cm $r = 4.147$ dkm $r = 13.58$ m

$r = 16.52$ ft $q = 11.81$ cm $s = 14.60$ dkm $s = 15.76$ m

3 **21.** 28.6° **22.** 4.075 m **23.** 5.979 ft **24.** 23.5° **25.** 0.04663 cm

26. 0.01212 m **27.** $D = 34.8°$ **28.** $E = 48°$ **29.** $D = 16.5°$ **30.** $D = 54.0°$

$E = 55.2°$ $d = 6.303$ ft $d = 5.963$ in. $E = 36.0°$

$f = 5.604$ m $f = 9.419$ ft $f = 20.99$ in. $f = 13.60$ ft

Self-Study Exercises 23–2

1 **1.** 150.1 m **2.** 11.91 ft **3.** 9.950 cm **4.** 46.71 cm **5.** 21.09 in.

2 **6.** 22.38 in. **7.** 53.62 ft **8.** 59.0° **9.** 39.23 ft **10.** 15.5° **11.** 23.25 ft **12.** 10 ohms

13. 17.32 ohms **14.** 7.654 cm **15.** 35.8°

Assignment Exercises, Chapter 23

1. $A = 38.1°$ **3.** $h = 2.537$ mm **5.** $y = 5.469$ m **7.** $D = 28.3°, E = 61.7°, f = 14.76$ ft
9. $G = 53°, g = 25.21$ m, $k = 31.57$ m **11.** 32.2° **13.** 11.57 in. **15.** $B = 60.5°, b = 3.641$ m, $c = 4.183$ m
17. 48.6° and 41.4° **19.** $B = 56°, a = 88.91$ ft, $b = 131.8$ ft **21.** 44.7° **23.** $X_L = 9.64$ ohms; $R = 11.49$ ohms
25. 56.54 cm

Trial Test, Chapter 23

1. 48.8° **3.** 15.41 ft **5.** 42.9° **7.** 6.420 cm **9.** 62.9°
11. $B = 65.5°, b = 30.72$ m, $c = 33.76$ m **13.** 38.7° **15.** 48 in. **17.** 40.06 A **19.** 84.3°

Self-Study Exercises 24–1

1 **1.** 13 **2.** 15 **3.** 7.280 **4.** 8.544 **5.** 2.746

2 **6.** 67.38° **7.** 53.13° **8.** 20.56° **9.** 68.20° **10.** 75.96°

3 **11.** 19; 42° **12.** 8; 72° **13.** 1; 225° **14.** 6; 75° **15.** $12 + 5i$; 13; 22.62° **16.** $6 + 4i$; 7.211; 33.69°

Self-Study Exercises 24–2

1 **1.** 60° **2.** 15° **3.** 70° **4.** 15° **5.** 32° **6.** 70° **7.** 32° **8.** 62° **9.** 0.96 rad **10.** 0.44 rad

2 **11.** all positive **12.** sin, csc positive; others negative **13.** tan, cot positive; others negative
14. cos, sec positive; others negative **15.** cos, sec positive; tan, sin zero; cot, csc undefined

3 **16.** −0.5000 **17.** −0.8391 **18.** −0.8011 **19.** −0.6536 **20.** −0.8660 **21.** −0.2910
22. −0.1736 **23.** 0.5000 **24.** 146.3° **25.** 3.61 rad

Self-Study Exercises 24–3

1 **1.** $C = 63°; b = 40.00; c = 45.22$ **2.** $A = 50°; a = 6.504; c = 8.075$ **3.** $C = 33°; b = 56.37; c = 35.45$
4. $B = 30°; a = 5.543; b = 3.200$ **5.** $C = 50°; a = 20.84; c = 16.03$

2 **6.** $B = 21.5°; C = 118.5°; c = 57.42$ **7.** $C = 10.8°; A = 99.2°; a = 15.76$ **8.** $A = 43.2°;$
$B = 116.8°; b = 23.48$ or $A = 136.8°; B = 23.2°; b = 10.35$ **9.** $B = 14.5°; C = 135.5°; c = 11.21$

3 **10.**   **11.** 805.9 ft

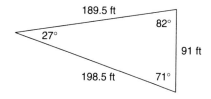

Self-Study Exercises 24–4

1 **1.** $A = 30.8°$ **2.** $D = 71.7°$ 2 **3.** $r = 32.42$ **4.** $j = 18.68$ 3 **5.** 49.27 ft
      $B = 125.1°$ $E = 62.9°$ $S = 49.7°$ $K = 37.5°$
      $C = 24.1°$ $F = 45.4°$ $T = 58.3°$ $L = 34.0°$

6. 60.8°
    44.1°
    75.1°

Self-Study Exercises 24–5

1 **1.** 27.93 cm² **2.** 90.42 m² **3.** 73.21 ft² **4.** 21.22 in² **5.** 122.6 mm²
6. 76.31 m² **7.** 192.8 dm² **8.** 453.6 dkm²

Selected Answers to Student Exercise Material

2   **9.** 14.97 m²   **10.** 3.712 in²   **11.** 24.38 cm²   **12.** 26.72 cm²

3   **13.** 3.32 cm²   **14.** 70.4 ft²   **15.** 5,108 ft²

Assignment Exercises, Chapter 24

1. 7.211; 56.3°   **3.** 6.083; 9.5°   **5.** 13.89; 59.7°   **7.** 0.1016 rad   **9.** 41°   **11.** 0.8832 rad   **13.** 32°15'10"
15. 0.8632   **17.** −0.3420   **19.** 0.3420   **21.** 1.000   **23.** 2.83; 2.36 rad   **25.** $A = 40°$; $b = 10.8$; $c = 4.3$
27. $A = 30.3°$; $B = 104.7°$; $b = 9.6$ **29.** 1st; $A = 39.4°$; $C = 112.6°$; $c = 13.4$
                                                    2nd; $A = 140.6°$; $C = 11.4°$; $c = 2.9$

31. 42.0 ft

33.

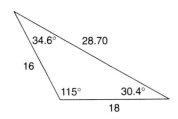

35.

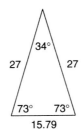

37.

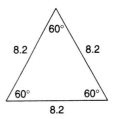

39. 97.98 ft   **41.** 343.3 ft; 665.4 ft   **43.** 99.68 m²   **45.** 710.4 cm²   **47.** 185.4 km²   **49.** 27.71 ft²
51. 708.6 in²   **53.** 12,046 ft²

Trial Test, Chapter 24

1. 0.8192   **3.** −0.9397   **5.** 135°   **7.** 101.3°   **9.** 8.991 ft   **11.** 131.8°   **13.** 6.830 m
15. 36.98 m²   **17.** 150.9 ft   **19.** 66.88 ft²   **21.** 6.4 milliamps   **23.** 30 ohms

Glossary and Index

Base angles, 802
Base in exponential notation: the value used in repeated multiplication when raising to a power, 49, 91–92
Base in a problem involving percent: the number (B) that represents the original or total amount, 175–176, 189
Base of any polygon: the horizontal side of any polygon or a side that would be horizontal if the polygon's orientation were modified, 277, 279, 280
Basic principle of equality: To preserve equality if we perform an operation on one side of an equation, we must perform the same operation on the other side, 362
Binary operation: an operation that involves working with two numbers at a time, 19, 81, 89
Binomial: a polynomial containing two terms. Examples include $2x + 4$ and $7xy - 21a$, 448, 541
Binomial square, 545
Borrowing, 26
Boundary, 653, 665, 701–703

Calculator
    accuracy, 181
    addition of whole numbers and decimals, 22
    angle calculations, 794
    angle measure given a trigonometric value, 855
    checking solutions, 612
    continuous sequence, 294, 628
    decimal part of degree, 794
    decimal points and zeros, 22
    decimals in equations, 415
    degrees, minutes, and seconds, 794
    division of whole numbers and decimals, 41
    exponential terms, 627–628
    exponents, 53, 295, 478
    fraction key, 476
    graphing linear equations in two variables, 697–698
    graphing quadratic equations and inequalities, 709–711
    graphing systems of equations, 763
    integers, 97–98
    logarithms, 640–643
    multiplication of whole numbers and decimals, 38
    negative key, 97
    natural exponential e, 631
    order of operations, 29, 55–56, 97–98
    order of operations with integers, 56
    parentheses, 295
    percents, 179, 180
    pi π, 293
    powers and roots, 53, 295, 476, 627
    proportions, 179–180
    radians, 898
    reciprocals, 855–856
    repeating decimals, 181
    roots, 53, 476
    roots of linear equations, 699–700
    roots of quadratic equations, 709–710
    scientific notation, 464–465

sign-change key, 97
signed numbers, 97–98
solving equations from a graph, 699–700, 709–710
solving systems of equations, 763
square roots, 53
subtraction of whole numbers and decimals, 29
toggle keys, 97
trigonometric values, 854–856
trigonometric values of an angle more than 90 degrees, 898
value of pi (π), 292, 293
verify concepts, 22
vertex of a parabola, 709
zero, 22
Canceling: dividing a numerator and denominator by the same amount to reduce the terms of the fraction, 138
Capacity
    metric, 236
    U. S. customary, 216
Cardinal number, 7
Carry, 14, 20
Cell, 378
Celsius, 260
Center: the point that is the same distance from every point on the circumference of the circle, 291
Centi-: the prefix used for a unit that is 1/100 of the standard unit, 235
Centimeter: the most common unit used to measure objects less than 1 meter long; it is 1/100 of a meter, 235
Central angle, 843
Change, 730
Check
    addition, 21
    division, 46
    multiplication, 36–37
    solution, 366
    subtraction, 28
Central tendancy, 329
Chord: a line segment joining two points on the circumference of a circle, 813
Circle: a closed curved line whose points lie in a plane and are the same distance from the center of the figure, 291
    arc length, 812, 843
    area, 291, 294–295
    center, 291
    circumference, 291–293
    diameter, 291
    graph, 316
    inscribed and circumscribed polygons, 815–821
    radius, 291
    sector, 812
    segment, 813
    semicircle, 291
    semicircular-sided figure, 298
    see also Geometry
Circle graph: the graph that uses a divided circle to show pictorially how a total amount is divided into parts, 316
Circumference: the perimeter or length of the closed curved of the line that forms the circle, 291–293

Circumscribed polygon: a polygon is circumscribed about a circle when the polygon is outside the circle and all its sides are tangent to (intersecting in exactly one point) the circumference. The circle is said to be inscribed in the polygon, 815
Class frequency: the number of values in each class interval, 321
Class intervals: the groupings of a large amount of data into smaller groups of data, 320
Class midpoint: the middle value of each class interval, 320
Clockwise, 788
Coefficient: each factor of a term containing more than one factor is the coefficient of the remaining factor(s). The number factor, also called the numerical coefficient, is often implied when the term coefficient is used, 356–357, 492
    leading, 450, 557, 599
Cofunction, 852, 856–858
Coincide, 744, 763, 786
Commission: a salary based on the amount of sales made. It is usually a certain percent of the total sales, 189–191
Common denominator: when comparing two or more fractions, a denominator that each denominator will divide into evenly, 127–128
Common denominator in decimals, 12
Common factor: a factor that appears in both terms of a fraction or in each of several terms in an expression, 117, 538
Common fraction: A common fraction is the division of one integer by another, 108
Common logarithm: a logarithm with a base of 10, 640
Commutative property of addition: the property that states that values being added may be added in any order, 18–19
Commutative property of multiplication: this property permits numbers to be multiplied in any order, 31
Compare
    decimals, 11–12, 129
    fractions, 128–129
    integers, 73
    whole numbers, 6
Complement of a percent: the difference between 100% and a percent, 200
Complement of a set: the complement of a set is a set that includes every element of the universal set that is not an element of the given set, 661
Complementary angles: two angles whose sum is one right angle or 90°, 789, 856
Complete quadratic equation: a quadratic equation and that has a quadratic term, a linear term, and a constant term, 599, 604–605
Complex fraction: A complex fraction has a fraction or mixed number in its numerator or denominator or both, 109, 145–146
Complex notation, 493, 891–892

uses parentheses to enclose
boundaries that are not included in
the set and brackets to enclose
boundaries that are included. [3,8)
means the set of values equal to or
greater than 3 and less than (but not
equal to) 8, 653

Inverse, 25
  additive, 75, 83
  multiplicative, 142
  of addition, 25
  of subtraction, 40
Inverse operations: a pair of operations that
    "do" and "undo" an operation
    addition/subtraction, and
    multiplication/division are pairs of
    inverse operations, 25, 40
  trigonometric, 855
Inverse proportion or variation, 423–426
Inverting: interchanging the numerator and
    denominator of a fraction, 143
Irrational number: roots of non-perfect
    powers or other non-terminating,
    non-repeating decimals, 474–475
Irrational numbers, 474–475
  estimating values of, 475
  positioned on number line, 475
Isosceles triangle: a triangle with exactly two
    equal sides. The angles opposite the
    equal sides are also equal, 798, 800

j-factor, 491, 892

Kelvin, 260
Key words
  addition, 58
  division, 58
  equality, 58
  multiplication, 58
  negative, 88
  percent, 88
  positive, 88
  subtraction, 26, 58
Kilo-, 261
Kilogram: the most often used unit for
    measuring weights in the metric
    system, 236
Kilometer: 100 meters; the kilometer is used
    to measure long distances, 234

Lateral surface area: for a three-dimensional
    figure the area of the sides only (bases
    excluded), 522
  see also Surface area
Latitude, 248
Law of cosines, 906–910
  ambiguous case, 902–905
Law of sines, 899–905
Laws of exponents, 441–447, 480
LCD, 127–128
  see also least common denominator
LCM, 117–118, 127–128, 414, 585
  see also least common multiple
Leading coefficient: the coefficient of the
    leading term of a polynomial, 450,
    557, 599
Leading term: the first term of a polynomial
    arranged in descending order, 450, 599

Least common denominator (LCD): the
    smallest common denominator
    among two or more fractions, 127–128
  rational expressions, 573
Least common multiple: The least common
    multiple (LCM) of two or more
    natural numbers is the smallest
    number that is a multiple each
    number. The LCM is divisible by each
    number, 117–118, 585
Leg: one of the two sides of a right triangle
    that is not the hypotenuse. Or, one of
    the two sides that forms the right
    angle, 497
Leg of an angle, 787
Length
  metric, 234–235
  U.S. Customary, 215
Less than (<), 6, 27
Letter term or variable term: a term that
    contains one letter, several letters used
    as factors, or a combination of letters
    and numbers used as factors, 355
Like bases, 441
Like denominators, 127
Like fractions, 127
Like measures: measures with the same
    units, 224
Like radicals: when the radicands are
    identical and the radicals have the
    same order or index, the radicals are
    like radicals, 484
Like signs
  addition, 80
  multiplication, 89
  division, 93
Like terms: terms are like terms if they are
    number terms or if they are letter
    terms with exactly the same letter
    factors, 359–360
Limit dimensions: the largest and smallest
    possible size of an object with a plus
    or minus (6) tolerance value on a
    blueprint specification, 27
Line, 257, 785
  equation, 735–738
  standard form, 737
Line graph: a graph that uses one or more
    lines to show changes in data, 317–318
Line segment or segment: a line segment, or
    segment, consists of all points on the
    line between and including two points
    that are called end points, 257, 786
Linear equation: an equation whose graph is
    a straight line. It can be written in the
    standard form $ax + by = c$ where a, b,
    and c are real numbers and a and b are
    not both zero, 361
Linear equation in two variables, 684–700
Linear equations
  point-slope form, 735
  slope-intercept form, 694
  standard form, 688
Linear function, 627, 686
Linear inequality in one variable, 655–659
Linear inequality in two variables, 701–703
Linear polynomial: a polynomial that has
    degree 1, 449

Linear term: a term that has degree 1, 599
Liter: the standard metric unit of capacity or
    volume, 232, 236
Litre, *see* liter
Logarithmic expression: an expression that
    contains at least one term with a
    logarithm, 638
Logarithms
  base other than 10 or *e*, 641
  common, 640
  equivalent exponential expressions,
    638–641
  evaluating formulas, 641–643
  evaluation, 640–643
  natural, 640
  powers, 638
  properties, 643
Long division, 41
Longitude, 249
Loss, 88
Lowest terms, 120–121

Magnitude, 888–889
Mass
  metric, 235–236
  U.S. customary, 215–216
Mean, arithmetic, 46–47, 327–329
Measures of central tendency: measures
    that show typical or central
    characteristics of a set of data.
    Examples are the mean, the median
    and, and, and a mode of a data set, 329
  see also Data, interpreting and analyzing
Measures of variation or dispersion:
    measures that show the dispersion
    or spread of a set of data. Examples
    are the range and the standard
    deviation, 330
Median: the middle quantity of the data set
    when the given quantities are
    arranged in order of size, 329–330
Member or element of a set, 652
Meridians of longitude, 249
Meter: the standard unit for measuring
    length in the metric system, 284
Metre, *see* meter
Metric rule: the tool for measuring length
    that uses centimeters as a standard
    unit, 256–257
Metric system (SI): An international system
    of measurement that uses standard
    units and prefixes to indicate their
    powers of 10, 232
  addition, 240–241
  capacity, 236
  conversions within system, 237–240
  division, 242
  length, 234–235
  metric rule, 256–257
  multiplication, 241
  prefixes, 232–236
  subtraction, 240
  U.S. customary conversions, 244–246
  weight or mass, 235–236
Midpoint, 258, 751–752
Mile, 215
Milli-: the prefix used for a unit that is
    1/1000 of the standard unit, 232